VOLUME EIGHTY-EIGHT

VITAMINS AND HORMONES

Hedgehog Signaling

VOLUME EIGHTY-EIGHT

VITAMINS AND HORMONES
Hedgehog Signaling

Editor-in-Chief

GERALD LITWACK, PhD
Toluca Lake, North Hollywood, California

ELSEVIER

AMSTERDAM • BOSTON • HEIDELBERG • LONDON
NEW YORK • OXFORD • PARIS • SAN DIEGO
SAN FRANCISCO • SINGAPORE • SYDNEY • TOKYO
Academic Press is an imprint of Elsevier

Cover photo credit:
Hatayama, M., and Aruga, J.
Gli Protein Nuclear Localization Signal
Vitamins and Hormones (2012) **88**, pp. 73-90

Academic Press is an imprint of Elsevier
32 Jamestown Road, London, NW1 7BY, UK
Radarweg 29, PO Box 211, 1000 AE Amsterdam, The Netherlands
Linacre House, Jordan Hill, Oxford OX2 8DP, UK
225 Wyman Street, Waltham, MA 02451, USA
525 B Street, Suite 1900, San Diego, CA 92101-4495, USA

First edition 2012

ISBN: 978-0-12-394622-5
ISSN: 0083-6729

Printed and bound by CPI Group (UK) Ltd, Croydon, CR0 4YY

Transferred to Digital Print 2012

Contents

Contributors

Fritz Aberger
Division of Molecular Tumor Biology, Department of Molecular Biology, University of Salzburg, Salzburg, Austria

Jun Aruga
Laboratory for Behavioral and Developmental Disorders, RIKEN Brain Science Institute, Wako-shi, and Saitama University Brain Science Institute, Saitama-shi, Saitama, Japan

Yoshinari Asaoka
Department of Gastroenterology, Graduate School of Medicine, The University of Tokyo, Tokyo, Japan

Nicholas C. Bambakidis
Department of Neurological Surgery, University Hospitals Case Medical Center, Cleveland, Ohio, USA

Elspeth M. Beauchamp
Lombardi Comprehensive Cancer Center, Georgetown University Medical Center, Washington, District of Columbia, USA

Donna Brennan
Department of Biochemistry and Molecular Biology, and Department of Dermatology, Thomas Jefferson University, Philadelphia, Pennsylvania, USA

John A. Buglino
Cell Biology Program, Memorial Sloan-Kettering Cancer Center, New York, USA

Gianluca Canettieri
Department of Molecular Medicine, Sapienza University, Rome, Italy

Richard L. Carpenter
Division of Surgical Sciences, Department of Surgery, Duke University School of Medicine, Durham, North Carolina, USA

Arunabha Chakrabarti
Structural Genomics Division, Saha Institute of Nuclear Physics, Kolkata, West Bengal, India

Xiaole Chen
Department of Biochemistry and Molecular Biology, Thomas Jefferson University, Philadelphia, Pennsylvania, USA

Lan Cheng
Department of Biochemistry and Molecular Biology, Thomas Jefferson University, Philadelphia, Pennsylvania, USA

Shuk Han Cheng
Department of Biology and Chemistry, City University of Hong Kong, Kowloon, Hong Kong, China

Siu Wah Choy
Department of Biology and Chemistry, City University of Hong Kong, Kowloon, Hong Kong, China

Enrico De Smaele
Department of Experimental Medicine, Sapienza University, Rome, Italy

Lucia Di Marcotullio
Department of Molecular Medicine, Sapienza University, Rome, Italy

Eva Dohle
Institute of Pathology, Langenbeckstr. 1, Johannes Gutenberg University, Mainz, Germany

Rui Feng
Department of Molecular and Cellular Physiology, University of Cincinnati College of Medicine, Cincinnati, Ohio, USA

Sabine Fuchs
Institute of Pathology, Langenbeckstr. 1, Johannes Gutenberg University, Mainz, Germany

Sayantani Ghosh
Structural Genomics Division, Saha Institute of Nuclear Physics, Kolkata, West Bengal, India

Richard Greil
Laboratory for Immunological and Molecular Cancer Research, Third Medical Department with Hematology, Oncology, Hemostaseology, Infectiology, and Rheumatology, Paracelsus Medical University, Salzburg, Austria

Alberto Gulino
Department of Molecular Medicine, Sapienza University, Rome, and Neuromed Institute, Pozzilli, Italy

Tanja Nicole Hartmann
Laboratory for Immunological and Molecular Cancer Research, Third Medical Department with Hematology, Oncology, Hemostaseology, Infectiology, and Rheumatology, Paracelsus Medical University, Salzburg, Austria

Minoru Hatayama
Laboratory for Behavioral and Developmental Disorders, RIKEN Brain Science Institute, Wako-shi, Saitama, Japan

Zhenhua He
Division of Dermatology, Department of Medicine, Washington University School of Medicine, St. Louis, Missouri, USA

G. Jason Huang
Division of Dermatology, Department of Medicine, Washington University School of Medicine, St. Louis, Missouri, USA

Anton M. Jetten
Cell Biology Section, Division of Intramural Research, National Institute of Environmental Health Sciences, National Institutes of Health, Research Triangle Park, North Carolina, USA

Jianhang Jia
Department of Molecular and Cellular Biochemistry, Markey Cancer Center, University of Kentucky College of Medicine, Lexington, Kentucky, USA

Hong Soon Kang
Cell Biology Section, Division of Intramural Research, National Institute of Environmental Health Sciences, National Institutes of Health, Research Triangle Park, North Carolina, USA

Fergal C. Kelleher
Department of Medical Oncology, St. Vincent's University Hospital, Dublin, Ireland, and Peter Mac Callum Cancer Centre, Melbourne, Victoria, Australia

Daniela Kern
Division of Molecular Tumor Biology, Department of Molecular Biology, University of Salzburg, Salzburg, Austria

Charles James Kirkpatrick
Institute of Pathology, Langenbeckstr. 1, Johannes Gutenberg University, Mainz, Germany

Tomoya Kotani
Laboratory of Reproductive and Developmental Biology, Department of Biological Sciences, Faculty of Science, Hokkaido University, Sapporo, Japan

Kristin Lichti-Kaiser
Cell Biology Section, Division of Intramural Research, National Institute of Environmental Health Sciences, National Institutes of Health, Research Triangle Park, North Carolina, USA

Hui-Wen Lo
Division of Surgical Sciences, Department of Surgery, Duke University School of Medicine, and Duke Cancer Institute, Durham, North Carolina, USA

Liang Ma
Division of Dermatology, Department of Medicine, Washington University School of Medicine, St. Louis, Missouri, USA

My Mahoney
Department of Dermatology, Thomas Jefferson University, Philadelphia, Pennsylvania, USA

Alla Maloverjan
Protobios LCC, Akadeemia tee 15a, Tallinn, Estonia

Maria Carmen Martinez
INSERM U1063, Université d'Angers, Angers, France

Raymond McDermott
St. Vincent's University Hospital, Dublin, Ireland and Adelaide and Meath Hospital, Dublin, Ireland

Debashis Mukhopadhyay
Structural Genomics Division, Saha Institute of Nuclear Physics, Kolkata, West Bengal, India

Kaine Onwuzulike
Department of Neurological Surgery, University Hospitals Case Medical Center, Cleveland, Ohio, USA

Diane Pham
Department of Pediatrics, Division of Child Neurology, Weill Cornell Medical College, New York, USA

Marko Piirsoo
Tallinn University of Technology, Akadeemia tee 15, Tallinn, Estonia

G. Praveen Raju
Department of Pediatrics, Division of Child Neurology, Weill Cornell Medical College, New York, USA

Marilyn D. Resh
Cell Biology Program, Memorial Sloan-Kettering Cancer Center, New York, USA

Natalia A. Riobo
Department of Biochemistry and Molecular Biology, Thomas Jefferson University, Philadelphia, Pennsylvania, USA

Isabella Screpanti
Department of Molecular Medicine, Sapienza University, Rome, Italy

Raffaella Soleti
INSERM U1063, Université d'Angers, Angers, France

Esther T. Stoeckli
Institute of Molecular Life Sciences, University of Zurich, Zurich, Switzerland

Xiulan Su
Clinical Research Center of the Affiliated Hospital, Inner Mongolia Medical College, Hohhot, Inner Mongolia, China

Aykut Üren
Lombardi Comprehensive Cancer Center, Georgetown University Medical Center, Washington, District of Columbia, USA

Shivakumar Vasanth
Cell Biology Section, Division of Intramural Research, National Institute of Environmental Health Sciences, National Institutes of Health, Research Triangle Park, North Carolina, USA

Nicole H. Wilson
Institute of Molecular Life Sciences, University of Zurich, Zurich, Switzerland

Chang Xiao
Department of Molecular and Cellular Physiology, University of Cincinnati College of Medicine, Cincinnati, Ohio, USA

Jingwu Xie
Wells Center for Pediatric Research, Department of Pediatrics and The Simon Cancer Center, Indiana University, Indianapolis, Indiana, USA

Ling Yang
Clinical Research Center of the Affiliated Hospital, Inner Mongolia Medical College, Hohhot, Inner Mongolia, China

Yana Zavros
Department of Molecular and Cellular Physiology, University of Cincinnati College of Medicine, Cincinnati, Ohio, USA

Gary ZeRuth
Cell Biology Section, Division of Intramural Research, National Institute of Environmental Health Sciences, National Institutes of Health, Research Triangle Park, North Carolina, USA

PREFACE

The Hedgehog signaling pathway is vital to human development and differentiation. The pathway was first described in *Drosophila* and the human version of the pathway is similar. There are three human Hedgehog genes: Sonic Hedgehog, Desert Hedgehog, and Indian Hedgehog. The genes encode Hedgehog proteins that bind to the Patch receptor (there are two Patch genes), an important receptor in the pathway. Ultimately, the Smoothened receptor, when it becomes activated, generates the activation of Gli proteins that are transported to the cell nucleus. Two Gli transactivating proteins bring about gene expression, while the third Gli protein is a repressor of transactivation. Although this pathway is critical for development, it is a vital pathway for stem cell generation and differentiation and its failure to function properly can lead to various cancers. The 23 chapters in this book elaborate all aspects of this pathway, its relation to disease, and its utility as a therapeutic target.

Chapter 1 has the broad title of "Hedgehog signaling" prepared by S. W. Choy and S. H. Cheng. Chapter 2 by F. Aberger, D. Kern, R. Greil, and T. N. Hartmann is entitled "Canonical and noncanonical Hedgehog/Gli signaling in hematological malignancies." This is followed by Chapter 3, "Noncanonical Hedgehog signaling," by D. Brennan, X. Chen, L. Cheng, M. Mahoney, and N. A. Riobo. M. Hatayama and J. Aruga report in Chapter 4 on "Gli protein nuclear localization signal." "Mammalian homologues of *Drosophila* fused kinase" is the topic of Chapter 5, written by A. Maloverjan and M. Piirsoo. R. L. Carpenter and H.-W. Lo in Chapter 6 describe "Identification, functional characterization, and pathological significance of Gli1 isoforms in human cancer." Chapter 7 "Gli-similar (Glis) proteins: their mechanisms of action, physiological functions, and roles in disease" is a review by K. Licht-Kaiser, G. ZeRuth, H. S. Kang, S. Vasanth, and A. M. Jetten. N. H. Wilson and E.T. Stoeckli write on "Sonic Hedgehog regulates Wnt activity during neural circuit formation" in Chapter 8. "Hedgehog/Gli control by ubiquitination/acetylation interplay" is reviewed by A. Gulino, L. D. Marcotullio, G. Canettieri, E. Desmaele, and I. Screpanti in Chapter 9. J. A. Buglino and M. D. Resh cover "Palmitoylation of Hedgehog proteins" in Chapter 10 and J. Jia's interest is "Phosphorylation regulation of Hedgehog signaling" (Chapter 11). In a similar vein, T. Kotani, in Chapter 12, reports on "Protein kinase A activity and Hedgehog signaling pathway." A further posttranslational aspect involves "Phosphorylation of Gli by cAMP-dependent protein kinase" by Y. Asaoka (Chapter 13).

On the topic of effective drugs, G. J. Huang, Z. He, and L.Ma, in Chapter 14, describe "ZFP932 suppresses cellular Hedgehog response and Patched1 transcription." Another drug, arsenic trioxide, is the subject of E. M. Beauchamp and A. Üren's Chapter 15: "A new era for an ancient drug: arsenic trioxide and Hedgehog signaling." F. C. Kelleher and R. McDermott report on "Aberrations and therapeutics involving the developmental pathway Hedgehog in pancreatic cancer" in Chapter 16, while N. C. Bambakidis and K. Onwuzulike have written on "Sonic Hedgehog signaling and potential therapeutic indications" in Chapter 17.

"Sonic Hedgehog on microparticles and neovascularization" is the subject of Chapter 18, written by R. Soleti and M. C. Martinez. S. Ghosh, A. Chakrabarti, and D. Mukhopadhyay, in Chapter 19, contribute ""Patch"-ing up the neurons: revival or enervation?" L. Yang, X. Su, and J. Xie rendered "Activation of Hedgehog pathway in gastrointestinal cancers" (Chapter 20), and "The role of Sonic Hedgehog as a regulator of gastric function and differentiation" is the work of R. Feng, C. Xiao, and Y. Zavros (Chapter 21). S. Fuchs, E. Dohle, and C. J. Kirkpatrick are the authors of "Sonic Hedgehog-mediated synergistic effects guiding angiogenesis and osteogenesis" (Chapter 22). Chapter 23, by G. P. Raju and D. Pham, is entitled "Hedgehog inhibition as an anti-cancer strategy."

The illustration on the book cover is a reproduction of Fig. 4.4 from Chapter 4.

The transition from manuscripts to assembled volume was facilitated by Sarah Latham of Elsevier and, initially, by Mary Ann Zimmerman.

Gerald Litwack
North Hollywood, CA

CHAPTER ONE

Hedgehog Signaling

Siu Wah Choy *and* Shuk Han Cheng

Contents

Abstract

The original *hedgehog* (*hh*) gene was found in *Drosophila* and named for the appearance of a mutant phenotype which causes an embryo to be covered with pointy denticles, thus resembling a hedgehog. The hedgehog family consists of sonic hedgehog (Shh), desert hedgehog (Dhh), and Indian hedgehog (Ihh). Shh is found in vertebrates and the best studied ligand of the hedgehog signaling pathway (Gilbert, 2000). It plays an important role in regulating vertebrate organogenesis, such as in the growth of digits on limbs and organization of the brain, and earlier studies also show that it is important during retinal development (for a review, see Amato *et al.*, 2004). Hedgehog expression drives waves of neurogenesis in animal retina, although genetic programs that control its expression are poorly elucidated. Recently, a novel transcriptional cascade which involves the atonal and Iroquois gene family was proposed in the regulation of hedgehog waves during vertebrate retinal development (Choy *et al.*, 2010). This chapter will focus on Shh by addressing its signaling mechanisms and roles in vertebrate eye development, as well as a novel finding in retinogenesis. © 2012 Elsevier Inc.

Department of Biology and Chemistry, City University of Hong Kong, Kowloon, Hong Kong, China

Vitamins and Hormones, Volume 88
ISSN 0083-6729, DOI: 10.1016/B978-0-12-394622-5.00001-8

I. Hedgehog Ligands

Cell–cell signaling is a crucial aspect of development and five well-known intercellular signal transduction pathways, including Wnt, TGF-β, Notch, receptor tyrosine kinase, and Hedgehog pathways, are well known in controlling the early development of most animals (Gerhart, 1999). To determine embryonic patterning and cell fate decisions, different signaling pathways activate or repress different target genes at distinct times and places in the embryo (Gerhart, 1999). Analysis of *Drosophila* development has been vital in elucidating the components and functions of these signaling pathways.

Sonic hedgehog (SHH) is the best established example of a morphogen—a molecule that diffuses to form a concentration gradient and has different effects on the cells of a developing embryo depending on its concentration. It is the most extensively characterized vertebrate homolog that is not only found in most animals, but often its developmental role is similar to in *Drosophila*. Shh is involved in a wide variety of embryonic events in both vertebrates and invertebrates. It plays a key role in regulating vertebrate organogenesis, such as in the growth of digits on limbs and organization of the brain. It can act as both a short-range, contact-dependent factor, and a long-range, diffusible morphogen (Chuang and McMahon, 1999; Johnson and Tabin, 1995). To qualify as a morphogen, a signal must fulfill two criteria: form a concentration gradient and elicit distinct responses at different concentrations. Cells encounter different concentrations of a morphogen according to their distance from the organizer that secretes it. Different transcription factors are therefore induced, committing the cells to different fates (Vincent and Briscoe, 2001). As molecular details have been elucidated, the idea has been well established in explaining a number of important embryonic events, from the initiation of anterior–posterior polarity in the *Drosophila* egg by a gradient of bicoid protein (Driever and Nusslein-Volhard, 1988) to the patterning of the ventral neural tube by different concentrations of SHH (reviewed in McMahon *et al.*, 2003).

What is SHH and how does it signal? SHH is proteolytically cleaved to produce two secreted proteins, a 19kDa N-terminal protein (N-SHH) that mediates all signaling activities and a 25kDa C-terminal protein (C-SHH) that possesses protease activity (Porter *et al.*, 1995) (Fig. 1.1). Autoprocessing of the Hedgehog protein induces the covalent attachment of cholesterol onto the carboxy terminus of the N-terminal domain. The cholesterol moiety is thought to direct Hedgehog protein traffic in the secretory cell. The N-terminal domain retains all known signaling capabilities while the C-terminal domain is responsible for intramolecular precursor processing, acting as a cholesterol transferase. The attachment of cholesterol provides

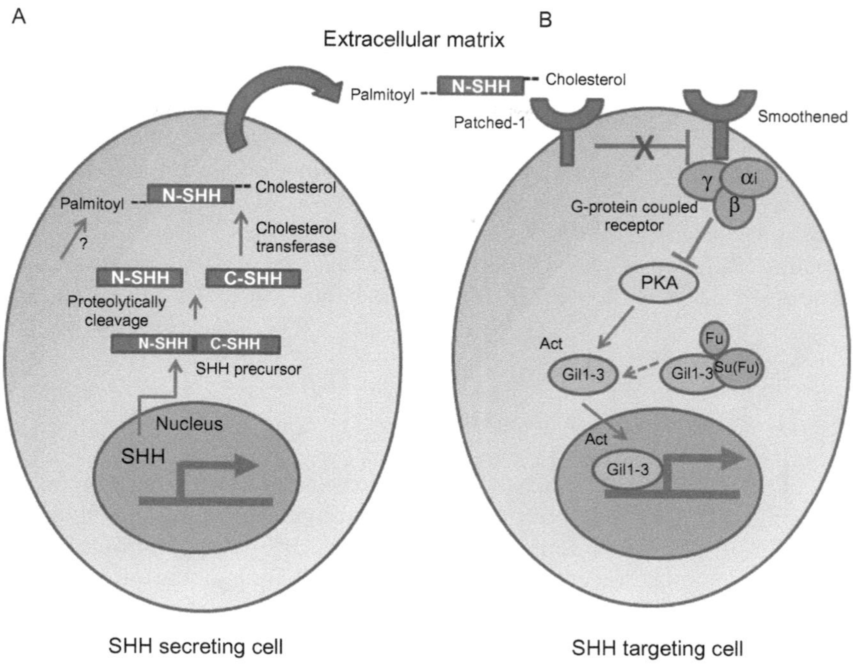

Figure 1.1 Schematic representation of the sonic hedgehog (SHH) signaling pathway. Only components identified in vertebrates are shown. (A) the SHH precursor is processed to proteolytically cleavage to produce two secreted proteins, N-SHH that mediates all signaling activities, and C-SHH, which possesses cholesterol transferase activity. Two covalent modifications of the N-SHH morphogen are introduced: a cholesterol group is added by C-SHH at the C-terminal and a palmitoyl moiety is added by an unknown component (?) at the N-terminal. The modified N-SHH is secreted into the extracellular matrix and bind to the target cells. (B) In the presence of the N-SHH ligand, binding of SHH to Patch-1 receptor prevents the inhibition of Smo, which activates the transduction pathway. Smo is no longer inhibited and activates downstream events through the inhibition of PKA. Zinc-finger transcription factors of the Gli (Glis 1–3) family is released from the intracellular complex and translocated to the nucleus as a transcriptional activator. (For color version of this figure, the reader is referred to the Web version of this chapter.)

specific properties on Hedgehogs that are essential for their complex signaling activities. For example, Hedgehogs in *Drosophila* show polarized distribution within the cell and present in rafts (Rietveld *et al.*, 1999), and cholesterol-rich rafts have been implicated in directed protein trafficking within the secretory pathway (Simons and Ikonen, 1997). Thus, the cholesterol moiety is involved in directing the intracellular transport of SHH within epithelia. Apart from the addition of a cholesterol group to the C-terminal of N-SHH, it is further modified by the addition of a palmitoyl group to the N-terminal of N-SHH (Pepinsky *et al.*, 1998).

This N-terminal palmitate modification, introduced in *Drosophila* by a different enzyme (Chamoun *et al.*, 2001), increases the inductive potency of SHH, both *in vitro* and *in vivo* (Kohtz *et al.*, 2001).

N-SHH has been described as a morphogen responsible for the key early patterning processes, such as controlling of left–right asymmetry, dorso–ventral patterning of the central nervous system (CNS) and somites, patterning of the limbs, as well as in some aspects of inducing organogenesis (Hammerschmidt *et al.*, 1997). Despite these various roles of SHH, the signaling pathway is well conserved from invertebrates to mammals, and progress in defining the SHH pathway has been extensively reviewed.

II. Hedgehog Signaling

The Shh signaling pathway involves two transmembrane proteins on receiving cell, Patched (Ptc), a 12-transmembrane protein, and Smoothened (Smo), a 7-transmembrane protein with a topology reminiscent of G-protein-coupled receptors, which is the signaling component of the SHH-receptor complex (Murone *et al.*, 1999). Ptc binds Shh, whereas Smo acts as a signal transducer. In the absence of ligands, Ptc interacts with and inhibits Smo. This inhibition activates a transcriptional repressor (e.g., Gli in vertebrates). In the presence of ligands, the binding of Shh to Ptc prevents the inhibition of Smo, activating the transduction pathway. The interaction of Ptc and Smo is altered and Smo is no longer inhibited. The Gli protein may then enter the nucleus and function as a transcriptional activator (Fig. 1.1). In mammals, two isoforms of Ptc are encoded by *Ptc1* and *Ptc2*, although only the former is definitely involved in Hh signaling. It is confined to target cells and upregulated in response to Hh. Ptc-2 is coexpressed with Hh, but its transcription is independent of pathway activation (St-Jacques *et al.*, 1998). As expected from the model, mice that lack Ptc-1 are lacking the inhibition of Smo activity and have constitutively activated Shh response genes in target tissues (Goodrich *et al.*, 1997). To regulate the SHH signaling pathway during the development of the vertebrate CNS, either inhibition of Gli proteins (Hammerschmidt and McMahon, 1998) or expression of a constitutively active form of Smo (Hynes *et al.*, 2000) is sufficient to trigger some actions of SHH. Analysis of Smo null mutant mice reveals phenotypes identical to Shh mutants and additional phenotypes not associated with Shh. These results indicate that Smo activity during development is likely not limited to activation of the Shh pathway (Zhang and Yang, 2001).

In the nucleus of a responding cell, zinc-finger transcription factors of the Gli family (GLI1–3) act at the last step of the SHH-signal-transduction pathway (Hynes *et al.*, 1997), although it is still unclear whether GLI proteins mediate all aspects of SHH signaling during development (Litintung and Chiang, 2000). GLI-mediated transcription of SHH target genes is regulated, at least in part by protein kinase A (PKA) (Fig. 1.1) (Kaesler *et al.*, 2000), an inhibitor of SHH activity (Hammerschmidt *et al.*, 1996). Hedgehog signaling is aberrantly activated via GLI in cancer cells, such as medulloblastoma; basal cell carcinoma; lung, esophageal, gastric, pancreatic, and breast cancers; and other tumors (Katoh and Katoh, 2009). However, other studies have demonstrated that mouse *Gli1* is not required for embryonic development and normal Shh signaling, unless one copy of *Gli2* is defective, which indicates that the GLI1-3 complex is required in the Shh signaling pathway (Bai *et al.*, 2002). Cytoplasmic–nuclear shuttling of GLI-3 proteins also depends on the complex formed by the kinesin-like cytoplasmic protein, serine–threonine kinase Fused (Murone *et al.*, 1999), and Suppressor of Fused (Pearse *et al.*, 1999) (Fig. 1.1), although this mechanism is not yet fully understood. Further research is also required to link Smo activation to the release of transcriptionally active forms of the GLI proteins. Interestingly, in *Drosophila*, Hedgehog-independent Ptc and Ptc-independent Hedgehog functions have also been observed (Goodrich and Scott, 1998). Thus, the Hedgehog signaling pathway may be more complicated than the canonical model suggests, as it is likely that pathway components interact with additional ligands and receptors that have yet to be identified.

III. Hedgehog Signaling in Zebrafish Eye Vesicle Patterning

The Shh signaling pathway functions throughout development. Shh is involved in the determination of cell fate and embryonic patterning during early vertebrate development. An example of this activity is the patterning and formation of the retinal cells during eye development in zebrafish. A role for Hedgehog (Hh) signaling in the development of the vertebrate eye was predicted based upon its known roles in *Drosophila* eye development, and the principle of evolutionary conservation of developmental function (Cho *et al.*, 2000; Heberlein *et al.*, 1995). Extensive experimental work has proven that this prediction is correct and it turns out that the Hh signaling system has been used for a variety of developmental purposes, such as the formation and differentiation of the vertebrate eye. Recently, many of the key experiments which have uncovered these roles had used the zebrafish as a genetic and organismal tool. The zebrafish has emerged as an outstanding model for the understanding of eye development, with its large, rapidly developing eyes, amenability to pharmacological and genetic manipulation, and suitability for a variety of imaging and gene expression studies (Easter and Malicki, 2002; Neumann, 2001).

In vertebrates, the eye is a bilateral organ that originates from a single field positioned in the anterior portion of the neural plate. Anterior migration of diencephalic precursor cells along the midline separates the primordial eye field into two regions (Varga *et al.*, 1999); the proliferation and evagination of which give rise to two visible optic primordia, the optic vesicles. This early event establishes the proximodistal (PD) axis of the optic vesicle: the distal-most region will invaginate, forming the optic cup and then the retina, while the proximal region will differentiate to the optic stalk. Proliferation of the multipotential retinal progenitor cells, as well as the generation and differentiation of all these diverse cell types, is regulated by a combination of cell-intrinsic factors and cell-extrinsic (extracellular) cues (Livesey and Cepko, 2001). The Hh protein is demonstrated as one of these very important extracellular cues in controlling vertebrate retinogenesis. The requirement for a midline source of Hh at early neurulation for separating and patterning the eye fields along the PD axis provides the first indications about Hh signaling pathways in vertebrate eye development (Ekker *et al.*, 1995; Macdonald *et al.*, 1995).

Genetic ablation of Shh in mice leads to severe defects in the anterior neural tube and the development of cyclopia (i.e., the presence of an unseparated optic vesicle) (Chiang *et al.*, 1996). Zebrafish embryos that lack the Hh-expressing cells in midline tissues also develop a single anterior eye rather than two lateral eyes, whereas overexpressing the shh gene in zebrafish results in the development of lateral optical structures with excess in the optic stalk at the expense of the neural retina (Cohen and Shiota, 2002; Incardona and Roelink, 2000; Ingham and McMahon, 2001; Roessler and Muenke, 2001). The cyclopic-eye phenotype is explained by the observation that Shh activity regulates the spatial expression of genes that encode Pax6 and Pax2, which normally demarcate the distal (retina and lens) and proximal (optic disk and optic stalk) optic primordium, respectively (Fig. 1.2). Gain-of-function experiments led to the conclusion that *Shh* promotes proximal and represses distal fates by regulating the expression of *Pax* genes. In zebrafish and *Xenopus*, *Shh* overexpression promotes the expression of *Pax2* and represses the expression of *Pax6* (Fig. 1.2). In contrast, the absence of SHH and tiggy-winkle hedgehog (twhh) in cyclops zebrafish mutants correlates with a severe reduction in *Pax2* concentration and a corresponding increase in *Pax6* across the midline, which leads to the fusion of the two retinal fields (Ekker *et al.*, 1995; Macdonald *et al.*, 1995; Perron *et al.*, 2003). By studying mouse embryos at early developmental stages, an expansion of the Pax2 expression domain was observed in the *Pax6−/−* mutant, whereas an expansion of the Pax6 expression domain was observed in the *Pax2−/−* embryo. As *Pax2* and *Pax6* transcriptionally repress each other, a precise boundary between the retina and the optic stalk is maintained by the Shh signaling pathway which controls the expression boundary of *Pax2* and *Pax6* (Schwarz *et al.*, 2000) (Fig. 1.2). Loss of Pax2

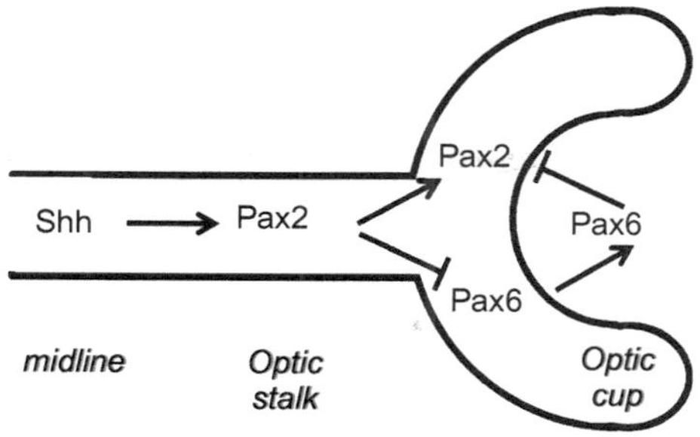

Figure 1.2 Molecular cascade involved in the partitioning of the visual system in the eye vesicle cup. Pax2 is induced in the optic stalk by midline signals, such as Shh. The reciprocal inhibition between Pax2 and Pax6 is required in maintaining the optic stalk/optic cup boundary during eye development. In this model, Pax2 in the optic stalk can repress Pax6 expression and enhance its own expression. Conversely, Pax6 in the optic cup can repress Pax2 expression and enhance its own expression.

function leads to optic nerve hypoplasia and colobomas in human patients, as well as in a number of animal model systems (Macdonald *et al.*, 1997; Otteson *et al.*, 1998; Sanyanusin *et al.*, 1995; Schwarz *et al.*, 2000; Torres *et al.*, 1996). In mice, the Shh pathway cooperates with the bone morphogenetic protein (BMP) pathway to regulate Pax2 expression (Sehgal *et al.*, 2009). By utilizing the Smo conditional knockout (CKO) mouse line, it was found that Bmp4 expression is greatly upregulated in the optic vesicle, which suggests that Shh is involved in the formation of the ventral optic cup by limiting Bmp4 expression to the dorsal domain (Zhao *et al.*, 2010).

IV. Hedgehog Signaling in Zebrafish Photoreceptor Differentiation

Zebrafish eyes, like the complex structure of vertebrate camera eyes, are composed of a thin stratified neural tissue retina and a single optic lens. The retina contains three nuclear layers and two plexiform layers. The inner most nuclear layer is called the ganglion cell layer (GCL), which is composed of retinal ganglion cells (RGCs) and a small number of amacrine cells. The middle nuclear layer is called the inner nuclear layer (INL), which consists of interneurons and glial cells, named the amacrine, bipolar, horizontal, and Müller glial cells. The bipolar cells can also be categorized into cone bipolar and rod bipolar. The outermost layer is called the outer nuclear layer (ONL), which consists of cone and rod photoreceptors. The retinal pigment epithelium (RPE) is a thin monolayer of cells that underlies the retina and maintains the function and development of photoreceptors

(Strauss, 2005). The RPE is also important for the normal development of photoreceptor cells and regulates the transport of various nutrients, ions, and metabolic waste between the subretinal space and choroid, thus maintaining a proper physiological environment for the functioning of photoreceptors (Jablonski *et al.*, 2000; Stenkamp *et al.*, 2000).

In *Drosophila*, the expression of Hh coordinates retinal development by propagating a wave of photoreceptor differentiation across the eye primordium. In zebrafish, two of the three *hh* genes, that is, *shh* and *twhh*, perform similar functions during eye development. Stenkamp *et al.* (2000) intensively examined their expression profiles in zebrafish embryos. The results showed that both *shh* and *twhh* are expressed in embryonic zebrafish RPE, initially in a discrete ventral patch which then expands outward in advance of an expanding wave of photoreceptor recruitment in the subjacent neural retina. The expression of *shh* and *twhh* in the RPE begins at about 45 hours postfertilization (hpf). Both genes in the RPE are restrictively expressed to a small region in the ventronasal quadrant of the developing eye. The expression wave later centrifugally spreads at 54 hpf, embryos in a larger patch of RPE, but not throughout the eye. By 81 hpf, *shh* and *twhh* are expressed throughout the RPE. Their spatiotemporal expression patterns in the RPE are similar, and slightly beyond the limit of photoreceptor lamination where the photoreceptor layer has not yet differentiated. Expression of *shh* and *twhh* therefore occurs immediately prior to, or concomitant with, the first morphological manifestation of photoreceptor differentiation. Hh protein can be immunocytochemically detected in the RPE and subretinal space, which suggests that it is secreted toward the retina. It is proposed that Hh genes may play a role in propagating photoreceptor differentiation across the developing eye of a zebrafish (Stenkamp *et al.*, 2000).

Zebrafish strains known to have mutations in Hh signaling pathway genes exhibit retardation of photoreceptor differentiation. *You-too* (*yotty17*) is a mutation in the *gli-2* gene (Karlstrom *et al.*, 1999), which encodes for a transcription factor known to mediate Hh signaling. *sonic-you* (*syu14*) is a deletion of the *shh* (Schauerte *et al.*, 1998), and the slow muscle-omitted *(smu)* mutant, which is a functional null mutation in the Smo gene (Varga *et al.*, 2001). Smo encodes a critical component of the Hh signal transduction pathway. When Hh binds to the Ptc receptor, inhibition from Ptc on Smo is relaxed, and intracellular signals are generated (Fig. 1.1). Both *yotty17* and *syu14* mutants showed differences in photoreceptor differentiation compared to wild-type embryos at 75 hpf while the *syu14* phenotype was particularly striking (Stenkamp *et al.*, 2000). Recently, it was found that Shh activity is necessary for the maintenance and survival of cone photoreceptor cells in adult zebrafish retina (Stenkamp *et al.*, 2008). *Syu14* mutants even lost the lamination layers in retinas; these retinas contained small, dense cells that appeared to be apoptotic. The *yotty17* mutants showed more variable and less extreme phenotypes with normal lamination in the retina

(Stenkamp *et al.*, 2000). Similarly, a large fraction of the *smu* mutants also failed to develop normal retinal lamination and showed reduced expression of photoreceptor markers (Stenkamp and Frey, 2003).

As expected, co-injection of a cocktail of antisense oligonucleotides of zebrafish shh and twhh arrests the progression of rod and cone photoreceptor differentiation. However, neither anti-shh nor anti-twhh alone is sufficient to mimic the effects of a combined oligo injection, which suggests the functional similarity of *shh* and *twhh* in the RPE (Stenkamp *et al.*, 2000). Conversely, overexpression of Shh to rat retinal increases the proportion of photoreceptors (Levine *et al.*, 1997), which also suggests a possible role for Hh in mammalian photoreceptor differentiation. The question has been further analyzed by Stenkamp *et al.* (2002), who have undertaken a careful analysis of retinal gene expression in zebrafish Shh mutants (*syu*). No abnormalities were found in the expression of *neuroD*, *crx*, and *rx2*. All of these genes are implicated in photoreceptor differentiation (Stenkamp *et al.*, 2002). Instead, the expression of *rx1*, which is expressed in the photoreceptor layer and thought to be involved in photoreceptor differentiation (Chen and Cepko, 2002), was clearly reduced in *syu* mutants (Stenkamp *et al.*, 2002). This gene is thus a good candidate to mediate the effect of reduced Hh signaling in photoreceptors.

Aside from being important for photoreceptor differentiation, Hh signaling in the RPE is essential for differentiation of the RPE itself. Treatment with anti-Shh antibodies in chick embryos showed that Shh is required for ventral RPE formation, since there is a lack of pigmentation in the ventral eye (Zhang and Yang, 2001). The same conclusion was drawn for *Xenopus*. Blocking the Hh cascade in developing embryos with cyclopamine has a severe consequence on RPE formation, which causes a reduction in pigmentation and loss of expression of RPE differentiation markers. Further, reduced Hh signaling resulted in cell death within the developing retina, which indicates that it is required in retinal cell proliferation and retinal cell survival (Stenkamp *et al.*, 2002). The timing of this cell death is consistent with an important role for the Hh signal, specifically from the RPE, in promoting retinal cell survival. Cell death is initially the highest in neuroepithelial cells and photoreceptors and then spreads to other retinal layers (Stenkamp *et al.*, 2002). A more recent finding showed that loss of Shh signaling leads to p53-dependent apoptosis in the zebrafish developing nervous system and retina. Moreover, Shh-mediated control of p53 activity is required for proliferation and cell cycle exit of retinal cells as well as differentiation of amacrine cells and photoreceptors (Prykhozhij, 2010).

With the conserved functions of Hh signaling between *Drosophila* and zebrafish in eye development, the differentiation of photoreceptors may be conserved. In *Drosophila*, Hh expression drives the progression of the morphogenetic furrow across the developing eye by synchronizing terminal mitoses in a linear array of neuroepithelial cells and stimulating subsequent photoreceptor differentiation. These newly differentiated photoreceptors then express Hh, and the entire process is repeated for the array of undifferentiated cells thus propagating the morphogenetic furrow (Heberlein and Moses, 1995). In the zebrafish, *shh* and *twhh* similarly coordinate the progression of photoreceptor differentiation in the developing eye by driving retinal progenitor cells through a terminal differentiation (Levine *et al.*, 1997) and stimulating subsequent photoreceptor differentiation (Jensen and Wallace, 1997; Levine *et al.*, 1997; Stenkamp *et al.*, 2000).

The importance of the RPE for photoreceptor development and survival has been recognized for a long time. The absence of RPE *in vivo* results in failure of retinal development followed by degeneration of the eye (Raymond and Jackson, 1995). It was proposed that identification of soluble factors in RPE-conditioned medium or the interphotoreceptor matrix that have photoreceptor survival and/or differentiation-promoting activities in *in vitro* models could help in photoreceptor development (Sheedlo and Turner, 1996; Watanabe and Raff, 1992). Hh proteins may now be considered as candidates for some of these activities.

V. Hedgehog Signaling in Zebrafish Inner Nuclear Layer

In transgenic zebrafish, green fluorescent protein under the control of the *shh* promoter *(shh-GFP)* reveals that in addition to the wave in the GCL, *shh* expression also spreads into the amacrine cells of the INL of the zebrafish retina (Stenkamp and Frey, 2003). This expression has been verified by *in situ* hybridization, and the spatiotemporal pattern of expression mimics that of the slightly earlier wave of *hh* gene expression in the GCL. Like expression in the RPE, there is a spatiotemporal gradient of expression that resembles the pattern of retinal cellular differentiation.

Besides the importance of *shh* signaling in retina development, the basic helix-loop-helix (bHLH) transcription factor *atonal homolog 5* (*ath5*; *atoh7*—Zebrafish Information Network) is also required for the generation of RGCs in the mouse and the zebrafish retinas (Brown *et al.*, 2001; Kay *et al.*, 2001). In the zebrafish, *ath5* mutants completely lack RGCs but have a thin GCL that is composed of "misplaced" amacrine cells from the INL (Kay *et al.*, 2001). *ath5* is expressed in a wave that precedes the wave of RGC differentiation (Masai *et al.*, 2000), reminiscent of the way in which *Drosophila atonal* is expressed during eye development (Dominguez, 1999). Interestingly, the onset and spread of Hh signaling in the amacrine cell of the INL is independent of that in the ganglion cells (GCs), and the expression of the retinal transcription factor *ath5*. The wave of Hh in amacrine cells could still be observed in the *lakritz/ath5* mutant, which lacks GCs (Kay *et al.*, 2001; Shkumatava *et al.*, 2004).

It is proven that INL expressed Hh is important for photoreceptor differentiation. Shkumatava *et al.* (2004) created mosaic embryos, which consisted of a combination of *syu−/−* cells and wild-type *shh-GFP* cells. The goal was to determine whether failed photoreceptor differentiation was associated with the absence of *shh* expression in nearby RPE or amacrine cells from the INL. The retinal regions that displayed normal photoreceptor differentiation were radially contiguous with the regions that contained wild-type *shh-GFP* amacrine cells (Shkumatava *et al.*, 2004). However, in regions where wild-type *shh-GFP* RPE cells were located, photoreceptors were not found. This suggests that the amacrine cell Hh signal, rather than the RPE Hh signal, promoted photoreceptor differentiation. The authors proposed several alternative explanations: (1) the amacrine cell Hh signal is required for retinal cell survival, while the RPE signal is needed for photoreceptor differentiation, (2) Hh signals from a basal (amacrine) as well as an apical (RPE) source are needed for photoreceptor differentiation, and (3) the total amount of Hh signal available to the photoreceptor layer must exceed a certain threshold before differentiation can take place. The significance of gradients and thresholds for Hh signaling is exemplified during cell determination and differentiation events within the embryonic spinal cord (Ericson *et al.*, 1997). These latter possibilities would be difficult to address by using mosaic embryos because the twhh gene is expressed in both wild-type and *syu−/−* RPEs, and its functions overlap with those of *shh* (Neumann and Nuesslein-Volhard, 2000; Stenkamp *et al.*, 2000).

Finally, it was concluded that Hh signaling is clearly required for photoreceptor differentiation in zebrafish; however, the cellular source of this important signal remains unclear. Possible sources include the RPE, amacrine cells from the INL, or a combination of the two.

VI. Hedgehog Signaling in Zebrafish Ganglion Cell Differentiation

A. Shh induces its own expression in the GCL and drives the wave of GC differentiation

So far, we have described the role of Hh signaling in the differentiation of the ONL and INL in zebrafish. A wealth of data suggests that Hh signaling also controls glial specification throughout the retinogenesis in vertebrates. RGCs are the first neurons to be born, followed by interneurons and photoreceptors, and lastly, by Müller glial cells. RGC differentiation roughly begins in the central part of the retina and proceeds toward the periphery along a central to peripheral pattern (Peters, 2002). Both *shh* and *twhh* are also found to be expressed in GCs through the use of a transgenic zebrafish, *shh-GFP* reporter line, and was subsequently verified by *in situ* hybridization (Stenkamp and Frey, 2003). Immunocytochemical techniques also revealed Hh protein expression in GCs at the time of GC differentiation (Stenkamp and Frey, 2003). Like expression in the RPE, there is a wave-like expression pattern that resembles the GC differentiation pattern. However, hh genes express in GCs at a much earlier time, which is at about 28 hpf, just after the first GCs withdraw from the cell cycle (Neumann and Nuesslein-Volhard, 2000).

The wave of GC differentiation is reminiscent of the neurogenic wave that occurs during *Drosophila* eye differentiation. It has been shown that Hh signaling controls this process in a similar way in both *Drosophila* and zebrafish (Neumann and Nuesslein-Volhard, 2000). It was found that *shh-GFP* expression is initiated in the first RGCs, but then fails to spread further in zebrafish *syu* mutants, in which *shh* gene is disrupted (Neumann and Nuesslein-Volhard, 2000). This is very similar to the *Drosophila* eye, where Hh signaling is required for the spread, but not the induction, of the first Hh-expressing neurons, which instead requires *decapentaplegic* (*dpp*) signaling (Treisman and Heberlein, 1998). This study further showed that Shh is both necessary and sufficient to control a wave of its own expression that sweeps through the GCL. This is strikingly similar to the function of Hh in controlling the morphogenetic furrow of the *Drosophila* eye (Heberlein *et al.*, 1993, 1995; Zhou *et al.*, 1993). In contrast to *Drosophila*, neurogenesis in the eye is only partially dependent on Shh in the zebrafish retina. As in several other Hh genes, such as *twhh* which is expressed in the GCL, and that this expression is detectable in *syu* eyes in the zebrafish (Neumann and Nuesslein-Volhard, 2000), it is possible that *twhh* might be responsible for the Shh-independent neurogenesis. To further address this issue, Neumann *et al.* (1999) treated zebrafish embryos with cyclopamine, which inhibits signaling by both Shh and other Hh family members (Cooper *et al.*, 1998; Incardona *et al.*, 1998; Neumann *et al.*, 1999). Treatment of embryos with cyclopamine from 26 to 52 hpf completely blocked both the spread of *shh-GFP* and also neurogenesis, which indicated that several Hh genes cooperated to drive the wave of neurogenesis in the zebrafish retina. In *Drosophila*, Hh is continuously required for furrow progression. Further treatment of embryos at a later stage from 30 to 52 hpf also resulted in the blockage of *shh-GFP* and neurogenesis, which indicated a continuous requirement for Hh signaling for the neurogenic wave in the zebrafish retina. With all the evidence that show the conserved role played by Hh signaling in retinal differentiation between flies and fish, this suggests that Hh is already used to pattern a primordial eye structure before vertebrate and invertebrate lineages diverge, and thus supports a common evolutionary origin of the animal eye.

B. Conserved molecular genetics in zebrafish and *Drosophila* retinogenesis

1. Shh interacts with Atonal in ganglion cell neurogenesis

The conserved wave-like retinogenesis progression in both zebrafish and *Drosophila* is driven by *sonic hedgehog*. In *Drosophila*, Hh is secreted by differentiating ommatidial cells and induces ommatidial assembly in more anterior cells. Some of the newly formed ommatidial cells become Hh secreting cells and hence, the furrow advances (Ma *et al.*, 1993; Strutt and Mlodzik, 1997). Hh controls the progression of furrow by induction of at least 2 signals. The downstream of *hedgehog* pathway *dpp* may act as a long-range signaling molecule on anterior undifferentiated cells and induces the expression of Hairy (Chanut and Heberlein, 1997; Heberlein *et al.*, 1993). The second signal is the short-range activation of the proneural gene *atonal* (*ato*) via the activation of serine–threonine kinase *Raf* (Greenwood and Struhl, 1999). As the neurogenesis wave moves, rows of evenly spaced cells are selected to become photoreceptor cells (Cagan, 1997; Freeman, 1996; Tomlinson and Ready, 1987). In the eye imaginal disk, *ato* is activated along the anterior edge of the furrow in response to *hedgehog* signaling (reviewed in Treisman and Heberlein, 1998). *Ato* ortholog in zebrafish, *ath5*, is expressed as a wave-like manner during retinogenesis and necessary for the propagation of neurogenesis. Complete loss of *ath5* in the zebrafish *lakritz* mutant embryos disrupts retinal ganglion cell development (Kay *et al.*, 2001). *ath5* transiently sweeps across the differentiating retinoblasts, and its expression is maintained in the periphery of the retina, where cells keep differentiating (Masai *et al.*, 2000). The waves of *ath5* and *shh* expression, which mirror the earliest wave of neurogenesis, have been observed in zebrafish retina, which suggest that the molecular mechanisms that drive the neurogenic wave have been conserved from flies to fish (Masai *et al.*, 2000; Neumann and Nuesslein-Volhard, 2000).

In zebrafish, since the *shh* wave normally propagates in the INL but not in the GCL of the *ath5* mutant retina which completely lacks RGCs (Kay *et al.*, 2001), the wave of *shh* expression in the INL is independent of the *ath5* expression in the GCL (Shkumatava *et al.*, 2004). Interestingly, neuronal differentiation is initiated at the ventronasal patch in the *shh* mutant retina whereas the spread of neuronal differentiation does not occur, which suggests that the onset and propagation of *shh* expression and neurogenesis employ different mechanisms (Stenkamp *et al.*, 2002). However, the ability of vertebrate *shh* and *ath5* to affect each other's expression has not been well defined and recently challenged, although it was defined in *Drosophila* that *hh* signaling is required in *ato* activation in eye development.

In flies, Hh triggers expression of *ato* in progenitor cells ahead of the furrow; *ato*, in turn, induces the formation of the next group of photoreceptors, which secrete their own Hh, thus a self-propagating wave spreads by "sequential induction." Loss of *ato* function blocks photoreceptor formation, thereby removing the cellular source of Hh and turning to case the neurogenesis wave (Jarman *et al.*, 1995). By analogy with the mechanism in *Drosophila*, it has been hypothesized that Shh signals derived from newborn RGCs might drive the *ath5* wave. In support of this idea, *shh* is expressed by newborn RGCs, and *shh* expression spreads across the zebrafish retina in what appears to be a self-propagating wave (Neumann and Nuesslein-Volhard, 2000; Zhang and Yang, 2001). Together, these findings led to a "sequential-induction" model that Hh molecules derived from RGCs should drive progression of the *ath5* wave and hence, the wave of RGC differentiation. However, Kay *et al.* (2005) found that neither RGCs nor the *ath5* gene itself are essential for driving the *ath5* wave in zebrafish. They demonstrated that the spread of *ath5* expression is normal in *lak* mutant which causes complete elimination of RGCs. This indicates that the induction of *ath5* expression is independent of the *hh* signaling derived from RGCs. Later on, they found that the midline-derived, but not the retinal-derived Shh, acts before neurogenesis regulates the timing of *ath5* expression and RGC differentiation. It was shown that the spatial pattern of the *ath5* wave-like expression was unaffected in the *shh* null mutant *syut4*, but the timing of wave progression was dramatically delayed by almost 10h. Thus, loss of *shh* function did not block the *ath5* wave but had a substantial effect on the time of neurogenic activation (Kay *et al.*, 2005). In order to determine whether this phenotype results from loss of midline- or retina-derived Shh, the time window during which *shh* is necessary for timely wave progression was defined by employing the drug cyclopamine, a small molecule inhibitor of the Hh receptor Smo. The early treated cyclopamine at 13hpf (before *shh* was expressed in the retina) delayed *ath5* expression similar to that seen in *syu* mutants, whereas blocking retina-derived Hh signaling by the drug treatment at a later stage indicated no effect on the progression of the *ath5* wave. Together, these results suggest that early, midline-derived *shh* signaling is required to ensure the subsequent timely expression of *ath5* by retinoblasts (Kay *et al.*, 2005).

2. A cascade of Iroquois genes controls *shh* expression during retinogenesis

a. Functional roles of Iroquois genes in eye development The Iroquois proteins belong to the three aminoacid loop extension (TALE) superclass of homeodomain proteins and are characterized by the presence, in addition to the homeodomain, of a conserved specific domain of an unknown function, named "IRO box" (Mukherjee and Bürglin, 2007). In addition to the protein structure, the genomic organization of the *Iroquois* genes is also apparently conserved. In *Drosophila*, three *Irx* genes—*araucan*, *caupolican*, and *mirror*—have been identified, and form a gene complex that spans about 130 kb in the genome, Iro-C. In humans and mice, a total of six members of the *Irx* genes were found and they are organized into two clusters. Orthologs of these genes have been found in zebrafish in which 11 *Irx* genes are organized into four clusters (Feijóo *et al.*, 2004).

In *Drosophila*, all of the *Iroquois* have essentially identical expression patterns during eye development, which suggest that they may share the same *cis* element within the *Iro-C*. In mice and zebrafish, the clustered *Irx* genes are expressed in GCL, but with different temporal expression patterns. We previously showed that zebrafish *irx1a* is first found in the ventronasal retina adjacent to the choroid fissure at 24hpf and gradually spreads to the dorsal parts of the retina, which is reminiscent of the differentiation of RGCs (Cheng *et al.*, 2006). The expression pattern of *irx2a* was found to be similar to that of *irx1a* during retinogenesis except that the *irx2a* expression commenced at 30hpf, which is later than that of *irx1a*. The expression of *irx4a* as well as the other *Iroquois* genes was not detectable in early retinogenesis before 36hpf, but these genes are highly expressed at 48 hpf during the later stages of RGC neurogenesis (Choy *et al.*, 2010). The divergence in the temporal expression pattern between *Irx* clustered genes indicates that in contrast to the *Drosophila Iro* genes, the zebrafish *Irx* genes are not completely redundant in their functions in eye development.

In both *Drosophila* and vertebrates, the *Iro* genes are required at early stages of development in order to define large territories. Later in development, the *Iro* genes function again to subdivide those territories into smaller domains (to review, see Cavodeassi *et al.*, 2001). The function of *Iroquois* genes in defining the equator in the dorsal eye disk is revealed by *mirrorP1* insertion in the *Drosophila* mutant. It was found that *mirrorP1* is expressed in the dorsal half of the eye and *mirr*-clones induce ectopic equators at the boundary of *mirr*-expressing and non-expressing cells (McNeill *et al.*, 1997). Moreover, *mirror, ara*, and *caup* are also expressed in the dorsal cells of the eye disk. Generalized misexpression of *ara, caup*, or *mirr* throughout the eye disk leads to the loss-of-eye phenotypes (Cavodeassi *et al.*, 1999; Dominguez and de Celis, 1998; McNeill *et al.*, 1997). Genetic studies revealed that *Iroquois* genes are required for the specification of the ommatidia at the dorsal rim (Tomlinson, 2003; Wernet *et al.*, 2003) and the photoreceptor cells (Mazzoni *et al.*, 2008) at later stages of eye development. The *Iroquois* expressing cells border in the eye disk is also required in the activation of the Hh morphogenetic furrow. The Hh expression was initiated normally, but its levels fail to increase at the most posterior domain in the eye disk (Cavodeassi *et al.*, 1999) (Fig. 1.3).

Although *Iroquois* genes were also identified in vertebrates soon after characterization in *Drosophila*, their functions in vertebrate eye development remains largely unclear at that moment. This was the case until 2003, when a study showed that *Irx4* may contribute to the definition of axonal path finding in chick retinas by regulating the *Slit1* expression (Jin *et al.*, 2003). In retinas, Slit1 positively acts on the growth of the retinal axons. Misexpression of *Irx4* represses the expression of *Slit1*, while the dominant negative form of Irx4 protein could upregulate the *Slit1* expression. We previously demonstrated that mouse *Irx5* plays a unique role in the differentiation of specific bipolar neurons (Cheng *et al.*, 2005). In Irx5-deficient mice, defects were observed in the expression of some, but not all, immunohistological markers that define mature Types 2 and 3 OFF cone bipolar

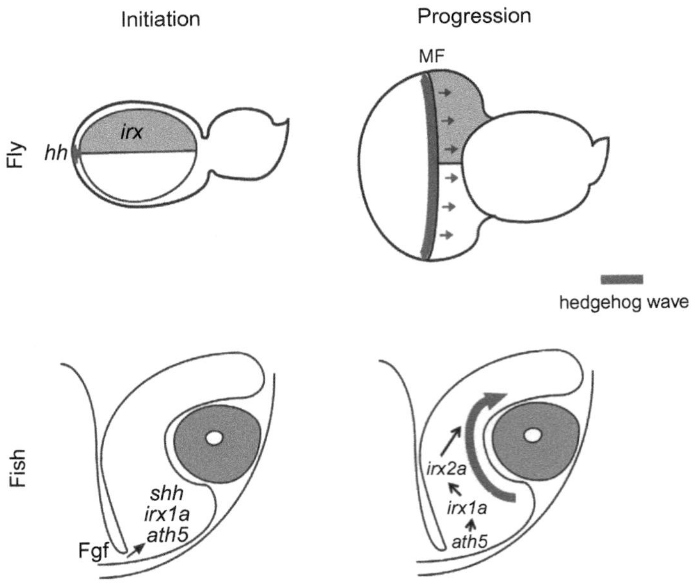

Figure 1.3 The schematic model for the roles of Iroquois genes in Hedgehog wave propagation. In the *Drosophila* eye disk, *Irx* genes expression border is required for the upregulation of Hh in the posterior region, which is sufficient to drive the progression of morphogenetic furrow. In zebrafish retinas, expression of *irx1a*, *ath5*, and *shh* are controlled by *Fgf* signaling. *ath5* and *irx* genes are not required for the initial expression of *shh*, but for the progression of *shh* from the ventronasal to the dorsal during retinogenesis. (For color version of this figure, the reader is referred to the Web version of this chapter.)

cells, which indicate a role for Irx5 in bipolar cell differentiation (Cheng *et al.*, 2005). In contrast, *Drosophila Iro* genes define the photoreceptor cell identity as controlling the receptor expression, a type of *rhodopsin*, in photoreceptor cells (Mazzoni *et al.*, 2008). Although we found that the terminal differentiation of bipolar cells is affected in mice with mutant *Irx5* (Cheng *et al.*, 2005), it will be of interest to study the cone *opsin* expression in this and other *Irx* mutants to test whether these genes are also involved in regulating the expression of opsins in vertebrates.

In zebrafish, all the *irxAa* clustered genes are expressed in the retina (Choy *et al.*, 2010). We recently demonstrated that *irx1a* and *irx2a* are required for the propagation of neurogenesis in developing zebrafish retinas. Knockdown of *irx1a* (*irx1a* morphants) resulted in small eyes with ectopic cell death, defective optic nerve and loss of lamination in the zebrafish retina. *irx1a* is also required for the propagation, but not the onset of neurogenesis. In all three retinal nuclear layers of the *irx1a* morphant retinas, a small number of differentiated neurons could be found in the ventronasal region, where retinogenesis is initiated (Cheng *et al.*, 2006). *irx1a* morphants exhibit retinal defects strikingly similar to those of *shh* mutants and that propagation of *shh* waves is abolished in the retina of *irx1a* morphants. The mosaic studies showed that knockdown of *irx1a* inhibits *shh*-GFP expression in a non-cell-autonomous manner. These observations indicate that *Iroquois* genes play evolutionarily conserved functions in the control of Hh expression during fly and zebrafish retinogenesis (Cheng *et al.*, 2006). Meanwhile, we showed that *irx1a*, but not *irx1b*, is expressed in the 5HT neurons. Knockdown of *irx1a* by antisense morpholino nucleotides revealed that it is a critical determinant for the differentiation of 5HT neurons in the hindbrain. However, *irx1a* morphants did not show a reduction of the progenitors of 5HT neurons. Hence, *irx1a* is not required for the initial specification, but it is required for the complete differentiation of 5HT neurons (Cheng *et al.*, 2007).

b. Conserved genetic components in retinal morphogenesis:* Sonic hedgehog, atonal, *and* Iroquois *gene family Like those of *ath5* and *shh* in zebrafish, *irx1a* expression is initiated at the ventronasal aspect of the nascent retina. Subsequently, the expression of all three genes spreads through the retina in a similar manner in GCL although *irx1a* expression appears later than that of *ath5*. Also similar to *ath5* and *shh*, the expression of *irx1a* in the retina is dependent on fibroblast growth factor (FGF) signaling (Cheng *et al.*, 2006; Martinez-Morales *et al.*, 2005) (Fig. 1.3).

Recently, it was demonstrated that *irx2a* morphants exhibit retinal defects strikingly similar to those of *irx1a* morphants, such as lacking GCL and INL differentiation, blindness, as well as lamination defects (Choy *et al.*, 2010). *irx1a*, *irx2a*, and *ath5* are all expressed in GCL, but with different temporal profiles: *ath5* was expressed the earliest, then sequentially followed by *irx1a* and *irx2a* (Choy *et al.*, 2010). The sequential expression of all these retinogenesis key factors implicates their interaction in zebrafish retinogenesis. Interestingly, we demonstrated the novel hierarchy genetic order in GCL retinogenesis: Ath5→Irx1a→Irx2a→Shh cascade. *irx2a* expression is reduced in both *irx1a* and *ath5* morphant retinas. Overexpression of *irx2a* in *irx1a* knockdown embryos could significantly rescue eye defects and *shh* propagation defects in *irx1a* morphant retinas. Meanwhile, *irx1a* expression is also *ath5*-dependent. A synergistic effect of *ath5* and *irx2a* was observed in knockdown experiments, which resulted in a dramatic increase of the retinogenesis phenotype. These all suggest that *irx1a* and *ath5* regulate *irx2a* expression in the developing retina and ensure a proper *shh* wave-propagation during retinogenesis (Choy *et al.*, 2010) (Fig. 1.3).

Similar to zebrafish retinogenesis, the border of *IRO-C*$^{+}$ and *IRO-C*$^{-}$ expressing cells in the *Drosophilia* eye disk is sufficient to promote the upregulation of *hh* and morphogenetic furrow progression, but not essential for *hh* initiation (Fig. 1.3) (Cavodeassi *et al.*, 1999). Hh controls the progression of furrow by induction of *ato* via the activation of serine–threonine kinase *Raf* (Greenwood and Struhl, 1999). In contrast, zebrafish *ath5* is one of the earliest transcription factors specifically expressed in differentiating RGCs, which coincides with the ontogeny of GCs (Kay *et al.*, 2005; Masai *et al.*, 2000). *ath5* controls *shh* wave propagation in RGCs in part through the control of *irx1a* and *irx2a* expressions (Choy *et al.*, 2010).

In the mouse retina, the transcriptional relationship of *Atonal* and *Iroquois* genes was identified recently. A microarray analysis indicated that *Irx2* and *Irx5* are potential downstream target genes of the mouse ortholog of *ato*, Math5 (Mu *et al.*, 2005). Meanwhile, *irx5a* was identified as a direct target of *ath5* in the zebrafish retina, Ath5 binds to the promoter region of *irx5a*, and *irx5a* expression is lost in the *ath5* mutant (Del Bene *et al.*, 2007). We found that the expression of *irx5a* is initiated at 48hpf in the developing zebrafish RGCs whereas *irx1a* and *irx2a* were initiated in the early stages of retinogenesis (Choy *et al.*, 2010). Future analyses will be needed to establish whether *irx5* can cooperate with *irx1a* and *irx2a* to maintain *shh* wave progression in zebrafish.

In summary, while zebrafish *irx2a* regulates *shh* expression in the GCL, its action on *shh* expression in the INL is indirect and involves a yet unknown cell signaling pathway. We speculate that this *irx2a*-dependent GCL-derived signal induces the expression of *shh* in the INL through a yet to be undefined factor(s). Identification of the unknown factor(s) as well as other downstream targets of *irx1a* and *irx2a* should greatly enhance our understanding of vertebrate retinal development.

REFERENCES

Amato, M. A., Boy, S., and Perron, M. (2004). Hedgehog signaling in vertebrate eye development: A growing puzzle. *Cell. Mol. Life Sci.* **61,** 899–910.

Bai, C. B., Auerbach, W., Lee, J. S., Stephen, D., and Joyner, A. L. (2002). *Gli2*, but not *Gli1*, is required for initial Shh signaling and ectopic activation of the Shh pathway. *Development* **129,** 4753–4761.

Brown, N. L., Patel, S., Brzezinski, J., and Glaser, T. (2001). Math5 is required for retinal ganglion cell and optic nerve formation. *Development* **128,** 2497–2508.

Cagan, R. (1997). Patterning in the *Drosophila* retina: The role of local cell interactions and programmed cell death. *Dev. Biol.* **186,** S29.

Cavodeassi, F., Diez del Corral, R., Campuzano, S., and Domínguez, M. (1999). Compartments and organising boundaries in the *Drosophila* eye: The role of the homeodomain Iroquois proteins. *Development* **126,** 4933–4942.

Cavodeassi, F., Modolell, J., and Gómez-Skarmeta, J. L. (2001). The Iroquois family of genes: From body building to neural patterning. *Development* **128,** 2847–2855.

Chamoun, Z., Mann, R. K., Nellen, D., von Kessler, D. P., and Bellotto, M. (2001). Skinny hedgehog, an acyltransferase required for palmitoylation and activity of the hedgehog signal. *Science* **293,** 2080–2084.

Chanut, F., and Heberlein, U. (1997). Retinal morphogenesis in Drosophila: Hints from an eye-specific decapentaplegic allele. *Dev. Genet.* **20,** 197–207.

Chen, C. M., and Cepko, C. L. (2002). The chicken RaxL gene plays a role in the initiation of photoreceptor differentiation. *Development* **129,** 5363–5375.

Cheng, C. W., Chow, R. L., Lebel, M., Sakuma, R., Cheung, H. O.-L., Thanabalasingham, V., Zhang, X., Bruneau, B. G., Birch, D. G., Hui, C.-C., McInnes, R. R., and Cheng, S. H. (2005). The Iroquois homeobox gene, *Irx5*, is required for retinal cone bipolar cell development. *Dev. Biol.* **287,** 48–60.

Cheng, C. W., Yan, C. H. M., Hui, C.-C., Strahle, U., and Cheng, S. H. (2006). The homeobox gene *irx1a* is required for the propagation of the neurogenic waves in the zebrafish retina. *Mech. Dev.* **123,** 252–263.

Cheng, C. W., Yan, C. H., Choy, S. W., Hui, M. N., Hui, C. C., and Cheng, S. H. (2007). Zebrafish homologue *irx1a* is required for the differentiation of serotonergic neurons. *Dev. Dyn.* **236,** 2661–2667.

Chiang, C., Litingtung, Y., Lee, E., Young, K. E., Corden, J. L., Westphal, H., and Beachy, P. A. (1996). Cyclopia and defective axial patterning in mice lacking *Sonic hedgehog* gene function. *Nature* **383,** 407–413.

Cho, K. O., Chern, J., and Izzadoost, S. (2000). Novel signaling from the peripodial membrane is essential for eye disc patterning in Drosophiia. *Cell* **103,** 331–342.

Choy, S. W., Cheng, C. W., Lee, S. T., Li, V. W., Hui, M. N., Hui, C. C., Liu, D., and Cheng, S. H. (2010). A cascade of *irx1a* and *irx2a* controls *shh* expression during retinogenesis. *Dev. Dyn.* **239,** 3204–3214.

Chuang, P.-T., and McMahon, A. P. (1999). Vertebrate Hedgehog signalling modulated by induction of a Hedgehog-binding protein. *Nature* **397,** 617–621.

Cohen, M. M., and Shiota, K. (2002). Teratogenesis of holoprosencephaly. *Am. J. Med. Genet.* **109,** 1–15.

Cooper, M. K., Porter, J. A., Young, K. E., and Beachy, P. A. (1998). Plant-derived and synthetic teratogens inhibit the ability of target tissues to respond to Sonic hedgehog signaling. *Science* **280,** 1603.

Del Bene, F., Ettwiller, L., Skowronska-Krawczyk, D., Baier, H., Matter, J. M., Birney, E., and Wittbrodt, J. (2007). In vivo validation of a computationally predicted conserved Ath5 target gene set. *PLoS Genet.* **3,** 1661–1671.

Dominguez, M. (1999). Dual role for Hedgehog in the regulation of the proneural gene atonal during ommatidia development. *Development* **126,** 2345–2353.

Dominguez, M., and de Celis, J. F. (1998). A dorsal/ventral boundary established by Notch controls growth and polarity in the Drosophila eye. *Nature* **396,** 276–278.

Driever, W., and Nusslein-Volhard, C. (1988). The bicoid protein determines position in the *Drosphila* embryo in a concentration-dependent manner. *Cell* **54,** 95–104.

Easter, S. S., Jr., and Malicki, J. J. (2002). The zebrafish eye: Developmental and genetic analysis. *Results Probl. Cell Differ.* **40,** 346–370.

Ekker, S. C., Ungar, A. R., Greenstein, P., von Kessler, D. P., Porter, J. A., and Moon, R. T. (1995). Patterning activities of vertebrate hedgehog proteins in the developing eye and brain. *Curr. Biol.* **5,** 944–955.

Ericson, J., Briscoe, J., and Rashbass, P. (1997). Graded sonic hedgehog signaling and the specification of cell fate in the ventral neural tube. *Cold Spring Harb. Symp. Quant. Biol.* **62,** 451–466.

Feijóo, C. G., Manzanares, M., de la Calle-Mustienes, E., Gómez-Skarmeta, J. L., and Allende, M. L. (2004). The Irx gene family in zebrafish: Genomic structure, evolution and initial characterization of irx5b. *Dev. Genes Evol.* **214,** 277–284.

Freeman, M. (1996). Reiterative use of the EGF receptor triggers differentiation of all cell types in the *Drosophila* eye. *Cell* **87,** 651–660.

Gerhart, J. (1999). Warkany lecture: Signaling pathways in development. *Teratology* **60,** 226–239.

Gilbert, S. F. (2000). *Developmental Biology* (6th ed.). Sinauer Associates, Inc., Sunderland, MA.

Goodrich, L. V., and Scott, M. P. (1998). Hedgehog and patched in neural development and disease. *Neuron* **21,** 1243–1257.

Goodrich, L. V., Milenkovic, L., and Higgins, K. M. (1997). Altered neural cell fates and medulloblastoma in mouse patched mutants. *Science* **277,** 1109–1113.

Greenwood, S., and Struhl, G. (1999). Progression of the morphogenetic furrow in the Drosophila eye: The roles of Hedgehog, Decapentaplegic and the Raf pathway. *Development* **126,** 5795–5808.

Hammerschmidt, M., and McMahon, A. P. (1998). The effect of pertussis toxin on zebrafish development: a possible role for inhibitory G-proteins in hedgehog signaling. *Dev. Biol.* **194,** 166–171.

Hammerschmidt, M., Bitgood, M. J., and McMahon, A. P. (1996). Protein kinase A is a common negative regulator of Hedgehog signalling in the vertebrate embryo. *Genes Dev.* **10,** 647–658.

Hammerschmidt, M., Brook, A., and McMahon, A. P. (1997). The world according to Hedgehog. *Trends Genet.* **13,** 14–21.

Heberlein, U., and Moses, K. (1995). Mechanisms of Drosophila retinal morphogenesis: The virtues of being progressive. *Cell* **30,** 987–990.

Heberlein, U., Wolff, T., and Rubin, G. M. (1993). The TGF beta homolog dpp and the segment polarity gene hedgehog are required for propagation of a morphogenetic wave in the Drosophila retina. *Cell* **75,** 913–926.

Heberlein, U., Singh, C. M., and Luk, A. Y. (1995). Growth and differentiation in the Drosophiia eye coordinated by hedgehog. *Nature* **373,** 709–711.

Hynes, M., Stone, D. M., Dowd, M., Pitts-Meek, S., Goddard, A., Gurney, A., and Rosenthal, A. (1997). Control of cell pattern in the neural tube by the zinc finger transcription factor and oncogene Gli1. *Neuron* **19,** 15–26.

Hynes, M., Ye, W., Wang, K., Stone, D., Murone, M., Sauvage, F., and Rosenthal, A. (2000). The seven-transmembrane receptor smoothened cell-autonomously induces multiple ventral cell types. *Nat. Neurosci.* **3,** 41–46.

Incardona, J. P., and Roelink, H. (2000). The role of cholesterol in Shh signaling and teratogen-induced holoprosencephaly. *Cell. Mol. Life Sci.* **57,** 1709–1719.

Incardona, J. P., Gaffield, W., Kapur, R. P., and Roelink, H. (1998). The teratogenic Veratrum alkaloid cyclopamine inhibits sonic hedgehog signal transduction. *Development* **125,** 3553.

Ingham, P. W., and McMahon, A. P. (2001). Hedgehog signaling in animal development: Paradigms and principles. *Genes Dev.* **15,** 3059–3087.

Jablonski, M. M., Tombran-Tink, J., Mrazek, D. A., and Iannaccone, A. (2000). Pigment epithelium-derived factor supports normal development of photoreceptor neurons and opsin expression after retinal pigment epithelium removal. *J. Neurosci.* **20,** 7149–7157.

Jarman, A. P., Sun, Y., Jan, L. Y., and Jan, Y. N. (1995). Role of the proneural gene, atonal, in formation of Drosophila chordotonal organs and photoreceptors. *Development* **121,** 2019–2030.

Jensen, A. M., and Wallace, V. A. (1997). Expression of Sonic hedgehog and its putative role as a precursor cell mitogen in the developing mouse retina. *Development* **124,** 363–371.

Jin, Z., Zhang, J., Klar, A., Chedotal, A., Rao, Y., Cepko, C. L., and Bao, Z. Z. (2003). Irx4-mediated regulation of Slit1 expression contributes to the definition of early axonal paths inside the retina. *Development* **130,** 1037–1048.

Johnson, R., and Tabin, C. (1995). The long and short of *hedgehog* signaling. *Cell* **81,** 313–316.

Kaesler, S., Lüscher, B., and Rüther, U. (2000). Transcriptional activity of GLI1 is negatively regulated by protein kinase A. *Biol. Chem.* **381,** 545–551.

Karlstrom, R. O., Talbot, W. S., and Schier, A. F. (1999). Comparative synteny cloning of zebrafish you-too: Mutations in the Hedgehog target gli2 affect ventral forebrain patterning. *Genes Dev.* **13,** 388–393.
Katoh, Y., and Katoh, M. (2009). Hedgehog target genes: Mechanisms of carcinogenesis induced by aberrant hedgehog signaling activation. *Curr. Mol. Med.* **9,** b873–b886.
Kay, J. N., Finger, K. C., Roeser, T., Staub, W., and Baier, H. (2001). Retinal ganglion cell determination requires *lakritz*, a Zebrafish homolog. *Neuron* **30,** 725–736.
Kay, J. N., Link, B. A., and Baier, H. (2005). Staggered cell-intrinsic timing of ath5 expression underlies the wave of ganglion cell neurogenesis in the zebrafish retina. *Development* **132,** 2573–2585.
Kohtz, J. D., Lee, H. Y., Gaiano, N., Segal, J. D., Ng, E., Larson, T. A., Baker, D. P., Garber, E. A., Williams, K. P., and Fishell, G. (2001). N-terminal fatty-acylation of Sonic Hedgehog enhances the induction of rodent ventral forebrain neurons. *Development* **128,** 2351–2363.
Levine, E. M., Roelink, H., Turner, J., and Reh, T. A. (1997). Sonic hedgehog promotes rod photoreceptor differentiation in mammalian retinal cells in vitro. *J. Neurosci.* **17,** 6277–6788.
Litintung, Y., and Chiang, C. (2000). Specification of ventral neuron types is mediated by an antagonistic interaction between Shh and Gli3. *Nat. Neurosci.* **3,** 979–985.
Livesey, F. J., and Cepko, C. L. (2001). Vertebrate neural cell-fate determination: Lessons from the retina. *Nat. Rev. Neurosci.* **2,** 109–118.
Ma, C., Zhou, Y., Beachy, P. A., and Moses, K. (1993). The segment polarity gene hedgehog is required for progression of the morphogenetic furrow in the developing Drosophila eye. *Cell* **75,** 927–938.
Macdonald, R., Barth, K. A., Xu, Q., Holder, N., Mikkola, I., and Wilson, S. W. (1995). Midline signalling is required for Pax gene regulation and patterning of the eyes. *Development* **121,** 3267–3278.
Macdonald, R., Scholes, J., Strahle, U., Brennan, C., Holder, N., Brand, M., and Wilson, S. W. (1997). The Pax protein Noi is required for commissural axon pathway formation in the rostral forebrain. *Development* **124,** 2397–2408.
Martinez-Morales, J. R., Del Bene, F., Nica, G., Hammerschmidt, M., Bovolenta, P., and Wittbrodt, J. (2005). Differentiation of the vertebrate retina is coordinated by an FGF signaling center. *Dev. Cell* **8,** 565–574.
Masai, I., Stemple, D. L., Okamoto, H., and Wilson, S. W. (2000). Midline signals regulate retinal neurogenesis in zebrafish. *Neuron* **27,** 251–263.
Mazzoni, E. O., Celik, A., Wernet, M. F., Vasiliauskas, D., and Johnston, R. J. (2008). *Iroquois Complex* Genes Induce Co-Expression of *rhodopsins* in *Drosophila. PLoS Biol.* **6,** e97.
McMahon, A. P., Ingham, P. W., and Tabin, C. J. (2003). Developmental roles and clinical significance of hedgehog signaling. *Curr. Top. Dev. Biol.* **53,** 1–114.
McNeill, H., Yang, C. H., Brodsky, M., Ungos, J., and Simon, M. A. (1997). Mirror encodes a novel PBX-class homeoprotein that functions in the definition of the dorsal-ventral border in the Drosophila eye. *Genes Dev.* **11,** 1073–1082.
Mu, X., Fu, X., Sun, H., Beremand, P. D., and Thomas, T. L. (2005). A gene network downstream of transcription factor Math5 regulates retinal progenitor cell competence and ganglion cell fate. *Dev. Biol.* **280,** 467–481.
Mukherjee, K., and Bürglin, T. R. (2007). Comprehensive analysis of animal TALE homeobox genes: New conserved motifs and cases of accelerated evolution. *J. Mol. Evol.* **65,** 137–153.
Murone, M., Rosenthal, A., and de Sauvage, F. J. (1999). Sonic hedgehog signaling by the patched-smoothened receptor complex. *Curr. Biol.* **9,** 76–84.
Neumann, C. J. (2001). Pattern formation in the zebrafish retina. *Semin. Cell Dev. Biol.* **12,** 485–490.

Neumann, C. J., and Nuesslein-Volhard, C. (2000). Patterning of the zebrafish retina by a wave of sonic hedgehog activity. *Science* **289,** 2137–2139.

Neumann, C. J., Grandel, H., Gaffield, W., Schulte-Merker, S., and Nusslein-Volhard, C. (1999). Transient establishment of anteroposterior polarity in the zebrafish pectoral fin bud in the absence of sonic hedgehog activity. *Development* **126,** 4817–4826.

Otteson, D. C., Shelden, E., Jones, J. M., Kameoka, J., and Hitchcock, P. F. (1998). Pax2 expression and retinal morphogenesis in the normal and Krd mouse. *Dev. Biol.* **193,** 209–224.

Pearse, R. V., II, Collier, L. S., Scott, M. P., and Tabin, C. J. (1999). Vertebrate homologs of *Drosophila* suppressor of fused interact with the gli family of transcriptional regulators. *Dev. Biol.* **212,** 323–336.

Pepinsky, R. B., Zeng, C., Wen, D., Rayhorn, P., Baker, D. P., Williams, K. P., Bixler, S. A., Ambrose, C. M., Garber, E. A., and Miatkowski, K. (1998). Identification of a palmitic acid-modified form of human Sonic hedgehog. *J. Biol. Chem.* **273,** 14037–14045.

Perron, M., Boy, S., Amato, M. A., Viczian, A., Koebernick, K., and Pieler, T. (2003). A novel function for Hedgehog signaling in retinal pigment epithelium differentiation. *Development* **130,** 1565–1577.

Peters, M. A. (2002). Patterning the neural retina. *Curr. Opin. Neurobiol.* **12,** 43–48.

Porter, J. A., von Kessler, D. P., Ekker, S. C., Young, K. E., Lee, J. J., Moses, K., and Beachy, P. A. (1995). The product of hedgehog auto-proteolytic cleavage active in local and long-range signaling. *Nature* **374,** 363–366.

Prykhozhij, S. V. (2010). In the absence of Sonic hedgehog, p53 induces apoptosis and inhibits retinal cell proliferation, cell-cycle exit and differentiation in zebrafish. *PLoS One* **5**(10), e13549.

Raymond, S. M., and Jackson, I. J. (1995). The retinal pigmented epithelium is required for development and maintenance of the mouse neural retina. *Curr. Biol.* **5,** 1286–1295.

Rietveld, A., Neutz, S., Simons, K., and Eaton, S. (1999). Association of sterol- and glycosylphosphatidylinositol-linked proteins with *Drosophila* raft lipid microdomains. *J. Biol. Chem.* **274,** 12049–12054.

Roessler, E., and Muenke, M. (2001). MidUne and laterality defects: Left and right meet in the middle. *Bioessays* **23,** 888–900.

Sanyanusin, P., Schimmenti, L. A., McNoe, L. A., Ward, T. A., Pierpont, M. E., Sullivan, M. J., Dobyns, W. B., and Eccles, M. R. (1995). Mutation of the PAX2 gene in a family with optic nerve colobomas, renal anomalies and vesicoureteral reflux. *Nat. Genet.* **9,** 358–364.

Schauerte, H. E., van Eeden, F. J., Fricke, C., Odenthal, J., Strahle, U., and Haffter, P. (1998). Sonic hedgehog is not required for the induction of medial floor plate cells in the zebrafish. *Development* **125,** 2983–2993.

Schwarz, M., Cecconi, F., Bernier, G., Andrejewski, N., Kammandel, B., and Wagner, M. (2000). Spatial specification of mammalian eye territories by reciprocal transcriptional repression of Pax2 and Pax6. *Development* **127,** 4325–4334.

Sehgal, R., Sheibani, N., Rhodes, S. J., and Belecky Adams, T. L. (2009). BMP7 and SHH regulate Pax2 in mouse retinal astrocytes by relieving TLX repression. *Dev. Biol.* **332,** 429–443.

Sheedlo, H. J., and Turner, J. E. (1996). Effects of retinal pigment epithelial cell-secreted factors on neonatal rat retinal explant progenitor cells. *J. Neurosci. Res.* **15,** 519–531.

Shkumatava, A., Fischer, S., and Muller, F. (2004). Sonic hedgehog, secreted by amacrine cells, acts as a short-range signal to direct differentiation and lamination in the zebrafish retina. *Development* **131,** 3849–3858.

Simons, K., and Ikonen, E. (1997). Functional rafts in cell membranes. *Nature* **387,** 569–572.

Stenkamp, D. L., and Frey, R. A. (2003). Extraretinal and retinal hedgehog signaling sequentially regulate retinal differentiation in zebrafish. *Dev. Biol.* **15,** 349–363.
Stenkamp, D. L., Frey, R. A., Prabhudesai, S. N., and Raymond, P. A. (2000). Function for Hedgehog genes in zebrafish retinal development. *Dev. Biol.* **220,** 238–252.
Stenkamp, D. L., Frey, R. A., Mallory, D. E., and Shupe, E. E. (2002). Embryonic retinal gene expression in sonic-you mutant zebrafish. *Dev. Dyn.* **225,** 344–350.
Stenkamp, D. L., Satterfield, R., Muhunthan, K., Sherpa, T., Vihtelic, T. S., and Cameron, D. A. (2008). Age-related cone abnormalities in zebrafish with genetic lesions in sonic hedgehog. *Invest. Ophthalmol. Vis. Sci.* **49,** 4631–4640.
St-Jacques, B., Dassule, H. R., and Karavanova, I. (1998). Sonic hedgehog signaling is essential for hair development. *Curr. Biol.* **8,** 1058–1068.
Strauss, O. (2005). The retinal pigment epithelium in visual function. *Physiol. Rev.* **85,** 845–881.
Strutt, D. I., and Mlodzik, M. (1997). Hedgehog is an indirect regulator of morphogenetic furrow progression in the Drosophila eye disc. *Development* **124,** 3233–3240.
Tomlinson, A., and Ready, D. F. (1987). Neuronal differentiation in the *Drosophila* ommatidium. *Dev. Biol.* **120,** 366–376.
Tomlinson, A. (2003). Patterning the peripheral retina of the fly: decoding a gradient. *Dev. Cell* **5**(5), 799--809.
Torres, M., Gomez-Pardo, E., and Gruss, P. (1996). Pax2 contributes to inner ear patterning and optic nerve trajectory. *Development* **122,** 3381–3391.
Treisman, J. E., and Heberlein, U. (1998). Eye development in Drosophila: Formation of the eye-field and control of differentiation. *Curr. Top. Dev. Biol.* **39,** 119–158.
Varga, Z. M., Wegner, J., and Westerfield, M. (1999). Anterior movement of ventral diencephalic precursors separates the primordial eye field in the neural plate and requires cyclops. *Development* **126,** 5533–5546.
Varga, Z. M., Amores, A., Lewis, K. E., Yan, Y. L., Postlethwait, J. H., Eisen, J. S., and Westerfield, M. (2001). Zebrafish smoothened functions in ventral neural tube specification and axon tract formation. *Development* **128,** 3497–3509.
Vincent, J. P., and Briscoe, J. (2001). Morphogens. *Curr. Biol.* **11,** 851–854.
Watanabe, T., and Raff, M. C. (1992). Diffusible rod-promoting signals in the developing rat retina. *Development* **114,** 899–906.
Wernet, M. F., Labhart, T., Baumann, F., Mazzoni, E. O., Pichaud, F., and Desplan, C. (2003). Homothorax switches function of Drosophila photoreceptors from color to polarized light sensors. *Cell* **115,** 267–279.
Zhang, X. M., and Yang, X. J. (2001). Temporal and spatial effects of Sonic hedgehog signaling in chick eye morphogenesis. *Dev. Biol.* **233,** 271–290.
Zhao, L., Saitsu, H., Sun, X., Shiota, K., and Ishibashi, M. (2010). Sonic hedgehog is involved in formation of the ventral optic cup by limiting Bmp4 expression to the dorsal domain. *Mech. Dev.* **127,** 62–72.
Zhou, M., Lambert, H., and Landry, J. (1993). Transient activation of a distinct serine protein kinase is responsible for 27-kDa heat shock protein phosphorylation in mitogen-stimulated and heat-shocked cells. *J. Biol.Chem.* **268,** 35–43.

CHAPTER TWO

Canonical and Noncanonical Hedgehog/GLI Signaling in Hematological Malignancies

Fritz Aberger,* Daniela Kern,* Richard Greil,† *and* Tanja Nicole Hartmann†

Contents

Abstract

The highly conserved Hedgehog/GLI signaling pathway regulates multiple aspects of embryonic development and plays a decisive role in tissue homeostasis and the hematopoietic system by controlling cell fate decisions, stem cell self-renewal, and activation. Loss of negative control of Hedgehog signaling contributes to tumor pathogenesis and progression. In the classical view of canonical Hedgehog signaling, Hedgehog ligand binding to its receptor Patched

* Division of Molecular Tumor Biology, Department of Molecular Biology, University of Salzburg, Salzburg, Austria

† Laboratory for Immunological and Molecular Cancer Research, Third Medical Department with Hematology, Oncology, Hemostaseology, Infectiology, and Rheumatology, Paracelsus Medical University, Salzburg, Austria

Vitamins and Hormones, Volume 88

ISSN 0083-6729, DOI: 10.1016/B978-0-12-394622-5.00002-X

culminates in the activation of the key pathway activator Smoothened, followed by activation of the GLI transcription factors. Its essential function and druggability render Smoothened well suited to therapeutic intervention. However, recent evidence suggests a critical role of Smoothened-independent regulation of GLI activity by several other signaling pathways including the PI3K/AKT and RAS/RAF/MEK/ERK axes. In addition, the contribution of canonical Hedgehog signaling via Patched and Smoothened to normal and malignant hematopoiesis has been the subject of recent controversies. In this review, we discuss the current understanding and controversial findings of canonical and noncanonical GLI activation in hematological malignancies in light of the current therapeutic strategies targeting the Hedgehog pathway.

I. Hedgehog Introduction

The Hedgehog (HH)/GLI pathway was originally identified as a key signaling system that controls multiple processes during invertebrate and vertebrate embryonic development (Ingham and McMahon, 2001; Nusslein-Volhard and Wieschaus, 1980; Teglund and Toftgard, 2010). More recently, the pathway has also been implicated in the control of tissue homeostasis, regeneration, and healing in adult organisms where signaling contributes to the regulation of stem cell activation and self-renewal (Beachy *et al.*, 2004; Ruiz i Altaba *et al.*, 2007). Precise control of the onset, strength, and termination of HH/GLI signaling is a critical requirement for proper pattern formation, cell proliferation, survival, differentiation, and morphogenesis. In line with its pivotal regulatory role in these processes, aberrant activation of HH/GLI signaling by either genetic alterations or uncontrolled expression of selected pathway effectors turned out to be an etiological factor in the initiation and growth of numerous human cancer entities including a wide spectrum of solid tumors and hematological malignancies (reviewed in Barakat *et al.*, 2010; Ruiz i Altaba *et al.*, 2002). Targeted inhibition of inappropriate HH/GLI signaling in cancer patients has therefore become a major interest of biotech and pharmaceutical companies with more than half a dozen novel Hedgehog pathway inhibitors (HPIs) currently evaluated for their therapeutic benefit in several clinical trials (for details, see Table 2.1; Epstein, 2008; Scales and de Sauvage, 2009; Takebe *et al.*, 2011).

II. Regulation of Canonical HH Signaling and Its Implication in Cancer

Regulation of Hedgehog signaling is a complex process and a detailed review of the mechanisms of pathway regulation would be beyond the scope of this chapter (for detailed reviews, see Huangfu and Anderson,

Table 2.1 Clinical and preclinical Hedgehog pathway inhibitors

Inhibitor	Target	Type	Indication and experimental validation	Company	References	Clinical trials[a]
GDC-0449 (Vismodegib)	SMO	SMI	BCC, metastatic colon cancer, gastric cancer, GBM, MB, MM, pancreatic, prostate, ovarian cancer, SCLC	Genentech, Roche	Rudin *et al.* (2009), Von Hoff *et al.* (2009)	Phase I/II
Cur-61414	SMO	SMO	BCC	Curis	Williams *et al.* (2003)	Failed in phase I
Cyclopamine	SMO	Natural SMI	BCC, MB	NA	Taipale *et al.* (2000), Thayer *et al.* (2003), Berman *et al.* (2003), Sanchez and Ruiz i Altaba (2005)	Limited use in human
SANT1–4	SMO	SMI	*In vitro*	NA	Chen *et al.* (2002a,b)	No
HhAntag	SMO	SMI	MB	Curis	Romer *et al.* (2004)	No
NVP-LDE225	SMO	SMI	BCC, MB, advanced solid tumors	Novartis	Buonamici *et al.* (2010), Skvara *et al.* (2011)	Phase I/II
NVP-LEQ506	SMO	SMI	Advanced/metastatic BCC, refractory MB	Novartis	NA	Phase I
IPI-926	SMO	SMI	MB, chondrosarcoma, MF, pancreatic cancer	Infinity	Tremblay *et al.* (2009)	Phase I/II
BMS-833923	SMO	SMI	BCC, BCNS, CML, esophageal, gastric, SCLC, solid tumors	Exelixis, Bristol–Myers Squibb	NA	Phase I/II

(*Continued*)

Table 2.1 *(Continued)*

Inhibitor	Target	Type	Indication and experimental validation	Company	References	Clinical trials[a]
PF-04449913	SMO	SMI	CML, hematologic malignancies, solid tumors	Pfizer	NA	Phase I
TAK-441	SMO	SMI	BCC	Millennium Pharma.	NA	Phase I
Itraconazol	SMO	SMI (antifungal)	BCC, MB	NA	Kim *et al.* (2010b)	Phase I
Arsenic trioxide	GLI	SMI	*In vitro, in vivo* xenografts	Various	Kim *et al.* (2010a), Beauchamp *et al.* (2011)	No
HPI 1–4	ND	SMI	*In vitro*	NA	Hyman *et al.* (2009)	No
GANT61	GLI	SMI	*In vivo, in vitro* (various)	NA	Lauth *et al.* (2007a,b)	No
GANT58	GLI	SMI	*In vitro*	NA	Lauth *et al.* (2007a,b)	No
Robotnikinin	Hh	SMI	*In vitro*	NA	Stanton *et al.* (2009)	No
Anti-Hh (5E1)	Hh	Antibody	*In vivo* (GI tumors, HF growth), *in vitro*	NA	Ericson *et al.* (1996), Yauch *et al.* (2008), Bailey *et al.* (2008)	No

Abbreviations: BCC, basal cell carcinoma; BCNS, basal cell nevous syndrome; CML, chronic myeloid leukemia; EWS, Ewing's Sarcoma; GBM, glioblastoma multiforme; GI, gastrointestinal; MB, medulloblastoma; MF, myelofibrosis; MM, multiple myeloma; PMF, primary myelofibrosis; SCLC, small cell lung cancer; NA, not applicable; SMI, small molecule inhibitor; ND, not determined in detail.

[a] Clinical trials as listed at http://clinicaltrials.gov on June 21, 2011.

2006; Ingham *et al.*, 2011; Varjosalo and Taipale, 2007). We will provide in the following an overview of the key steps and players in canonical HH signaling relayed via the Patched/Smoothened/GLI route and subsequently elaborate on the noncanonical Smoothened-independent signals that have been implicated in the control of the GLI zinc finger transcription factors controlling Hedgehog target gene expression.

In the absence of HH ligand, HH signaling is actively maintained in a repressed state by the activity of the HH receptor Patched (PTCH). PTCH is a 12-transmembrane domain protein whose intracellular localization is concentrated at the base of the primary cilium, a single antenna-like structure that protrudes from the cell surface of many adherent cell types and functions as a signal transduction compartment (Rohatgi *et al.*, 2007). Although the detailed mechanism of pathway repression by PTCH has not been elucidated, recent studies have shown that unliganded PTCH, which can act catalytically (Taipale *et al.*, 2002), prevents the translocation of the seven-transmembrane domain protein and essential pathway effector Smoothened (SMO) into the primary cilium (Rohatgi *et al.*, 2007). This leads to proteolytic cleavage of the latent zinc finger transcription factors GLI3—and to some extent also of GLI2—into C-terminally truncated repressor forms (GLI^R) (Wang *et al.*, 2000; Wen *et al.*, 2010). GLI^R formation involves preceding and sequential phosphorylation by protein kinase A (PKA), glycogen synthase kinase 3-beta (GSK), and casein kinase I (CKI) (Price and Kalderon, 2002) as well as a functional primary cilium (Huangfu *et al.*, 2003; Wong *et al.*, 2009). Following processing, GLI^R translocates to the nucleus to bind to HH target gene promoters and repress target gene expression (Aza-Blanc *et al.*, 1997). GLI signals are also negatively regulated by proteasome-mediated degradation of GLI and by binding to Suppressor of Fused (SUFU), which sequesters GLI proteins in the cytoplasm (Fig. 2.1, left) (reviewed in Ruiz i Altaba *et al.*, 2007; Teglund and Toftgard, 2010).

Paracrine or autocrine activation of HH signaling is initiated by binding of secreted and posttranslationally modified HH proteins, that is, Sonic (SHH), Indian (IHH), or Desert (DHH) Hedgehog, to their receptor PTCH (Gallet, 2011; Ingham *et al.*, 2011). The interaction of HH protein with PTCH removes PTCH from the primary cilium, thus allowing SMO to enter the cilium via lateral transport and activate downstream signaling events (Milenkovic *et al.*, 2009; Rohatgi *et al.*, 2007). The translocation and activation of SMO involve association of SMO with β-arrestins and the GPCR kinase GKR2, respectively (Chen *et al.*, 2011; Kovacs *et al.*, 2008; Meloni *et al.*, 2006; Philipp *et al.*, 2008). Active SMO localized in the primary cilium (i) interferes with GLI repressor formation and (ii) triggers release of GLI from SUFU (Humke *et al.*, 2010; Tukachinsky *et al.*, 2010). As a result of these complex regulatory steps, the full-length activator form of GLI, hitherto referred to as GLI^A, can translocate to the nucleus, where it binds to target gene promoters and activates transcription of HH target

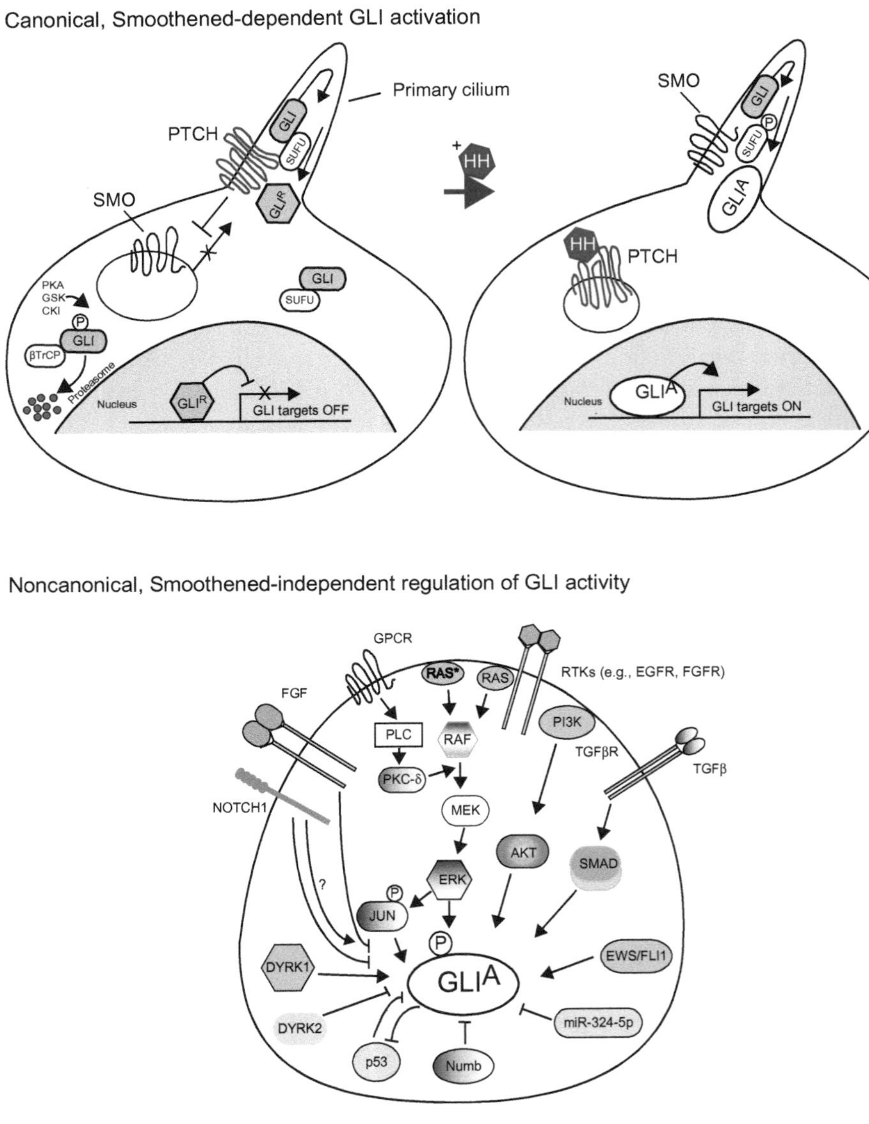

Figure 2.1 Smoothened-dependent and -independent regulation of GLI activity. (A) Active repression of HH target gene expression in the absence of ligand stimulation by GLI repressor (GLI^R) formation, proteasome-mediated GLI protein degradation and SUFU-dependent cytoplasmic retention of GLI proteins. (B) Activation of HH signaling and target gene expression in response to binding of HH protein to its receptor PTCH, leading to removal of PTCH from the primary cilium and entrance of SMO into the primary cilium via lateral transport mechanisms. In the primary cilium, active SMO prevents cleavage of the full-length GLI into the truncated repressor form and also triggers the dissociation of GLI from SUFU, allowing the GLI activator form (GLI^A) to translocate to the nucleus and turn on target gene expression. (C) Smoothened-

genes including GLI1. Being a strong transcriptional activator, induction of GLI1 in response to canonical PTCH/SMO-dependent pathway activation causes amplification of the GLI^{A} signal (Ikram *et al.*, 2004; Regl *et al.*, 2002; Ruiz i Altaba *et al.*, 2002).

In essence, one can therefore state that the precise regulation of the GLI^{A}/GLI^{R} ratio is the most critical parameter for proper development and homeostasis. This is supported by the fact that the balance of GLI^{A}/GLI^{R} forms determines not only the strength and output of HH signaling but also the fate of a cell. High GLI^{A}/GLI^{R} ratios are mainly associated with proliferation, increased survival, and stem cell self-renewal, while low ratios favor differentiation and quiescence (Stecca *et al.*, 2007).

During the past years, numerous studies have unraveled a fundamental role of the HH/GLI pathway in a wide variety of human cancers including solid tumors and hematological malignancies. Current estimates suggest that up to 25% of human cancers display aberrant HH/GLI signaling (Teglund and Toftgard, 2010). These cancers are characterized by uncontrolled and persistent activation of the HH/GLI pathway leading to an increase in GLI^{A} forms, mostly GLI1, at the expense of GLI^{R} forms. High GLI^{A}/GLI^{R} ratios can be the result of (i) loss-of-function mutations in pathway repressors such as PTCH and SUFU, (ii) gain-of-function mutations and amplifications in SMO and GLI2/GLI1, respectively, or (iii) overexpression of and sustained exposure to HH ligands (Kasper *et al.*, 2006a; Stecca and Ruiz, 2010). While it is clear that aberrant activation of HH/GLI signaling is an etiologic factor in many cancers, the identity of the HH signal-receiving cell is more complex as there is evidence for both the tumor cell itself and the tumor-associated stromal cell being the target of HH pathway activation (for detailed reviews, see Scales and de Sauvage, 2009; Teglund and Toftgard, 2010).

The widespread involvement of HH/GLI in human malignancies has initiated a remarkable effort to identify selective HPIs. As shown in Table 2.1, most of these small molecule inhibitors target the essential effector protein SMO, which should lead to pathway abrogation by eventually decreasing the GLI^{A}/GLI^{R} ratio. This has already been successfully demonstrated in patients with basal cell carcinoma (BCC) and medulloblastoma showing aberrant SMO-dependent HH/GLI signaling (Rudin *et al.*, 2009; Skvara *et al.*, 2011; Von Hoff *et al.*, 2009). However, there is also

independent regulation of GLI activity. The illustration summarizes recent data on diverse signaling cascades and regulatory proteins impinging on the GLI transcription factors to regulate their transcriptional activity, stability, and expression. Arrows indicate a positive impact on the activity of GLI, inhibitory symbols a negative regulatory interaction including protein destabilization (e.g., numb) or repression of GLI expression (e.g., NOTCH1). RAS*, oncogenic RAS; encircled P, phosphorylation; TGFβR, TGFβ receptor; RTK, receptor tyrosine kinase; PLC, phospholipase C; GPCR, G-protein-coupled receptor.

increasing evidence for SMO-independent stimulation of GLI activity in cancer cells, raising the question of whether SMO antagonists will prove efficient therapeutic drugs in such settings. As SMO function is associated with the primary cilium (Huangfu and Anderson, 2006), this limitation is likely to be relevant for GLI-dependent cancer cells lacking this single cellular protrusion, as is the case for most myeloid and lymphoid cells (Pazour and Witman, 2003). In such cases, targeting SMO-independent mechanisms that impinge upon the GLI transcription factors to modulate their activity becomes more relevant.

III. SMO-Independent Modulation of GLI Activity

A. RAS–RAF–MEK and PI3K–AKT signaling as modulators of GLI activity

Constitutive activation of RAS signaling is frequently found in human cancers such as pancreatic ductal adenocarcinoma (PDAC) of which the vast majority displays activating K-RAS mutations. Aside from activated RAS, PDAC frequently expresses high levels of HH ligands, pointing to a possible interaction of these two pathways (Lau *et al.*, 2006). Notably, coexpression of a dominant active form of GLI2 and oncogenic K-RAS in the pancreas of transgenic mice led to synergistically enhanced disease progression (Pasca di Magliano *et al.*, 2006). A possible explanation for this cooperative effect came from studies of melanoma, prostate cancer, and glioma cells showing that oncogenic N- and H-RAS as well as active AKT can stimulate the transcriptional activity and nuclear import of GLI1. Further, endogenous GLI activity depends on RAS and PI3K/AKT function and AKT has been shown to enhance GLI protein stability (Riobo *et al.*, 2006b; Stecca *et al.*, 2007). Studies in zebrafish provided additional *in vivo* evidence for HH/GLI–AKT cooperation. Expression of zebrafish Smoothened in combination with constitutively active human Akt1 induced various tumor types resembling rhabdomyoma, melanoma, and astrocytoma (Ju *et al.*, 2009).

Oncogenic RAS signaling has also been implicated in transcriptional activation and protein stabilization of GLI (Ji *et al.*, 2007). Aside from its direct effect on GLI, RAS has been shown to promote paracrine tumor-to-stroma HH signaling in pancreatic cancer cells by activating HH ligand expression and inhibiting autocrine HH pathway activation in the cancer cells (Lauth *et al.*, 2010). Activation of the MEK/ERK cascade appears to be a central player mediating the GLI activity promoting effect of RAS and also of PKC-δ, the latter of which may involve G-protein-coupled receptor signaling including that of SMO (Lauth *et al.*, 2007b; Riobo *et al.*, 2006a;

Stecca *et al.*, 2007). However, another study has implicated PKC-δ in GLI activation downstream of SUFU (Lauth *et al.*, 2007b). Coexpression of GLI1 and dominant active MEK has been shown to drive xenograft growth of human nontumorigenic keratinocytes, while neither factor alone is able to transform these cells, suggesting cooperative interactions between MEK and GLI1 (Schnidar *et al.*, 2009). MEK stabilizes GLI proteins and generally enhances the transcriptional activity of GLI1, which requires the presence of the GLI1 N-terminus (Riobo *et al.*, 2006a; Schnidar *et al.*, 2009; Stecca *et al.*, 2007). Of note, a recent study provided evidence for GLI3 and GLI1 as novel substrate of MAP-kinases (MAPKs) including the MEK substrate ERK2 (Whisenant *et al.*, 2010). Although *in vivo* evidence is still lacking, it is likely that ERK-mediated phosphorylation of the N-terminus of GLI1 accounts for its increased stability and transcriptional activity, which would also be in line with the notion that deletion of the GLI1 N-terminus abrogates the enhanced transcriptional activity of GLI1 observed upon concomitant basic fibroblast growth factor (bFGF) treatment (Riobo *et al.*, 2006a).

Other kinases affecting the activity of GLI proteins in different ways include the dual-specificity tyrosine-(Y)-phosphorylation-regulated kinases DYRK1 and DYRK2. While DYRK1 enhances the transcriptional activity of GLI1 by promoting its nuclear localization (Mao *et al.*, 2002), DYRK2 negatively regulates GLI2 by phosphorylation thereby targeting GLI2 to proteasomal degradation (Varjosalo *et al.*, 2008).

B. Modulation of GLI activity by growth factor signaling pathways and oncogenes

Given the well-documented role of RAS, MAPK, and PI3K/AKT signaling in the modulation of GLI proteins, it may at first glance not be surprising that growth factor receptor tyrosine kinase (RTK) signaling also affects the activity of GLI in an SMO-independent manner. For instance, treatment with epidermal growth factor (EGF) synergizes with HH signaling in the activation of neural stem cell proliferation (Palma and Ruiz i Altaba, 2004) and bFGF enhances the transcriptional activity of GLI1 (Riobo *et al.*, 2006a). Both interactions could simply be explained by the RTK-mediated activation of intracellular PI3K/AKT and RAS/MAPK signaling branches. However, this interaction appears to be more complex, as in HH/GLI-dependent medulloblastoma cells, FGF signaling actually counteracts the proliferative effect of HH signaling (Fogarty *et al.*, 2007), suggesting context-dependent integration of distinct signal transduction pathways yielding very different outcomes. Our own group has intensely studied the mechanisms of HH/GLI and EGF-receptor (EGFR) signal cooperation in skin cancer. The key findings are that coactivation of EGFR and GLI1 or GLI2 results in selective modulation of GLI target gene expression rather than in a general activation of GLI activity. The integration of both pathways relies on EGFR-mediated activation of

MEK/ERK signaling. This, in turn, leads to phosphorylation and activation of the JUN/AP1 transcription factor, which itself is a direct GLI target. EGFR-activated JUN/AP1 then together with GLI binds to selected GLI target genes resulting in synergistic transcriptional activation (Kasper *et al.*, 2006b; Laner-Plamberger *et al.*, 2009; Schnidar *et al.*, 2009). Thus, the integration of EGFR and HH/GLI signal cooperation at the level of selected GLI target genes constitutes another mechanism by which growth factor signaling pathways can modulate the activity of GLI proteins (Fig. 2.1).

Activation of GLI transcription accounts for another mode of SMO-independent modulation of GLI activity. In frog embryos, bFGF induces Gli2 expression in the ventro-posterior mesoderm through an as-yet unidentified mechanism (Brewster *et al.*, 2000). Gli2 expression is also controlled by transforming growth factor β (TGFβ)/SMAD signaling via direct binding of activated SMAD3 to the Gli2 promoter. Intriguingly, the same study also identified an adjacent T-cell factor/lymphoid enhancer binding factor (TCF/LEF) binding site required for efficient activation of Gli2 in response to TGFβ, suggesting cooperative interactions of TGFβ/SMAD and Wnt/β-catenin/TCF signaling in the control of Gli2 expression (Dennler *et al.*, 2009). Likewise, TGFβ signaling and oncogenic K-RAS signaling induce GLI1 expression in pancreatic cancer cells, and inhibition of GLI1, but not SMO, decreases pancreatic cancer cell survival and proliferation (Nolan-Stevaux *et al.*, 2009). In agreement with these studies, TGFβ/SMAD signaling is required for HH/GLI-induced BCC development (Fan *et al.*, 2010).

Insulin-like growth factor 2 (IGF2) signaling represents an essential pathway in HH/GLI-dependent medulloblastoma and rhabdomyosarcoma (Hahn *et al.*, 2000). Notably, IGF2 activates AKT and increases transcription of GLI1, suggesting a dual mode for noncanonical GLI activation (Hartmann *et al.*, 2005; Rao *et al.*, 2004).

Transcription of GLI1 is also regulated by the EWS–FLI1 fusion oncogene in Ewing sarcoma cells. EWS–FLI1 induces GLI1 transcription via direct binding to the GLI1 promoter, which is a critical molecular mechanism in Ewing sarcoma development, as interference with GLI1 function blocks EWS cell proliferation and tumor growth *in vitro* and *in vivo* (Beauchamp *et al.*, 2009, 2011).

C. Noncanonical negative regulators of GLI activity

Aside from the numerous oncogenic signals that positively affect GLI activity, there is increasing evidence for noncanonical PTCH-independent repression of GLI activity by tumor suppressor genes (Fig. 2.1). Several studies have demonstrated a negative regulatory interaction between GLI and the p53 tumor suppressor. Stecca and Ruiz i Altaba (2009) found that p53 negatively controls GLI1 transcriptional activity, prevents nuclear GLI1 localization, and reduces GLI1 expression levels. Conversely, GLI1 expression reduces p53 levels via upregulation of the p53 inhibitor HMD2/MDM2, suggesting a mutual negative interaction loop between GLI1 and p53 (Abe *et al.*, 2008; Stecca and Ruiz i Altaba, 2009). This would also explain the dramatic enhancement in the tumor phenotype of patched heterozygous mice upon concomitant deletion of p53 (Romer *et al.*, 2004).

In neural cells, Numb, a tumor suppressor involved in asymmetric division of stem cells, encodes another PTCH-independent negative regulator of GLI1. Numb inhibits GLI function by targeting GLI1 to Itch-dependent proteasomal degradation (Di Marcotullio *et al.*, 2006). A different negative regulatory mechanism of GLI activity applies to the Notch signaling pathway, which in skin displays tumor suppressor function (Dotto, 2008). Analysis of mice with epidermal-specific deletion of Notch1 revealed upregulation of Gli2 expression and BCC development, respectively, suggesting that Notch1-mediated negative control mechanisms actively prevent epidermal Gli2 expression in normal skin (Nicolas *et al.*, 2003). The negative interaction of Notch–HH/GLI may, however, be context- and cell-type-specific as Notch signaling may cooperate with HH/GLI in medulloblastoma development, although data from different laboratories have provided somewhat controversial results (Hallahan *et al.*, 2004; Hatton *et al.*, 2010; Julian *et al.*, 2010).

Finally, recent reports have implicated microRNAs as regulators of HH/GLI signaling. For instance, miR125b, miR326, and miR324-5p all suppress SMO expression, while miR324-5p also reduces GLI1 expression (Ferretti *et al.*, 2008).

In light of these studies, it is obvious that the regulation of GLI activity is not only subject to canonical HH/PTCH/SMO-dependent signaling but also controlled by the integration of multiple non-HH signals. This concept has led to the funnel hypothesis, which views GLI proteins as an information nexus in the regulation of cell fate, stemness, and cancer (Ruiz i Altaba *et al.*, 2007; Stecca and Ruiz, 2010). The SMO-independent activation of GLI activator forms therefore challenges the efficacy of SMO inhibitors as therapeutic drugs in GLI-dependent cancers lacking classical HH/GLI signaling. This may apply to some cancers of the hematopoietic system where canonical HH/GLI signaling may be compromised by the possible lack of a primary cilium (Pazour and Witman, 2003).

IV. Hedgehog in Hematopoiesis and Hematopoietic Stem Cells

Adult hematopoiesis begins in the bone marrow with the differentiation of multipotent hematopoietic stem cells into progenitors, and further into all blood cell types of the myeloid and lymphoid lineages. As with other developmental signaling pathways such as Wnt or Notch signaling, the role of HH signaling in normal and malignant hematopoiesis and in the regulation of adult stem and progenitor cell pools has been the subject of intense studies. While HH signaling was initially reported to be of great importance for these processes, recent studies have challenged this view and have led to an ongoing discussion on the contribution of HH to the different aspects of hematopoiesis.

A role of HH in human hematopoiesis was first described in 2001 when Bhardwaj *et al.* (2001) observed that primitive $CD34^+$ human blood cells express SHH, PTCH, and SMO. Addition of exogenous SHH to *in vitro* expansion cultures increased the number of functional progenitor cells and promoted their differentiation. HH inhibition preserved the primitive state of the human progenitors and their capability to repopulate the murine bone marrow when adoptively transferred to NOD/SCID mice. Further, high levels of SHH stimulation caused proliferation of hematopoietic stem and progenitor cells (HSC/Ps) via downstream signaling cascades involving the hematopoietic inducer bone morphogenetic protein 4 (Bhardwaj *et al.*, 2001).

In a later study, IHH but not SHH was found to be expressed by bone marrow stromal cells (Kobune *et al.*, 2004). Consistently, IHH gene transfer into stromal cells enhanced their hematopoietic supporting potential. Expansion of repopulating cells on these IHH-overexpressing stromal cells resulted in increased engraftment in NOD/SCID mice. While in these assays IHH was presented by the stromal cells, it is important to note that it is also intrinsically present on $CD34^+$ cells and may also support the proliferative capacity of the HSC/Ps in an autocrine manner (Kobune *et al.*, 2004).

Subsequently, several mouse models with modulation of HH signaling pathway members were used to elucidate HH signaling activity in the hematopoietic system. HH signaling activity is increased with impaired activity of the pathway suppressor PTCH1, encoded by the tumor suppressor gene *PTCH1.* Mice carrying only one allele of *Ptch* ($Ptch^{+/-}$ mice) exhibited a higher expression of GLI1 and cyclin-D1 in primitive Lin- Sca-1+ c-Kit+ (LSK) hematopoietic cells along with an increase of their total number and frequency (Trowbridge *et al.*, 2006). In addition, the heterozygous mice demonstrated an enhanced ability to regenerate the hematopoietic system following its ablation with 5-fluorouracil (5-FU) in both primary and transplant models. However, these phenomena turned out to be transient and the regeneration capacity of the HSCs could not be sustained beyond 8 weeks posttransplantation despite continued HH activity (Trowbridge *et al.*, 2006). Uhmann *et al.* (2007) used a tamoxifen-inducible Ptch1 deletion to demonstrate that Ptch1 is necessary for commitment to the B- and T-cell lineage fate. They did not find any influence of Ptch ablation on the myeloid lineage (Uhmann *et al.*, 2007). However, transfer assays of Lin- bone marrow cells from $tPtch^{-/-}$ mice into $Rag2^{-/-}$ $\gamma^{-/-}$ mice lacking mature lymphocytes suggested that the defects in lymphoid lineage commitments were caused by the Ptch defect in the stromal compartment rather than in the progenitors (Uhmann *et al.*, 2007). To further determine cell-intrinsic versus -extrinsic hematopoietic effects of Hh signaling activation, Siggins *et al.* (2009) used inducible, cell-type-specific deletion models of Ptch1. Hematopoietic-specific deletion of Ptch1 using a tamoxifen-inducible Cre-ER recombinase under the control of the stem cell leukemia enhancer did not lead to Hh signaling activation in hematopoietic cells or to any phenotypic effects. In contrast, deletion of Ptch1 in epithelial cells led to the apoptosis of lymphoid progenitors, whereas deletion of Ptch1 in the bone marrow cell niche led to increased numbers in LKS cells and increased mobilization of myeloid progenitors (Siggins *et al.*, 2009).

Whether or not SMO and SMO-dependent HH signaling play a role in hematopoiesis is a subject of strong controversy. Obviously, the effects of SMO deletion are highly contextual and dependent on the timing of deletion. Dierks *et al.* (2008) transplanted murine $Smo^{-/-}$ fetal liver cells into sublethally irradiated mice and observed no significant effect of the Smo defect on long-term reconstitution of the bone marrow. However, they noted that the short-term repopulation of the bone marrow by HSCs was impaired (Dierks *et al.*, 2008). When challenging the hematopoietic regeneration capacity of these mice by injecting 5-FU, they found reduced regeneration 10 days but not 40 days after the treatment. Thus, loss of Smo seems to affect short-term progenitors rather than long-term repopulating HSCs (Dierks *et al.*, 2008). In contrary, Zhao *et al.* (2009), using a Vav-driven Cre-mediated Smo deletion model, observed a clear defect in long-term HSC function in this mouse model including decreased reconstitution of the bone marrow in transplantation assays.

While the above-mentioned reports mostly supported the by-then consensus view that HH signaling is needed in adult hematopoiesis, two seminal studies published back-to-back challenged this view in 2009 when they both suggested its complete dispensability (Gao *et al.*, 2009; Hofmann *et al.*, 2009). Both studies were based on use of an inducible conditional Smo knockout model in adult mice. A Cre recombinase under the control of myxovirus-resistance 1 (Mx1) gene promoter (Mx1-Cre) allowed for interferon-inducible Smo deletion by stimulation with polyI:polyC. This deletion of Smo in adult mice did not change their hematopoiesis. Consistently, no differences in peripheral blood counts, colony formation *in vitro*, and HSC/Ps numbers upon Smo deletion could be detected by the authors of both groups (Gao *et al.*, 2009; Hofmann *et al.*, 2009). Strikingly, HSC-specific gene expression signature was preserved in the Smo-deficient HSCs (Gao *et al.*, 2009) and pharmacological inhibition of Hh signaling did not affect murine hematopoiesis (Hofmann *et al.*, 2009). One explanation for these observations might be that despite the expression of Hh upstream elements, Hh signaling activity is shut off in the adult murine hematopoietic system. Indeed, expression of the downstream elements Gli1, Gli2, or Gli3, which serve as a sign for pathway activity, was not detectable in these LSKs and myeloid progenitors (Gao *et al.*, 2009).

Focusing on the positive downstream effector Gli1, Merchant *et al.* (2010) found that mice with a homozygous LacZ insertion in the first exon of Gli1 displayed an increase in the long-term (LT)-HSC compartment. Gli^{null} LT-HSCs were more quiescent and had a higher engraftment potential upon transplantation. However, in the proliferative progenitor compartment, impaired myeloid differentiation and defective hematopoiesis in response to stress was observed (Merchant *et al.*, 2010).

V. From HSC to CML

Chronic myeloid leukemia (CML) is a myeloproliferative clonal disorder that originates from a transformed hematopoietic stem or multipotent progenitor cell. CML is characterized by the Philadelphia chromosome, resulting from the translocation t(9;22) between chromosomes 22 and 9. The resulting Bcr–Abl fusion gene encodes a constitutively active tyrosine kinase and results in strong malignant hematopoiesis and gross disturbances in the normal hematopoietic bone marrow environment (Sawyers, 1992). Looking at GLI1 and PTCH in $CD34^{+}$ cells from CML patients, Dierks *et al.* (2008) observed higher transcript levels in samples from CML patients compared to those from healthy donors. To further evaluate this finding, they induced a CML-like disorder in mice by introducing a Bcr–Abl retrovirus in fetal liver cells that were either Smo negative ($Smo^{-/-}$) or heterozygous for Ptch ($Ptch^{+/-}$). These cells were then transplanted into lethally irradiated recipient mice. The retroviral transduction caused upregulation of Smo in the respective Bcr–Abl-positive cells, with higher efficiency in the $Ptch^{+/-}$ model than in WT. $Smo^{-/-}$ Bcr-Abl cells had a reduced potential for expansion, and the disease developed with longer latency and at lower frequency. Treatment with the Smo inhibitor cyclopamine-induced apoptosis in the leukemic cells *ex vivo* and reduced their clonogenic potential (Dierks *et al.*, 2008). The therapeutic potential of Smo inhibition was then explored in combination with the Abl inhibitor nilotinib, both *in vitro* and *in vivo*. *In vitro*, the combination of both agents induced apoptosis of CML cells while sparing those from healthy donors. *In vivo*, combination treatment of leukemic mice reduced their tumor burden and prolonged their overall survival (Dierks *et al.*, 2008). Analogous findings were made by Zhao *et al.* (2009). When they transplanted Bcr–Abl-transduced hematopoietic $Smo^{-/-}$ or WT progenitors into irradiated mice, the $Smo^{-/-}$ cells caused CML in about half of the recipients and with increased latency, whereas transduced WT cells caused leukemia in almost all recipients. Vice versa, transduction of cells harboring constitutively active Smo strongly accelerated the disease progression. Treatment of the diseased mice with cyclopamine extended their overall survival, and leukemic cells from these treated mice were unable to induce the disease upon subsequent transplantation (Zhao *et al.*, 2009).

The regulation of the stem cell compartment in CML is dynamic, in particular, during disease progression to a blast crisis and might involve a complex cross-regulation of Hh, Wnt, Notch, and hox signaling pathways (Sengupta *et al.*, 2007). Significant upregulation of Ptch1 and cyclin D1, upon other proteins, marks the blast crisis with Shh-dependent Stat3 activation hypothetically regulating the interconnection between the signaling pathways (Sengupta *et al.*, 2007).

In light of this complex time- and tissue-dependent regulation, it is not difficult to see why inconsistent data on the role of Smo in malignant hematopoiesis exist. For example, Hofmann *et al.* (2009) induced AML by transducing whole bone marrow from Smo-null or WT animals with retrovirus harboring the leukemia-associated disease allele MLL-AF9 and could not detect any influence of Smo deficiency on the replating potential of the transduced cells on methylcellulose or on AML development when transplanting these cells into mice (Hofmann *et al.*, 2009).

VI. T- and B-Cell Malignancies

A. Hedgehog role in T cells and T-ALL

The complexity of the Hh signaling pathway also becomes apparent when dissecting its contribution to T-lymphocyte differentiation in the thymus. Hh signaling regulates distinct stages of thymocyte development (Drakopoulou *et al.*, 2010; El Andaloussi *et al.*, 2006; Outram *et al.*, 2000; Rowbotham *et al.*, 2007). Using several experimental models to ablate Smo function in a T-cell specific manner, El Andaloussi *et al.* (2006) found that Smo-dependent Hh signaling is necessary for early T-cell progenitor survival, proliferation, and differentiation (El Andaloussi *et al.*, 2006). However, recent data from Uhmann *et al.* (Ji, 2011), who analyzed cell-type-specific deletion of Ptch in mice, suggested that T-cell development in thymus is independent of T-cell intrinsic Ptch expression but that Ptch is involved in either the homing of thymic progenitors or their further development. Expression of constitutively active Gli2 or activation of the Hh signaling pathway with Shh shifts the balance in the thymocyte subpopulations toward the $CD8^+$ population suggesting that Hh signaling affects CD4/8 lineage commitment (Rowbotham *et al.*, 2007). As TCR signal strength during repertoire selection controls decision toward the single positive lineage (Kappes and He, 2005), this could point to an influence of Hh signaling on TCR signals (Rowbotham *et al.*, 2007).

Evidence for proliferative and/or antiapoptotic effects of Hh ligands on T lymphocytes were first reported by the group around Lamb and Howie

(Lowrey *et al.*, 2002; Stewart *et al.*, 2002). While Shh addition did not have any effect on resting $CD4^+$ T cells, it significantly enhanced the proliferation of activated ones. In addition, when activated T lymphocytes were exposed to Shh, increased transcription of Ptch, which is known as a classical downstream target of canonical Hh signaling, was found in about half of the investigated samples (Stewart *et al.*, 2002). Lack of Smo can result in loss of the $CD8^+$ T-cell population (Dierks *et al.*, 2008). However, as conditional Smo deletion in the T-cell lineage did not affect the activation of T cells (El Andaloussi *et al.*, 2006), the requirement of T-cell intrinsic Smo remains undetermined. In line with this, Smo was found to be dispensable for the induction of T-acute lymphoblastic leukemia (T-ALL) (Gao *et al.*, 2009). When the authors transformed $Smo^{-/-}$ or WT adult murine cells with an activated form of Notch and transplanted them to recipients to induce a Notch-dependent T-ALL, they did not find any role of Smo in the onset or progression of this disease. In addition, treatment of several T-ALL lines with cyclopamine did not reveal any role of Smo in leukemic cell survival or proliferation (Gao *et al.*, 2009). However, other signaling pathways involved in survival of malignant T-lymphocyte might regulate Gli in noncanonical ways. For example, Singh *et al.* (2009) found that the SHH/GLI axis is activated in anaplastic large cell lymphoma with expression of anaplastic lymphoma kinase (ALK^+) and described a cross talk of the PI3K/Akt axis and GLI in these cells.

B. Hedgehog role in B cells and B-cell malignancies

The influence of HH signals on differentiation and malignant transformation of B cells is still poorly understood although several studies have looked into the role of hedgehog signaling in B-cell malignancies. Sacedon *et al.* (2005) suggested that the production of Shh by follicular dendritic cells provides antiapoptotic effects to B cells within germinal centers (GCs). Further, blockade of Hh signaling by addition of cyclopamine or anti-Shh antibody to proliferating GC B-cell cultures reduced their proliferation and their ability to produce different immunoglobulin isotypes (Sacedon *et al.*, 2005). Later, a comparative approach of immunohistochemical detection of SHH expression in the microenvironment of several B-cell malignancies was pursued by Kim *et al.* (2009). The authors confirmed SHH expression by follicular dendritic cells in nonmalignant lymph nodes and further investigated its expression in lymph nodes from diffused large B-cell lymphoma (DLBCL), follicular lymphoma, and chronic lymphocytic leukemia (CLL) patients. They found SHH positivity in 91% of investigated cases of DLBCL. It is important to note, however, that SHH was expressed by the tumor cells rather than by the tumor microenvironment indicating an autocrine loop of SHH signaling in this leukemia (Kim *et al.*, 2009). Low expression of SHH was also found in a subset of centroblasts in 29% of investigated lymph nodes from follicular lymphoma patients. In contrast, none of the five investigated lymph nodes from CLL patients was clearly positive for SHH, albeit prolymphocytes and paraimmunoblasts inside proliferation centers displayed sporadic SHH expression (Kim *et al.*, 2009) (see also below).

With regard to the major HH receptors and downstream transcription factors, most studies have reported positive transcriptional expression data and a reduction in tumor cell viability *in vitro* following treatment with the SMO inhibitor cyclopamine (see below). However, the fact that cyclopamine exhibits significant off-target effects at high doses (Yauch *et al.*, 2008) complicates our understanding of the existing data. Dierks *et al.* (2007) established a variety of Myc-positive primary lymphoma cell cultures from Eμ-Myc mice, a well-established model for mouse B-cell lymphomas. Activation of Hh signaling activity in these cells by either stimulation with recombinant Shh or Ihh or treatment with the Smo agonist purmorphamine increased lymphoma cell survival, which could be reversed by Hh antagonism with 5E1 or cyclopamine. In addition, cyclopamine treatment of mice with fully developed disease effectively inhibited lymphoma growth (Dierks *et al.*, 2007). In DLBCL, canonical SMO-dependent signaling via stroma-secreted SHH seems to be involved in development of chemoresistance of the malignant B lymphocytes (Kim *et al.*, 2009; Singh *et al.*, 2010, 2011). Such multidrug resistance may—at least in part—arise from high expression of the adenosine triphosphate-binding cassette drug transporter ABCG2 on the malignant cells (Kim *et al.*, 2009), which enables efflux of chemotherapeutic drugs such as doxorubicin (Gottesman *et al.*, 2002). The majority of investigated DLBCL cell lines turned out to be SMO positive and sensitive toward SMO inhibition with cyclopamine (compared to the negative control compound tomatidine). The same group of authors linked HH signaling to ABCG2 expression in DLBCL (Singh *et al.*, 2011). Activation of HH signaling by SHH addition or coculture of DLBCL with a human stromal cell line increased ABCG2 expression accompanied by increased chemotolerance. Vice versa, cyclopamine treatment of the DLBCL cell lines resulted in decreased ABCG2 mRNA levels compared to treatment with tomatidine, consistent with the presence of a GLI transcription binding site within the ABCG2 promoter (Singh *et al.*, 2011).

Hegde *et al.* demonstrated that GLI1 and GLI2 were expressed in several mantle cell lymphoma cell lines and primary MCL cells from human patients (Hegde *et al.*, 2008a). In one of the investigated cell lines, JVM2, HH signaling activity upon SHH treatment could be confirmed by increased GLI1 transcription; and SHH significantly influenced cell proliferation in an SMO-dependent manner which could be antagonized by cyclopamine. On the contrary, SHH-mediated HH signaling or effects by SMO inhibition could not be demonstrated in several other investigated cell lines. From these data, the authors deduced subset-specific differences based on the classical (JVM2) versus blastoid cytological nature of the malignant B cells (Hegde *et al.*, 2008a). However, an additional explanation might be that canonical signaling via SMO takes place only in JVM2 cells, which express higher levels of SMO as compared to the other cell lines. In contrast to the differential dependence on SMO, functional intrinsic GLI was shown to be indispensable for proliferation in all investigated cell lines, as demonstrated by antisense experiments, and is also involved in chemoresistance of the cells (Hegde *et al.*, 2008a).

In CLL, canonical Hh signaling via SMO could not be definitely proven (Fig. 2.2). CLL is a non-Hodgkin lymphoma characterized by leukemic accumulation of malignant B lymphocytes in peripheral blood, invariable infiltration of bone marrow, and progressive infiltration of secondary lymphoid organs. *In vitro*, CLL cells undergo rapid spontaneous apoptosis if not cocultured with stromal cells or with cells of the monocyte lineage such as "nurse-like" cells (Burger *et al.*, 2000; Kurtova *et al.*, 2009). Also, it is thought that proliferation of CLL takes place in specific proliferation centers inside lymph nodes and bone marrow (Schmid and Isaacson, 1994) and this is governed by microenvironmental stimuli. This high dependence on cell–cell contact or soluble factors secreted by the microenvironment prompted an investigation into the contribution of stroma-derived HH ligands on tumor cell survival in CLL. Based on the finding that GLI1 transcription is increased upon incubation of CLL cells with conditioned medium derived from stromal cells and that cyclopamine induces a decrease in cell viability, Hegde *et al.* (2008b) suggested that HH ligands secreted by the stromal cells contribute to CLL cell protection in the coculture system via canonical SMO-dependent HH signaling. While we could confirm a cyclopamine-mediated decrease in CLL cell viability when using the same dosage, we also observed a reduction in viability by tomatidine, the negative control compound (Desch *et al.*, 2010). Further, SMO expression is strongly reduced in CLL as compared to normal B lymphocytes, and siRNA-mediated knockdown of SMO in a CLL cell line exhibiting higher SMO levels did not decrease its cell viability. Moreover, CLL cells did not specifically react to SHH (Desch *et al.*, 2010).

In contrast, we could define an intrinsic, SMO-independent role of GLI in CLL cell survival (Fig. 2.2). Treatment of CLL cells with the GLI antagonist GANT61 massively reduced their cell viability without affecting the viability of normal B lymphocytes and this could be verified by knockdown experiments of GLI and transcriptional reduction of the cognate HH–GLI target gene PTCH. Notably, we observed a strong *in vitro* response of fludarabine-resistant CLL samples toward GLI inhibition, suggesting a therapeutic potential of GLI targeting especially in this difficult-to-treat group. We could not confirm a robust role of HH ligands secreted by stromal cells in supporting the survival of CLL cells and suggest that factors other than HH ligands from these accessory cells are beneficial for CLL cell

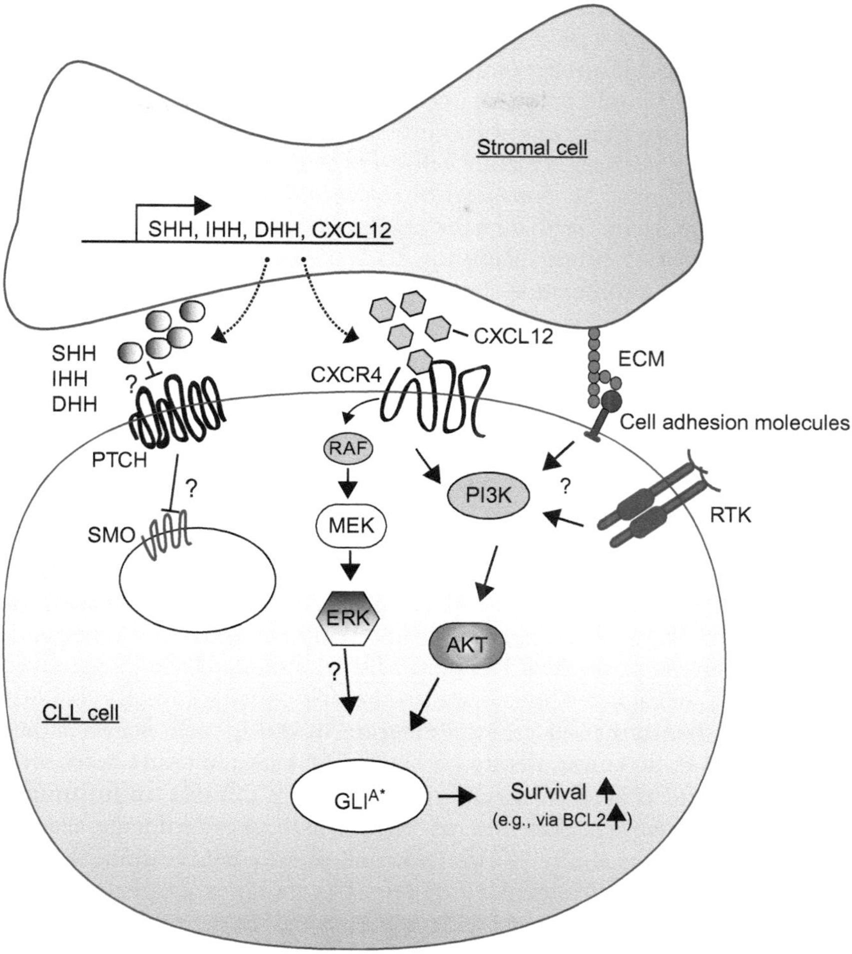

Figure 2.2 Model of SMO-independent GLI activity regulation in CLL. Stromal cells support the survival of CLL cells by cell contact-dependent and -independent signals. Stromal cells express secreted factors including HH ligands (SHH, IHH, and DHH) and chemokines such as the CXCR4 ligand CXCL12. As we have shown that genetic and pharmacological targeting of GLI—but not of SMO—reduces survival of CLL cells, we propose SMO-independent regulation of GLI^{A} forms in CLL. Possible GLI activating signals could be provided by RAF/MEK/ERK and/or by PI3K/AKT signaling in response to CXCL12-mediated CXCR4 activation or downstream of adhesion molecules and B-cell antigen receptor signaling. In either scenario, the signals would be funneled through GLI^{A} forms, resulting in increased expression of GLI target genes such as the prominent prosurvival factor BCL2 (Regl *et al.*, 2004).

survival and/or proliferation. However, it is also possible that microenvironment-induced signals converge on GLI by noncanonical mechanisms. For example, we recently observed that the presence of protective stromal cells can activate Akt pathways in CLL (Hofbauer *et al.*, 2010). Convergence of the GLI and PI3K/Akt and MEK pathways have already been demonstrated in solid tumors (Schnidar *et al.*, 2009; Stecca and Ruiz, 2010; Stecca *et al.*, 2007) and could also occur in CLL.

Nevertheless, as our study was mostly focused on the peripheral blood pool of CLL cells, which is mainly cell-cycle arrested and nonproliferative, we do not discount the possibility that canonical SMO-dependent Hh signaling is important for malignant cell proliferation or for maintenance of a tumor-initiating subpopulation in CLL. Prolymphocytes and paraimmunoblasts inside proliferation centers displayed sporadic SHH expression (Kim *et al.*, 2009). Thus, HH pathway activity may occur in a small subpopulation, which is responsible for CLL leukemogenesis. In multiple myeloma, a subset of cells with active HH signaling is markedly concentrated in the tumor stem cell compartment (Peacock *et al.*, 2007). Accordingly, inhibition of HH signaling blocks clonal expansion of the disease but hardly affects the bulk of terminally differentiated plasma cells (Peacock *et al.*, 2007). In a study of pre-B acute lymphocytic leukemia (B-ALL), all cells of investigated cell lines and primary samples expressed the major HH pathway components PTCH1, PTCH2, SMO, and GLI1. Inhibition of HH signaling primarily impacted a highly clonogenic cell population expressing aldehyde dehydrogenase (ALDH) (Lin *et al.*, 2010). Treatment with cyclopamine or the SMO inhibitor IPI-926 reduced the frequency of ALDH+ cells in each cell line, as well as their clonogenic capacity. IPI-926 also mainly affected the self-renewal potential of B-ALL cells *in vivo*. When the cell line Reh was injected into NOD/SCID mice and then treated with IPI-926, no effects were observed in the primary transplantation model. However, in subsequent bone marrow transplantation experiments, a loss of serial transplantation ability of the leukemia due to SMO inhibition was observed (Lin *et al.*, 2010). Although in CLL, such a cancer stem cell or clonogenic subpopulation with self-renewal capacity has not yet been identified, it is possible that during proliferation HH pathway could be switched on by environmental factors, thus restoring the renewal potential of the cells.

VII. Outlook

The complexity of HH/GLI signal transduction including its context dependence and the various noncanonical variations of GLI regulation by pathway cross talk and signal integration pose challenges in defining the precise role of HH and GLI in cancer, particularly in hematopoietic diseases. The impressive list of highly specific HH pathway inhibitors currently tested in clinical trials (see Table 2.1) raises the hope that targeting HH signaling will mark a leap forward in molecular medicine and oncology, eventually providing a therapeutic benefit to many cancer patients. Results of the first clinical trials with BCC and medulloblastoma patients treated with SMO inhibitors are highly promising while others have been disappointing (Rudin *et al.*, 2009; Skvara *et al.*, 2011; Von Hoff *et al.*, 2009). Whether the failure of SMO inhibition in ovarian or metastatic colorectal cancer has to do with SMO-independent regulation of GLI activator forms remains to be addressed. Also, in HH-dependent medulloblastoma, treatment with SMO inhibitors can cause a dramatic, yet transient response that is rapidly compensated by the development of drug resistance (Rudin *et al.*, 2009). SMO inhibitor resistance has been ascribed to the selection for SMO mutants with reduced drug affinity, to amplification of GLI2 and/or parallel activation of PI3K/AKT signaling (Buonamici *et al.*, 2010; Yauch *et al.*, 2009).

In summary, and despite the first successful trials with SMO inhibitors in BCC patients, we propose that the proper combination of therapeutic compounds simultaneously targeting SMO and positively interacting pathways such as PI3K/AKT, RAS/RAF/MEK/ERK, or RTK signaling will eventually turn out to be the most successful strategy. In fact, a number of studies have provided evidence for an increased anticancer effect of such combinations at least *in vitro* and in preclinical mouse models (Buonamici *et al.*, 2010; Mimeault *et al.*, 2007; Schnidar *et al.*, 2009; Stecca *et al.*, 2007).

The design of proper therapeutic strategies for GLI-dependent yet SMO-independent malignant diseases such as CLL is likely to be more challenging as transcription factors are generally considered poorly druggable targets. However, the identification of HH pathway inhibitors acting downstream of SMO including GLI antagonists (Hyman *et al.*, 2009; Lauth *et al.*, 2007a) may open new avenues for novel therapeutic opportunities, although their *in vivo* anticancer activity still needs rigorous testing. In addition, the discovery that arsenic trioxide (ATO), which is successfully used for the treatment of anaplastic promyelocytic leukemia (de The and Chen, 2010), efficiently reduces survival of CLL cells from patients with poor prognosis (Merkel *et al.*, 2008), and directly affects GLI activation and stability (Beauchamp *et al.*, 2011; Kim *et al.*, 2010a,b), represents an important step toward anti-GLI activator-based strategies. In the future, it will therefore be necessary to critically evaluate the therapeutic efficacy of single agent therapies compared to combination regimes with compounds that target SMO, GLI, and GLI activity enhancing signaling pathways. Targeting HH signaling has developed into an exciting field of research, with a realistic chance of identifying novel rationale-based drug combinations that are likely to exceed the therapeutic benefit of current regimens in a variety of human malignancies.

ACKNOWLEDGMENTS

Work of the authors has been supported by the Austrian Science Fund FWF (projects P20652 to F. A., W1213 to F. A. and R. G., SFB program P021 to R. G.), the Austrian Genome Research Program GEN-AU (project MedSys MOGLI to F. A.), the priority program "Life Sciences and Health" of the University of Salzburg to F. A., the Austrian National Bank (project 13420 to T. N. H.), the Paracelsus Medical University Salzburg (project E-10/11/058-HAR to T. N. H.), the "Klinische Malignom und Zytokinforschung Salzburg-Innsbruck GmbH," and the province of Salzburg.

REFERENCES

Abe, Y., Oda-Sato, E., Tobiume, K., Kawauchi, K., Taya, Y., Okamoto, K., Oren, M., and Tanaka, N. (2008). Hedgehog signaling overrides p53-mediated tumor suppression by activating Mdm2. *Proc. Natl. Acad. Sci. USA* **105,** 4838–4843.

Aza-Blanc, P., Ramirez-Weber, F. A., Laget, M. P., Schwartz, C., and Kornberg, T. B. (1997). Proteolysis that is inhibited by hedgehog targets Cubitus interruptus protein to the nucleus and converts it to a repressor. *Cell* **89,** 1043–1053.

Bailey, J. M., Swanson, B. J., Hamada, T., Eggers, J. P., Singh, P. K., Caffery, T., Ouellette, M. M., and Hollingsworth, M. A. (2008). Sonic hedgehog promotes desmoplasia in pancreatic cancer. *Clin. Cancer Res.* **14,** 5995–6004.

Barakat, M. T., Humke, E. W., and Scott, M. P. (2010). Learning from Jekyll to control Hyde: Hedgehog signaling in development and cancer. *Trends Mol. Med.* **16,** 337–348.

Beachy, P. A., Karhadkar, S. S., and Berman, D. M. (2004). Tissue repair and stem cell renewal in carcinogenesis. *Nature* **432,** 324–331.

Beauchamp, E., Bulut, G., Abaan, O., Chen, K., Merchant, A., Matsui, W., Endo, Y., Rubin, J. S., Toretsky, J., and Uren, A. (2009). GLI1 is a direct transcriptional target of EWS–FLI1 oncoprotein. *J. Biol. Chem.* **284,** 9074–9082.

Beauchamp, E. M., Ringer, L., Bulut, G., Sajwan, K. P., Hall, M. D., Lee, Y. C., Peaceman, D., Ozdemirli, M., Rodriguez, O., Macdonald, T. J., Albanese, C., Toretsky, J. A., *et al.* (2011). Arsenic trioxide inhibits human cancer cell growth and tumor development in mice by blocking Hedgehog/GLI pathway. *J. Clin. Invest.* **121,** 148–160.

Berman, D. M., Karhadkar, S. S., Maitra, A., Montes De Oca, R., Gerstenblith, M. R., Briggs, K., Parker, A. R., Shimada, Y., Eshleman, J. R., Watkins, D. N., and Beachy, P. A. (2003). Widespread requirement for Hedgehog ligand stimulation in growth of digestive tract tumours. *Nature* **425,** 846–851.

Bhardwaj, G., Murdoch, B., Wu, D., Baker, D. P., Williams, K. P., Chadwick, K., Ling, L. E., Karanu, F. N., and Bhatia, M. (2001). Sonic hedgehog induces the proliferation of primitive human hematopoietic cells via BMP regulation. *Nat. Immunol.* **2,** 172–180.

Brewster, R., Mullor, J. L., and Ruiz i Altaba, A. (2000). Gli2 functions in FGF signaling during antero-posterior patterning. *Development* **127,** 4395–4405.

Buonamici, S., Williams, J., Morrissey, M., Wang, A., Guo, R., Vattay, A., Hsiao, K., Yuan, J., Green, J., Ospina, B., Yu, Q., Ostrom, L., *et al.* (2010). Interfering with resistance to smoothened antagonists by inhibition of the PI3K pathway in medulloblastoma. *Sci. Transl. Med.* **2,** 51ra70.

Burger, J. A., Tsukada, N., Burger, M., Zvaifler, N. J., Dell'Aquila, M., and Kipps, T. J. (2000). Blood-derived nurse-like cells protect chronic lymphocytic leukemia B cells from spontaneous apoptosis through stromal cell-derived factor-1. *Blood* **96,** 2655–2663.

Chen, J. K., Taipale, J., Cooper, M. K., and Beachy, P. A. (2002a). Inhibition of Hedgehog signaling by direct binding of cyclopamine to Smoothened. *Genes Dev.* **16,** 2743–2748.

Chen, J. K., Taipale, J., Young, K. E., Maiti, T., and Beachy, P. A. (2002b). Small molecule modulation of Smoothened activity. *Proc. Natl. Acad. Sci. USA* **99,** 14071–14076.

Chen, Y., Sasai, N., Ma, G., Yue, T., Jia, J., Briscoe, J., and Jiang, J. (2011). Sonic Hedgehog dependent phosphorylation by CK1alpha and GRK2 is required for ciliary accumulation and activation of Smoothened. *PLoS Biol.* **9,** e1001083.

de The, H., and Chen, Z. (2010). Acute promyelocytic leukaemia: Novel insights into the mechanisms of cure. *Nat. Rev. Cancer* **10,** 775–783.

Dennler, S., Andre, J., Verrecchia, F., and Mauviel, A. (2009). Cloning of the human GLI2 Promoter: Transcriptional activation by transforming growth factor-beta via SMAD3/beta-catenin cooperation. *J. Biol. Chem.* **284,** 31523–31531.

Desch, P., Asslaber, D., Kern, D., Schnidar, H., Mangelberger, D., Alinger, B., Stoecher, M., Hofbauer, S. W., Neureiter, D., Tinhofer, I., Aberger, F., Hartmann, T. N., *et al.* (2010). Inhibition of GLI, but not Smoothened, induces apoptosis in chronic lymphocytic leukemia cells. *Oncogene* **29,** 4885–4895.

Di Marcotullio, L., Ferretti, E., Greco, A., De Smaele, E., Po, A., Sico, M. A., Alimandi, M., Giannini, G., Maroder, M., Screpanti, I., and Gulino, A. (2006). Numb is a suppressor of Hedgehog signalling and targets Gli1 for Itch-dependent ubiquitination. *Nat. Cell Biol.* **8,** 1415–1423.

Dierks, C., Grbic, J., Zirlik, K., Beigi, R., Englund, N. P., Guo, G. R., Veelken, H., Engelhardt, M., Mertelsmann, R., Kelleher, J. F., Schultz, P., and Warmuth, M. (2007). Essential role of stromally induced hedgehog signaling in B-cell malignancies. *Nat. Med.* **13,** 944–951.

Dierks, C., Beigi, R., Guo, G. R., Zirlik, K., Stegert, M. R., Manley, P., Trussell, C., Schmitt-Graeff, A., Landwerlin, K., Veelken, H., and Warmuth, M. (2008). Expansion of Bcr–Abl-positive leukemic stem cells is dependent on Hedgehog pathway activation. *Cancer Cell* **14,** 238–249.

Dotto, G. P. (2008). Notch tumor suppressor function. *Oncogene* **27,** 5115–5123.

Drakopoulou, E., Outram, S. V., Rowbotham, N. J., Ross, S. E., Furmanski, A. L., Saldana, J. I., Hager-Theodorides, A. L., and Crompton, T. (2010). Non-redundant role for the transcription factor Gli1 at multiple stages of thymocyte development. *Cell Cycle* **9,** 4144–4152.

El Andaloussi, A., Graves, S., Meng, F., Mandal, M., Mashayekhi, M., and Aifantis, I. (2006). Hedgehog signaling controls thymocyte progenitor homeostasis and differentiation in the thymus. *Nat. Immunol.* **7,** 418–426.

Epstein, E. H. (2008). Basal cell carcinomas: Attack of the hedgehog. *Nat. Rev. Cancer* **8,** 743–754.

Ericson, J., Morton, S., Kawakami, A., Roelink, H., and Jessell, T. M. (1996). Two critical periods of Sonic Hedgehog signaling required for the specification of motor neuron identity. *Cell* **87,** 661–673.

Fan, Q., He, M., Sheng, T., Zhang, X., Sinha, M., Luxon, B., Zhao, X., and Xie, J. (2010). Requirement of TGFbeta signaling for SMO-mediated carcinogenesis. *J. Biol. Chem.* **285,** 36570–36576.

Ferretti, E., De Smaele, E., Miele, E., Laneve, P., Po, A., Pelloni, M., Paganelli, A., Di Marcotullio, L., Caffarelli, E., Screpanti, I., Bozzoni, I., and Gulino, A. (2008). Concerted microRNA control of Hedgehog signalling in cerebellar neuronal progenitor and tumour cells. *EMBO J.* **27,** 2616–2627.

Fogarty, M. P., Emmenegger, B. A., Grasfeder, L. L., Oliver, T. G., and Wechsler-Reya, R. J. (2007). Fibroblast growth factor blocks Sonic hedgehog signaling in neuronal precursors and tumor cells. *Proc. Natl. Acad. Sci. USA* **104,** 2973–2978.

Gallet, A. (2011). Hedgehog morphogen: From secretion to reception. *Trends Cell Biol.* **21,** 238–246.

Gao, J., Graves, S., Koch, U., Liu, S., Jankovic, V., Buonamici, S., El Andaloussi, A., Nimer, S. D., Kee, B. L., Taichman, R., Radtke, F., and Aifantis, I. (2009). Hedgehog signaling is dispensable for adult hematopoietic stem cell function. *Cell Stem Cell* **4,** 548–558.

Gottesman, M. M., Fojo, T., and Bates, S. E. (2002). Multidrug resistance in cancer: Role of ATP-dependent transporters. *Nat. Rev. Cancer* **2,** 48–58.

Hahn, H., Wojnowski, L., Specht, K., Kappler, R., Calzada-Wack, J., Potter, D., Zimmer, A., Muller, U., Samson, E., and Quintanilla-Martinez, L. (2000). Patched target Igf2 is indispensable for the formation of medulloblastoma and rhabdomyosarcoma. *J. Biol. Chem.* **275,** 28341–28344.

Hallahan, A. R., Pritchard, J. I., Hansen, S., Benson, M., Stoeck, J., Hatton, B. A., Russell, T. L., Ellenbogen, R. G., Bernstein, I. D., Beachy, P. A., and Olson, J. M. (2004). The SmoA1 mouse model reveals that notch signaling is critical for the growth and survival of sonic hedgehog-induced medulloblastomas. *Cancer Res.* **64,** 7794–7800.

Hartmann, W., Koch, A., Brune, H., Waha, A., Schuller, U., Dani, I., Denkhaus, D., Langmann, W., Bode, U., Wiestler, O. D., Schilling, K., and Pietsch, T. (2005). Insulin-like growth factor II is involved in the proliferation control of medulloblastoma and its cerebellar precursor cells. *Am. J. Pathol.* **166,** 1153–1162.

Hatton, B. A., Villavicencio, E. H., Pritchard, J., LeBlanc, M., Hansen, S., Ulrich, M., Ditzler, S., Pullar, B., Stroud, M. R., and Olson, J. M. (2010). Notch signaling is not essential in sonic hedgehog-activated medulloblastoma. *Oncogene* **29,** 3865–3872.

Hegde, G. V., Munger, C. M., Emanuel, K., Joshi, A. D., Greiner, T. C., Weisenburger, D. D., Vose, J. M., and Joshi, S. S. (2008a). Targeting of sonic hedgehog-GLI signaling: A potential strategy to improve therapy for mantle cell lymphoma. *Mol. Cancer Ther.* **7,** 1450–1460.

Hegde, G. V., Peterson, K. J., Emanuel, K., Mittal, A. K., Joshi, A. D., Dickinson, J. D., Kollessery, G. J., Bociek, R. G., Bierman, P., Vose, J. M., Weisenburger, D. D., and Joshi, S. S. (2008b). Hedgehog-induced survival of B-cell chronic lymphocytic leukemia cells in a stromal cell microenvironment: A potential new therapeutic target. *Mol. Cancer Res.* **6,** 1928–1936.

Hofbauer, S. W., Pinon, J. D., Brachtl, G., Haginger, L., Wang, W., Johrer, K., Tinhofer, I., Hartmann, T. N., and Greil, R. (2010). Modifying akt signaling in B-cell chronic lymphocytic leukemia cells. *Cancer Res.* **70,** 7336–7344.

Hofmann, I., Stover, E. H., Cullen, D. E., Mao, J., Morgan, K. J., Lee, B. H., Kharas, M. G., Miller, P. G., Cornejo, M. G., Okabe, R., Armstrong, S. A., Ghilardi, N., *et al.* (2009). Hedgehog signaling is dispensable for adult murine hematopoietic stem cell function and hematopoiesis. *Cell Stem Cell* **4,** 559–567.

Huangfu, D., and Anderson, K. V. (2006). Signaling from Smo to Ci/Gli: Conservation and divergence of Hedgehog pathways from Drosophila to vertebrates. *Development* **133,** 3–14.

Huangfu, D., Liu, A., Rakeman, A. S., Murcia, N. S., Niswander, L., and Anderson, K. V. (2003). Hedgehog signalling in the mouse requires intraflagellar transport proteins. *Nature* **426,** 83–87.

Humke, E. W., Dorn, K. V., Milenkovic, L., Scott, M. P., and Rohatgi, R. (2010). The output of Hedgehog signaling is controlled by the dynamic association between Suppressor of Fused and the Gli proteins. *Genes Dev.* **24,** 670–682.

Hyman, J. M., Firestone, A. J., Heine, V. M., Zhao, Y., Ocasio, C. A., Han, K., Sun, M., Rack, P. G., Sinha, S., Wu, J. J., Solow-Cordero, D. E., Jiang, J., *et al.* (2009). Small-molecule inhibitors reveal multiple strategies for Hedgehog pathway blockade. *Proc. Natl. Acad. Sci. USA* **106,** 14132–14137.

Ikram, M. S., Neill, G. W., Regl, G., Eichberger, T., Frischauf, A. M., Aberger, F., Quinn, A., and Philpott, M. (2004). GLI2 is expressed in normal human epidermis and BCC and induces GLI1 expression by binding to its promoter. *J. Invest. Dermatol.* **122,** 1503–1509.

Ingham, P. W., and McMahon, A. P. (2001). Hedgehog signaling in animal development: Paradigms and principles. *Genes Dev.* **15,** 3059–3087.

Ingham, P. W., Nakano, Y., and Seger, C. (2011). Mechanisms and functions of Hedgehog signalling across the metazoa. *Nat. Rev. Genet.* **12,** 393–406.

Ji, Z., Mei, F. C., Xie, J., and Cheng, X. (2007). Oncogenic KRAS activates hedgehog signaling pathway in pancreatic cancer cells. *J. Biol. Chem.* **282,** 14048–14055.

Ju, B., Spitsbergen, J., Eden, C. J., Taylor, M. R., and Chen, W. (2009). Co-activation of hedgehog and AKT pathways promote tumorigenesis in zebrafish. *Mol. Cancer* **8,** 40.

Julian, E., Dave, R. K., Robson, J. P., Hallahan, A. R., and Wainwright, B. J. (2010). Canonical Notch signaling is not required for the growth of Hedgehog pathway-induced medulloblastoma. *Oncogene* **29,** 3465–3476.

Kappes, D. J., and He, X. (2005). CD4–CD8 lineage commitment: An inside view. *Nat. Immunol.* **6,** 761–766.

Kasper, M., Regl, G., Frischauf, A. M., and Aberger, F. (2006a). GLI transcription factors: Mediators of oncogenic Hedgehog signalling. *Eur. J. Cancer* **42,** 437–445.

Kasper, M., Schnidar, H., Neill, G. W., Hanneder, M., Klingler, S., Blaas, L., Schmid, C., Hauser-Kronberger, C., Regl, G., Philpott, M. P., and Aberger, F. (2006b). Selective modulation of Hedgehog/GLI target gene expression by epidermal growth factor signaling in human keratinocytes. *Mol. Cell. Biol.* **26,** 6283–6298.

Kim, J. E., Singh, R. R., Cho-Vega, J. H., Drakos, E., Davuluri, Y., Khokhar, F. A., Fayad, L., Medeiros, L. J., and Vega, F. (2009). Sonic hedgehog signaling proteins and ATP-binding cassette G2 are aberrantly expressed in diffuse large B-cell lymphoma. *Mod. Pathol.* **22,** 1312–1320.

Kim, J., Lee, J. J., Gardner, D., and Beachy, P. A. (2010a). Arsenic antagonizes the Hedgehog pathway by preventing ciliary accumulation and reducing stability of the Gli2 transcriptional effector. *Proc. Natl. Acad. Sci. USA* **107,** 13432–13437.

Kim, J., Tang, J. Y., Gong, R., Lee, J. J., Clemons, K. V., Chong, C. R., Chang, K. S., Fereshteh, M., Gardner, D., Reya, T., Liu, J. O., Epstein, E. H., *et al.* (2010b). Itraconazole, a commonly used antifungal that inhibits Hedgehog pathway activity and cancer growth. *Cancer Cell* **17,** 388–399.

Kobune, M., Ito, Y., Kawano, Y., Sasaki, K., Uchida, H., Nakamura, K., Dehari, H., Chiba, H., Takimoto, R., Matsunaga, T., Terui, T., Kato, J., *et al.* (2004). Indian hedgehog gene transfer augments hematopoietic support of human stromal cells including NOD/SCID-beta2m$^{-/-}$ repopulating cells. *Blood* **104,** 1002–1009.

Kovacs, J. J., Whalen, E. J., Liu, R., Xiao, K., Kim, J., Chen, M., Wang, J., Chen, W., and Lefkowitz, R. J. (2008). Beta-arrestin-mediated localization of smoothened to the primary cilium. *Science* **320,** 1777–1781.

Kurtova, A. V., Balakrishnan, K., Chen, R., Ding, W., Schnabl, S., Quiroga, M. P., Sivina, M., Wierda, W. G., Estrov, Z., Keating, M. J., Shehata, M., Jager, U., *et al.* (2009). Diverse marrow stromal cells protect CLL cells from spontaneous and drug-induced apoptosis: Development of a reliable and reproducible system to assess stromal cell adhesion-mediated drug resistance. *Blood* **114,** 4441–4450.

Laner-Plamberger, S., Kaser, A., Paulischta, M., Hauser-Kronberger, C., Eichberger, T., and Frischauf, A. M. (2009). Cooperation between GLI and JUN enhances transcription of JUN and selected GLI target genes. *Oncogene* **28,** 1639–1651.

Lau, J., Kawahira, H., and Hebrok, M. (2006). Hedgehog signaling in pancreas development and disease. *Cell. Mol. Life Sci.* **63,** 642–652.

Lauth, M., Bergstrom, A., Shimokawa, T., and Toftgard, R. (2007a). Inhibition of GLI-mediated transcription and tumor cell growth by small-molecule antagonists. *Proc. Natl. Acad. Sci. USA* **104,** 8455–8460.

Lauth, M., Bergstrom, A., and Toftgard, R. (2007b). Phorbol esters inhibit the Hedgehog signalling pathway downstream of Suppressor of Fused, but upstream of Gli. *Oncogene* **26,** 5163–5168.

Lauth, M., Bergstrom, A., Shimokawa, T., Tostar, U., Jin, Q., Fendrich, V., Guerra, C., Barbacid, M., and Toftgard, R. (2010). DYRK1B-dependent autocrine-to-paracrine shift of Hedgehog signaling by mutant RAS. *Nat. Struct. Mol. Biol.* **17,** 718–725.

Lin, T. L., Wang, Q. H., Brown, P., Peacock, C., Merchant, A. A., Brennan, S., Jones, E., McGovern, K., Watkins, D. N., Sakamoto, K. M., and Matsui, W. (2010). Self-renewal of acute lymphocytic leukemia cells is limited by the Hedgehog pathway inhibitors cyclopamine and IPI-926. *PLoS One* **5,** e15262.

Lowrey, J. A., Stewart, G. A., Lindey, S., Hoyne, G. F., Dallman, M. J., Howie, S. E., and Lamb, J. R. (2002). Sonic hedgehog promotes cell cycle progression in activated peripheral CD4(+) T lymphocytes. *J. Immunol.* **169,** 1869–1875.

Mao, J., Maye, P., Kogerman, P., Tejedor, F. J., Toftgard, R., Xie, W., Wu, G., and Wu, D. (2002). Regulation of Gli1 transcriptional activity in the nucleus by Dyrk1. *J. Biol. Chem.* **277,** 35156–35161.

Meloni, A. R., Fralish, G. B., Kelly, P., Salahpour, A., Chen, J. K., Wechsler-Reya, R. J., Lefkowitz, R. J., and Caron, M. G. (2006). Smoothened signal transduction is promoted by G protein-coupled receptor kinase 2. *Mol. Cell. Biol.* **26,** 7550–7560.

Merchant, A., Joseph, G., Wang, Q., Brennan, S., and Matsui, W. (2010). Gli1 regulates the proliferation and differentiation of HSCs and myeloid progenitors. *Blood* **115,** 2391–2396.

Merkel, O., Heyder, C., Asslaber, D., Hamacher, F., Tinhofer, I., Holler, C., Stocher, M., Prokesch, A., Papak, C., Scheideler, M., Trajanoski, Z., and Greil, R. (2008). Arsenic trioxide induces apoptosis preferentially in B-CLL cells of patients with unfavourable prognostic factors including del17p13. *J. Mol. Med. (Berl)* **86,** 541–552.

Milenkovic, L., Scott, M. P., and Rohatgi, R. (2009). Lateral transport of Smoothened from the plasma membrane to the membrane of the cilium. *J. Cell Biol.* **187,** 365–374.

Mimeault, M., Johansson, S. L., Vankatraman, G., Moore, E., Henichart, J. P., Depreux, P., Lin, M. F., and Batra, S. K. (2007). Combined targeting of epidermal growth factor receptor and hedgehog signaling by gefitinib and cyclopamine cooperatively improves the cytotoxic effects of docetaxel on metastatic prostate cancer cells. *Mol. Cancer Ther.* **6,** 967–978.

Nicolas, M., Wolfer, A., Raj, K., Kummer, J. A., Mill, P., van Noort, M., Hui, C. C., Clevers, H., Dotto, G. P., and Radtke, F. (2003). Notch1 functions as a tumor suppressor in mouse skin. *Nat. Genet.* **33,** 416–421.

Nolan-Stevaux, O., Lau, J., Truitt, M. L., Chu, G. C., Hebrok, M., Fernandez-Zapico, M. E., and Hanahan, D. (2009). GLI1 is regulated through Smoothened-independent mechanisms in neoplastic pancreatic ducts and mediates PDAC cell survival and transformation. *Genes Dev.* **23,** 24–36.

Nusslein-Volhard, C., and Wieschaus, E. (1980). Mutations affecting segment number and polarity in Drosophila. *Nature* **287,** 795–801.

Outram, S. V., Varas, A., Pepicelli, C. V., and Crompton, T. (2000). Hedgehog signaling regulates differentiation from double-negative to double-positive thymocyte. *Immunity* **13,** 187–197.

Palma, V., and Ruiz i Altaba, A. (2004). Hedgehog-GLI signaling regulates the behavior of cells with stem cell properties in the developing neocortex. *Development* **131,** 337–345.

Pasca di Magliano, M., Sekine, S., Ermilov, A., Ferris, J., Dlugosz, A. A., and Hebrok, M. (2006). Hedgehog/Ras interactions regulate early stages of pancreatic cancer. *Genes Dev.* **20,** 3161–3173.

Pazour, G. J., and Witman, G. B. (2003). The vertebrate primary cilium is a sensory organelle. *Curr. Opin. Cell Biol.* **15,** 105–110.

Peacock, C. D., Wang, Q., Gesell, G. S., Corcoran-Schwartz, I. M., Jones, E., Kim, J., Devereux, W. L., Rhodes, J. T., Huff, C. A., Beachy, P. A., Watkins, D. N., and Matsui, W. (2007). Hedgehog signaling maintains a tumor stem cell compartment in multiple myeloma. *Proc. Natl. Acad. Sci. USA* **104,** 4048–4053.

Philipp, M., Fralish, G. B., Meloni, A. R., Chen, W., MacInnes, A. W., Barak, L. S., and Caron, M. G. (2008). Smoothened signaling in vertebrates is facilitated by a G protein-coupled receptor kinase. *Mol. Biol. Cell* **19,** 5478–5489.

Price, M. A., and Kalderon, D. (2002). Proteolysis of the Hedgehog signaling effector Cubitus interruptus requires phosphorylation by Glycogen Synthase Kinase 3 and Casein Kinase 1. *Cell* **108,** 823–835.

Rao, G., Pedone, C. A., Del Valle, L., Reiss, K., Holland, E. C., and Fults, D. W. (2004). Sonic hedgehog and insulin-like growth factor signaling synergize to induce medulloblastoma formation from nestin-expressing neural progenitors in mice. *Oncogene* **23,** 6156–6162.

Regl, G., Neill, G. W., Eichberger, T., Kasper, M., Ikram, M. S., Koller, J., Hintner, H., Quinn, A. G., Frischauf, A. M., and Aberger, F. (2002). Human GLI2 and GLI1 are part of a positive feedback mechanism in Basal Cell Carcinoma. *Oncogene* **21,** 5529–5539.

Regl, G., Kasper, M., Schnidar, H., Eichberger, T., Neill, G. W., Philpott, M. P., Esterbauer, H., Hauser-Kronberger, C., Frischauf, A. M., and Aberger, F. (2004). Activation of the BCL2 promoter in response to Hedgehog/GLI signal transduction is predominantly mediated by GLI2. *Cancer Res.* **64,** 7724–7731.

Riobo, N. A., Haines, G. M., and Emerson, C. P., Jr. (2006a). Protein kinase C-delta and mitogen-activated protein/extracellular signal-regulated kinase-1 control GLI activation in hedgehog signaling. *Cancer Res.* **66,** 839–845.

Riobo, N. A., Lu, K., Ai, X., Haines, G. M., and Emerson, C. P., Jr. (2006b). Phosphoinositide 3-kinase and Akt are essential for Sonic Hedgehog signaling. *Proc. Natl. Acad. Sci. USA* **103,** 4505–4510.

Rohatgi, R., Milenkovic, L., and Scott, M. P. (2007). Patched1 regulates hedgehog signaling at the primary cilium. *Science* **317,** 372–376.

Romer, J. T., Kimura, H., Magdaleno, S., Sasai, K., Fuller, C., Baines, H., Connelly, M., Stewart, C. F., Gould, S., Rubin, L. L., and Curran, T. (2004). Suppression of the Shh pathway using a small molecule inhibitor eliminates medulloblastoma in Ptc1(+/−)p53 (−/−) mice. *Cancer Cell* **6,** 229–240.

Rowbotham, N. J., Hager-Theodorides, A. L., Cebecauer, M., Shah, D. K., Drakopoulou, E., Dyson, J., Outram, S. V., and Crompton, T. (2007). Activation of the Hedgehog signaling pathway in T-lineage cells inhibits TCR repertoire selection in the thymus and peripheral T-cell activation. *Blood* **109,** 3757–3766.

Rudin, C. M., Hann, C. L., Laterra, J., Yauch, R. L., Callahan, C. A., Fu, L., Holcomb, T., Stinson, J., Gould, S. E., Coleman, B., LoRusso, P. M., Von Hoff, D. D., *et al.* (2009). Treatment of medulloblastoma with hedgehog pathway inhibitor GDC-0449. *N. Engl. J. Med.* **361,** 1173–1178.

Ruiz i Altaba, A., Sanchez, P., and Dahmane, N. (2002). Gli and hedgehog in cancer: Tumours, embryos and stem cells. *Nat. Rev. Cancer* **2,** 361–372.

Ruiz i Altaba, A., Mas, C., and Stecca, B. (2007). The Gli code: An information nexus regulating cell fate, stemness and cancer. *Trends Cell Biol.* **17,** 438–447.

Sacedon, R., Diez, B., Nunez, V., Hernandez-Lopez, C., Gutierrez-Frias, C., Cejalvo, T., Outram, S. V., Crompton, T., Zapata, A. G., Vicente, A., and Varas, A. (2005). Sonic hedgehog is produced by follicular dendritic cells and protects germinal center B cells from apoptosis. *J. Immunol.* **174,** 1456–1461.

Sanchez, P., and Ruiz i Altaba, A. (2005). In vivo inhibition of endogenous brain tumors through systemic interference of Hedgehog signaling in mice. *Mech. Dev.* **122,** 223–230.
Sawyers, C. L. (1992). The Bcr–Abl gene in chronic myelogenous leukaemia. *Cancer Surv.* **15,** 37–51.
Scales, S. J., and de Sauvage, F. J. (2009). Mechanisms of Hedgehog pathway activation in cancer and implications for therapy. *Trends Pharmacol. Sci.* **30,** 303–312.
Schmid, C., and Isaacson, P. G. (1994). Proliferation centres in B-cell malignant lymphoma, lymphocytic (B-CLL): An immunophenotypic study. *Histopathology* **24,** 445–451.
Schnidar, H., Eberl, M., Klingler, S., Mangelberger, D., Kasper, M., Hauser-Kronberger, C., Regl, G., Kroismayr, R., Moriggl, R., Sibilia, M., and Aberger, F. (2009). Epidermal growth factor receptor signaling synergizes with Hedgehog/GLI in oncogenic transformation via activation of the MEK/ERK/JUN pathway. *Cancer Res.* **69,** 1284–1292.
Sengupta, A., Banerjee, D., Chandra, S., Banerji, S. K., Ghosh, R., Roy, R., and Banerjee, S. (2007). Deregulation and cross talk among Sonic hedgehog, Wnt, Hox and Notch signaling in chronic myeloid leukemia progression. *Leukemia* **21,** 949–955.
Siggins, S. L., Nguyen, N. Y., McCormack, M. P., Vasudevan, S., Villani, R., Jane, S. M., Wainwright, B. J., and Curtis, D. J. (2009). The Hedgehog receptor Patched1 regulates myeloid and lymphoid progenitors by distinct cell-extrinsic mechanisms. *Blood* **114,** 995–1004.
Singh, R. R., Cho-Vega, J. H., Davuluri, Y., Ma, S., Kasbidi, F., Milito, C., Lennon, P. A., Drakos, E., Medeiros, L. J., Luthra, R., and Vega, F. (2009). Sonic hedgehog signaling pathway is activated in ALK-positive anaplastic large cell lymphoma. *Cancer Res.* **69,** 2550–2558.
Singh, R. R., Kim, J. E., Davuluri, Y., Drakos, E., Cho-Vega, J. H., Amin, H. M., and Vega, F. (2010). Hedgehog signaling pathway is activated in diffuse large B-cell lymphoma and contributes to tumor cell survival and proliferation. *Leukemia* **24,** 1025–1036.
Singh, R. R., Kunkalla, K., Qu, C., Schlette, E., Neelapu, S. S., Samaniego, F., and Vega, F. (2011). ABCG2 is a direct transcriptional target of hedgehog signaling and involved in stroma-induced drug tolerance in diffuse large B-cell lymphoma. *Oncogene* **30**(49), 4874–4886.
Skvara, H., Kalthoff, F., Meingassner, J. G., Wolff-Winiski, B., Aschauer, H., Kelleher, J. F., Wu, X., Pan, S., Mickel, L., Schuster, C., Stary, G., Jalili, A., *et al.* (2011). Topical treatment of Basal cell carcinomas in nevoid Basal cell carcinoma syndrome with a smoothened inhibitor. *J. Invest. Dermatol.* **131,** 1735–1744.
Stanton, B. Z., Peng, L. F., Maloof, N., Nakai, K., Wang, X., Duffner, J. L., Taveras, K. M., Hyman, J. M., Lee, S. W., Koehler, A. N., Chen, J. K., Fox, J. L., *et al.* (2009). A small molecule that binds Hedgehog and blocks its signaling in human cells. *Nat. Chem. Biol.* **5,** 154–156.
Stecca, B., and Ruiz, I. A. A. (2010). Context-dependent regulation of the GLI code in cancer by HEDGEHOG and non-HEDGEHOG signals. *J. Mol. Cell Biol.* **2,** 84–95.
Stecca, B., and Ruiz i Altaba, A. (2009). A GLI1-p53 inhibitory loop controls neural stem cell and tumour cell numbers. *EMBO J.* **28,** 663–676.
Stecca, B., Mas, C., Clement, V., Zbinden, M., Correa, R., Piguet, V., Beermann, F., and Ruiz, I. A. A. (2007). Melanomas require HEDGEHOG-GLI signaling regulated by interactions between GLI1 and the RAS-MEK/AKT pathways. *Proc. Natl. Acad. Sci. USA* **104,** 5895–5900.
Stewart, G. A., Lowrey, J. A., Wakelin, S. J., Fitch, P. M., Lindey, S., Dallman, M. J., Lamb, J. R., and Howie, S. E. (2002). Sonic hedgehog signaling modulates activation of and cytokine production by human peripheral $CD4^+$ T cells. *J. Immunol.* **169,** 5451–5457.
Taipale, J., Chen, J. K., Cooper, M. K., Wang, B., Mann, R. K., Milenkovic, L., Scott, M. P., and Beachy, P. A. (2000). Effects of oncogenic mutations in Smoothened and Patched can be reversed by cyclopamine. *Nature* **406,** 1005–1009.

Taipale, J., Cooper, M. K., Maiti, T., and Beachy, P. A. (2002). Patched acts catalytically to suppress the activity of Smoothened. *Nature* **418,** 892–897.

Takebe, N., Harris, P. J., Warren, R. Q., and Ivy, S. P. (2011). Targeting cancer stem cells by inhibiting Wnt, Notch, and Hedgehog pathways. *Nat. Rev. Clin. Oncol.* **8,** 97–106.

Teglund, S., and Toftgard, R. (2010). Hedgehog beyond medulloblastoma and basal cell carcinoma. *Biochim. Biophys. Acta* **1805,** 181–208.

Thayer, S. P., di Magliano, M. P., Heiser, P. W., Nielsen, C. M., Roberts, D. J., Lauwers, G. Y., Qi, Y. P., Gysin, S., Fernandez-del Castillo, C., Yajnik, V., Antoniu, B., McMahon, M., *et al.* (2003). Hedgehog is an early and late mediator of pancreatic cancer tumorigenesis. *Nature* **425,** 851–856.

Tremblay, M. R., Lescarbeau, A., Grogan, M. J., Tan, E., Lin, G., Austad, B. C., Yu, L. C., Behnke, M. L., Nair, S. J., Hagel, M., White, K., Conley, J., *et al.* (2009). Discovery of a potent and orally active hedgehog pathway antagonist (IPI-926). *J. Med. Chem.* **52,** 4400–4418.

Trowbridge, J. J., Scott, M. P., and Bhatia, M. (2006). Hedgehog modulates cell cycle regulators in stem cells to control hematopoietic regeneration. *Proc. Natl. Acad. Sci. USA* **103,** 14134–14139.

Tukachinsky, H., Lopez, L. V., and Salic, A. (2010). A mechanism for vertebrate Hedgehog signaling: Recruitment to cilia and dissociation of SuFu-Gli protein complexes. *J. Cell Biol.* **191,** 415–428.

Uhmann, A., Dittmann, K., Nitzki, F., Dressel, R., Koleva, M., Frommhold, A., Zibat, A., Binder, C., Adham, I., Nitsche, M., Heller, T., Armstrong, V., *et al.* (2007). The Hedgehog receptor Patched controls lymphoid lineage commitment. *Blood* **110,** 1814–1823.

Uhmann, A., van den Brandt, J., Dittmann, K., Hess, I., Dressel, R., Binder, C., Luhder, F., Christiansen, H., Fassnacht, M., Bhandoola, A., Wienands, J., Reichardt, H. M., and Hahn, H. (2011). T cell development critically depends on prethymic stromal patched expression. *J. Immunol.* **186,** 3383–3391.

Varjosalo, M., and Taipale, J. (2007). Hedgehog signaling. *J. Cell Sci.* **120,** 3–6.

Varjosalo, M., Bjorklund, M., Cheng, F., Syvanen, H., Kivioja, T., Kilpinen, S., Sun, Z., Kallioniemi, O., Stunnenberg, H. G., He, W. W., Ojala, P., and Taipale, J. (2008). Application of active and kinase-deficient kinome collection for identification of kinases regulating hedgehog signaling. *Cell* **133,** 537–548.

Von Hoff, D. D., LoRusso, P. M., Rudin, C. M., Reddy, J. C., Yauch, R. L., Tibes, R., Weiss, G. J., Borad, M. J., Hann, C. L., Brahmer, J. R., Mackey, H. M., Lum, B. L., *et al.* (2009). Inhibition of the hedgehog pathway in advanced basal-cell carcinoma. *N. Engl. J. Med.* **361,** 1164–1172.

Wang, B., Fallon, J. F., and Beachy, P. A. (2000). Hedgehog-regulated processing of Gli3 produces an anterior/posterior repressor gradient in the developing vertebrate limb. *Cell* **100,** 423–434.

Wen, X., Lai, C. K., Evangelista, M., Hongo, J. A., de Sauvage, F. J., and Scales, S. J. (2010). Kinetics of hedgehog-dependent full-length Gli3 accumulation in primary cilia and subsequent degradation. *Mol. Cell. Biol.* **30,** 1910–1922.

Whisenant, T. C., Ho, D. T., Benz, R. W., Rogers, J. S., Kaake, R. M., Gordon, E. A., Huang, L., Baldi, P., and Bardwell, L. (2010). Computational prediction and experimental verification of new MAP kinase docking sites and substrates including Gli transcription factors. *PLoS Comput. Biol.* **6,** e1000908.

Williams, J. A., Guicherit, O. M., Zaharian, B. I., Xu, Y., Chai, L., Wichterle, H., Kon, C., Gatchalian, C., Porter, J. A., Rubin, L. L., and Wang, F. Y. (2003). Identification of a small molecule inhibitor of the hedgehog signaling pathway: Effects on basal cell carcinoma-like lesions. *Proc. Natl. Acad. Sci. USA* **100,** 4616–4621.

Wong, S. Y., Seol, A. D., So, P. L., Ermilov, A. N., Bichakjian, C. K., Epstein, E. H., Jr., Dlugosz, A. A., and Reiter, J. F. (2009). Primary cilia can both mediate and suppress Hedgehog pathway-dependent tumorigenesis. *Nat. Med.* **15,** 1055–1061.

Yauch, R. L., Gould, S. E., Scales, S. J., Tang, T., Tian, H., Ahn, C. P., Marshall, D., Fu, L., Januario, T., Kallop, D., Nannini-Pepe, M., Kotkow, K., *et al.* (2008). A paracrine requirement for hedgehog signalling in cancer. *Nature* **455,** 406–410.

Yauch, R. L., Dijkgraaf, G. J., Alicke, B., Januario, T., Ahn, C. P., Holcomb, T., Pujara, K., Stinson, J., Callahan, C. A., Tang, T., Bazan, J. F., Kan, Z., *et al.* (2009). Smoothened mutation confers resistance to a Hedgehog pathway inhibitor in medulloblastoma. *Science* **326,** 572–574.

Zhao, C., Chen, A., Jamieson, C. H., Fereshteh, M., Abrahamsson, A., Blum, J., Kwon, H. Y., Kim, J., Chute, J. P., Rizzieri, D., Munchhof, M., VanArsdale, T., *et al.* (2009). Hedgehog signalling is essential for maintenance of cancer stem cells in myeloid leukaemia. *Nature* **458,** 776–779.

CHAPTER THREE

Noncanonical Hedgehog Signaling

Donna Brennan,*,† Xiaole Chen,* Lan Cheng,*
My Mahoney,† *and* Natalia A. Riobo*

Contents

Abstract

The notion of noncanonical hedgehog (Hh) signaling in mammals has started to receive support from numerous observations. By noncanonical, we refer to all those cellular and tissue responses to any of the Hh isoforms that are independent of transcriptional changes mediated by the Gli family of transcription factors. In this chapter, we discuss the most recent findings that suggest that Patched1 can regulate cell proliferation and apoptosis independently of Smoothened (Smo) and Gli and the reports that Smo modulates actin cytoskeleton-dependent processes such as fibroblast migration, endothelial cell tubulogenesis, axonal extension, and neurite formation by diverse mechanisms that exclude any involvement of Gli-dependent transcription. We also acknowledge the existence of less stronger evidence of noncanonical signaling in *Drosophila*. © 2012 Elsevier Inc.

* Department of Biochemistry and Molecular Biology, Thomas Jefferson University, Philadelphia, Pennsylvania, USA
† Department of Dermatology, Thomas Jefferson University, Philadelphia, Pennsylvania, USA

Vitamins and Hormones, Volume 88
ISSN 0083-6729, DOI: 10.1016/B978-0-12-394622-5.00003-1

Abbreviations

ARA	arachidonic acid
Ci	Cubitus interruptus
CTD	C-terminal domain
Hh	hedgehog
Ptc1	Patched1
Shh	sonic hedgehog
Smo	Smoothened

I. Definition

The main players of the so-called "canonical" hedgehog (Hh) signaling pathway in mammals have been the objects of intense study in the last decade. We have learnt that in the absence of any Hh ligand (Sonic (Shh), Indian (Ihh), or Desert (Dhh)), the receptor Patched1 (Ptc1) prevents activation of the seven-transmembrane protein Smoothened (Smo) by inhibiting its translocation into the primary cilium. In the presence of Hh, Smo accumulates in the cilium in an active conformation that initiates a complex signaling network that results in activation of the Gli family of transcription factors (Riobo and Manning, 2007). However, a vast number of recent studies have demonstrated that not all Hh signaling proceeds through Gli activation, and we have named this subset of Gli-independent responses "noncanonical" Hh signaling (Jenkins, 2009).

In this chapter, we review the current evidence for the existence of at least two distinct classes of noncanonical Hh signaling (Fig. 3.1): a Type I, which works through functions of Ptch1 that are unrelated to its inhibitory activity on Smo, and a Type II, which operates through Smo functions (or activities) beyond Gli regulation.

II. Type I Noncanonical Signaling: Pathways Engaged Exclusively by Ptc1

A. Apoptosis

Apoptosis, or programmed cell death, is an essential process for development, tissue homeostasis, and tumor suppression. Hh opposes apoptosis by at least three mechanisms: increased expression of pro-survival genes (such

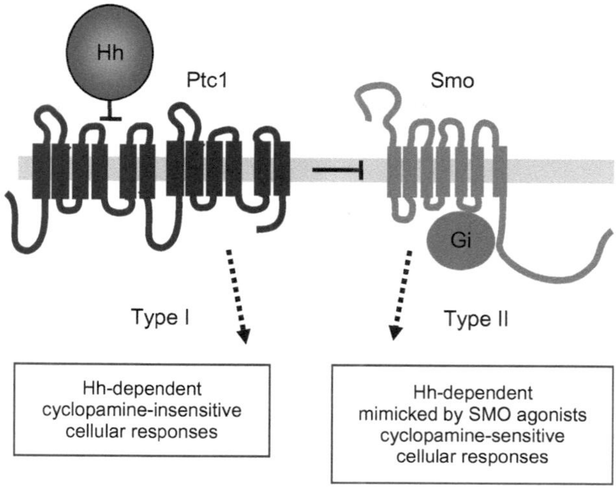

Figure 3.1 Schematic representation of the two types of noncanonical Hh signaling. Type I requires only binding of an Hh isoform to Ptc1 and is mediated by novel functions of Ptc1 unrelated to Smo repression, and it is by definition insensitive to Smo modulators. Type II is dependent on Smo and in some cases it has been shown to rely on signaling through G_i proteins, and it is both mimicked by Smo agonists and inhibited by Smo antagonists. ((See Color Insert.)

as Bcl-2) through the modulation of Gli transcription (Cayuso *et al.*, 2006; Katoh and Katoh, 2009), activation of the survival kinase Akt (Riobo *et al.*, 2006a), and Type I noncanonical pathway.

Growing evidence suggests that Ptc1 can promote or induce apoptosis independently of the canonical pathway (Chang *et al.*, 2010; Chinchilla *et al.*, 2010; Mille *et al.*, 2009; Thibert *et al.*, 2003). In the absence of an Hh ligand, ectopic expression of Ptc1 induces apoptosis in 293T and neuroepithelial cells (Thibert *et al.*, 2003). This process is independent of the downstream elements of the canonical Hh pathway, as overexpression of Smo cannot prevent cell death. Soft-agar assays also showed that Ptc1 function as a tumor suppressor might be related to its capacity to induce apoptosis. In line with those findings, we found that siRNA-mediated depletion of Ptc1 in endothelial cells extends cell survival in the absence of serum and to limit caspase-3 activation (Chinchilla *et al.*, 2010). We proposed that the pro-apoptotic effect of Ptc1 is exerted through a Type I noncanonical pathway for the following reasons: (1) lack of detectable canonical Hh signaling in endothelial cells both *in vitro* and *in vivo* (Chinchilla *et al.*, 2010; Lavine *et al.*, 2008; Pola *et al.*, 2001); (2) lack of effect of the Smo antagonists SANT-1, cyclopamine, and KAAD-cyclopamine to prevent the Shh-dependent reduction in caspase-3 activity (Fig. 3.2); and (3) inability of the Smo agonist (SAG) to mimic the antiapoptotic effect of Shh (Fig. 3.2).

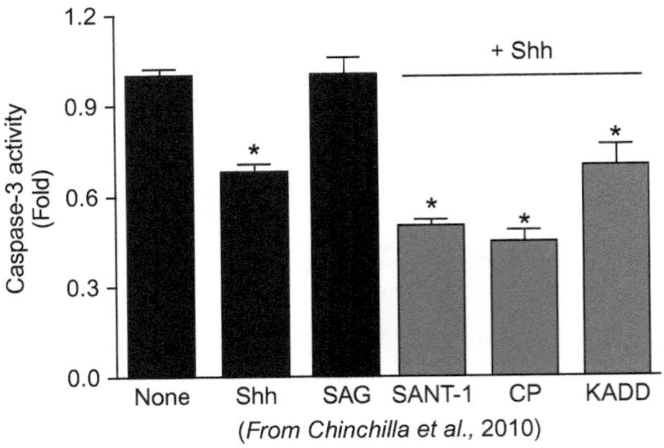

Figure 3.2 Apoptosis was assessed by caspase-3 activity measurement of serum starved HUVECs in the absence (control) or in the presence of 2.5 μg/ml Shh (Shh), 0.5 μM Smo agonist (SAG), or Shh plus the Smo inhibitors SANT-1, cyclopamine (CP), or KAAD-cyclopamine (KAAD) (Chinchilla *et al.*, 2010).

Ptc1 has been proposed to function as a dependence receptor, meaning that cell survival is dependent on the presence of the ligand when the receptor is expressed (Bredesen *et al.*, 2004; Mehlen and Thibert, 2004). There are more than 10 dependence receptors, most of which are involved in neural development and tumor suppression. Ptc1 shares the common features for the dependent receptors: (i) in the presence of ligand, receptors transduce signal, leading to the activation of the canonical signaling transcription pathway; (ii) in the absence of ligand, dependence receptors activate a different (noncanonical) pathway and induce apoptosis, which can be reverted by reintroducing the respective ligand; (iii) all dependence receptors contain the DART (dubbed dependence-associated receptor transmembrane) motif (del Rio *et al.*, 2007); and (iv) in order to induce cell death, the receptor needs to be cleaved by a caspase to expose a pro-apoptotic domain. Thibert *et al.* showed that the C-terminal cytoplasmic domain of Ptc1 is a substrate for caspase-3, -7, and -8 (Thibert *et al.*, 2003). Cleavage of Ptc1 is essential for induction of apoptosis because mutation of the caspase site (D1392N) abolishes Ptc1-induced cell death. Moreover, a truncated mutant constitutively exposing the caspase site Ptc1 (1–1392) induces cell death regardless of the presence of Shh, while full-length Ptc1 pro-apoptotic activity is exquisitely sensitive to Shh, demonstrating the functional importance of caspase cleavage. Further studies shed light on this mechanism, revealing that Ptc1 induces apoptosis by recruitment of a pro-apoptotic complex that includes caspase-9, DRAL, and TUCAN-1 to the C-terminal tail (Mille *et al.*, 2009). However, it is still unknown whether caspase-9 is involved in the generation of the pro-apoptotic Ptc1 (1–1392) and whether recruitment of the complex is a requisite for or a consequence of the cleavage at D1392.

It is worth noting the importance of the C-terminal domain (CTD) of Ptc1, containing 273 aa, in the regulation of cell death. The CTD alone (soluble) or a fragment truncated at the caspase cleavage site, Ptc1 (1165–1392), induces cell death in 293T and neuroepithelial cells at similar extend as the full-length Ptc1 (Thibert *et al.*, 2003). Mutation of the cleavage site (D1392N) acts as a dominant negative inhibitor for Ptc1-induced apoptosis. The CTD was suggested to also be a critical regulator of degradation and localization of Ptc1 (Chang *et al.*, 2010; Kawamura *et al.*, 2008; Lu *et al.*, 2006). In addition, it was recently reported that Ptc1 undergoes proteolytic processing at the C-terminus and the soluble CTD translocates to the nucleus and mediates a new form of signal transduction (Kagawa *et al.*, 2011). In *Drosophila*, the CTD is indispensible for repression of the canonical pathway when Hh is absent (Johnson *et al.*, 2000). However, studies using mammalian cells and mouse models suggest that in mammals, the CTD is not required for the canonical signal transduction but most likely has a distinct function in apoptosis and/or regulation of proliferation (Makino *et al.*, 2001; Mille *et al.*, 2009; Nieuwenhuis *et al.*, 2007; Sweet *et al.*, 1996; Thibert *et al.*, 2003). A spontaneous Ptc1 mouse mutant has a frameshift in the CTD that gives rise to retention of only the first 53 aa after the last transmembrane domain and addition of 63 nonsense residues. Named mesenchymal dysplasia ($Ptc1^{mes}$), the mice exhibit epidermal hyperplasia, increased musculature and preaxial polydactyly (Makino *et al.*, 2001; Nieuwenhuis *et al.*, 2007; Sweet *et al.*, 1996). Compared with $\text{Ptc1}^{-/-}$ homozygotes, which die between embryonic days 9 and 10.5 with severe neural tube defects as a result of hyperactivation of the canonical Hh signaling (Goodrich *et al.*, 1997), the $\text{Ptc1}^{mes/mes}$ embryos exhibit normal spinal cord development, suggesting that most of the CTD is absolutely dispensable for canonical signaling in mammalian embryogenesis. Studies using $Ptc1^{-/mes}$ mice (Makino *et al.*, 2001) also demonstrated that $Ptc1^{mes}$ can substitute a wild-type allele with regard to the canonical Hh pathway (Nieuwenhuis *et al.*, 2007). However, the same report also indicates that the CTD plays an important role in body size control and skin development since the $Ptc1^{mes/mess}$ animals show a marked increase in proliferation of the basal cell layer of the skin (Nieuwenhuis *et al.*, 2007). While they did not observe changes in cell death, the result is inconclusive due to the technical limitations on measuring the apoptosis in that system.

A different mouse model to study the function of Ptc-1 CTD in tumor suppression is a natural polymorphism (T1267N) responsible for a 600-fold change in susceptibility to develop Ras-induced skin squamous cell carcinomas in FVB versus C57BL/6 mice. The polymorphism was shown to reduce the binding of Ptc1 to Tid1, a tumor suppressor protein (Wakabayashi *et al.*, 2007). Direct interaction of Ptc1 with other potential apoptosis-related proteins has also been reported, as mentioned above (Mille *et al.*, 2009). Shh also activates the ERK pathway independently of Smo, probably through modulating the interaction of the SH3 interacting motif of Ptc1 CTD with SH3-containing proteins like Grb2 (Chang *et al.*, 2010). Taken together, these findings strongly support the notion that the mammalian Ptc1 CTD functions in the induction of cell death independently of the canonical Hh signaling pathway.

B. Cyclin B1 and cell cycle regulation

A large number of studies have demonstrated that the canonical Hh signaling pathway plays an essential role in proliferation of some cell types, the best-characterized example being postnatal proliferation of cerebellar granule precursor cells (Dahmane and Ruiz i Altaba, 1999; Wallace, 1999; Wechsler-Reya and Scott, 1999). Induction of *n-myc* and *cyclin D1* by Gli-dependent transcription mediates the mitogenic effect of Shh in those cells (Kenney and Rowitch, 2000; Kenney *et al.*, 2003). Cyclins are a family of proteins and act as regulatory subunits for cyclin-dependent kinases, and these complexes are responsible for regulating cell cycle progression through the cell cycle by directly affecting a cell's transcriptional regime (Murray, 2004). Dynamic modulation of the expression, phosphorylation, interactions, and subcellular localization of these complexes and their individual subunits all influence the activity of these complexes, and thus cell cycle progression. The Cyclin B1–Cdk1 complex (also known as M-phase promoting factor (MPF)) is specifically involved in the G2/M checkpoint, and its activation is essential for mitotic progression (Takizawa and Morgan, 2000). The activated form of this complex contains a series of phosphorylated serine residues on Cyclin B1. Interestingly, noncanonical Hh signaling also regulates the cell cycle at the level of Cyclin B1. An early report detected Ptc1 in a yeast two-hybrid assay designed to identify binding partners of phosphorylated Cyclin B1 (Barnes *et al.*, 2001). Biochemical analysis revealed that Ptc1 interacts with phospho-Cyclin B1, but not with the unphosphorylated protein, through the large intracellular loop between transmembrane domains 6 and 7. Overexpression of Ptc1 in 293T cells resulted in a redistribution of Cyclin B1 from the nucleus to the cytoplasm and reduced cell proliferation, which was restored by addition of a Cyclin B1 mutant that cannot be phosphorylated. In a later study, the same group demonstrated that the Ptc1^{Q688X}, a truncation mutant found in some basal cell carcinomas (BCCs), is unable to interact with Cyclin B1 and, unlike wild-type Ptc1, did not reduce the mitotic index (Barnes *et al.*, 2005).

The identification of Ptc1 as a binding partner of Cyclin B1 suggested a new mode of cell cycle regulation by Hh signaling. Indeed, the presence of Shh disrupts Ptc1 interaction with Cyclin B1, allowing its nuclear translocation and promoting completion of mitosis. The regulation of Cyclin B1 by Shh/Ptc1 was not limited to overexpression experiments but was also evident with the endogenous proteins. These results support the notion that Shh regulates proliferation by modulating the interaction between Ptc1 and phosphorylated Cyclin B1 via a noncanonical signaling pathway.

While those studies were conducted *in vitro*, several histological observations are in line with those findings. Work characterizing Shh expression pattern in human urinary tract development revealed a correlation between Shh expression and Cyclin B1 localization: higher levels of Shh correlated to cytoplasmic and nuclear Cyclin B1, while when Shh levels go down Cyclin B1 relocalizes to the apical membranes together with Ptc1 (Jenkins *et al.*, 2007). Another study showed that epidermal ablation of Ptc1 in mice coincided with an increase in nuclear Cyclin B1, when compared to normal epidermis (Adolphe *et al.*, 2006). Wang *et al.* demonstrated that epidermal tumors that arise in Ptc1$^{+/-}$ mice are derived from hair follicle progenitor cells and are characterized by increased nuclear Cyclin B1. This is different from tumors developed in the SmoM2 mice model, which express a constitutively active from of Smo (Wang *et al.*, 2011; Youssef *et al.*, 2010). They suggested that nuclear Cyclin B1 and not increased Smo activity results in hair follicle-derived tumors. These data allude to a possible mechanism by which loss of Ptc1 can lead to tumorigenesis independently of the canonical Hh pathway, though more work will need to be done to determine if this alternative pathway is sufficient to induce tumor formation or synergizes with enhanced Gli1 activity.

Finally, studies in zebrafish provided some mechanistic insight into the regulation of Ptc1–Cyclin B1 interaction (Jiang *et al.*, 2009). The authors reported that dissociation of the Ptc1–Cyclin B1 complex after Shh activation involves the formation of another complex between Ptc1 and G-protein receptor kinase-2 (GRK2). GRK2 silencing by siRNA abrogated the ability of Shh to redistribute Cyclin B1 to the nucleus, and exogenous GRK2 expression was able to rescue the antiproliferative phenotype associated with Ptc1 overexpression. Taken together, these results suggest that Shh stimulation may induce a conformational change in Ptc1 that increases its affinity for GRK2 at the expense of its interaction with Cyclin B1.

In summary, there is significant evidence for a noncanonical role for Ptc1 in cell cycle regulation through Cyclin B1 in an apparently Smo- and Gli-independent manner (Type I noncanonical signaling). The relative contribution of this noncanonical pathway on overall cell growth and survival mediated by Hh signaling remains to be established.

III. Type II Noncanonical Signaling: Pathways Engaged by Smoothened

A. Small GTPases

The discovery that Smo is a functional G-protein-coupled receptor (GPCR) with selectivity toward heterotrimeric G_i proteins had profound implications on Hh signaling (Riobo *et al.*, 2006b). Several GPCRs have been shown to activate small GTPases, which play key roles in cytoskeletal reorganization (Ridley, 2006). Small GTPases are monomeric G proteins that act as molecular switches and, thus, rapidly regulate cellular processes (Vetter and Wittinghofer, 2001). Activated by guanine exchange factors (GEFs), they bound GTP in their "on" state, when they are able to interact with and activate their molecular targets. GTPases are then inactivated by nucleotide hydrolysis to GDP, an intrinsically slow process, enhanced by GTPase-activating proteins (GAPs). Small GTPases can be divided into four families: Ras, Rho, Arf, and Rab. The Rho family, which has been shown to mediate cytoskeletal reorganization, is further divided into three subfamilies: Rho, Rac, and Cdc42. These subfamilies differentially modulate distinct cytoskeletal rearrangements through polymerization of actin filaments, and these unique rearrangements can then regulate coordinated movement of the cell.

Recent work demonstrated that Hh proteins promote actin stress fiber formation and endothelial cell tubulogenesis in a Smo-dependent manner in endothelial cells (Chinchilla *et al.*, 2010). Interestingly, while endothelial cells express all key proteins of the Hh signaling pathway, they are unable to activate the canonical pathway in response to Hh ligands, as measured by activation of the marker Gli-target genes *gli1* or *ptch1*. However, the human umbilical vein endothelial cells (HUVECs) and human cardiac microvascular endothelial cells (HMVECs) utilized in this study exhibit strong morphological changes in response to Shh, Ihh, or Dhh stimulation. In this system, stimulation by Hh ligands results in the formation of actin stress fibers within minutes of treatment, thus suggesting a noncanonical pathway based on the time course and the lack of detectable Gli-dependent transcription. The acute cytoskeletal changes were shown to depend on Smo- and G_i-protein-mediated activation of the small GTPase RhoA (Fig. 3.3). In addition to short-term changes, stimulation of endothelial cells by the three Hh ligands resulted in a significantly increase in tubulogenesis after several hours by a process that was also the result of Smo-mediated, G_i-dependent activation of RhoA. Interestingly, this study also demonstrated that, unlike the canonical paradigm whereby the different Hh ligands exhibit distinct potencies for Gli activation (Shh≫Ihh>Dhh) (Pathi *et al.*, 2001), endothelial cells responded equally to all three ligands in terms of stress fiber formation, RhoA activation, and tubulogenesis.

Another study demonstrated that Shh stimulates fibroblast migration via activation of small Rho GTPases (Polizio *et al.*, 2011). In this system, both Shh and purmorphamine (PUR, a potent SAG) enhance migration of NIH-3T3 fibroblasts but have no effect on $Smo^{-/-}$ MEFs. Inhibition experiments demonstrated that activation of both G_i and PI3K by Smo was necessary for cell migration and activation of RhoA and Rac1. To confirm that these effects were indeed noncanonical, the canonical pathway was inhibited by the expression of Gli3R, the repressor form of the Gli3 transcription factor. Effective inhibition of the canonical pathway was

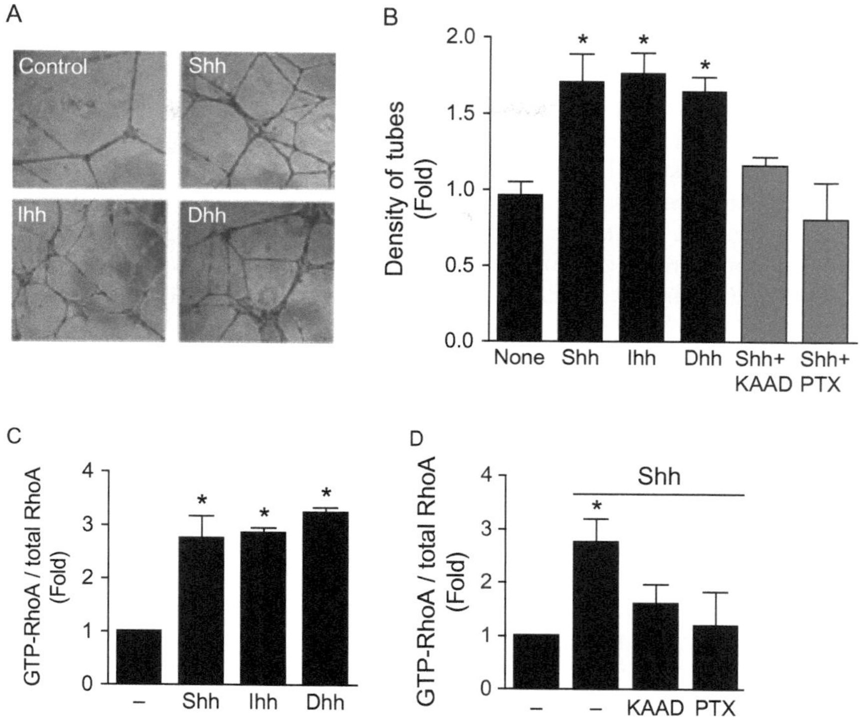

(*From Chinchilla et al.*, 2010)

Figure 3.3 The three Hh isoforms promote tubulogenesis in endothelial cells by a G_i protein and RhoA-dependent mechanism. (A) HUVECs were cultured in a 3D collagen Type I matrix for 24h in the absence (control) or presence of Shh, Ihh, or Dhh (all at 2.5μg/ml). Photographs are representative tube densities. (B) Quantification of tube density in the absence (control) or presence of Shh, Ihh, and Dhh, and in the presence of Shh after preincubation with 0.5μM KAAD-cyclopamine (KAAD, Smo inhibitor) or 100ng/ml Pertussis toxin (PTX; inhibitor of G_i protein activation). (C) HUVECs were serum staved for 24h and stimulated with Shh, Ihh, or Dhh (all at 2.5μg/ml) during 15 min. Active RhoA (RhoA-GTP) was pulled down from whole cell lysates with a Rhotekin-binding domain–GST fusion protein coupled to GSH-sepharose beads. Active and total RhoA were evaluated by Western blot, and the densitometric values are shown in the bar graph (n=3). (D) RhoA pull-down assay as in (C) in cells preincubated with 0.5μM KAAD-cyclopamine (KAAD) or 100ng/ml PTX (n=3) (Chinchilla *et al.*, 2010).

confirmed, but expression of Gli3R did not perturb Shh- or PUR-mediated migration, indicating that fibroblast migration is affected by a noncanonical Hh signaling pathway.

While these studies provide solid evidence that Smo-dependent, noncanonical Hh signaling elicits specific cellular responses via the activation of small GTPases, the mechanism by which Smo or G_i activates RhoA and

Rac1 has not been yet established. A recent report looking at Hh signaling during dendritic spine formation sheds some light on this issue (Sasaki *et al.*, 2010). Tiam1 (T-lymphoma invasion and metastasis) is a GEF specific for Rac1 that localizes to dendritic spines via interactions with spine-specific proteins. The Tiam1–Rac1 cascade is activated in response to Shh stimulation, without parallel activation of the canonical Hh pathway. Results indicated that both stimulation by Hh ligand and depletion of Smo via siRNA resulted in an increase in spine formation. Subsequent binding studies demonstrated that Smo interacts directly with Tiam1 and sequesters it, thus preventing Rac1 activation. This complex dissociates upon Smo activation, which in turn allows Tiam1 to activate Rac1. While these studies do not ascertain the requirement of G_i activity, they do provide a direct mechanism by which Smo activation leads to Rac1 activation. It will be interesting to see if these types of interactions are common to Smo-mediated activation of small GTPases and whether or not these events are G_i-dependent.

In summary, there is considerable evidence for Smo-dependent, noncanonical Hh signaling that is mediated through the activation of small GTPases. Future studies designed to elucidate the exact mechanisms by which Smo activates these molecular switches have the potential to identify specific molecular targets for modulating noncanonical Hh signaling responses as well as provide insights into the diverse phenotypes observed when Hh signaling is dysregulated.

B. Src activation in the axonal growth cone

The identification of a noncanonical Hh signaling pathway playing a role in cytoskeleton remodeling is reminiscent of the identification, over the past few years, of noncanonical Wnt and BMP signaling pathways. Before the identification of these noncanonical signaling pathways, all of the functions of Hh isoforms were thought to be executed by alteration of gene expression by the Gli family of transcription factors. Recently, it has been demonstrated that Shh can signal through stimulating the activity of Src family kinase (SFK) members in a Smo-dependent manner to control axon guidance (Yam *et al.*, 2009). In this study, axons of dissociated commissural neurons placed in a Shh gradient turned rapidly toward increasing concentrations of Shh in a transcriptional-independent manner. Instead, Shh stimulates the activity of SFK members (Src and Fyn) in a Smo-dependent fashion. This study established the existence of a transcription-independent Shh signaling pathway, which acts locally at the axonal growth cone. Through this pathway, Shh gradients can elicit a rapid and spatially polarized response within the growth cone, without retrograde signaling to elicit transcription in the nucleus to induce global gene expression changes in the cell. It has been proposed that a graded activation of SFKs in response to a Shh gradient mediates changes in the actin cytoskeleton (Liu *et al.*, 2007; Robles *et al.*, 2005; Suter and Forscher, 2001). However, the regulation of Src by a noncanonical Hh pathway appears to be even more complex. In a screening for Ptc1-interacting proteins containing SH3 domains, it was found that the C-terminal tail of Ptc1 (1163–1435) binds to Src (Chang *et al.*, 2010). These finding define a novel signaling cascade operating through SH3 domain-containing proteins that could be directly stimulated by the Hh ligands. Together, both studies suggest that Shh is able to stimulate Src activity via both Type I and Type II noncanonical signaling pathways, and that the latter is preferred in axonal guidance.

Contributions of the canonical Hh pathway to Src activation are unlikely since (1) Hh-responsive transcriptional reporters have failed to identify Src or Fyn as components of the canonical Hh signaling pathway (Lum *et al.*, 2003; Nybakken *et al.*, 2005; Varjosalo *et al.*, 2008) and (2) SFK mutant mice lack phenotypes reminiscent of dysregulated canonical Shh signaling (Kuo *et al.*, 2005).

C. Arachidonic acid metabolites

Arachidonic acid (ARA) is an important signaling molecule that mediates many cellular functions including inflammation and actin remodeling (Glenn and Jacobson, 2002; Peppelenbosch *et al.*, 1993). ARA is synthesized in the cell upon receptor stimulation and can act directly as a second messenger or can be further metabolized into leukotrienes and prostaglandins by the action of lipoxygenase or cyclooxygenase, respectively (Funk, 2001).

Recent studies have found that ARA metabolites, particularly leukotrienes synthesized by the lipoxygenase pathway, are involved in Gli-independent Hh signal transduction. In fibroblasts, Shh induces the formation of lamellipodia and promotes migration (Bijlsma *et al.*, 2007; Polizio *et al.*, 2011). A role of leukotrienes was proposed since these responses were partially inhibited by treatment with 5-lipoxygenase inhibitors, could be mimicked by addition of ARA to cell culture medium, and the ARA metabolite leukotriene was upregulated in the cells following Shh stimulation (Bijlsma *et al.*, 2007). These results strongly support the speculation that ARA, together with trimeric G-protein-dependent activation of RhoA and Rac1, is one of the signals that mediate Shh-induced cytoskeletal rearrangements. This study also demonstrated that Shh-induced cell migration is Smo dependent but Gli independent (Type II noncanonical pathway). The authors claimed the following reasons for their conclusion: (1) acute response (only minutes after addition of Shh) and (2) insensitivity to overexpression of Sufu, a negative regulator of the Gli transcription factors.

The same group also reported that leukotrienes mediate the formation of neurite projections in motor neurons differentiated from embryonic stem cells (ESCs) treated with Shh (Bijlsma *et al.*, 2008). Neurite formation is also Gli independent and mediated by Smo, since this response occurs in wild type and ESC deficient in various combinations of Gli transcription factors, but not in Smo-deficient ESCs. In conclusion, these studies point to ARA and leukotrienes as novel players in Type II noncanonical Hh signaling in fibroblast migration and neuronal development.

D. Calcium transients

Calcium ions regulate a vast number of neuronal cell functions, including proliferation, differentiation, apoptosis, and migration (Berridge, 1998; Komuro and Rakic, 1996; LoTurco *et al.*, 1995). Stimulation of transient elevations of intracellular calcium ($Ca^{2+}i$) activates protein kinases, regulates transcription, and influences motility and morphology (Catterall, 2010). It is not surprising that Ca^{2+} has also been shown both to influence expression of some Hh pathway genes and to mediate some cellular responses to Hh signaling.

Calcium has repeatedly reported to induce expression of Hh ligands. It was reported that Ca^{2+} chelation inhibits *ihh* gene expression in chick chondrocytes (Zuscik *et al.*, 2002). Also, classical protein kinase C (PKC) isoforms, which are activated by Ca^{2+} release, sustain *shh* gene expression in the chick wing bud (Lu *et al.*, 2001). In mouse primary gastric cultures, gadolinium-, thapsigargin-, and carbachol-mediated release of intracellular Ca^{2+} induces Shh expression as well (El-Zaatari *et al.*, 2010). In contrast, Ca^{2+}-chelation with BAPTA–EGTA reduced Shh expression. Thus, during gastric acid secretion, intracellular calcium release and PKC activation stimulate *shh* gene expression.

On the other hand, since Smo was proved to act as a GPCR, and stimulation of GPCRs often engages second messengers such as Ca^{2+}, the interplay between Ca^{2+} transient and Shh signaling was recently investigated. Belgacem and Borodinsky reported that recombinant N-terminal Shh peptide acutely increases Ca^{2+} spike activity in a dose-dependent manner in the developing spinal cord (Belgacem and Borodinsky, 2011). This effect is mimicked by a SAG and is prevented by cyclopamine, a Smo antagonist. They further demonstrated that this effect depends on both extracellular Ca^{2+} and intracellular Ca^{2+} stores and on Pertussis toxin (PTX)-sensitive heterotrimeric G_i proteins. While the exact mechanisms are not yet clear, they proposed that activation Smo, resulting in the activation of a PTX-sensitive G_i protein, leads to activation of Phospholipase C-γ (PLC) and increases IP_3 generation. Opening of IP_3R-operated stores will deplete intracellular Ca^{2+} stores, which leads to the subsequent activation of transient receptor potential channel 1 (TRPC1) and voltage-gated channels (Cav), resulting in increased Ca^{2+} spike activity. This hypothesis is supported by the observed inhibition of Ca^{2+} spike activity by inhibitors of PLC, IP_3R, and TRCP1 channels. Also in support, the authors found that acute stimulation of Smo with SAG results in sequential IP_3 and Ca^{2+} transients in the primary cilia of neurons, and that this effect is abolished by the Smo inhibitor cyclopamine.

This provoking finding suggests that Hh signaling might regulate a cohort of physiological processes involving fluctuations of Ca^{2+} that acutely affect membrane potential and Ca^{2+}-dependent signaling pathways.

IV. Noncanonical Hh Signaling in *Drosophila*

The concept of noncanonical Hh signaling has not been formally extended to *Drosophila*, despite the existence of an old controversy about a subset of Hh target genes that seem to be independent of the Gli ortholog Cubitus interruptus (Ci). A large number of studies on Hh signaling in *Drosophila* focus on the effect of mutations that alter the body segment pattern at the larva stage. Each segment has a denticle-cover anterior part and a naked cuticle posterior part. *Drosophila* Hh is expressed and secreted from two rows of cells in the posterior compartment, from where it patterns the expression of *wingless* (*wg*), the Wnt ortholog, and of *rhomboid* (*rho*), ortholog of the mammalian EGFR protein. Absence of Hh results in lack of expression of *wg* and *rho* resulting in abnormal segmentation and lack of specification of the naked cuticle, leading to an "Hh-like" larva. If all Hh signaling in *Drosophila* is channeled through Ci activation, thus absence of Ci should result in a phenotype identical to that of the *hh* mutant. However, Gallet *et al.* showed that a null allele of Ci (*ci94*) has a much weaker phenotype than the Hh null (*hhAC*) (Gallet *et al.*, 2000). Moreover, over-expression of Hh in a Ci-deficient background induces early stage expression of *wg* and a stronger expression of *rho* than in the presence of Ci, indicating that these are noncanonical effects. This group further demonstrated that expression of *rho* is mediated by the C2H2-zinc finger transcription factor Teashirt (*tsh*), which would act redundantly with Ci to transduce Hh signaling in the fly.

Another group presented a different point of view: instead of activating the pathway by Hh overexpression, they used two ptc-null alleles (*ptcIIW* and *ptcS2*) in the Ci-null background and analyzed the cuticle phenotype but not *wg* and *rho* expression (Méthot and Basler, 2001). While they agreed that the cuticle phenotype of *ci94* is different from *hhAC*, they showed that an allele encoding just the repressor form of Ci (*ciCell*) resembles more closely the Hh-null phenotype. Notwithstanding, some stripes of naked cuticle are still present, suggesting that Ci-independent signaling may occur. In addition, this group analyzed the requirement for Ci in Hh-dependent imaginal wing disc patterning using somatic recombination techniques.

In this model, clones that lack Ci cannot induce expression of *engrailed* (*en*) or *ptc*, when the Hh pathway was activated by loss of function of Ptc. The authors concluded that all Hh signaling in *Drosophila* requires Ci activity, either the activator or the repressor. However, an alternative explanation is that, as demonstrated in vertebrates, some Ci-independent signaling could be independent of Smo, for example, mediated by other Hh-interacting proteins (iHog, Boi, Dsp, etc.) and thus cannot be induced by removal of Ptc but is evidenced by Hh overexpression.

Another aspect of fly development that appears to utilize a noncanonical Hh pathway is the formation of the Bolwig's organ, a light-sensing organ at the larval stage. Hh induces expression of *atonal* (*ato*) in the precursors of the Bolwig's organ and that *ato* expression can be induced by loss of Ptc and blocked in a *smo* mutant (Susuki and Saigo, 2000). Elegant genetic analysis revealed that (1) loss of Ci does not affect Bolwig's organ formation or *ato* expression, (2) an activator form of Ci (*ciZnC*) cannot rescue *ato* expression in an Hh mutant, and (3) the Ci repressor (*ciCell*) also fails to rescue the Hh mutants. These observations strongly argue in favor of the existence of at least some form of Ci-independent Hh signaling in *Drosophila*.

V. Concluding Remarks

We have presented ample evidence that noncanonical Hh signaling regulates key physiological functions in mammals during development and after birth. The recognition of those Gli-independent functions has been delayed due to the earlier consensual notion in the research community that true Hh-dependent effects are those requiring Smo and Gli activity, which was initially assessed using genetic models and later with pharmacological and molecular biology tools. The availability of recombinant active Hh isoforms and more specific inhibitors leads to discredit such belief and opened up a new area of research that teaches us that Hh signaling is more complex and regulates more physiological processes than previously thought.

REFERENCES

Adolphe, C., Hetherington, R., Ellis, T., and Wainwright, B. (2006). Patched1 functions as a gatekeeper by promoting cell cycle progression. *Cancer Res.* **66,** 2081–2088.

Barnes, E. A., Kong, M., Ollendorff, V., and Donoghue, D. J. (2001). Patched1 interacts with cyclin B1 to regulate cell cycle progression. *EMBO J.* **20,** 2214–2223.

Barnes, E. A., Heidtman, K. J., and Donoghue, D. J. (2005). Constitutive activation of the shh-ptc1 pathway by a patched mutation identified in BCC. *Oncogene* **24,** 902–915.

Belgacem, Y. H., and Borodinsky, L. N. (2011). Sonic hedgehog signaling is decoded by calcium spike activity in the developing spinal cord. *Proc. Natl. Acad. Sci. USA* **108,** 4482–4487.

Berridge, M. J. (1998). Neuronal calcium signaling. *Neuron* **21,** 13–26.

Bijlsma, M. F., Borensztajn, K. S., Roelink, H., Peppelenbosch, M. P., and Spek, C. A. (2007). Sonic hedgehog induces transcription-independent cytoskeletal rearrangement and migration regulated by arachidonate metabolites. *Cell. Signal.* **19,** 2596–2604.

Bijlsma, M. F., Peppelenbosch, M. P., Spek, C. A., and Roelink, H. (2008). Leukotriene synthesis is required for hedgehog-dependent neurite projection in neuralized embryoid bodies but not for motor neuron differentiation. *Stem Cells* **26,** 1138–1145.

Bredesen, D. E., Mehlen, P., and Rabizadeh, S. (2004). Apoptosis and dependence receptors: A molecular basis for cellular addiction. *Physiol. Rev.* **84,** 411–430.

Catterall, W. A. (2010). Signaling complexes of voltage-gated sodium and calcium channels. *Neurosci. Lett.* **486,** 107–116.

Cayuso, J., Ulloa, F., Cox, B., Briscoe, J., and Martí, E. (2006). The Sonic hedgehog pathway independently controls the patterning, proliferation and survival of neuroepithelial cells by regulating Gli activity. *Development* **133**(3), 517–528.

Chang, H., Li, Q., Moraes, R. C., Lewis, M. T., and Hamel, P. A. (2010). Activation of Erk by sonic hedgehog independent of canonical hedgehog signalling. *Int. J. Biochem. Cell Biol.* **42,** 1462–1471.

Chinchilla, P., Xiao, L., Kazanietz, M. G., and Riobo, N. A. (2010). Hedgehog proteins activate pro-angiogenic responses in endothelial cells through non-canonical signaling pathways. *Cell Cycle* **9,** 570–579.

Dahmane, N., and Ruiz i Altaba, A. (1999). Sonic hedgehog regulates the growth and patterning of the cerebellum. *Development* **126**(14), 3089–3100.

Del Rio, G., Kane, D. J., Ball, K. D., and Bredesen, D. E. (2007). A novel motif identified in dependence receptors. *PLoS One* **2,** e463.

El-Zaatari, M., Zavros, Y., Tessier, A., Waghray, M., Lentz, S., Gumucio, D., Todisco, A., and Merchant, J. L. (2010). Intracellular calcium release and protein kinase C activation stimulate sonic hedgehog gene expression during gastric acid secretion. *Gastroenterology* **139,** 2061–2071.

Funk, C. D. (2001). Prostaglandins and leukotrienes: Advances in eicosanoid biology. *Science* **294,** 1871–1875.

Gallet, A., Angelats, C., Kerridge, S., and Therond, P. P. (2000). Cubitus interruptus-independent transduction of the Hedgehog signal in *Drosophila. Development* **127,** 5509–5522.

Glenn, H. L., and Jacobson, B. S. (2002). Arachidonic acid signaling to the cytoskeleton: The role of cyclooxygenase and cyclic AMP-dependent protein kinase in actin bundling. *Cell Motil. Cytoskeleton* **53,** 239–250.

Goodrich, L. V., Milenkovic, L., Higgins, K. M., and Scott, M. P. (1997). Altered neural cell fates and medulloblastoma in mouse patched mutants. *Science* **277,** 1109–1113.

Jenkins, D. (2009). Hedgehog signalling: Emerging evidence for non-canonical pathways. *Cell. Signal.* **21,** 1023–1034.

Jenkins, D., Winyard, P. J., and Woolf, A. S. (2007). Immunohistochemical analysis of Sonic hedgehog signaling in normal human urinary tract development. *J. Anat.* **211,** 620–629.

Jiang, X., Yang, P., and Ma, L. (2009). Kinase activity-independent regulation of cyclin pathway by GRK2 is essential for zebrafish early development. *Proc. Natl. Acad. Sci. USA* **106,** 10183–10188.

Johnson, R. L., Milenkovic, L., and Scott, M. P. (2000). In vivo functions of the patched protein: Requirement of the C terminus for target gene inactivation but not Hedgehog sequestration. *Mol. Cell* **6,** 467–478.

Kagawa, H., Shino, Y., Kobayashi, D., Demizu, S., Shimada, M., Ariga, H., and Kawahara, H. (2011). A novel signaling pathway mediated by the nuclear targeting of C-terminal fragments of mammalian Patched 1. *PLoS One* **6,** e18638.

Katoh, Y., and Katoh, M. (2009). Hedgehog target genes: Mechanisms of carcinogenesis induced by aberrant hedgehog signaling activation. *Curr. Mol. Med.* **9,** 873–886.

Kawamura, S., Hervold, K., Ramirez-Weber, F. A., and Kornberg, T. B. (2008). Two patched protein subtypes and a conserved domain of group I proteins that regulates turnover. *J. Biol. Chem.* **283,** 30964–30969.

Kenney, A. M., and Rowitch, D. H. (2000). Sonic hedgehog promotes G(1) cyclin expression and sustained cell cycle progression in mammalian neuronal precursors. *Mol. Cell. Biol.* **20**(23), 9055–9067.

Kenney, A. M., Cole, M. D., and Rowitch, D. H. (2003). Nmyc upregulation by sonic hedgehog signaling promotes proliferation in developing cerebellar granule neuron precursors. *Development* **130**(1), 15–28.

Komuro, H., and Rakic, P. (1996). Intracellular Ca^{2+} fluctuations modulate the rate of neuronal migration. *Neuron* **17,** 275–285.

Kuo, G., Arnaud, L., Kronstad-O'Brien, P., and Cooper, J. A. (2005). Absence of Fyn and Src causes a reeler-like phenotype. *J. Neurosci.* **25,** 8578–8586.

Lavine, K. J., Kovacs, A., and Ornitz, D. M. (2008). Hedgehog signaling is critical for maintenance of the adult coronary vasculature in mice. *J. Clin. Invest.* **118**(7), 2404–2414.

Liu, G., Li, W., Gao, X., Li, X., Jurgensen, C., Park, H. T., Shin, N. Y., Yu, J., He, M. L., Hanks, S. K., Wu, J. Y., Guan, K. L., *et al.* (2007). p130CAS is required for netrin signaling and commissural axon guidance. *J. Neurosci.* **27,** 957–968.

LoTurco, J. J., Owens, D. F., Heath, M. J., Davis, M. B., and Kriegstein, A. R. (1995). GABA and glutamate depolarize cortical progenitor cells and inhibit DNA synthesis. *Neuron* **15,** 1287–1298.

Lu, H. C., Swindell, E. C., Sierralta, W. D., Eichele, G., and Thaller, C. (2001). Evidence for a role of protein kinase C in FGF signal transduction in the developing chick limb bud. *Development* **128,** 2451–2460.

Lu, X., Liu, S., and Kornberg, T. B. (2006). The C-terminal tail of the Hedgehog receptor Patched regulates both localization and turnover. *Genes Dev.* **20,** 2539–2551.

Lum, L., Yao, S., Mozer, B., Rovescalli, A., Von Kessler, D., Nirenberg, M., and Beachy, P. A. (2003). Identification of Hedgehog pathway components by RNAi in Drosophila cultured cells. *Science* **299,** 2039–2045.

Makino, S., Masuya, H., Ishijima, J., Yada, Y., and Shiroishi, T. (2001). A spontaneous mouse mutation, mesenchymal dysplasia (mes), is caused by a deletion of the most C-terminal cytoplasmic domain of patched (ptc). *Dev. Biol.* **239,** 95–106.

Mehlen, P., and Thibert, C. (2004). Dependence receptors: Between life and death. *Cell. Mol. Life Sci.* **61,** 1854–1866.

Méthot, N., and Basler, K. (2001). An absolute requirement for Cubitus interruptus in Hedgehog signaling. *Development* **128,** 733–742.

Mille, F., Thibert, C., Fombonne, J., Rama, N., Guix, C., Hayashi, H., Corset, V., Reed, J. C., and Mehlen, P. (2009). The Patched dependence receptor triggers apoptosis through a DRAL-caspase-9 complex. *Nat. Cell Biol.* **11,** 739–746.

Murray, A. W. (2004). Recycling the cell cycle: Cyclins revisited. *Cell* **116,** 221–234.

Nieuwenhuis, E., Barnfield, P. C., Makino, S., and Hui, C. C. (2007). Epidermal hyperplasia and expansion of the interfollicular stem cell compartment in mutant mice with a C-terminal truncation of Patched1. *Dev. Biol.* **308,** 547–560.

Nybakken, K., Vokes, S. A., Lin, T. Y., McMahon, A. P., and Perrimon, N. (2005). A genome-wide RNA interference screen in Drosophila melanogaster cells for new components of the Hh signaling pathway. *Nat. Genet.* **37,** 1323–1332.

Pathi, S., Pagan-Westphal, S., Baker, D. P., Garber, E. A., Rayhorn, P., Bumcrot, D., Tabin, C. J., Blake Pepinsky, R., and Williams, K. P. (2001). Comparative biological responses to human Sonic, Indian, and Desert hedgehog. *Mech. Dev.* **106,** 107–117.

Peppelenbosch, M. P., Tertoolen, L. G., Hage, W. J., and de Laat, S. W. (1993). Epidermal growth factor-induced actin remodeling is regulated by 5-lipoxygenase and cyclooxygenase products. *Cell* **74,** 565–575.

Pola, R., Ling, L. E., Silver, M., Corbley, M. J., Kearney, M., Blake Pepinsky, R., Shapiro, R., Taylor, F. R., Baker, D. P., Asahara, T., and Isner, J. M. (2001). The morphogen Sonic hedgehog is an indirect angiogenic agent upregulating two families of angiogenic growth factors. *Nat. Med.* **7**(6), 706–711.

Polizio, A. H., Chinchilla, P., Chen, X., Kim, S., Manning, D. R., and Riobo, N. A. (2011). Heterotrimeric Gi proteins link Hedgehog signaling to activation of Rho small GTPases to promote fibroblast migration. *J. Biol. Chem.* **286,** 19589–19596.

Ridley, A. J. (2006). Rho GTPases and actin dynamics in membrane protrusions and vesicle trafficking. *Trends Cell Biol.* **16,** 522–529.

Riobo, N. A., and Manning, D. R. (2007). Pathways of signal transduction employed by vertebrate Hedgehogs. *Biochem. J.* **403**(3), 369–379.

Riobo, N. A., Lu, K., Ai, X., Haines, G. M., and Emerson, C. P., Jr. (2006a). Phosphoinositide 3-kinase and Akt are essential for Sonic Hedgehog signaling. *Proc. Natl. Acad. Sci. USA* **103**(12), 4505–4510.

Riobo, N. A., Saucy, B., Dilizio, C., and Manning, D. R. (2006b). Activation of heterotrimeric G proteins by Smoothened. *Proc. Natl. Acad. Sci. USA* **103**(33), 12607–12612.

Robles, E., Woo, S., and Gomez, T. M. (2005). Src-dependent tyrosine phosphorylation at the tips of growth cone filopodia promotes extension. *J. Neurosci.* **25,** 7669–7681.

Sasaki, N., Kuriso, J., and Kengaku, M. (2010). Sonic hedgehog signaling regulates actin cytoskeleton via Tiam1-Rac1 cascade during spine formation. *Mol. Cell. Neurosci.* **45,** 335–344.

Susuki, T., and Saigo, K. (2000). Transcriptional regulation of atonal required for *Drosophila* larval eye development by concerted action of eyes absent, sine oculis and hedgehog signaling independent of fused kinase and cubitus interruptus. *Development* **127,** 1531–1540.

Suter, D. M., and Forscher, P. (2001). Transmission of growth cone traction force through apCAM-cytoskeletal linkages is regulated by Src family tyrosine kinase activity. *J. Cell Biol.* **155,** 427–438.

Sweet, H. O., Bronson, R. T., Donahue, L. R., and Davisson, M. T. (1996). Mesenchymal dysplasia: A recessive mutation on chromosome 13 of the mouse. *J. Hered.* **87,** 87–95.

Takizawa, C. G., and Morgan, D. O. (2000). Control of mitosis by changes in the subcellular location of cyclin-B1-Cdk1 and Cdc25C. *Curr. Opin. Cell Biol.* **12,** 658–665.

Thibert, C., Teillet, M. A., Lapointe, F., Mazelin, L., Le Douarin, N. M., and Mehlen, P. (2003). Inhibition of neuroepithelial patched-induced apoptosis by sonic hedgehog. *Science* **301,** 843–846.

Varjosalo, M., Bjorklund, M., Cheng, F., Syvanen, H., Kivioja, T., Kilpinen, S., Sun, Z., Kallioniemi, O., Stunnenberg, H. G., He, W. W., Ojala, P., and Taipale, J. (2008). Application of active and kinase-deficient kinome collection for identification of kinases regulating hedgehog signaling. *Cell* **133,** 537–548.

Vetter, I. R., and Wittinghofer, A. (2001). The guanine nucleotide-binding switch in three dimensions. *Science* **294,** 1299–1304.

Wakabayashi, Y., Mao, J. H., Brown, K., Girardi, M., and Balmain, A. (2007). Promotion of Hras-induced squamous carcinomas by a polymorphic variant of the Patched gene in FVB mice. *Nature* **445**(7129), 761–765.

Wallace, V. A. (1999). Purkinje-cell-derived Sonic hedgehog regulates granule neuron precursor cell proliferation in the developing mouse cerebellum. *Curr. Biol.* **9**(8), 445–448.

Wang, G. Y., Wang, J., Mancianti, M. L., and Epstein, E. H., Jr. (2011). Basal cell carcinomas arise from hair follicle stem cells in Ptch1+/− mice. *Cancer Cell* **19,** 114–124.

Wechsler-Reya, R. J., and Scott, M. P. (1999). Control of neuronal precursor proliferation in the cerebellum by Sonic Hedgehog. *Neuron* **22**(1), 103–114.

Yam, P. T., Langlois, S. D., Morin, S., and Charron, F. (2009). Sonic hedgehog guides axons through a noncanonical, Src-family-kinase-dependent signaling pathway. *Neuron* **62,** 349–362.

Youssef, K. K., Van Keymeulen, A., Lapouge, G., Beck, B., Michaux, C., Achouri, Y., Sotiropoulou, P. A., and Blanpain, C. (2010). Identification of the cell lineage at the origin of basal cell carcinoma. *Nat. Cell Biol.* **12,** 299–307.

Zuscik, M. J., D'Souza, M., Ionescu, A. M., Gunter, K. K., Gunter, T. E., O'Keefe, R. J., Schwarz, E. M., Puzas, J. E., and Rosier, R. N. (2002). Growth plate chondrocyte maturation is regulated by basal intracellular calcium. *Exp. Cell Res.* **276,** 310–319.

CHAPTER FOUR

Gli Protein Nuclear Localization Signal

Minoru Hatayama* *and* Jun Aruga*,†

Contents

Abstract

Drosophila cubitus interruptus (Ci) and its vertebrate homologues, the glioblastoma (Gli) protein family, are the transcription factors belonging to the metazoan Gli/Glis/Zic ZF protein superfamily that shares similar five tandemly repeated C2H2-type zinc finger (ZF) motifs. Nuclear transport of Gli/Ci proteins is regulated by hedgehog (Hh) signaling and is an essential part of the Hh signal transduction pathway. Gli/Ci proteins possess a nuclear localization signal (NLS) and a nuclear export signal (NES), both of which are key signatures for controlling nucleocytoplasmic shuttling. The NLS of the Gli/Ci proteins has been mapped to the fifth ZF domain and its C-terminal side. It contains two clusters of basic residues (classical bipartite-type), which are conserved in metazoan Gli/Ci homologues, but which partially deviate from the intra-ZF domain NLSs in the

* Laboratory for Behavioral and Developmental Disorders, RIKEN Brain Science Institute, Wako-shi, Saitama, Japan
† Saitama University Brain Science Institute, Saitama-shi, Saitama, Japan

Vitamins and Hormones, Volume 88
ISSN 0083-6729, DOI: 10.1016/B978-0-12-394622-5.00004-3

Glis and Zic proteins. Recently, Importin α3 was identified as a nuclear transport protein for Ci. When we modeled the 3D structure of the Gli NLS–Importin α complex, the two basic clusters were predicted to fit in the two binding interfaces of Importin α. The mechanisms controlling the function of NLSs and NESs involve the elimination of the NES by Hh signaling-dependent protein cleavage in the Ci and the Gli3 proteins, and the phosphorylation of a threonine residue close to the NLS in Gli1. Both processes depend on the activity of protein kinase A, which has a critical role in Hh signaling in fly wing discs. In addition, the Roadkill protein, a substrate recognition component of E3 ubiquitin ligase, competes with the Ci protein to interact with Importin α3 resulting in inhibition of Ci protein nuclear import.

I. Introduction

Drosophila cubitus interruptus (Ci) and its vertebrate homologues, the glioblastoma (Gli) protein family (Gli1, Gli2, and Gli3), are critical regulators in animal development (Huangfu and Anderson, 2006; Ingham *et al.*, 2011; Jiang and Hui, 2008; Lum and Beachy, 2004; Stecca and Ruiz i Altaba, 2010). They are key transcription factors (TFs) in the hedgehog (Hh) signal transduction pathway. Structurally, Gli1–3 and Ci are 75–150kDa zinc finger (ZF) proteins with a single ZF domain composed of five tandemly repeated C2H2 ZFs. The ZF domain of Gli1 is known to bind DNA (consensus sequence 5′-TGGGTGGTC-3′) in a sequence-specific manner (Kinzler and Vogelstein, 1990) and is similar to the ZF domains of the Glis and Zic families, which are also animal development-controlling TFs. Together, these proteins comprise the Gli/Glis/Zic ZF protein superfamily.

Recent studies revealed that subcellular localization of Gli/Ci is dynamically controlled in the signal transduction process, and that transport of these proteins is an essential part of the signal transduction process. In particular, the nucleocytoplasmic shuttling of Ci, Gli2, and Gli3 is closely related to the signal-on and signal-off states of Hh signaling. Because of that, many studies have examined the control of the nucleocytoplasmic localization of the Gli/Ci proteins. In this chapter, we discuss the nucleocytoplasmic localization-determining elements contained within the Gli/Ci proteins themselves and consider their significance from structural, functional, and evolutionary perspectives.

II. Nuclear Localization Signal and Nuclear Export Signal

In eukaryotes, trafficking between the cytoplasm and the nucleus is restricted by the nuclear membrane. Macromolecules are transported into and out of the nucleus through large proteinaceous structures called nuclear pore complexes (NPCs) (Lange *et al.*, 2007; Marfori *et al.*, 2010). Although small macromolecules (<40kDa) can passively diffuse through NPCs, larger molecules must be imported via active carrier-mediated transport. The Importin α-mediated pathway is one of the most well-characterized pathways for nuclear import (Goldfarb *et al.*, 2004). Importin α binds to the nuclear localization signal (NLS) of target cargo proteins (Lange *et al.*, 2007; Marfori *et al.*, 2010). The Importin α–NLS cargo protein complex then binds an additional component, Importin β, and the resultant heterotrimeric protein is transported into the nucleus with the support of the RanGDP–RanGTP gradient from the cytoplasm to the nucleus (Marfori *et al.*, 2010). After import to the nucleus, the complex is disassembled and the cargo protein is released (Fig. 4.1) (Goldfarb *et al.*, 2004; Gorlich and Mattaj, 1996; Marfori *et al.*, 2010). Protein localization is also affected by nuclear export signals (NESs), which mediate the binding of adaptor proteins (e.g., CRM1) involved in the export of proteins from the nucleus through NPCs. Thus, NLSs and NESs are major intrinsic determinants of nucleus–cytoplasm protein distribution.

Previous studies have addressed the conserved features of NLS-constituting amino acid sequences. Some NLSs contain one or two clusters of basic amino acid residues (Lange *et al.*, 2007). These NLSs were initially identified in SV40 large T-antigen (PKKKRKV) and Xenopus nucleoplasmin (KRPAATKKAGQAKKKK), and they are collectively referred to as classical NLSs (Lange *et al.*, 2007). Single cluster (monopartite) classical NLSs contain a three to four basic amino acid stretch as is found in SV40 large T, whereas bipartite classical NLSs contain two clusters of basic amino acids that are separated by 10–12 intervening amino acids as is found in Xenopus nucleoplasmin (Dingwall and Laskey, 1991; Fontes *et al.*, 2000). The binding of classical NLSs to Importin α has been well studied (Conti and Kuriyan, 2000; Conti *et al.*, 1998; Fontes *et al.*, 2000); however, recent studies indicate the presence of another pathway that employs proline–tyrosine (PY)-NLSs and the receptor transportin-1/karyopherin-β2 besides the Importin α/karyopherin-α and Importin β/karyopherin-β1 that may target classical NLSs (Marfori *et al.*, 2010). Thus, the study of the mechanisms and molecular determinants of nuclear import is still a developing research field.

III. NLS and NES Mapping

A. NLS and NES mapping in the Ci protein

The Ci protein has two forms (Figs. 4.1 and 4.2). The full-length form (Ci-155; 155kDa) acts as a transcriptional activator. In the absence of Hh signals, Ci-155 is proteolytically cleaved to generate a carboxy-terminally

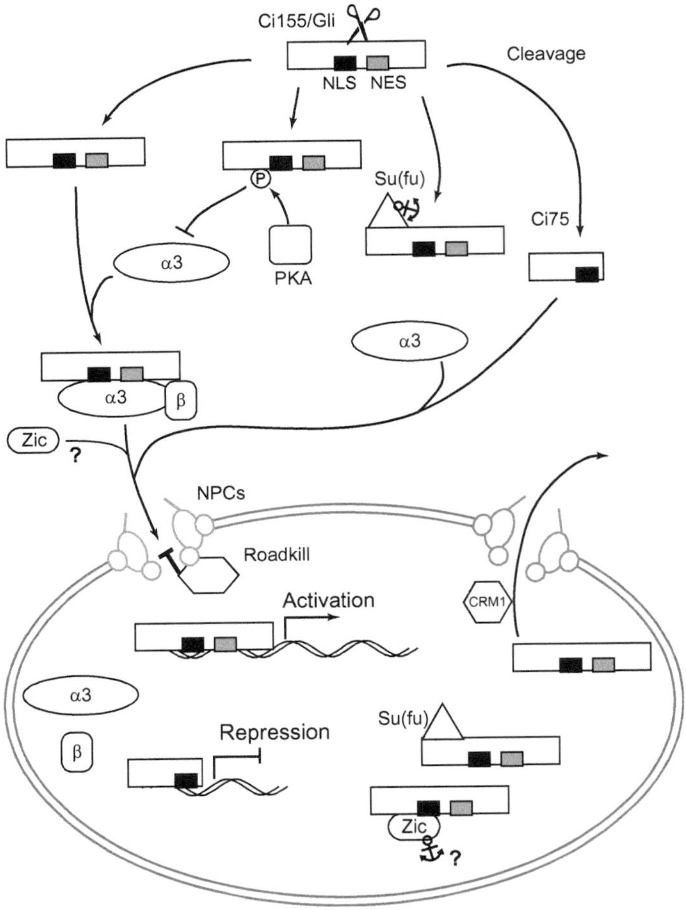

Figure 4.1 Regulation of nucleocytoplasmic shuttling of Gli/Ci by NLS/NES. The Gli/Ci NLS (black box) is a binding target of Importin α3 (α3). Importin β (β) binds to the Gli/Ci–Importin α3 complex via the Importin β binding domain of Importin α. The Gli–Importin α3–Importin β heterotrimer is transported into the nucleus through nuclear pore complexes (NPCs). Phosphorylation of residues close to the NLS inhibits nuclear localization, and Suppressor of fused [Su(fu)] also inhibits nuclear transport. In addition, the Roadkill protein, which localizes around NPCs, inhibits Gli/Ci–Importin α3 binding and accelerates degradation. After nuclear import, the complex is disassembled and Importin α and β are exported from the nucleus. Full-length Gli/Ci that includes an NES (gray box) is exported from the nucleus by the CRM1-mediated export pathway, whereas truncated Gli/Ci without an NES may accumulate in the nucleus. Full-length Gli/Ci transcripts target genes to enhance Hh signal, and cleavaged suppressor form Ci75 act as a transcriptional repressor for the target genes. Zic proteins enhance Gli nuclear localization presumably by modulating the function of the NLS or NES or both or by tethering Gli in the nucleus by physical interaction.

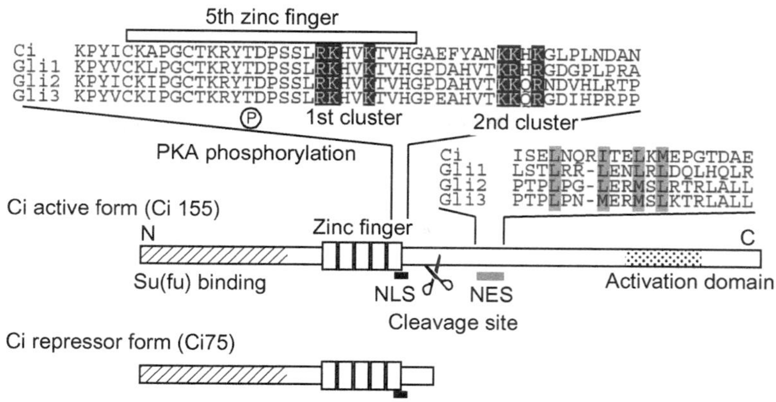

Figure 4.2 Positioning of the NLS and NES and other functional domains in Gli/Ci proteins. The ZF domain, comprising five C2H2 ZFs, is shown as five open boxes. The NLS (black box) is mapped in the fifth ZF domain and its C-terminal flanking region. The NES (gray box) is mapped in the C-terminal region. The NLS and NES amino acid sequence alignments are shown at the top (NLS) and bottom (NES), where fly (Ci) and human (Gli1–3) amino acid sequences are aligned. Top open box indicates the position of the fifth ZF, and NLS-forming basic residues (the first and second clusters) are shown in black. Phosphorylation site near the NLS is indicated as P with a circle. Gray background indicates NES-forming hydrophobic residues. Su(fu) binding site and activation domain are also indicated as diagonal lines and condensed dots, respectively. The symbol of scissors indicates the cleavage site location. The cleavaged suppressor form Ci75 is shown in bottom.

truncated form, Ci-75 (75kDa), which is thought to act as a transcriptional repressor (Aza-Blanc *et al.*, 1997). Ci-155 is primarily located in the cytoplasm (Motzny and Holmgren, 1995), despite being required for the transactivation of target genes, whereas Ci-75 is primarily located in the nucleus (Aza-Blanc *et al.*, 1997).

In a pioneering study, Wang and Holmgren (1999) determined NLS and NES of the Ci protein. They identified a bipartite classical NLS-like sequence in the fifth ZF and its carboxy terminal flanking region (596–614). The two basic clusters reside one in the fifth ZF (R596–K600) and one in the region outside the ZFs (K611–K614) (Fig. 4.2, black box). β-galactosidase fused with the NLS-candidate region (581–616) resulted in strong nuclear localization in cultured insect cells, while the unmodified β-galactosidase was restricted entirely to the cytoplasm. They also tested the necessity of the candidate NLS using a carboxy-terminally truncated Ci (1–616) that shows predominant nuclear localization. When the second cluster was removed from the truncated Ci by further seven amino acids deletion (1–609), nuclear localization was substantially impaired *in vivo*. Thus, they concluded that the 596–614 region acts as an NLS in the Ci protein.

Wang and Holmgren (1999) also reported on the NES of the Ci protein. By fusing the 703–835 region of the Ci protein to the β-galactosidase NLS, they showed that the 703–835 region works in opposition to the NLS; however, mutants lacking the 685–836 region still localized in the cytoplasm. Therefore, while the 703–835 region is sufficient for the cytoplasmic localization, further C-terminal region may have supportive NES activity. The positions of NLS and NES in the Ci protein are consistent with the findings that the absence of the ZF C-terminal sequence results in nuclear localization (Aza-Blanc *et al.*, 1997; Hepker *et al.*, 1997). This provides the structural basis of Ci protein nuclear transport in the transcriptional regulation of Hh target genes (Chen *et al.*, 1999).

B. NLS and NES mapping in Gli proteins

Among the vertebrate Gli proteins, Gli1 shows the most predominant nuclear localization in certain cell types (Barnfield *et al.*, 2005; Ding *et al.*, 1999; Kinzler and Vogelstein, 1990; Koyabu *et al.*, 2001; Merchant *et al.*, 2004; Ruiz i Altaba, 1999). NLS and NES function analyses have been carried out for Gli1. Barnfield *et al.* (2005) predicted two candidate NLSs based on sequence analysis of mouse Gli1. The first was a monopartite NLS candidate in region 79–84, and the second was a bipartite NLS candidate in region 383–401. The second, bipartite NLS candidate was in the homologous region of the Ci protein NLS (596–614). Cytoplasmic localization of Gli1 was markedly increased by substitution mutations in the 596–614 region. Mutation of the 79–84 region had minimal effect on Gli1 localization, but mutations of both candidate regions at the same time resulted in a drastic increase in cytoplasmic localization. Therefore, both NLSs may play a functional role in Gli1. Additional information on the Gli1 NLS came from the analysis of the Zic family NLSs (Hatayama *et al.*, 2008). When the fifth ZF of human Gli1 and a C-terminal flanking region (Gli1 ZF5+C-region; 364–410) were placed between glutathione *S*-transferase and enhanced green fluorescent protein, the Gli1 ZF5+C-region exhibited strong NLS activity, which was comparable with that of the SV40 large T NLS measured in the same assay. Collectively, these results indicate that the Gli1 ZF5+C-region NLS strongly directs nuclear protein sorting.

The NES of human Gli1 was examined by Kogerman *et al.* (1999). In previous studies, deletion of the carboxy terminal regions of Gli3 resulted in nuclear localization (Ruiz i Altaba, 1999; Shin *et al.*, 1999). They found a leucine-rich sequence in the 496–504 region of Gli1, the leucine residues of which were conserved in Gli2 and Gli3. When they substituted two of the conserved leucine residues (495 and 501; human Gli1) with alanine, the mutant Gli1 showed absolute nuclear localization. This indicates that this leucine-rich sequence acts as an NES. Leucine-rich NESs are known to mediate the nuclear export of proteins that are dependent on the NES receptor protein, CRM1. Accordingly, when Kogerman *et al.* (1999) added a CRM1 inhibitor (leptomycin B), Gli1 accumulated in the nucleus. The function of the NES was also confirmed in mouse Gli1 (Barnfield *et al.*, 2005), in which alanine-substitution mutations were introduced into the residues corresponding to the final two leucine residues in the human GLI1 NES. Barnfield *et al.* (2005) also showed that a leucine-rich sequence motif in mouse Gli2 (887–895) with sequence similarity to Gli1 NES acts as an NES by mutating it (Fig. 4.2).

As shown above, both NLSs and NESs endow the Gli/Ci proteins with a nucleocytoplasmic shuttling capacity. The importance of NLSs and NESs in controlling subcellular localization of the Gli/Ci proteins is discussed in Sections VI and VII.

IV. Sequence Comparison of Gli NLSs—NLSs in ZF Domains

Alignment of the Gli NLSs of representative metazoan animals suggests that the amino acid residues in the ZF domains are strongly conserved (Figs. 4.2 and 4.3B). However, the length of the intervening amino acid sequence between the first and second basic clusters is shorter in sea anemone, *Nematostella vectensis* (11 amino acid residues), and sponge, *Amphimedon queenslandica* (8 amino acid residues), than in bilateria. Bipartite classical NLSs generally have 10–12 residues between the sequences that separate the two basic clusters (Dingwall and Laskey, 1991). Whether or not the sponge sequence acts as an NLS is not clear at this point.

Gli/Ci NLSs are partly conserved in Glis3, but not in Glis2 and Zic3 (Fig. 4.3C). Comparison of the Gli/Ci NLSs with the Glis and Zic NLSs is intriguing because Gli, Glis, and Zic proteins are likely derived from a common metazoan ancestor (Hatayama and Aruga, 2010). These gene families are widely distributed in metazoan animals, except that placozoa lack Gli but possess Zic (Layden *et al.*, 2010; Srivastava *et al.*, 2010) (Fig. 4.3A). This fact suggested that metazoan common ancestor had already acquired Gli/Glis, and Zic may have been innovated in eumetazoan ancestor. The divergence of NLS sequences may partly reflect this evolutional history. In a comparison between the eumetazoan Gli/Ci proteins, the two basic residue clusters in the NLS are strongly conserved (Fig. 4.3C), suggesting a biological significance of the bipartite Gli/Ci NLS motif.

The Zic NLS sequence is not a classical NLS (dispersed NLS where NLS-forming basic residues does not form a cluster), but in its 3D conformation, the basic residues are positioned similar to those in the bipartite nucleoplasmin NLS (Hatayama *et al.*, 2008). The corresponding region in the Gli ZF5+C-region NLS lacks the second cluster and acts as a

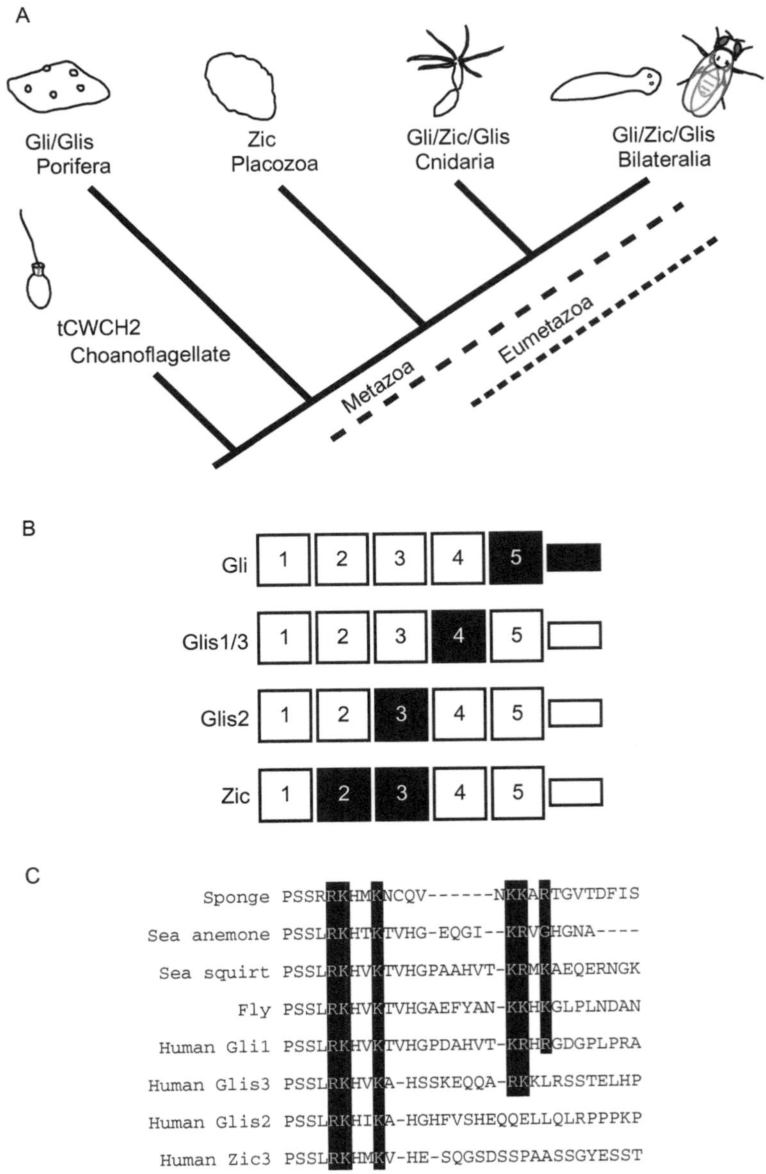

Figure 4.3 Phylogeny of NLSs in the Gli/Glis/Zic ZF family proteins. (A) Distribution of Gli, Glis, and Zic families in animal phylogeny. (B) Location of NLSs in Gli, Glis, and Zic ZF domains. Each ZF is depicted as a white box and the NLS-coding ZF is depicted as a black box. (C) Sequence alignment of the Gli/Ci NLS region. Sponge (*Amphimedon queenslandica*), sea anemone (*Nematostella vectensis*), sea squirt (*Ciona intestinalis*), fly (*Drosophila melanogaster*), and human (*Homo sapiens*) Gli1 sequences are aligned. Human Glis3, Glis2, and Zic3 are also aligned. We aligned Glis3 as a representative for Glis1/3 group.

supplementary NLS (Bedard *et al.*, 2007; Hatayama *et al.*, 2008). Thus, the position of the NLS clearly differs between Gli and Zic proteins.

Glis ZF sequences can be divided into two groups (data not shown): Glis1/3 and Glis2. Our sequence alignment suggests that basic amino acid residues in the Gli/Ci NLS region are partly conserved in Glis1/3, but not in Glis2 (Fig. 4.3C). Beak *et al.* (2008) reported that the Glis3 NLS was located in ZF4 and that removal of the Gli NLS-corresponding region did not affect nuclear localization. The Glis2 NLS is located in the ZF3 region (237–264) (Vasanth *et al.*, 2011). Therefore, it can be concluded that the NLSs in Gli/Ci, Glis2, Glis3, and Zic3 are not identical. It is intriguing that the NLSs in the closely related ZF domains are divergent. This presumably reflects the structural restriction of ZFs due to their multifunctional properties.

C2H2-type ZF domain NLSs have recently been mapped in the snail (Mingot *et al.*, 2009; Yamasaki *et al.*, 2005) and in Sp1 proteins (Ito *et al.*, 2009). In the snail, the NLS was mapped to a ZF domain that directly binds Importin β (Yamasaki *et al.*, 2005). Further, mutational analysis of the basic residues in the snail ZF domain showed that five out of six basic residues in defined positions in the three consecutive ZFs are essential for nuclear localization (Mingot *et al.*, 2009). Interestingly, the positions of six basic residues are conserved in C2H2 ZF proteins, including the Gli, Glis, and Zic proteins. The ZF C-terminal flanking basic cluster in the bipartite Gli NLS may have been uniquely acquired over the course of evolution, while the other ZF NLSs may have diversified using basic residues scattered within the ZF.

V. Predicted 3D Structure of Gli NLS

The crystallographic 3D structure was reported for the GLI1 ZF (Pavletich and Pabo, 1993). To visualize the structure of the NLS, we modeled the 3D structure of the human GLI1 bipartite NLS using Modeller software (version 9.9; http://salilab.org/modeller/) (Eswar *et al.*, 2006). In our model, the fifth ZF domain folded into a solid structure, but the flanking C-terminal region had an indeterminate structure (data not shown). This result suggests that the bipartite Gli1 NLS outside of ZF5 is flexible. Interestingly, one out of ten Gli1 NLS models fitted well with the cap-binding protein 80 NLS (CBP80; PDBID 3FEX) in terms of side-chain position of the NLS-constituting basic amino acid residues (Fig. 4.4A).

In accordance with its similarity to classical NLSs, Importin α3 was recently reported as a nuclear transport receptor for Ci (Seong *et al.*, 2010). We therefore superimposed the Gli1 NLS model with the Importin α2 structure (3EFX), which has a structure similar to *Drosophila* Importin α3 (Mason

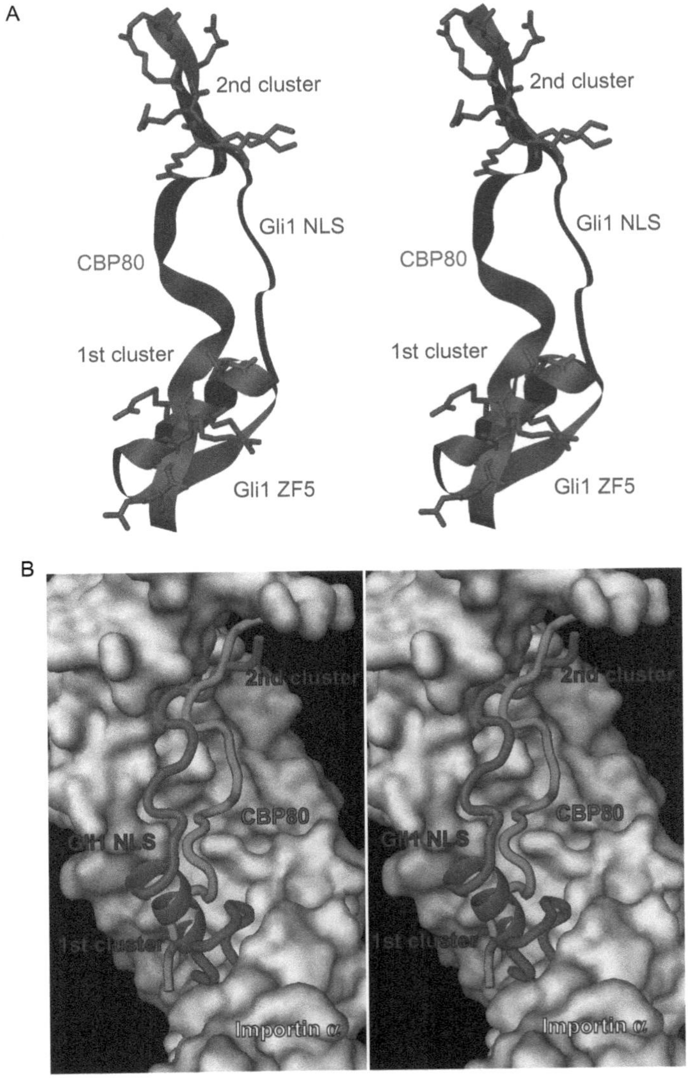

Figure 4.4 Three-dimensional model of Gli1 NLS. The results are shown as stereograms. (A) Superimposition of the Gli NLS structure (thin ribbon) and a known bipartite NLS (thick ribbon) structure from CBP80 (PDB ID, 3FEX). We calculated and created 10 models of the Gli1 NLS, referring to known bipartite NLS structures. NLS-forming side chains are shown as a stick model and the C-terminus is at the top. (B) Superimposition of the Gli NLS and Importin α. Critical steric hindrance was not found. Importin α is shown as a surface model, and the Gli NLS and CBP80 are shown as thick wire models. (See Color Insert.)

et al., 2009) (Fig. 4.4B). Gli1 NLS fitted well with the Importin α2 interaction groove without serious steric collision. Previous studies showed that Importin α has two binding interfaces in the side chains of the NLS-forming residues (Conti *et al.*, 1998; Fontes *et al.*, 2000, 2003). The Gli1 NLS-forming residues were located in these two binding interfaces. The result of the modeling was consistent with the interaction between Importin α3 and Gli1, and suggested that the length of the intervening sequence between the two basic clusters in the Gli NLS was appropriate for the distance between the two binding pockets in Importin α.

VI. NLS Control of Gli/Ci Subcellular Localization in Hh Signaling

A. Gli/Ci processing and the function of the NLS/NES

As described in Section III, proteolytic cleavage of Ci-155, the full-length activator form of the Ci protein, produces the Ci-75 repressor form (Aza-Blanc *et al.*, 1997). Because the cleavage point is mapped between NLS and NES, Ci-75 contains only an NLS (Figs. 4.1 and 4.2). This cleavage-based elimination of the NES is the primary mechanism for the nuclear localization of Ci-75. In the vertebrate Gli protein family, Gli2 and Gli3 are the vertebrate counterparts of the Ci protein (Ingham *et al.*, 2011; Sasaki *et al.*, 1999). Both mouse Gli2 and Gli3 function as transcriptional activators in full-length form but function as transcriptional repressors after removal of the C-terminal repressor domain (Pan *et al.*, 2006; Wang *et al.*, 2000). Therefore, the nuclear transport accompanying the C-terminal processing (NES removal) may proceed in a manner similar to that of the Ci protein. However, in Gli2, the processing is very inefficient compared to that of Gli3 (Pan *et al.*, 2006). The actual contribution of processing-based nuclear transport needs to be further examined. Gli1 lacks an N-terminal repression domain and has been known to act as a transcriptional activator. Recently, it was reported that an N-terminally processed isoform and a C-terminally processed isoform exist for human Gli1 (Stecca and Ruiz i Altaba, 2010). The latter isoform possibly coincides with that of the Gli2 and Gli3 processed form, but there are no details regarding its subcellular localization.

Processing-based removal of the NES partly explains the nuclear import control of Gli/Ci proteins. However, understanding of the control of the nuclear import of full-length Gli/Ci proteins remains elusive. Sisson *et al.* (2006) addressed the localization of NLS-mutated Ci *in vivo* by using an antibody that recognizes only the full-length Ci. For this purpose, Sisson *et al.* carried out transgenic expression of a Ci NLS mutant in a ci^{94} null mutation background. While the wild-type Ci cDNA completely rescued the segmental defects that appeared in the larval cuticles, mutation of the second cluster in the Ci NLS (KKHK) to GGLE resulted in partial loss of naked cuticles, indicating downregulation of the Hh signal. In the same experimental model, they also found that mutant Ci with SV40 large T NLS expression resulted in impaired cuticle patterns indicative of ectopic Hh signaling. These results indicate that the Ci NLS actually contributes to the determination of the extent of Ci nuclear localization. However, endogenous Ci NLS is not required for appropriate nuclear import in salivary glands or in wing discs. This may be explained either by the presence of additional NLSs other than the ZF5 bipartite NLS or may be the remaining first cluster has a residual NLS function (Sisson *et al.*, 2006). Alternatively, the functionality of NLSs and NESs is modified by other proteins (Sisson *et al.*, 2006). This seems to be the case for NLSs as we discuss below.

B. Phosphorylation-mediated control of NLS function

One potential mechanism of Gli/Ci NLS functional regulation is protein phosphorylation because a threonine residue, Thr374, adjacent to the first basic cluster of the Gli1 NLS can be phosphorylated by protein kinase A (PKA) (Sheng *et al.*, 2006). Mutation of Thr374 to aspartate (T374D) results in more cytoplasmic Gli1, whereas mutation to lysine (T374K) leads to more nuclear Gli1 (Sheng *et al.*, 2006). Thr374 is located close to the Gli1 NLS in the 3D structure. Therefore, phosphorylation of Thr374 is predicted to change the local charge and therefore NLS function (Sheng *et al.*, 2006). Generally, phosphorylation within or adjacent to classic NLSs is associated with cytoplasmic protein localization in many proteins, and phosphorylation near NLSs decreases the affinity of cargo to Importin α (Harreman *et al.*, 2004). The sequence around Thr374 is conserved among the Gli1, Gli2, Gli3, and Ci proteins (Sheng *et al.*, 2006). Therefore, the same mechanism is plausible for the functional regulation of the other Gli/Ci NLSs.

Interestingly, PKA signaling is also involved in the regulation of Gli/Ci protein processing. Therefore, PKA signaling functionally controls Gli/Ci NLSs and NESs in two different ways. PKA activation and phosphorylation are triggered by Hh signals. This is consistent with the fact that PKA negatively regulates Gli nuclear accumulation. Together with the enhanced proteasomal degradation of the Gli2 protein by PKA (Pan *et al.*, 2009), PKA phosphorylation seems to regulate Gli dynamics at multiple points.

C. Importin α3-mediated nuclear transport of Ci and its inhibition by Roadkill

Accumulating evidence indicates the importance of NLS-mediated control of Gli proteins in Hh signaling. Importin α3 was identified as a Ci-binding protein in yeast two-hybrid assays (Formstecher *et al.*, 2005; Seong *et al.*,

2010), and the Ci–Importin α3 interaction was confirmed by immunoprecipitation assay in cultured insect cells (Seong *et al.*, 2010). Knockdown of Importin α3 inhibited nuclear import of Ci (Seong *et al.*, 2010), indicating that Importin α3 acts as a nuclear import receptor for Ci.

In an emerging regulatory mechanism of Ci NLS function, the Roadkill protein has been shown to compete with Importin α3 for binding to Ci and to inhibit the nuclear import of Ci-155 (Seong *et al.*, 2010). These results suggest that Gli NLSs serve as the platform for Roadkill–Importin α3 competition. Interestingly, Roadkill is known as the substrate recognition component of Cul3 E3 ubiquitin ligase, which targets Ci for degradation (Kent *et al.*, 2006; Zhang *et al.*, 2006). These two modes of Hh signal negative-regulation may play an important role in the fine-tuning of Hh responses (Seong *et al.*, 2010).

D. Other potential factors modulating NLS functionality

Suppressor of fused [Su(fu)] is a binding partner of Gli/Ci (Ding *et al.*, 1999; Kogerman *et al.*, 1999; Monnier *et al.*, 1998) and a negative regulator of the nuclear accumulation of Gli/Ci (Ding *et al.*, 1999; Kogerman *et al.*, 1999; Methot and Basler, 2000; Ohlmeyer and Kalderon, 1998). It has been proposed that NLS masking is involved in the cytoplasmic retention of Ci-155 by Su(fu) (Wang and Jiang, 2004). This idea is based on the fact that SV40 large T NLS-attached Ci in wing discs is predominantly localized in the nucleus (Wang and Jiang, 2004). However, Su(fu)–Ci complexes can be transported into nuclei in the salivary glands in the presence of the CRM1-mediated nuclear export inhibitor leptomycin B (Sisson *et al.*, 2006), suggesting that NLS masking is not the only mechanism for Ci cytoplasmic restriction by Su(fu). Another potential explanation that was proposed by Sisson *et al.* (2006) is that the Su(fu)–Ci complex has an affinity for a cytoplasmic tether away from the Hh signaling pathway and therefore a low probability of being imported into the nucleus. The modulation of Ci NLS functionality by Su(fu) is still controversial.

Other potential candidates for the altering of Gli/Ci NLS functionality are the Zic proteins. Zic1 increases the nuclear localization of both Gli1 and Gli3 in transfected cells (Koyabu *et al.*, 2001), and Zic2 increases the nuclear localization of the Gli1–3 proteins (Chan *et al.*, 2011; Nguyen *et al.*, 2005). Zic proteins and Gli proteins physically interact through their ZF domains, and the presence of Zic proteins affects the transcriptional activity of the Gli proteins in a context-dependent manner (Chan *et al.*, 2011; Koyabu *et al.*, 2001; Mizugishi *et al.*, 2001). Because the Zic1–Gli3 interaction occurs through the ZF3–5 domain, which partly includes the NLS (Koyabu *et al.*, 2001), Zic proteins might affect Gli NLS/NES functionality or may otherwise keep Gli in the nuclei through physical interaction. Further, functional modulation of Gli protein or Hh signaling by Zic proteins has been shown in the development of Xenopus and zebrafish embryos (Brewster *et al.*, 1998; Maurus and Harris, 2009). Also, in the context of cervical cancer development, Zic2 enhances Hh signaling through nuclear retention of Gli1 (Chan *et al.*, 2011). Not only Zic–Gli interactions, but also Gli–Gli interactions, are known to occur through their ZF domains (Nguyen *et al.*, 2005). Gli1 enhanced the nuclear localization of both Gli2 and Gli3 (Nguyen *et al.*, 2005). Whether or not the NLS functionality is altered remains to be elucidated in these cases. The mechanism underlying the enhanced nuclear localization through the ZF-mediated complex formation has attracted our attention.

VII. Perspectives

Having reviewed the current understanding of the Gli/Ci NLS, it is clear that the NLS is one of the critical platforms for Hh signaling and its modulation. Since the nucleocytoplasmic distribution of Gli/Ci proteins is a useful parameter in Hh signaling, we can expect experimental manipulation of, or therapeutics targeting, the nuclear import/export processes. Considering the involvement of Gli proteins in tumorigenesis (Jiang and Hui, 2008; Stecca and Ruiz i Altaba, 2010), clinical importance is high. Gli NLS studies have significantly contributed to the basic understanding of the nuclear import/export mechanism from the point of view of general cell biology. Further clarification of the function of the Gli NLS, or more broadly, the NLSs in ZF proteins, is an important challenge for the biological sciences.

REFERENCES

Aza-Blanc, P., Ramirez-Weber, F. A., Laget, M. P., Schwartz, C., and Kornberg, T. B. (1997). Proteolysis that is inhibited by hedgehog targets Cubitus interruptus protein to the nucleus and converts it to a repressor. *Cell* **89,** 1043–1053.

Barnfield, P. C., Zhang, X., Thanabalasingham, V., Yoshida, M., and Hui, C. C. (2005). Negative regulation of Gli1 and Gli2 activator function by suppressor of fused through multiple mechanisms. *Differentiation* **73,** 397–405.

Beak, J. Y., Kang, H. S., Kim, Y. S., and Jetten, A. M. (2008). Functional analysis of the zinc finger and activation domains of Glis3 and mutant Glis3(NDH1). *Nucleic Acids Res.* **36,** 1690–1702.

Bedard, J. E., Purnell, J. D., and Ware, S. M. (2007). Nuclear import and export signals are essential for proper cellular trafficking and function of ZIC3. *Hum. Mol. Genet.* **16,** 187–198.

Brewster, R., Lee, J., and Ruiz i Altaba, A. (1998). Gli/Zic factors pattern the neural plate by defining domains of cell differentiation. *Nature* **393,** 579–583.

Chan, D. W., Liu, V. W., Leung, L. Y., Yao, K. M., Chan, K. K., Cheung, A. N., and Ngan, H. Y. (2011). Zic2 synergistically enhances Hedgehog signalling through nuclear retention of Gli1 in cervical cancer cells. *J. Pathol.* **225,** 525–534.

Chen, C. H., von Kessler, D. P., Park, W., Wang, B., Ma, Y., and Beachy, P. A. (1999). Nuclear trafficking of Cubitus interruptus in the transcriptional regulation of Hedgehog target gene expression. *Cell* **98,** 305–316.

Conti, E., and Kuriyan, J. (2000). Crystallographic analysis of the specific yet versatile recognition of distinct nuclear localization signals by karyopherin alpha. *Structure* **8,** 329–338.

Conti, E., Uy, M., Leighton, L., Blobel, G., and Kuriyan, J. (1998). Crystallographic analysis of the recognition of a nuclear localization signal by the nuclear import factor karyopherin alpha. *Cell* **94,** 193–204.

Ding, Q., Fukami, S., Meng, X., Nishizaki, Y., Zhang, X., Sasaki, H., Dlugosz, A., Nakafuku, M., and Hui, C. (1999). Mouse suppressor of fused is a negative regulator of sonic hedgehog signaling and alters the subcellular distribution of Gli1. *Curr. Biol.* **9,** 1119–1122.

Dingwall, C., and Laskey, R. A. (1991). Nuclear targeting sequences—A consensus? *Trends Biochem. Sci.* **16,** 478–481.

Eswar, N., Webb, B., Marti-Renom, M. A., Madhusudhan, M. S., Eramian, D., Shen, M. Y., Pieper, U., and Sali, A. (2006). Comparative protein structure modeling using Modeller. *Curr. Protoc. Bioinform.* Chapter 5, Unit 5 6.

Fontes, M. R., Teh, T., and Kobe, B. (2000). Structural basis of recognition of monopartite and bipartite nuclear localization sequences by mammalian importin-alpha. *J. Mol. Biol.* **297,** 1183–1194.

Fontes, M. R., Teh, T., Jans, D., Brinkworth, R. I., and Kobe, B. (2003). Structural basis for the specificity of bipartite nuclear localization sequence binding by importin-alpha. *J. Biol. Chem.* **278,** 27981–27987.

Formstecher, E., Aresta, S., Collura, V., Hamburger, A., Meil, A., Trehin, A., Reverdy, C., Betin, V., Maire, S., Brun, C., Jacq, B., Arpin, M., *et al.* (2005). Protein interaction mapping: A Drosophila case study. *Genome Res.* **15,** 376–384.

Goldfarb, D. S., Corbett, A. H., Mason, D. A., Harreman, M. T., and Adam, S. A. (2004). Importin alpha: A multipurpose nuclear-transport receptor. *Trends Cell Biol.* **14,** 505–514.

Gorlich, D., and Mattaj, I. W. (1996). Nucleocytoplasmic transport. *Science* **271,** 1513–1518.

Harreman, M. T., Kline, T. M., Milford, H. G., Harben, M. B., Hodel, A. E., and Corbett, A. H. (2004). Regulation of nuclear import by phosphorylation adjacent to nuclear localization signals. *J. Biol. Chem.* **279,** 20613–20621.

Hatayama, M., and Aruga, J. (2010). Characterization of the tandem CWCH2 sequence motif: A hallmark of inter-zinc finger interactions. *BMC Evol. Biol.* **10,** 53.

Hatayama, M., Tomizawa, T., Sakai-Kato, K., Bouvagnet, P., Kose, S., Imamoto, N., Yokoyama, S., Utsunomiya-Tate, N., Mikoshiba, K., Kigawa, T., and Aruga, J. (2008). Functional and structural basis of the nuclear localization signal in the ZIC3 zinc finger domain. *Hum. Mol. Genet.* **17,** 3459–3473.

Hepker, J., Wang, Q. T., Motzny, C. K., Holmgren, R., and Orenic, T. V. (1997). Drosophila cubitus interruptus forms a negative feedback loop with patched and regulates expression of Hedgehog target genes. *Development* **124,** 549–558.

Huangfu, D., and Anderson, K. V. (2006). Signaling from Smo to Ci/Gli: Conservation and divergence of Hedgehog pathways from Drosophila to vertebrates. *Development* **133,** 3–14.

Ingham, P. W., Nakano, Y., and Seger, C. (2011). Mechanisms and functions of Hedgehog signalling across the metazoa. *Nat. Rev. Genet.* **12,** 393–406.

Ito, T., Azumano, M., Uwatoko, C., Itoh, K., and Kuwahara, J. (2009). Role of zinc finger structure in nuclear localization of transcription factor Sp1. *Biochem. Biophys. Res. Commun.* **380,** 28–32.

Jiang, J., and Hui, C. C. (2008). Hedgehog signaling in development and cancer. *Dev. Cell* **15,** 801–812.

Kent, D., Bush, E. W., and Hooper, J. E. (2006). Roadkill attenuates Hedgehog responses through degradation of Cubitus interruptus. *Development* **133,** 2001–2010.

Kinzler, K. W., and Vogelstein, B. (1990). The GLI gene encodes a nuclear protein which binds specific sequences in the human genome. *Mol. Cell. Biol.* **10,** 634–642.

Kogerman, P., Grimm, T., Kogerman, L., Krause, D., Unden, A. B., Sandstedt, B., Toftgard, R., and Zaphiropoulos, P. G. (1999). Mammalian suppressor-of-fused modulates nuclear-cytoplasmic shuttling of Gli-1. *Nat. Cell Biol.* **1,** 312–319.

Koyabu, Y., Nakata, K., Mizugishi, K., Aruga, J., and Mikoshiba, K. (2001). Physical and functional interactions between Zic and Gli proteins. *J. Biol. Chem.* **276,** 6889–6892.

Lange, A., Mills, R. E., Lange, C. J., Stewart, M., Devine, S. E., and Corbett, A. H. (2007). Classical nuclear localization signals: Definition, function, and interaction with importin alpha. *J. Biol. Chem.* **282,** 5101–5105.

Layden, M. J., Meyer, N. P., Pang, K., Seaver, E. C., and Martindale, M. Q. (2010). Expression and phylogenetic analysis of the zic gene family in the evolution and development of metazoans. *Evodevo* **1,** 12.

Lum, L., and Beachy, P. A. (2004). The Hedgehog response network: Sensors, switches, and routers. *Science* **304,** 1755–1759.

Marfori, M., Mynott, A., Ellis, J. J., Mehdi, A. M., Saunders, N. F., Curmi, P. M., Forwood, J. K., Boden, M., and Kobe, B. (2010). Molecular basis for specificity of nuclear import and prediction of nuclear localization. *Biochim. Biophys. Acta.*

Mason, D. A., Stage, D. E., and Goldfarb, D. S. (2009). Evolution of the metazoan-specific importin alpha gene family. *J. Mol. Evol.* **68,** 351–365.

Maurus, D., and Harris, W. A. (2009). Zic-associated holoprosencephaly: Zebrafish Zic1 controls midline formation and forebrain patterning by regulating Nodal, Hedgehog, and retinoic acid signaling. *Genes Dev.* **23,** 1461–1473.

Merchant, M., Vajdos, F. F., Ultsch, M., Maun, H. R., Wendt, U., Cannon, J., Desmarais, W., Lazarus, R. A., de Vos, A. M., and de Sauvage, F. J. (2004). Suppressor of fused regulates Gli activity through a dual binding mechanism. *Mol. Cell. Biol.* **24,** 8627–8641.

Methot, N., and Basler, K. (2000). Suppressor of fused opposes hedgehog signal transduction by impeding nuclear accumulation of the activator form of Cubitus interruptus. *Development* **127,** 4001–4010.

Mingot, J. M., Vega, S., Maestro, B., Sanz, J. M., and Nieto, M. A. (2009). Characterization of Snail nuclear import pathways as representatives of C2H2 zinc finger transcription factors. *J. Cell Sci.* **122,** 1452–1460.

Mizugishi, K., Aruga, J., Nakata, K., and Mikoshiba, K. (2001). Molecular properties of Zic proteins as transcriptional regulators and their relationship to GLI proteins. *J. Biol. Chem.* **276,** 2180–2188.

Monnier, V., Dussillol, F., Alves, G., Lamour-Isnard, C., and Plessis, A. (1998). Suppressor of fused links fused and Cubitus interruptus on the hedgehog signalling pathway. *Curr. Biol.* **8,** 583–586.

Motzny, C. K., and Holmgren, R. (1995). The Drosophila cubitus interruptus protein and its role in the wingless and hedgehog signal transduction pathways. *Mech. Dev.* **52,** 137–150.

Nguyen, V., Chokas, A. L., Stecca, B., and Ruiz i Altaba, A. (2005). Cooperative requirement of the Gli proteins in neurogenesis. *Development* **132,** 3267–3279.

Ohlmeyer, J. T., and Kalderon, D. (1998). Hedgehog stimulates maturation of Cubitus interruptus into a labile transcriptional activator. *Nature* **396,** 749–753.

Pan, Y., Bai, C. B., Joyner, A. L., and Wang, B. (2006). Sonic hedgehog signaling regulates Gli2 transcriptional activity by suppressing its processing and degradation. *Mol. Cell. Biol.* **26,** 3365–3377.

Pan, Y., Wang, C., and Wang, B. (2009). Phosphorylation of Gli2 by protein kinase A is required for Gli2 processing and degradation and the Sonic Hedgehog-regulated mouse development. *Dev. Biol.* **326,** 177–189.

Pavletich, N. P., and Pabo, C. O. (1993). Crystal structure of a five-finger GLI-DNA complex: New perspectives on zinc fingers. *Science* **261,** 1701–1707.

Ruiz i Altaba, A. (1999). Gli proteins encode context-dependent positive and negative functions: Implications for development and disease. *Development* **126,** 3205–3216.

Sasaki, H., Nishizaki, Y., Hui, C., Nakafuku, M., and Kondoh, H. (1999). Regulation of Gli2 and Gli3 activities by an amino-terminal repression domain: Implication of Gli2 and Gli3 as primary mediators of Shh signaling. *Development* **126,** 3915–3924.

Seong, K. H., Akimaru, H., Dai, P., Nomura, T., Okada, M., and Ishii, S. (2010). Inhibition of the nuclear import of cubitus interruptus by Roadkill in the presence of strong hedgehog signal. *PLoS One* **5,** e15365.

Sheng, T., Chi, S., Zhang, X., and Xie, J. (2006). Regulation of Gli1 localization by the cAMP/protein kinase A signaling axis through a site near the nuclear localization signal. *J. Biol. Chem.* **281,** 9–12.

Shin, S. H., Kogerman, P., Lindstrom, E., Toftgard, R., and Biesecker, L. G. (1999). GLI3 mutations in human disorders mimic Drosophila cubitus interruptus protein functions and localization. *Proc. Natl. Acad. Sci. USA* **96,** 2880–2884.

Sisson, B. E., Ziegenhorn, S. L., and Holmgren, R. A. (2006). Regulation of Ci and Su(fu) nuclear import in Drosophila. *Dev. Biol.* **294,** 258–270.

Srivastava, M., Simakov, O., Chapman, J., Fahey, B., Gauthier, M. E., Mitros, T., Richards, G. S., Conaco, C., Dacre, M., Hellsten, U., Larroux, C., Putnam, N. H., *et al.* (2010). The Amphimedon queenslandica genome and the evolution of animal complexity. *Nature* **466,** 720–726.

Stecca, B., and Ruiz i Altaba, A. (2010). Context-dependent regulation of the GLI code in cancer by Hedgehog and non-Hedgehog signals. *J. Mol. Cell Biol.* **2,** 84–95.

Vasanth, S., ZeRuth, G., Kang, H. S., and Jetten, A. M. (2011). Identification of nuclear localization, DNA binding, and transactivating mechanisms of Kruppel-like zinc finger protein Gli-similar 2 (Glis2). *J. Biol. Chem.* **286,** 4749–4759.

Wang, Q. T., and Holmgren, R. A. (1999). The subcellular localization and activity of Drosophila cubitus interruptus are regulated at multiple levels. *Development* **126,** 5097–5106.

Wang, G., and Jiang, J. (2004). Multiple Cos2/Ci interactions regulate Ci subcellular localization through microtubule dependent and independent mechanisms. *Dev. Biol.* **268,** 493–505.

Wang, B., Fallon, J. F., and Beachy, P. A. (2000). Hedgehog-regulated processing of Gli3 produces an anterior/posterior repressor gradient in the developing vertebrate limb. *Cell* **100,** 423–434.

Yamasaki, H., Sekimoto, T., Ohkubo, T., Douchi, T., Nagata, Y., Ozawa, M., and Yoneda, Y. (2005). Zinc finger domain of Snail functions as a nuclear localization signal for importin beta-mediated nuclear import pathway. *Genes Cells* **10,** 455–464.

Zhang, Q., Zhang, L., Wang, B., Ou, C. Y., Chien, C. T., and Jiang, J. (2006). A hedgehog-induced BTB protein modulates hedgehog signaling by degrading Ci/Gli transcription factor. *Dev. Cell* **10,** 719–729.

CHAPTER FIVE

Mammalian Homologues of *Drosophila* Fused Kinase

Alla Maloverjan* *and* Marko Piirsoo†

Contents

Abstract

Sonic Hedgehog (Shh) signaling pathway is implicated in various developmental and postnatal processes. Much of the current knowledge about the mechanisms of Shh signal transduction in vertebrates comes from the investigations of the respective pathway in fruit fly *Drosophila melanogaster*. In *Drosophila*, serine/threonine kinase fused is involved in all aspects of regulation of the Hh-dependent transcription factor *cubitus interruptus* possessing both catalytic and regulatory functions. Two proteins, Stk36 and Ulk3, share similarity with fu and have been suggested as mammalian fu homologues. However, *in vivo* data clarify that Stk36 is not required for embryonic development in mice and participates in Shh-independent genesis of motile cilia. Even if Stk36 is associated with any pathological or physiological aspect of postnatal Shh signaling in mammals, it has perhaps only regulatory functions since its catalytic activity seems to be lost during evolution. In contrast to Stk36, Ulk3 is an active kinase.

* Protobios LCC, Akadeemia tee 15a, Tallinn, Estonia
† Tallinn University of Technology, Akadeemia tee 15, Tallinn, Estonia

Vitamins and Hormones, Volume 88
ISSN 0083-6729, DOI: 10.1016/B978-0-12-394622-5.00005-5

In non-stimulated cells, Ulk3 catalytic activity is blocked, and it is involved in negative control of Gli proteins, mediators of Shh signaling. In response to Shh, Ulk3 positively regulates Gli proteins by directly phosphorylating them. Thus, Ulk3 is able to recapitulate both positive and negative roles of fu *in vitro*. However, Ulk3 functioning *in vivo* remains to be investigated.

Abbreviations

ACT	activator
ci	cubitus interruptus
CTD	C-terminal regulatory domain
Fu	fused
HSC	hedgehog signaling complex
KD	kinase domain
REP	repressor
Shh	sonic hedgehog
Stk36	serine/threonine kinase 36
Sufu	suppressor of fused
Ulk3	unc51-like kinase 3

I. Introduction

Reductionistic view of pattern formation during embryonic development argues that a limited number of signaling pathways represents the core regulators, termed as "developmental toolkit," and others are involved in fine-tuning of the developmental processes (Canestro *et al.*, 2007). "Developmental toolkit" includes evolutionally conserved BMP, FGF, Notch, Hedgehog (Hh), and Wnt signaling pathways. All of these signaling pathways start with a ligand binding on the cell surface, followed by interpretation of the signal in the cytoplasm and end up in converting the signal to differential transcription in the nucleus. In principle, these signaling pathways can be divided into two groups. First group, including BMP and FGF pathways, utilizes protein kinases as transmembrane receptors, a serine/threonine kinase in case of BMP and a tyrosine kinase in case of FGF pathway. Kinase activity of these receptors is activated upon ligand binding and the receptor phosphorylation is the rate limiting protein kinase involved step in the pathway regulation. Notch, Hh, and Wnt signaling pathways utilize cell surface receptors that are not protein kinases. Regulation of the pathways by protein kinases occurs downstream of the receptors in the cytoplasm or nucleus and usually does not involve phosphorylation of the receptors.

The human genome codes for about 500 protein kinases, and the signaling complexes in the cytoplasm are exposed to multiple kinases simultaneously. This variety has hampered the understanding of which kinases are important in regulating specifically the Hh, Notch, or Wnt pathways.

Hh signaling molecules trigger one of the fundamental developmental pathways required for proper formation of body plan. *Hh* was first discovered as a gene responsible for the segment polarity phenotype in fruit fly *Drosophila melanogaster.* Sonic Hedgehog (Shh) is the most widely expressed and best-studied mammalian homologue of Hh proteins and is vitally required for numerous developmental and postnatal processes. At present, it is evident that the Hh signaling pathway is conserved in many aspects from flies to humans. However, severe variations in the Hh signaling cascade generated during evolution by mutations or gene duplication events illustrate divergences of Hh pathway between different species.

One of the most intriguing molecules in the Hh pathway is serine/threonine kinase Fused (Fu). *Fu* was discovered as a segment polarity gene in *Drosophila.* In *Drosophila,* fu is maternally required for proper segmentation during embryonic development and for development of wings and legs during the larval stage. Numerous studies have demonstrated that fu participates in Hh signal transduction as an obligatory component of multiprotein complex involved in regulation of *cubitus interruptus* (ci), a sole transcription factor mediating the Hh activities in *Drosophila.* Ci is a homologue of Gli transcription factors (Gli1, Gli2, and Gli3), the differential regulation of which is the final step of Hh signaling in mammals. In *Drosophila,* fu kinase activity is required for positive regulation of ci, while fu kinase activity independent regulatory properties are essential for negative regulation of ci in the absence of Hh. Such an absolute requirement of fu homologues for Hh signal transduction has not been described in mammals. This review is aimed at describing the function of two proteins, suggested as mammalian homologues of *Drosophila* fu kinase, serine/threonine kinase Stk36 (also known as Fu) and unc51-like kinase 3 (Ulk3), and comparing their roles in Hh signal transduction to that of fu.

II. The Major Steps of Hh Signaling

Major stages of Hh signal transduction are rather conserved between flies and vertebrates. The principle difference between mammalian and *Drosophila* cells in reception and intracellular transduction of Hh signal is requirement of primary cilia for the mammalian signaling (Huangfu and Anderson, 2005). The Hh pathway may be divided into three separate steps. The first one is the signal recognition which occurs on plasma membrane and involves an Hh ligand binding by its receptor Patched (Ptc) and rescue of Ptc inhibitory effect on intrinsic signaling activity of other transmembrane protein Smoothened (Smo) (Incardona *et al.*, 2002; Milenkovic *et al.*, 2009; Taipale *et al.*, 2002). The signal execution involves activation of ci/Gli transcription factor target genes expression that often represent genes belonging to one family (e.g., *ptc* and *Ptc1* or *dpp* and *Bmp*-s) (Drossopoulou *et al.*, 2000; Goodrich *et al.*, 1996; te Welscher *et al.*, 2002). The signal regulation in the cytoplasm, involving regulation of ci/Gli transcription factor activity, occurs in Hh stimulated cells (positive regulation) as well as in Hh non-stimulated cells (negative regulation).

III. Ci/Gli Proteins are Transcription Factors Mediating Hh Activities

Proteins ci in *Drosophila* and its three vertebrate homologues Gli1, Gli2, and Gli3 belong to a family of zinc finger (ZnF) transcription factors. These proteins are known to mediate all identified Hh signaling activities. Gli proteins have partly overlapping expression domains and possess distinct but partly redundant functions (Bai and Joyner, 2001; Ding *et al.*, 1998; Park *et al.*, 2000). All of the differentiated functions of Gli proteins are performed by ci in insects.

Ci/Gli proteins are regulated at three levels: stability of the proteins and nuclear transport and activation. Besides that, unlike *ci*, *Gli* genes are transcriptionally regulated by Shh (Grindley *et al.*, 1997; Lee *et al.*, 1997; Marigo *et al.*, 1996; Ruiz i Altaba, 1998; Sasaki *et al.*, 1999).

Gli2 and Gli3 proteins are more similar to ci than Gli1. Mechanism of ci/Gli2/3 regulation has several similar features. These proteins possess N-terminal repressor domain in addition to highly conserved Sufu-binding and C-terminal activator domains present also in Gli1 (Fig. 5.1). Sufu (Suppressor of Fused) is one of the most important negative regulators of Gli proteins, and disruption of Sufu binding domain leads to abrogation of Sufu-mediated control (Dunaeva *et al.*, 2003). Further, ci and Gli2/3 are known to be subjected to posttranslational modifications in an Hh signaling dependent manner, yielding three distinct forms of the proteins: the C-terminally truncated forms possessing the repressor and DNA-binding domains ci/Gli2/3REP, full-length ci/Gli2/3, and activated form of full-length proteins ci/Gli2/3ACT.

The full-length ci/Gli2/3 and truncated proteins ci/Gli2/3REP are present in the non-stimulated cells. In this case, the full-length proteins are tethered in cytoplasm and proteosomally processed. Processing of full-length ci and Gli2/3 occurs due to hyperphosphorylation by protein kinase A (pka), glycogen synthase kinase 3β (Gsk3β or shaggy), and casein kinase-1 (ck1) at the serine residues adjacent to ZnF domain (Chen *et al.*, 1998; Jia *et al.*, 2002;

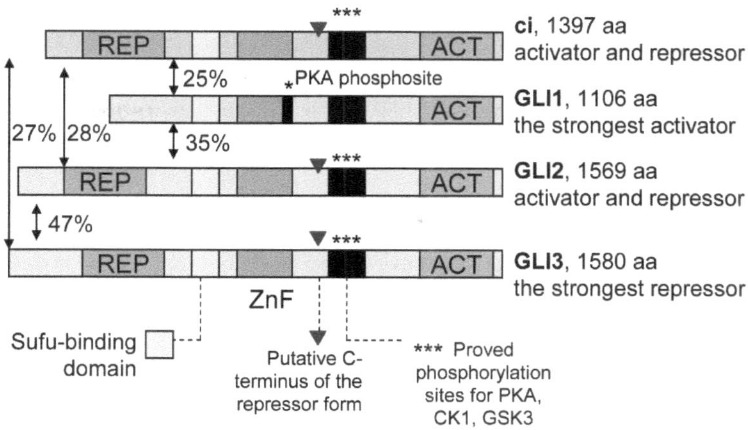

Figure 5.1 Schematic representation of ci/GLI proteins. Regardless of low overall identity between the proteins, they belong to one family of ci/GLI transcription factors containing DNA-binding zinc finger (ZnF) domain. The percents of identity between the respective proteins have been calculated using pairwise alignment (ClustalW program). Ci, GLI2, and GLI3 contain repressor (REP) and activator (ACT) domains and may serve as full-length transcriptional activators or C-terminally truncated transcriptional repressors. Ci/Gli proteins contain a highly conserved Sufu binding domain and sites for PKA, GSK3, and CK1 phosphorylation (showed as asterisks) situated in the phosphorylation cluster (colored with black). The cluster is less conserved in Gli1 that is phosphorylated by PKA at Threonine-374 situated in the end of ZnF domain (Sheng *et al.*, 2006).

Price and Kalderon, 2002). Hyperphosphorylation induces ci/Gli2/3 degradation via the SPOP/Cul3 and SCF/βTrCP ubiquitin-proteasomal pathway (Jiang and Struhl, 1998; Smelkinson *et al.*, 2007; Zhang *et al.*, 2005). Degradation may be complete or partial, until generation of ci/Gli2/3REP. The resulted forms ci/Gli2/3REP lack activator domains and are localized in the nucleus mediating repression of their target genes expression.

Full-length transcriptional activators ci/Gli2/3ACT are generated in response to Hh signal in cytosol, transported to the nucleus, where they activate their target gene expression and are degraded. The process of ciACT generation from full-length ci is pka-dependent and requires high levels of Hh and additional, yet unknown, activation step (Ohlmeyer and Kalderon, 1998; Wang and Holmgren, 2000; Wang *et al.*, 1999). Transcriptional activator forms of Gli2/3 have not been described yet in molecular terms and the mechanism of Gli2/3ACT generation is unknown. However, activation of full-length Gli2/3 is believed to occur via their phosphorylation in the primary cilia in response to Hh binding and requires the presence of Smo there (Wen *et al.*, 2010).

Gli3 is the most potent transcriptional repressor among Gli proteins. Besides, Gli3 as well as ci in *Drosophila* is very efficiently and almost fully processed in the absence of Hh, whereas activation of the Hh signaling blocks their processing (Wang *et al.*, 2000a). In the absence of Gli2, full-length Gli3 can function as a weak mediator of Shh signaling *in vivo* and *in vitro* inducing *Gli1* expression (Dai *et al.*, 1999; Motoyama *et al.*, 1998). Comparing with Gli3, the processing of Gli2 is relatively inefficient. Gli2 exists presumably in a full-length form even in non-stimulated cells, representing a tool for the immediate response to Hh and acting as a primary activator of Hh target genes including *Gli1* (Aza-Blanc *et al.*, 2000; Bai *et al.*, 2002; Ikram *et al.*, 2004; Sasaki *et al.*, 1999).

Gli1 is the most divergent member of the family. It lacks an N-terminal repressor domain and functions as a strong transcriptional activator (Fig. 5.1). However, recently the C-terminally truncated form of GLI1 has been described as a weak transcriptional repressor (Stecca and Ruiz i Altaba, 2009). The *Gli1* promoter possesses Gli-binding consensus sequence, and Gli2 and Gli3 are able to bind to these and directly activate *Gli1* expression (Dai *et al.*, 1999; Ikram *et al.*, 2004). Normally, *Gli1* is not expressed in the absence of Hh; however, its expression is highly activated upon Hh stimulation. Therefore, Gli1 serves as a marker of Hh activity and is very important protein for the diagnostic of the Hh-associated tumors.

IV. Hedgehog Signaling Complex is Responsible for Regulation of ci Protein

In *Drosophila*, interactions between ci and cytoplasmic proteins including serine/threonine fu, kinesin-related protein costal 2 (cos2), and PEST motif containing protein suppressor of fused (sufu) play a central role in the intracellular Hh signal transduction (Fig. 5.2).

Fu, cos2, and sufu physically interact with each other and bind full-length ci through cos2 and sufu. These proteins form a complex named Hedgehog signaling complex (HSC) (Robbins *et al.*, 1997; Stegman *et al.*, 2000). In the absence of Hh, HSC is responsible for negative regulation of ci that comprises tethering of full-length ci in the cytoplasm and generation of ciREP (Fig. 5.2A). The complex is bound to microtubules through cos2 and also directly associated with the cytoplasmic tail of smo (Lum *et al.*, 2003; Malpel *et al.*, 2007; Sisson *et al.*, 1997). Additionally, pka, ck1, and gsk3, responsible for ci hyperphosphorylation, are sequestered by cos2 (Zhang *et al.*, 2005). It has been proposed that smo-HSC shuttles between the membrane and vesicular compartment in the resting cells (Jia *et al.*, 2003; Ruel *et al.*, 2003).

An alternative model suggests that sufu is not a member of HSC, since even in the absence of sufu, a competent complex containing full-length ci, fu, and cos2 is formed (Stegman *et al.*, 2000). Moreover, the presence of

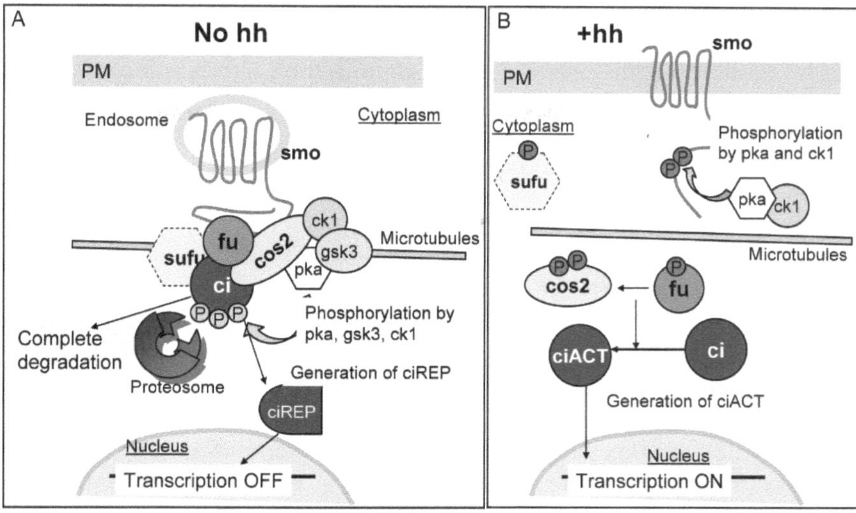

Figure 5.2 Fu, cos2, probably sufu, and ci proteins form HSC that is needed for the mediation of Hh signal from the cell membrane to nucleus. (A) In the absence of Hh, the complex is bound to smo cytoplasmic tail and is responsible for the negative regulation of ci—full-length ci degradation or/and ciREP generation. Proteosomal degradation of ci is primed by phosphorylation by pka, gsk3, and ck1 kinases sequestered by cos2. The multiprotein complex, bound to smo, may be tethered in the cytoplasm and bound to microtubules via cos2. Sufu is showed by dashed lines since it may not be a member of HSC. (B) In the presence of Hh, fu is (auto?) phosphorylated. Fu phosphorylates cos2, and the complex dissociates. Full-length ci escapes the negative phosphorylation and subsequent degradation. By an unknown mechanism ci is converted to ciACT that moves to the nucleus and activates transcription of its target genes. Generation of ciACT and full activation of the pathway requires the presence of fu and cos2. PM—plasma membrane.

those proteins is necessary and sufficient for ciREP generation. At the same time, sufu is thought to interact with the fraction of full-length ci that is not associated with fu and cos2 (Lum *et al.*, 2003; Monnier *et al.*, 1998).

Upon Hh binding, fu becomes activated and phosphorylates cos2 thus inducing dissociation of HSC, release of full-length ci from its cytoplasmic anchor, and generation of ciACT (Ruel *et al.*, 2007; Therond *et al.*, 1996) (Fig. 5.2B). Although the precise mechanism of ciACT generation is not clear yet, it is to some extent also dependent on the members of HSC. For instance, cos2 is required for the full activation of the Hh pathway and mediation of high levels of Hh signal (Wang *et al.*, 2000b). Loss of sufu is associated with a decrease in levels of full-length ci in wing disks and consequently weaker immediate response to Hh (Lefers *et al.*, 2001; Ohlmeyer and Kalderon, 1998). Besides, sufu has been shown to elicit stabilizing effects on ci because it competes with an ubiquitin–ligase adaptor proteins for binding of ci (Zhang *et al.*, 2006). Pka and ck1, instead of ci phosphorylation, phosphorylate specific sites in the cytoplasmic tail of smo that is required for the pathway activation in response to binding of Hh by ptc (Jia *et al.*, 2003; Zhang *et al.*, 2004; Zhao *et al.*, 2007). Thus, all the components of HSC have a dual role since they contribute to generation and regulation of ciREP and ciACT.

V. Regulation of Gli Proteins

In contrast to *Drosophila*, a similar Hh-dependent signaling complex has not been described in vertebrates. Genetic studies suggest that Sufu is a major negative regulator of Gli proteins in mammals (Cooper *et al.*, 2005; Svärd *et al.*, 2006). Recently, the role of Kif7, a mammalian homologue of cos2, in transduction of Shh signaling has been described (Endoh-Yamagami *et al.*, 2009). The function of Kif7 appears to be corresponding to the dual function of cos2 in *Drosophila* Hh signaling pathway.

The mechanism of Gli regulation by Sufu and Kif7 is complex and seems to occur in different cellular compartments: primary cilia, cytoplasm, and nucleus. First, in the absence of Hh, Sufu controls Gli proteins by direct binding and sequestering them in cytoplasm (Ding *et al.*, 1999). Cytoplasmic tethering of Gli proteins by Sufu seems to be primary cilium-independent, since Sufu is able to inhibit signaling in cells lacking components of intraflagellar transport (Chen *et al.*, 2009; Jia *et al.*, 2009). Kif7 physically interacts with Sufu and all Gli proteins. In the absence of Shh, Kif7 cooperates with Sufu in negative regulation of Gli proteins and is needed for efficient processing of full-length Gli3 (Endoh-Yamagami *et al.*, 2009). This interaction on one hand stabilizes Gli2/3 proteins, protects them from complete proteosomal degradation, since Sufu antagonizes SPOP that promotes Gli2/3FL ubiquitination, and contributes to the accumulation of a pool of Gli2/3 proteins ready to be converted to transcriptional activators (Chen *et al.*, 2009). On the other hand, it has been shown that Sufu contributes to the generation of Gli2/3REP (Humke *et al.*, 2010).

In response to the pathway activation, full-length Gli2/3 proteins dissociate from Sufu and accumulate in the primary cilia (Haycraft *et al.*, 2005; Humke *et al.*, 2010). The Gli/Sufu complex dissociation and ciliar translocation of Gli proteins depend on the ciliary motor Kif3a. Kif7 also moves to the cilia and is shown to be required for efficient ciliar accumulation of Gli3 (Endoh-Yamagami *et al.*, 2009). Gli2/3 activation is believed to occur via direct phosphorylation in the primary cilia (Wen *et al.*, 2010). Activated Gli2/3 proteins move to the nucleus in a Kif3a-dependent manner (Humke *et al.*, 2010).

VI. Fu is a Protein Kinase Playing Catalytic and Regulatory Roles in ci Regulation

Segment polarity gene *fu* encoding a serine/threonine kinase is maternally required for proper pattern formation within embryonic segments and imaginal discs in the cells responsive to the Hh signal. Homozygous *fu* embryos derived from *fu*−/− females are not viable. *Fu*−/− embryos derived from heterozygous females develop normally probably due to accumulation of *fu* product (mRNA or protein) in oocytes. However, adults display several anomalies such as reduced female fertility, tumorous ovaries, expansion of wing vein 3 and disappearance of vein 4, and a posterior expansion of the double-row wing margin bristles (Preat *et al.*, 1990; Therond *et al.*, 1993).

The product of *fu* is an 805-amino-acid-long protein. It comprises an N-terminal catalytic domain (KD) and a C-terminal regulatory domain (CTD). CTD shows no significant homology with other proteins (Therond *et al.*, 1993).

Genetic and biochemical studies show that fu has two functional roles—catalytic and noncatalytic (here referred as regulatory). Through genetic interactions with *sufu*, three classes of *fu* alleles have been distinguished: *fu0* (amorphic alleles; fu protein is disrupted), *fuI* (the protein is altered in the N-terminal KD), and *fuII* (the protein bears deletions in the CTD). Analysis of the ci protein distribution in *fuI* and *fuII* mutants showed that, in contrast to *fuI* mutant disks, *fuII* mutant wing disks demonstrated increased level of full-length ci protein and expansion of the ci expression domain abnormally far anteriorly from the source of Hh (Alves *et al.*, 1998; Ohlmeyer and Kalderon, 1998). Besides that, deletion of the fu CTD in *fuII* mutants resulted in a reduced amount of the ciREP. In contrast, *fuI* mutants, deficient in fu KD, did not exhibit reduction of ciREP in imaginal wing disks (Methot and Basler, 2000). The ci processing was completely abolished in *fuII* mutants deficient also for *sufu* (mutants with *cos2*-similar phenotype) (Lefers *et al.*, 2001). These data indicate that cos2, sufu, and fu are required for efficient ci processing. Biochemical studies suggest that in non-stimulated cells fu is bound by cos2 and probably sufu. In this case, fu exists in a nonphosphorylated and presumably inactive state, which is mediated by autoinhibition through its CTD (Ascano and Robbins, 2004).

In response to Hh signaling, fu becomes phosphorylated (Robbins *et al.*, 1997; Therond *et al.*, 1996). Recently, it has been demonstrated that fu is activated by fu-dependent phosphorylation (Zhou and Kalderon, 2011). The possible site of Hh-dependent phosphorylation has been mapped to threonine-158 that is identified as a crucial residue for regulation of fu kinase catalytic activity (Fukumoto *et al.*, 2001). In response to Hh, fu phosphorylates another component of Hh signaling, cos2, at serine residues 572 and 931 (Nybakken *et al.*, 2002; Ruel *et al.*, 2007). Phosphorylation of cos2 by fu occurs in all Hh-responsive cells and its intensity matches up the Hh gradient (Raisin *et al.*, 2010). Besides that, reception of Hh induces phosphorylation of sufu. It is not clear yet, if fu directly phosphorylates sufu, but sufu phosphorylation depends on fu kinase activity (Dussillol-Godar *et al.*, 2006; Lum *et al.*, 2003). Genetic studies have shown that fu KD is absolutely required for positive regulation of ci that is restricted by Hh signaling region (Lefers *et al.*, 2001; Methot and Basler, 2000; Therond *et al.*, 1999). Moreover, deletion of fu CTD results in stronger positive effects elicited by fu KD on different components of the Hh pathway. These findings have been confirmed by data showing that the functional KD of fu is needed for the positive regulation of ci transcriptional activity in Hh-responsive cultured cells, whereas deletion of fu CTD even enhanced fu coactivator properties (Fukumoto *et al.*, 2001). Fu kinase activity is also required for smo phosphorylation and accumulation on the cell surface (Claret *et al.*, 2007). It has been suggested that in response to Hh, fu may be anchored to the plasma membrane which is sufficient to make it a potent smo-dependent and ptc-resistant activator of the pathway.

VII. Protein Kinases Participating in the Regulation of Gli Proteins

A number of protein kinases regulate activity of Gli proteins. The role of Pka, Gsk3, and Ck1 in the regulation of Gli2/3 has been discussed above. Gli1 is known to be negatively regulated by Pka that directly phosphorylates Gli1 at threonine residue 374 and thus prevents its nuclear localization (Sheng *et al.*, 2006). Nuclear kinase DYRK1A directly phophorylates Gli1 and promotes sustaining of Gli1 in the nucleus thus contributing to enhancement of transcriptional response (Mao *et al.*, 2002). MAP3K10 and DYRK2 have been identified in a large-scale kinome screening as protein kinases required for Shh signal transduction *in vitro* (Varjosalo *et al.*, 2008). Both kinases are able to modulate Gli transcriptional activity, DYRK2 negatively and MAP3K10 positively. DYRK2 phosphorylates directly Gli2 protein at serine residues 385 and 1011 and promotes its degradation by ubiquitin/proteosome pathway. MAP3K10 modulates the activity of Gli2 via regulation of DYRK2 and GSK3. Cdc2l1 regulates positively Gli transcriptional activity by competing with Sufu for Gli-binding site (Evangelista *et al.*, 2008). Shh induces phosphoinositide 3-kinase (PI3K)-dependent phosphorylation of Akt kinase that in turn positively regulates Gli/Shh signaling antagonizing Pka-dependent degradation of Gli2 (Riobo *et al.*, 2006b). Hh signaling requires also PKCδ and MEK1 that both positively regulate Gli transcriptional activity utilizing phorbol esters (Riobo *et al.*, 2006a). The role of Fu kinase in mammalian Hh signaling has remained enigmatic.

VIII. Controversial Roles of Stk36

In parallel with *Drosophila* studies, vertebrate homologues of fu have been identified (Murone *et al.*, 2000; Wolff *et al.*, 2003). First, a serine/threonine kinase 36 (Stk36) has been suggested as a *Drosophila* fu homologue, because among mammalian proteins, it displays the highest sequence similarity with fu. STK36 or FU is 1315-amino-acid-long protein, possessing an N-terminal KD and a C-terminal regulatory domain. The latter is approximately 500 amino acids longer than the respective domain of fu. The KDs of fu and STK36 share 54% identity and 73% similarity, whereas their CTDs are considerably less conserved. Nevertheless, six short regions of fu and STK36 C-termini demonstrate limited homology (23–47% of similarity) (Daoud and Blanchet-Tournier, 2005).

Mouse and human *Stk36* are broadly expressed during development. Expression patterns of *Stk36* and *Shh* overlap in such tissues as developing limbs, neural tube, and somites suggesting a potential role for Stk36 in the Shh pathway (Chen *et al.*, 2005). Postnatally, mouse *Stk36* is highly expressed in choroid plexus, ependymal lining of the ventricles in the brain, respiratory epithelium, oviduct, and testis (Merchant *et al.*, 2005; Wilson *et al.*, 2009). Human *STK36* is expressed in adult brain, testis, pancreas, and kidney (Osterlund *et al.*, 2004).

In vitro experiments have demonstrated that Stk36 is able to activate the transcriptional activity of Gli1 and Gli2 proteins in a cell type specific way. STK36 synergizes with Gli1 in SHH signal responsive C3H10T½, NIH3T3, and SW480 cells, but not in primary cilia deficient HEK293 cells where SHH signaling pathway is known to be incomplete. However, enhancement of the transcriptional activity of Gli2 by STK36 has been detected in all cell lines tested (Daoud and Blanchet-Tournier, 2005; Kise *et al.*, 2006; Murone *et al.*, 2000; Osterlund *et al.*, 2004). It has been shown that STK36 is able to interact physically with GLI proteins protecting them from degradation and prolonging their half-lives (Kise *et al.*, 2006; Murone *et al.*, 2000). Therefore, the positive effect of Stk36 on GLI1/2 transcriptional activity may be caused by binding and stabilization of GLI proteins. Further cell culture experiments show that Stk36 kinase activity is dispensable for its function. Catalytically dead versions of Stk36 carrying the appropriate mutation in the ATP binging pocket or partly lacking the KD as well as Stk36 CTD alone show wt activities *in vitro* (Murone *et al.*, 2000; Osterlund *et al.*, 2004). These data suggest that Stk36 binds Gli proteins through its CTD.

It is interesting that human STK36 fails to induce GLI1 but induces GLI2 transcriptional activity in NIH3T3-derived Shh-L2 cells (Maloverjan *et al.*, 2010b). However, mouse Stk36 is able to synergize with both proteins in the same cell line (Maloveryan *et al.*, 2007). One possible explanation for this phenomenon arises from comparative analysis of human and mouse Stk36 sequences. According to ClustalW alignment results, those proteins share 84% of overall identity, and the most diverse region sharing only 60% identity is situated in the CTD between amino acids 305–402 (our unpublished observations). It may be speculated that this particularly region is highly important for the regulation (e.g., binding) of GLI1 but not essential for the regulation of GLI2.

Two independent groups have demonstrated that the functions of fly and vertebrate Fu homologues are not conserved also *in vivo* because Stk36 appeared to be dispensable for mouse embryonic development. If *Shh* knockout mice die at very early stages of development, Stk36-deficient mice develop normally until birth and do not display any Hh-related phenotype (Chen *et al.*, 2005; Merchant *et al.*, 2005). All organs affected in Shh-deficient mice develop normally in Stk36 knockouts; the expression pattern of Hh target genes (*Ptc1*, *Gli1*, and *Hip1*) is similar to wt. These data indicate that embryonic Hh signaling is not perturbed in Stk36-deficient mice. However, postnatally *Stk36* knockout mice display growth retardation, hydrocephalus, and rhinitis, a massive inflammation in the nasal cavity, and die within 2–3 weeks after birth. Besides those defects, *Stk36−/−* mice display atrophy of thymus and spleen with massive apoptosis of lymphocytes (Merchant *et al.*, 2005). Contrary to fu and sufu genetic interaction taking place in *Drosophila*, mammalian *Sufu* and *Stk36* do not cooperate genetically and loss of Stk36 cannot rescue Hh defects in *Sufu−/−* mutants (Chen *et al.*, 2009).

As discussed above, Shh signaling requires primary cilia that are not disrupted in *Stk36−/−* mouse embryonic fibroblasts. However, *Stk36−/−* mice are deficient in (9+2) motile cilia. It has been shown that Stk36 is able to physically interact with Kif27, a mammalian orthologue of Cos2, and Spag16 (sperm associated antigen 16), a protein found in the central apparatus of the axoneme. Thus, it seems that Stk36 is not required for Hh signaling during embryonic development in mammals but is needed for the generation of the central pair apparatus of motile cilia (Wilson *et al.*, 2009). This fact gives a reasonable explanation to hydrocephalus displayed by *Stk36−/−* mice. Hydrocephalus is a disorder affecting the flow, synthesis, and absorption of cerebrospinal fluid (CSF) in the brain that lead to the accumulation of CSF and, as a result, to hydrocephalus. One of the factors influencing normal CSF circulation is the presence of functional motile cilia. Since Stk36 is involved in motile ciliogenesis, hydrocephalus in *Stk36*-deficient mice may be linked to disturbance of motile cilia in the brain.

Knockout of Fu in zebrafish results in mild Hh-related phenotype and, additionally, in impaired motile cilia genesis. It is interesting that mouse

Stk36 is able to rescue zebrafish Fu−/− deficiencies (Wilson *et al.*, 2009; Wolff *et al.*, 2003). However, overexpression of human STK36 in *fu* mutant flies shows that STK36 is not able to rescue *fu* phenotype (Daoud and Blanchet-Tournier, 2005). It is not yet explored whether zebrafish Fu could substitute *Drosophila* fu.

Regardless of the fact that Stk36 is dispensable for embryonic development in mouse, it may play a role in postnatal developmental events controlled by Shh in cooperation with other regulatory networks. Several groups have reported the role of Shh signaling in the immune system. It has been shown that, during the early postnatal period, Shh and/or its downstream effectors are expressed in the adult thymus, spleen, and lymph nodes, where they are required for survival of thymocyte progenitors and germinal-center B cells and formation of mature T and B lymphocytes (Dierks *et al.*, 2007; El Andaloussi *et al.*, 2006; Sacedon *et al.*, 2003). Atrophy of spleen and thymus and apoptosis of lymphocytes in Stk36 knockout mice may be linked to defects in Hh signal transduction in those organs. Besides that, a potential oncogenic role of Stk36 in malignant B-cell lymphomas has been described (Dierks *et al.*, 2007). It has been shown that Shh, secreted by bone-marrow, nodal, and splenic stromal cells, functions as a survival factor for lymphoma cells and ectopic Shh signaling induces lymphoma growth. Treatment of the lymphoma cells with cyclopamine, an inhibitor of Smo, leads to their apoptosis. In this model, forced expression of STK36 overcomes the inhibitory effects of cyclopamine preventing cyclopamine-induced apoptosis of lymphoma cells. These data suggest that STK36 may act downstream of Smo as a positive regulator of oncogenic Shh pathway.

High expression of human and mouse *Stk36* in testis suggests a prominent role of Stk36 in this tissue. It has been shown that *Gli1* and *Ptc1* are expressed in adult mouse testis suggesting a potential role for Hh pathway (Turner *et al.*, 2004). Indeed, an important role for Desert Hh (Dhh, a homologue of Shh) in testicular development has been elucidated by analysis of the phenotype of *Dhh* knockout mice showing that Dhh is required for the development of Leydig cells (Clark *et al.*, 2000). However, the question of Stk36 function in testis development remains to be addressed.

IX. The Role of Ulk3 in Shh Signaling Pathway

In the light of the fact that Stk36 has lost its kinase activity-dependent function during evolution, other serine/threonine kinases are thought to substitute fu in the mammalian Hh pathway. A family of unc51-like protein kinases, particularly Ulk3 and Ulk2, have been hypothetically suggested as candidates for recapitulation of fu catalytic function (Wilson *et al.*, 2009).

The reason for such a speculation is that among other serine/threonine kinases Ulk2 and Ulk3 share higher identity with Stk36 and fu in KD (37% and 38%, respectively) (Chan *et al.*, 2009; Wilson *et al.*, 2009).

Unc51 (uncoordinated movement)-51 is an ancient family of serine/threonine kinases conserved from yeast to human. Yeast, *Drosophila,* and *Caenorhabditis elegance* genomes have a single *unc51* gene, also called *atg1* or *autophagy-related 1* or *pegarn*. In vertebrates, four *Ulk* genes (unc51-like kinases 1, 2, 3, and 4) are annotated to belong to *unc51* family. Among Ulk proteins, Ulk1/2 proteins display the highest homology with each other and unc51/atg1 proteins and represent the functionally conserved members of the atg1/unc51 family (Chan and Tooze, 2009).

Several studies have reported that yeast, *Drosophila* and *C. elegance* atg1 kinases, as well as their closest mammalian homologues Ulk1 and Ulk2, are implicated in autophagocytosis (Chan and Tooze, 2009; Lee and Tournier, 2011). Autophagy is a lysosome-dependent cytoplasmic degradation of damaged or unwanted cell's own components. It involves generation of autophagosome—a structure that envelops the degradation target into a membrane and afterward fuses with lysosomes. Additionally, atg1/unc51/Ulk1/2 proteins are shown to be critical for neuronal development, where they are involved in neuronal vesicular trafficking and axonal elongation and guidance (Ahantarig *et al.*, 2009; Ogura *et al.*, 1994; Tomoda *et al.*, 1999, 2004).

Ulk3 diverges from the members of the family sharing comparable similarity with Ulk1/2 kinases (39% and 37%) and Stk36 (38%) in the KD. ULK3 is widely expressed, showing the highest expression in hippocampus, olfactory bulb, cerebellum, and optic nerve (Maloverjan *et al.*, 2010b).

In contrast to Stk36 and similarly to fu, ULK3 acts as an active kinase in biological assays, demonstrating requirement of its functional KD for positive regulation of GLI proteins. ULK3 enhances GLI1 and GLI2 transcriptional activity and promotes nuclear localization of GLI1in a kinase activity-dependent way (Maloverjan *et al.*, 2010b). These data indicate that ULK3 may act as a positive regulator of Shh pathway at least in Shh responsive cancer cell lines.

Moreover, ULK3 is able to phosphorylate itself and GLI proteins *in vitro*. Mouse Ulk1 and Ulk2 are also shown to be autophosphorylated mainly in the central proline/serine rich domains (Dorsey *et al.*, 2009; Yan *et al.*, 1999). However, ULK3 lacks such a domain. Instead, ULK3 possesses a domain contained within Microtubule Interacting and Trafficking molecules, the function of which is yet unclear. ULK3 autophosphorylation occurs at four serine residues (Ser-300, Ser-350, Ser-384, and Ser-464) situated outside the KD. ULK3 is not able to autophosphorylate in *trans*. The sites of ULK3 autophosphorylation are hidden in catalytically inactive ULK3 and unexposed to other ULK3 molecules. It is interesting that autophosphorylation of ULK3 seems to involve conformational changes and resulted in exposure of its CTD to KD and consequently in generation of the catalytically active kinase (Maloverjan *et al.*, 2010a). Such mechanism of self-regulation control is observed also in the autophosphorylated kinases Ulk1/2 as well as *Drosophila* fu kinases (Ascano and Robbins, 2004; Chan *et al.*, 2009). Phosphorylation of GLI1 by ULK3 occurs at least at two serine or threonine residues situated in GLI1 amino- and carboxy-termini (amino acids 1–426 and 754–1106). One of the putative phosphorylated by ULK3 residues in GLI1 has been identified by LC-ESI-MS/MS analysis as serine-882 (our unpublished data).

In *Drosophila*, fu is known to bind directly several proteins—sufu, cos2, and smo. Being in HSC, fu is catalytically inactive. Those interactions are highly important for the negative regulation of ci in the absence of Hh (Malpel *et al.*, 2007; Monnier *et al.*, 1998, 2002; Robbins *et al.*, 1997). In response to Hh, fu becomes an active kinase and HCS dissociates releasing full-length ci ready to be converted to ciACT (Lum *et al.*, 2003; Ruel *et al.*, 2007; Stegman *et al.*, 2000). Biochemical studies have shown that *Drosophila* fu may interact with sufu through both KD and CTD, but the region between aa 306 and aa 436 has been defined as a minimal domain responsible for sufu binding (Monnier *et al.*, 1998, 2002). However, genetic studies strongly suggest that fu bind sufu through its KD (Preat *et al.*, 1993). ULK3 is also able to interact with Sufu via its KD (Maloverjan *et al.*, 2010a). The biological role of ULK3-SUFU interaction involves blocking of ULK3 catalytic domain by SUFU, the negative regulator of SHH pathway, thus preventing the positive functioning of ULK3. This effect may be inhibited by SHH signaling that induces dissociation of the SUFU/ULK3 complex and release of catalytically active ULK3.

Another possible function of ULK3-SUFU complex may be linked to generation of GLI2REP. It has been shown that Sufu contributes to the generation of Gli2/3REP (Humke *et al.*, 2010). In overexpression studies, neither SUFU nor ULK3 alone can induce the generation of GLI2REP, but acting in concert, they promote C-terminal processing of full-length GLI2 (Maloverjan *et al.*, 2010a). As it can be expected, the ULK3 kinase activity is not required for GLI2REP generation or SUFU binding indicating that ULK3 plays also a regulatory role. Results of RNAi experiments support a hypothesis that suppression of *Ulk3* expression in Shh responsive cells results in accumulation of full-length Gli proteins, since more intensive activation of *Gli1* expression occurs in response to Shh in the cells with reduced level of *Ulk3* mRNA (Maloverjan *et al.*, 2010a).

Additionally, it has been reported that involvement of ULK3 in a special form of autophagy associated with oncogene-induced cell senescence, indicating that ULK3 may have multiple roles in cells (Young *et al.*, 2009). However, the common opinion considers a role of Ulk3 in autophagy as controversial and requiring additional investigations (Chan and Tooze, 2009; Narita *et al.*, 2009).

X. An Emerging Shh Signaling Model Involving Ulk3

From the presented data it can be speculated that Ulk3 has at least two distinct roles in the Hh signaling pathway eliciting kinase activity-dependent and -independent effects on Gli transcription factors. The proposed model is depicted in Fig. 5.3. In the absence of Shh, Kif7, and Sufu interact with Gli2/3 proteins and dock them in the cytoplasm (Fig. 5.3A). Sufu binds the KD of Ulk3 thus inhibiting its catalytic activity. The Gli2/3FL proteins are sequestered by the Ulk3/Sufu/Kif7 complex and cannot be converted to active forms. Moreover, the complex participates in the generation of C-terminally truncated transcriptional repressors Gli2/3Rep. Through its putative MIT domain (domain contained within Microtubule Interacting and Trafficking molecules), Ulk3 may be a link between Sufu/Kif7/GliFL complex and microtubules. In the presence of Shh, the Ulk3/Sufu/Kif7 complex dissociates (Fig. 5.3B). Ulk3 becomes an active

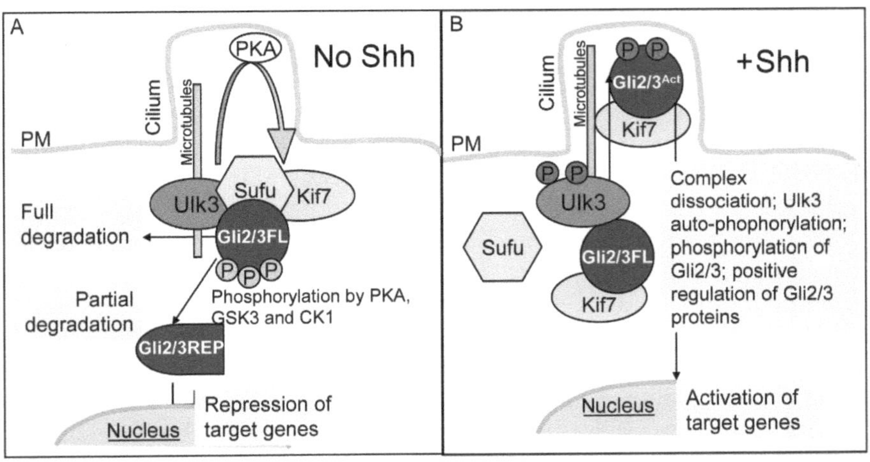

Figure 5.3 A model for the actions of Ulk3 in the Shh pathway. (A) In the absence of Shh, Sufu forms a complex with Kif 7 and Ulk3 that possibly interacts with cytoskeleton through its MIT domain. Ulk3 is not catalytically active but plays a regulatory role. The complex binds full-length Gli2/3 through Sufu and Kif 7. Ulk3/Sufu/Kif 7 complex contributes to C-terminal processing of Gli2/3 probably through recruiting PKA, GSK3, and CK1 kinases to Gli2/3. This results in generation of Gli2/3REP that may enter the nucleus and inhibit the target gene expression. (B) In the presence of Shh, Ulk3/Sufu/Kif 7/Gli2/3 complex dissociates and Ulk3 activates itself by autophosphorylation and phosphorylates Gli2/3. This contributes to generation of Gli2/3ACT that translocate to the nucleus and activate their target genes. For simplification, the other proteins participating in Shh signal transduction and Gli proteins regulation (Ptc1, Smo, Kif3, etc.) are not shown. PM—plasma membrane.

kinase that autophosphorylates and is stabilized in an active conformation. Ulk3 is now able to phosphorylate the released Gli2/3FL and contribute to Gli2/3ACT formation. Gli2/3ACT enter the nucleus and activate the expression of their target genes including *Ptc1* and *Gli1*. Other molecule(s), for instance, Stk36, may participate in the regulation of Gli2/3 activity depending on cellular context and strength of Shh signaling. Whether Ulk3 also interacts with and phosphorylates Kif7 and thereby contributes to the complex dissociation remains to be investigated. It also remains to be investigated if Ulk3 is able to phosphorylate other signaling components like Smo and Ptc further influencing the pathway. However, it is clear from these analyses that Ulk3 plays an important role in regulation of Gli proteins and is likely to be the "missing link" in the Sufu/Gli regulatory complex in Shh signaling.

ACKNOWLEDGMENTS

This material is based upon work supported by the Estonian Science Foundation (grants 7279, and MJD266) and Estonian Ministry of Education and Research Grant 0140143.

REFERENCES

Ahantarig, A., Chadwell, L. V., Terrazas, I. B., Garcia, C. T., Nazarian, J. J., Lee, H. K., Lundell, M. J., and Cassill, J. A. (2009). Molecular characterization of Pegarn: A Drosophila homolog of UNC-51 kinase. *Mol. Biol. Rep.* **36,** 1311–1321.

Alves, G., Limbourg-Bouchon, B., Tricoire, H., Brissard-Zahraoui, J., Lamour-Isnard, C., and Busson, D. (1998). Modulation of Hedgehog target gene expression by the Fused serine-threonine kinase in wing imaginal discs. *Mech. Dev.* **78,** 17–31.

Ascano, M., Jr., and Robbins, D. J. (2004). An intramolecular association between two domains of the protein kinase Fused is necessary for Hedgehog signaling. *Mol. Cell. Biol.* **24,** 10397–10405.

Aza-Blanc, P., Lin, H. Y., Ruiz i Altaba, A., and Kornberg, T. B. (2000). Expression of the vertebrate Gli proteins in Drosophila reveals a distribution of activator and repressor activities. *Development* **127,** 4293–4301.

Bai, C. B., and Joyner, A. L. (2001). Gli1 can rescue the *in vivo* function of Gli2. *Development* **128,** 5161–5172.

Bai, C. B., Auerbach, W., Lee, J. S., Stephen, D., and Joyner, A. L. (2002). Gli2, but not Gli1, is required for initial Shh signaling and ectopic activation of the Shh pathway. *Development* **129,** 4753–4761.

Canestro, C., Yokoi, H., and Postlethwait, J. H. (2007). Evolutionary developmental biology and genomics. *Nat. Rev. Genet.* **8,** 932–942.

Chan, E. Y., and Tooze, S. A. (2009). Evolution of Atg1 function and regulation. *Autophagy* **5,** 758–765.

Chan, E. Y., Longatti, A., McKnight, N. C., and Tooze, S. A. (2009). Kinase-inactivated ULK proteins inhibit autophagy via their conserved C-terminal domains using an Atg13-independent mechanism. *Mol. Cell. Biol.* **29,** 157–171.

Chen, Y., Gallaher, N., Goodman, R. H., and Smolik, S. M. (1998). Protein kinase A directly regulates the activity and proteolysis of cubitus interruptus. *Proc. Natl. Acad. Sci. USA* **95,** 2349–2354.
Chen, M. H., Gao, N., Kawakami, T., and Chuang, P. T. (2005). Mice deficient in the fused homolog do not exhibit phenotypes indicative of perturbed hedgehog signaling during embryonic development. *Mol. Cell. Biol.* **25,** 7042–7053.
Chen, M. H., Wilson, C. W., Li, Y. J., Law, K. K., Lu, C. S., Gacayan, R., Zhang, X., Hui, C. C., and Chuang, P. T. (2009). Cilium-independent regulation of Gli protein function by Sufu in Hedgehog signaling is evolutionarily conserved. *Genes Dev.* **23,** 1910–1928.
Claret, S., Sanial, M., and Plessis, A. (2007). Evidence for a novel feedback loop in the Hedgehog pathway involving Smoothened and Fused. *Curr. Biol.* **17,** 1326–1333.
Clark, A. M., Garland, K. K., and Russell, L. D. (2000). Desert hedgehog (Dhh) gene is required in the mouse testis for formation of adult-type Leydig cells and normal development of peritubular cells and seminiferous tubules. *Biol. Reprod.* **63,** 1825–1838.
Cooper, A. F., Yu, K. P., Brueckner, M., Brailey, L. L., Johnson, L., McGrath, J. M., and Bale, A. E. (2005). Cardiac and CNS defects in a mouse with targeted disruption of suppressor of fused. *Development* **132,** 4407–4417.
Dai, P., Akimaru, H., Tanaka, Y., Maekawa, T., Nakafuku, M., and Ishii, S. (1999). Sonic Hedgehog-induced activation of the Gli1 promoter is mediated by GLI3. *J. Biol. Chem.* **274,** 8143–8152.
Daoud, F., and Blanchet-Tournier, M. F. (2005). Expression of the human FUSED protein in Drosophila. *Dev. Genes Evol.* **215,** 230–237.
Dierks, C., Grbic, J., Zirlik, K., Beigi, R., Englund, N. P., Guo, G. R., Veelken, H., Engelhardt, M., Mertelsmann, R., Kelleher, J. F., Schultz, P., and Warmuth, M. (2007). Essential role of stromally induced hedgehog signaling in B-cell malignancies. *Nat. Med.* **13,** 944–951.
Ding, Q., Motoyama, J., Gasca, S., Mo, R., Sasaki, H., Rossant, J., and Hui, C. C. (1998). Diminished Sonic hedgehog signaling and lack of floor plate differentiation in Gli2 mutant mice. *Development* **125,** 2533–2543.
Ding, Q., Fukami, S., Meng, X., Nishizaki, Y., Zhang, X., Sasaki, H., Dlugosz, A., Nakafuku, M., and Hui, C. (1999). Mouse suppressor of fused is a negative regulator of sonic hedgehog signaling and alters the subcellular distribution of Gli1. *Curr. Biol.* **9,** 1119–1122.
Dorsey, F. C., Rose, K. L., Coenen, S., Prater, S. M., Cavett, V., Cleveland, J. L., and Caldwell-Busby, J. (2009). Mapping the phosphorylation sites of Ulk1. *J. Proteome Res.* **8,** 5253–5263.
Drossopoulou, G., Lewis, K. E., Sanz-Ezquerro, J. J., Nikbakht, N., McMahon, A. P., Hofmann, C., and Tickle, C. (2000). A model for anteroposterior patterning of the vertebrate limb based on sequential long- and short-range Shh signalling and Bmp signalling. *Development* **127,** 1337–1348.
Dunaeva, M., Michelson, P., Kogerman, P., and Toftgard, R. (2003). Characterization of the physical interaction of Gli proteins with SUFU proteins. *J. Biol. Chem.* **278,** 5116–5122.
Dussillol-Godar, F., Brissard-Zahraoui, J., Limbourg-Bouchon, B., Boucher, D., Fouix, S., Lamour-Isnard, C., Plessis, A., and Busson, D. (2006). Modulation of the Suppressor of fused protein regulates the Hedgehog signaling pathway in Drosophila embryo and imaginal discs. *Dev. Biol.* **291,** 53–66.
El Andaloussi, A., Graves, S., Meng, F., Mandal, M., Mashayekhi, M., and Aifantis, I. (2006). Hedgehog signaling controls thymocyte progenitor homeostasis and differentiation in the thymus. *Nat. Immunol.* **7,** 418–426.
Endoh-Yamagami, S., Evangelista, M., Wilson, D., Wen, X., Theunissen, J. W., Phamluong, K., Davis, M., Scales, S. J., Solloway, M. J., de Sauvage, F. J., and Peterson, A. S. (2009). The mammalian Cos2 homolog Kif7 plays an essential role in modulating Hh signal transduction during development. *Curr. Biol.* **19,** 1320–1326.

Evangelista, M., Lim, T. Y., Lee, J., Parker, L., Ashique, A., Peterson, A. S., Ye, W., Davis, D. P., and de Sauvage, F. J. (2008). Kinome siRNA screen identifies regulators of ciliogenesis and hedgehog signal transduction. *Sci. Signal.* **1,** ra7.
Fukumoto, T., Watanabe-Fukunaga, R., Fujisawa, K., Nagata, S., and Fukunaga, R. (2001). The fused protein kinase regulates Hedgehog-stimulated transcriptional activation in Drosophila Schneider 2 cells. *J. Biol. Chem.* **276,** 38441–38448.
Goodrich, L. V., Johnson, R. L., Milenkovic, L., McMahon, J. A., and Scott, M. P. (1996). Conservation of the hedgehog/patched signaling pathway from flies to mice: Induction of a mouse patched gene by Hedgehog. *Genes Dev.* **10,** 301–312.
Grindley, J. C., Bellusci, S., Perkins, D., and Hogan, B. L. (1997). Evidence for the involvement of the Gli gene family in embryonic mouse lung development. *Dev. Biol.* **188,** 337–348.
Haycraft, C. J., Banizs, B., Aydin-Son, Y., Zhang, Q., Michaud, E. J., and Yoder, B. K. (2005). Gli2 and Gli3 localize to cilia and require the intraflagellar transport protein polaris for processing and function. *PLoS Genet.* **1,** e53.
Huangfu, D., and Anderson, K. V. (2005). Cilia and Hedgehog responsiveness in the mouse. *Proc. Natl. Acad. Sci. USA* **102,** 11325–11330.
Humke, E. W., Dorn, K. V., Milenkovic, L., Scott, M. P., and Rohatgi, R. (2010). The output of Hedgehog signaling is controlled by the dynamic association between Suppressor of Fused and the Gli proteins. *Genes Dev.* **24,** 670–682.
Ikram, M. S., Neill, G. W., Regl, G., Eichberger, T., Frischauf, A. M., Aberger, F., Quinn, A., and Philpott, M. (2004). GLI2 is expressed in normal human epidermis and BCC and induces GLI1 expression by binding to its promoter. *J. Invest. Dermatol.* **122,** 1503–1509.
Incardona, J. P., Gruenberg, J., and Roelink, H. (2002). Sonic hedgehog induces the segregation of patched and smoothened in endosomes. *Curr. Biol.* **12,** 983–995.
Jia, J., Amanai, K., Wang, G., Tang, J., Wang, B., and Jiang, J. (2002). Shaggy/GSK3 antagonizes Hedgehog signalling by regulating Cubitus interruptus. *Nature* **416,** 548–552.
Jia, J., Tong, C., and Jiang, J. (2003). Smoothened transduces Hedgehog signal by physically interacting with Costal2/Fused complex through its C-terminal tail. *Genes Dev.* **17,** 2709–2720.
Jia, J., Kolterud, A., Zeng, H., Hoover, A., Teglund, S., Toftgard, R., and Liu, A. (2009). Suppressor of Fused inhibits mammalian Hedgehog signaling in the absence of cilia. *Dev. Biol.* **330,** 452–460.
Jiang, J., and Struhl, G. (1998). Regulation of the Hedgehog and Wingless signalling pathways by the F-box/WD40-repeat protein Slimb. *Nature* **391,** 493–496.
Kise, Y., Takenaka, K., Tezuka, T., Yamamoto, T., and Miki, H. (2006). Fused kinase is stabilized by Cdc37/Hsp90 and enhances Gli protein levels. *Biochem. Biophys. Res. Commun.* **351,** 78–84.
Lee, E. J., and Tournier, C. (2011). The requirement of uncoordinated 51-like kinase 1 (ULK1) and ULK2 in the regulation of autophagy. *Autophagy* **7,** 689–695.
Lee, J., Platt, K. A., Censullo, P., and Ruiz i Altaba, A. (1997). Gli1 is a target of Sonic hedgehog that induces ventral neural tube development. *Development* **124,** 2537–2552.
Lefers, M. A., Wang, Q. T., and Holmgren, R. A. (2001). Genetic dissection of the Drosophila Cubitus interruptus signaling complex. *Dev. Biol.* **236,** 411–420.
Lum, L., Zhang, C., Oh, S., Mann, R. K., von Kessler, D. P., Taipale, J., Weis-Garcia, F., Gong, R., Wang, B., and Beachy, P. A. (2003). Hedgehog signal transduction via Smoothened association with a cytoplasmic complex scaffolded by the atypical kinesin, Costal-2. *Mol. Cell* **12,** 1261–1274.
Maloverjan, A., Piirsoo, M., Kasak, L., Peil, L., Osterlund, T., and Kogerman, P. (2010a). Dual function of UNC-51-like kinase 3 (Ulk3) in the Sonic hedgehog signaling pathway. *J. Biol. Chem.* **285,** 30079–30090.

Maloverjan, A., Piirsoo, M., Michelson, P., Kogerman, P., and Osterlund, T. (2010b). Identification of a novel serine/threonine kinase ULK3 as a positive regulator of Hedgehog pathway. *Exp. Cell Res.* **316,** 627–637.

Maloveryan, A., Finta, C., Osterlund, T., and Kogerman, P. (2007). A possible role of mouse Fused (STK36) in Hedgehog signaling and Gli transcription factor regulation. *J. Cell Commun. Signal.* **1,** 165–173.

Malpel, S., Claret, S., Sanial, M., Brigui, A., Piolot, T., Daviet, L., Martin-Lanneree, S., and Plessis, A. (2007). The last 59 amino acids of Smoothened cytoplasmic tail directly bind the protein kinase Fused and negatively regulate the Hedgehog pathway. *Dev. Biol.* **303,** 121–133.

Mao, J., Maye, P., Kogerman, P., Tejedor, F. J., Toftgard, R., Xie, W., Wu, G., and Wu, D. (2002). Regulation of Gli1 transcriptional activity in the nucleus by Dyrk1. *J. Biol. Chem.* **277,** 35156–35161.

Marigo, V., Johnson, R. L., Vortkamp, A., and Tabin, C. J. (1996). Sonic hedgehog differentially regulates expression of GLI and GLI3 during limb development. *Dev. Biol.* **180,** 273–283.

Merchant, M., Evangelista, M., Luoh, S. M., Frantz, G. D., Chalasani, S., Carano, R. A., van Hoy, M., Ramirez, J., Ogasawara, A. K., McFarland, L. M., Filvaroff, E. H., French, D. M., *et al.* (2005). Loss of the serine/threonine kinase fused results in postnatal growth defects and lethality due to progressive hydrocephalus. *Mol. Cell. Biol.* **25,** 7054–7068.

Methot, N., and Basler, K. (2000). Suppressor of fused opposes hedgehog signal transduction by impeding nuclear accumulation of the activator form of Cubitus interruptus. *Development* **127,** 4001–4010.

Milenkovic, L., Scott, M. P., and Rohatgi, R. (2009). Lateral transport of Smoothened from the plasma membrane to the membrane of the cilium. *J. Cell Biol.* **187,** 365–374.

Monnier, V., Dussillol, F., Alves, G., Lamour-Isnard, C., and Plessis, A. (1998). Suppressor of fused links fused and Cubitus interruptus on the hedgehog signalling pathway. *Curr. Biol.* **8,** 583–586.

Monnier, V., Ho, K. S., Sanial, M., Scott, M. P., and Plessis, A. (2002). Hedgehog signal transduction proteins: Contacts of the Fused kinase and Ci transcription factor with the kinesin-related protein Costal2. *BMC Dev. Biol.* **2,** 4.

Motoyama, J., Liu, J., Mo, R., Ding, Q., Post, M., and Hui, C. C. (1998). Essential function of Gli2 and Gli3 in the formation of lung, trachea, and oesophagus. *Nat. Genet.* **20,** 54–57.

Murone, M., Luoh, S. M., Stone, D., Li, W., Gurney, A., Armanini, M., Grey, C., Rosenthal, A., and de Sauvage, F. J. (2000). Gli regulation by the opposing activities of fused and suppressor of fused. *Nat. Cell Biol.* **2,** 310–312.

Narita, M., Young, A. R., and Narita, M. (2009). Autophagy facilitates oncogene-induced senescence. *Autophagy* **5,** 1046–1047.

Nybakken, K. E., Turck, C. W., Robbins, D. J., and Bishop, J. M. (2002). Hedgehog-stimulated phosphorylation of the kinesin-related protein Costal2 is mediated by the serine/threonine kinase fused. *J. Biol. Chem.* **277,** 24638–24647.

Ogura, K., Wicky, C., Magnenat, L., Tobler, H., Mori, I., Muller, F., and Ohshima, Y. (1994). Caenorhabditis elegans unc-51 gene required for axonal elongation encodes a novel serine/threonine kinase. *Genes Dev.* **8,** 2389–2400.

Ohlmeyer, J. T., and Kalderon, D. (1998). Hedgehog stimulates maturation of Cubitus interruptus into a labile transcriptional activator. *Nature* **396,** 749–753.

Osterlund, T., Everman, D. B., Betz, R. C., Mosca, M., Nothen, M. M., Schwartz, C. E., Zaphiropoulos, P. G., and Toftgard, R. (2004). The FU gene and its possible protein isoforms. *BMC Genomics* **5,** 49.

Park, H. L., Bai, C., Platt, K. A., Matise, M. P., Beeghly, A., Hui, C. C., Nakashima, M., and Joyner, A. L. (2000). Mouse Gli1 mutants are viable but have defects in SHH signaling in combination with a Gli2 mutation. *Development* **127,** 1593–1605.

Preat, T., Therond, P., Lamour-Isnard, C., Limbourg-Bouchon, B., Tricoire, H., Erk, I., Mariol, M. C., and Busson, D. (1990). A putative serine/threonine protein kinase encoded by the segment-polarity fused gene of Drosophila. *Nature* **347,** 87–89.

Preat, T., Therond, P., Limbourg-Bouchon, B., Pham, A., Tricoire, H., Busson, D., and Lamour-Isnard, C. (1993). Segmental polarity in Drosophila melanogaster: Genetic dissection of fused in a Suppressor of fused background reveals interaction with costal-2. *Genetics* **135,** 1047–1062.

Price, M. A., and Kalderon, D. (2002). Proteolysis of the Hedgehog signaling effector Cubitus interruptus requires phosphorylation by Glycogen Synthase Kinase 3 and Casein Kinase 1. *Cell* **108,** 823–835.

Raisin, S., Ruel, L., Ranieri, N., Staccini-Lavenant, L., and Therond, P. P. (2010). Dynamic phosphorylation of the kinesin Costal-2 *in vivo* reveals requirement of fused kinase activity for all levels of hedgehog signalling. *Dev. Biol.* **344,** 119–128.

Riobo, N. A., Haines, G. M., and Emerson, C. P., Jr. (2006a). Protein kinase C-delta and mitogen-activated protein/extracellular signal-regulated kinase-1 control GLI activation in hedgehog signaling. *Cancer Res.* **66,** 839–845.

Riobo, N. A., Lu, K., Ai, X., Haines, G. M., and Emerson, C. P., Jr. (2006b). Phosphoinositide 3-kinase and Akt are essential for Sonic Hedgehog signaling. *Proc. Natl. Acad. Sci. USA* **103,** 4505–4510.

Robbins, D. J., Nybakken, K. E., Kobayashi, R., Sisson, J. C., Bishop, J. M., and Therond, P. P. (1997). Hedgehog elicits signal transduction by means of a large complex containing the kinesin-related protein costal2. *Cell* **90,** 225–234.

Ruel, L., Rodriguez, R., Gallet, A., Lavenant-Staccini, L., and Therond, P. P. (2003). Stability and association of Smoothened, Costal2, and Fused with Cubitus interruptus are regulated by Hedgehog. *Nat. Cell Biol.* **5,** 907–913.

Ruel, L., Gallet, A., Raisin, S., Truchi, A., Staccini-Lavenant, L., Cervantes, A., and Therond, P. P. (2007). Phosphorylation of the atypical kinesin Costal2 by the kinase Fused induces the partial disassembly of the Smoothened-Fused-Costal2-Cubitus interruptus complex in Hedgehog signalling. *Development* **134,** 3677–3689.

Ruiz i Altaba, A. (1998). Combinatorial Gli gene function in floor plate and neuronal inductions by Sonic hedgehog. *Development* **125,** 2203–2212.

Sacedon, R., Varas, A., Hernandez-Lopez, C., Gutierrez-deFrias, C., Crompton, T., Zapata, A. G., and Vicente, A. (2003). Expression of hedgehog proteins in the human thymus. *J. Histochem. Cytochem.* **51,** 1557–1566.

Sasaki, H., Nishizaki, Y., Hui, C., Nakafuku, M., and Kondoh, H. (1999). Regulation of Gli2 and Gli3 activities by an amino-terminal repression domain: Implication of Gli2 and Gli3 as primary mediators of Shh signaling. *Development* **126,** 3915–3924.

Sheng, T., Chi, S., Zhang, X., and Xie, J. (2006). Regulation of Gli1 localization by the cAMP/protein kinase A signaling axis through a site near the nuclear localization signal. *J. Biol. Chem.* **281,** 9–12.

Sisson, J. C., Ho, K. S., Suyama, K., and Scott, M. P. (1997). Costal2, a novel kinesin-related protein in the Hedgehog signaling pathway. *Cell* **90,** 235–245.

Smelkinson, M. G., Zhou, Q., and Kalderon, D. (2007). Regulation of Ci-SCFSlimb binding, Ci proteolysis, and hedgehog pathway activity by Ci phosphorylation. *Dev. Cell* **13,** 481–495.

Stecca, B., and Ruiz i Altaba, A. (2009). A GLI1-p53 inhibitory loop controls neural stem cell and tumour cell numbers. *EMBO J.* **28,** 663–676.

Stegman, M. A., Vallance, J. E., Elangovan, G., Sosinski, J., Cheng, Y., and Robbins, D. J. (2000). Identification of a tetrameric hedgehog signaling complex. *J. Biol. Chem.* **275,** 21809–21812.

Svärd, J., Heby-Henricson, K., Persson-Lek, M., Rozell, B., Lauth, M., Bergström, A., Ericson, J., Toftgård, R., and Teglund, S. (2006). Genetic elimination of Suppressor of

fused reveals an essential repressor function in the mammalian Hedgehog signaling pathway. *Dev. Cell* **10,** 187–197.
Taipale, J., Cooper, M. K., Maiti, T., and Beachy, P. A. (2002). Patched acts catalytically to suppress the activity of Smoothened. *Nature* **418,** 892–897.
te Welscher, P., Zuniga, A., Kuijper, S., Drenth, T., Goedemans, H. J., Meijlink, F., and Zeller, R. (2002). Progression of vertebrate limb development through SHH-mediated counteraction of GLI3. *Science* **298,** 827–830.
Therond, P., Busson, D., Guillemet, E., Limbourg-Bouchon, B., Preat, T., Terracol, R., Tricoire, H., and Lamour-Isnard, C. (1993). Molecular organisation and expression pattern of the segment polarity gene fused of Drosophila melanogaster. *Mech. Dev.* **44,** 65–80.
Therond, P. P., Knight, J. D., Kornberg, T. B., and Bishop, J. M. (1996). Phosphorylation of the fused protein kinase in response to signaling from hedgehog. *Proc. Natl. Acad. Sci. USA* **93,** 4224–4228.
Therond, P. P., Limbourg Bouchon, B., Gallet, A., Dussilol, F., Pietri, T., van den Heuvel, M., and Tricoire, H. (1999). Differential requirements of the fused kinase for hedgehog signalling in the Drosophila embryo. *Development* **126,** 4039–4051.
Tomoda, T., Bhatt, R. S., Kuroyanagi, H., Shirasawa, T., and Hatten, M. E. (1999). A mouse serine/threonine kinase homologous to C. elegans UNC51 functions in parallel fiber formation of cerebellar granule neurons. *Neuron* **24,** 833–846.
Tomoda, T., Kim, J. H., Zhan, C., and Hatten, M. E. (2004). Role of Unc51.1 and its binding partners in CNS axon outgrowth. *Genes Dev.* **18,** 541–558.
Turner, T. T., Bomgardner, D., and Jacobs, J. P. (2004). Sonic hedgehog pathway genes are expressed and transcribed in the adult mouse epididymis. *J. Androl.* **25,** 514–522.
Varjosalo, M., Bjorklund, M., Cheng, F., Syvanen, H., Kivioja, T., Kilpinen, S., Sun, Z., Kallioniemi, O., Stunnenberg, H. G., He, W. W., Ojala, P., and Taipale, J. (2008). Application of active and kinase-deficient kinome collection for identification of kinases regulating hedgehog signaling. *Cell* **133,** 537–548.
Wang, Q. T., and Holmgren, R. A. (2000). Nuclear import of cubitus interruptus is regulated by hedgehog via a mechanism distinct from Ci stabilization and Ci activation. *Development* **127,** 3131–3139.
Wang, G., Wang, B., and Jiang, J. (1999). Protein kinase A antagonizes Hedgehog signaling by regulating both the activator and repressor forms of Cubitus interruptus. *Genes Dev.* **13,** 2828–2837.
Wang, B., Fallon, J. F., and Beachy, P. A. (2000a). Hedgehog-regulated processing of Gli3 produces an anterior/posterior repressor gradient in the developing vertebrate limb. *Cell* **100,** 423–434.
Wang, G., Amanai, K., Wang, B., and Jiang, J. (2000b). Interactions with Costal2 and suppressor of fused regulate nuclear translocation and activity of cubitus interruptus. *Genes Dev.* **14,** 2893–2905.
Wen, X., Lai, C. K., Evangelista, M., Hongo, J. A., de Sauvage, F. J., and Scales, S. J. (2010). Kinetics of hedgehog-dependent full-length Gli3 accumulation in primary cilia and subsequent degradation. *Mol. Cell. Biol.* **30,** 1910–1922.
Wilson, C. W., Nguyen, C. T., Chen, M. H., Yang, J. H., Gacayan, R., Huang, J., Chen, J. N., and Chuang, P. T. (2009). Fused has evolved divergent roles in vertebrate Hedgehog signalling and motile ciliogenesis. *Nature* **459,** 98–102.
Wolff, C., Roy, S., and Ingham, P. W. (2003). Multiple muscle cell identities induced by distinct levels and timing of hedgehog activity in the zebrafish embryo. *Curr. Biol.* **13,** 1169–1181.
Yan, J., Kuroyanagi, H., Tomemori, T., Okazaki, N., Asato, K., Matsuda, Y., Suzuki, Y., Ohshima, Y., Mitani, S., Masuho, Y., Shirasawa, T., and Muramatsu, M. (1999). Mouse ULK2, a novel member of the UNC-51-like protein kinases: Unique features of functional domains. *Oncogene* **18,** 5850–5859.

Young, A. R., Narita, M., Ferreira, M., Kirschner, K., Sadaie, M., Darot, J. F., Tavare, S., Arakawa, S., Shimizu, S., Watt, F. M., and Narita, M. (2009). Autophagy mediates the mitotic senescence transition. *Genes Dev.* **23,** 798–803.

Zhang, C., Williams, E. H., Guo, Y., Lum, L., and Beachy, P. A. (2004). Extensive phosphorylation of Smoothened in Hedgehog pathway activation. *Proc. Natl. Acad. Sci. USA* **101,** 17900–17907.

Zhang, W., Zhao, Y., Tong, C., Wang, G., Wang, B., Jia, J., and Jiang, J. (2005). Hedgehog-regulated Costal2-kinase complexes control phosphorylation and proteolytic processing of Cubitus interruptus. *Dev. Cell* **8,** 267–278.

Zhang, Q., Zhang, L., Wang, B., Ou, C. Y., Chien, C. T., and Jiang, J. (2006). A hedgehog-induced BTB protein modulates hedgehog signaling by degrading Ci/Gli transcription factor. *Dev. Cell* **10,** 719–729.

Zhao, Y., Tong, C., and Jiang, J. (2007). Hedgehog regulates smoothened activity by inducing a conformational switch. *Nature* **450,** 252–258.

Zhou, Q., and Kalderon, D. (2011). Hedgehog activates fused through phosphorylation to elicit a full spectrum of pathway responses. *Dev. Cell* **20,** 802–814.

CHAPTER SIX

Identification, Functional Characterization, and Pathobiological Significance of GLI1 Isoforms in Human Cancers

Richard L. Carpenter* *and* Hui-Wen Lo*,†

Contents

Abstract

Glioma-associated oncogene homolog 1 (GLI1) is the nuclear mediator of Hedgehog signaling that activates gene transcription via its zinc finger DNA-binding and transactivation domains. GLI1 plays a critical role in several cellular processes, including embryonic development, tumorigenesis, and

* Division of Surgical Sciences, Department of Surgery, Duke University School of Medicine, Durham, North Carolina, USA
† Duke Cancer Institute, Durham, North Carolina, USA

Vitamins and Hormones, Volume 88
ISSN 0083-6729, DOI: 10.1016/B978-0-12-394622-5.00006-7

tumor growth and progression. The human GLI1 gene was identified in 1987 as an amplified gene in glioblastoma. Somatic mutations have never been reported in the GLI1 gene in any cell or tumor type. Very recently in 2008–2009, the full-length GLI1 transcript was discovered to undergo alternative splicing to form two shorter isoforms, namely N-terminal deletion variant (GLI1ΔN) and truncated GLI1 (tGLI1). Emerging evidence suggests that the three structurally different GLI1 isoforms are distinctly different in their expression patterns and functions in the context of human cancers. The tGLI1 isoform, in particular, has been shown to gain the ability to modulate expression of the genes that are not regulated by GLI1 and to support the biology of more aggressive cancer. Consequently, a key focus of this chapter is to summarize and compare the properties of the three GLI1 isoforms and their relations to malignant biology of human cancers.

I. Introduction

The hedgehog signaling pathway is conserved in both vertebrates and invertebrates and is linked to a broad array of biological processes most prominently embryonic development (Ingham *et al.*, 2011), tumorigenesis, and cancer progression (Yang *et al.*, 2010; Zhu and Lo, 2010). Mammalian hedgehog signaling is initiated by three ligands: Sonic Hedgehog (Shh), Indian Hedghog (Ihh), and Desert Hedgehog (Dhh). Shh is the most potent of the three ligands (Pathi *et al.*, 2001) and is expressed most widely in embryos and adult tissues (Ingham and McMahon, 2001). Aberrant Shh signaling is implicated in endodermally derived human cancers that account for up to 25% of human cancer deaths (Lum and Beachy, 2004). The main effectors of hedgehog signaling in Drosophila and mammals are Cubitus interruptus (Ci) and the GLI (glioma-associated oncogene homolog) proteins (GLI1, GLI2, GLI3), respectively. Hedgehog induces GLI protein translocation to the nucleus where transcription can be induced primarily by GLI1 and/or GLI2 or repressed primarily by GLI2 and/or GLI3. Via GLI-mediated gene regulation, Hedgehog plays an essential role in many important cellular processes in normal and cancerous cells. Consequent to the pivotal role that the Hedgehog pathway plays in tumor growth and progression, extensive efforts have been invested in the development of Hedgehog-targeted therapy (Ng and Curran, 2011; Yang *et al.*, 2010). Notably, a small molecule inhibitor of smoothened (SMO), Vismodegib/GDC-0449, is being evaluated in several clinical trials for cancers (LoRusso *et al.*, 2008; Rudin *et al.*, 2009; Von Hoff *et al.*, 2009).

The human GLI1 gene was discovered in 1987 by Vogelstein and colleagues, as an amplified gene in human glioblastoma multiforme, GBM (Kinzler *et al.*, 1987). It was later found to encode for a zinc finger transcription factor belonging to the Kruppel family of zinc finger proteins (Kinzler *et al.*, 1988). The clinical importance of GLI1 gene amplification is still unclear as early studies found low rates of GLI1 amplification ranging from 1.6% to 3.3% occurrence in gliomas (Bigner *et al.*, 1987; Forus *et al.*, 1993; Wong *et al.*, 1987) whereas a more recent study reported a 22.6% occurrence in gliomas (Rao *et al.*, 2010). Other cancers were observed to have varying ranges of occurrences for GLI1 amplification with the highest occurrence of 28% in rhabdomyosarcomas (Gordon *et al.*, 2000). Amplification of the GLI1 gene has been shown to result in increased GLI1 gene expression (Reifenberger *et al.*, 1994).

Somatic mutations have never been reported in the GLI1 gene in any cell or tumor type, unlike other components of the Hedgehog pathway that contain somatic mutations such as Shh, patched (PTCH), SMO, suppressor of fused (SUFU), GLI2, and GLI3 (Kang *et al.*, 1997; Radhakrishna *et al.*, 1997; Raffel *et al.*, 1997; Roessler *et al.*, 1996; Xie *et al.*, 1998). Very recently in 2008 and 2009, the full-length GLI1 transcript was found to undergo alternative splicing to form two shorter isoforms, namely N-terminal deletion variant (GLI1ΔN) identified by Shimokawa *et al.* (2008) and truncated GLI1 (tGLI1) identified in our laboratory (Lo *et al.*, 2009). While both GLI1 variants are under intensive investigations for their properties and functions, emerging evidence indicates that the three GLI1 isoforms differ in their expression patterns and their ability to regulate gene activation. In comparing tGLI1 to GLI1, our results indicated that although they display some overlapping properties, the two isoforms demonstrate distinctly different functions by regulating different sets of target genes (Cao *et al.*, 2012; Lo *et al.*, 2009). These differences may lead to their unique differences in mediating cellular processes important to tumor behaviors such as proliferation, migration, invasion, angiogenesis, and metastasis. To provide a comprehensive review of the GLI1 isoforms in relation to human cancers, we will summarize the structures and biochemical properties of the three GLI1 isoforms (Section II), provide an overview of the regulation of the GLI1 isoforms by the canonical and noncanonical signaling pathways (Section III), and compare the malignant phenotypes of human cancers that can be modulated by the GLI1 isoforms (Section IV).

II. Structures and Properties of GLI1 Isoforms

A. GLI1

The human GLI1 gene was identified in 1987 by Vogelstein and colleagues (Kinzler *et al.*, 1987). In this study, the investigators found evidence of gene amplification in a human GBM cell line and mapped the gene to chromosome 12 (12q13-14.3). The gene locus did not correspond to already well-known oncogenes such as MYC or RAS, or other oncogenes known to be located on chromosome 12. Subsequently, the amplified gene was named as GLI for the glioma tumor in which it was discovered. The GLI gene was later mapped, more specifically, to 12q13.3-14.1 (Arheden *et al.*, 1989). The GLI gene was found to contain 3318 bp as part of the open reading frame that predicted a 1106-residue protein with a calculated molecular mass of 118kDa (Kinzler *et al.*, 1988). The encoded protein was detected by immunoblotting soon afterward and observed to migrate as a 150kDa protein (Kinzler and Vogelstein, 1990). It was later concluded that the discrepancy between calculated and apparent molecular mass was likely due to the intrinsic amino acid sequence as the GLI1 protein synthesized from *in vitro* translation migrated similarly as the 150kDa GLI1 protein isolated from the cells.

The GLI1 protein contains five successive repeats of a zinc finger DNA-binding consensus sequence (Fig. 6.1A). It was determined that the GLI1 protein belongs to the Kruppel family of zinc finger proteins as it contained the H–C link between zinc fingers common to this family of transcription factors. This finding defined GLI1 as the first human member of the Kruppel family of proteins (Kinzler *et al.*, 1988). GLI1 was detected in the nuclei of the cells of cancer cell lines, xenografts, and primary tumor tissue, which is consistent with its structure being that of a zinc finger transcription factor (Kinzler and Vogelstein, 1990). In the nucleus, the GLI protein binds to a 9-bp DNA sequence of 5′-GACCACCCA-3′. Later analysis of GLI-DNA binding found that only fingers 2–5 bind to the major groove of DNA wrapping around for a full helical turn with fingers 4–5 having the most contact (Pavletich and Pabo, 1993). It was also reported that finger one does not even contact DNA but rather has protein–protein interaction with finger two. These early studies characterized the GLI gene to encode for a zinc finger transcription factor and upon discovery of GLI2 and GLI3, this original GLI gene was termed GLI1.

B. GLI1ΔN

In 2008, Shimokawa *et al.* reported the existence of a splice variant of GLI1 that has 128 amino acids deleted from the N-terminus as a result of splicing exon 1 directly to exon 4 (Shimokawa *et al.*, 2008). Due to the deletion of this region from the N-terminus, this variant was termed GLI1ΔN (Fig. 6.1B). Primary functional domains remained intact in the GLI1ΔN variant include the zinc finger domains, the transactivation domain, the nuclear localization signal (NLS), and the nuclear export signal. The SUFU-binding site was lost in GLI1ΔN. Notably, the

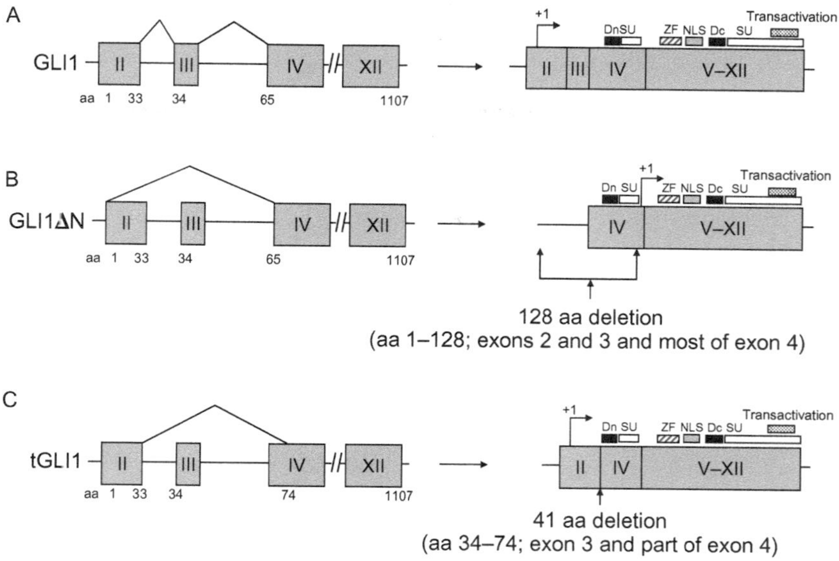

Figure 6.1 Structures of the human GLI1 gene and the encoded full-length GLI1, GLI1ΔN and tGLI1 isoforms. (A) The full-length GLI1 gene comprises 12 exons, including the 5′-untranslated exon 1. The GLI1 coding region spans nt +79 to +3399 with the initiating methionine codon, ATG, at +79 in exon 2 (arrows). Exons are indicated as gray boxes while introns are shown by lines. The known functional domains of full-length GLI1 include the degron degradation signals (Dn and Dc; aa 77–116; 464–469), SUFU-binding domains (SU; aa 111–125 and C-terminus) (Dunaeva *et al.*, 2003), zinc finger domains (ZF; aa 235–387), the nuclear localization signal (NLS; aa 380–420), and the transactivation domain (aa 1020–1091). (B) Alternative splicing of GLI1 RNA can lead to the deletion of exons 1–3 totaling 128 amino acids in the N-terminus, forming the GLI1ΔN variant. (C) The deletion of the entire exon 3 and part of exon 4 totaling 41 amino acids yields the tGLI1 isoform. Notably, tGLI1 retains all the known functional domains of the full-length GLI1. (See Color Insert.)

SUFU-binding domain in the GLI1 protein interacts with cytoplasmic SUFU complex and consequently, is sequestered in the cytoplasm (Kogerman *et al.*, 1999). In contrast to the prediction that GLI1ΔN would be enriched in the nucleus because of the loss of SUFU-binding domain, GLI1ΔN showed a weakened ability to translocate into the nucleus, which leads to weaker activation of GLI1 target genes such as GLI1 and PTCH. Further, GLI1ΔN was detected in several cancer cell lines derived from human tumors as well as in human embryonic kidney cells indicating it may have relevance in human tissues (Shimokawa *et al.*, 2008). Our study showed that GLI1ΔN is undetectable in GBMs (Lo *et al.*, 2009). The significance of GLI1ΔN in normal physiology and cancer biology is still largely unknown and remains to be defined.

C. tGLI1

In 2009, our laboratory reported the discovery of tGLI1 when analyzing cell lines and tumor tissue of GBM (Lo *et al.*, 2009). As summarized in Fig. 6.1C, tGLI1 is a product of alternative splicing that lacks 41 amino acids corresponding to the entire exon 3 and part of exon 4 of the GLI1 gene. Analysis of the GLI1 gene in GBM cell lines and normal peripheral lymphocytes found no somatic abnormality in the GLI1 gene. tGLI1 retains all the known functional domains present in full-length GLI1, including the zinc finger domains, transactivation domain, NLS, degrons signals, and the SUFU-binding domains. In line with the preservation of these functional domains, tGLI1 retains its ability to translocalize into the nucleus, transactivate GLI1 binding sites and activates expression of PTCH1, a GLI1-regulated gene. Like GLI1, tGLI1 responds to Shh stimulation (Lo *et al.*, 2009).

The expression pattern of tGLI1 is significant in the context of human cancers because currently available evidence suggests that tGLI1 is expressed in a tumor-specific fashion (Cao *et al.*, 2012; Lo *et al.*, 2009). For example, tGLI1 is highly expressed in cell lines and primary specimens of GBM but not in normal brain tissues and other normal tissues (Lo *et al.*, 2009). More recently, we further reported that tGLI1 is frequently expressed in breast cancer cell lines and primary tumors but rarely in normal mammary tissues (Cao *et al.*, 2012). This pattern of expression for tGLI1 is distinctly different than those of GLI1 and GLI1ΔN, which are expressed in both normal and cancer cells (Cao *et al.*, 2012; Lo *et al.*, 2009; Shimokawa *et al.*, 2008).

Notably, the molecular weight difference between the tGLI1 and GLI1 proteins is only about 4.5kDa, which makes it difficult to distinguish the two isoforms using immunoblotting (146kDa for tGLI1 and 150kDa for GLI1). Commercially available antibodies appear to recognize both isoforms (Cao *et al.*, 2012; Lo *et al.*, 2009). These technical limitations may have contributed to the inability to uncover the existence of tGLI1 for more than 20years after the 1987 discovery of the GLI1 gene. These notions also suggest that some of the known GLI1 functions may be attributed to tGLI1 but not GLI1. Evidence supporting this speculation will be discussed in detail in Section IV. Importantly, we have recently optimized a gel electrophoresis condition that can effectively resolve the GLI1 and tGLI1 proteins (Cao *et al.*, 2012). Although the combination of transcript analysis (by RT-PCR and DNA sequencing of the PCR product) with protein analysis (by gel electrophoresis followed by immunoblotting) adequately and accurately distinguish tGLI1 from GLI1, it is recognized that a tGLI1-specific antibody is needed to selectively detect the tGLI1 protein but not GLI1, in order to help further decipher the functions of the two isoforms and to gain a greater understanding of the long overlooked tGLI1 isoform. In Section IV, we will discuss in detail the transcriptional functions of tGLI1 and its ability to modulate several important malignant phenotypes of human cancers, including growth, migration, invasion, and angiogenesis.

III. Regulation of GLI1 Isoforms by Canonical and Noncanonical Pathways

A. Canonical pathway

Canonical Hedgehog signaling is initiated by binding of the ligand Shh to the 12-transmembrane receptors PTCH 1 and 2 (Fig. 6.2A). Shh binding to PTCH proteins relieves constitutive PTCH repression of the 7-transmembrane protein receptor SMO. Activation of SMO leads to the release of GLI1 from SUFU-mediated cytoplasmic sequestration and to GLI1 nuclear translocalization (Kogerman *et al.*, 1999). SMO also increases levels of full-length GLI2/3, which proceeds to induce transcription of GLI1. GLI1 acts on many target genes to regulate several physiological processes and its aberrant activity causes several pathologies including tumor progression.

GLI1 nuclear translocation and induction of transcription is further controlled by Shh pathway proteins. SUFU was cloned in 1999 as a 484 residue protein (Kogerman *et al.*, 1999). SUFU physically interacts with GLI1 and the interaction traps GLI1 in the cytoplasm and prevents it from translocating into the nucleus. Expression of a GLI1 mutant that is constitutively localized to the nucleus also showed SUFU directly interacted with GLI1 while in the nucleus and reduced GLI1-mediated transcriptional activity (Kogerman *et al.*, 1999). It was later discovered that SUFU forms a complex with SAP18 and mSin3, a transcription corepressor, and this complex directly binds GLI1 on DNA to repress GLI1-mediated transcription (Chen *et al.*, 2002). While SUFU negatively regulates GLI1, the dual specificity Yak1-related kinase Dyrk1 appears to potentiate GLI1 activity. Dyrk1 phosphorylates GLI1 at multiple serine/threonine sites and induce nuclear accumulation of GLI1 (Mao *et al.*, 2002). This study also showed that Dyrk1 enhances transcription by a mutant GLI1, which is constitutively localized to the nucleus, indicating Dyrk1 can enhance the process of GLI1-mediated transcription and not only nuclear localization.

Canonical hedgehog signaling is critical for Shh-induced patterning during embryonic development. For example, Shh-GLI1 signaling mediates the induction of floor plate cells in the posterial neural tube (Lee *et al.*, 1997). Shh-GLI1 signaling also directly induced transcription of HNF-3β, a winged-helix transcription factor that mediated expression of floor plate differentiation in mice (Sasaki *et al.*, 1997). Shh-GLI1 signaling also regulates pathological processes, most notably several cancer types (Kasper *et al.*, 2006). Mutations in SMO, PTCH1, and SUFU have been reported and were associated with formation of cancer (Taylor *et al.*, 2002; Villavicencio

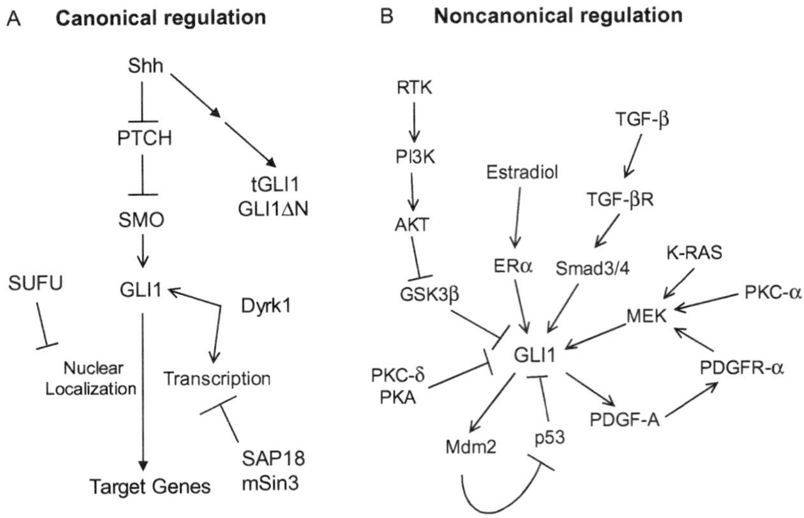

Figure 6.2 Regulation of GLI1 the isoforms by the canonical and noncanonical pathways. (A) Canonical activation of GLI1 is initiated by Shh, which binds and inhibits PTCH1 relieving PTCH1-mediated suppression of SMO. Derepressed SMO releases GLI1 from SUFU-mediated cytoplasmic sequestration. In turn, GLI1 translocalizes to the nucleus to induce transcription of target genes. Dyrk1 enhances GLI1 activity. Both GLI1ΔN and tGLI1 respond to Shh; however, it is unknown whether they are subjected to the regulation by other components of the Hedgehog pathway. (B) GLI1 can be affected by several noncanonical pathways. Activators of GLI1 include TGF-β, K-Ras, PKC-α, PDGF-A, PI3K-AKT, and estradiol. PKC-δ, PKA, and p53 suppress GLI1. The response of tGLI1 or GLI1ΔN to noncanonical pathway activation is currently unknown.

et al., 2000). The role of GLI1 in hedgehog-associated tumorigenesis is still not well understood with mixed results being reported (Kimura *et al.*, 2005; Villavicencio *et al.*, 2000).

Limited information is available in the regulation of GLI1ΔN and tGLI1. What is known is that Shh upregulates GLI1ΔN to a similar level as GLI1 (Shimokawa *et al.*, 2008). It is also known that tGLI1 responds to Shh stimulation to a similar extent as GLI1, as reported in our study (Lo *et al.*, 2009). However, the role of SUFU in regulating the cytoplasmic-nuclear shuttling of GLI1ΔN and tGLI1 is still not well understood. GLI1ΔN does not retain the N-terminal SUFU-binding domain, suggesting that its nuclear import may be regulated by the C-terminal SUFU-binding domain or by other means (Shimokawa *et al.*, 2008). It is yet to be investigated whether tGLI1 interacts with SUFU, although it is expected that tGLI1 will associate with SUFU since tGLI1 retains the SUFU-binding domains. Also uninvestigated are the stability of GLI1ΔN and tGLI1 and their relationships to other components of the Hedgehog pathway such as

SMO, GLI2, and GLI3. Taken together, future efforts are required to elucidate the regulatory role that the Hedgehog-PTCH-SMO signaling axis plays in GLI1ΔN and tGLI1.

B. Noncanonical pathway

Canonical activation of the Shh–PTCH–SMO pathway had been thought to be the primary mechanism for control of GLI1. However, evidence continues to accumulate that GLI1 can also be regulated by other signaling pathways and proteins outside of the canonical Shh–PTCH–SMO pathway (Fig. 6.2B). Elucidation of these pathways could potentially be important for treatment strategies as redundant regulation of GLI1 could lead to cancers that develop resistance to Shh-targeted therapies.

Gastric cancer patients showed a positive correlation between activation of ERK and PTCH expression, a GLI1 gene target (Seto *et al.*, 2009). Expression of K-Ras increased GLI-induced transcription, which was suppressed by a MEK inhibitor. The effect of K-Ras-MEK signaling on GLI1 is independent of SUFU as knockdown of SUFU did not alter K-Ras induction of GLI1 (Seto *et al.*, 2009). Further, suppression of GLI1 inhibits cell transformation induced by K-Ras overexpression (Ji *et al.*, 2007). Perhaps the strongest evidence that Ras can control GLI1 is that mice overexpressing K-Ras, which induces pancreatic adenocarcinoma, strongly expressed GLI1 and PTCH despite the fact that these mice lack the SMO gene (Nolan-Stevaux *et al.*, 2009). Suppression of GLI1 in cancer cells isolated from these mice showed reduced indices of cell transformation and decreased K-Ras indicating positive feedback between GLI1 and K-Ras. Thus, the Ras–MEK–ERK pathway has the ability to regulate GLI1. Despite clear evidence for this regulation, the mechanism by which Ras–MEK regulates GLI1 levels has still not been discovered.

Another pathway that has recently gained attention for GLI1 regulation is the TGF-β-Smad pathway. Mice overexpressing TGF-β in the epidermis also overexpress GLI1 (Dennler *et al.*, 2007). TGF-β treatment induces cellular expression of GLI1 that was independent of the SMO inhibitor cyclopamine (Dennler *et al.*, 2007). Further, knockdown of Smad4 or Smad3 in these cells prevented GLI protein induction by TGF-β. The mechanism of GLI1 induction by TGF-β was further elucidated when loss of GLI2 prevented induction of GLI1 by TGF-β (Dennler *et al.*, 2007). Adding to this, the GLI2 gene was found to have a TGF-β-responsive region (Dennler *et al.*, 2009). Thus, it appears that TGF-β directly induces GLI2, which then promotes GLI1 production. Results also show pancreatic adenocarcinoma cells with enhanced GLI expression, but are resistant to cyclopamine, are sensitive to TGF-β inhibitors or GLI2 inhibition (Dennler *et al.*, 2007). Since GLI1 expression can be controlled by GLI2 in response to TGF-β, it stands to reason that the tGLI1 and GLI1ΔN variants should be inducible by GLI2 in order to be responsive to TGF-β. However, exactly what drives expression of these variants is still unknown.

There is recent evidence to suggest the p53 tumor suppressor also can regulate GLI1. Intracranial tumor development in mice was seen to increase with p53 knockdown or with increased GLI1 expression but tumors growth was synergistic when both p53 was suppressed and GLI1 expression increased (Stecca and Ruiz i Altaba, 2009). Increasing p53 levels in these tumor cells by oxaliplatin treatment decreased cell proliferation, which was rescued by enhancement of GLI1 levels. Similarly, expression of a p53 dominant negative in frog embryos induced tumor formation that could be rescued with GLI1 suppression (Stecca and Ruiz i Altaba, 2009). Further, p53 knockdown decreased GLI1-induced transcription while p53 over-expression reduced GLI1-induced transcription. These data clearly indicate an inhibitory loop between GLI1 and p53. Another study found that GLI1 enhanced MDM2 interaction with p53, which led to p53 degradation indicating an indirect regulation by GLI1 (Abe *et al.*, 2008). The mechanism of GLI1 regulation by p53 was not assessed and could be direct or indirect. Since GLI1 regulates p53 via its physical association with MDM2, it is difficult to predict whether tGLI1 and/or GLI1ΔN have influence on MDM2-p53 because it is not known whether the deletions within the two isoforms have an impact on their ability to interact with MDM2. In the case of tGLI1, we speculate that it may interact with proteins and promoter sequences in a different fashion compared to GLI1. This speculation is built on our findings showing that tGLI1, but not GLI1, binds to the promoters of CD24 and VEGF-A genes (Cao *et al.*, 2012; Lo *et al.*, 2009).

Signaling through the PI3K–AKT pathway appears to participate in GLI1 regulation, but perhaps not independent of Shh signaling. GLI1-induced transcription can be enhanced by IGF-1, which activates PI3K–AKT signaling, and Shh-induced GLI1 can be inhibited with PI3K inhibitors (Riobo *et al.*, 2006). However, activation of PI3K signaling by IGF-1 could not induce GLI-mediated transcription independent of Shh. AKT knockdown reduced Shh-induced transcription by GLI1. Another study found that inhibition of PI3K signaling reduced cell proliferation induced by ectopic GLI1 expression (Stecca *et al.*, 2007). This study showed PTEN, an endogenous inhibitor to PI3K signaling, could inhibit GLI1 nuclear localization and transcription. Evidence indicates a clear role for PI3K-AKT signaling in Shh-induced GLI1 but the precise mechanism for this regulation has not been made clear. Regulation of tGLI1 or GLI1ΔN by PI3K-AKT signaling is unknown.

Protein kinase A (PKA) negatively regulates GLI1. The first evidence of PKA-mediated negative regulation of GLI1 came from expression of dominant-negative PKA in mouse embryos leading to effects that mimic GLI1 hyperactivation (Epstein *et al.*, 1996). More direct evidence showed that activation of PKA by forskolin resulted in an alteration of GLI1 localization from the nucleus to the cytoplasm (Sheng *et al.*, 2006). PKA phosphorylates GLI1 at Thr374, which when mutated to valine prevented PKA-induced cytoplasmic localization (Sheng *et al.*, 2006). Threonine 374 is adjacent to the NLS on GLI1. The phosphorylation-mimicking Thr374Asp mutant is sequestered in the cytoplasm. Consistent with these findings, activation of PKA decreased GLI1-mediated transcription indicative of GLI1 cytoplasmic localization. PKA negative regulation of GLI1 has also been shown previously in the context of embryonic development in which PKA inhibition led to hyperproliferation in several brain areas resulting in malformations (Epstein *et al.*, 1996). Although PKA likely retains the ability to phosphorylate both tGLI1 and GLI1ΔN as Thr374 is present in both variants, this has not been established by experimental evidence.

PKC-α was found to enhance GLI1-mediated transcription in a MEK-dependent manner whereas PKC-δ inhibited GLI1-mediated transcription (Cai *et al.*, 2009). GBM neurospheres that are resistant to Notch inhibitors were found to have increased GLI1 levels as the Notch target gene Hes1 directly interacted with the GLI1 promoter suggesting these cells become resistant by Notch-induced activation of GLI1 (Schreck *et al.*, 2010). GLI1 expression can also induce phosphorylation of ERK2 that is dependent on PDGFR-α, which is a direct GLI1 target (Xie *et al.*, 2001). In these cells, inhibition of MEK-ERK2 signaling or neutralizing antibodies to PDGF-A reduced enhanced proliferation rates due to GLI1 expression (Xie *et al.*, 2001). Breast cancer cells exposed to 17β-estradiol displayed increased GLI1 levels that could be prevented with an estrogen receptor-α (ER-α) inhibitor (Xu *et al.*, 2010). The potential for these pathways to regulate tGLI1 or GLI1ΔN levels, nuclear localization, or transcriptional activity is unclear as there is not currently a mechanism for how many of these noncanonical pathways influence GLI1 expression or activity.

GLI1 protein stability can be regulated by the noncanonical pathways. The GLI1 protein has two degron degradation signals with one on the N-terminus and the other on the C-terminus. The C-terminal degron signal within GLI1 had the motif of DSGVEM (Huntzicker *et al.*, 2006), which is a consensus sequence recognized by the E3 ubiquitin ligase βTrCP (Jiang and Struhl, 1998). Increased βTrCP expression resulted in decreased GLI1 protein levels (Huntzicker *et al.*, 2006). GLI1 was seen to immunoprecipitate with βTrCP in a manner dependent on the presence of the DSGVEM motif. In addition, ubiquitinylated GLI1 was immunoprecipitated with βTrCP whereas mutation of the DSGVEM motif resulted in no detectable ubiqutinylated GLI1 (Huntzicker *et al.*, 2006). Further deletion of residues 51–116 on the N-terminus of GLI1 resulted in further stabilization beyond that provided by deletion of the C-terminal degrons signal indicating an N-terminal degron was likely present. However, the N-terminal degrons signals were not affected by βTrCP or by SUFU suggesting an alternative method of degradation by this degron signal (Huntzicker *et al.*, 2006). The importance of GLI1 protein stability controlled by these degron signals was illustrated in this same study by mice expressing GLI1 mutants. Mice expressing double mutants where both degrons were inactivated displayed a more severe phenotype of basal cell carcinoma compared to single mutants. The tGLI1 variant, although lacking several residues, is not missing residues in either degron signal and would not be expected to have altered degradation compared to GLI1. However, GLI1ΔN has lost the N-terminal degron signal but protein stability was not assessed in the one study examining this variant (Shimokawa *et al.*, 2008).

IV. GLI1 Isoforms and Malignant Phenotypes of Cancer

GLI1 overexpression has been reported in normal tissues and also cancerous tissues, including those of GBM (Cui *et al.*, 2010; Lo *et al.*, 2009), hepatocellular carcinoma (Huang *et al.*, 2006), gastric adenocarcinoma (Ma *et al.*, 2005), esophageal cancer (Ma *et al.*, 2006), breast cancer (Cao *et al.*, 2012; ten Haaf *et al.*, 2009), melanoma (Das *et al.*, 2009), rhabdomyosarcoma (Roberts *et al.*, 1989), and pancreatic adenocarcinoma (Nolan-Stevaux *et al.*, 2009). Similar to GLI1, GLI1ΔN appears to be expressed in both normal and cancerous cells (Shimokawa *et al.*, 2008). Currently available evidence suggests that tGLI1 is expressed in a tumor-specific fashion, at least in the two tumor types (GBM and breast carcinoma) examined to date (Cao *et al.*, 2012; Lo *et al.*, 2009). In line with the difference in expression patterns, the GLI1 and tGLI1 isoforms have been found to display overlapping and also distinctly different tissue- and tumor-specific functions (Fig. 6.3). In some instances, they differ in their ability to regulate expression of the genes that are important modulators of malignant phenotypes of human cancers, including proliferation, migration and invasiveness, progression, angiogenesis, cancer stem cell renewal, and drug resistance.

A. Proliferation

A major characteristic of cancer cells is an enhanced ability for cell division leading to uncontrolled proliferation. GLI1 has a proliferative-promoting effect when expressed in neural cells, evidenced by studies of GLI1 in the context of both cancer and embryonic development. For example, mice with Shh mutants leading to decreased GLI1 levels have brains with decreased proliferation (Dahmane *et al.*, 2001). This same study found GLI1 could induce formation of several hyperplasias throughout the neural

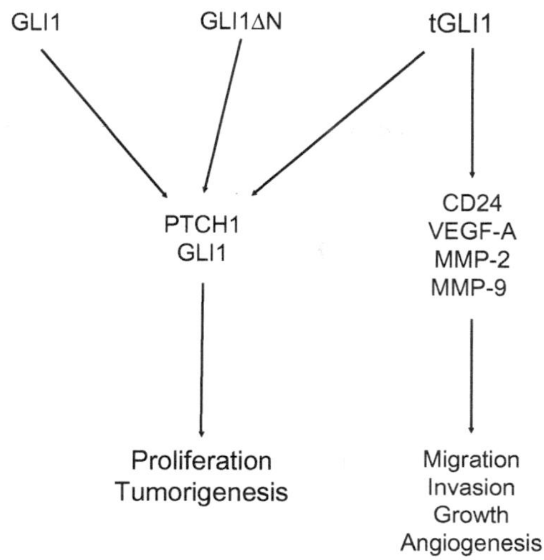

Figure 6.3 Regulation of malignant phenotypes of cancer by GLI1 and tGLI1. Expression of all three GLI1 variants results in increased expression of PTCH1, an established GLI1 target gene. Expression of GLI1 leads to increased cell proliferation rates and other cellular processes. Expression of tGLI1 activates additional genes, namely VEGF-A, CD24, MMP-2, and MMP-9, leading to its ability to promote cancer cell growth, migration, invasion, and angiogenesis.

tube when GLI1 RNA was injected into tadpoles. Similarly, ectopic expression of GLI1 in embryonic mouse brain regions resulted in outgrowths of neural tissue (Hynes *et al.*, 1997; Stecca and Ruiz i Altaba, 2009). The opposite also holds true in that inhibition of GLI1, either by cyclopamine or siRNA targeting, greatly reduces proliferation and tumor size (Clement *et al.*, 2007; Wang *et al.*, 2010). GLI1 has a similar effect on brain progenitor cells as GLI1 expression caused formation of tumor-like hyperplasia in the thalamus and cerebellum (Stecca and Ruiz i Altaba, 2009).

GLI1 expression in epithelial cells can induce cell transformation characterized by increased proliferation and anchorage-independent proliferation (Kimura *et al.*, 2005; Li *et al.*, 2006; Yoon *et al.*, 2002). This is specific to GLI1 or GLI2 as GLI3 had no such effect (Kimura *et al.*, 2005). These effects have been recapitulated *in vivo* as GLI1 induction in mouse epidermis led to hyperproliferative skin lesions (Li *et al.*, 2006). In other epithelial cell types, GLI1 targeted expression in mammary glands led to mammary hyperplasia and tumor formation (Fiaschi *et al.*, 2009). Interestingly, GLI1 was driven by a doxycycline promoter in this study and removal of doxycycline after GLI1 induction did not reduce mammary tumors, suggesting other cellular factors could replace GLI1 in sustaining tumor growth. In contrast, specific knockdown of GLI1 in inflammatory metastatic breast cancer cells resulted in decreased proliferation and increased apoptosis (Thomas *et al.*, 2011).

These data indicate GLI1 can induce a proliferative cell phenotype that can lead to tumor formation but may not be required for maintenance of a cancer phenotype in mammary tumors.

However, it is worth noting that siRNA-targeted expression knockdown would have silenced expression of both GLI1 and tGLI1, thus making it difficult to conclude that the growth inhibition is attributed to GLI1 suppression. This is a valid point as our studies have shown that tGLI1 may have stronger growth-stimulating effects on breast cancer cells (Cao *et al.*, 2012). Using isogenic MDA-MB-231 cell lines stably expressing GLI1, tGLI1, or control vector, we observed that the cells with tGLI1 expression underwent significant anchorage-dependent and -independent growth than those with GLI1 and control cells (Cao *et al.*, 2012). In the two GBM cell lines, we had examined to date, tGLI1 and GLI1 appeared to have similar effects on cell proliferation (Lo *et al.*, 2009). Our most recent, unpublished results indicate that the *in vivo* growth of a GBM cell line, U87MG, was greatly enhanced by tGLI1 expression and to a lesser extent, by GLI1. The discrepancy between the *in vitro* and *in vivo* effects of tGLI1 on GBM growth is likely due to tumor microenvironment such as tumor vasculature.

GLI1ΔN did not have any significant impact on cell proliferation despite its ability to activate a similar gene set to that of GLI1 (Shimokawa *et al.*, 2008). However, the induction of target genes by GLI1ΔN was significantly weaker than that of GLI1, indicating that GLI1ΔN may not be a strong promoter of cell proliferation. The ability of GLI1ΔN to regulate *in vivo* tumor growth is presently unknown.

B. Migration and invasiveness

Melanomas have been shown to express increased levels of GLI1 and osteopontin, along with enhanced ability to migrate, invade, and develop tumor xenografts and lung metastases (Das *et al.*, 2009). This study found that osteopontin was a direct target of GLI1 and loss of either GLI1 or osteopontin resulted in decreased cell migration, invasion, and tumor growth, and reduced numbers of lung metastases. In breast cancer, increased GLI1 expression promotes cell motility (Thomas *et al.*, 2011). Conversely, cyclopamine decreased the ability of melanoma cells to form lung metastases (Stecca *et al.*, 2007), and also reduced invasiveness of prostate cancer cells (Sheng *et al.*, 2004). In line with these findings, GLI1 has been shown to positively correlate with tumor grade and even lymph node status indicating it may play a role in progression of tumors toward metastasis (Stein *et al.*, 1999; ten Haaf *et al.*, 2009; Wang *et al.*, 2010). Further, high GLI1 expression resulted in a less favorable prognosis (ten Haaf *et al.*, 2009).

In light of the high homology between GLI1 and tGLI1, and the inability of antibodies and siRNA to selectively recognize tGLI1 versus GLI1, the reported positive correlations between GLI1 and tumor cell migration, invasion, and metastasis could be due to both GLI1 and tGLI1, or to a greater extent to tGLI1. This speculation is based on our results showing that tGLI1-expressing GBM and breast cancer cells migrated and invaded in significantly higher rates than those with GLI1 and control cells (Cao *et al.*, 2012; Lo *et al.*, 2009). When implanted in the flanks of nude mice, tGLI1-expressing GBM xenografts underwent a greater extent of infiltration into the skeletal muscles than those with GLI1 (Lo *et al.*, 2009). In elucidating the molecular mechanisms underlying tGLI1-mediated migration and invasion, we conducted DNA microarray to compare the gene expression profiles of three isogenic U87MG GBM cell lines with control vector, GLI1 and tGLI1 (Lo *et al.*, 2009). The results showed 75 genes to be expressed at a significantly higher level and 26 genes to be more suppressed in U87MG-tGLI1 cells compared to U87MG-vector and U87MG-GLI1 cells. Interestingly, the levels of well-known GLI1 target genes such as PTCH1 were higher in both U87MG-GLI1 and U87MG-tGLI1 cells compared to U87MG-vector cells. Importantly, the results of DNA microarray led us to identify and subsequently validate CD24 as a tGLI1-targeted gene and an important gene essential for tGLI1-associated migration and invasiveness in GBM cells. Notably, CD24 has been shown to recruit adhesion molecules to lipid rafts, thereby contributing to tumor cell migration, dissemination, and metastasis (Baumann *et al.*, 2005; Runz *et al.*, 2008). In addition to CD24, our results further indicated that tGLI1 induces expression of two invasion-associated genes, namely MMP-2 and MMP-9 (Cao *et al.*, 2012). Ongoing efforts are investigating the role of tGLI1 in tumor metastasis, the leading cause of death for most cancers.

C. Epithelial–mesenchymal transition

Development of an invasive cell phenotype with enhanced ability for migration is necessary for progression to metastasis. In epithelial tumors, this is often accompanied by a dedifferentiaion program termed epithelial–mesenchymal transition (EMT), most notably characterized by loss of E-cadherin and increases in Snail, Twist, Vimentin, and other mesenchymal markers (Lo *et al.*, 2007). There is also evidence that GLI1 may mediate EMT by increasing expression of Snail and repression of E-cadherin (Li *et al.*, 2006; Louro *et al.*, 2002). Expression of a dominant negative of Snail in GLI1-transformed epithelial cells decreased the ability of these cells to grow independent of attachment and led to increased E-cadherin levels (Li *et al.*, 2006). This likely indicates GLI1 regulates EMT by induction of Snail and results were confirmed *in vivo* as targeted GLI1 expression in mouse epidermis caused hyperproliferative skin lesions along with increased Snail expression (Li *et al.*, 2006). GLI1 has shown differing effects on E-cadherin as one study found GLI1 promoted E-cadherin redistribution toward the cell membrane (Neill *et al.*, 2008) and another found increased E-cadherin expression induced by GLI1 (Liao *et al.*, 2009). Despite showing increased E-cadherin, this last study also found increased expression of vimentin and Snail along with cells that had a greater capacity for proliferation, migration, and invasion (Liao *et al.*, 2009). Inhibition of GLI1 in these cells with cyclopamine decreased proliferation, invasion, migration, E-cadherin expression, and Snail expression. This suggests GLI1 had a profound effect on the phenotype but perhaps did not induce a full EMT change. The impact of tGLI1 on EMT is still unknown.

D. Cancer stem cells

Cancer stem cells are thought to control tumor development and progression. In addition, cancer stem cells are thought to have limitless proliferation potential and mediate drug resistance. Stem cells are characterized by the presence of undifferentiated cell markers such as CD133 or Nestin. Considering that GLI proteins play a significant role in embryonic development and differentiation, it has been investigated whether GLI1 expression is associated with cancer stem cells.

Cells derived from GBM are found to express NANOG, a transcription factor that participates in regulation of self-renewal in stem cells (Zbinden *et al.*, 2010). These cells show the ability to form gliomaspheres, which is exclusive to non-differentiated cells. NANOG appears to control the "stemness" of these cells as downregulation of NANOG reduced proliferation and the ability of these cells to form gliomaspheres and tumors. NANOG was found to be enriched in $CD133^{+}$ cells along with GLI1, GLI2, Nestin, and OCT4. Loss of NANOG also resulted in decreased GLI1, Nestin, and OCT4 indicating a loss of stem cell markers. Enhancement of GLI1 levels by PTCH knockdown results in increased NANOG levels, which control the stemness and tumor formation abilities of these cells (Zbinden *et al.*, 2010). Overexpression of GLI1 in normal mice brains also induces expression of stem cell markers such as SOX2, Nestin, Prominin1, and Bmi-1 that is accompanied by hyperproliferation and brain outgrowths (Stecca and Ruiz i Altaba, 2009). GLI1 is also involved in maintaining the stem cell phenotype in other cell types. Mammary stem cells form mammospheres and are enriched with greater than 20-fold higher levels of GLI1 compared to differentiated mammary cells (Liu *et al.*, 2006). Shh enhanced mammosphere formation whereas cyclopamine suppresses mammosphere formation, further suggesting Shh signaling plays a role in the stem cell phenotype. For direct evidence for a role of GLI1, overexpression of GLI1 or GLI2 enhances the ability of these cells to form mammospheres (Liu *et al.*, 2006). These data indicate the Shh pathway, and GLI1 in particular, play a significant role in cancer cells with stem-like characteristics. Future investigations are required to determine if some of the reported GLI1 activity is attributed to both tGLI1 and GLI1, or exclusively to tGLI1 or GLI1 alone.

The role of tGLI1 or GLI1ΔN in cancer stem cell marker expression or stem cell renewal has not been assessed. GLI1ΔN appears to have weaker transcriptional activity than GLI1 (Shimokawa *et al.*, 2008) and likely will contribute to a stem cell phenotype to a lesser extent than GLI1. However, tGLI1 has a stronger ability to activate multiple genes compared to GLI1 including VEGF-A (Cao *et al.*, 2012) that has been linked to stem cell phenotype (Oka *et al.*, 2007; Ricci-Vitiani *et al.*, 2010). In addition, cells expressing tGLI1 appear to have greater capabilities for migration and invasion, characteristics cancer stem cells purportedly have.

Our results indicate that CD24 is a transcriptional target of tGLI1, but not GLI1, in GBM and breast cancer (Cao *et al.*, 2012; Lo *et al.*, 2009). This finding is in agreement with our results showing that tGLI1 enhanced the propensity of cancer cells to migrate and invade both *in vitro* and *in vivo*. In line with our observations, CD24 has been shown to be involved in cancer cell motility, invasion, and metastasis (Baumann *et al.*, 2005; Kristiansen *et al.*, 2003; Runz *et al.*, 2008) and shortened patient survival (Kristiansen *et al.*, 2003; Lee *et al.*, 2009). Also consistent with these reports are the facts that CD24 recruits adhesion molecules to lipid rafts (Runz *et al.*, 2008) and that CD24 indirectly stimulates cell adhesion to fibronectin, collagens I/IV, and laminin through the activation of integrins (Baumann *et al.*, 2005). Interestingly, several studies have shown that some breast cancer-initiating stem cells display the characteristic of being low/negative for CD24 expression but high for CD44 (Al-Hajj *et al.*, 2003; Kim *et al.*, 2007). This notion is not in agreement with our results showing that tGLI1 expression enables MDA-MB-231 cells to grow in an anchorage-independent fashion, a property commonly displayed by cancer stem cells. In support of our results, a recent report, however, showed that CD24 does not define breast cancer stem cells (Meyer *et al.*, 2010). Also in agreement with our results, another recent study reported that pancreatic cancer stem cells demonstrate the CD44+CD24+ESA+phenotype (Li *et al.*, 2007). In support of these notions, it has been suggested that the role of CD24 in cancer stemness may be tissue-specific (Keysar and Jimeno, 2010). Clearly, more work is needed to define the role that CD24 plays in both cancer progression and stem cell biology. Also needed is future research into the role of tGLI1 in mediating cancer stem cell properties will be interesting considering tGLI1 is expressed in multiple human cancers including GBM and breast cancers (Cao *et al.*, 2012; Lo *et al.*, 2009). More significantly, tGLI1 could be mistakenly detected as GLI1 by RT-PCR or western blotting.

E. Angiogenesis

Tumor growth beyond 2mm in diameter requires angiogenesis in order for tumors to receive nutrients and remove wastes to survive (Bergers and Benjamin, 2003). Angiogenesis is regulated by the balance of proangiogenic factors (e.g., VEGFs) and anti-angiogenic factors (e.g., Tsp-1). Tumor cells alter this balance by secreting proangiogenic factors, which draws growing microvessels toward the tumor in order to retrieve nutrients and discard wastes. Turning on the angiogenic switch is a crucial point in the progression of tumors toward metastasis as angiogenic tumors greatly increase the pace of tumor growth. In addition to increasing the rate of tumor growth, angiogenesis increases vascular density around the tumor providing a door for metastatic cells to enter the general circulation where they can find a new tissue for seeding secondary tumors. Thus, angiogenesis has attracted much attention in recent decades as a potential drug target for cancer.

Pancreatic adenocarcinoma cells have increased GLI1 levels whereas cyclopamine decreases GLI1 levels, xenograft tumor growth, and reduced vascularity within the tumor (Nakamura *et al.*, 2010). Conditioned medium made from these cancer cells was mixed with matrigel and implanted in nude mice causing bone marrow-derived cells to be recruited and increasing vascular density. Cyclopamine decreased the number of bone marrow-derived cells that were recruited to the matrigel plug (Nakamura *et al.*, 2010). Cyclopamine treatment to growing xenograft tumors decreased tumor angiogenesis and decreased tumor expression of several proangiogenic factors including VEGF-A, SDF-1, Ang-1, IGF-1, and PDGF-B (Nakamura *et al.*, 2010). GLI1, specifically, can mediate these effects as glioma stem cells, which express GLI1, have increased production and release of VEGF in response to Shh and IGF-1 (Hsieh *et al.*, 2011). However, knockdown of GLI1 significantly reduces VEGF production by these cells and reduces the ability of these cells to promote angiogenesis *in vitro*. Thus, GLI1 can have a proangiogenic effect with increased expression.

tGLI1 also induces a proangiogenic response. Expression of tGLI1 in breast cancer cells increased VEGF-A mRNA and protein production by these cells compared to cells expressing GLI1 (Cao *et al.*, 2012). As mentioned above, GLI1 can promote an angiogenic response by induction of VEGF-A (Hsieh *et al.*, 2011; Nakamura *et al.*, 2010). The ability of tGLI1 to induce a much stronger level of VEGF-A production was found to be due to direct interaction with the VEGF-A promoter by tGLI1 whereas GLI1 had minimal binding (Cao *et al.*, 2012). Medium from tGLI1-expressing cells also induced stronger endothelial cell tube formation compared to GLI1-expressing cells consistent with increased VEGF induction. This suggests that although GLI1 has a proangiogenic effect, tGLI1 may promote a stronger angiogenic response. Induction of angiogenesis by tGLI1 is consistent with its role in promoting a more aggressive cancer phenotype. There are currently no published data regarding the ability of GLI1ΔN to alter angiogenesis genes.

F. Drug resistance

Cancer cell resistance to chemotherapy is usually associated with more aggressive cell phenotypes that give a poor patient prognosis. Certain cancer cells adapt to the environment created by chemotherapy to escape toxicity induced by cancer killing therapies. GLI1 appears to have some association with the development of cellular resistance to chemotherapy. Patients with gliomas displayed a positive correlation between GLI1 expression and recurring tumors (Cui *et al.*, 2010). Concurrent cell treatment with Shh and chemotherapy led to an increase in clonogenic survival of these cells whereas cells receiving chemotherapy alone had decreased survival (Cui *et al.*, 2010). Adding to this, treatment of cancer cells with both chemotherapy and cyclopamine significantly increased cell death and apoptosis compared to either treatment alone (Cui *et al.*, 2010). These data suggest upregulation of the Shh pathway, likely mediated by GLI1, provides protection for cells and resistance to chemotherapy. The precise mechanism for this protection was not elucidated. Considering tGLI1 is associated with a more aggressive cell phenotype, it is logical that tGLI1 could provide similar, if not greater, resistance to a chemotherapy challenge. Experiments to test this hypothesis have not been published. It will be interesting to see if the tGLI1 variant continues to show a more aggressive cell phenotype and what characteristics this splice variant can provide for cells. It will be equally important to determine if tGLI1-expressing tumor cells respond to Shh-targeted therapy that typically inhibits the Shh-SMO-GLI1 signaling axis.

V. Conclusion and Future Directions

Significance advances have been made in our understanding of the GLI1 family of proteins in the past few years. Most notably, the full-length GLI1 transcript is alternatively spliced to yield two shorter isoforms, namely GLI1ΔN and tGLI1. While the expression pattern of GLI1ΔN appears to be similar to GLI1 that is widely expressed in both normal and cancerous cells, tGLI1 has been shown to be expressed in a tumor-specific fashion in GBM and breast cancer. The unique pattern expression of tGLI1 is significant in the context of human cancers, in particular, tumorigenesis, tumor progression, and therapeutics. In line with these implications, tGLI1 has gained the ability to transcriptionally regulate several genes that are not regulated by GLI1 such as CD24, VEGF-A, MMP-2, and MMP-9, and potentially other genes identified via gene expression profiling. Interestingly, tGLI1 has retained its ability to undergo nuclear translocalization, activate GLI1-binding sequence, and increase PTCH1 expression.

Consistent with the gained ability to upregulate these important cancer-related genes, tGLI1-expressing tumor cells display an increased propensity to migrate, invade, growth, and stimulate the proliferation of vascular endothelial cells. When implanted in the flanks of nude mice, tGLI1-expressing cells formed bigger xenografts compared to those with GLI1. These emerging evidences suggest that tGLI1 may be a stronger mediator of several malignant phenotypes of human cancer. They also present a new paradigm in which some of the known GLI1 functions may be attributed to both GLI1 and tGLI1, or to a greater extent to tGLI1. Future efforts are urgently needed to generate experimental evidence to establish this exciting new paradigm.

An important future direction is to further decipher the roles that the three GLI1 isoforms in human cancers. Given the minute difference between the GLI1 and tGLI1 transcripts and proteins, previous detection of GLI1 may have masked the existence of tGLI1. Also very likely, the siRNA knockdown studies may have downregulated expression of both GLI1 and tGLI1 rather than exclusively, that of GLI1. Further complicating the issue is the facts that the molecular difference between the tGLI1 and GLI1 is only about 4.5 kDa (146 kDa for tGLI1 and 150 kDa for GLI1) and that commercially available antibodies appear to recognize both isoforms. Collectively, these notions and technical limitations have contributed to our inability to uncover the existence of tGLI1 for more than 20 years and also call for a need to develop tools to selectively detect/silence tGLI1 or GLI1. For this, we have optimized two complementary methods to distinguish the two isoforms: (a) RNA extraction followed by RT-PCR to amplify the exons 1–4 of the GLI1 and tGLI1 transcripts, and then by DNA sequencing of the PCR product to confirm the deleted region within tGLI1 (Cao *et al.*, 2012; Lo *et al.*, 2009), and (b) a SDS-PAGE condition that can effectively resolve the mobility of the two GLI1 and tGLI1 proteins (Cao *et al.*, 2012). Although these methods effectively and accurately detect tGLI1 versus GLI1 mRNAs and proteins, it is recognized that tGLI1-specific and GLI1-specific antibodies will be helpful as we decipher the functions of the two isoforms and the contributions of them to malignant biology of human cancers. This effort will also help determine the clinical correlations between tGLI1 and GLI1 expression profiles in patient tumors with patient prognosis and response to treatments.

The exciting discoveries of GLI1ΔN and tGLI1 also call for future investigations on several aspects of the two novel variants that are presently unknown. First, it is important to gain a greater understanding of the regulation of GLI1ΔN and tGLI1 by various components of the Hedgehog signaling pathway including SMO, SUFU, GLI2, and GLI3. Second, it is essential to characterize the extent to which the two new isoforms are regulated by the noncanonical pathways, in comparison to GLI1. Third, it is imperative to determine the molecular mechanisms by which tGLI1 regulates its target genes since the available evidence indicates that it can bind to the consensus GLI1-binding sequence and additionally bind to other gene promoters that do not have the sequence and are not regulated by GLI1. This effort will help identify additional molecular pathways and cellular processes that are subjected to tGLI1 regulation. Fourth, given the established facts that GLI1 is associated with tumor development and other important aspects of tumor biology, a critical future task is to determine the role that tGLI1 plays in these processes, in particular, its involvement in comparison to GLI1 and GLI1ΔN. Finally, in light of the emerging importance of the Hedgehog–GLI1 pathway as a target of anti-cancer therapy, it will be highly relevant to investigate whether tumors with different expression profiles of the three GLI1 isoforms differ in their sensitivity to Hedgehog-targeted therapy, in particular, those that are being evaluated in the clinical trials in cancer patients. The derived information will help stratify cancer patients for more effective treatments and also aid in identifying rationales for overcoming therapeutic resistance.

ACKNOWLEDGMENTS

This study was supported by the NIH grant K01-CA118423, DOD grants W81XWH-07-1-0390 and W81XWH-11-1-0600, the Beez Foundation, the Pediatric Brain Tumor Foundation, and Intramural Division of Surgical Sciences Dani P. Bolognesi, Ph.D. Award (to H. W. L).

REFERENCES

Abe, Y., Oda-Sato, E., Tobiume, K., Kawauchi, K., Taya, Y., Okamoto, K., Oren, M., and Tanaka, N. (2008). Hedgehog signaling overrides p53-mediated tumor suppression by activating Mdm2. *Proc. Natl. Acad. Sci. USA* **105,** 4838–4843.

Al-Hajj, M., Wicha, M. S., Benito-Hernandez, A., Morrison, S. J., and Clarke, M. F. (2003). Prospective identification of tumorigenic breast cancer cells. *Proc. Natl. Acad. Sci. USA* **100,** 3983–3988.

Arheden, K., Ronne, M., Mandahl, N., Heim, S., Kinzler, K. W., Vogelstein, B., and Mitelman, F. (1989). In situ hybridization localizes the human putative oncogene GLI to chromosome subbands 12q13.3-14.1. *Hum. Genet.* **82,** 1–2.

Baumann, P., Cremers, N., Kroese, F., Orend, G., Chiquet-Ehrismann, R., Uede, T., Yagita, H., and Sleeman, J. P. (2005). CD24 expression causes the acquisition of multiple cellular properties associated with tumor growth and metastasis. *Cancer Res.* **65,** 10783–10793.

Bergers, G., and Benjamin, L. E. (2003). Tumorigenesis and the angiogenic switch. *Nat. Rev. Cancer* **3,** 401–410.

Bigner, S. H., Wong, A. J., Mark, J., Muhlbaier, L. H., Kinzler, K. W., Vogelstein, B., and Bigner, D. D. (1987). Relationship between gene amplification and chromosomal deviations in malignant human gliomas. *Cancer Genet. Cytogenet.* **29,** 165–170.

Cai, Q., Li, J., Gao, T., Xie, J., and Evers, B. M. (2009). Protein kinase Cδ negatively regulates Hedgehog signaling by inhibition of Gli1 activity. *J. Biol. Chem.* **284,** 2150–2158.

Cao, X., Geradts, J., Dewhirst, M. W., and Lo, H. W. (2012). Upregulation of VEGF-A and CD24 gene expression by the tGLI1 transcription factor contributes to the aggressive behavior of breast cancer cells. *Oncogene* **31,** 104–115.

Chen, J. K., Taipale, J., Young, K. E., Maiti, T., and Beachy, P. A. (2002). Small molecule modulation of Smoothened activity. *Proc. Natl. Acad. Sci. USA* **99,** 14071–14076.

Clement, V., Sanchez, P., de Tribolet, N., Radovanovic, I., and Ruiz i Altaba, A. (2007). HEDGEHOG-GLI1 signaling regulates human glioma growth, cancer stem cell self-renewal, and tumorigenicity. *Curr. Biol.* **17,** 165–172.

Cui, D., Xu, Q., Wang, K., and Che, X. (2010). Gli1 is a potential target for alleviating multidrug resistance of gliomas. *J. Neurol. Sci.* **288,** 156–166.

Dahmane, N., Sanchez, P., Gitton, Y., Palma, V., Sun, T., Beyna, M., Weiner, H., and Ruiz i Altaba, A. (2001). The Sonic Hedgehog-Gli pathway regulates dorsal brain growth and tumorigenesis. *Development* **128,** 5201–5212.

Das, S., Harris, L. G., Metge, B. J., Liu, S., Riker, A. I., Samant, R. S., and Shevde, L. A. (2009). The hedgehog pathway transcription factor GLI1 promotes malignant behavior of cancer cells by up-regulating osteopontin. *J. Biol. Chem.* **284,** 22888–22897.

Dennler, S., Andre, J., Alexaki, I., Li, A., Magnaldo, T., ten Dijke, P., Wang, X. J., Verrecchia, F., and Mauviel, A. (2007). Induction of sonic hedgehog mediators by transforming growth factor-beta: Smad3-dependent activation of Gli2 and Gli1 expression in vitro and in vivo. *Cancer Res.* **67,** 6981–6986.

Dennler, S., André, J., Verrecchia, F., and Mauviel, A. (2009). Cloning of the human GLI2 promoter. *J. Biol. Chem.* **284,** 31523–31531.

Dunaeva, M., Michelson, P., Kogerman, P., and Toftgard, R. (2003). Characterization of the physical interaction of Gli proteins with SUFU proteins. *J. Biol. Chem.* **278,** 5116–5122.

Epstein, D. J., Marti, E., Scott, M. P., and McMahon, A. P. (1996). Antagonizing cAMP-dependent protein kinase A in the dorsal CNS activates a conserved Sonic hedgehog signaling pathway. *Development* **122,** 2885–2894.

Fiaschi, M., Rozell, B., Bergstrom, A., and Toftgard, R. (2009). Development of mammary tumors by conditional expression of GLI1. *Cancer Res.* **69,** 4810–4817.

Forus, A., Florenes, V. A., Maelandsmo, G. M., Meltzer, P. S., Fodstad, O., and Myklebost, O. (1993). Mapping of amplification units in the q13-14 region of chromosome 12 in human sarcomas: Some amplica do not include MDM2. *Cell Growth Differ.* **4,** 1065–1070.

Gordon, A. T., Brinkschmidt, C., Anderson, J., Coleman, N., Dockhorn-Dworniczak, B., Pritchard-Jones, K., and Shipley, J. (2000). A novel and consistent amplicon at 13q31 associated with alveolar rhabdomyosarcoma. *Genes Chromosomes Cancer* **28,** 220–226.

Hsieh, A., Ellsworth, R., and Hsieh, D. (2011). Hedgehog/GLI1 regulates IGF dependent malignant behaviors in glioma stem cells. *J. Cell. Physiol.* **226,** 1118–1127.

Huang, S., He, J., Zhang, X., Bian, Y., Yang, L., Xie, G., Zhang, K., Tang, W., Stelter, A. A., Wang, Q., Zhang, H., and Xie, J. (2006). Activation of the hedgehog pathway in human hepatocellular carcinomas. *Carcinogenesis* **27,** 1334–1340.

Huntzicker, E. G., Estay, I. S., Zhen, H., Lokteva, L. A., Jackson, P. K., and Oro, A. E. (2006). Dual degradation signals control Gli protein stability and tumor formation. *Genes Dev.* **20,** 276–281.

Hynes, M., Stone, D. M., Dowd, M., Pitts-Meek, S., Goddard, A., Gurney, A., and Rosenthal, A. (1997). Control of cell pattern in the neural tube by the zinc finger transcription factor and oncogene Gli-1. *Neuron* **19,** 15–26.

Ingham, P. W., and McMahon, A. P. (2001). Hedgehog signaling in animal development: Paradigms and principles. *Genes Dev.* **15,** 3059–3087.

Ingham, P. W., Nakano, Y., and Seger, C. (2011). Mechanisms and functions of Hedgehog signalling across the metazoa. *Nat. Rev. Genet.* **12,** 393–406.

Ji, Z., Mei, F. C., Xie, J., and Cheng, X. (2007). Oncogenic KRAS activates Hedgehog signaling pathway in pancreatic cancer cells. *J. Biol. Chem.* **282,** 14048–14055.

Jiang, J., and Struhl, G. (1998). Regulation of the Hedgehog and Wingless signalling pathways by the F-box/WD40-repeat protein Slimb. *Nature* **391,** 493–496.

Kang, S., Graham, J. M., Jr., Olney, A. H., and Biesecker, L. G. (1997). GLI3 frameshift mutations cause autosomal dominant Pallister-Hall syndrome. *Nat. Genet.* **15,** 266–268.

Kasper, M., Regl, G., Frischauf, A. M., and Aberger, F. (2006). GLI transcription factors: Mediators of oncogenic Hedgehog signalling. *Eur. J. Cancer* **42,** 437–445.

Keysar, S. B., and Jimeno, A. (2010). More than markers: Biological significance of cancer stem cell-defining molecules. *Mol. Cancer Ther.* **9,** 2450–2457.

Kim, H. J., Kim, J. B., Lee, K. M., Shin, I., Han, W., Ko, E., Bae, J. Y., and Noh, D. Y. (2007). Isolation of CD24(high) and CD24(low/-) cells from MCF-7: CD24 expression is positively related with proliferation, adhesion and invasion in MCF-7. *Cancer Lett.* **258,** 98–108.

Kimura, H., Stephen, D., Joyner, A., and Curran, T. (2005). Gli1 is important for medulloblastoma formation in Ptc1+/− mice. *Oncogene* **24,** 4026–4036.

Kinzler, K. W., and Vogelstein, B. (1990). The GLI gene encodes a nuclear protein which binds specific sequences in the human genome. *Mol. Cell. Biol.* **10,** 634–642.

Kinzler, K. W., Bigner, S. H., Bigner, D. D., Trent, J. M., Law, M. L., O'Brien, S. J., Wong, A. J., and Vogelstein, B. (1987). Identification of an amplified, highly expressed gene in a human glioma. *Science* **236,** 70–73.

Kinzler, K. W., Ruppert, J. M., Bigner, S. H., and Vogelstein, B. (1988). The GLI gene is a member of the Kruppel family of zinc finger proteins. *Nature* **332,** 371–374.

Kogerman, P., Grimm, T., Kogerman, L., Krause, D., Unden, A. B., Sandstedt, B., Toftgard, R., and Zaphiropoulos, P. G. (1999). Mammalian suppressor-of-fused modulates nuclear-cytoplasmic shuttling of Gli-1. *Nat. Cell Biol.* **1,** 312–319.

Kristiansen, G., Winzer, K. J., Mayordomo, E., Bellach, J., Schluns, K., Denkert, C., Dahl, E., Pilarsky, C., Altevogt, P., Guski, H., and Dietel, M. (2003). CD24 expression is a new prognostic marker in breast cancer. *Clin. Cancer Res.* **9,** 4906–4913.

Lee, J., Platt, K. A., Censullo, P., and Ruiz i Altaba, A. (1997). Gli1 is a target of Sonic hedgehog that induces ventral neural tube development. *Development* **124,** 2537–2552.

Lee, J. H., Kim, S. H., Lee, E. S., and Kim, Y. S. (2009). CD24 overexpression in cancer development and progression: A meta-analysis. *Oncol. Rep.* **22,** 1149–1156.

Li, X., Deng, W., Nail, C. D., Bailey, S. K., Kraus, M. H., Ruppert, J. M., and Lobo-Ruppert, S. M. (2006). Snail induction is an early response to Gli1 that determines the efficiency of epithelial transformation. *Oncogene* **25,** 609–621.

Li, C., Heidt, D. G., Dalerba, P., Burant, C. F., Zhang, L., Adsay, V., Wicha, M., Clarke, M. F., and Simeone, D. M. (2007). Identification of pancreatic cancer stem cells. *Cancer Res.* **67,** 1030–1037.

Liao, X., Siu, M. K., Au, C. W., Wong, E. S., Chan, H. Y., Ip, P. P., Ngan, H. Y., and Cheung, A. N. (2009). Aberrant activation of hedgehog signaling pathway in ovarian cancers: Effect on prognosis, cell invasion and differentiation. *Carcinogenesis* **30,** 131–140.

Liu, S., Dontu, G., Mantle, I. D., Patel, S., Ahn, N. S., Jackson, K. W., Suri, P., and Wicha, M. S. (2006). Hedgehog signaling and Bmi-1 regulate self-renewal of normal and malignant human mammary stem cells. *Cancer Res.* **66,** 6063–6071.

Lo, H. W., Hsu, S. C., Xia, W., Cao, X., Shih, J. Y., Wei, Y., Abbruzzese, J. L., Hortobagyi, G. N., and Hung, M. C. (2007). Epidermal growth factor receptor cooperates with signal transducer and activator of transcription 3 to induce epithelial-mesenchymal transition in cancer cells via up-regulation of TWIST gene expression. *Cancer Res.* **67,** 9066–9076.

Lo, H. W., Zhu, H., Cao, X., Aldrich, A., and Ali-Osman, F. (2009). A novel splice variant of GLI1 that promotes glioblastoma cell migration and invasion. *Cancer Res.* **69,** 6790–6798.

LoRusso, P. M., Rudin, C. M., Borad, M. J., Vernillet, L., Darbonne, W. C., Mackey, H., DiMartino, J. F., de Sauvage, F., Low, J. A., and Von Hoff, D. D. (2008). A first-in-human, first-in-class, phase (ph) I study of systemic Hedgehog (Hh) pathway antagonist, GDC-0449, in patients (pts) with advanced solid tumors. *J. Clin. Oncol.* **26,** 3516, (Meeting Abstracts).

Louro, I. D., Bailey, E. C., Li, X., South, L. S., McKie-Bell, P. R., Yoder, B. K., Huang, C. C., Johnson, M. R., Hill, A. E., Johnson, R. L., and Ruppert, J. M. (2002). Comparative gene expression profile analysis of GLI and c-MYC in an epithelial model of malignant transformation. *Cancer Res.* **62,** 5867–5873.

Lum, L., and Beachy, P. A. (2004). The Hedgehog response network: Sensors, switches, and routers. *Science* **304,** 1755–1759.

Ma, X., Chen, K., Huang, S., Zhang, X., Adegboyega, P. A., Evers, B. M., Zhang, H., and Xie, J. (2005). Frequent activation of the hedgehog pathway in advanced gastric adenocarcinomas. *Carcinogenesis* **26,** 1698–1705.

Ma, X., Sheng, T., Zhang, Y., Zhang, X., He, J., Huang, S., Chen, K., Sultz, J., Adegboyega, P. A., Zhang, H., and Xie, J. (2006). Hedgehog signaling is activated in subsets of esophageal cancers. *Int. J. Cancer* **118,** 139–148.

Mao, J., Maye, P., Kogerman, P., Tejedor, F. J., Toftgard, R., Xie, W., Wu, G., and Wu, D. (2002). Regulation of Gli1 transcriptional activity in the nucleus by Dyrk1. *J. Biol. Chem.* **277,** 35156–35161.

Meyer, M. J., Fleming, J. M., Lin, A. F., Hussnain, S. A., Ginsburg, E., and Vonderhaar, B. K. (2010). CD44posCD49fhiCD133/2hi defines xenograft-initiating cells in estrogen receptor-negative breast cancer. *Cancer Res.* **70,** 4624–4633.

Nakamura, K., Sasajima, J., Mizukami, Y., Sugiyama, Y., Yamazaki, M., Fujii, R., Kawamoto, T., Koizumi, K., Sato, K., Fujiya, M., Sasaki, K., Tanno, S., *et al.* (2010). Hedgehog promotes neovascularization in pancreatic cancers by regulating Ang-1 and IGF-1 expression in bone-marrow derived pro-angiogenic cells. *PLoS One* **5,** e8824.

Neill, G. W., Harrison, W. J., Ikram, M. S., Williams, T. D., Bianchi, L. S., Nadendla, S. K., Green, J. L., Ghali, L., Frischauf, A. M., O'Toole, E. A., Aberger, F., and Philpott, M. P. (2008). GLI1 repression of ERK activity correlates with colony formation and impaired migration in human epidermal keratinocytes. *Carcinogenesis* **29,** 738–746.

Ng, J. M., and Curran, T. (2011). The Hedgehog's tale: Developing strategies for targeting cancer. *Nat. Rev. Cancer* **11,** 493–501.

Nolan-Stevaux, O., Lau, J., Truitt, M. L., Chu, G. C., Hebrok, M., Fernandez-Zapico, M. E., and Hanahan, D. (2009). GLI1 is regulated through Smoothened-independent mechanisms in neoplastic pancreatic ducts and mediates PDAC cell survival and transformation. *Genes Dev.* **23,** 24–36.

Oka, N., Soeda, A., Inagaki, A., Onodera, M., Maruyama, H., Hara, A., Kunisada, T., Mori, H., and Iwama, T. (2007). VEGF promotes tumorigenesis and angiogenesis of human glioblastoma stem cells. *Biochem. Biophys. Res. Commun.* **360,** 553–559.

Pathi, S., Pagan-Westphal, S., Baker, D. P., Garber, E. A., Rayhorn, P., Bumcrot, D., Tabin, C. J., Blake Pepinsky, R., and Williams, K. P. (2001). Comparative biological responses to human Sonic, Indian, and Desert hedgehog. *Mech. Dev.* **106,** 107–117.

Pavletich, N. P., and Pabo, C. O. (1993). Crystal structure of a five-finger GLI-DNA complex: New perspectives on zinc fingers. *Science* **261,** 1701–1707.

Radhakrishna, U., Wild, A., Grzeschik, K. H., and Antonarakis, S. E. (1997). Mutation in GLI3 in postaxial polydactyly type A. *Nat. Genet.* **17,** 269–271.

Raffel, C., Jenkins, R. B., Frederick, L., Hebrink, D., Alderete, B., Fults, D. W., and James, C. D. (1997). Sporadic medulloblastomas contain PTCH mutations. *Cancer Res.* **57,** 842–845.

Rao, S. K., Edwards, J., Joshi, A. D., Siu, I. M., and Riggins, G. J. (2010). A survey of glioblastoma genomic amplifications and deletions. *J. Neurooncol.* **96**(2), 169–179.

Reifenberger, G., Reifenberger, J., Ichimura, K., Meltzer, P. S., and Collins, V. P. (1994). Amplification of multiple genes from chromosomal region 12q13-14 in human malignant gliomas: Preliminary mapping of the amplicons shows preferential involvement of CDK4, SAS, and MDM2. *Cancer Res.* **54,** 4299–4303.

Ricci-Vitiani, L., Pallini, R., Biffoni, M., Todaro, M., Invernici, G., Cenci, T., Maira, G., Parati, E. A., Stassi, G., Larocca, L. M., and De Maria, R. (2010). Tumour vascularization via endothelial differentiation of glioblastoma stem-like cells. *Nature* **468,** 824–828.

Riobo, N. A., Lu, K., Ai, X., Haines, G. M., and Emerson, C. P., Jr. (2006). Phosphoinositide 3-kinase and Akt are essential for Sonic Hedgehog signaling. *Proc. Natl. Acad. Sci. USA* **103,** 4505–4510.

Roberts, W. M., Douglass, E. C., Peiper, S. C., Houghton, P. J., and Look, A. T. (1989). Amplification of the gli gene in childhood sarcomas. *Cancer Res.* **49,** 5407–5413.

Roessler, E., Belloni, E., Gaudenz, K., Jay, P., Berta, P., Scherer, S. W., Tsui, L. C., and Muenke, M. (1996). Mutations in the human Sonic Hedgehog gene cause holoprosencephaly. *Nat. Genet.* **14,** 357–360.

Rudin, C. M., Hann, C. L., Laterra, J., Yauch, R. L., Callahan, C. A., Fu, L., Holcomb, T., Stinson, J., Gould, S. E., Coleman, B., LoRusso, P. M., Von Hoff, D. D., *et al.* (2009). Treatment of medulloblastoma with hedgehog pathway inhibitor GDC-0449. *N. Engl. J. Med.* **361,** 1173–1178.

Runz, S., Mierke, C. T., Joumaa, S., Behrens, J., Fabry, B., and Altevogt, P. (2008). CD24 induces localization of beta1 integrin to lipid raft domains. *Biochem. Biophys. Res. Commun.* **365,** 35–41.

Sasaki, H., Hui, C., Nakafuku, M., and Kondoh, H. (1997). A binding site for Gli proteins is essential for HNF-3beta floor plate enhancer activity in transgenics and can respond to Shh in vitro. *Development* **124,** 1313–1322.

Schreck, K. C., Taylor, P., Marchionni, L., Gopalakrishnan, V., Bar, E. E., Gaiano, N., and Eberhart, C. G. (2010). The Notch target Hes1 directly modulates Gli1 expression and Hedgehog signaling: A potential mechanism of therapeutic resistance. *Clin. Cancer Res.* **16,** 6060–6070.

Seto, M., Ohta, M., Asaoka, Y., Ikenoue, T., Tada, M., Miyabayashi, K., Mohri, D., Tanaka, Y., Ijichi, H., Tateishi, K., Kanai, F., Kawabe, T., *et al.* (2009). Regulation of the hedgehog signaling by the mitogen-activated protein kinase cascade in gastric cancer. *Mol. Carcinog.* **48,** 703–712.

Sheng, T., Li, C., Zhang, X., Chi, S., He, N., Chen, K., McCormick, F., Gatalica, Z., and Xie, J. (2004). Activation of the hedgehog pathway in advanced prostate cancer. *Mol. Cancer* **3,** 29.

Sheng, T., Chi, S., Zhang, X., and Xie, J. (2006). Regulation of Gli1 localization by the cAMP/protein kinase A signaling axis through a site near the nuclear localization signal. *J. Biol. Chem.* **281,** 9–12.

Shimokawa, T., Tostar, U., Lauth, M., Palaniswamy, R., Kasper, M., Toftgard, R., and Zaphiropoulos, P. G. (2008). Novel human glioma-associated oncogene 1 (GLI1) splice variants reveal distinct mechanisms in the terminal transduction of the hedgehog signal. *J. Biol. Chem.* **283,** 14345–14354.

Stecca, B., and Ruiz i Altaba, A. (2009). A GLI1-p53 inhibitory loop controls neural stem cell and tumour cell numbers. *EMBO J.* **28,** 663–676.

Stecca, B., Mas, C., Clement, V., Zbinden, M., Correa, R., Piguet, V., Beermann, F., and Ruiz, I. A. A. (2007). Melanomas require HEDGEHOG-GLI signaling regulated by interactions between GLI1 and the RAS-MEK/AKT pathways. *Proc. Natl. Acad. Sci. USA* **104,** 5895–5900.

Stein, U., Eder, C., Karsten, U., Haensch, W., Walther, W., and Schlag, P. M. (1999). GLI gene expression in bone and soft tissue sarcomas of adult patients correlates with tumor grade. *Cancer Res.* **59,** 1890–1895.

Taylor, M. D., Liu, L., Raffel, C., Hui, C. C., Mainprize, T. G., Zhang, X., Agatep, R., Chiappa, S., Gao, L., Lowrance, A., Hao, A., Goldstein, A. M., *et al.* (2002). Mutations in SUFU predispose to medulloblastoma. *Nat. Genet.* **31,** 306–310.

ten Haaf, A., Bektas, N., von Serenyi, S., Losen, I., Arweiler, E., Hartmann, A., Knuchel, R., and Dahl, E. (2009). Expression of the glioma-associated oncogene homolog (GLI) 1 in human breast cancer is associated with unfavourable overall survival. *BMC Cancer* **9,** 298.

Thomas, Z. I., Gibson, W., Sexton, J. Z., Aird, K. M., Ingram, S. M., Aldrich, A., Lyerly, H. K., Devi, G. R., and Williams, K. P. (2011). Targeting GLI1 expression in human inflammatory breast cancer cells enhances apoptosis and attenuates migration. *Br. J. Cancer* **104,** 1575–1586.

Villavicencio, E. H., Walterhouse, D. O., and Iannaccone, P. M. (2000). The sonic hedgehog-patched-gli pathway in human development and disease. *Am. J. Hum. Genet.* **67,** 1047–1054.

Von Hoff, D. D., LoRusso, P. M., Rudin, C. M., Reddy, J. C., Yauch, R. L., Tibes, R., Weiss, G. J., Borad, M. J., Hann, C. L., Brahmer, J. R., Mackey, H. M., Lum, B. L., *et al.* (2009). Inhibition of the hedgehog pathway in advanced basal-cell carcinoma. *N. Engl. J. Med.* **361,** 1164–1172.

Wang, G., Zhang, Z., Xu, Z., Yin, H., Bai, L., Ma, Z., Decoster, M. A., Qian, G., and Wu, G. (2010). Activation of the sonic hedgehog signaling controls human pulmonary arterial smooth muscle cell proliferation in response to hypoxia. *Biochim. Biophys. Acta* **1803,** 1359–1367.

Wong, A. J., Bigner, S. H., Bigner, D. D., Kinzler, K. W., Hamilton, S. R., and Vogelstein, B. (1987). Increased expression of the epidermal growth factor receptor gene in malignant gliomas is invariably associated with gene amplification. *Proc. Natl. Acad. Sci. USA* **84,** 6899–6903.

Xie, J., Murone, M., Luoh, S. M., Ryan, A., Gu, Q., Zhang, C., Bonifas, J. M., Lam, C. W., Hynes, M., Goddard, A., Rosenthal, A., Epstein, E. H., Jr., *et al.* (1998). Activating Smoothened mutations in sporadic basal-cell carcinoma. *Nature* **391,** 90–92.

Xie, J., Aszterbaum, M., Zhang, X., Bonifas, J. M., Zachary, C., Epstein, E., and McCormick, F. (2001). A role of PDGFRα in basal cell carcinoma proliferation. *Proc. Natl. Acad. Sci. USA* **98,** 9255–9259.

Xu, L., Kwon, Y.-J., Frolova, N., Steg, A., Yuan, K., Johnson, M., Grizzle, W., Desmond, R., and Frost, A. (2010). Gli1 promotes cell survival and is predictive of a poor outcome in ERα-negative breast cancer. *Breast Cancer Res. Treat.* **123,** 59–71.

Yang, L., Xie, G., Fan, Q., and Xie, J. (2010). Activation of the hedgehog-signaling pathway in human cancer and the clinical implications. *Oncogene* **29,** 469–481.

Yoon, J. W., Kita, Y., Frank, D. J., Majewski, R. R., Konicek, B. A., Nobrega, M. A., Jacob, H., Walterhouse, D., and Iannaccone, P. (2002). Gene expression profiling leads to identification of GLI1-binding elements in target genes and a role for multiple downstream pathways in GLI1-induced cell transformation. *J. Biol. Chem.* **277,** 5548–5555.

Zbinden, M., Duquet, A., Lorente-Trigos, A., Ngwabyt, S. N., Borges, I., and Ruiz i Altaba, A. (2010). NANOG regulates glioma stem cells and is essential in vivo acting in a cross-functional network with GLI1 and p53. *EMBO J.* **29,** 2659–2674.

Zhu, H., and Lo, H.-W. (2010). The human glioma-associated oncogene homolog 1 (GLI1) family of transcription factors in gene regulation and diseases. *Curr. Genomics* **11,** 238–245.

CHAPTER SEVEN

Gli-Similar Proteins: Their Mechanisms of Action, Physiological Functions, and Roles in Disease

Kristin Lichti-Kaiser, Gary ZeRuth, Hong Soon Kang, Shivakumar Vasanth, *and* Anton M. Jetten

Contents

Cell Biology Section, Division of Intramural Research, National Institute of Environmental Health Sciences, National Institutes of Health, Research Triangle Park, North Carolina, USA

Vitamins and Hormones, Volume 88
ISSN 0083-6729, DOI: 10.1016/B978-0-12-394622-5.00007-9

Abstract

Gli-similar (Glis) 1–3 proteins constitute a subfamily of Krüppel-like zinc-finger proteins that are closely related to members of the Gli family. Glis proteins have been implicated in several pathologies, including cystic kidney disease, diabetes, hypothyroidism, fibrosis, osteoporosis, psoriasis, and cancer. In humans, a mutation in the *Glis2* gene has been linked to the development of nephronophthisis (NPHP), a recessive cystic kidney disease, while mutations in *Glis3* lead to an extended multisystem phenotype that includes the development of neonatal diabetes, polycystic kidneys, congenital hypothyroidism, and facial dysmorphism. *Glis3* has also been identified as a risk locus for type-1 and type-2 diabetes and additional studies have revealed a role for Glis3 in pancreatic endocrine development, β-cell maintenance, and insulin regulation. Similar to Gli1–3, Glis2 and 3 have been reported to localize to the primary cilium. These studies appear to suggest that Glis proteins are part of a primary cilium-associated signaling pathway(s). It has been hypothesized that Glis proteins are activated through posttranslational modifications and subsequently translocate to the nucleus where they regulate transcription by interacting with Glis-binding sites in the promoter regions of target genes. This chapter summarizes the current state of knowledge regarding mechanisms of action of the Glis family of proteins, their physiological functions, as well as their roles in disease.

I. Introduction

Krüppel-like zinc-finger proteins, so named for their similarity to the *Drosophila* segmentation gene product Krüppel (Preiss *et al.*, 1985), belong to a large and evolutionarily conserved family of transcriptional regulators implicated in a broad range of cellular processes including proliferation, apoptosis, differentiation, and development (Pearson *et al.*, 2008). The characteristic feature of the Krüppel-like family is the presence of two or more classical Cys_2/His_2 zinc fingers that are separated by the conserved consensus sequence TGEKP(Y/F)X (Dang *et al.*, 2000). The zinc fingers constitute a DNA-binding domain that recognizes specific DNA elements in the promoter/regulatory regions of target genes (Kaczynski *et al.*, 2003). Outside of the zinc-finger region, there is relatively little homology among family members. Krüppel-like zinc-finger proteins exhibit diverse roles during embryonic development and in the adult are implicated in a variety of pathologies.

Gli and Zic constitute two closely related subfamilies of Krüppel-like zinc-finger proteins that contain five Cys_2/His_2 zinc-finger motifs and function as activators and/or repressors of gene transcription. Gli and Zic proteins are essential for normal embryonic development and have been implicated in a number of human diseases and cancers (Kasper *et al.*, 2006; Merzdorf, 2007; Mo *et al.*, 1997; Ruiz i Altaba, 1999; Sasaki *et al.*, 1999). Gli-similar (Glis) proteins, the focus of this review, are closely related to the Gli and Zic subfamilies (Kang *et al.*, 2010).

Glis1 was independently identified by two laboratories employing a yeast two-hybrid screen against a mouse kidney cDNA library using the ligand-binding domain of the retinoid-related orphan receptor, RORγ, as bait (Kim *et al.*, 2002) or by sequence homology to the Gli zinc-finger domain (Nakashima *et al.*, 2002). Subsequently, two additional family members exhibiting a high degree of homology to the Glis1 zinc-finger domain were identified and referred to as Glis2 and Glis3 (Kim *et al.*, 2003; Lamar *et al.*, 2001; Zhang and Jetten, 2001; Zhang *et al.*, 2002). Like the members of the Gli and Zic subfamilies, Glis1–3 share a highly homologous zinc-finger domain consisting of five Cys_2/His_2 motifs. The zinc-finger domain of Glis1 is 58% and 93% identical to that of Glis2 and Glis3, respectively (Kim *et al.*, 2002, 2003); however, Glis family members exhibit little sequence homology outside of their zinc-finger domains. The Glis proteins are conserved across species with homologues for Glis3 identified in *Drosophila* and *Oryzias latipes* and a high degree of homology existing between the human and mouse proteins (Furlong *et al.*, 2001; Hashimoto *et al.*, 2009).

The human *GLIS1–3* genes are located on chromosomes 1p32.3, 16p13.3, and 9p24.2 and encode for proteins that are approximately 65.9, 55.7, and 90kDa in size, respectively. *Glis1–3* are expressed in a temporal and spatial manner during embryonic development and in the adult are expressed in a tissue- and cell type-specific manner. Under normal circumstances, all three Glis family members are most abundantly expressed in the kidney. *Glis1* is further expressed in the placenta, brown adipose tissue, brain, thymus, colon, and testis (Kim *et al.*, 2002), while *Glis2* expression was also detected in the lung, prostate, colon, brain, liver, heart, and intestine (Zhang and Jetten, 2001; Zhang *et al.*, 2002). In addition to the kidney, *Glis3* expression was detected in the pancreas, thyroid, thymus, uterus, ovary, brain, and lung (Kim *et al.*, 2003; Senee *et al.*, 2006).

Characterization of Glis-null mouse models has identified several biological functions associated with the Glis proteins. For example, mice deficient in Glis2 develop nephronophthisis (NPHP), characterized by renal atrophy and fibrosis involving defects in epithelial–mesenchymal transition (EMT) within renal tubules (Attanasio *et al.*, 2007; Kang *et al.*, 2010; Kim *et al.*, 2008). Glis3-null mice develop neonatal diabetes, hypothyroidism, and polycystic kidney disease (PKD) (Kang *et al.*, 2009b; Watanabe *et al.*, 2009).

This chapter reviews what is currently known about the mechanisms of action of Glis proteins, their physiological function, and their roles in disease.

II. Mechanism of Action of Glis Proteins

A. Subcellular localization

Examination of fluorescent-tagged Glis proteins by confocal microscopy revealed that these proteins were predominantly localized to the nuclear compartment (Kim *et al.*, 2002, 2003; Zhang *et al.*, 2002). However, the mechanisms by which Glis proteins are directed to the nucleus are not yet fully understood. Analyses of several putative bipartite nuclear localization sequences within the Glis proteins, including a motif within Glis3 overlapping ZF5 that shares homology with the *bona fide* nuclear localization signal (NLS) of the *Drosophila* homologue of Gli, Cubitus interruptus (Ci) (Wang and Holmgren, 1999), demonstrated that these sequences are not required for nuclear localization (Beak *et al.*, 2008; Kim *et al.*, 2002; Zhang *et al.*, 2002). Deletion analyses suggested that the region containing ZF4 is essential for the nuclear localization of Glis3, while ZF3 is essential for Glis2 nuclear localization (Beak *et al.*, 2008; Vasanth *et al.*, 2011). Further, while disruption of the tetrahedral configuration of ZF4 ablated Glis3 nuclear localization, Glis2 localization was unaffected by corresponding mutations. These observations indicated that the nuclear localization of Glis2 and Glis3 seems to be dependent upon their zinc-finger domains. Future studies are required to understand the mechanisms by which these regions mediate nuclear localization.

B. DNA binding

In addition to their role in nuclear localization, the zinc-finger domains of the Glis proteins are essential for the interaction of Glis proteins with DNA. The Glis zinc-finger domains comprise five tandem Cys_2His_2 zinc fingers. Cys_2His_2 zinc fingers, the most common DNA-binding domain found in eukaryotes, each form a $\beta\beta\alpha$ structure and maintain a tetrahedral configuration stabilized by a zinc ion that interacts with the four cysteine and histidine residues (Brayer and Segal, 2008; Frankel *et al.*, 1987; Lee *et al.*, 1989). The specificity of the zinc fingers for defined DNA sequences is mediated by the alpha helices of each finger, which fits within the major groove of the DNA and recognizes overlapping short stretches of nucleotides (Elrod-Erickson *et al.*, 1996). Thus, modularly composed zinc fingers can be capable of recognizing a relatively diverse array of DNA elements. The sequence 5′-(G/C)TGGGGGG(A/C) was identified as the optimal Glis-binding site (GlisBS) for Glis3 *in vitro* (Beak *et al.*, 2008).

The tetrahedral configuration of each ZF was required for DNA binding. Given the great degree of homology between the zinc-finger domains of Gli and Glis proteins, it came as little surprise that the Glis proteins were also capable of interacting with the consensus Gli-binding site (GBS), 5′-GACCACCCA *in vitro* (Kim *et al.*, 2002, 2003; Lamar *et al.*, 2001; Nakashima *et al.*, 2002; Ruppert *et al.*, 1988; Vasanth *et al.*, 2011). However, both Glis3 and Gli1 bound the GlisBS with a higher affinity than the GBS (Beak *et al.*, 2008). The fact that Gli and Glis proteins can both bind the same elements allows for the possibility of cross talk between the Gli and Glis signaling pathways.

Binding of Glis proteins to specific DNA elements is likely not only determined by the specificity of individual residues within the alpha-helices of the zinc fingers but also affected by the cofactors that are part of the Glis transcription regulatory complex and the promoter context of the GlisBS. In addition, posttranslational modifications of Glis proteins might influence their binding specificity for GlisBS. The latter is supported by a recent study showing that the binding of Glis2 to GlisBS was abrogated by a phosphomimetic mutation of Ser^{245} within the loop of ZF3 both *in vitro* and in cultured cells (Vasanth *et al.*, 2011). The specific kinase(s) involved in the potential phosphorylation of Glis2 and its relevance to the physiological function of Glis2 *in vivo* have yet to be determined.

C. Transcriptional regulation by Glis proteins

The ability of Glis proteins to function as transcription factors requires not only the DNA-binding domain but also a transcriptional activation or repressor domain (Kang *et al.*, 2010). In that regard, several distinct regions outside the zinc-finger domain that mediate transcriptional regulation by Glis1–3 have been characterized and are depicted in Fig. 7.1 (Beak *et al.*, 2007, 2008; Kang *et al.*, 2009a, 2010; Kim *et al.*, 2002, 2003; Nakashima *et al.*, 2002). Glis3 was able to induce transactivation of a reporter gene under the control of the consensus GlisBS or GBS, whereas Glis1 and Glis2 were unable to activate the reporter gene very effectively. A potent transactivation domain (TAD) was identified within the C-terminus of Glis1 (Kim *et al.*, 2002). However, the transactivation by Glis1 required the removal of the N-terminus, suggesting the presumed presence of an N-terminal repressor domain. In contrast, full-length Glis1 was found to potently activate a reporter under the control of the mouse insulin 2 (m*Ins2*) promoter in both HEK293 cells and INS1 (832/13) cells (ZeRuth *et al.*, 2011), suggesting that Glis1 activity is dependent on the cell type and the promoter context. Further, Ca^{2+}-dependent calmodulin kinase IV (CaM-KIV), which has been reported to phosphorylate transcription factors such as serum response factor (SRF) and CAATT enhancer binding protein (CEBPα) enhancing their transcriptional activity (Marshall *et al.*, 2003;

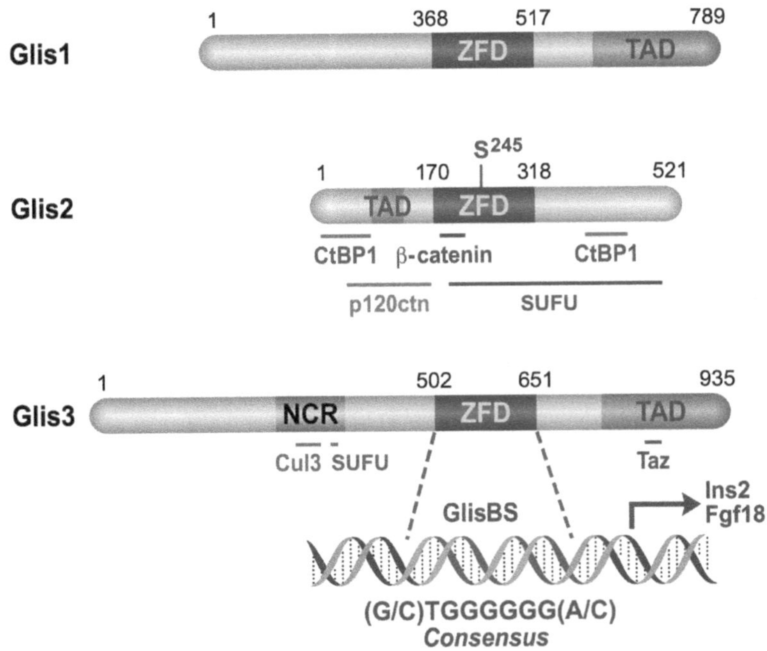

Figure 7.1 Schematic representation of Glis1–3 protein structure. The primary structure of Glis proteins and their identified domains are shown (ZFD, zinc-finger domain; TAD, transactivation domain; NCR, N-terminal conserved region that is shared with GLI proteins). Bold lines represent regions within Glis proteins required for their interaction with C-terminal-binding protein 1 (CtBP1), p120 catenin (p120ctn), Suppressor of Fused (SUFU), Cullin 3 (Cul3), and transcriptional coactivator with PDZ-binding motif (Taz) as indicated. The potential phosphorylation site, Serine^{245}, is specified in Glis2. Numbers indicate the amino acid position. (For color version of this figure, the reader is referred to the Web version of this chapter.)

Miranti *et al.*, 1995), increased the activation capabilities of Glis1 approximately fourfold (Kim *et al.*, 2002). The significance of CaMKIV to the transcriptional activity of Glis1 *in vivo* is unknown and elicits further study; however, given the noted ability of CaMKIV to phosphorylate and activate the ubiquitous coactivator, CREB-binding protein (CBP), it is possible that the enhancement of Glis1 transactivation function by CaMKIV involves the activation of CBP or another cofactor (Soderling, 1999).

A putative TAD and a potential repressor domain were also identified by monohybrid analyses within the N-terminus of Glis2 between amino acids 71 and 137 and extending into ZF1, respectively (Zhang *et al.*, 2002). While Glis2 was incapable of activating a $(GlisBS)_6$-controlled reporter, exogenous full-length Glis2 was capable of activating a reporter controlled by the m*Ins2* promoter in HEK293 cells (Vasanth *et al.*, 2011; ZeRuth *et al.*, 2011).

Maximal induction of the m*Ins2* promoter required the N- and C-terminus of Glis2, suggesting that both regions may be involved in the binding and recruitment of transcriptional mediators. In contrast, Glis2 was able to repress GLI1-mediated activation of a $(GliBS)_6$-driven reporter. This was likely due to competition between Glis2 and GLI1 for GliBS binding rather than Glis2 repressor activity given that the DNA-binding domain of Glis2 was sufficient to repress GLI1-mediated activation, while a DNA-binding mutant of Glis2 had no effect.

Glis3 is able to induce transactivation of a $(GlisBS)_6$-driven reporter as well as reporters under control of the m*Ins2* and *FGF18* promoters in various cellular contexts (Beak *et al.*, 2007; Kang *et al.*, 2009b; Yang *et al.*, 2009; ZeRuth *et al.*, 2011). A potent TAD was identified within the C-terminus of Glis3, which is indispensable for Glis3-mediated transactivation (Kang *et al.*, 2010; Kim *et al.*, 2003). N-terminal deletions of Glis3 increase its activation of the m*Ins2* promoter that attained a maximum with the ΔN302 mutant and decreased with subsequent deletions, suggesting the presence of a regulatory motif(s) within the N-terminus (Kang *et al.*, 2009b; Kim *et al.*, 2003; ZeRuth *et al.*, 2011). The molecular basis by which the N-terminus of Glis3 regulates Glis3 transactivation function is currently unknown. It is possible that deletions within the N-terminus eliminate a repressor or protein interaction domain. Alternatively, truncation of the N-terminus may lead to a conformational change that enhances interactions with cofactors and/or increases DNA-binding affinity.

D. Interacting partners

Transcriptional regulation by transcription factors is mediated through their interaction with coactivator and corepressor complexes. A number of proteins that interact with Glis proteins and mediate or modulate their transcriptional activity have been identified (Fig. 7.1). Among these, C-terminal-binding protein 1 (CtBP1) has been reported to interact with both the N- and the C-terminus of Glis2 (Kim *et al.*, 2005). CtBP1 interacts with and functions as a corepressor for a number of transcription factors. It mediates transcriptional repression by recruiting proteins with various histone modifying enzymatic activities, including histone deacetylases (HDACs) and histone lysine methyl transferases (Chinnadurai, 2007). In fact, HDAC3 has been identified as part of a Glis2–CtBP1 complex and may contribute to the ability of Glis2 to act as a transcriptional repressor (Kim *et al.*, 2005).

Glis2 has also been reported to interact with p120 catenin (p120ctn) and to promote its nuclear localization (Hosking *et al.*, 2007). Interaction with p120ctn was found to induce a Src kinase-dependent cleavage of Glis2 between zinc fingers 4 and 5. p120ctn has been reported to be bound to E-cadherin at cell–cell contacts and can also be associated with microtubules.

Overexpression of E-cadherin resulted in a reduction of Glis2 cleavage, while the induction of microtubule depolymerization enhanced Glis2 cleavage. Taken together, this suggests that p120ctn must be free in the cytosol to interact with Glis2 (Hosking *et al.*, 2007). While the physiological function of its interaction with Glis2 is currently unknown, p120ctn, a member of the Armadillo family of proteins, has emerged as a regulator of RhoGTPases and E-cadherin stability as well as a regulator of the transcriptional activity of the transcription factor Kaiso (Daniel, 2007; Reynolds, 2007; Xiao *et al.*, 2007). Because of its association with Kaiso, p120ctn has also been implicated as a modulator of the Wnt signaling pathway (Kim *et al.*, 2004; Na *et al.*, 2007; Park *et al.*, 2005) and therefore raises the possibility that Glis2 may be linked to Wnt signaling. The connection between Glis2 and p120ctn is interesting in the light that both proteins have been implicated in the control of EMT.

Further reinforcing the possibility of Glis2 cross talk with the Wnt pathway, Glis2 interacts with the armadillo repeats of β-catenin *via* its first zinc finger (Kim *et al.*, 2007). β-Catenin is an integral component of the canonical Wnt signaling pathway and, in combination with T-cell factor/lymphoid enhancer factor (TCF/LEF), positively regulates Wnt target genes (Akiyama, 2000). β-Catenin also regulates cell adhesion and migration through interactions with the cytoplasmic domains of cadherins and is therefore implicated in a number of cancers (Gavert and Ben-Ze'ev, 2007). Glis2 acts as a negative regulator of β-catenin and subsequently inhibits TCF/LEF signaling and the β-catenin-TCF/LEF-mediated activation of cyclin D1. Because of the essential role β-catenin plays in Wnt signaling, its interaction with Glis2 suggests cross talk between the Wnt and Glis signaling pathways.

Recently, Glis3 was reported to interact with the tumor suppressor and negative regulator of Hedgehog (Hh) signaling, Suppressor of Fused (SUFU), via a YGH motif (ZeRuth *et al.*, 2011). This motif is part of a 58-amino-acid region within the Glis3 N-terminus that exhibits high levels of homology to a corresponding region in the Ci/Gli proteins. A putative degron was identified within this conserved region of the Glis3 N-terminus upstream of the SUFU interaction motif, a loss of which resulted in a significant increase in Glis3 protein levels. Glis3 protein levels are stabilized by the proteasome inhibitor, MG132, suggesting that Glis3 is targeted for proteolytic degradation by the 26S proteasome. In fact, the E3-ubiquitin ligase scaffolding protein, Cullin 3 (Cul3), was shown to associate with the Glis3 N-terminus and its overexpression enhanced Glis3 polyubiquitination in cultured cells. SUFU, conversely, inhibited Glis3–Cul3 interaction and decreased the level of Glis3 polyubiquitination, thereby stabilizing Glis3 protein levels (ZeRuth *et al.*, 2011). Cul3 is typically associated with ubiquitin ligases by means of BTB-domain-containing proteins that target specific proteins for degradation. To date, a BTB-domain-containing protein that mediates interaction of Glis3 and Cul3 has not yet been identified.

In addition to stabilizing Glis3, SUFU has been shown to modulate Glis3-mediated transactivation of the *Ins2* promoter. SUFU also interacts with the C-terminus of Glis2 through a disparate mechanism, although the effects of the interaction are unknown and do not appear to influence Glis2 protein stability or transactivation function. SUFU has been reported to interact with members of the GLI family and shown to regulate the stability and processing of GLI2 and GLI3 into repressor or activator forms (Barnfield *et al.*, 2005; Cheng and Bishop, 2002; Humke *et al.*, 2010; Svard *et al.*, 2006; Wang *et al.*, 2010). In the absence of signaling, SUFU restrains Gli3 in the cytoplasm, promoting its processing into a repressor, while initiation of hedgehog signaling triggers the dissociation of SUFU and promotes the translocation of the activated form into the nucleus. SUFU has also been reported to interact with β-catenin to negatively regulate TCF/LEF signaling (Meng *et al.*, 2001). The interaction of Glis2–3 with SUFU yields the possibility of cross talk between the Glis, hedgehog, and Wnt signaling pathways.

Glis3 has further been reported to interact with the key insulin transcriptional regulatory factors, pancreatic duodenal homeobox 1 (Pdx1), v-maf musculoaponeurotic fibrosarcoma oncogene homolog A (MafA), and neuronal differentiation 1 (NeuroD1) resulting in synergistic activation of *Ins2* (Yang *et al.*, 2009). Transcriptional regulation and β-cell specific expression of the *insulin 2* gene occur via a 600bp promoter region upstream of the transcriptional start site (Edlund *et al.*, 1985; German *et al.*, 1992; Hanahan, 1985; Melloul *et al.*, 2002; Ohneda *et al.*, 2000a; Walker *et al.*, 1983). The most notable elements within this regulatory region are the A, E, and C-boxes, which bind Pdx1, β2/NeuroD1, and MafA, respectively (Melloul *et al.*, 2002; Ohneda *et al.*, 2000a). These transcription factors interact to form a stable DNA-binding complex capable of recruiting additional cofactors and transcriptional machinery (Ohneda *et al.*, 2000a,b; Peshavaria *et al.*, 1997; Petersen *et al.*, 1994). The precise dynamics of Glis3 interaction with insulin transcriptional regulators and the role Glis3 plays in the maintenance of adult pancreatic endocrine cells is not fully understood and requires further investigation.

Recent studies have demonstrated that Glis3 is able to interact with the transcriptional coregulator, transcriptional coactivator with PDZ-binding motif (TAZ, also known as WWTR1) through a PPXY site within the Glis3 C-terminus (Kang *et al.*, 2009a). TAZ interaction with Glis3 modestly enhances Glis3-mediated activation of a $(\text{GlisBS})_6$-controlled reporter. The interaction between Glis3 and TAZ and the observation that loss of either TAZ or Glis3 expression lead to PKD suggest a possible common link in the development of this phenotype (Chan *et al.*, 2008; Hossain *et al.*, 2007; Kang *et al.*, 2009a; Makita *et al.*, 2008; Senee *et al.*, 2006). TAZ is part of the Hippo signaling pathway (Lei *et al.*, 2008) that has been implicated in the regulation of cell proliferation, EMT, and planar cell polarity (PCP), cellular processes that have been implicated in the formation of polycystic kidneys.

Thus, disruption in Glis3–TAZ interaction might result in defects in these functions and be causally involved in the development of polycystic kidneys.

E. Glis signaling and the primary cilium

In addition to the nucleus, both Glis2 and Glis3 have been reported to localize to the primary cilium (Attanasio *et al.*, 2007; Hashimoto *et al.*, 2009; Kang *et al.*, 2009a, 2010). Thus far, no evidence has emerged linking Glis1 to the primary cilium. The primary cilium, a microtubule-based, hair-like projection present on nearly all mammalian cells, extends from the apical surface and functions as a sensory organelle (Goetz and Anderson, 2010). It serves as a signaling hub for an increasing number of distinct signaling pathways such as phototransduction, mechano-, osmo-, and chemosensation. Key components of the canonical hedgehog, Wnt, platelet-derived growth factor, and PCP pathways have also been shown to localize within the primary cilium, which has been demonstrated as indispensable for their proper signaling (Berbari *et al.*, 2009). Moreover, certain G-protein-coupled receptors (GPCRs), including somatostatin receptor 3 (Sstr3), melanin-concentrating hormone receptor 1 (Mchr1), and serotonin receptor 6 (Htr6), have been reported to localize to the ciliary membrane in various cell types (Berbari *et al.*, 2008a; Iwanaga *et al.*, 2011; Stanic *et al.*, 2009). The trafficking of many signaling receptors to the ciliary membrane appears to be mediated by the BBsome, a large protein complex containing specific Bardet–Biedl syndrome (BBS) proteins (Berbari *et al.*, 2008b; Domire *et al.*, 2010; Goetz and Anderson, 2010; Nachury *et al.*, 2010). Once inside the ciliary compartment, many essential ciliary proteins utilize a modified system of microtubule associated transport termed intraflagellar transport (IFT). Anterograde transport of proteins toward the cilium tip requires the kinesin II motor proteins, while retrograde transport back toward the cilium base requires dynein motor proteins (Blacque *et al.*, 2008; Goetz and Anderson, 2010; Rosenbaum and Witman, 2002). Since many of the proteins necessary for cilia construction utilize IFT, disruption of IFT components inhibits ciliogenesis or results in the deformation of cilia (Pedersen and Rosenbaum, 2008). Further, disruption of IFT proteins affects cilia-related signaling pathways such as several aspects of sonic hedgehog (Shh)–Gli signaling, including the movement of the downstream effectors, Gli2–3, suggesting that they depend on a functional IFT (Haycraft *et al.*, 2005; Liu *et al.*, 2005; May *et al.*, 2005; Qin *et al.*, 2011). Whether or not Glis proteins associate with IFT motors or whether IFT is required for proper Glis signaling is currently unknown and requires further investigation.

It appears likely that Glis proteins are part of a primary cilium-associated signaling pathway, and although the functional significance of Glis protein accumulation in the primary cilium has not yet been established, the role that the primary cilium plays in Shh/Gli signaling has been fairly well characterized (Goetz and Anderson, 2010). In the absence of Shh, its receptor Patched (Ptch1) is localized to the primary cilium and prevents ciliary localization of Smoothened (Smo). Shh binding to Ptch1 results in its exit from the ciliary compartment and relieves repression of Smo, allowing it to enter the cilium (Corbit *et al.*, 2005; Rohatgi *et al.*, 2007). The mutually exclusive presence of Ptch1 or Smo in the cilium regulates the processing of Gli3 into its activator or repressor forms (Haycraft *et al.*, 2005; Huangfu *et al.*, 2003; Liu *et al.*, 2005). Modified Gli proteins then exit the cilium and translocate to the nucleus where they regulate the transcription of target genes. We proposed previously that the primary cilium might play a similar role in the regulation of Glis. The activity of Glis proteins might be regulated by external signals received through receptors (e.g., GPCRs) localized to the primary ciliary membrane that control Glis activity by promoting their proteolytic cleavage or regulating other posttranslational modifications. After activation, Glis proteins might then translocate to the nucleus where they regulate the transcription of target genes and subsequently influence various physiological processes (Fig. 7.2). Defects in the structure or synthesis of the primary cilium and abnormalities in primary cilium-associated signaling pathways have been implicated in several pathologies collectively referred to as ciliopathies, including disorders with various sensory defects and cystic renal diseases (Fliegauf *et al.*, 2007). Because of their connection with the primary cilium, it is not surprising that defective Glis signaling is associated with the development of several ciliopathies.

III. Role of Glis Proteins in Renal Physiology and Pathology

A. Glis2, nephronophthisis, and mesenchymal–epithelial transition

All three Glis proteins are highly expressed in adult kidney and exhibit a temporal and spatial pattern of expression during embryonic kidney development. At E9.0 of murine development, the ureteric bud develops from budding of the caudal portion of the mesonephric Wolffian duct and invades the metanephric mesenchyme (Michos, 2009; Wilson, 2008). This is followed by extensive branching of the ureteric bud, giving rise to the renal collecting tubules. At E11.5, the mesenchymal cells condense and undergo mesenchymal–epithelial transition (MET) to form comma and S-shaped bodies that by E13.5 begin to develop into the glomeruli and renal tubules. Reciprocal cross talk between the metanephric mesenchyme and the ureteric bud plays an important role in the control of MET and ureteric bud branching. During murine metanephric development,

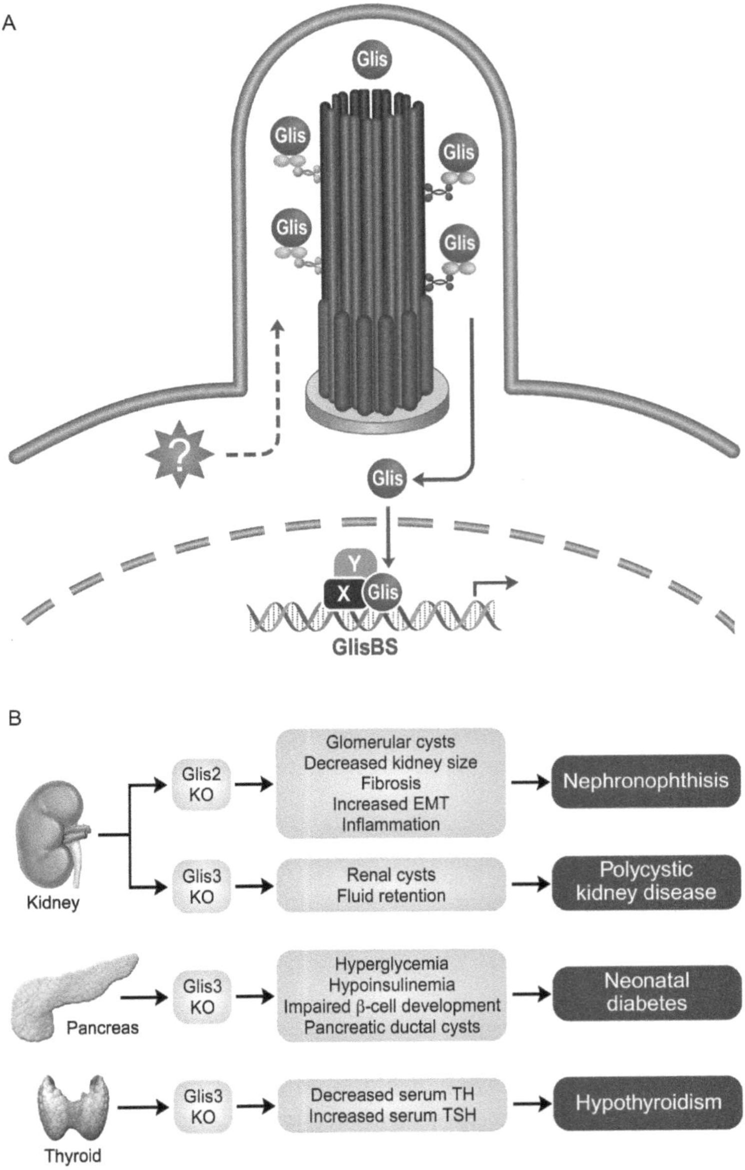

Figure 7.2 Schematic representation of Glis signaling and its role in disease. Glis2 and 3 localize to the primary cilium and defective Glis signaling is associated with the development of several ciliopathies. (A) An as-of-yet unidentified upstream signal may regulate the subcellular localization and activation of Glis proteins. During activation in the primary cilium, Glis proteins may undergo posttranslational modifications or proteolytic processing, and upon translocation to the nucleus, regulate the transcription of target genes through interactions with Glis-binding sites (GlisBSs) and transcriptional coregulators (X and Y). (B) Disruption of Glis signaling in the kidney, pancreas, and thyroid leads to the development of nephronophthisis and polycystic kidney disease, neonatal diabetes, and hypothyroidism, respectively. (For color version of this figure, the reader is referred to the Web version of this chapter.)

Glis2 is most highly expressed in the ureteric bud, whereas mesenchymal tissue, comma- and S-shaped bodies expressed at low levels or were devoid of Glis2 (Attanasio *et al.*, 2007; Zhang *et al.*, 2002). In adult kidneys, Glis2 was detected in the epithelial cells of all segments of the renal tubule and in the epithelial cells of Bowman's capsule, but not in glomeruli, endothelial, or mesenchymal cells (Attanasio *et al.*, 2007). Recent studies reported that loss of Glis2 function in humans and mice leads to the development of NPHP, a recessive cystic kidney disease that is the most common genetic cause of end-stage renal disease in children and young adults (Hildebrandt and Zhou, 2007). NPHP is characterized by tubular atrophy, fibrosis, and glomerular cyst formation which progressively leads to dramatic changes in renal architecture and ultimately end-stage kidney disease. Through positional cloning, mutations in nine distinct genes (*NPHP1–9*) have been linked to NPHP (Hildebrandt and Zhou, 2007; Simms *et al.*, 2011). Mutations in the NPHP1 gene account for approximately 21% of all NPHP cases, whereas mutations in NPHP2–9 each account for less than 9% of cases. Mutations that abolish the splice site within exon 5 of Glis2, which is also referred to as NPHP7, were linked to NPHP in three patients from a single family (Attanasio *et al.*, 2007). Similarly, Glis2-null mice develop renal atrophy and tubulointerstitial fibrosis characteristic of NPHP at an early age that is accompanied by increased inflammation. The severity of the symptoms progresses with age as evidenced by increased blood urea nitrogen (BUN) and creatinine levels and the development of proteinuria, which ultimately lead to renal failure and premature death (Attanasio *et al.*, 2007; Kim *et al.*, 2008).

The progressive tubulointerstitial fibrosis associated with NPHP7 is accompanied by increased accumulation of extracellular matrix (ECM) components produced by myofibroblasts. The increase in myofibroblasts during fibrosis may occur through several mechanisms, including differentiation of fibroblasts into myofibroblast, EMT, infiltration of fibrocytes, or a combination of the three processes. In EMT, epithelial cells transdifferentiate into myofibroblasts that migrate into the interstitium (Liu, 2009; Wada *et al.*, 2007; Wynn, 2007; Zeisberg and Kalluri, 2004). Gene expression profile analysis of kidneys from Glis2-null and wild-type mice revealed increased expression of a number of genes that play a critical role in EMT, including transforming growth factor β (Tgfβ), vimentin, matrix metallopeptidase 14 (Mmp14), connective tissue growth factor (Ctgf), Snail, and Slug (Attanasio *et al.*, 2007; Kim *et al.*, 2008). These data suggested that the induction of renal fibrosis in Glis2-null mice is mediated through EMT in renal tubule epithelial cells. These observations further implied that Glis2 may act as a repressor of EMT and EMT-related gene expression. Additional research is required to determine the precise molecular mechanism by which Glis2 regulates the maintenance of normal renal structure and function and to determine its role in controlling EMT.

B. Glis3 and polycystic kidney disease

Glis3 deficiency in humans and mice is associated with an extended multi-system phenotype that includes the development of PKD (Dimitri *et al.*, 2011; Kang *et al.*, 2009b; Senee *et al.*, 2006; Taha *et al.*, 2003; Watanabe *et al.*, 2009). Further, a mutation in the *pc* gene, an ortholog of Glis3, has been associated with the development of polycystic kidneys in medaka (Hashimoto *et al.*, 2009; Kang *et al.*, 2010).

PKD is a common heritable kidney disease characterized by the formation of large fluid-filled cysts in the glomerulus, renal tubules, and collecting ducts. Autosomal dominant PKD, the most common PKD in humans, results from mutations in *PKD1* and *PKD2*, which encode polycystin-1 and -2, respectively (Gallagher *et al.*, 2010; Harris and Torres, 2009). Autosomal recessive PKD, which is mostly associated with young children, is caused by mutations in PKHD1, which encodes polyductin/fibrocystin (Al-Bhalal and Akhtar, 2008; Bergmann *et al.*, 2004). At E14.5 of kidney development, Glis3 mRNA is highly expressed in the branches of the ureteric bud of the metanephros (Kim *et al.*, 2003), while in the adult mouse kidney, it is expressed in the epithelia of the collecting ducts, renal tubules, and Bowman's capsule (Hashimoto *et al.*, 2009; Kang *et al.*, 2009a). Histological analyses of Glis3-null mouse kidneys showed dilation of Bowman's spaces as early as E14.5. Loss of Glis3 does not appear to affect the extent of ureteric bud branching, and the renal phenotype became more pronounced at later stages of mouse development. By postnatal day 3 (PND3), major glomerular cysts were observed along with the formation of cystic renal tubules and dilation of collecting ducts (Kang *et al.*, 2009a). These observations indicate that Glis3 plays a critical role in the maintenance of normal renal architecture and function (Kang *et al.*, 2010).

C. Links between Glis signaling, the primary cilium, and renal disease

The molecular mechanisms that lead to cyst formation are still not fully understood; however, it has become evident that there is a strong causal relationship between the primary cilium and cystic renal diseases, as defects in the formation of the primary cilium or in primary cilium-associated signaling pathways result in the development of cystic kidney diseases, including PKD and NPHP (Hildebrandt and Zhou, 2007; Hildebrandt *et al.*, 2011; Yoder, 2007). Moreover, many proteins involved in NPHP and PKD, including NPHP1–5 and polycystins 1 and 2, have been localized to the primary cilium. Thus, the development of PKD and NPHP due to Glis3 and Glis2 deficiency, respectively, is consistent with an association between Glis signaling pathways and the primary cilium (Fig. 7.2). Although a reduction in the percentage of primary cilium-containing cells was observed in renal cysts, Glis3 deficiency does not result in the loss of the primary cilium in renal tubule epithelial cells, implying that Glis3 is not required for ciliogenesis (Hashimoto *et al.*, 2009; Kang *et al.*, 2009a; Kim *et al.*, 2008). The reduction in primary cilium-containing cells is likely a consequence rather than a cause of renal cyst formation.

Although the primary cilium plays a critical role in the regulation of many developmental processes (Bisgrove and Yost, 2006; Eggenschwiler and Anderson, 2007), the function of the primary cilium and its relationship to the development of cystic kidney disease are not fully understood. A role for the primary cilia in sensing fluid flow has been suggested by a number of laboratories (Nauli and Zhou, 2004; Praetorius and Spring, 2001; Schwartz *et al.*, 1997), and mechanosensation by the primary cilium has been shown to result in increased levels of intracellular Ca^{2+} concentration both in cultured cells and in collecting ducts of kidney (Nauli *et al.*, 2003). Alternatively, the primary cilium contributes to the control of proper kidney development and maintenance by regulating cell proliferation, PCP, and oriented cell division (OCD). Several protein kinases, including the mTOR, MAPK, Akt pathways, and Wnt signaling, have been implicated in these processes (McNeill, 2009; Menezes and Germino, 2009).

PCP, which is defined as the polarization of a field of cells within the plane of a sheet, has been implicated in postnatal development of nephrons as well as in cystic kidney diseases (Fischer and Pontoglio, 2009). Disruption of PCP genes, including Vang-like 2 (Vangl2) and four-jointed box 1 (Fjx1), has been shown to result in the development of tubular cysts. Kidney tubules undergo extensive proliferation in the developing kidney. The elongation of the tubule in the longitudinal direction is driven by cell division along the axis of the tubule, a process referred to as OCD. Non-directional (misoriented) cell division has been proposed as a mechanism of cyst formation; however, cyst formation has been observed at stages of renal development before cells undergo OCD, suggesting that additional mechanisms may be involved (Karner *et al.*, 2009; Nishio *et al.*, 2010). Defects in non-canonical Wnt signaling have been reported to result in cystic kidney disease (Lancaster and Gleeson, 2010). Whether defects in PCP, OCD, and Wnt signaling play a role in the cystogenic phenotype associated with deficiencies Glis2 and Glis3 requires further study.

IV. Role of Glis3 in Pancreas Physiology and Pathology

A. Development of diabetes

In humans, mutations in the gene encoding GLIS3 are associated with a rare syndrome characterized by neonatal diabetes and congenital hypothyroidism with eight affected cases from five families reported to date (Dimitri *et al.*, 2011; Kang *et al.*, 2010; Senee *et al.*, 2006). Depending on the nature of the mutation, additional features include hepatic fibrosis, congenital glaucoma, PKD, facial dysmorphism, bilateral sensorineural deafness, and osteopenia (Dimitri *et al.*, 2011; Senee *et al.*, 2006; Taha *et al.*, 2003). According to several human genome-wide association studies (GWAS), *GLIS3* has also been linked to aberrant glucose regulation and reduced β-cell function, and was identified as a risk locus for both type-1 and type-2 diabetes (Barker *et al.*, 2011; Barrett *et al.*, 2009; Boesgaard *et al.*, 2010; Dupuis *et al.*, 2010; Hu *et al.*, 2010). In accordance with the human studies, two independent laboratories have reported that Glis3-null mice developed neonatal diabetes characterized by hyperglycemia and hypoinsulinemia (Kang *et al.*, 2009b; Watanabe *et al.*, 2009). These findings suggest that GLIS3 contributes to the maintenance of endocrine functions in the pancreas, while aberrant GLIS3 function is associated with the development of diabetes (including neonatal, type-1, and type-2). Thus, Glis3-null mice provide an excellent functional model system in which to study the biological role of Glis3 in diabetes.

B. Pancreatic β-cell development and maintenance

The pancreas exhibits important exocrine and endocrine functions. It contains three major cell compartments: the acini, which produce digestive enzymes; the ducts through which the secretory enzymes are transported into the gastrointestinal tract; and the islets of Langerhans (Leung, 2010). The islets produce and secrete various hormones, including glucagon, insulin, pancreatic polypeptide, somatostatin, and ghrelin from α, β, γ, δ, and ε cells, respectively. Although islets compose a small portion (roughly 5–10%) of the pancreas, they are crucial for the maintenance of pancreas function and endocrine hormone signaling, including the control of glucose homeostasis. Glis2 and Glis3 are highly expressed in pancreatic islets, specifically in the insulin-producing β-cells. Glis3 is also highly expressed in the pancreatic ducts (Kang *et al.*, 2009b; Senee *et al.*, 2006; Watanabe *et al.*, 2009).

The mouse pancreas develops from two buds of the foregut into the dorsal and ventral pancreas at approximately embryonic day 9 (E9) and then forms a single organ after rotation and fusion between E13 and E14 (Guney and Gannon, 2009; Pan and Wright, 2011). The major components of the mature pancreas are not discernible until E16.5. For example, islet-like structures are formed at E16.5 and continue to develop and mature until 2–3 weeks after birth. Several genetically modified mouse models have been used to identify the hierarchy of the signaling pathways and transcription factors critical for pancreatic lineage determination (Jorgensen *et al.*, 2007; Murtaugh and Melton, 2003). At E9, epithelial cells of the dorsal and ventral pancreas, regarded as pancreatic progenitors, express Pdx1, Nkx2.2, pancreatic transcription factor 1a (Ptfla), and carboxypeptidase A1 (Cpa1)

(Burlison *et al.*, 2008; Zhou *et al.*, 2007). In fact, Pdx1-null mice fail to develop a pancreas and die shortly after birth (Jonsson *et al.*, 1994). Notch signaling is also crucial for pancreas development at multiple stages. Under normal circumstances, reduced Notch signaling induces the expression of neurogenin 3 (Ngn3)-positive endocrine progenitor cells from pancreatic progenitors, which subsequently become committed to the various endocrine lineages (Apelqvist *et al.*, 1999; Guney and Gannon, 2009; Pan and Wright, 2011). Ngn3-deficient embryos expressed Pdx1^{+} pancreatic progenitor cells at E11.5 but lack endocrine cells at birth (Gradwohl *et al.*, 2000). Finally, transcription factors including NeuroD1, Pax6, Pax4, Nkx2.2, and Nxk6.1 play a critical role in the further development of endocrine progenitors into pancreatic β-cells (Bernardo *et al.*, 2008; Guney and Gannon, 2009; Pan and Wright, 2011).

Glis3-null mice die within 1 week after birth likely due to neonatal diabetes. Islet size and the expression of several β-cell markers, including insulin, glucose transporter type 2 (Glut2), MafA, and Nkx6.1, were dramatically decreased in Glis3-null mice. Other hormones including somatostatin and pancreatic polypeptide were also significantly reduced in pancreas of Glis3-null mice, while the expression of acini-specific genes was not changed (Kang *et al.*, 2009b). The depletion of β-cells was not due to increased apoptosis, as the number of apoptotic cells was not changed in the pancreas of Glis3-null compared to wild-type mice (Kang *et al.*, 2009b; Watanabe *et al.*, 2009). Given that the expression of Pdx1 and Cpa1 was not changed in the pancreas of Glis3-null compared to wild-type embryos at E13.5 and E14.5, Glis3 does not appear to play a critical role in early pancreatic progenitor cell maintenance or development. However, the expression of Ngn3 was significantly decreased in Glis3-null compared to wild-type pancreas at E15.5, PND0, and PND3 (Kang *et al.*, 2009b). As Ngn3 is known to be a key transcription factor for the development of endocrine progenitor cells, these data suggest that Glis3 plays a critical role in the maintenance or proliferation of endocrine progenitor cells and/or in the specification or development of endocrine cells, particularly β-cells (Fig. 7.3; Kang *et al.*, 2010).

Hh signaling molecules including Indian hedgehog (Ihh), Desert hedgehog (Dhh), and Ptch1 are expressed in developing pancreas of embryos. A recent study showed that Hh signaling molecules including Smo, Ptch1, and the downstream effector, Gli, localized to the primary cilia in pancreatic epithelium, and that activated Hh signaling induced the expansion of ductal and endocrine progenitor cells, suggesting that a certain level of Hh signaling is required for the development of endocrine cells (Cervantes *et al.*, 2010; Chen *et al.*, 2009; Tukachinsky *et al.*, 2010). In fact, activation of Hh signaling induced Pdx1-dependent insulin expression (Thomas *et al.*, 2000, 2001), and the expression of insulin was decreased in pancreas-specific Smo-null mice (Lau and Hebrok, 2010). Given that genes encoding Hh signaling proteins including Ptch1, Smo, Dhh, and Ihh showed similar expression

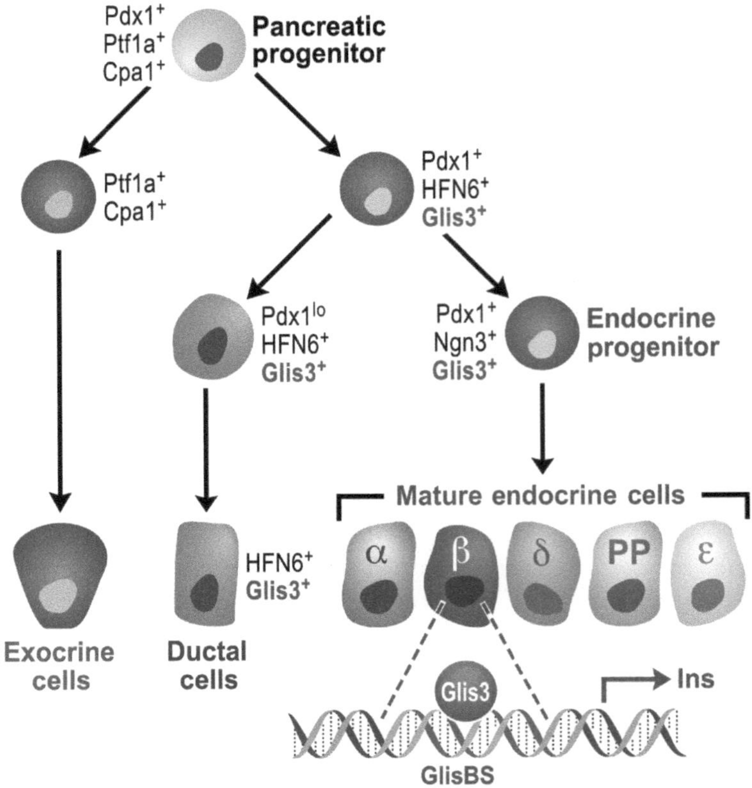

Figure 7.3 Schematic representation of the role of Glis3 in pancreatic cell lineage determination. Pancreatic progenitor cells, expressing Pdx1, Ptf1a, and Cpa1, can differentiate into exocrine, ductal, and proendocrine cell lineages. The expression of Glis3 is induced during the second transition of pancreas development and remains highly expressed in endocrine progenitor cells, ductal cells, and in mature β-cells where Glis3 can regulate insulin gene expression. In mice, loss of Glis3 inhibits the generation of β-cells and leads to the development of neonatal diabetes. (For color version of this figure, the reader is referred to the Web version of this chapter.)

pattern with Glis3 in islets and ducts of adult pancreas, there could be a possibility for cross talk between Hh and Glis3.

C. Glis-Mediated Regulation of Insulin

Many transcription factors involved in pancreatic development also play a critical role in the regulation of insulin gene expression. For example, Pdx1, NeuroD, MafA, and Pax4/6 regulate the expression of the insulin gene through their binding the A-box, E-box, C-box, and C2 element of the insulin promoter, respectively (Cerf, 2006; Melloul *et al.*, 2002). Given the role of Glis3 in β-cell development and that the expression of *Ins2* was dramatically decreased in pancreas of Glis3-null mice at PND3 and in Glis3-null embryos as early as E15.5, Glis3 may also directly regulate insulin gene expression in mature β-cells (Fig. 7.3).

Ins2 expression in rat insulinoma 832/13 cells is markedly increased by Glis3 overexpression and decreased when Glis3 is knocked down by siRNA. Further, Glis3 binds to the m*Ins2* promoter in these cells as indicated by chromatin immunoprecipitation (Yang *et al.*, 2009). Indeed, promoter analysis showed that the human *INS* and mouse *Ins2* genes contain two well-conserved GlisBS within 600bp upstream of the transcription start site. *In vitro* DNA-binding assays and cell-based reporter gene assays provided further support that Glis3 directly regulates the expression of human and mouse insulin genes through these two GlisBS sequences (Kang *et al.*, 2009b; Yang *et al.*, 2009). Glis2 can bind a similar GlisBS as Glis3. Glis2 activated the mouse m*Ins2* promoter *in vitro* in non-beta cells through the same two GlisBS as Glis3 (Vasanth *et al.*, 2011). Thus, Glis2, which is expressed in mouse pancreatic islets, may also play a role in insulin regulation in the mouse pancreas. Glis2-null mice did not show any diabetic or abnormal pancreatic phenotype; however, it is possible that Glis3 may compensate for the loss of Glis2 function in pancreas of Glis2-null mice.

Glis3 is also reported to physically and functionally interact with other known regulators of insulin gene expression, including MafA, NeuroD, and Pdx1. Glis3 coimmunoprecipitated with MafA, NeuroD, and Pdx1 in insulinoma 832/13 cells, and cell-based reporter assays indicated that Pdx1 and, to a lesser degree, MafA had synergistic effects with Glis3 on insulin promoter activity (Yang *et al.*, 2009). Taken together, these results reveal potential for cross talk between Glis2, Glis3, and other known regulators of insulin gene expression. Further examination is required to determine the mechanism by which Glis2 and Glis3 may directly regulate insulin gene expression.

Even though insulin has been identified as a direct Glis3 target gene, Glis3 may also regulate insulin expression by indirect mechanisms. The expression of MafA was significantly decreased in the pancreas of Glis3-null mice at PND3 and in insulinoma 832/13 cells when Glis3 is knocked down by siRNA (Kang *et al.*, 2009b; Yang *et al.*, 2009). Although MafA is known to regulate the expression of *Ins2*, MafA-null mice do not exhibit neonatal diabetes. At an early age, no difference was observed in insulin content in the pancreas of MafA-null compared to wild-type mice, but MafA-null mice were intolerant to glucose challenge and developed diabetes with age. In addition, glucose-stimulated insulin secretion (GSIS) was impaired in the islets of MafA-null compared to wild-type mice (Zhang *et al.*, 2005). These results suggest that MafA may be more critical for insulin secretion than insulin production. *Abcc8*, another gene decreased in PND3 pancreas of Glis3-null mice, encodes the sulfonylurea receptor 1 (Sur1), which plays a role in the regulation of insulin secretion (Aguilar-Bryan *et al.*, 1995; Kang *et al.*, 2009b; Koehn *et al.*, 2008). These results indicate that in addition to its function in the control of insulin gene expression, Glis3 may also play a role, either directly or indirectly, in regulating insulin secretion. In addition to the continual investigation into the role that Glis3 plays in insulin regulation, further research is needed to identify additional Glis3 target genes in order to elucidate the full scope of Glis3 biological function in the pancreas.

V. Glis Functions in Other Tissues

A. Glis1 function in the epidermis

Although Glis1 is expressed in several tissues, the physiological functions of this Glis family member are still poorly understood (Kang *et al.*, 2010). The characterization of Glis1-null mice did not reveal any obvious phenotype. In the skin, Glis1 mRNA expression was detected in the dermal papilla, but not in normal human epidermis (Nakanishi *et al.*, 2006; Nakashima *et al.*, 2002). However, Glis1 expression was significantly induced in psoriatic epidermis where its expression was associated with the suprabasal, differentiated layers. Glis1 is also expressed in mouse skin treated with the tumor promoter phorbol-12-myristate-13-acetate (PMA) and induced in normal human epidermal keratinocytes (NHEK) cells in culture by the addition of PMA and IFNγ, both inflammatory mediators and inducers of epidermal differentiation. The overexpression of Glis1 in NHEK cells resulted in the increased expression of several markers of epidermal differentiation, including S100 calcium binding protein A9 (S100A9), kallikrein 7 (KLK7), small proline-rich protein (SPRR), involucrin, and transglutaminase 1 (Nakanishi *et al.*, 2006). The induction of differentiation markers by Glis1 is consistent with its expression in the suprabasal, differentiated layers of hyperplastic epidermis. Taken together, these observations suggest that Glis1 promotes differentiation of epidermal keratinocytes and may play a role in aberrant epidermal differentiation associated with psoriatic skin.

B. Promotion of iPS cell generation by Glis1

A recent study showed that Glis1 is highly expressed in the unfertilized egg and in one-cell embryos but is rapidly downregulated in two-cell embryos and blastocysts (Maekawa *et al.*, 2011). Moreover, expression of Glis1 was very low in embryonic stem (ES) cells. Maekawa *et al.* further demonstrated that expression of Glis1 significantly promoted the generation of induced pluripotent stem (iPS) cells. The generation of iPS cells has been achieved by the transgenic expression of the transcription factors Oct3/4, Sox2, and Klf4; however, conversion has been inefficient. Co-expression of Glis1 greatly enhanced the generation of iPS cells from both mouse and human fibroblasts. Glis1 was shown to associate with Klf4, Oct3/4, and Sox2 protein complexes. Gene expression profiling showed that Glis1 induced the expression of several genes that have been reported to enhance iPS generation, including estrogen-related receptor β (Esrrb), several Wnt ligands, lin-28 homologue A (Lin28a), the homeobox transcription factor Nanog, Mycn, and Mycl1, while the expression of Myc was suppressed. Interestingly, Glis1 also enhanced expression of forkhead box A2 (Foxa2), which has been reported to antagonize EMT. MET is required for the reprogramming of fibroblasts into iPS cells (Li *et al.*, 2010). Thus, Glis1 might stimulate somatic cell reprogramming by promoting MET. These observations indicate that Glis1 enhances iPS generation by inducing several pro-reprogramming pathways.

C. Role for Glis2 in neurogenesis

At embryonic day E9.5, Glis2 is expressed in the neural tube, cranial ganglia, and dorsal root ganglia (Lamar *et al.*, 2001). By E10.5, Glis2 is expressed in the intermediate zone of the hindbrain and spinal cord and in the dorsal root ganglia. A similar pattern of Glis2 expression was observed in the chick embryo in the intermediate zone where newly postmitotic neurons are located. In *Xenopus*, Glis2 expression corresponds to two midline regions that contain precursors of primary neurons. Overexpression of Glis2 in neuronal progenitors in the spinal cord of chick embryos was shown to promote the neuronal differentiation. Likewise, ectopic expression of Glis2 in *Xenopus* induced the expression of several neuron-specific markers. Expression of the neurogenin 1 (Ngn1) was shown to induce Glis2, suggesting that Glis2 functions downstream of Ngn1 (Lamar *et al.*, 2001). Together, these observations suggest that Glis2 plays a role in the regulation of neurogenesis.

D. Glis3 and osteogenesis

A reduction in bone formation leads to reduced bone density and an increased risk of fractures. Several studies have indicated a role for Glis3 in osteogenesis. Human patients with aberrant Glis3 expression develop osteopenia with thoracolumbar lordosis and multiple rib fractures (Dimitri *et al.*, 2011). Moreover, Glis3 was highly expressed in human osteoblasts and induced during osteoblast differentiation of mesenchymal stem cells (Beak *et al.*, 2007). Ectopic expression of Glis3 in the multipotent cell line C3H10T1/2 acts synergistically with bone morphogenic protein 2 (BMP2) to promote osteoblast differentiation as measured by the increased levels of alkaline phosphatase activity and *osteopontin* and *osteocalcin* expression. In contrast,

Glis3 expression inhibited adipocyte differentiation in C3H10T1/2 cells. Gene expression profiling identified a number of additional genes that were induced or downregulated by Glis3 in this cell system. Interestingly, the expression of fibroblast growth factor 18 (*FGF18*), which has a positive role in osteogenesis, was significantly increased. EMSA and reporter gene analysis revealed that Glis3 regulates *FGF18* by binding a GlisBS in proximal promoter region (Beak *et al.*, 2007). These observations suggest that *FGF18* is a direct target gene of Glis3 and indicates that Glis3 is an important modulator of osteogenesis. Interestingly, the primary cilium also mediates important functions in osteoblast differentiation and normal bone development as well as in the generation of osteopenia (Malone *et al.*, 2007; Xiao and Quarles, 2010; Xiao *et al.*, 2008) Further insights into the role of the Glis3 signaling pathway in osteoblast differentiation might lead to novel approaches for the treatment of osteoporosis.

E. Glis3 and hypothyroidism

In addition to diabetes and PKD, patients with aberrant Glis3 expression also develop congenital hypothyroidism that is accompanied by reduced levels of T3 and T4 and elevated blood levels of thyroid-stimulating hormone (TSH) and thyroglobulin (Dimitri *et al.*, 2011; Senee *et al.*, 2006; Taha *et al.*, 2003). Similarly, hypothyroidism was observed in Glis3-null mice (Watanabe *et al.*, 2009). About 85% of this disorder is caused by abnormal thyroid gland development; however, histological examination of thyroid gland of Glis3-null mice suggested that Glis3 does not affect thyroid gland development. Some patients responded well to T4 treatment, while others did not. While all patients with Glis3 mutations profiled thus far display thyroid dysfunction, the inconsistent clinical features make it difficult to establish a causative mechanism. Further analysis is required to determine whether the development of hypothyroidism is related to misregulation of the hypothalamus–pituitary–thyroid axis and/or involves changes in iodine or thyroid hormone metabolism.

F. Glis3 and cancer

Increased Glis3 expression has been detected in several different cancer cell types, while no links have yet been made between tumorigenesis and Glis1 or Glis2 (Kang *et al.*, 2010). Increased expression of Glis3 has been reported in ependymomas with high proliferation indices and poor patient prognosis (Lukashova-v Zangen *et al.*, 2007). Amplification of Glis3 was observed in proneural glioblastomas (Cooper *et al.*, 2010). Glis3 was also highly expressed in chromophobe renal cell carcinomas (Yusenko and Kovacs, 2009). Analysis of the role of Glis3 in cell fate and tumor metastasis could provide additional information regarding a link between Glis3 expression and cancer progression.

VI. Conclusion

It is clear that Glis transcription factors play a critical role in the regulation of several physiological processes and are implicated in various pathologies. Mutations in Glis2 have been linked to NPHP, an end-stage renal disease characterized by renal atrophy and fibrosis. This appears to involve induction of EMT in tubule epithelial cells. Mutations in Glis3 lead to the development of diabetes, hypothyroidism, PKD, and several other abnormalities. Glis3 was shown to play a key role in endocrine lineage determination in the pancreas and is required for β-cell development. In addition, Glis3 has a role in the regulation of insulin gene expression. Although the consensus GlisBS and several target genes have been identified, little is known about the proteins that mediate transcriptional regulation by Glis proteins. Moreover, the mechanisms by which Glis activity is controlled are not well understood. Several reports have indicated that Glis2 and Glis3 localize to the primary cilium, suggesting that Glis proteins are part of a primary cilium-mediated signaling pathway. This is consistent with studies showing that loss of Glis2 or Glis3 is linked to several ciliopathies, such as NPHP and PKD. The activities of Glis proteins may be regulated by external signals that interact with membrane-bound receptors in the primary cilium. Reception of such signals may result in a posttranslational modification of Glis proteins and their subsequent translocation into the nucleus where they regulate the transcription of target genes. Elucidation of the different stages of the Glis signaling pathway will be crucial to understanding the biological functions of Glis proteins and to the discovery of therapeutic opportunities for the treatment of diabetes and cystic kidney disease.

ACKNOWLEDGMENTS

The authors would like to thank Drs. Zhengyu Yin and Christina Teng (NIEHS) for their comments on this chapter. This research was supported by the Intramural Research Program of the NIEHS, NIH (Z01-ES-100485).

REFERENCES

Aguilar-Bryan, L., Nichols, C. G., Wechsler, S. W., Clement, J. P., Boyd, A. E., Gonzalez, G., Herrera-Sosa, H., Nguy, K., Bryan, J., and Nelson, D. A. (1995). Cloning of the beta cell high-affinity sulfonylurea receptor: A regulator of insulin secretion. *Science* **268,** 423–426.

Akiyama, T. (2000). Wnt/beta-catenin signaling. *Cytokine Growth Factor Rev.* **11,** 273–282.

Al-Bhalal, L., and Akhtar, M. (2008). Molecular basis of autosomal recessive polycystic kidney disease (ARPKD). *Adv. Anat. Pathol.* **15,** 54–58.

Apelqvist, A., Li, H., Sommer, L., Beatus, P., Anderson, D. J., Honjo, T., Hrabe de Angelis, M., Lendahl, U., and Edlund, H. (1999). Notch signalling controls pancreatic cell differentiation. *Nature* **400,** 877–881.

Attanasio, M., Uhlenhaut, N. H., Sousa, V. H., O'Toole, J. F., Otto, E., Anlag, K., Klugmann, C., Treier, A. C., Helou, J., Sayer, J. A., Seelow, D., Nurnberg, G., *et al.* (2007). Loss of GLIS2 causes nephronophthisis in humans and mice by increased apoptosis and fibrosis. *Nat. Genet.* **39,** 1018–1024.

Barker, A., Sharp, S. J., Timpson, N. J., Bouatia-Naji, N., Warrington, N. M., Kanoni, S., Beilin, L. J., Brage, S., Deloukas, P., Evans, D. M., Grontved, A., Hassanali, N., *et al.* (2011). Association of genetic Loci with glucose levels in childhood and adolescence: A meta-analysis of over 6,000 children. *Diabetes* **60,** 1805–1812.

Barnfield, P. C., Zhang, X., Thanabalasingham, V., Yoshida, M., and Hui, C. C. (2005). Negative regulation of Gli1 and Gli2 activator function by Suppressor of fused through multiple mechanisms. *Differentiation* **73,** 397–405.

Barrett, J. C., Clayton, D. G., Concannon, P., Akolkar, B., Cooper, J. D., Erlich, H. A., Julier, C., Morahan, G., Nerup, J., Nierras, C., Plagnol, V., Pociot, F., *et al.* (2009). Genome-wide association study and meta-analysis find that over 40 loci affect risk of type 1 diabetes. *Nat. Genet.* **41,** 703–707.

Beak, J. Y., Kang, H. S., Kim, Y. S., and Jetten, A. M. (2007). Kruppel-like zinc finger protein Glis3 promotes osteoblast differentiation by regulating FGF18 expression. *J. Bone Miner. Res.* **22,** 1234–1244.

Beak, J. Y., Kang, H. S., Kim, Y. S., and Jetten, A. M. (2008). Functional analysis of the zinc finger and activation domains of Glis3 and mutant Glis3(NDH1). *Nucleic Acids Res.* **36,** 1690–1702.

Berbari, N. F., Johnson, A. D., Lewis, J. S., Askwith, C. C., and Mykytyn, K. (2008a). Identification of ciliary localization sequences within the third intracellular loop of G protein-coupled receptors. *Mol. Biol. Cell* **19,** 1540–1547.

Berbari, N. F., Lewis, J. S., Bishop, G. A., Askwith, C. C., and Mykytyn, K. (2008b). Bardet-Biedl syndrome proteins are required for the localization of G protein-coupled receptors to primary cilia. *Proc. Natl. Acad. Sci. USA* **105,** 4242–4246.

Berbari, N. F., O'Connor, A. K., Haycraft, C. J., and Yoder, B. K. (2009). The primary cilium as a complex signaling center. *Curr. Biol.* **19,** R526–R535.

Bergmann, C., Senderek, J., Kupper, F., Schneider, F., Dornia, C., Windelen, E., Eggermann, T., Rudnik-Schoneborn, S., Kirfel, J., Furu, L., Onuchic, L. F., Rossetti, S., *et al.* (2004). PKHD1 mutations in autosomal recessive polycystic kidney disease (ARPKD). *Hum. Mutat.* **23,** 453–463.

Bernardo, A. S., Hay, C. W., and Docherty, K. (2008). Pancreatic transcription factors and their role in the birth, life and survival of the pancreatic beta cell. *Mol. Cell. Endocrinol.* **294,** 1–9.

Bisgrove, B. W., and Yost, H. J. (2006). The roles of cilia in developmental disorders and disease. *Development* **133,** 4131–4143.

Blacque, O. E., Cevik, S., and Kaplan, O. I. (2008). Intraflagellar transport: From molecular characterisation to mechanism. *Front. Biosci.* **13,** 2633–2652.

Boesgaard, T. W., Grarup, N., Jorgensen, T., Borch-Johnsen, K., Hansen, T., and Pedersen, O. (2010). Variants at DGKB/TMEM195, ADRA2A, GLIS3 and C2CD4B loci are associated with reduced glucose-stimulated beta cell function in middle-aged Danish people. *Diabetologia* **53,** 1647–1655.

Brayer, K. J., and Segal, D. J. (2008). Keep your fingers off my DNA: Protein-protein interactions mediated by C_2H_2 zinc finger domains. *Cell Biochem. Biophys.* **50,** 111–131.

Burlison, J. S., Long, Q., Fujitani, Y., Wright, C. V., and Magnuson, M. A. (2008). Pdx-1 and Ptf1a concurrently determine fate specification of pancreatic multipotent progenitor cells. *Dev. Biol.* **316,** 74–86.

Cerf, M. E. (2006). Transcription factors regulating beta-cell function. *Eur. J. Endocrinol.* **155,** 671–679.

Cervantes, S., Lau, J., Cano, D. A., Borromeo-Austin, C., and Hebrok, M. (2010). Primary cilia regulate Gli/Hedgehog activation in pancreas. *Proc. Natl. Acad. Sci. USA* **107,** 10109–10114.

Chan, S. W., Lim, C. J., Guo, K., Ng, C. P., Lee, I., Hunziker, W., Zeng, Q., and Hong, W. (2008). A role for TAZ in migration, invasion, and tumorigenesis of breast cancer cells. *Cancer Res.* **68,** 2592–2598.

Chen, M. H., Wilson, C. W., Li, Y. J., Law, K. K., Lu, C. S., Gacayan, R., Zhang, X., Hui, C. C., and Chuang, P. T. (2009). Cilium-independent regulation of Gli protein function by Sufu in Hedgehog signaling is evolutionarily conserved. *Genes Dev.* **23,** 1910–1928.

Cheng, S. Y., and Bishop, J. M. (2002). Suppressor of Fused represses Gli-mediated transcription by recruiting the SAP18-mSin3 corepressor complex. *Proc. Natl. Acad. Sci. USA* **99,** 5442–5447.

Chinnadurai, G. (2007). Transcriptional regulation by C-terminal binding proteins. *Int. J. Biochem. Cell Biol.* **39,** 1593–1607.

Cooper, L. A., Gutman, D. A., Long, Q., Johnson, B. A., Cholleti, S. R., Kurc, T., Saltz, J. H., Brat, D. J., and Moreno, C. S. (2010). The proneural molecular signature is enriched in oligodendrogliomas and predicts improved survival among diffuse gliomas. *PLoS One* **5,** e12548.

Corbit, K. C., Aanstad, P., Singla, V., Norman, A. R., Stainier, D. Y., and Reiter, J. F. (2005). Vertebrate Smoothened functions at the primary cilium. *Nature* **437,** 1018–1021.

Dang, D. T., Pevsner, J., and Yang, V. W. (2000). The biology of the mammalian Kruppel-like family of transcription factors. *Int. J. Biochem. Cell Biol.* **32,** 1103–1121.

Daniel, J. M. (2007). Dancing in and out of the nucleus: p120(ctn) and the transcription factor Kaiso. *Biochim. Biophys. Acta* **1773,** 59–68.

Dimitri, P., Warner, J. T., Minton, J. A., Patch, A. M., Ellard, S., Hattersley, A. T., Barr, S., Hawkes, D., Wales, J. K., and Gregory, J. W. (2011). Novel GLIS3 mutations demonstrate an extended multisystem phenotype. *Eur. J. Endocrinol.* **164,** 437–443.

Domire, J. S., Green, J. A., Lee, K. G., Johnson, A. D., Askwith, C. C., and Mykytyn, K. (2011). Dopamine receptor 1 localizes to neuronal cilia in a dynamic process that requires the Bardet-Biedl syndrome proteins. *Cell. Mol. Life Sci.* **68,** 2951–2960.

Dupuis, J., Langenberg, C., Prokopenko, I., Saxena, R., Soranzo, N., Jackson, A. U., Wheeler, E., Glazer, N. L., Bouatia-Naji, N., Gloyn, A. L., Lindgren, C. M., Magi, R., *et al.* (2010). New genetic loci implicated in fasting glucose homeostasis and their impact on type 2 diabetes risk. *Nat. Genet.* **42,** 105–116.

Edlund, T., Walker, M. D., Barr, P. J., and Rutter, W. J. (1985). Cell-specific expression of the rat insulin gene: Evidence for role of two distinct 5′ flanking elements. *Science* **230,** 912–916.

Eggenschwiler, J. T., and Anderson, K. V. (2007). Cilia and developmental signaling. *Annu. Rev. Cell Dev. Biol.* **23,** 345–373.

Elrod-Erickson, M., Rould, M. A., Nekludova, L., and Pabo, C. O. (1996). Zif268 protein-DNA complex refined at 1.6 A: A model system for understanding zinc finger-DNA interactions. *Structure* **4,** 1171–1180.

Fischer, E., and Pontoglio, M. (2009). Planar cell polarity and cilia. *Semin. Cell Dev. Biol.* **20,** 998–1005.

Fliegauf, M., Benzing, T., and Omran, H. (2007). When cilia go bad: Cilia defects and ciliopathies. *Nat. Rev. Mol. Cell Biol.* **8,** 880–893.

Frankel, A. D., Berg, J. M., and Pabo, C. O. (1987). Metal-dependent folding of a single zinc finger from transcription factor IIIA. *Proc. Natl. Acad. Sci. USA* **84,** 4841–4845.

Furlong, E. E., Andersen, E. C., Null, B., White, K. P., and Scott, M. P. (2001). Patterns of gene expression during Drosophila mesoderm development. *Science* **293,** 1629–1633.

Gallagher, A. R., Germino, G. G., and Somlo, S. (2010). Molecular advances in autosomal dominant polycystic kidney disease. *Adv. Chronic Kidney Dis.* **17,** 118–130.

Gavert, N., and Ben-Ze'ev, A. (2007). beta-Catenin signaling in biological control and cancer. *J. Cell. Biochem.* **102,** 820–828.

German, M. S., Moss, L. G., Wang, J., and Rutter, W. J. (1992). The insulin and islet amyloid polypeptide genes contain similar cell-specific promoter elements that bind identical beta-cell nuclear complexes. *Mol. Cell. Biol.* **12,** 1777–1788.

Goetz, S. C., and Anderson, K. V. (2010). The primary cilium: A signalling centre during vertebrate development. *Nat. Rev. Genet.* **11,** 331–344.

Gradwohl, G., Dierich, A., LeMeur, M., and Guillemot, F. (2000). Neurogenin3 is required for the development of the four endocrine cell lineages of the pancreas. *Proc. Natl. Acad. Sci. USA* **97,** 1607–1611.

Guney, M. A., and Gannon, M. (2009). Pancreas cell fate. *Birth Defects Res. C Embryo Today* **87,** 232–248.

Hanahan, D. (1985). Heritable formation of pancreatic beta-cell tumours in transgenic mice expressing recombinant insulin/simian virus 40 oncogenes. *Nature* **315,** 115–122.

Harris, P. C., and Torres, V. E. (2009). Polycystic kidney disease. *Annu. Rev. Med.* **60,** 321–337.

Hashimoto, H., Miyamoto, R., Watanabe, N., Shiba, D., Ozato, K., Inoue, C., Kubo, Y., Koga, A., Jindo, T., Narita, T., Naruse, K., Ohishi, K., *et al.* (2009). Polycystic kidney disease in the medaka (Oryzias latipes) pc mutant caused by a mutation in the Gli-Similar3 (glis3) gene. *PLoS One* **4,** e6299.

Haycraft, C. J., Banizs, B., Aydin-Son, Y., Zhang, Q., Michaud, E. J., and Yoder, B. K. (2005). Gli2 and Gli3 localize to cilia and require the intraflagellar transport protein polaris for processing and function. *PLoS Genet.* **1,** e53.

Hildebrandt, F., and Zhou, W. (2007). Nephronophthisis-associated ciliopathies. *J. Am. Soc. Nephrol.* **18,** 1855–1871.

Hildebrandt, F., Benzing, T., and Katsanis, N. (2011). Ciliopathies. *N. Engl. J. Med.* **364,** 1533–1543.

Hosking, C. R., Ulloa, F., Hogan, C., Ferber, E., Figueroa, A., Gevaert, K., Birchmeier, W., Briscoe, J., and Fujita, Y. (2007). The transcriptional repressor Glis2 is a novel binding partner for p120 catenin. *Mol. Biol. Cell* **18,** 1918–1927.

Hossain, Z., Ali, S. M., Ko, H. L., Xu, J., Ng, C. P., Guo, K., Qi, Z., Ponniah, S., Hong, W., and Hunziker, W. (2007). Glomerulocystic kidney disease in mice with a targeted inactivation of Wwtr1. *Proc. Natl. Acad. Sci. USA* **104,** 1631–1636.

Hu, C., Zhang, R., Wang, C., Wang, J., Ma, X., Hou, X., Lu, J., Yu, W., Jiang, F., Bao, Y., Xiang, K., and Jia, W. (2010). Variants from GIPR, TCF7L2, DGKB, MADD, CRY2, GLIS3, PROX1, SLC30A8 and IGF1 are associated with glucose metabolism in the Chinese. *PLoS One* **5,** e15542.

Huangfu, D., Liu, A., Rakeman, A. S., Murcia, N. S., Niswander, L., and Anderson, K. V. (2003). Hedgehog signalling in the mouse requires intraflagellar transport proteins. *Nature* **426,** 83–87.

Humke, E. W., Dorn, K. V., Milenkovic, L., Scott, M. P., and Rohatgi, R. (2010). The output of Hedgehog signaling is controlled by the dynamic association between Suppressor of Fused and the Gli proteins. *Genes Dev.* **24,** 670–682.

Iwanaga, T., Miki, T., and Takahashi-Iwanaga, H. (2011). Restricted expression of somatostatin receptor 3 to primary cilia in the pancreatic islets and adenohypophysis of mice. *Biomed. Res.* **32,** 73–81.

Jonsson, J., Carlsson, L., Edlund, T., and Edlund, H. (1994). Insulin-promoter-factor 1 is required for pancreas development in mice. *Nature* **371,** 606–609.

Jorgensen, M. C., Ahnfelt-Ronne, J., Hald, J., Madsen, O. D., Serup, P., and Heckshe-Sorensen, J. (2007). An illustrated review of early pancreas development in the mouse. *Endocr. Rev.* **28,** 685–705.

Kaczynski, J., Cook, T., and Urrutia, R. (2003). Sp1- and Kruppel-like transcription factors. *Genome Biol.* **4,** 206.

Kang, H. S., Beak, J. Y., Kim, Y. S., Herbert, R., and Jetten, A. M. (2009a). Glis3 is associated with primary cilia and Wwtr1/TAZ and implicated in polycystic kidney disease. *Mol. Cell. Biol.* **29,** 2556–2569.

Kang, H. S., Kim, Y. S., ZeRuth, G., Beak, J. Y., Gerrish, K., Kilic, G., Sosa-Pineda, B., Jensen, J., Foley, J., and Jetten, A. M. (2009b). Transcription factor Glis3, a novel critical player in the regulation of pancreatic beta-cell development and insulin gene expression. *Mol. Cell. Biol.* **29,** 6366–6379.

Kang, H. S., ZeRuth, G., Lichti-Kaiser, K., Vasanth, S., Yin, Z., Kim, Y.-S., and Jetten, A. M. (2010). Gli-similar (Glis) Krüppel-like zinc finger proteins: Insights into their physiological functions and critical roles in neonatal diabetes and cystic renal disease. *Histol. Histopathol.* **25,** 1481–1496.

Karner, C. M., Chirumamilla, R., Aoki, S., Igarashi, P., Wallingford, J. B., and Carroll, T. J. (2009). Wnt9b signaling regulates planar cell polarity and kidney tubule morphogenesis. *Nat. Genet.* **41,** 793–799.

Kasper, M., Regl, G., Frischauf, A. M., and Aberger, F. (2006). GLI transcription factors: Mediators of oncogenic Hedgehog signalling. *Eur. J. Cancer* **42,** 437–445.

Kim, Y. S., Lewandoski, M., Perantoni, A. O., Kurebayashi, S., Nakanishi, G., and Jetten, A. M. (2002). Identification of Glis1, a novel Gli-related, Kruppel-like zinc finger protein containing transactivation and repressor functions. *J. Biol. Chem.* **277,** 30901–30913.

Kim, Y. S., Nakanishi, G., Lewandoski, M., and Jetten, A. M. (2003). GLIS3, a novel member of the GLIS subfamily of Kruppel-like zinc finger proteins with repressor and activation functions. *Nucleic Acids Res.* **31,** 5513–5525.

Kim, S. W., Park, J. I., Spring, C. M., Sater, A. K., Ji, H., Otchere, A. A., Daniel, J. M., and McCrea, P. D. (2004). Non-canonical Wnt signals are modulated by the Kaiso transcriptional repressor and p120-catenin. *Nat. Cell Biol.* **6,** 1212–1220.

Kim, S. C., Kim, Y. S., and Jetten, A. M. (2005). Kruppel-like zinc finger protein Gli-similar 2 (Glis2) represses transcription through interaction with C-terminal binding protein 1 (CtBP1). *Nucleic Acids Res.* **33,** 6805–6815.

Kim, Y. S., Kang, H. S., and Jetten, A. M. (2007). The Kruppel-like zinc finger protein Glis2 functions as a negative modulator of the Wnt/beta-catenin signaling pathway. *FEBS Lett.* **581,** 858–864.

Kim, Y. S., Kang, H. S., Herbert, R., Beak, J. Y., Collins, J. B., Grissom, S. F., and Jetten, A. M. (2008). Kruppel-like zinc finger protein Glis2 is essential for the maintenance of normal renal functions. *Mol. Cell. Biol.* **28,** 2358–2367.

Koehn, J., Fountoulakis, M., and Krapfenbauer, K. (2008). Multiple drug resistance associated with function of ABC-transporters in diabetes mellitus: Molecular mechanism and clinical relevance. *Infect. Disord. Drug Targets* **8,** 109–118.

Lamar, E., Kintner, C., and Goulding, M. (2001). Identification of NKL, a novel Gli-Kruppel zinc-finger protein that promotes neuronal differentiation. *Development* **128,** 1335–1346.

Lancaster, M. A., and Gleeson, J. G. (2010). Cystic kidney disease: The role of Wnt signaling. *Trends Mol. Med.* **16,** 349–360.

Lau, J., and Hebrok, M. (2010). Hedgehog signaling in pancreas epithelium regulates embryonic organ formation and adult beta-cell function. *Diabetes* **59,** 1211–1221.

Lee, M. S., Gippert, G. P., Soman, K. V., Case, D. A., and Wright, P. E. (1989). Three-dimensional solution structure of a single zinc finger DNA-binding domain. *Science* **245,** 635–637.

Lei, Q. Y., Zhang, H., Zhao, B., Zha, Z. Y., Bai, F., Pei, X. H., Zhao, S., Xiong, Y., and Guan, K. L. (2008). TAZ promotes cell proliferation and epithelial-mesenchymal transition and is inhibited by the hippo pathway. *Mol. Cell. Biol.* **28,** 2426–2436.

Leung, P. S. (2010). Physiology of the pancreas. *Adv. Exp. Med. Biol.* **690,** 13–27.

Li, R., Liang, J., Ni, S., Zhou, T., Qing, X., Li, H., He, W., Chen, J., Li, F., Zhuang, Q., Qin, B., Xu, J., *et al.* (2010). A mesenchymal-to-epithelial transition initiates and is required for the nuclear reprogramming of mouse fibroblasts. *Cell Stem Cell* **7,** 51–63.

Liu, Y. (2009). New insights into epithelial-mesenchymal transition in kidney fibrosis. *J. Am. Soc. Nephrol.* **21,** 212–222.

Liu, A., Wang, B., and Niswander, L. A. (2005). Mouse intraflagellar transport proteins regulate both the activator and repressor functions of Gli transcription factors. *Development* **132,** 3103–3111.

Lukashova-v Zangen, I., Kneitz, S., Monoranu, C. M., Rutkowski, S., Hinkes, B., Vince, G. H., Huang, B., and Roggendorf, W. (2007). Ependymoma gene expression profiles associated with histological subtype, proliferation, and patient survival. *Acta Neuropathol.* **113,** 325–337.

Maekawa, M., Yamaguchi, K., Nakamura, T., Shibukawa, R., Kodanaka, I., Ichisaka, T., Kawamura, Y., Mochizuki, H., Goshima, N., and Yamanaka, S. (2011). Direct reprogramming of somatic cells is promoted by maternal transcription factor Glis1. *Nature* **474,** 225–229.

Makita, R., Uchijima, Y., Nishiyama, K., Amano, T., Chen, Q., Takeuchi, T., Mitani, A., Nagase, T., Yatomi, Y., Aburatani, H., Nakagawa, O., Small, E. V., *et al.* (2008). Multiple renal cysts, urinary concentration defects, and pulmonary emphysematous changes in mice lacking TAZ. *Am. J. Physiol. Renal Physiol.* **294,** F542–F553.

Malone, A. M., Anderson, C. T., Stearns, T., and Jacobs, C. R. (2007). Primary cilia in bone. *J. Musculoskelet. Neuronal Interact.* **7,** 301.

Marshall, J., Dolan, B. M., Garcia, E. P., Sathe, S., Tang, X., Mao, Z., and Blair, L. A. (2003). Calcium channel and NMDA receptor activities differentially regulate nuclear C/EBPbeta levels to control neuronal survival. *Neuron* **39,** 625–639.

May, S. R., Ashique, A. M., Karlen, M., Wang, B., Shen, Y., Zarbalis, K., Reiter, J., Ericson, J., and Peterson, A. S. (2005). Loss of the retrograde motor for IFT disrupts localization of Smo to cilia and prevents the expression of both activator and repressor functions of Gli. *Dev. Biol.* **287,** 378–389.

McNeill, H. (2009). Planar cell polarity and the kidney. *J. Am. Soc. Nephrol.* **20,** 2104–2111.

Melloul, D., Marshak, S., and Cerasi, E. (2002). Regulation of insulin gene transcription. *Diabetologia* **45,** 309–326.

Menezes, L. F., and Germino, G. G. (2009). Polycystic kidney disease, cilia, and planar polarity. *Methods Cell Biol.* **94,** 273–297.

Meng, X., Poon, R., Zhang, X., Cheah, A., Ding, Q., Hui, C. C., and Alman, B. (2001). Suppressor of fused negatively regulates beta-catenin signaling. *J. Biol. Chem.* **276,** 40113–40119.

Merzdorf, C. S. (2007). Emerging roles for zic genes in early development. *Dev. Dyn.* **236,** 922–940.

Michos, O. (2009). Kidney development: From ureteric bud formation to branching morphogenesis. *Curr. Opin. Genet. Dev.* **19,** 484–490.

Miranti, C. K., Ginty, D. D., Huang, G., Chatila, T., and Greenberg, M. E. (1995). Calcium activates serum response factor-dependent transcription by a Ras- and Elk-1-independent mechanism that involves a Ca2+/calmodulin-dependent kinase. *Mol. Cell. Biol.* **15,** 3672–3684.

Mo, R., Freer, A. M., Zinyk, D. L., Crackower, M. A., Michaud, J., Heng, H. H., Chik, K. W., Shi, X. M., Tsui, L. C., Cheng, S. H., Joyner, A. L., and Hui, C. (1997). Specific and redundant functions of Gli2 and Gli3 zinc finger genes in skeletal patterning and development. *Development* **124,** 113–123.

Murtaugh, L. C., and Melton, D. A. (2003). Genes, signals, and lineages in pancreas development. *Annu. Rev. Cell Dev. Biol.* **19,** 71–89.

Na, J., Lykke-Andersen, K., Torres Padilla, M. E., and Zernicka-Goetz, M. (2007). Dishevelled proteins regulate cell adhesion in mouse blastocyst and serve to monitor changes in Wnt signaling. *Dev. Biol.* **302,** 40–49.

Nachury, M. V., Seeley, E. S., and Jin, H. (2010). Trafficking to the ciliary membrane: How to get across the periciliary diffusion barrier? *Annu. Rev. Cell Dev. Biol.* **26,** 59–87.

Nakanishi, G., Kim, Y. S., Nakajima, T., and Jetten, A. M. (2006). Regulatory role for Kruppel-like zinc-finger protein Gli-similar 1 (Glis1) in PMA-treated and psoriatic epidermis. *J. Invest. Dermatol.* **126,** 49–60.

Nakashima, M., Tanese, N., Ito, M., Auerbach, W., Bai, C., Furukawa, T., Toyono, T., Akamine, A., and Joyner, A. L. (2002). A novel gene, GliH1, with homology to the Gli zinc finger domain not required for mouse development. *Mech. Dev.* **119,** 21–34.

Nauli, S. M., and Zhou, J. (2004). Polycystins and mechanosensation in renal and nodal cilia. *Bioessays* **26,** 844–856.

Nauli, S. M., Alenghat, F. J., Luo, Y., Williams, E., Vassilev, P., Li, X., Elia, A. E., Lu, W., Brown, E. M., Quinn, S. J., Ingber, D. E., and Zhou, J. (2003). Polycystins 1 and 2 mediate mechanosensation in the primary cilium of kidney cells. *Nat. Genet.* **33,** 129–137.

Nishio, S., Tian, X., Gallagher, A. R., Yu, Z., Patel, V., Igarashi, P., and Somlo, S. (2010). Loss of oriented cell division does not initiate cyst formation. *J. Am. Soc. Nephrol.* **21,** 295–302.

Ohneda, K., Ee, H., and German, M. (2000a). Regulation of insulin gene transcription. *Semin. Cell Dev. Biol.* **11,** 227–233.

Ohneda, K., Mirmira, R. G., Wang, J., Johnson, J. D., and German, M. S. (2000b). The homeodomain of PDX-1 mediates multiple protein-protein interactions in the formation of a transcriptional activation complex on the insulin promoter. *Mol. Cell. Biol.* **20,** 900–911.

Pan, F. C., and Wright, C. (2011). Pancreas organogenesis: From bud to plexus to gland. *Dev. Dyn.* **240,** 530–565.

Park, J. I., Kim, S. W., Lyons, J. P., Ji, H., Nguyen, T. T., Cho, K., Barton, M. C., Deroo, T., Vleminckx, K., Moon, R. T., and McCrea, P. D. (2005). Kaiso/p120-catenin and TCF/beta-catenin complexes coordinately regulate canonical Wnt gene targets. *Dev. Cell* **8,** 843–854.

Pearson, R., Fleetwood, J., Eaton, S., Crossley, M., and Bao, S. (2008). Kruppel-like transcription factors: A functional family. *Int. J. Biochem. Cell Biol.* **40,** 1996–2001.

Pedersen, L. B., and Rosenbaum, J. L. (2008). Intraflagellar transport (IFT) role in ciliary assembly, resorption and signalling. *Curr. Top. Dev. Biol.* **85,** 23–61.

Peshavaria, M., Henderson, E., Sharma, A., Wright, C. V., and Stein, R. (1997). Functional characterization of the transactivation properties of the PDX-1 homeodomain protein. *Mol. Cell. Biol.* **17,** 3987–3996.

Petersen, H. V., Serup, P., Leonard, J., Michelsen, B. K., and Madsen, O. D. (1994). Transcriptional regulation of the human insulin gene is dependent on the homeodomain protein STF1/IPF1 acting through the CT boxes. *Proc. Natl. Acad. Sci. USA* **91,** 10465–10469.

Praetorius, H. A., and Spring, K. R. (2001). Bending the MDCK cell primary cilium increases intracellular calcium. *J. Membr. Biol.* **184,** 71–79.

Preiss, A., Rosenberg, U. B., Kienlin, A., Seifert, E., and Jackle, H. (1985). Molecular genetics of Kruppel, a gene required for segmentation of the Drosophila embryo. *Nature* **313,** 27–32.

Qin, J., Lin, Y., Norman, R. X., Ko, H. W., and Eggenschwiler, J. T. (2011). Intraflagellar transport protein 122 antagonizes Sonic Hedgehog signaling and controls ciliary localization of pathway components. *Proc. Natl. Acad. Sci. USA* **108,** 1456–1461.

Reynolds, A. B. (2007). p120-catenin: Past and present. *Biochim. Biophys. Acta* **1773,** 2–7.

Rohatgi, R., Milenkovic, L., and Scott, M. P. (2007). Patched1 regulates hedgehog signaling at the primary cilium. *Science* **317,** 372–376.

Rosenbaum, J. L., and Witman, G. B. (2002). Intraflagellar transport. *Nat. Rev. Mol. Cell Biol.* **3,** 813–825.

Ruiz i Altaba, A. (1999). Gli proteins encode context-dependent positive and negative functions: Implications for development and disease. *Development* **126,** 3205–3216.

Ruppert, J. M., Kinzler, K. W., Wong, A. J., Bigner, S. H., Kao, F. T., Law, M. L., Seuanez, H. N., O'Brien, S. J., and Vogelstein, B. (1988). The GLI-Kruppel family of human genes. *Mol. Cell. Biol.* **8,** 3104–3113.

Sasaki, H., Nishizaki, Y., Hui, C., Nakafuku, M., and Kondoh, H. (1999). Regulation of Gli2 and Gli3 activities by an amino-terminal repression domain: Implication of Gli2 and Gli3 as primary mediators of Shh signaling. *Development* **126,** 3915–3924.

Schwartz, E. A., Leonard, M. L., Bizios, R., and Bowser, S. S. (1997). Analysis and modeling of the primary cilium bending response to fluid shear. *Am. J. Physiol.* **272,** F132–F138.

Senee, V., Chelala, C., Duchatelet, S., Feng, D., Blanc, H., Cossec, J. C., Charon, C., Nicolino, M., Boileau, P., Cavener, D. R., Bougneres, P., Taha, D., *et al.* (2006). Mutations in GLIS3 are responsible for a rare syndrome with neonatal diabetes mellitus and congenital hypothyroidism. *Nat. Genet.* **38,** 682–687.

Simms, R. J., Hynes, A. M., Eley, L., and Sayer, J. A. (2011). Nephronophthisis: A genetically diverse ciliopathy. *Int. J. Nephrol.* **2011,** 527137.

Soderling, T. R. (1999). The Ca-calmodulin-dependent protein kinase cascade. *Trends Biochem. Sci.* **24,** 232–236.

Stanic, D., Malmgren, H., He, H., Scott, L., Aperia, A., and Hokfelt, T. (2009). Developmental changes in frequency of the ciliary somatostatin receptor 3 protein. *Brain Res.* **1249,** 101–112.

Svard, J., Heby-Henricson, K., Persson-Lek, M., Rozell, B., Lauth, M., Bergstrom, A., Ericson, J., Toftgard, R., and Teglund, S. (2006). Genetic elimination of Suppressor of fused reveals an essential repressor function in the mammalian Hedgehog signaling pathway. *Dev. Cell* **10,** 187–197.

Taha, D., Barbar, M., Kanaan, H., and Williamson Balfe, J. (2003). Neonatal diabetes mellitus, congenital hypothyroidism, hepatic fibrosis, polycystic kidneys, and congenital glaucoma: A new autosomal recessive syndrome? *Am. J. Med. Genet. A* **122A,** 269–273.

Thomas, M. K., Rastalsky, N., Lee, J. H., and Habener, J. F. (2000). Hedgehog signaling regulation of insulin production by pancreatic beta-cells. *Diabetes* **49,** 2039–2047.

Thomas, M. K., Lee, J. H., Rastalsky, N., and Habener, J. F. (2001). Hedgehog signaling regulation of homeodomain protein islet duodenum homeobox-1 expression in pancreatic beta-cells. *Endocrinology* **142,** 1033–1040.

Tukachinsky, H., Lopez, L. V., and Salic, A. (2010). A mechanism for vertebrate Hedgehog signaling: Recruitment to cilia and dissociation of SuFu-Gli protein complexes. *J. Cell Biol.* **191,** 415–428.

Vasanth, S., ZeRuth, G., Kang, H. S., and Jetten, A. M. (2011). Identification of nuclear localization, DNA binding, and transactivating mechanisms of Kruppel-like zinc finger protein Gli-similar 2 (Glis2). *J. Biol. Chem.* **286,** 4749–4759.

Wada, T., Sakai, N., Matsushima, K., and Kaneko, S. (2007). Fibrocytes: A new insight into kidney fibrosis. *Kidney Int.* **72,** 269–273.

Walker, M. D., Edlund, T., Boulet, A. M., and Rutter, W. J. (1983). Cell-specific expression controlled by the 5′-flanking region of insulin and chymotrypsin genes. *Nature* **306,** 557–561.

Wang, Q. T., and Holmgren, R. A. (1999). The subcellular localization and activity of Drosophila cubitus interruptus are regulated at multiple levels. *Development* **126,** 5097–5106.

Wang, C., Pan, Y., and Wang, B. (2010). Suppressor of fused and Spop regulate the stability, processing and function of Gli2 and Gli3 full-length activators but not their repressors. *Development* **137,** 2001–2009.
Watanabe, N., Hiramatsu, K., Miyamoto, R., Yasuda, K., Suzuki, N., Oshima, N., Kiyonari, H., Shiba, D., Nishio, S., Mochizuki, T., Yokoyama, T., Maruyama, S., *et al.* (2009). A murine model of neonatal diabetes mellitus in Glis3-deficient mice. *FEBS Lett.* **583,** 2108–2113.
Wilson, P. D. (2008). Mouse models of polycystic kidney disease. *Curr. Top. Dev. Biol.* **84,** 311–350.
Wynn, T. A. (2007). Common and unique mechanisms regulate fibrosis in various fibroproliferative diseases. *J. Clin. Invest.* **117,** 524–529.
Xiao, Z. S., and Quarles, L. D. (2010). Role of the polycytin-primary cilia complex in bone development and mechanosensing. *Ann. N. Y. Acad. Sci.* **1192,** 410–421.
Xiao, K., Oas, R. G., Chiasson, C. M., and Kowalczyk, A. P. (2007). Role of p120-catenin in cadherin trafficking. *Biochim. Biophys. Acta* **1773,** 8–16.
Xiao, Z., Zhang, S., Magenheimer, B. S., Luo, J., and Quarles, L. D. (2008). Polycystin-1 regulates skeletogenesis through stimulation of the osteoblast-specific transcription factor RUNX2-II. *J. Biol. Chem.* **283,** 12624–12634.
Yang, Y., Chang, B. H., Samson, S. L., Li, M. V., and Chan, L. (2009). The Kruppel-like zinc finger protein Glis3 directly and indirectly activates insulin gene transcription. *Nucleic Acids Res.* **37,** 2529–2538.
Yoder, B. K. (2007). Role of primary cilia in the pathogenesis of polycystic kidney disease. *J. Am. Soc. Nephrol.* **18,** 1381–1388.
Yusenko, M. V., and Kovacs, G. (2009). Identifying CD82 (KAI1) as a marker for human chromophobe renal cell carcinoma. *Histopathology* **55,** 687–695.
Zeisberg, M., and Kalluri, R. (2004). The role of epithelial-to-mesenchymal transition in renal fibrosis. *J. Mol. Med.* **82,** 175–181.
ZeRuth, G. T., Yang, X. P., and Jetten, A. M. (2011). Modulation of the transactivation function and stability of Kruppel-like Zinc finger protein Gli-similar 3 (Glis3) by suppressor of fused. *J. Biol. Chem.* **24,** 22077–22089.
Zhang, F., and Jetten, A. M. (2001). Genomic structure of the gene encoding the human GLI-related, Kruppel-like zinc finger protein GLIS2. *Gene* **280,** 49–57.
Zhang, F., Nakanishi, G., Kurebayashi, S., Yoshino, K., Perantoni, A., Kim, Y. S., and Jetten, A. M. (2002). Characterization of Glis2, a novel gene encoding a Gli-related, Kruppel-like transcription factor with transactivation and repressor functions. Roles in kidney development and neurogenesis. *J. Biol. Chem.* **277,** 10139–10149.
Zhang, C., Moriguchi, T., Kajihara, M., Esaki, R., Harada, A., Shimohata, H., Oishi, H., Hamada, M., Morito, N., Hasegawa, K., Kudo, T., Engel, J. D., *et al.* (2005). MafA is a key regulator of glucose-stimulated insulin secretion. *Mol. Cell. Biol.* **25,** 4969–4976.
Zhou, Q., Law, A. C., Rajagopal, J., Anderson, W. J., Gray, P. A., and Melton, D. A. (2007). A multipotent progenitor domain guides pancreatic organogenesis. *Dev. Cell* **13,** 103–114.

CHAPTER EIGHT

Sonic Hedgehog Regulates Wnt Activity During Neural Circuit Formation

Nicole H. Wilson *and* Esther T. Stoeckli

Contents

Institute of Molecular Life Sciences, University of Zurich, Zurich, Switzerland

Vitamins and Hormones, Volume 88
ISSN 0083-6729, DOI: 10.1016/B978-0-12-394622-5.00008-0

Abstract

Gradients of secreted morphogens, such as Sonic hedgehog (Shh), Wnt, and TGFβ/Bmp, have classically been shown to control many aspects of early development by regulating cell proliferation and determining cell fate. However, recent studies demonstrate that these molecules also play important and evolutionarily conserved roles in later aspects of neural development. Depending on the context, these molecules can elicit gene transcription in the nucleus, or alternatively can provide instructive signals at the growth cone that induce local and rapid changes in cytoskeletal organization. Shh can activate different cellular transduction pathways via its binding to alternative coreceptor complexes or simply by adaptation of its "classical" signaling pathway. However, in most of its activities during neural development, Shh does not act alone but rather in concert with other morphogens, particularly the Wnts. This review provides an overview of the mechanisms by which Shh signaling acts in concert with Wnts to mediate a myriad of cellular processes that are required for neural circuit formation.

I. Introduction

The establishment of neural circuits during development is a well-ordered series of events: the proliferation and differentiation of neuronal subtypes in the early neural tube; the polarization and outgrowth of axons and dendrites from differentiated neurons; the navigation of axons along defined, stereotypical pathways; and finally, the formation and maintenance of synaptic connections with appropriate target cells. Shh signaling has been implicated in many of these processes, but it never acts in isolation. Rather, signaling by another family of morphogens, the Wnts, often accompanies Shh activity to ensure correct cellular specification, outgrowth, guidance, and targeting during the development of the nervous system. Wnts may function in parallel to Shh in either an antagonistic or cooperative manner, or the pathways may be more closely linked by the cross regulation of various signaling modulators or the activation of common downstream signaling components. Here, we review the cooperative and antagonistic roles of Wnt and Shh during neural circuit formation, focusing primarily on the developing spinal cord as a model of neural development.

II. Shh and Wnt Signaling: Canonical Signaling Pathways

The Sonic hedgehog (Shh) and Wnt/Wingless (Wg) families of secreted signaling molecules play crucial roles in many aspects of embryonic development and adult tissue homeostasis (reviewed by Ingham and

McMahon, 2001; Logan and Nusse, 2004). The canonical signaling pathways activated by these molecules, and a summary of their expression patterns in the developing neural tube, are briefly described below.

A. Canonical Shh signaling

Canonical Shh signaling (Fig. 8.1) occurs via a derepression mechanism involving the multiple-pass transmembrane receptor Patched (Ptch) and the G-protein-coupled receptor-like molecule Smoothened (Smo). In the absence of Shh, Ptch represses Smo activity. Shh binding to Ptch relieves this repression, allowing Smo to translocate to the primary cilium where it activates intracellular signaling pathways. This activation of Smo culminates

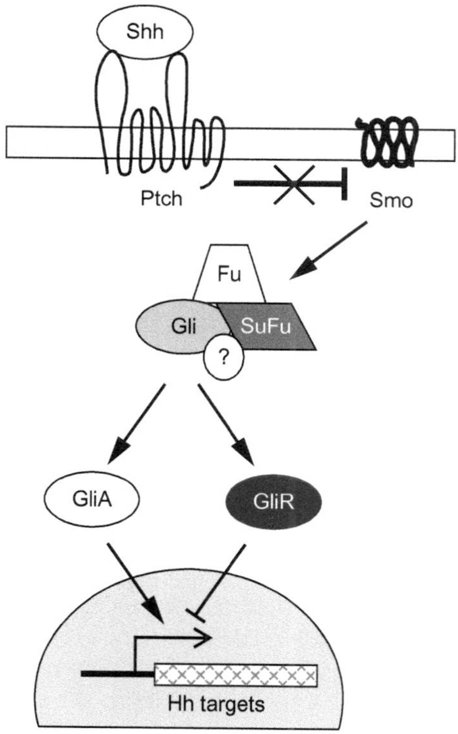

Figure 8.1 Canonical Shh signaling pathway. Binding of Shh to Ptch relieves Ptch inhibition of Smo. In the absence of ligand, transcription of target genes is blocked because Gli transcription factors are proteolytically processed to repressor forms (GliR). Shh binding suppresses GliR production and promotes the accumulation of the activator forms of Gli (GliA). The modulation of Gli activity occurs via an intracellular complex composed of Fused (Fu), Suppressor of Fused (SuFu), and possibly other components (?). In vertebrates, Gli1 is primarily an activator, Gli2 is bifunctional, and Gli3 is primarily a repressor.

in transcriptional activity, via its regulation of Gli transcription factors that can activate or repress Shh target genes (Martí and Bovolenta, 2002). During neural development, Shh is first expressed by the notochord, which induces the formation of the floor plate at the ventral midline of the neural tube. Subsequently, the floor plate itself begins to express Shh, thereby becoming a source from which this morphogen diffuses to form a ventral^{high} to dorsal^{low} expression gradient (Martí *et al.*, 1995a,b; Roelink *et al.*, 1995).

In vertebrates, three Gli transcription factors (Gli1–Gli3) have been described, all of which are expressed in the developing neural tube (Bai *et al.*, 2002; Hui *et al.*, 1994; Jacob and Briscoe, 2003; Lee *et al.*, 1997; Ruiz i Altaba, 1998). While Gli2 and Gli3 both have transcriptional activator and repressor activities, Gli1 is solely a transcriptional activator (reviewed by Ruiz i Altaba *et al.*, 2007). Each Gli family member responds differently to Shh. In the absence of Shh, Gli2 is fully degraded, while Gli3 is converted (via proteolytic processing of its C-terminus) to a repressor of transcription (Gli3R). When Shh is present, Gli1 expression is induced and proteolysis of Gli2 and Gli3 is inhibited, allowing full-length activator forms of the Gli proteins to accumulate. The Gli transcriptional activators then induce the expression of Shh pathway target genes (Aza-Blanc *et al.*, 2000; Dai *et al.*, 1999; Pan *et al.*, 2006; Ruiz i Altaba, 1999).

B. Canonical Wnt signaling

Current models of Wnt signaling describe three cellular transduction pathways: the canonical/β-catenin pathway, the planar cell polarity (PCP) pathway, and the calcium (Ca^{2+}) pathway (Komiya and Habas, 2008). In the canonical Wnt pathway (Fig. 8.2), transcriptional activity is regulated by β-catenin, a bifunctional protein that either regulates gene expression, when translocated to the nucleus, or acts as a linker between surface receptors and the cytoskeleton. When Wnt is not present, Glycogen synthase kinase 3 (GSK3β) constitutively phosphorylates β-catenin, leading to its degradation. When Wnt binds to the receptor complex formed by Frizzled (Fzd) and LDL receptor-related protein (LRP)5/6, the cytoplasmic protein Dishevelled (Dvl) is recruited to the membrane, where it inhibits GSK3β. This allows β-catenin to accumulate in the cytosol (Aberle *et al.*, 1997; Bhanot *et al.*, 1996; Komiya and Habas, 2008; Tamai *et al.*, 2000). β-Catenin is subsequently translocated to the nucleus to activate the transcription of specific target genes, by interacting with members of the T-cell factor/Lymphoid enhancer factor (Tcf/Lef) family of DNA-binding proteins (Eastman and Grosschedl, 1999).

Multiple Wnts are expressed both in the dorsal and ventral aspects of the neural tube (Hollyday *et al.*, 1995; Megason and McMahon, 2002;

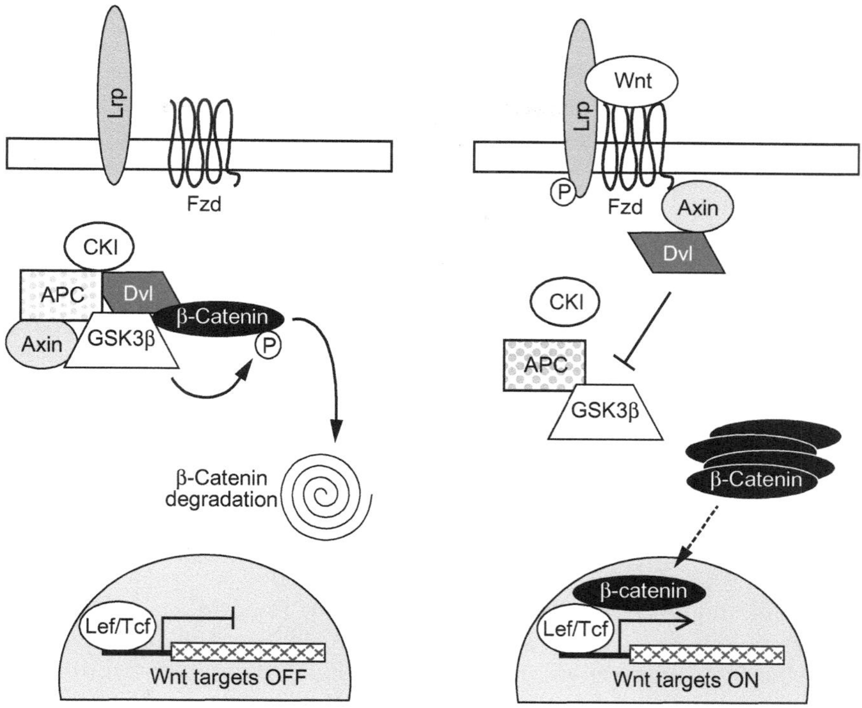

Figure 8.2 Canonical Wnt/β-catenin signaling pathway. In the absence of Wnt, β-catenin is recruited by a "destruction complex" in which it is phosphorylated by GSK3β and targeted for degradation. Wnt target genes are not transcribed. When Wnt interacts with Fzd and Lrp5/6 receptors, Dvl is activated, leading to the disintegration of the GSK3β/APC/CKI/Axin complex and accumulation of β-catenin. β-Catenin translocates to the nucleus, where it interacts with Tcf/Lef transcription factors to promote transcription of Wnt target genes. APC, Adenomatous polyposis coli; CKI, Casein kinase I.

Parr *et al.*, 1993), as are transcriptional effectors of Wnt/β-catenin activity, such as Tcf1, Tcf3, Tcf4, but not Lef1 (Alvarez-Medina *et al.*, 2008).

III. Patterning and Morphogenesis of the Neural Tube

The first step in neural circuit formation is the differentiation of distinct neuronal subtypes from the developing neuroepithelium. The developing spinal cord has commonly been used as a model to investigate the mechanisms governing tissue proliferation and patterning. The simple, early neural tube contains a lumen lined with mitotically active neuroepithelial cells.

As the cells become postmitotic and begin to differentiate, they migrate and distribute laterally. The neural progenitors are then arranged in 11 different subtypes along the dorsoventral axis, each identifiable by the expression of a combinatorial code of transcription factors (Helms and Johnson, 2003; Jessell, 2000). This dorsoventral patterning occurs due to the activities of several secreted molecules, including Shh, Wnts, Bone morphogenetic proteins (Bmps), Fibroblast growths factors (FGFs), and retinoic acid, which provide positional information to the neural progenitors (reviewed by Helms and Johnson, 2003; Shirasaki and Pfaff, 2002). The main players, however, are Shh which is both necessary and sufficient to induce the five ventral-most neural subsets (V3, MN, V2, V1, V0), as well as Bmps and Wnts, which induce the differentiation of the dorsal neural tube (dI1–dI6). Recent evidence suggests that both proliferation and patterning of neural progenitors involve interactions between Wnt and Shh along the dorsoventral axis (Alvarez-Medina *et al.*, 2009; Ulloa and Martí, 2010).

A. Shh and Wnt activity coordinately regulate neural progenitor proliferation

Both Shh and Wnt have been demonstrated to regulate the growth and proliferation of neural tissue. In the neural tube, Shh regulates the expression of cyclin D1 (Cayuso *et al.*, 2006; Lobjois *et al.*, 2004), which controls the G1/S transition phase of the cell cycle. Cyclin D1 is also a transcriptional target of the canonical Wnt pathway (Shtutman *et al.*, 1999; Tetsu and McCormick, 1999), which may account for the neural growth-regulatory effects of Wnt/β-catenin (Megason and McMahon, 2002; Zechner *et al.*, 2003).

In the developing neural tube, the Wnt family members that regulate mitogenic activity (Wnt1 and Wnt3a) are mainly expressed dorsally (Megason and McMahon, 2002) while Shh is expressed ventrally (Martí *et al.*, 1995b). These opposing expression patterns suggest a simplistic model of proliferation control, in which these pathways act in parallel on distinct subpopulations of neural precursor cells (i.e., Wnt dorsally and Shh ventrally). However, the loss of either Wnt or Shh leads to cell cycle arrest and a uniform tissue reduction throughout the dorsoventral axis of the neural tube (Alvarez-Medina *et al.*, 2009; Cayuso *et al.*, 2006), arguing against this simplistic model. Another possibility is that the two pathways act in the same cells, by controlling the activity or transcription of different regulators of cell cycle progression.

Recent studies in chicken support this idea (Alvarez-Medina *et al.*, 2009; Ulloa *et al.*, 2007). By *in vivo* gain- and loss-of-function experiments, Alvarez-Medina and colleagues (2009) showed that the expression of Tcf3/4 in the neural tube depends on Shh activity, thus linking Shh signaling to the canonical Wnt/β-catenin transcription pathway. In a Shh loss-of-function background, the expression of Wnt1/Wnt3a or constitutively active β-catenin did not restore proliferation in the neural tube, whereas expression of a β-catenin-independent Tcf activator was able to rescue the BrdU incorporation rate and reestablish the normal size of the neural tube. The authors found that an upstream Shh signal was required for the transcription of *Tcf3/4*. In the absence of Tcf3/4, canonical Wnt signaling could not activate mitogenic target genes, including *cyclin D1*, leading to arrest of the cell cycle at G1. Thus, by controlling the levels of Tcf transcription factors, Shh activity is required for canonical Wnt/β-catenin activity during neural tube proliferation. Wnt and Shh activities are thus integrated to control cyclin D1 expression and the subsequent progression through the G1 phase of the cell cycle (Fig. 8.3).

Another level of interaction between the Wnt and Shh signaling pathways in the regulation of proliferation has been proposed by Ulloa and colleagues (2007). They found that the loss of Shh signaling reduced the responsiveness of neural cells to canonical Wnt signaling. In this case, the molecule responsible for the inhibition of Wnt signaling in the absence of Shh was identified as Gli3R, the transcriptional repressor form of Gli3 that is generated when Shh is absent. Gli3R was demonstrated to physically interact with β-catenin *in vitro* and to inhibit the transcriptional activity of β-catenin in many different cell types *in vitro* and *in vivo*. Interestingly, Gli3R is not expected to directly regulate the expression of Tcf3/4 (since *Tcf3* and *Tcf4* genes both appear to lack core consensus Gli binding domains; Alvarez-Medina *et al.*, 2009; Hallikas *et al.*, 2006), thus the precise mechanism underlying its inhibition of the canonical Wnt signaling pathway is currently unknown.

While Shh and Wnt apparently regulate some aspects of cell cycle progression in a coordinated manner, they may also have independent roles in cell proliferation. For example, Shh has been shown to activate cell cycle genes promoting the transition from G1 to G2 (including *cyclin A*, *cyclin B*, *and cyclin E*), suggesting that Shh signaling, but not Wnt/β-catenin signaling, also regulates late cell cycle progression (Alvarez-Medina *et al.*, 2009; Megason and McMahon, 2002). Thus, in addition to their integrated activities, the modulation of different transcription factors also allows Wnt and Shh to regulate distinct mechanisms that control cell proliferation in the neural tube.

B. Shh defines cell fate in the ventral spinal cord

Following proliferation, neural precursors must adopt a particular cell fate. In the neural tube, distinct types of neurons are generated along the dorsoventral axis (Fig. 8.4). Developmental studies over the past two decades have identified the dorsal and ventral midlines as the major signaling centers that regulate the differentiation of the appropriate types and numbers of cells along the dorsoventral axis of the neural tube.

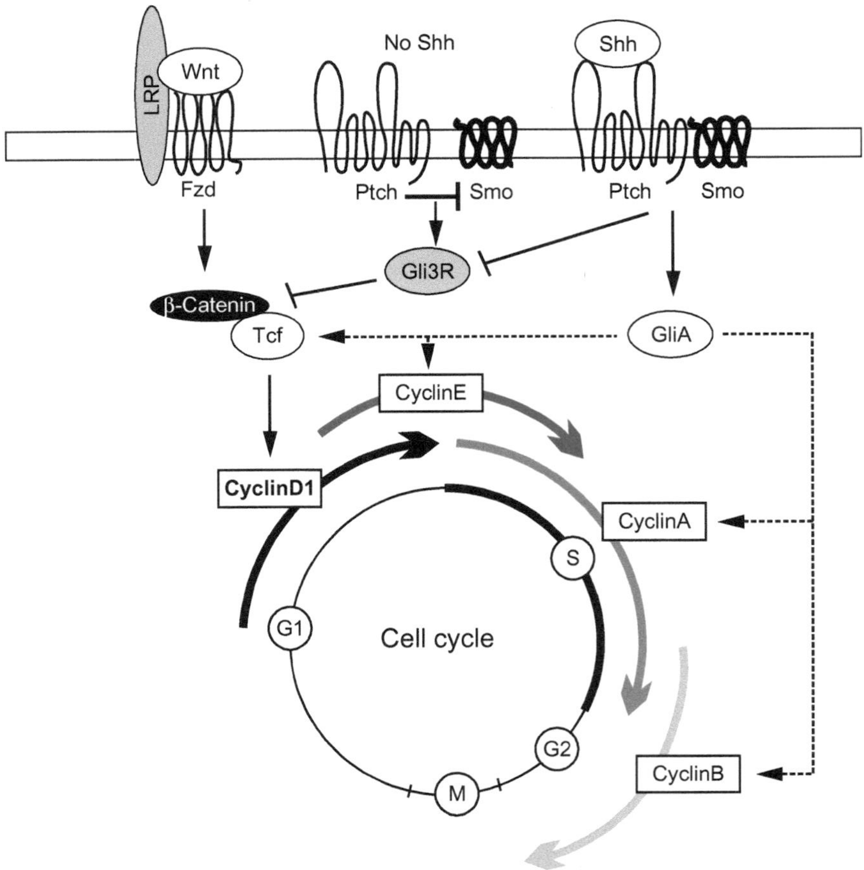

Figure 8.3 Summary of Wnt and Shh activities in the regulation of cell cycle progression in neuroepithelial cells. CyclinD1 is induced downstream of Wnt and Shh signaling, promoting the G1/S transition of the cell cycle. However, the Wnt signaling response relies on Shh-induced Tcf expression, thus linking the two pathways in the control of neural progenitor proliferation. In the absence of Shh, Gli3R also inhibits the activity of β-catenin. Shh can also induce other cyclin genes independently of Wnt to control later stages of cell cycle progression. Figure modified from Alvarez-Medina and colleagues (2009).

In the ventral neural tube, the gradient of secreted Shh emanating from the floor plate represents the primary morphogenetic signal that regulates cellular identity. Shh controls the specification of ventral cell types in a dose-dependent manner (Ericson *et al.*, 1996; Jessell, 2000; Roelink *et al.*, 1995; Tanabe and Jessell, 1996). Current models suggest that the morphogenetic activity of Shh is attributable to a gradient of intracellular Gli activity. Incremental changes in Shh concentration, Gli activation levels,

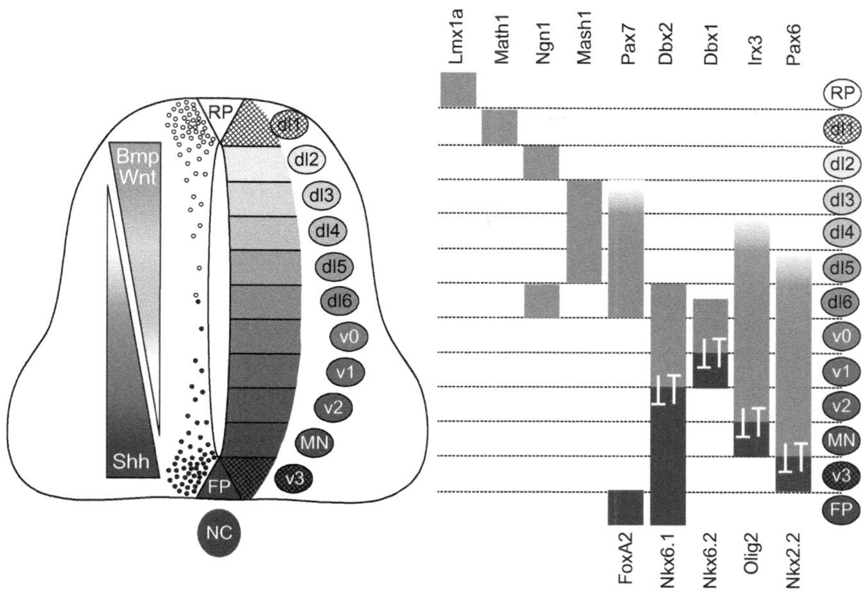

Figure 8.4 Model by which Shh and Wnt/Bmp pattern the developing spinal cord. The lumen of the neural tube is lined with mitotically active progenitor cells. As these cells exit the cell cycle, they distribute laterally in a specific order along the dorsoventral axis (dI1–v3). This patterning is established via the activities of counteracting gradients of Wnt/Bmp, secreted from the roof plate (RP) and Shh, secreted from the notochord (NC) and floor plate (FP). The Wnt/Bmp and Shh gradients are transduced intracellularly by the generation of transcriptional repressors and activators in a concentration-dependent manner. The specific combinations of transcription factors induced generate a cell identity code that specifies the neural progenitor subtypes. Some pairs of transcription factors show cross-inhibitory interactions that define their boundary of expression along the dorsoventral axis.

or in the duration of Shh/Gli signaling determine different neuronal subtypes, by regulating the expression of transcription factors of the basic helix-loop-helix (bHLH) or homeodomain (HD) families (Briscoe and Ericson, 2001; Dessaud *et al.*, 2010; Jessell, 2000; Stamataki *et al.*, 2005). The resulting profiles of HD and bHLH transcription factors expressed by the precursor cells act to specify the particular neural identity arising from each progenitor domain (Jessell, 2000).

Most of the ventral cell types are missing in the neural tube of *Shh*$^{-/-}$ mice (Chiang *et al.*, 1996). However, *Shh*$^{-/-}$ and *Gli3*$^{-/-}$ double mutants exhibit some ventral cell types, including motor neurons (Litingtung and Chiang, 2000), suggesting that Shh is normally required to antagonize Gli3, which would otherwise repress ventral fates. In accordance with this, the repressor activity of Gli3 (which functions in the absence of Shh) has been demonstrated to be important for many of its patterning activities (Meyer and Roelink, 2003; Persson *et al.*, 2002). Thus, it seems that Shh not only induces Gli activator but also restricts Gli3 repressor activities during neural tube patterning.

C. Wnt canonical signaling regulates neural proliferation and patterning

The Wnt/β-catenin signaling pathway has clearly been demonstrated, by gain- and loss-of-function approaches in chick, mice, and zebrafish, to play a proliferative role in neural tissue. This function is mainly attributable to the activities of Wnt3a and Wnt1, both of which are expressed at the dorsal midline of the spinal cord (Bonner *et al.*, 2008; Chesnutt *et al.*, 2004; Dickinson *et al.*, 1994; Ille *et al.*, 2007; Megason and McMahon, 2002; Zechner *et al.*, 2007). However, an independent role of Wnt in specifying cellular identity has, until recently, been controversial (Chesnutt *et al.*, 2004; Ille *et al.*, 2007; Megason and McMahon, 2002). Several lines of evidence now support the idea that Wnts are involved in dorsoventral patterning of the neural tube.

Muroyama and colleagues (2002) demonstrated that *Wnt1/Wnt3a* double mutant mice show reduced numbers of the dorsal-most (dI1–dI3) interneurons, and a compensatory increase in intermediate interneurons. Similarly, expression of the secreted Wnt antagonist Dickkopf in zebrafish resulted in the loss or reduction of dorsal progenitor cell types (Bonner *et al.*, 2008). Disruption of *β-catenin* also results in patterning defects in the dorsal neural tube: expression of a constitutively active form of β-catenin results in ectopic expression of Olig3, a bHLH transcription factor that is essential for the correct specification of dI1–dI3 neurons (Müller *et al.*, 2005), while the conditional loss of β-catenin in neural progenitors prevents Olig3 expression (Zechner *et al.*, 2007).

Further experiments in chick embryos suggested that Wnt/β-catenin not only induces dorsal cell types but also inhibits ventral cell types (Alvarez-Medina *et al.*, 2008). Ectopic activation of Wnt/β-catenin in the chick neural tube expanded the expression domains of dorsal markers such as Pax6 and Pax7, but also inhibited the expression of ventral makers such as Olig2 and Nkx2.2. Conversely, inhibition of the Wnt/β-catenin pathway by expression of dominant-negative Tcf not only repressed the expression of dorsal cell subtypes but also resulted in an expansion of the ventral markers (Alvarez-Medina *et al.*, 2008). In agreement with the results from chicken electroporation assays, Yu and colleagues (2008b) used transgenic mice to induce the expression of constitutively active β-catenin in patches of the ventral neural tube, and found that dorsal identities were ectopically induced in a cell-autonomous, time-dependent manner, while ventral cell identities were repressed. Taken together, the results suggest a model in which the secretion of Wnt1/Wnt3a from the dorsal neural tube simultaneously induces dorsal cell types while repressing ventral cell types. The activation of β-catenin appears to antagonize the activity of Shh in neural tube patterning.

D. Wnt antagonizes Shh signaling via Gli3 during neural tube patterning

How does canonical Wnt signaling antagonize the patterning activity of Shh? Evidence supports at least two models that have been investigated to date: (i) Wnt/β-catenin acts indirectly on patterning by affecting other signaling molecules downstream of Shh, such as the Glis; and (ii) Wnt/β-catenin directly regulates patterning molecules that oppose those of Shh/Gli, acting in parallel to Bmp-mediated patterning of the dorsal neural tube (Ulloa and Martí, 2010).

Alvarez-Medina and colleagues (2008) used luciferase reporter constructs containing Gli binding sites to assess activation of the canonical Shh pathway *in ovo*. The misexpression of Wnt1/Wnt3a inhibited the transcriptional activity of Gli3, demonstrating that Wnt/β-catenin activity inhibits the Shh/Gli pathway in the chicken spinal cord. When endogenous Gli3 activator and repressor functions were blocked by the electroporation of the DNA-binding domain of *Gli3* (Gli3–Znf), the misexpression of Wnt1/Wnt3a no longer expanded the dorsal expression domains of Pax6 or Pax7. Conversely, Pax7 expression was not inhibited following the coexpression of *Gli3–Znf* and a dominant-negative form of Tcf. Taken together, the data indicates that the patterning activity of Wnt/β-catenin depends, at least in part, on its direct regulation of *Gli3* transcription, which in turn restricts Shh signaling (Alvarez-Medina *et al.*, 2008).

In line with its role as a potential mediator of Wnt/β-catenin patterning activity, Gli3 is expressed in an appropriate gradient in the dorsal spinal cord, and is a potent inhibitor of the Shh pathway (Jacob and Briscoe, 2003). Further support for this model comes from a study in mice (Yu *et al.*, 2008b), in which stabilized β-catenin was expressed in the ventral Olig1 domain. This resulted in the induction of Gli3 expression and a ventral shift in cell identities: V2 cells were ectopically specified at the expense of motor neurons. The concomitant loss of Gli2 had no effect on this phenotype, whereas the loss of Gli3 elicited a partial rescue, showing that the ventralizing activity of Wnt/β-catenin was dependent on Gli3. The idea that Wnt/β-catenin can directly regulate Gli3 expression is supported by the identification of several Tcf binding domains in highly conserved noncoding regions of the human *Gli3* locus (Abbasi *et al.*, 2007). Two of these Tcf binding sites were shown to respond to Wnt signaling in neural cells and are capable of driving Gli3 expression in the dorsal neural tube (Alvarez-

Medina *et al.*, 2008). However, analysis of the direct binding of Tcf/β-catenin to *Gli3* regulatory domains is yet to be performed.

Does Gli3 account for all of the Wnt/β-catenin patterning effects? Experimental evidence suggests that this is not the case, since the loss of Gli3 does not completely reproduce the patterning defects seen in Wnt/β-catenin loss-of-function models (Yu *et al.*, 2008b). Gli3-deficient mice still possess the dI1–dI3 interneurons that are lacking in *Wnt1/Wnt3a* double mutants, showing that Wnt/β-catenin can act independently of Gli3 to specify the dorsal-most cell types. While the precise mechanism by which this occurs remains unknown, one possibility is that Wnt/β-catenin directly controls the expression of bHLH and HD proteins that act in parallel to Bmps in dorsal spinal cord patterning. There may also be a differential requirement for Gli3-dependent versus Gli3-independent activities along the dorsoventral axis of the neural tube. The Gli3-independent activities of Wnt/β-catenin in neural tube patterning remain to be clarified.

Taken together, the patterning of the spinal cord in the dorsoventral axis is achieved by the concerted activities of several families of signaling molecules, and especially by counteracting gradients of Wnts, Bmps, and Shh. One molecule at the heart of this activity is Gli3, whose relative transcriptional repressor versus activator functions appear to be crucial for correct dorsoventral patterning. Wnt canonical signaling acts to induce the expression of Gli3, thus tipping the balance toward Gli3 repressor activity, which antagonizes the patterning activity of Shh. On the other hand, Shh signaling favors the production of Gli3 activator, thus limiting the patterning effects of Gli3 repressor activity. These antagonistic interactions between the canonical Wnt and Shh signaling pathways help to define the specific cell types along the dorsoventral axis of the developing neural tube (Fig. 8.5).

IV. Establishment of Neuronal Polarization

Mature neurons are segregated into two distinct morphological and functional domains: (i) a single axon that navigates to its target cell, forms synaptic contacts, and releases neurotransmitters at the presynaptic terminal; and (ii) multiple dendrites that receive and integrate synaptic inputs. This establishment of neuronal polarity is essential for appropriate flow of information and connectivity in neural circuits.

A number of molecules have been demonstrated to play important roles in neuronal polarization, most of which act by modulating the cytoskeleton or mediating the vesicular trafficking of proteins (Arimura and Kaibuchi, 2007). These molecules act to produce asymmetries in growth and responsiveness in the developing neuron, thus facilitating the establishment of structurally and functionally distinct compartments—axons and

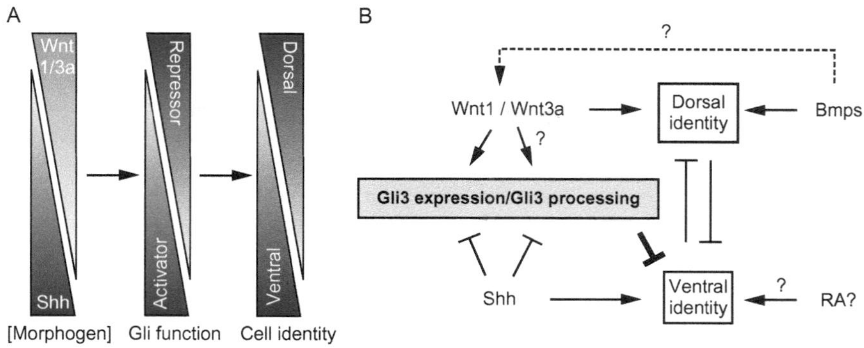

Figure 8.5 The relationships between patterning signals in the spinal cord. (A) The graded signaling of Wnt1/Wnt3a and Shh is transduced into an intracellular gradient of Gli activity. Shh promotes the formation of Gli activator forms, while Wnt1/3a directly induces the expression of Gli3, which acts as a transcriptional repressor and a potent antagonist of Shh signaling. In turn, the ventral and dorsal cell identities are dependent on the combinatorial expression of transcription factors that are activated or repressed by Gli. (B) Gli3 expression and its processing to the repressor form is central to the control of neural identity by Shh and Wnt. Wnt1/Wnt3a promote the expression of Gli3, which antagonizes Shh signaling and suppresses the expression of genes that specify ventral cell identities. On the other hand, Shh suppresses both the expression and proteolysis of Gli3, thus favoring the production of ventral cell identities. Both Shh and Wnts may have Gli3-independent activities in spinal cord patterning, and other signaling molecules such as retinoic acid (RA) and Bmps also contribute to the production of dorsal and ventral subtypes. Figure modified from Ulloa and Martí (2010).

dendrites—within one cell. Extracellular growth and guidance cues lead to a polarized intracellular response that allows neurons to determine axon and dendrite fate during development. While several studies now suggest that Wnt signaling is required in the establishment of axon–dendrite polarity in developing neurons (recently reviewed by Yang and Luo, 2011), there is currently no evidence that Shh influences neuronal polarization. Interestingly, however, Shh has been demonstrated to induce the expression of genes implicated in cell polarity, including *Dishevelled-associated activator of morphogenesis 1* (*Daam1*) (Fournier-Thibault *et al.*, 2009). Daam1 is a formin-homology (FH) protein that mediates reorganization of the actin cytoskeleton, and has previously been demonstrated to act downstream of Wnt signaling. During PCP, Daam1 is thought to control cell polarity and movement by acting as a scaffolding protein for the small GTPase Rho, a key regulator of cytoskeleton architecture (Govek *et al.*, 2005). The cooperation of Daam1 and Rho facilitates the formation of a Dvl–Rho complex, downstream of Wnt/Fzd signaling (Habas *et al.*, 2001). Interestingly, Daam1 has also been implicated in the orientation of the centrosome (Ang *et al.*, 2010), which is important for axon determination by influencing where the nascent axon will eventually protrude (de Anda *et al.*, 2005; Lefcort and Bentley, 1989; Zmuda and Rivas, 1998). Daam1 is also involved in the endocytosis of EphB receptor tyrosine kinases (Kida *et al.*, 2007), which may in turn influence the polarity of neurons by modulating neurite extension and/or dendritic branching (Chumley *et al.*, 2007; Witze *et al.*, 2008). Together (in a speculative view), these data suggest that Shh may be involved in the establishment of neuronal polarity by regulating Daam1 expression. Since Daam1 is also a downstream effector of Wnt/Fzd signaling, the Shh and Wnt pathways may interact during neuronal polarization. However, whether Shh is required for neuronal polarization and whether this activity is mediated by Daam1 remains to be tested.

V. Axon Guidance

Outgrowing axons are steered toward their targets by the expression of chemoattractive and chemorepulsive cues (Dickson, 2002; Tessier-Lavigne and Goodman, 1996). The growth cone, the sensory organelle at the leading edge of the outgrowing axon, contains receptors that transduce the environmental cues into signals that ultimately influence the actin cytoskeleton to cause growth cone turning, extension, retraction, or collapse (reviewed by Bashaw and Klein, 2010; Geraldo and Gordon-Weeks, 2009). The molecular mechanisms underlying axon guidance have been well characterized, and classically involve many signaling cues of the IgCAM, netrin, semaphorin, slit, and ephrin families, and receptors such as Dcc, neuropilin, plexin, roundabout, and Ephs (for a recent review, see Dickson and Zou, 2010). However, studies in recent years have shown that the morphogenetic gradients of Shh and Wnt that are established during tissue patterning are reused later in development as axon guidance cues. How is this molecular multitasking achieved, and how do Wnt and Shh cooperate in axon guidance?

A. Wnt and Shh signaling in commissural axon guidance

The developing spinal cord has been used as an accessible, informative *in vivo* model to elucidate the mechanisms of axon guidance. Commissural neurons in the dorsolateral spinal cord extend axons that grow toward and across the floor plate, at the ventral midline. After crossing the midline, the axons make an abrupt rostral turn and extend along the contralateral floor plate border (Fig. 8.6). The simplicity of this stereotypical trajectory belies the fact that the axons must make many complex pathfinding decisions en route to their target: they first seek out the floor plate, but later must escape from the midline and make a precise postcrossing turning decision in order

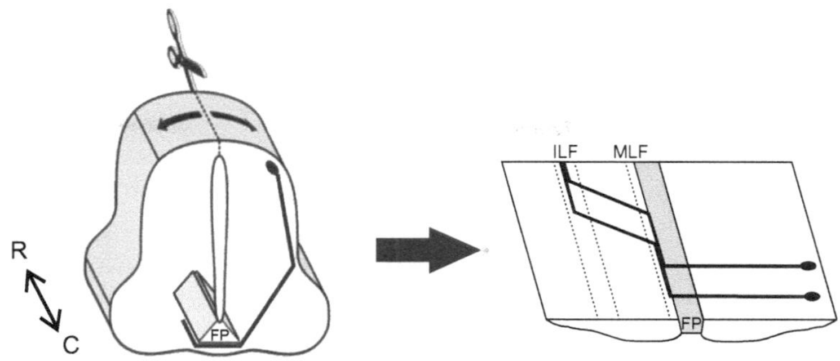

Figure 8.6 Schematic representation of the trajectory of commissural axons of the developing spinal cord. The cell bodies are located in the dorsolateral spinal cord. The axons project ventrally toward and across the floor plate (FP), before making a sharp rostral turn at the contralateral floor plate border and extending toward the head. An "open-book" preparation (right) is generated by cutting the spinal cord open at the roof plate and lying the tissue flat, thus facilitating analysis of the axon trajectories. Dorsal is to the top, the rostral–caudal axis is indicated. MLF, medial longitudinal fascicle; ILF, intermediate longitudinal fascicle.

to continue their trajectory on the opposite side. The floor plate is an intermediate target for these axons, and is a major source of attractive and repulsive, long- and short-range cues. The correct interpretation of these signals by the navigating axons is only possible by very precise spatiotemporal control over signaling activities, which in turn may be achieved by local translational control (reviewed by Yoon *et al.*, 2009), receptor trafficking, sorting, and cleavage (Keleman *et al.*, 2002; Zisman *et al.*, 2007), and hierarchical ordering amongst different guidance systems (Stein and Tessier-Lavigne, 2001). Both Wnt and Shh have been demonstrated as axon guidance cues for commissural axons. In axon guidance, both molecules seem to activate noncanonical pathways that permit rapid, local changes in growth cone dynamics.

1. Shh is both an attractant and a repellant cue in commissural axon guidance

Shh has been demonstrated to play two roles during the guidance of commissural axons in the spinal cord: first, it collaborates with netrin-1 in the dorsoventral axis to attract precrossing axons toward the floor plate (Charron *et al.*, 2003); and second, just a few hours later, it acts as a repulsive guidance cue in the longitudinal axis to guide postcrossing axons toward the brain (Bourikas *et al.*, 2005). This switch from attractant to repellant is due to a change in the growth cone receptors that detect and transduce the Shh signal. Precrossing axons express Smo (Charron *et al.*, 2003; Yam *et al.*, 2009) and Brother of Cdo (Boc) (Okada *et al.*, 2006), and are attracted to an increasing gradient of Shh in the floor plate via a noncanonical pathway. Rather than signaling retrogradely to the nucleus and eliciting Gli-dependent transcriptional changes, Shh attracts commissural axons by activating Src family kinases (SFKs) in the growth cone, in a Smo- and Boc-dependent manner (Yam *et al.*, 2009). In the presence of a Shh gradient, the SFKs Src and Fyn are rapidly phosphorylated and asymmetrically distributed in the growth cone of commissural neurons. This local, polarized response elicits growth cone turning, since SFKs can regulate cytoskeletal rearrangement and filopodial dynamics (Liu *et al.*, 2007; Robles *et al.*, 2005; Suter and Forscher, 2001). Thus, in attractive axon guidance, Shh signals via a rapid, local, transcription-independent mechanism (Fig. 8.7).

The signaling mechanism governing the repulsive response of postcrossing commissural axons to Shh is less clear. Consistent with its graded expression in the longitudinal axis of the chick floor plate (caudal^{high} to

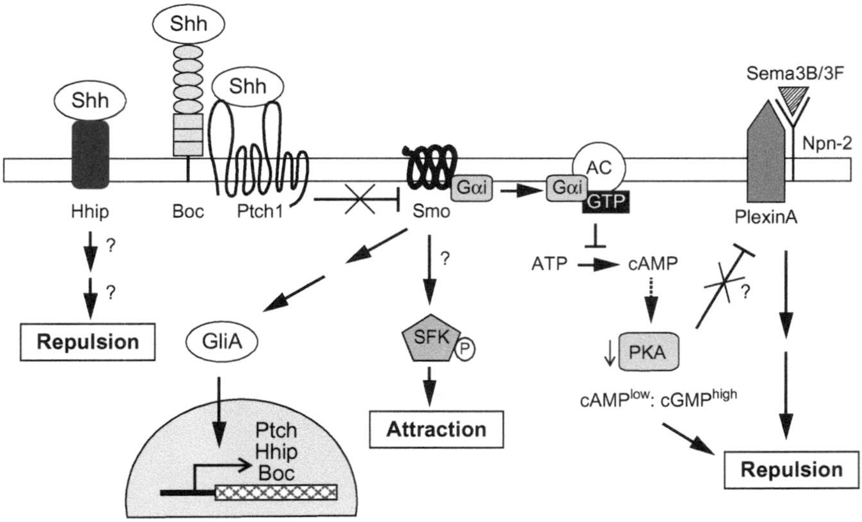

Figure 8.7 Shh signaling pathways in axon guidance. In repulsive axon guidance, Shh binds to Hhip, which activates signaling mechanisms that are so far uncharacterized. Alternatively, Shh binding to Ptch1 can induce repulsive axon guidance by reducing the intracellular cAMP concentration, which leads to reduced PKA activity. The change in intracellular cyclic nucleotides promotes axon repulsion. By an unknown mechanism, the low PKA/cAMP also acts as a "switch" to promote the activity of a PlexinA/Npn-2 complex, which can then transduce a repulsive axon guidance response to Sema3B/3F after midline crossing. In attractive axon guidance, the activation of Smo downstream of Ptch1 and Boc leads to phosphorylation of SFKs, which induce cytoskeletal rearrangements in the growth cone. The Gli-mediated transcription of Shh target genes likely also influences axon guidance decisions. AC, adenylate cyclase; Gαi, inhibitory G-protein α; Npn-2, Neuropilin-2.

rostral^{low}), Bourikas and colleagues (2005) demonstrated that Shh is a direct repellant for postcrossing commissural axons *in vitro*. However, Smo and Ptch1 are no longer expressed in commissural neurons once their axons have reached the midline. Rather, Hedgehog-interacting protein (Hhip) is transiently upregulated at the time when the commissural axons turn longitudinally. Knock down of Hhip by *in ovo* RNAi phenocopied the loss of Shh in longitudinal axon guidance, suggesting that Hhip is required to mediate the repulsive response of postcrossing commissural axons to Shh (Bourikas *et al.*, 2005). At present, it is not known if or how Shh acts as a longitudinal repellent in mammals. It is also unclear how Hhip transmits the Shh signal that results in repulsive axon guidance. Although Hhip is a type I transmembrane glycoprotein that can bind directly to all vertebrate hedgehog proteins (Chuang and McMahon, 1999), it does not contain an intracellular domain. However, Hhip could mediate repulsive axon guidance by acting as a Shh-binding unit in a receptor complex containing an as-yet unidentified signal-transducing component. Alternatively, Hhip might modulate the Shh response by sequestering it away from other receptors that transduce the attractive guidance signal, thus facilitating the switch in axon responsiveness from attraction to repulsion that occurs at the midline. Consistent with this idea, Hhip has been reported as a negative regulator of Shh signaling (Chuang and McMahon, 1999), and may compete with other Hedgehog receptors for Shh binding, including Cdo (McLellan *et al.*, 2008) and possibly Ptch (Bosanac *et al.*, 2009). The requirement for Shh-induced transcriptional activity in postcrossing axon guidance has not been formally tested, but the lack of dependence on Smo and Ptch1 suggests that repulsive axon guidance, like the attractive guidance response, occurs via a noncanonical Shh signaling pathway.

2. Wnt is an attractive cue in postcrossing commissural axon guidance

In the mouse floor plate, several Wnts (particularly *Wnt4* and *Wnt7b*) are expressed in a rostral^{high} to caudal^{low} gradient. In agreement with this expression pattern, Wnts are chemoattractive cues in the longitudinal axis of the spinal cord (Lyuksyutova *et al.*, 2003). This attractive response is mediated by the Wnt receptor Frizzled3 (Fzd3), since commissural axons of *Fzd3*$^{-/-}$ mice (Wang *et al.*, 2002) display random turning after midline crossing, which phenocopies the loss of *Wnt4*. Reminiscent of the Shh response, Wnt-induced axon guidance activity is mediated via a noncanonical, transcription-independent Wnt signaling pathway (Wolf *et al.*, 2008) (Fig. 8.8). Using pharmacological inhibitors and dominant-negative approaches in open-book spinal cord explant assays, Wolf and colleagues (2008) demonstrated that atypical protein kinase C (aPKC) activity was required for correct commissural axon guidance in the anterior–posterior axis. The authors propose that PKCζ is downstream of heterotrimeric

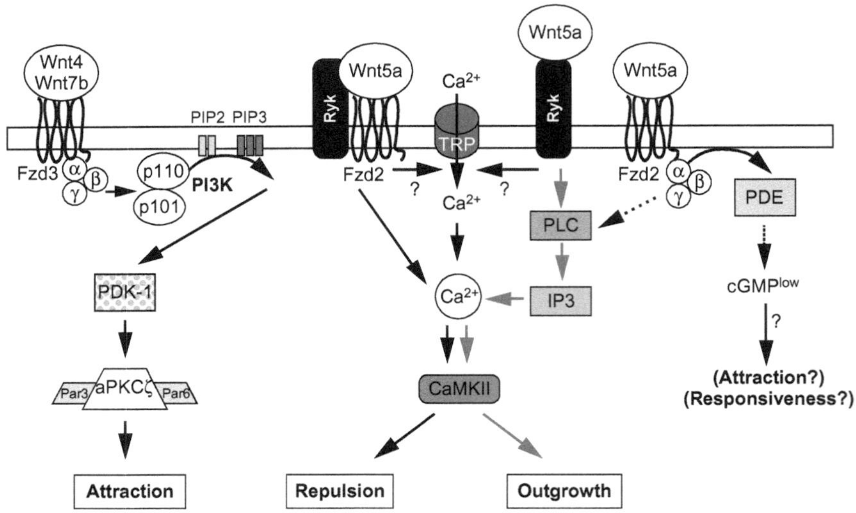

Figure 8.8 Wnt signaling pathways in axon guidance. In attractive axon guidance, a noncanonical pathway is activated in response to Wnt4/Wnt7b. Ligand binding to Fzd3 activates a Par3/Par6/aPKCζ complex through the p110 subunit of PI3K (figure modified from Sanchez-Camacho and Bovolenta, 2009). Noncanonical pathways also operate through Ca^{2+}, which activates CaMKII. While Wnt5a binding to Ryk receptors leads to axon outgrowth via PLC/IP3 and activation of membrane TRP channels, the binding of Wnt5a to Ryk and Fzd2 leads to axon repulsion. Wnt5a might also modulate growth cone behavior by modifying intracellular cGMP levels, via Gα and PDE.

G-proteins and phosphatidylinositol-3-kinases (PI3Ks). Atypical PKC has previously been linked to cell motility and is best known for its function in establishing apical–basal polarity (Karner *et al.*, 2006). In an extended parallel to the acquisition of polarity, Wnt4/Fzd3-mediated axon guidance requires the p110γ catalytic domain of PI3K, upstream of aPKC (Wolf *et al.*, 2008).

3. Wnt and Shh cooperate in postcrossing commissural axon guidance

The question arose whether mouse and chick use different morphogens to guide postcrossing commissural axons, or whether Shh and Wnt cooperate in this process (Stoeckli, 2006). To address this issue, Domanitskaya and colleagues (2010) assessed the expression patterns of *Wnts* in the developing chicken spinal cord. In contrast to findings in mice, *Wnt4* was not expressed in the floor plate in chick. Rather, *Wnt5a* and *Wnt7a* were good candidate guidance cues for postcrossing commissural axons, based on their spatiotemporal expression patterns. Indeed, knockdown of these molecules by *in ovo* RNAi demonstrated that Wnt5a/Wnt7a were required for correct longitudinal axon guidance. Surprisingly, the *Wnts* were not found to be expressed in a gradient in the chicken spinal cord. Instead, the graded expression of Wnt antagonists appears to generate a functional gradient of Wnt activity (Domanitskaya *et al.*, 2010). *Secreted frizzled-related protein 1* (*Sfrp1*), a demonstrated antagonist of Wnt activity on postcrossing commissural axons (Lyuksyutova *et al.*, 2003), was expressed strongly in the chicken floor plate and exhibited a clear caudal^{high} to rostral^{low} gradient along the longitudinal axis. Gain- and loss-of-function approaches demonstrated that the graded expression of Sfrp1 in turn shaped a gradient of Wnt activity that attracted postcrossing commissural axons rostrally.

Taken together, longitudinal axon guidance in the chick is achieved by a cooperation between the graded signaling activities of Shh and Wnt. On one hand, Shh acts as a repellent to "push" postcrossing axons toward the brain in a Hhip-dependent manner (Bourikas *et al.*, 2005). On the other hand, Wnts act as attractants to "pull" postcrossing axons rostrally. The Wnt activity gradient in chick is not directly transcriptional (as occurs for *Wnt4* in mouse), but is functionally generated via the graded expression of the Wnt antagonist, Sfrp1 (Domanitskaya *et al.*, 2010).

4. Shh indirectly regulates Wnt activity in postcrossing commissural axon guidance

Intriguingly, Sfrp is induced by Shh in mesodermal tissue (Lee *et al.*, 2000), and Shh is expressed in the same longitudinal gradient in the chicken floor plate as Sfrp1 (Bourikas *et al.*, 2005; Domanitskaya *et al.*, 2010). Therefore, Domanitskaya and colleagues (2010) asked whether Shh could also control Sfrp expression in the developing neural tube. When Shh was ectopically expressed in the chicken spinal cord, both Sfrp1 and Sfrp2 were induced. Interestingly, these findings demonstrate that the two guidance cues for postcrossing commissural axons in chicken (Shh and Wnt) do not act independently. Shh not only repels postcrossing commissural axons directly (Bourikas *et al.*, 2005), but also affects Wnt activity by inducing the expression of the Wnt antagonists Sfrp1 and Sfrp2. Thus, Shh also has an indirect role in postcommissural axon guidance by shaping the functional gradient of Wnt activity (Fig. 8.9). However, the signaling mechanism by which Shh induces Sfrps is currently unknown, as are the signaling mechanisms downstream of Wnt in postcrossing commissural axon guidance in the chicken.

B. Shh can modulate the response of commissural axons to other guidance cues

In addition to acting directly as an axon guidance cue and indirectly as a modulator of the axon guidance activity of Wnts, Shh may also regulate growth cone responsiveness via local intracellular mechanisms. One intracellular, noncanonical target of Shh signaling is the cyclic nucleotides,

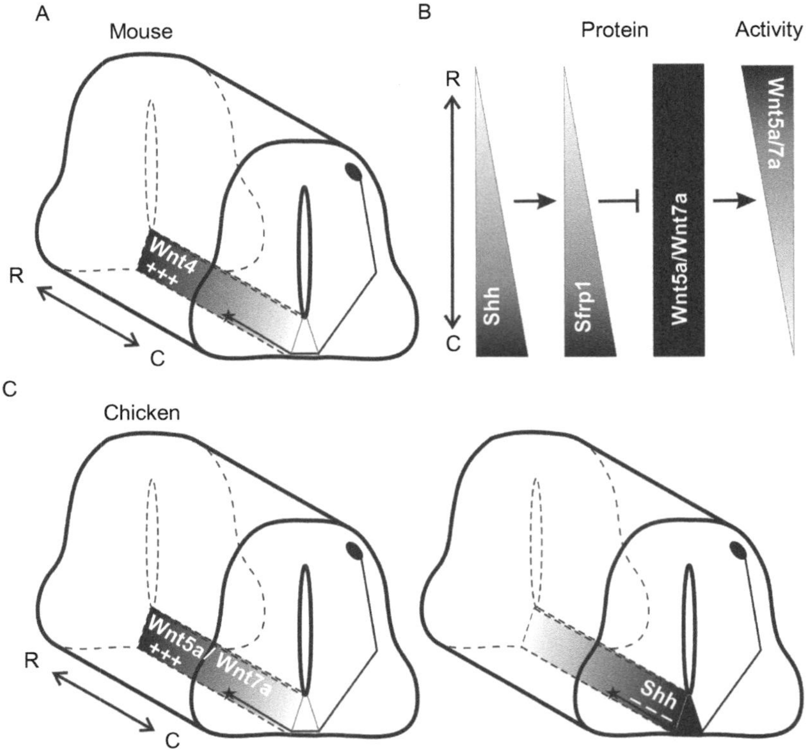

Figure 8.9 Control of postcrossing commissural axon guidance in mouse and chicken spinal cords. (A) In mouse, a transcriptional gradient of *Wnt4* in the floor plate attracts axons rostrally. (B) In the chicken spinal cord, Wnt5a and Wnt7a are not expressed in a gradient in the floor plate. Rather, an activity gradient of Wnts is generated along the rostrocaudal axis by the graded expression of the Wnt antagonist, Sfrp1, which is in turn induced by the graded expression of Shh. (C) In the chicken, commissural axons are directed to grow along the longitudinal axis via a cooperation between Wnt and Shh signaling. The rostral^{high} to caudal^{low} Wnt activity gradient, generated as a result of Shh activation of Sfrps, acts to attract axons rostrally. At the same time, the graded expression of Shh (caudal^{high} to rostral^{low}) repels axons from the caudal areas.

which are important for signaling in response to guidance cues (Song *et al.*, 1997). The ratio of cAMP/cGMP determines attractive or repulsive axonal responses (Song *et al.*, 1998), with cAMP^{high}/cGMP^{low} favoring attraction and cAMP^{low}/cGMP^{high} favoring repulsion. Since Shh can influence cyclic nucleotide levels by reducing the activity of protein kinase A (PKA), it may influence the responsiveness of axons to guidance cues (Parra and Zou, 2010).

This hypothesis is supported from several studies. Chick retinal ganglion cells (RGCs), for example, display a strong reduction in cytoplasmic cAMP and a subsequent lack of outgrowth following exposure to Shh (Trousse *et al.*, 2001). More directly, Parra and Zou (2010) demonstrated that the modulation of cAMP levels by Shh can act as a molecular "switch" at the floor plate, allowing commissural axons to acquire responsiveness to midline repellants and successfully exit to the contralateral side. In open-book explants, the authors found that disruption of Shh signaling (via function-blocking antibodies, expression of a dominant-negative form of Ptch, or shRNA-mediated knock down of Smo) resulted in severe guidance defects in commissural axons. The abnormalities that they saw (including stalling/knotting in the floor plate, recrossing, randomized turning, and overshooting of postcrossing axons) were reminiscent of those found in embryos deficient for Neuropilin-2, a receptor component for the secreted class 3 Semaphorins, Sema3B and Sema3F (Zou *et al.*, 2000). Together with the Slits, Sema3B/3F provide important chemorepulsive cues that expel commissural axons out of the midline during floor plate crossing. Commissural axons only acquire responsiveness to these repellants upon reaching the floor plate, suggesting that a floor plate-derived signal, such as Shh, might mediate a switch in guidance responses. Indeed, the authors found that exposure to Shh was required to induce a repellant response to Sema3B/3F in precrossing commissural axons *in vitro*. Finally, Parra and Zou (2010) tested whether cyclic nucleotides are important for this phenomenon. They first examined whether cAMP/PKA activity was needed for proper midline axon pathfinding. PKA activity had previously been shown to be coupled with Semaphorin–PlexinA-mediated repulsion *in vivo* (Terman and Kolodkin, 2004). Indeed, treatment of the explant cultures with the adenyl cyclase activator forskolin (which increases the level of cAMP) caused similar defects on commissural axons as seen with disrupted Shh signaling. Further, the Shh-induced repulsive response to Sema3B/3F by precrossing commissural axons was reduced *in vitro* in the presence of forskolin. Together, these findings suggest that Shh regulates cAMP levels, which in turn modulates the sensitivity of growth cones to Semaphorins (Fig. 8.7). Thus, Shh is not only a diffusible morphogen/chemoattractant, but can also act as a switch to activate growth cone responses to other axon guidance molecules.

1. Can Wnt modulate the response of commissural axons to other guidance cues?

Wnt signaling can also modulate levels of cyclic nucleotides (Wang, 2004; Wang and Malbon, 2003). In the noncanonical Wnt-Ca^{2+} pathway, the binding of Wnt5a to Fzd2 leads to an activation of several G-proteins, one of which (Gα) activates phosphodiesterase (PDE), which leads to a decreased concentration of cGMP (Ahumada *et al.*, 2002) (Fig. 8.8). However, a role for Wnt in modulating axon guidance responses via its regulation of cGMP levels is yet to be reported.

Interestingly, a similar pathway, acting through Gβ/γ and activation of phospholipase C (PLC), leads to increased concentrations of intracellular Ca^{2+} (Ahumada *et al.*, 2002). Like cyclic nucleotides, the cytosolic concentration of Ca^{2+} can influence the attractive or repulsive behavior of axons (Wen *et al.*, 2004; Zheng, 2000). Several lines of recent evidence now suggest that Wnt-mediated regulation of Ca^{2+} levels can influence both axon outgrowth and guidance, via the activation of distinct Wnt receptors (Ryk and Ryk/Fzd, respectively) (Hutchins *et al.*, 2011; Li *et al.*, 2009) (Fig. 8.8). Whether this modulation of Ca^{2+} levels by Wnt can also alter the growth cone responsiveness to other guidance cues (in an analogous manner to Shh modulation of cAMP) is yet to be investigated.

C. Do the canonical signaling pathways contribute to the axon guidance activities of Shh and Wnt?

The mechanisms described above delineate several noncanonical mechanisms by which Shh and Wnt can guide axons or modulate growth cone responsiveness. However, the transcriptional activities of these molecules are also likely to contribute to correct axon pathfinding, by regulating the levels and types of guidance receptors, coreceptor components, cell adhesion molecules, and/or modulators of signaling that are expressed. Together, such changes would facilitate alterations in growth cone responsiveness as axons navigate intermediate targets. However, the role of canonical signaling in regulating axon guidance behavior has not been extensively studied to date. In fact, the modulation of Sfrp1 levels by Shh, as described above, is the first report to directly support the idea that Shh-induced transcriptional activation affects axon guidance.

In a speculative view, commissural axons that approach the midline express molecules that promote Shh signaling (such as Boc, Smo, and low levels of Ptch). This would activate both the noncanonical attractive guidance response, as well as initiating transcriptional activity in the nucleus. As the axons approach the midline and encounter increasing levels of Shh expression, there is a progressive upregulation of Shh-induced genes. Some of these directly affect the signaling response to Shh, including *Ptch1/2* and *Hhip* (negative regulators), while others could influence or activate other signaling pathways that mediate axon guidance (such as *EphB4, ephrinB2, PlexinA2, or Adamts1*; Oliver *et al.*, 2003; Yu *et al.*, 2008a). Shh expression may also induce the expression of different molecules that affect axon guidance in the floor plate (e.g., Sfrp1; Domanitskaya *et al.*, 2010). In sum, Shh-induced transcriptional activity would modify the growth cone sensitivity to numerous guidance factors, thus making an important contribution to axon pathfinding. This activity, and that of Wnt-induced transcription, is of interest for future investigation.

VI. Synapse Formation

Upon reaching their appropriate target cell, navigating axons must initiate and stabilize contacts, remodel into synaptic boutons, and assemble presynaptic machinery. Similarly, the postsynaptic cell must recruit and arrange receptive components that detect and respond to neurotransmitters. These processes must be coordinated during synaptogenesis, which requires precise communication between the pre- and postsynaptic cells. Recent studies have shown that this transsynaptic signaling depends crucially on the Wnts. The role of Shh in synapse formation, in contrast, has been largely neglected. However, recent reports of the subcellular localization of Shh and its receptors suggest that, like Wnts, Shh may be involved in target recognition, synapse formation, and/or maintenance of circuits.

A. Wnt signaling has diverse presynaptic and postsynaptic effects

Wnts have recently been demonstrated to promote synaptic assembly, plasticity, and maintenance by signaling to both pre- and postsynaptic cells, in both vertebrates and invertebrates (reviewed by Budnik and Salinas, 2011; Marqués, 2005; Salinas and Zou, 2008). An extensive review of the role of Wnts in synaptic development and plasticity is beyond the scope of this chapter. Below, we focus on the central synapse to provide some brief examples of selected activities of Wnt signaling, with the intention of demonstrating the general mechanisms by which these signaling molecules can direct the coordinated assembly and function of synapses.

1. Wnts direct remodeling and protein assembly in presynaptic and postsynaptic cells

Wnt signaling was first implicated in synapse formation in the vertebrate nervous system. Briefly, Wnt7a was demonstrated to act via its effector Dishevelled-1 (Dvl1) to regulate the remodeling of the axon terminal and the accumulation of presynaptic components (Ahmad-Annuar *et al.*, 2006; Hall *et al.*, 2000). In presynaptic hippocampal neurons, this activity of Wnt7a was mediated by the receptor Frizzled5 (Fzd5) (Sahores *et al.*, 2010). The expression of Dvl1 has been demonstrated to increase the stability of microtubules through a divergent canonical pathway that inhibits GSK3β but does not require transcription (Ciani *et al.*, 2004). Rather, inhibition of GSK3β reduces phosphorylation of the microtubule-associated protein MAP1B, thus linking Wnt signaling to the regulation of microtubule stability and growth cone remodeling (Lucas *et al.*, 1998). Interestingly, Wnt signaling also regulates presynaptic protein assembly at the *Drosophila* neuromuscular junction. In *Drosophila*, Wnt/Wg is secreted from presynaptic fibers and has been proposed to act in an autocrine fashion to mediate microtubule dynamics and presynaptic protein assembly (Miech *et al.*, 2008; Packard *et al.*, 2002).

Wnt signaling does not act only in the presynaptic cell, however. Wnt5a, for example, preferentially stimulates postsynaptic assembly (Farías *et al.*, 2009), suggesting that different Wnts might promote protein assembly in either pre- or postsynaptic cells to regulate synapse formation. In the postsynaptic cell, Wnt5a does not act via GSK3β, but instead operates via an unknown receptor to activate a noncanonical pathway that leads to clustering of PSD95 through JNK. Thus, different Wnt-mediated signaling pathways are activated in pre- and postsynaptic cells, probably by the presence of distinct Wnt receptors.

2. Wnts contribute to activity-dependent plasticity of synapses

Wnt signaling has also been demonstrated to promote activity-dependent synaptic connectivity. The expression and release of Wnt proteins is enhanced by depolarization (Ataman *et al.*, 2008; Wayman *et al.*, 2006; Yu and Malenka, 2003). In hippocampal neurons, high-frequency stimulation increases the level of cell-surface Fzd5 and its localization to synapses, in a Wnt-dependent manner (Sahores *et al.*, 2010). In addition, remodeling of hippocampal circuits in response to an enriched environment was repressed when Wnt signaling was locally blocked (Gogolla *et al.*, 2009). Together, these studies link Wnt activity to synaptic remodeling in response to behavioral experience and activity.

B. Is Shh involved in synaptogenesis?

In contrast to the Wnts, there is currently no direct evidence that Shh signaling is involved in synaptogenesis. However, Shh and its receptors are appropriately expressed to support such activities. In the retina of adult hamster, Shh-N peptides are synthesized and transported axonally along the optic nerve to the superior colliculus *in vivo* (Traiffort *et al.*, 2001). Similarly, Shh is distributed intracellularly in regulated secretory vesicles and on the extracellular surface of the soma, dendrites, and axons of rat RGCs. The secretion of Shh can be induced upon depolarization of RGCs, suggesting a link between neuronal activity and Shh release (Beug *et al.*, 2011). In a recent study, Ptch and Smo were localized to dendrites and spines of mature rat hippocampal neurons, suggesting that Shh signaling may also influence synaptic transmission (Petralia *et al.*, 2011). Surprisingly, in an earlier study, a presynaptic localization of Smo was reported (Masdeu *et al.*, 2007). Taken together, the subcellular localizations of Shh and its receptors makes their involvement in synapse formation an attractive possibility to be examined in future studies.

VII. Other Factors Influencing Shh and Wnt Pathways

Evidence from outside the nervous system suggests a number of other molecules, pathways, or processes in which Shh and Wnt signaling could converge. At the membrane, both Shh and Wnt are modulated by heparan sulfate proteoglycans (HSPGs). Intracellularly, the signaling pathways rely (to a greater or lesser extent) on the function of primary cilia. How might these factors influence Shh and Wnt activity during neural development?

A. Heparan sulfate proteoglycans

HSPGs are comprised of a core protein, to which long heparan sulfate glycosaminoglycan (GAG) side chains are attached. They are crucial to many aspects of development, and are thought to act via two general mechanisms: (1) by controlling the availability or distribution of secreted ligands; and/or (2) by modulating the efficiency of receptor–ligand interactions (Rhiner and Hengartner, 2006). In the nervous system, they are found both on neurons and in the extracellular matrix, and are involved in proliferation, cell migration, axon guidance, and synaptogenesis (Chan *et al.*, 2009; Jen *et al.*, 2009; van Vactor *et al.*, 2006; Yamaguchi, 2001). Although Wnt and Shh both bind to HSPGs, the role of these proteoglycans in regulating gene function in the nervous system is only beginning to be understood, since classical genetic disruption of HSPG function typically perturbs development during early embryogenesis.

Of particular interest to the regulation of cell signaling are HSPGs belonging to the glypican and syndecan families. These molecules are associated with the cell membrane, either by a glycosylphosphatidylinositol (GPI)-anchor (glypican) or a transmembrane domain (syndecan), thus allowing them to influence the dispersal of secreted ligands away from their source (i.e., gradient formation), the clustering of coreceptor complexes, and/or the selective activation of intracellular signaling pathways. While syndecans have been implicated in the regulation of axon guidance (Smart *et al.*, 2011; Steigemann *et al.*, 2004) and in some aspects of Wnt activity (Muñoz *et al.*, 2006), there is limited evidence that they modulate the signaling or distribution of Shh. In contrast, the glypicans are absolutely required to regulate the dispersal, turnover, and signaling responses to both Shh and Wnt (reviewed by Filmus *et al.*, 2008).

Evidence for the functional roles of glypican HSPGs during development comes primarily from studies in *Drosophila*. While mammals have six glypican genes (Gpc1–6), there are only two in the fly: *Dally and Dally-like* (*Dlp*). Dally and Dlp play crucial roles in regulating the distributions of the secreted morphogens Wg/Wnt and Hh/Shh, by controlling their cell–cell movement across the wing imaginal discs (Han *et al.*, 2004, 2005). This regulation of morphogen distribution by Gpc is also evident in vertebrates: *Gpc3* mutant mice display overgrowth, renal cystic dysplasia, and limb defects, consistent with disruption of Shh, Wnt, and Bmp pathways (Cano-Gauci *et al.*, 1999; Paine-Saunders *et al.*, 2000; Song *et al.*, 2005). However, the way in which Gpc specifically controls these signaling activities is unknown. In fact, Gpc may act as positive or negative regulator of these cues, depending on the context. For example, in *Drosophila* Dlp plays dual roles in Wg distribution and signaling, suppressing high levels of Wg signaling but enhancing low levels (Wu *et al.*, 2010). While Dlp is a positive regulator of Hh signaling in *Drosophila* (Williams *et al.*, 2010), Gpc3 in mice acts as a competitive inhibitor of Shh signaling and induces the endocytosis and degradation of Shh from the cell surface (Capurro *et al.*, 2008). The picture emerging from these and other examples is that vertebrate and invertebrate HSPGs have diverse and specialized roles in the modulation of Wnt and Shh signaling, demonstrating both positive and negative activities (Filmus *et al.*, 2008). The relative contributions of the GAG side chains, the core protein, and GPI-anchor of glypican in regulating these behaviors are just beginning to be elucidated (Capurro *et al.*, 2008; Wu *et al.*, 2010; Yan *et al.*, 2009).

Further studies are essential to determine how glypican contributes to the regulation of Shh and Wnt signaling *in vivo*. These studies are complicated by the fact that different glypican family members display overlapping and distinct functions, and are promiscuous in their ligand binding. To overcome such complications, a recent study used an elegant method to perturb Shh–proteoglycan interactions in vertebrates: rather than mutating the proteoglycans themselves (which may variously affect the activities of Shh, Wnt, Bmp, and/or FGF), the authors took the alternative approach of mutating *Shh* (Chan *et al.*, 2009). Shh contains a Cardin–Weintraub motif (KRRHPKK) at its N-terminal that mediates interactions with proteoglycans. By mutating this motif, the interactions between Shh and proteoglycans were reduced without disturbing Shh's affinity for Ptch (Rubin *et al.*, 2002). By selectively inhibiting Shh–proteoglycan interactions, Chan and colleagues (2009) were able to demonstrate in a mouse model that proteoglycans were required for the proliferative activity of Shh, but not for tissue patterning. The mutants displayed a significantly reduced ratio of Gli2 activator to Gli2 repressor compared to wildtypes, suggesting that Shh–proteoglycan interactions regulate the timing and nature of Gli2-dependent transcription and subsequent gene expression, including that of *cyclinD1*. Further studies will elucidate additional mechanisms by which HSPGs (especially the glypicans) regulate the activities of Shh and Wnt during neural circuit formation.

B. Primary cilia

The primary cilium is a small microtubule-based organelle that projects from the plasma membrane of almost all nonmitotic cells throughout the body (Wheatley *et al.*, 1996). Although these nonmotile structures were initially regarded as vestigial and nonfunctional, a flurry of papers in the past decade has sparked renewed interest in primary cilia as crucial organelles for signal transduction (recently reviewed by Goetz and Anderson, 2010; Tasouri and Tucker, 2011). Defects in the structure or function of the primary cilia now define a new group of disorders known as the "ciliopathies" that are characterized by broad, partially overlapping phenotypes resulting from abnormalities in cell determination, differentiation, and maturation during organogenesis (reviewed by Hildebrandt *et al.*, 2011).

In particular, the Hedgehog signaling pathway is critically dependent on intact primary cilia during organogenesis (Haycraft *et al.*, 2005; Huangfu and Anderson, 2005; Huangfu *et al.*, 2003). Ptch1 is found at the base of the primary cilium and, upon Shh binding, becomes internalized into the cytoplasm (Rohatgi *et al.*, 2007). Smo is not constitutively associated with the cilium, but is translocated there in the presence of Shh (Corbit *et al.*, 2005) and activates cilium-localized Gli2 and Gli3 (Kim *et al.*, 2009; Wen *et al.*, 2010). Both the transcriptional activation of Gli2 and Gli3 and the processing of Gli3 to its transcriptional repressor form are dependent on intact ciliary function. In recent years, dozens of papers have linked primary cilia with Shh signaling in diverse models of organogenesis and cancer. Since Shh signaling is crucial to many aspects of neural development, it is not surprising that mutants affecting the structure and/or function of primary cilia display defects in the formation of the spinal cord, forebrain, hippocampus, cerebellum, and hindbrain (Lee and Gleeson, 2011; Tasouri and Tucker, 2011).

Although the role of cilia in Shh signaling is well established, it remains controversial whether Wnt signaling requires the primary cilium. Components of Wnt signaling pathways are localized to basal bodies and/or primary cilia (Lancaster *et al.*, 2011; Watanabe *et al.*, 2003), and many ciliary mutants display abnormalities in canonical or noncanonical Wnt signaling in different organs (e.g., Corbit *et al.*, 2008; Jonassen *et al.*, 2008; McDermott *et al.*, 2010; Willaredt *et al.*, 2008). However, ciliary mutants usually do not display more overt phenotypes that are classically correlated with disturbances in Wnt function (such as defects in gastrulation), as would be expected if Wnt signaling were dependent on primary cilia (Huang and Schier, 2009; Ocbina *et al.*, 2009). Reconciling these disparate findings is difficult, but a consensus may be that the primary cilium is not essential for Wnt signaling (as it is for Shh), but is required to modulate Wnt activity in some contexts. How could this be achieved? One study suggests that β-catenin is subjected to degradation at the ciliary basal body, and that the absence of primary cilia would then allow cytoplasmic β-catenin to accumulate and upregulate Wnt responses (Gerdes *et al.*, 2007). Another recent study suggests an alternative mechanism that is dependent on the subcellular localization of Jouberin (Jbn), a positive regulator of canonical Wnt signaling (Lancaster *et al.*, 2009). Upon stimulation of ciliated cells with Wnt3a, Jbn was found to sequester β-catenin into the primary cilium, thus keeping nuclear levels of β-catenin low and dampening the Wnt signaling response. In contrast, in the absence of cilium, β-catenin protein that escapes degradation in response to Wnt3a binding, translocates with Jbn to the nucleus and activates Wnt pathways (Lancaster *et al.*, 2011). Thus, the presence of primary cilia seems to modulate a cell's sensitivity to a Wnt ligand.

The definitive mechanisms by which the primary cilium regulates Shh and Wnt signaling are currently areas of active research. In neural development, we are still far from understanding the roles of primary cilia and many questions remain unsolved. During which developmental stages do neurons possess primary cilia? Do ciliary defects lead to aberrant neural circuit formation? Why do different cell types respond differently to the loss of cilia? Do mechanisms exist that bypass the requirement for cilia in specific signaling pathways? Future studies will provide further insights into the roles of this key signaling organelle in neuronal structure and function.

VIII. Conclusions and Future Directions

Signaling in response to Shh and Wnt is crucial for embryonic development in all animal species described to date. In recent years, studies have revealed that these molecules do not always signal in isolation, but rather via a number of cross-talk mechanisms that can integrate multiple inputs to influence biological processes. In the development of the neural tube, Wnt and Shh coordinate their activities to drive cell proliferation, antagonize each other to mediate cell patterning, and cooperate to guide axons toward their targets. Taken together, the integration of the Shh and Wnt signaling pathways enables numerous cell types to communicate with each other in a manner that is flexible, dynamic, and sensitive.

Future studies will reveal whether additional cross-talk occurs at the level of cell surface receptors, signaling modulators, or intracellular signaling components. The activity of these molecules and their interpretation by different cells may also be influenced by posttranslational modifications or the contributions of nonreceptor binding molecules such as HSPGs. Determining how individual cells integrate all this signal information is one of the most challenging questions to address in the future.

REFERENCES

Abbasi, A. A., Paparidis, Z., Malik, S., Goode, D. K., Callaway, H., Elgar, G., and Grzeschik, K.-H. (2007). Human GLI3 intragenic conserved non-coding sequences are tissue-specific enhancers. *PLoS One* **2,** e366.

Aberle, H., Bauer, A., Stappert, J., Kispert, A., and Kemler, R. (1997). Beta-catenin is a target for the ubiquitin–proteasome pathway. *EMBO J.* **16,** 3797–3804.

Ahmad-Annuar, A., Ciani, L., Simeonidis, I., Herreros, J., Fredj, N. B., Rosso, S. B., Hall, A., Brickley, S., and Salinas, P. C. (2006). Signaling across the synapse: A role for Wnt and Dishevelled in presynaptic assembly and neurotransmitter release. *J. Cell Biol.* **174,** 127–139.

Ahumada, A., Slusarski, D. C., Liu, X., Moon, R. T., Malbon, C. C., and Wang, H.-Y. (2002). Signaling of rat Frizzled-2 through phosphodiesterase and cyclic GMP. *Science* **298,** 2006–2010.

Alvarez-Medina, R., Cayuso, J., Okubo, T., Takada, S., and Martí, E. (2008). Wnt canonical pathway restricts graded Shh/Gli patterning activity through the regulation of Gli3 expression. *Development* **135,** 237–247.

Alvarez-Medina, R., Le Dreau, G., Ros, M., and Martí, E. (2009). Hedgehog activation is required upstream of Wnt signalling to control neural progenitor proliferation. *Development* **136,** 3301–3309.

Ang, S.-F., Zhao, Z.-S., Lim, L., and Manser, E. (2010). DAAM1 is a formin required for centrosome re-orientation during cell migration. *PLoS One* **5,** e13064.

Arimura, N., and Kaibuchi, K. (2007). Neuronal polarity: From extracellular signals to intracellular mechanisms. *Nat. Rev. Neurosci.* **8,** 194–205.

Ataman, B., Ashley, J., Gorczyca, M., Ramachandran, P., Fouquet, W., Sigrist, S. J., and Budnik, V. (2008). Rapid activity-dependent modifications in synaptic structure and function require bidirectional Wnt signaling. *Neuron* **57,** 705–718.

Aza-Blanc, P., Lin, H. Y., Ruiz i Altaba, A., and Kornberg, T. B. (2000). Expression of the vertebrate Gli proteins in Drosophila reveals a distribution of activator and repressor activities. *Development* **127,** 4293–4301.

Bai, C. B., Auerbach, W., Lee, J. S., Stephen, D., and Joyner, A. L. (2002). Gli2, but not Gli1, is required for initial Shh signaling and ectopic activation of the Shh pathway. *Development* **129,** 4753–4761.

Bashaw, G. J., and Klein, R. (2010). Signaling from axon guidance receptors. *Cold Spring Harb. Perspect. Biol.* **2,** a001941.

Beug, S. T., Parks, R. J., McBride, H. M., and Wallace, V. A. (2011). Processing-dependent trafficking of Sonic hedgehog to the regulated secretory pathway in neurons. *Mol. Cell. Neurosci.* **46,** 583–596.

Bhanot, P., Brink, M., Samos, C. H., Hsieh, J. C., Wang, Y., Macke, J. P., Andrew, D., Nathans, J., and Nusse, R. (1996). A new member of the frizzled family from Drosophila functions as a Wingless receptor. *Nature* **382,** 225–230.

Bonner, J., Gribble, S. L., Veien, E. S., Nikolaus, O. B., Weidinger, G., and Dorsky, R. I. (2008). Proliferation and patterning are mediated independently in the dorsal spinal cord downstream of canonical Wnt signaling. *Dev. Biol.* **313,** 398–407.

Bosanac, I., Maun, H. R., Scales, S. J., Wen, X., Lingel, A., Bazan, J. F., de Sauvage, F. J., Hymowitz, S. G., and Lazarus, R. A. (2009). The structure of SHH in complex with HHIP reveals a recognition role for the Shh pseudo active site in signaling. *Nat. Struct. Mol. Biol.* **16,** 691–697.

Bourikas, D., Pekarik, V., Baeriswyl, T., Grunditz, A., Sadhu, R., Nardó, M., and Stoeckli, E. T. (2005). Sonic hedgehog guides commissural axons along the longitudinal axis of the spinal cord. *Nat. Neurosci.* **8,** 297–304.

Briscoe, J., and Ericson, J. (2001). Specification of neuronal fates in the ventral neural tube. *Curr. Opin. Neurobiol.* **11,** 43–49.

Budnik, V., and Salinas, P. C. (2011). Wnt signaling during synaptic development and plasticity. *Curr. Opin. Neurobiol.* **21,** 151–159.

Cano-Gauci, D. F., Song, H. H., Yang, H., McKerlie, C., Choo, B., Shi, W., Pullano, R., Piscione, T. D., Grisaru, S., Soon, S., *et al.* (1999). Glypican-3-deficient mice exhibit developmental overgrowth and some of the abnormalities typical of Simpson–Golabi–Behmel syndrome. *J. Cell Biol.* **146,** 255–264.

Capurro, M. I., Xu, P., Shi, W., Li, F., Jia, A., and Filmus, J. (2008). Glypican-3 inhibits Hedgehog signaling during development by competing with patched for Hedgehog binding. *Dev. Cell* **14,** 700–711.

Cayuso, J., Ulloa, F., Cox, B., Briscoe, J., and Martí, E. (2006). The Sonic hedgehog pathway independently controls the patterning, proliferation and survival of neuroepithelial cells by regulating Gli activity. *Development* **133,** 517–528.

Chan, J. A., Balasubramanian, S., Witt, R. M., Nazemi, K. J., Choi, Y., Pazyra-Murphy, M. F., Walsh, C. O., Thompson, M., and Segal, R. A. (2009). Proteoglycan interactions with Sonic Hedgehog specify mitogenic responses. *Nat. Neurosci.* **12,** 409–417.

Charron, F., Stein, E., Jeong, J., McMahon, A. P., and Tessier-Lavigne, M. (2003). The morphogen sonic hedgehog is an axonal chemoattractant that collaborates with netrin-1 in midline axon guidance. *Cell* **113,** 11–23.

Chesnutt, C., Burrus, L. W., Brown, A. M. C., and Niswander, L. (2004). Coordinate regulation of neural tube patterning and proliferation by TGFbeta and WNT activity. *Dev. Biol.* **274,** 334–347.

Chiang, C., Litingtung, Y., Lee, E., Young, K. E., Corden, J. L., Westphal, H., and Beachy, P. A. (1996). Cyclopia and defective axial patterning in mice lacking Sonic hedgehog gene function. *Nature* **383,** 407–413.

Chuang, P.-T., and McMahon, A. P. (1999). Vertebrate Hedgehog signalling modulated by induction of a Hedgehog-binding protein. *Nature* **397,** 617–621.

Chumley, M. J., Catchpole, T., Silvany, R. E., Kernie, S. G., and Henkemeyer, M. (2007). EphB receptors regulate stem/progenitor cell proliferation, migration, and polarity during hippocampal neurogenesis. *J. Neurosci.* **27,** 13481–13490.

Ciani, L., Krylova, O., Smalley, M. J., Dale, T. C., and Salinas, P. C. (2004). A divergent canonical WNT-signaling pathway regulates microtubule dynamics: Dishevelled signals locally to stabilize microtubules. *J. Cell Biol.* **164,** 243–253.

Corbit, K. C., Aanstad, P., Singla, V., Norman, A. R., Stainier, D. Y. R., and Reiter, J. F. (2005). Vertebrate Smoothened functions at the primary cilium. *Nature* **437,** 1018–1021.

Corbit, K. C., Shyer, A. E., Dowdle, W. E., Gaulden, J., Singla, V., Chen, M.-H., Chuang, P.-T., and Reiter, J. F. (2008). Kif3a constrains beta-catenin-dependent Wnt signalling through dual ciliary and non-ciliary mechanisms. *Nat. Cell Biol.* **10,** 70–76.

Dai, P., Akimaru, H., Tanaka, Y., Maekawa, T., Nakafuku, M., and Ishii, S. (1999). Sonic hedgehog-induced activation of the Gli1 promoter is mediated by GLI3. *J. Biol. Chem.* **274,** 8143–8152.

de Anda, F. C., Pollarolo, G., Da Silva, J. S., Camoletto, P. G., Feiguin, F., and Dotti, C. G. (2005). Centrosome localization determines neuronal polarity. *Nature* **436,** 704–708.

Dessaud, E., Ribes, V., Balaskas, N., Yang, L. L., Pierani, A., Kicheva, A., Novitch, B. G., Briscoe, J., Sasai, N., and Scott, M. P. (2010). Dynamic assignment and maintenance of positional identity in the ventral neural tube by the morphogen sonic hedgehog. *PLoS Biol.* **8,** e1000382.

Dickinson, M. E., Krumlauf, R., and McMahon, A. P. (1994). Evidence for a mitogenic effect of Wnt-1 in the developing mammalian central nervous system. *Development* **120,** 1453–1471.

Dickson, B. J. (2002). Molecular mechanisms of axon guidance. *Science* **298,** 1959–1964.

Dickson, B. J., and Zou, Y. (2010). Navigating intermediate targets: The nervous system midline. *Cold Spring Harb. Perspect. Biol.* **2,** a002055.

Domanitskaya, E., Wacker, A., Mauti, O., Baeriswyl, T., Esteve, P., Bovolenta, P., and Stoeckli, E. T. (2010). Sonic hedgehog guides post-crossing commissural axons both directly and indirectly by regulating Wnt activity. *J. Neurosci.* **30,** 11167–11176.

Eastman, Q., and Grosschedl, R. (1999). Regulation of LEF-1/TCF transcription factors by Wnt and other signals. *Curr. Opin. Cell Biol.* **11,** 233–240.

Ericson, J., Morton, S., Kawakami, A., Roelink, H., and Jessell, T. M. (1996). Two critical periods of Sonic hedgehog signaling required for the specification of motor neuron identity. *Cell* **87,** 661–673.

Farías, G. G., Alfaro, I. E., Cerpa, W., Grabowski, C. P., Godoy, J. A., Bonansco, C., and Inestrosa, N. C. (2009). Wnt-5a/JNK signaling promotes the clustering of PSD-95 in hippocampal neurons. *J. Biol. Chem.* **284,** 15857–15866.

Filmus, J., Capurro, M., and Rast, J. (2008). Glypicans. *Genome Biol.* **9,** 224.

Fournier-Thibault, C., Blavet, C., Jarov, A., Bajanca, F., Thorsteinsdóttir, S., and Duband, J.-L. (2009). Sonic hedgehog regulates integrin activity, cadherin contacts, and cell polarity to orchestrate neural tube morphogenesis. *J. Neurosci.* **29,** 12506–12520.

Geraldo, S., and Gordon-Weeks, P. R. (2009). Cytoskeletal dynamics in growth-cone steering. *J. Cell Sci.* **122,** 3595–3604.

Gerdes, J. M., Liu, Y., Zaghloul, N. A., Leitch, C. C., Lawson, S. S., Kato, M., Beachy, P. A., Beales, P. L., DeMartino, G. N., Fisher, S., *et al.* (2007). Disruption of the basal body compromises proteasomal function and perturbs intracellular Wnt response. *Nat. Genet.* **39,** 1350–1360.

Goetz, S. C., and Anderson, K. V. (2010). The primary cilium: A signalling centre during vertebrate development. *Nat. Rev. Genet.* **11,** 331–344.

Gogolla, N., Galimberti, I., Deguchi, Y., and Caroni, P. (2009). Wnt signaling mediates experience-related regulation of synapse numbers and mossy fiber connectivities in the adult hippocampus. *Neuron* **62,** 510–525.

Govek, E.-E., Newey, S. E., and van Aelst, L. (2005). The role of the Rho GTPases in neuronal development. *Genes Dev.* **19,** 1–49.

Habas, R., Kato, Y., and He, X. (2001). Wnt/Frizzled activation of Rho regulates vertebrate gastrulation and requires a novel Formin homology protein Daam1. *Cell* **107,** 843–854.

Hall, A. C., Lucas, F. R., and Salinas, P. C. (2000). Axonal remodeling and synaptic differentiation in the cerebellum is regulated by WNT-7a signaling. *Cell* **100,** 525–535.

Hallikas, O., Palin, K., Sinjushina, N., Rautiainen, R., Partanen, J., Ukkonen, E., and Taipale, J. (2006). Genome-wide prediction of mammalian enhancers based on analysis of transcription-factor binding affinity. *Cell* **124,** 47–59.

Han, C., Belenkaya, T. Y., Wang, B., and Lin, X. (2004). Drosophila glypicans control the cell-to-cell movement of Hedgehog by a dynamin-independent process. *Development* **131,** 601–611.

Han, C., Yan, D., Belenkaya, T. Y., and Lin, X. (2005). Drosophila glypicans Dally and Dally-like shape the extracellular Wingless morphogen gradient in the wing disc. *Development* **132,** 667–679.

Haycraft, C. J., Banizs, B., Aydin-Son, Y., Zhang, Q., Michaud, E. J., and Yoder, B. K. (2005). Gli2 and Gli3 localize to cilia and require the intraflagellar transport protein Polaris for processing and function. *PLoS Genet.* **1,** e53.

Helms, A. W., and Johnson, J. E. (2003). Specification of dorsal spinal cord interneurons. *Curr. Opin. Neurobiol.* **13,** 42–49.

Hildebrandt, F., Benzing, T., and Katsanis, N. (2011). Ciliopathies. *N. Engl. J. Med.* **364,** 1533–1543.

Hollyday, M., McMahon, J. A., and McMahon, A. P. (1995). Wnt expression patterns in chick embryo nervous system. *Mech. Dev.* **52,** 9–25.

Huang, P., and Schier, A. F. (2009). Dampened Hedgehog signaling but normal Wnt signaling in zebrafish without cilia. *Development* **136,** 3089–3098.

Huangfu, D., and Anderson, K. V. (2005). Cilia and Hedgehog responsiveness in the mouse. *Proc. Natl. Acad. Sci. USA* **102,** 11325–11330.

Huangfu, D., Liu, A., Rakeman, A. S., Murcia, N. S., Niswander, L., and Anderson, K. V. (2003). Hedgehog signalling in the mouse requires intraflagellar transport proteins. *Nature* **426,** 83–87.

Hui, C. C., Slusarski, D., Platt, K. A., Holmgren, R., and Joyner, A. L. (1994). Expression of three mouse homologs of the Drosophila segment polarity gene cubitus interruptus, Gli, Gli-2, and Gli-3, in ectoderm- and mesoderm-derived tissues suggests multiple roles during postimplantation development. *Dev. Biol.* **162,** 402–413.

Hutchins, B. I., Li, L., and Kalil, K. (2011). Wnt/calcium signaling mediates axon growth and guidance in the developing corpus callosum. *Dev. Neurobiol.* **71,** 269–283.

Ille, F., Atanasoski, S., Falk, S., Ittner, L. M., Märki, D., Büchmann-Møller, S., Wurdak, H., Suter, U., Taketo, M. M., and Sommer, L. (2007). Wnt/BMP signal integration regulates the balance between proliferation and differentiation of neuroepithelial cells in the dorsal spinal cord. *Dev. Biol.* **304,** 394–408.

Ingham, P. W., and McMahon, A. P. (2001). Hedgehog signaling in animal development: Paradigms and principles. *Genes Dev.* **15,** 3059–3087.

Jacob, J., and Briscoe, J. (2003). Gli proteins and the control of spinal-cord patterning. *EMBO Rep.* **4,** 761–765.

Jen, Y.-H. L., Musacchio, M., and Lander, A. D. (2009). Glypican-1 controls brain size through regulation of fibroblast growth factor signaling in early neurogenesis. *Neural Dev.* **4,** 33.

Jessell, T. M. (2000). Neuronal specification in the spinal cord: Inductive signals and transcriptional codes. *Nat. Rev. Genet.* **1,** 20–29.

Jonassen, J. A., San Agustin, J., Follit, J. A., and Pazour, G. J. (2008). Deletion of IFT20 in the mouse kidney causes misorientation of the mitotic spindle and cystic kidney disease. *J. Cell Biol.* **183,** 377–384.

Karner, C., Wharton, K. A., and Carroll, T. J. (2006). Apical–basal polarity Wnt signaling and vertebrate organogenesis. *Semin. Cell Dev. Biol.* **17,** 214–222.

Keleman, K., Rajagopalan, S., Cleppien, D., Teis, D., Paiha, K., Huber, L. A., Technau, G. M., and Dickson, B. J. (2002). Comm sorts robo to control axon guidance at the Drosophila midline. *Cell* **110,** 415–427.

Kida, Y. S., Sato, T., Miyasaka, K. Y., Suto, A., and Ogura, T. (2007). Daam1 regulates the endocytosis of EphB during the convergent extension of the zebrafish notochord. *Proc. Natl. Acad. Sci. USA* **104,** 6708–6713.

Kim, J., Kato, M., and Beachy, P. A. (2009). Gli2 trafficking links Hedgehog-dependent activation of Smoothened in the primary cilium to transcriptional activation in the nucleus. *Proc. Natl. Acad. Sci. USA* **106,** 21666–21671.

Komiya, Y., and Habas, R. (2008). Wnt signal transduction pathways. *Organogenesis* **4,** 68–75.

Lancaster, M. A., Louie, C. M., Silhavy, J. L., Sintasath, L., Decambre, M., Nigam, S. K., Willert, K., and Gleeson, J. G. (2009). Impaired Wnt-beta-catenin signaling disrupts adult renal homeostasis and leads to cystic kidney ciliopathy. *Nat. Med.* **15,** 1046–1054.

Lancaster, M. A., Schroth, J., and Gleeson, J. G. (2011). Subcellular spatial regulation of canonical Wnt signalling at the primary cilium. *Nat. Cell Biol.* **13,** 700–707.

Lee, J. E., and Gleeson, J. G. (2011). Cilia in the nervous system: Linking cilia function and neurodevelopmental disorders. *Curr. Opin. Neurol.* **24,** 98–105.

Lee, J., Platt, K. A., Censullo, P., and Ruiz i Altaba, A. (1997). Gli1 is a target of Sonic hedgehog that induces ventral neural tube development. *Development* **124,** 2537–2552.

Lee, C. S., Buttitta, L. A., May, N. R., Kispert, A., and Fan, C. M. (2000). SHH-N upregulates Sfrp2 to mediate its competitive interaction with WNT1 and WNT4 in the somitic mesoderm. *Development* **127,** 109–118.

Lefcort, F., and Bentley, D. (1989). Organization of cytoskeletal elements and organelles preceding growth cone emergence from an identified neuron in situ. *J. Cell Biol.* **108,** 1737–1749.

Li, L., Hutchins, B. I., and Kalil, K. (2009). Wnt5a induces simultaneous cortical axon outgrowth and repulsive axon guidance through distinct signaling mechanisms. *J. Neurosci.* **29,** 5873–5883.

Litingtung, Y., and Chiang, C. (2000). Specification of ventral neuron types is mediated by an antagonistic interaction between Shh and Gli3. *Nat. Neurosci.* **3,** 979–985.

Liu, G., Li, W., Gao, X., Li, X., Jürgensen, C., Park, H.-T., Shin, N.-Y., Yu, J., He, M.-L., Hanks, S. K., *et al.* (2007). p130CAS is required for netrin signaling and commissural axon guidance. *J. Neurosci.* **27,** 957–968.

Lobjois, V., Benazeraf, B., Bertrand, N., Medevielle, F., and Pituello, F. (2004). Specific regulation of cyclins D1 and D2 by FGF and Shh signaling coordinates cell cycle progression, patterning, and differentiation during early steps of spinal cord development. *Dev. Biol.* **273,** 195–209.

Logan, C. Y., and Nusse, R. (2004). The Wnt signaling pathway in development and disease. *Annu. Rev. Cell Dev. Biol.* **20,** 781–810.

Lucas, F. R., Goold, R. G., Gordon-Weeks, P. R., and Salinas, P. C. (1998). Inhibition of GSK-3beta leading to the loss of phosphorylated MAP-1B is an early event in axonal remodelling induced by WNT-7a or lithium. *J. Cell Sci.* **111,** 1351–1361.

Lyuksyutova, A. I., Lu, C.-C., Milanesio, N., King, L. A., Guo, N., Wang, Y., Nathans, J., Tessier-Lavigne, M., and Zou, Y. (2003). Anterior–posterior guidance of commissural axons by Wnt-frizzled signaling. *Science* **302,** 1984–1988.

Marqués, G. (2005). Morphogens and synaptogenesis in Drosophila. *J. Neurobiol.* **64,** 417–434.

Martí, E., and Bovolenta, P. (2002). Sonic hedgehog in CNS development: One signal, multiple outputs. *Trends Neurosci.* **25,** 89–96.

Martí, E., Bumcrot, D. A., Takada, R., and McMahon, A. P. (1995a). Requirement of 19K form of Sonic hedgehog for induction of distinct ventral cell types in CNS explants. *Nature* **375,** 322–325.

Martí, E., Takada, R., Bumcrot, D. A., Sasaki, H., and McMahon, A. P. (1995b). Distribution of Sonic hedgehog peptides in the developing chick and mouse embryo. *Development* **121,** 2537–2547.

Masdeu, C., Bernard, V., Faure, H., Traiffort, E., and Ruat, M. (2007). Distribution of Smoothened at hippocampal mossy fiber synapses. *Neuroreport* **18,** 395–399.

McDermott, K. M., Liu, B. Y., Tlsty, T. D., and Pazour, G. J. (2010). Primary cilia regulate branching morphogenesis during mammary gland development. *Curr. Biol.* **20,** 731–737.

McLellan, J. S., Zheng, X., Hauk, G., Ghirlando, R., Beachy, P. A., and Leahy, D. J. (2008). The mode of Hedgehog binding to Ihog homologues is not conserved across different phyla. *Nature* **455,** 979–983.

Megason, S. G., and McMahon, A. P. (2002). A mitogen gradient of dorsal midline Wnts organizes growth in the CNS. *Development* **129,** 2087–2098.

Meyer, N. P., and Roelink, H. (2003). The amino-terminal region of Gli3 antagonizes the Shh response and acts in dorsoventral fate specification in the developing spinal cord. *Dev. Biol.* **257,** 343–355.

Miech, C., Pauer, H.-U., He, X., and Schwarz, T. L. (2008). Presynaptic local signaling by a canonical wingless pathway regulates development of the Drosophila neuromuscular junction. *J. Neurosci.* **28,** 10875–10884.

Müller, T., Anlag, K., Wildner, H., Britsch, S., Treier, M., and Birchmeier, C. (2005). The bHLH factor Olig3 coordinates the specification of dorsal neurons in the spinal cord. *Genes Dev.* **19,** 733–743.

Muñoz, R., Moreno, M., Oliva, C., Orbenes, C., and Larraín, J. (2006). Syndecan-4 regulates non-canonical Wnt signalling and is essential for convergent and extension movements in Xenopus embryos. *Nat. Cell Biol.* **8,** 492–500.

Muroyama, Y., Fujihara, M., Ikeya, M., Kondoh, H., and Takada, S. (2002). Wnt signaling plays an essential role in neuronal specification of the dorsal spinal cord. *Genes Dev.* **16,** 548–553.

Ocbina, P. J. R., Tuson, M., and Anderson, K. V. (2009). Primary cilia are not required for normal canonical Wnt signaling in the mouse embryo. *PLoS One* **4,** e6839.

Okada, A., Charron, F., Morin, S., Shin, D. S., Wong, K., Fabre, P. J., Tessier-Lavigne, M., and McConnell, S. K. (2006). Boc is a receptor for sonic hedgehog in the guidance of commissural axons. *Nature* **444,** 369–373.

Oliver, T. G., Grasfeder, L. L., Carroll, A. L., Kaiser, C., Gillingham, C. L., Lin, S. M., Wickramasinghe, R., Scott, M. P., and Wechsler-Reya, R. J. (2003). Transcriptional profiling of the Sonic hedgehog response: A critical role for N-myc in proliferation of neuronal precursors. *Proc. Natl. Acad. Sci. USA* **100,** 7331–7336.

Packard, M., Koo, E. S., Gorczyca, M., Sharpe, J., Cumberledge, S., and Budnik, V. (2002). The Drosophila Wnt, wingless, provides an essential signal for pre- and postsynaptic differentiation. *Cell* **111,** 319–330.

Paine-Saunders, S., Viviano, B. L., Zupicich, J., Skarnes, W. C., and Saunders, S. (2000). Glypican-3 controls cellular responses to Bmp4 in limb patterning and skeletal development. *Dev. Biol.* **225,** 179–187.

Pan, Y., Bai, C. B., Joyner, A. L., and Wang, B. (2006). Sonic hedgehog signaling regulates Gli2 transcriptional activity by suppressing its processing and degradation. *Mol. Cell. Biol.* **26,** 3365–3377.

Parr, B. A., Shea, M. J., Vassileva, G., and McMahon, A. P. (1993). Mouse Wnt genes exhibit discrete domains of expression in the early embryonic CNS and limb buds. *Development* **119,** 247–261.

Parra, L. M., and Zou, Y. (2010). Sonic hedgehog induces response of commissural axons to Semaphorin repulsion during midline crossing. *Nat. Neurosci.* **13,** 29–35.

Persson, M., Stamataki, D., te Welscher, P., Andersson, E., Böse, J., Rüther, U., Ericson, J., and Briscoe, J. (2002). Dorsal–ventral patterning of the spinal cord requires Gli3 transcriptional repressor activity. *Genes Dev.* **16,** 2865–2878.

Petralia, R. S., Schwartz, C. M., Wang, Y.-X., Mattson, M. P., and Yao, P. J. (2011). Subcellular localization of patched and smoothened, the receptors for sonic hedgehog signaling, in the hippocampal neuron. *J. Comp. Neurol.* **519,** 3684–3699.

Rhiner, C., and Hengartner, M. O. (2006). Sugar antennae for guidance signals: Syndecans and glypicans integrate directional cues for navigating neurons. *ScientificWorldJournal* **6,** 1024–1036.

Robles, E., Woo, S., and Gomez, T. M. (2005). Src-dependent tyrosine phosphorylation at the tips of growth cone filopodia promotes extension. *J. Neurosci.* **25,** 7669–7681.

Roelink, H., Porter, J. A., Chiang, C., Tanabe, Y., Chang, D. T., Beachy, P. A., and Jessell, T. M. (1995). Floor plate and motor neuron induction by different concentrations of the amino-terminal cleavage product of sonic hedgehog autoproteolysis. *Cell* **81,** 445–455.

Rohatgi, R., Milenkovic, L., and Scott, M. P. (2007). Patched1 regulates hedgehog signaling at the primary cilium. *Science* **317,** 372–376.

Rubin, J. B., Choi, Y., and Segal, R. A. (2002). Cerebellar proteoglycans regulate sonic hedgehog responses during development. *Development* **129,** 2223–2232.

Ruiz i Altaba, A. (1998). Combinatorial Gli gene function in floor plate and neuronal inductions by Sonic hedgehog. *Development* **125,** 2203–2212.

Ruiz i Altaba, A. (1999). Gli proteins encode context-dependent positive and negative functions: Implications for development and disease. *Development* **126,** 3205–3216.

Ruiz i Altaba, A., Mas, C., and Stecca, B. (2007). The Gli code: An information nexus regulating cell fate, stemness and cancer. *Trends Cell Biol.* **17,** 438–447.

Sahores, M., Gibb, A., and Salinas, P. C. (2010). Frizzled-5, a receptor for the synaptic organizer Wnt7a, regulates activity-mediated synaptogenesis. *Development* **137,** 2215–2225.

Salinas, P. C., and Zou, Y. (2008). Wnt signaling in neural circuit assembly. *Annu. Rev. Neurosci.* **31,** 339–358.

Sanchez-Camacho, C., and Bovolenta, P. (2009). Emerging mechanisms in morphogen-mediated axon guidance. *Bioessays* **31,** 1013–1025.

Shirasaki, R., and Pfaff, S. L. (2002). Transcriptional codes and the control of neuronal identity. *Annu. Rev. Neurosci.* **25,** 251–281.

Shtutman, M., Zhurinsky, J., Simcha, I., Albanese, C., D'Amico, M., Pestell, R., and Ben-Ze'ev, A. (1999). The cyclin D1 gene is a target of the beta-catenin/LEF-1 pathway. *Proc. Natl. Acad. Sci. USA* **96,** 5522–5527.

Smart, A. D., Course, M. M., Rawson, J., Selleck, S., van Vactor, D., and Johnson, K. G. (2011). Heparan sulfate proteoglycan specificity during axon pathway formation in the Drosophila embryo. *Dev. Neurobiol.* **71,** 608–618.

Song, H. J., Ming, G. L., and Poo, M. M. (1997). cAMP-induced switching in turning direction of nerve growth cones. *Nature* **388,** 275–279.

Song, H., Ming, G., He, Z., Lehmann, M., McKerracher, L., Tessier-Lavigne, M., and Poo, M. (1998). Conversion of neuronal growth cone responses from repulsion to attraction by cyclic nucleotides. *Science* **281,** 1515–1518.

Song, H. H., Shi, W., Xiang, Y.-Y., and Filmus, J. (2005). The loss of glypican-3 induces alterations in Wnt signaling. *J. Biol. Chem.* **280,** 2116–2125.

Stamataki, D., Ulloa, F., Tsoni, S. V., Mynett, A., and Briscoe, J. (2005). A gradient of Gli activity mediates graded Sonic Hedgehog signaling in the neural tube. *Genes Dev.* **19,** 626–641.

Steigemann, P., Molitor, A., Fellert, S., Jäckle, H., and Vorbrüggen, G. (2004). Heparan sulfate proteoglycan syndecan promotes axonal and myotube guidance by slit/robo signaling. *Curr. Biol.* **14,** 225–230.

Stein, E., and Tessier-Lavigne, M. (2001). Hierarchical organization of guidance receptors: Silencing of netrin attraction by slit through a Robo/DCC receptor complex. *Science* **291,** 1928–1938.

Stoeckli, E. T. (2006). Longitudinal axon guidance. *Curr. Opin. Neurobiol.* **16,** 35–39.

Suter, D. M., and Forscher, P. (2001). Transmission of growth cone traction force through apCAM–cytoskeletal linkages is regulated by Src family tyrosine kinase activity. *J. Cell Biol.* **155,** 427–438.

Tamai, K., Semenov, M., Kato, Y., Spokony, R., Liu, C., Katsuyama, Y., Hess, F., Saint-Jeannet, J. P., and He, X. (2000). LDL-receptor-related proteins in Wnt signal transduction. *Nature* **407,** 530–535.

Tanabe, Y., and Jessell, T. M. (1996). Diversity and pattern in the developing spinal cord. *Science* **274,** 1115–1123.

Tasouri, E., and Tucker, K. L. (2011). Primary cilia and organogenesis: Is Hedgehog the only sculptor? *Cell Tissue Res.* **345,** 21–40.

Terman, J. R., and Kolodkin, A. L. (2004). Nervy links protein kinase a to plexin-mediated semaphorin repulsion. *Science* **303,** 1204–1207.

Tessier-Lavigne, M., and Goodman, C. S. (1996). The molecular biology of axon guidance. *Science* **274,** 1123–1133.

Tetsu, O., and McCormick, F. (1999). Beta-catenin regulates expression of cyclin D1 in colon carcinoma cells. *Nature* **398,** 422–426.

Traiffort, E., Moya, K. L., Faure, H., Hässig, R., and Ruat, M. (2001). High expression and anterograde axonal transport of aminoterminal sonic hedgehog in the adult hamster brain. *Eur. J. Neurosci.* **14,** 839–850.

Trousse, F., Martí, E., Gruss, P., Torres, M., and Bovolenta, P. (2001). Control of retinal ganglion cell axon growth: A new role for Sonic hedgehog. *Development* **128,** 3927–3936.

Ulloa, F., and Martí, E. (2010). Wnt won the war: Antagonistic role of Wnt over Shh controls dorso-ventral patterning of the vertebrate neural tube. *Dev. Dyn.* **239,** 69–76.

Ulloa, F., Itasaki, N., and Briscoe, J. (2007). Inhibitory Gli3 activity negatively regulates Wnt/beta-catenin signaling. *Curr. Biol.* **17,** 545–550.

van Vactor, D., Wall, D. P., and Johnson, K. G. (2006). Heparan sulfate proteoglycans and the emergence of neuronal connectivity. *Curr. Opin. Neurobiol.* **16,** 40–51.

Wang, H.-Y. (2004). WNT-frizzled signaling via cyclic GMP. *Front. Biosci.* **9,** 1043–1047.

Wang, H.-Y., and Malbon, C. C. (2003). Wnt signaling, Ca^{2+}, and cyclic GMP: Visualizing Frizzled functions. *Science* **300,** 1529–1530.

Wang, Y., Thekdi, N., Smallwood, P. M., Macke, J. P., and Nathans, J. (2002). Frizzled-3 is required for the development of major fiber tracts in the rostral CNS. *J. Neurosci.* **22,** 8563–8573.

Watanabe, D., Saijoh, Y., Nonaka, S., Sasaki, G., Ikawa, Y., Yokoyama, T., and Hamada, H. (2003). The left–right determinant Inversin is a component of node monocilia and other 9+0 cilia. *Development* **130,** 1725–1734.

Wayman, G. A., Impey, S., Marks, D., Saneyoshi, T., Grant, W. F., Derkach, V., and Soderling, T. R. (2006). Activity-dependent dendritic arborization mediated by CaM-kinase I activation and enhanced CREB-dependent transcription of Wnt-2. *Neuron* **50,** 897–909.

Wen, Z., Guirland, C., Ming, G.-L., and Zheng, J. Q. (2004). A CaMKII/calcineurin switch controls the direction of Ca(2+)-dependent growth cone guidance. *Neuron* **43,** 835–846.

Wen, X., Lai, C. K., Evangelista, M., Hongo, J.-A., de Sauvage, F. J., and Scales, S. J. (2010). Kinetics of hedgehog-dependent full-length Gli3 accumulation in primary cilia and subsequent degradation. *Mol. Cell. Biol.* **30,** 1910–1922.

Wheatley, D. N., Wang, A. M., and Strugnell, G. E. (1996). Expression of primary cilia in mammalian cells. *Cell Biol. Int.* **20,** 73–81.

Willaredt, M. A., Hasenpusch-Theil, K., Gardner, H. A. R., Kitanovic, I., Hirschfeld-Warneken, V. C., Gojak, C. P., Gorgas, K., Bradford, C. L., Spatz, J., Wölfl, S., *et al.* (2008). A crucial role for primary cilia in cortical morphogenesis. *J. Neurosci.* **28,** 12887–12900.

Williams, E. H., Pappano, W. N., Saunders, A. M., Kim, M.-S., Leahy, D. J., and Beachy, P. A. (2010). Dally-like core protein and its mammalian homologues mediate stimulatory and inhibitory effects on Hedgehog signal response. *Proc. Natl. Acad. Sci. USA* **107,** 5869–5874.

Witze, E. S., Litman, E. S., Argast, G. M., Moon, R. T., and Ahn, N. G. (2008). Wnt5a control of cell polarity and directional movement by polarized redistribution of adhesion receptors. *Science* **320,** 365–369.

Wolf, A. M., Lyuksyutova, A. I., Fenstermaker, A. G., Shafer, B., Lo, C. G., and Zou, Y. (2008). Phosphatidylinositol-3-kinase-atypical protein kinase C signaling is required for Wnt attraction and anterior–posterior axon guidance. *J. Neurosci.* **28,** 3456–3467.

Wu, Y., Belenkaya, T. Y., and Lin, X. (2010). Dual roles of Drosophila glypican Dally-like in Wingless/Wnt signaling and distribution. *Methods Enzymol.* **480,** 33–50.

Yam, P. T., Langlois, S. D., Morin, S., and Charron, F. (2009). Sonic hedgehog guides axons through a noncanonical, Src-family-kinase-dependent signaling pathway. *Neuron* **62,** 349–362.

Yamaguchi, Y. (2001). Heparan sulfate proteoglycans in the nervous system: Their diverse roles in neurogenesis, axon guidance, and synaptogenesis. *Semin. Cell Dev. Biol.* **12,** 99–106.

Yan, D., Wu, Y., Feng, Y., Lin, S.-C., and Lin, X. (2009). The core protein of glypican Dally-like determines its biphasic activity in wingless morphogen signaling. *Dev. Cell* **17,** 470–481.

Yang, G.-Y., and Luo, Z.-G. (2011). Implication of Wnt signaling in neuronal polarization. *Dev. Neurobiol.* **71,** 495–507.

Yoon, B. C., Zivraj, K. H., and Holt, C. E. (2009). Local translation and mRNA trafficking in axon pathfinding. *Results Probl. Cell Differ.* **48,** 269–288.

Yu, X., and Malenka, R. C. (2003). Beta-catenin is critical for dendritic morphogenesis. *Nat. Neurosci.* **6,** 1169–1177.

Yu, M., Gipp, J., Yoon, J. W., Iannaccone, P., Walterhouse, D., and Bushman, W. (2008a). Sonic hedgehog-responsive genes in the fetal prostate. *J. Biol. Chem.* **284,** 5620–5629.

Yu, W., McDonnell, K., Taketo, M. M., and Bai, C. B. (2008b). Wnt signaling determines ventral spinal cord cell fates in a time-dependent manner. *Development* **135,** 3687–3696.

Zechner, D., Fujita, Y., Hülsken, J., Müller, T., Walther, I., Taketo, M. M., Crenshaw, E. B., Birchmeier, W., and Birchmeier, C. (2003). Beta-catenin signals regulate cell growth and the balance between progenitor cell expansion and differentiation in the nervous system. *Dev. Biol.* **258,** 406–418.

Zechner, D., Müller, T., Wende, H., Walther, I., Taketo, M. M., Crenshaw, E. B., Treier, M., Birchmeier, W., and Birchmeier, C. (2007). Bmp and Wnt/beta-catenin signals control expression of the transcription factor Olig3 and the specification of spinal cord neurons. *Dev. Biol.* **303,** 181–190.

Zheng, J. Q. (2000). Turning of nerve growth cones induced by localized increases in intracellular calcium ions. *Nature* **403,** 89–93.

Zisman, S., Marom, K., Avraham, O., Rinsky-Halivni, L., Gai, U., Kligun, G., Tzarfaty-Majar, V., Suzuki, T., and Klar, A. (2007). Proteolysis and membrane capture of F-spondin generates combinatorial guidance cues from a single molecule. *J. Cell Biol.* **178,** 1237–1249.

Zmuda, J. F., and Rivas, R. J. (1998). The Golgi apparatus and the centrosome are localized to the sites of newly emerging axons in cerebellar granule neurons in vitro. *Cell Motil. Cytoskeleton* **41,** 18–38.

Zou, Y., Stoeckli, E., Chen, H., and Tessier-Lavigne, M. (2000). Squeezing axons out of the gray matter: A role for slit and semaphorin proteins from midline and ventral spinal cord. *Cell* **102,** 363–375.

CHAPTER NINE

HEDGEHOG/GLI CONTROL BY UBIQUITINATION/ACETYLATION INTERPLAY

Alberto Gulino,*,‡ Lucia Di Marcotullio,* Gianluca Canettieri,* Enrico De Smaele,† *and* Isabella Screpanti*

Contents

Abstract

Hedgehog is a key morphogen regulating development and leading to tumorigenesis, when hyperactivated. Hedgehog signaling is mediated by transcriptional effectors belonging to the Gli family. Ubiquitination-related posttranslational modifications of the Gli transcription factors, leading to proteasome-dependent proteolytic cleavage or massive degradation, represent an important mechanism of regulation of the pathway. Gli ubiquitination is controlled by a number of E3 ligases belonging to the RING/Cullin and HECT families. These E3 ligases are regulated by several members of the Hh pathway itself (e.g., Smo-activated kinases) as well as by proteins belonging to other signaling cascades (i.e., Numb-activated Itch). These proteolytic signals

* Department of Molecular Medicine, Sapienza University, Rome, Italy
† Department of Experimental Medicine, Sapienza University, Rome, Italy
‡ Neuromed Institute, Pozzilli, Italy

Vitamins and Hormones, Volume 88
ISSN 0083-6729, DOI: 10.1016/B978-0-12-394622-5.00009-2

finally suppress Gli function either directly or indirectly (i.e., suppression of HDAC1-mediated Gli deacetylation). The complex of these regulatory circuitries finely tunes Hedgehog signaling providing a tight control of developmental processes, the subversion of which leads to tumorigenesis. To this regard, these ubiquitination processes represent promising targets for novel therapeutic strategies.

I. Introduction

Hedgehog (Hh) signaling is an evolutionary conserved pathway initially identified in *Drosophila* where it is a key regulator of developmental processes, determining proper embryonic patterning, cell fate, and body-segment polarity (Nusslein-Volhard and Wieschaus, 1980). In *Drosophila*, the secreted protein Hh binds the receptor Patched (Ptc), resulting in the activation of the Cubitus interruptus (Ci) transcription factor, whch is positively regulated by Fused (Fu) serine/threonine kinase and suppressed by Suppressor of Fused (Sufu) and the kinesin-like molecule Costal2 (Cos2) (Hooper and Scott, 2005).

The dual Ci function has expanded into three Gli proteins (Gli1, Gli2, and Gli3) in vertebrates. Gli1 is a strong constitutive transcriptional activator. In contrast, Gli2 and Gli3 display both positive (Gli2A and Gli3A) and negative (Gli2R and Gli3R) transcriptional functions. Indeed, they can be converted into cleaved forms provided of transcriptional repressor function (Gli2R and Gli3R). The shift from "A" toward "R" forms is favored for Gli3 with respect to Gli2. Therefore, Gli3 mainly behaves as a transcriptional repressor while Gli2 as an activator (reviewed in Teglund and Toftgard, 2010).

Both in *Drosophila* and mammals, Hh/Ci/Gli activates the expression of a number of genes involved in several functions related to cell development, such as cell proliferation and differentiation (e.g., CyclinD1 and D2, N-Myc, Wnts, PdgfRa, Igf2, FoxM1, FoxA2, Nkx2.2, FoxF1, Myf5, Hes1, Igfbp3), survival (Bcl2), self-renewal, cell fate determination (Bmi1, Nanog), angiogenesis (Vegf), epithelial–mesenchymal transition (Snail1, Sip1, Elk1, and Msx2), and invasiveness (osteopontin) in several tissues, in addition to the positive Gli1 and negative Ptch1 and Hhip regulators of the pathway, forming a feedback loops enhancing or diminishing the Hh response (Po *et al.*, 2010; reviewed in Teglund and Toftgard, 2010; Stecca ad Ruiz i Altaba, 2010). While several additional candidate transcriptional targets are being identified by ChIP-chip technology (Lee *et al.*, 2010), one important and poorly investigated question is the identification of target genes that are specifically regulated by Gli-A compared to Gli-R.

Remarkably, when hyperactivated by genetic and/or epigenetic alterations, Hh pathway promotes tumorigenesis in several tissues (e.g., medulloblastoma, glioblastoma, basal cell carcinoma, rhabdomyosarcoma, breast, prostate, ovarian, lung, colon, gastric, liver, pancreas cancers) implicating the pathological role of subversion of Hh-regulated developmental processes (Barakat *et al.*, 2010; Teglund and Toftgard, 2010; Kelleher, 2011). An example is represented by heterozygous Ptch1+/− mice which are prone to medulloblastoma, reproducing the human Gorlin's syndrome (Goodrich *et al.*, 1997).

II. Ubiquitin-Dependent Regulation of Hh Signaling

A critical regulation of Hh signaling is mediated by posttranslational modifications (i.e., ubiquitination) of the transcriptionl effectors.

Postsynthetic protein modifications by ubiquitin-dependent processes have emerged as crucial mechanisms by which protein stability, activity, or localization is controlled, thus regulating a number of cellular pathways (Welchman *et al.*, 2005). The ubiquitination process involves the action of different enzymes (E1, E2, E3, and E4) and terminates with the transfer of ubiquitin to lysine residues of substrate proteins, as polyubiquitin chains or monoubiquitination or multimonoubiquitination events. The major determinants of the ubiquitination mechanism are the E3 ubiquitin ligases. Based on the sequence homology of their E2-binding domains, E3s can be generally classified into subfamilies: the two main classes belonging to the really interesting new gene (RING) or homologous to E6-AP C-terminus (HECT) families and the recently identified U-box (Hatakeyama *et al.*, 2001), PHD (Boname and Stevenson, 2001), and HUL-1 (Van Sant *et al.*, 2001), all known to bind specific degron sequences of the target proteins (Pickart, 2001; Welchman *et al.*, 2005). Each one of these poly- or mono- or multimonoubiquitination processes results in a different fate of ubiquitinated target protein and functional outcome (Welchman *et al.*, 2005). Protein polyubiquitination frequently leads to proteolytic processing of the target, allowing recognition by the 26S proteasome system.

Regulation of Hh signaling by ubiquitin-dependent processing has been first identified in *Drosophila*. In the absence of Hh, the full-length Ci (Ci-FL) is phosphorylated first by protein kinase A (PKA) and subsequently by glycogen synthase kinase 3 (Gsk3/Shaggy) and casein kinase 1 (CK1) at multiple sites of the Ci C-terminus, thus driving the Slimb (Supernumerary limbs, an F-box protein of the SCF (Skip1-Cul1-F-box) E3 ligase complex)-mediated ubiquitination and proteasome-dependent proteolytic processing to generate a C-terminally truncated transcriptional repressor, Ci-Rep (Jia *et al.*, 2002, 2005; Price and Kalderon, 2002; Smelkinson and Kalderon, 2006). Ci-Rep is a repressor of Hh target genes (Aza-Blanc *et al.*, 1997). In contrast, the presence of Hh inhibits Ci processing leading to transcriptional activation (Chen *et al.*, 1999; Wang and Holmgren, 1999).

In addition to such an ubiquitination and proteasome-mediated Ci degradation is also induced by an additional E3 ligase complex formed by HIB (Hh-induced MATH and BTB domain protein) or Rdx, a substrate-binding adaptor for Cullin3 (Cul3)-based ligase (Kent *et al.*, 2006; Zhang *et al.*, 2006).

III. Increasing the Complexity of Hh Signaling in Vertebrates: The Cul1- and Cul3-Dependent Ubiquitination Processes

According to the expansion of Ci into three Gli proteins (Gli1, Gli2, and Gli3), the complexity of Gli regulation increases in vertebrates. The ubiquitin-dependent pathway, although conserved in vertebrates, has been adapted for controlling in multiple ways the three different Ci-homologs. In mammals, Gli2 and Gli3 receive still poorly characterized signals from Smo in the primary cilium, a critical platform for Hh signaling in mammals (reviewed in Teglund and Toftgard, 2010). Gli2 acts mainly as a strong activator with a weak repressor function, whereas Gli3 has weak activator activity and is processed to generate Gli3-R, a strong transcriptional repressor. In the absence of Hh, Ptc receptor is located at the base of the primary cilium, where it prevents Smo entry into the cilium, thus suppressing its function. Full-length Gli3 is converted to a C-terminally truncated Gli-R by sequential phosphorylation by PKA, followed by GSK3β and CK1, which creates binding sites for the adaptor protein β-TrCP, that subsequently associates with the RING-type Cul1-based E3 ligase, resulting in ubiquitination and partial Gli3 degradation by the 26S proteasome to form the Gli-R.

In the presence of Hh, the complex with Ptc becomes internalized in endosomes and later degraded, while Smo trafficking is promoted by GRK2 phosphorylation, thereby recruiting β-arrestin (Arrb2), which facilitates the movement of Smo along the cilia microtubule through interaction with the anterograde IFT motor kinesin-2, including Kif3a, Kif3b, and kinesin-2-associated proteins. In this way, Smo becomes activated and promotes the enhancement of transcriptionally active full-length Gli proteins, mainly Gli2 and to a lesser extent Gli3. Among the Hh target genes activated by Gli2 and Gli3, Gli1 plays a critical role being a strong constitutive transcriptional activator, thus representing an automaintenance loop of the pathway. Being synthetically regulated, Gli1 needs additional translational and posttranslational mechanisms to limit its function, including ubiquitination (see below).

The invertebrate HIB/Cul3 ligase-dependent events is also conserved in mammals. The vertebrate homolog of HIB is speckle-type POZ protein or Spop (Kwon *et al.*, 2006); by recruiting the RING-type Cul3, it promotes ubiquitination and proteasome-dependent degradation of Gli2-FL and Gli3-FL but not their repressor forms (Wang *et al.*, 2010). Such a Spop function is suppressed by Sufu, which protects Gli2-FL and Gli3-FL from degradation, thus preserving a pool of transcription factors that can be converted in activators (Gli2-FL and to a lesser extent Gli3-FL) or repressors (Gli3-R and to a lesser extent Gli2-R) (Chen *et al.*, 2009; Humke *et al.*, 2010). In this way, by balancing these ubiquitin-dependent functions, Sufu plays a dual positive and negative role on Hh signaling. These mechanisms add to the complexity of Sufu function, which is known to suppress the transcriptional activity of the pathway also by sequestering Gli1 in the cytoplasm thus preventing its nuclear transfer and transcription function, as well as SAP18/Sin3 corepressor-mediated transcriptional suppression (Cheng and Bishop, 2002; Kogerman *et al.*, 1999).

In summary, according to the complexity and heterogeneity of the three Gli effectors, the overall Hh inhibitory activity played by Sufu is mostly mediated by transcriptional repression mainly through Gli1 suppression and Gli3-R generation. The protective positive role of Sufu upon Gli-FL proteins (i.e., Gli2-FL) contributes to fine-tuning of the Hh signaling.

IV. An Alternative Gli1 Control by the HECT-Type Itch E3 Ubiquitin Ligase-Based Processing

Since Gli1 is a strong constitutive transcriptional activator and is not converted onto a “Repressor” form, like Gli2 and Gli3, but is transcriptionally induced by upstream Gli2-A and Gli3-A signals, an important question deals with the control of Gli1 protein once it has been produced.

Gli1 has been reported to be targeted for proteolysis by Itch (also called AIP4 and, in *Drosophila*, Suppressor of Deltex), a member of an additional E3 ubiquitin ligase family (Nedd4/Rsp5p-like HECT—Homologous to E6-associated protein C Terminus), containing an N-terminal lipid-interacting C2 region, four WW protein–protein interacting motifs, characterized by the presence of two conserved tryptophan residues and a C-terminal HECT ligase domain (Liu, 2004). Itch binds a number of proteins ranging from cell adhesion proteins or membrane receptors or associated proteins (e.g., occludin, Notch, Deltex, Cbl, c-FLIP, endophilin) or transcription factors (e.g., Jun and JunB, p73, NF-E2, Smad2) diverting most of them to a proteasome- or lysosome-dependent degradative pathway (Angers *et al.*, 2004; Bai *et al.*, 2004; Chang *et al.*, 2006; Chastagner *et al.*, 2006; Chen *et al.*, 2001; Ettenberg *et al.*, 2003; Gao *et al.*, 2004; Qiu *et al.*, 2000; Rossi *et al.*, 2005; Traweger *et al.*, 2002).

Itch binds to two distinct types of WW docking sites (PPXY and pSP) at positions Y859, Y872, and pS1060 of Gli1 C-terminus and promotes its polyubiquitination and proteasome-dependent proteolytic degradation. The reduced intracellular levels of the protein lead to the impairment of nuclear accumulation, of expression of Hh-target genes, and of Hh-induced cell proliferation (Di Marcotullio *et al.*, 2006, 2011). Therefore, the Itch-dependent ubiquitination of Gli1 extends the list of E3-ligase-related mechanisms that control the Hedgehog pathway signaling output.

Whether such a HECT-type E3 ligase-dependent processing of Gli protein is also conserved in *Drosophila* via targeting of Ci is unknown, although the HECT-type E3 ligase *Hyperplastic discs* has been described to be a negative regulator of Hh signaling downstream to Smo, and its loss-of-function mutants accumulate Ci protein (Lee *et al.*, 2002). However, the ability of *Hyperplastic discs* to promote ubiquitination and degradation of Ci has not been tested.

V. Activation of Itch Function by Numb

An important question deals with the regulatory mechanisms that control Hh signaling through ubiquitin-dependent events. Downregulation of the Cul1-based ubiquitination of Gli proteins is a part of an Hh-induced activator signal by which Hh maintains a low-degradative pathway to enable Gli function (Huntzicker *et al.*, 2006). Conversely, the adaptor protein HIB/SPOP is upregulated by Hh thus promoting Gli degradation by Cul3-based ubiquitination and representing an Hh-induced negative feedback loop to modulate the signaling response (Zhang *et al.*, 2006). These Cullin-based events, together with the role played by Sufu in SPOP-mediated control, provide a fine-tuning of Hh signaling output. A contribution to such a fine-tuning is also provided by the HECT ligase Itch-dependent mechanism suppressing Gli1 function. HECT E3 ligases, initially thought to be constitutively active, are subject to regulation by phosphorylation that induces their activity. Indeed, JNK-dependent phosphorylation on three sites within the proline-rich region has been reported to control Itch activity by disrupting the inhibitory conformation due to self-binding of the HECT domain to the proline-rich/WW region that would hamper the access of the substrate (Gallagher *et al.*, 2006). In contrast, in the context of Gli1, Itch activity has been reported to be enhanced by the developmental protein Numb (Di Marcotullio *et al.*, 2006), that by binding the WW2 site of Itch releases, by displacement, the HECT domain from its interaction with the WW region, thereby recruiting and favoring the association of Gli1 with Itch. Released from its self-inhibited conformation, the catalytic activity of Itch is strongly enhanced, thus promoting Gli1 ubiquitination (Di Marcotullio *et al.*, 2011). Such an auxiliary function of Numb appears to belong to a quite general mechanism of Itch activation. Indeed, recent data have shown that Itch-like E3 family can be activated by the binding of the NDFIP proteins to their WW domains (Mund and Pelham Hugh, 2009).

VI. Acetylation to Ubiquitination Connection in Hh Signaling

Acetylation has recently hit the scene of ubiquitination-dependent control of Gli1. Protein acetylation is one of the most critical posttranslational modifications by which crucial steps of cell functions are regulated at molecular level. Two main families of enzymes (histone acetyltransferases, HAT and histone deacetylases, HDACs) maintain the equilibrium of acetyl groups added to or removed from lysine residues, respectively (Gershey *et al.*, 1968; Jenuwein and Allis, 2001; Shahbazian and Grunstein, 2007). Initially identified as involved in transcriptional regulation by targeting histones, acetyl transferases may also target other regulatory proteins (e.g., transcription factors or cytoskeletal proteins; Glozak *et al.*, 2005). Acetylation of transcription factors regulates their transacting function and, consequently, the extent of gene transcription, whereas histone acetylation directly influences chromatin remodeling and its accessibility to coactivators and to the basal transcriptional machinery (Yang and Seto, 2007). Mammalian HDACs function by removing the acetyl groups from ε-*N*-acetyl lysines of histone and nonhistone proteins. There are 18 different HDACs, divided in two families, based on sequence similarity and cofactor dependency. The first family includes Class I, II, and IV HDACs and requires Zn2+ for enzymatic activity, while the second family of HDACs is composed by seven members, belonging to the Class III HDACs or sirtuins (SIRT1–7), that are provided of a characteristically NAD+-dependent enzymatic activity.

Gli1 and Gli2 have been recently reported to be acetylated proteins at Lysine 518 and 757, respectively (Canettieri *et al.*, 2010). Remarkably, their HDAC-mediated deacetylation represents a transcriptional switch in which deacetylation promotes transcriptional activation and sustains a positive autoregulatory loop through Hh-induced upregulation of HDAC1 (Canettieri *et al.*, 2010). This mechanism is turned off by HDAC1 degradation through a E3 ubiquitin ligase complex formed by Cullin3 and a family of adaptor proteins (KCASH1, KCASH2, and KCASH3, *K*CTD containing, *C*ullin3 *A*daptor, *S*uppressor of *H*edgehog) whose N-terminal BTB domain binds to Cul3 to form homo- or heterotetramers, thus recruiting C-terminal-bound HDAC1 to the enzyme (Canettieri *et al.*, 2010; Correale *et al.*, 2011; De Smaele *et al.*, 2011).

These data identify an integrated HDAC- and ubiquitin-mediated circuitry, where acetylation of Gli proteins functions as an unexpected key transcriptional checkpoint of Hedgehog signaling.

VII. Implications of Ubiquitin-Dependent Events for Cell Development and Tumorigenesis

Progenitor cells are the major targets of Hh signaling (Ahn and Joyner, 2005; Lai *et al.*, 2003; Machold *et al.*, 2003; Palma and Ruiz i Altaba, 2004; Palma *et al.*, 2005; Po *et al.*, 2010; Ruiz i Altaba *et al.*, 2002), and when this pathway is aberrantly activated, it is suggested to be responsible for their progression toward malignancy (Ruiz i Altaba *et al.*, 2002). A critical tumorigenic role of Hh is highlighted by its activity on stem cells (e.g., neural stem cells; Po *et al.*, 2010; Zbinden *et al.*, 2010). Transformation of stem cells to cancer stem cells is believed to be a critical mechanism of tumorigenesis in which cancer stem cells sustain and maintain a never ending reservoir of the tumor cell mass (Lobo *et al.*, 2007). Indeed, cancer stem cells are rare cells populating the tumor cell population and, when transplanted *in vivo*, are able to reconstitute the tumor mass (Morrison and Kimble, 2006). Remarkably, Hh signaling is active in cancer stem cells of various types of tumors (e.g., Grachtchouk *et al.*, 2011; Lim *et al.*, 2011; Liu *et al.*, 2006; Peacock *et al.*, 2007; Po *et al.*, 2010; Song *et al.*, 2011; Varnat *et al.*, 2009; Zbinden *et al.*, 2010; Zhao *et al.*, 2009).

The role of ubiquitination-related events in the Gli-dependent control of cell development and tumorigenesis is indicated by several observations.

Gli1 is a potent inducer of cell transformation *in vitro* (Canettieri *et al.*, 2010; Liu *et al.*, 2006; Ruppert *et al.*, 1991), while abrogation of Gli1 expression counteracts the development of MB in Ptch1$^{+/-}$ mutant mice (Kimura *et al.*, 2005). These reports highlight its critical role in promoting tumorigenesis and emphasize the relevance of the processes that control Gli1 activity.

Accordingly, Cul1/βTrCP-insensitive Gli1 mutants are able to accelerate tumor formation in transgenic mice (Huntzicker *et al.*, 2006). Furthermore, the newly identified role of Numb in regulating the Itch-dependent control of Gli1 provides novel insights in the link between ubiquitination-based Gli processing and tumorigenesis. Indeed, mutations of Numb/Itch-dependent degron, by inducing Gli1 stability, lead to an increased proliferation rate of medulloblastoma cells, anchorage-independent tumor growth, a higher capacity of migration and invasion, as well as a significantly higher neoplastic transformation of rat kidney fibroblasts *in vitro* (Di Marcotullio *et al.*, 2011).

The role of Numb in the control of Gli function is consistent with the low expression of this protein in Hh-dependent medulloblastomas and with its tumor suppressor activity (Colaluca *et al.*, 2008; Di Marcotullio *et al.*, 2006; Pece *et al.*, 2004).

Furthermore, Numb deregulation also suggests a link between cancer stem cell-related tumorigenesis and stem cell biology. Indeed, Numb is a major determinant of asymmetric cell division by which daughter cells acquire binary cell fates including self-renewal versus differentiation ability. Such a role has been described in *Drosophila* larval brain neuroblasts which divide asymmetrically to generate either self-renewing neuroblast or terminally differentiating ganglion mother cell (GMC) daughters. Numb asymmetrically distributes into the GMC to inhibit cell self-renewal (Cayouette and Raff, 2002). Interestingly, loss-of-function mutants of *Drosophila* Numb lead to a tumor-like phenotype characterized by ectopic self-renewing overproliferating neuroblasts (Caussinus and Gonzalez, 2005). These observations underscore a tumor suppressor function for Numb in *Drosophila* neuroblasts, likely due to impairment of asymmetric cell division and self-renewal of progenitor cells.

A similar asymmetric distribution and function in neural progenitor cells has also been described in mouse (Shen *et al.*, 2002), although, because of the complex functions and heterogeneity of murine Numb proteins, the mechanisms underlying its influence on neurogenesis remain controversial (Li *et al.*, 2003; Petersen *et al.*, 2002; reviewed in Gulino *et al.*, 2010). In keeping with its ability to promote neuronal differentiation (Li *et al.*, 2003; Shen *et al.*, 2002; Verdi *et al.*, 1999), Numb deficiency in mouse cerebellum prevents the differentiation of granule cell progenitors (GCPs) which accumulate in the external germinal layer (EGL) (Klein *et al.*, 2004). Accordingly, downregulated expression of Numb is observed in the outer layer of EGL where proliferating progenitor cells reside, while Numb expression is observed in inner EGL and internal granule layer (IGL) where GCPs stop proliferating and differentiate (Di Marcotullio *et al.*, 2006). According to these observations, the expression of Numb and Gli1 in developing cells of cerebellar EGL and IGL as well as in medulloblastomas are mutually exclusive (Di Marcotullio *et al.*, 2006). This is consistent with the ability of Numb to suppress Gli1 function through an ubiquitin-dependent event, therefore providing a plausible explanation for the impairment of cerebellar granule cell maturation described in Numb-deficient mice (Klein *et al.*, 2004) and by the antiproliferative and prodifferentiation effects of Numb on both GCPs (Di Marcotullio *et al.*, 2006). These observations also suggest that inappropriate Hh signaling responsible for the growth of GCP-derived brain tumors could be related to the failure of a Numb-mediated restraint mechanism. Indeed, the role of Hh signaling in the maintenance and self-renewal of neural stem cells derived from cerebellum and medulloblastoma stem cells is well established (Po *et al.*, 2010; Zbinden *et al.*, 2010). Furthermore, Purkinje cell-derived Shh is the major signal that promotes proliferation and keeps undifferentiated the cerebellar GCPs (Dahmane and Ruiz i Altaba, 1999; Wallace, 1999; Wechsler-Reya and Scott, 1999) which may further progress toward malignancy if aberrantly activated Hh signaling is maintained (Ruiz i Altaba *et al.*, 2002).

Numb is reported to control cell functions via regulation of additional molecules, besides Gli1, (e.g., Notch, the endocytic proteins α-adaptin and Eps15 and the adherens junction molecule E-cadherin), which are all involved in determining cell fate decisions (McGill and McGlade, 2003, Berdnick *et al.*, 2002; Kuo *et al.*, 2006; Santolini *et al.*, 2000). Some of these interactions involve the ability of Numb to regulate ubiquitination processes, as occurring for Gli1. For instance, the E3 ubiquitin ligase Itch is also recruited by Numb to negatively regulate the transmembrane receptor Notch (McGill and McGlade, 2003) that plays a critical role in determining binary cell fates and in controlling proliferation versus differentiation (Yoon and Gaiano, 2005). Remarkably, Notch has been described to suppress Numb (Chapman *et al.*, 2006) providing a cascade mechanism which integrates Notch and Hh signals. Therefore, Numb appears to behave as a scaffold protein, adapting ubiquitin-related molecular events to multiple targets that belong to distinct signaling pathways (i.e., Hh and Notch), in order to coordinate their influence upon cell development.

VIII. Implications for Ubiquitination/Acetylation Interplay in Cell Development and Cancer Control

The functional interaction between the ubiquitination and acetylation mechanisms regulating Hh pathway has also implications for cell development and tumorigenesis. Indeed, the KCASH family-dependent ubiquitination of HDAC, and the consequent regulation of Gli factors, appears to be developmentally regulated and associated with cancer.

A member of this family, KCASH1, initially termed REN^{KCTD11}, was identified a few years ago as a putative tumor suppressor playing also a role in neurogenesis. It is expressed early during embryonic development and restricted to differentiating rather than proliferating neuroblasts of the embryo cortical ventricular zone as well as to low-proliferating inner EGL GCPs and IGL of the developing cerebellum (Argenti *et al.*, 2005; Gallo *et al.*, 2002). These observations suggest a role for REN as an inhibitory signal required for withdrawing GCP expansion at the outer to inner EGL transition. Consistently, REN induces growth arrest and differentiation of neural progenitors (Gallo *et al.*, 2002).

Interestingly, REN maps to chromosome 17p and is deleted in about 50% of medulloblastomas, a brain tumor generated by hyperactivation of Hh pathway in EGL GCPs, whose proliferation depends on Hh signaling (Dahmane and Ruiz i Altaba, 1999; Wallace, 1999; Wechsler-Reya and Scott, 1999). Consistently, REN suppresses the growth of medulloblastoma both *in vitro* and *in vivo* (Di Marcotullio *et al.*, 2004). Similar expression profile and activity in medulloblastoma have been observed for the other KCASH members (De Smaele *et al.*, 2011). Therefore, the loss of such inhibitory signals, as a consequence of chromosome deletion (KCASH1 in 17p or KCASH2 in 11q14.1) or methylation (De Smaele *et al.*, 2011; Di Marcotullio *et al.*, 2004; Mancarelli *et al.*, 2010), may lead to the uncontrolled Hh-dependent GCP overgrowth linked to medulloblastoma development.

These findings highlight the relevance of the interplay between ubiquitination and acetylation events in which genetic and epigenetic loss of KCASH gene family enhances HDAC–Gli synergism in brain tumors, functioning as a native HDAC1 inhibitors and impairing cancer cell growth (Di Marcotullio *et al.*, 2011b).

Such an interplay may also be exploited as a part of a combination therapy aimed to prevent the resistance to the treatment against a primary target. An example is given by the targeting of the Hh pathway in brain tumors. Anti-Smo drugs (e.g., vismodegib, reviewed in De Smaele *et al.*, 2010) have proven to be active in suppressing the growth of medulloblastoma that, nevertheless, quickly rescues because of acquisition of Smo mutations (Rudin *et al.*, 2009; Yauch *et al.*, 2009). The emergence of this drug-resistant cell population is enhanced by mechanisms that activate the transcriptional function of Gli factors that are downstream to Smo. One of these mechanisms is represented by HDAC1 function, whose levels are, in turn, increased by both Smo-generated signals and loss of the HDAC1-specific REN^{KCTD11} and the other KCASH family members as adaptor of the Cul3 E3 ubiquitin ligase complex, observed in a consistent percentage of these tumors (Canettieri *et al.*, 2010; De Smaele *et al.*, 2011). Therefore, the therapeutic use of HDAC inhibitors provides a rationale for reverting the resistance to anti-Smo drugs by rescuing the physiological control of HDAC1 activity.

IX. Conclusions

Ubiquitination-related posttranslational modifications of the Gli transcriptional effectors of the Hh signaling represent an important mechanism of regulation of the pathway, which is, in turn, controlled by a number of E3 ligases (RING Cullin and HECT families). These E3 ligases are regulated by several members of the Hh pathway (i.e., Smo-activated kinases) as well as by proteins belonging to other signaling cascades (i.e., Numb-activated Itch). These proteolytic signals finally suppress Gli function either directly or indirectly (i.e., suppression of HDAC1-mediated Gli deacetylation). The complex of these regulatory circuitries finely tunes Hh signaling providing a tight control of developmental processes, a subversion of which may lead to tumorigenesis. To this regard, these ubiquitination processes represent promising targets for novel therapeutic strategies.

ACKNOWLEDGMENTS

This work was supported by AIRC (Associazione Italiana Ricerca Cancro), MIUR FIRB and PRIN projects, Ministry of Health, Fondazione Roma, Fondazione Mariani, EU Healing grant, IIT (Italian Institute of Technology), ASI (Agenzia Spaziale Italiana), and Pasteur Insitute/Cenci Bolognetti Foundation.

REFERENCES

Ahn, S., and Joyner, A. L. (2005). In vivo analysis of quiescent adult neural stem cells responding to Sonic hedgehog. *Nature* **437,** 894–897.

Angers, A., Ramjaun, A. R., and McPherson, P. S. (2004). The HECT domain ligase itch ubiquitinates endophilin and localizes to the trans-Golgi network and endosomal system. *J. Biol. Chem.* **279,** 11471–11479.

Argenti, B., Gallo, R., Di Marcotullio, L., Ferretti, E., Napolitano, M., Canterini, S., *et al.* (2005). Hedgehog antagonist REN(KCTD11) regulates proliferation and apoptosis of developing granule cell progenitors. *J. Neurosci.* **25,** 8338–8346.

Aza-Blanc, P., Ramirez-Weber, F. A., Laget, M. P., Schwartz, C., and Kornberg, T. B. (1997). Proteolysis that is inhibited by hedgehog targets Cubitus interruptus protein to the nucleus and converts it to a repressor. *Cell* **89,** 1043–1053.

Bai, Y., Yang, C., Hu, K., Elly, C., and Liu, Y. C. (2004). Itch E3 ligase-mediated regulation of TGF-beta signaling by modulating smad2 phosphorylation. *Mol. Cell* **15,** 825–831.

Barakat, M. T., Humke, E. W., and Scott, M. P. (2010). Learning from Jekyll to control Hyde: Hedgehog signaling in development and cancer. *Trend Mol. Med.* **16,** 1471–4914.

Berdnick, D., Torok, T., Gonzalez-Gaitan, M., and Knoblich, J. A. (2002). The endocytic protein alpha-Adaptin is required for numb-mediated asymmetric cell division in Drosophila. *Dev. Cell* **3,** 221–231.

Boname, J. M., and Stevenson, P. G. (2001). MHC class I ubiquitination by a viral PHD/LAP finger protein. *Immunity* **15,** 627–636.

Canettieri, G., Di Marcotullio, L., Greco, A., Coni, S., Antonucci, L., Infante, P., Pietrosanti, L., De Smaele, E., Ferretti, E., Miele, E., Pelloni, M., De Simone, G., *et al.* (2010). Histone deacetylase and Cullin3-REN(KCTD11) ubiquitin ligase interplay regulates Hedgehog signalling through Gli acetylation. *Nat. Cell Biol.* **12,** 132–142.

Caussinus, E., and Gonzalez, C. (2005). Induction of tumor growth by altered stem-cell asymmetric division in Drosophila melanogaster. *Nat. Genet.* **37,** 1125–1129.

Cayouette, M., and Raff, M. (2002). Asymmetric segregation of Numb: A mechanism for neural specification from Drosophila to mammals. *Nat. Neurosci.* **5,** 1265–1269.

Chang, L., Kamata, H., Solinas, G., Luo, J. L., Maeda, S., Venuprasad, K., Liu, Y. C., and Karin, M. (2006). The E3 ubiquitin ligase itch couples JNK activation to TNFalpha-induced cell death by inducing c-FLIP(L) turnover. *Cell* **124,** 601–613.

Chapman, G., Liu, L., Sahlgren, C., Dahlqvist, C., and Lendahl, U. (2006). High levels of Notch signaling down-regulate Numb and Numblike. *J. Cell Biol.* **175,** 535–540.

Chastagner, P., Israel, A., and Brou, C. (2006). Itch/AIP4 mediates Deltex degradation through the formation of K29-linked polyubiquitin chains. *EMBO Rep.* **7,** 1147–1153.

Chen, C. H., von Kessler, D. P., Park, W., Wang, B., Ma, Y., and Beachy, P. A. (1999). Nuclear trafficking of Cubitus interruptus in the transcriptional regulation of Hedgehog target gene expression. *Cell* **98,** 305–316.

Chen, X., Wen, S., Fukuda, M. N., Gavva, N. R., Hsu, D., Akama, T. O., Yang-Feng, T., and Shen, C. K. (2001). Human ITCH is a coregulator of the hematopoietic transcription factor NF-E2. *Genomics* **73,** 238–241.

Chen, M. H., Wilson, C. W., Li, Y. J., Law, K. K., Lu, C. S., Gacayan, R., Zhang, X., Hui, C. C., and Chuang, P. T. (2009). Cilium-independent regulation of Gli protein function by Sufu in Hedgehog signaling is evolutionarily conserved. *Genes Dev.* **23,** 1910–1928.

Cheng, S. Y., and Bishop, J. M. (2002). Suppressor of Fused represses Gli-mediated transcription by recruiting the SAP18-mSin3 corepressor complex. *Proc. Natl. Acad. Sci. USA* **99,** 5442–5447.

Colaluca, I. N., Tosoni, D., Nuciforo, P., Senic-Matuglia, F., Galimberti, V., Viale, G., Pece, S., and Di Fiore, P. P. (2008). NUMB controls p53 tumour suppressor activity. *Nature* **451,** 76–80.

Correale, S., Pirone, L., Di Marcotullio, L., De Smaele, E., Greco, A., Mazzà, D., Moretti, M., Alterio, V., Vitagliano, L., Di Gaetano, S., Gulino, A., and Pedone, E. M. (2011). Molecular organization of the cullin E3 ligase adaptor KCTD11. *Biochimie* **93,** 715–724.

Dahmane, N., and Ruiz I Altaba, A. (1999). Sonic hedgehog regulates the growth and patterning of the cerebellum. *Development* **126,** 3089–3100.

De Smaele, E., Ferretti, E., and Gulino, A. (2010). Vismodegib, a small-molecule inhibitor of the hedgehog pathway for the treatment of advanced cancers. *Curr. Opin. Investig. Drugs* **11,** 707–718.

De Smaele, E., Di Marcotullio, L., Moretti, M., Pelloni, M., Occhione, M., Infante, P., Cucchi, D., Greco, A., Pietrosanti, L., Todorovic, J., Coni, S., Canettieri, G., *et al.* (2011). Identification and characterization of KCASH2 and KCASH3, two novel Cullin3 adaptors suppressing histone deacetylase and Hedgehog activity in medulloblastoma. *Neoplasia* **13,** 374–385.

Di Marcotullio, L., Ferretti, E., De Smaele, E., Argenti, B., Mincione, C., Zazzeroni, F., Gallo, R., Masuelli, L., Napolitano, M., Maroder, M., Modesti, A., Giangaspero, F., *et al.* (2004). REN(KCTD11) is a suppressor of Hedgehog signaling and is deleted in human medulloblastoma. *Proc. Natl. Acad. Sci. USA* **101,** 10833–10838.

Di Marcotullio, L., Ferretti, E., Greco, A., De Smaele, E., Po, A., Sico, M. A., Alimandi, M., Giannini, G., Maroder, M., Screpanti, I., and Gulino, A. (2006). Numb is a suppressor of Hedgehog signalling and targets Gli1 for Itch-dependent ubiquitination. *Nat. Cell Biol.* **8,** 1415–1423.

Di Marcotullio, L., Greco, A., Mazzà, D., Canettieri, G., Pietrosanti, L., Infante, P., Coni, S., Moretti, M., De Smaele, E., Ferretti, E., Screpanti, I., and Gulino, A. (2011). Numb activates the E3 ligase Itch to control Gli1 function through a novel degradation signal. *Oncogene* **30,** 65–76.

Di Marcotullio, L., Canettieri, G., Infante, P., Greco, A., and Gulino, A. (2011b). Protected from the inside: endogenous histone deacetylase inhibitors and the road to cancer. *Biochim. Biophys. Acta.* **1815,** 241–252.

Ettenberg, S., Yang, C., Mariano, J., Tiwari, S., Fang, S., Lipkowitz, S., and Weissman, A. M. (2003). WW domain HECT E3s target Cbl RING finger E3s for proteasomal degradation. *J. Biol. Chem.* **278,** 43169–43177.

Gallagher, E., Gao, M., Liu, Y. C., and Karin, M. (2006). Activation of E3 ubiquitin ligase Itch through a phosphorylation-induced conformational change. *Proc. Natl. Acad. Sci. USA* **103,** 1717–1722.

Gallo, R., Zazzeroni, F., Alesse, E., Mincione, C., Borello, U., Buanne, P., D'Eugenio, R., Mackay, A. R., Argenti, B., Gradini, R., Russo, M. A., Maroder, M., *et al.* (2002). REN: A novel, developmentally regulated gene that promotes neural cell differentiation. *J. Cell Biol.* **158,** 731–740.

Gao, M., Labuda, T., Xia, Y., Gallagher, E., Fang, D., Liu, Y. C., and Karin, M. (2004). Jun turnover is controlled through JNK-dependent phosphorylation of the E3 ligase Itch. *Science* **306,** 271–275.

Gershey, E. L., Vidali, G., and Allfrey, V. G. (1968). Chemical studies of histone acetylation. The occurrence of epsilon-N-acetyllysine in the f2a1 histone. *J. Biol. Chem.* **243,** 5018–5022.

Glozak, M. A., Sengupta, N., Zhang, X., and Seto, E. (2005). Acetylation and deacetylation of nonhistone proteins. *Gene* **363,** 15–23.

Goodrich, L. V., Milenkovic, L., Higgins, K. M., and Scott, M. P. (1997). Altered neural cell fates and medulloblastoma in mouse patched mutants. *Science* **277,** 1109–1113.

Grachtchouk, M., Pero, J., Yang, S. H., Ermilov, A. N., Michael, L. E., Wang, A., Wilbert, D., Patel, R. M., Ferris, J., Diener, J., Allen, M., Lim, S., *et al.* (2011). Basal cell carcinomas in mice arise from hair follicle stem cells and multiple epithelial progenitor populations. *J. Clin. Invest.* **121,** 1768–1781.

Gulino, A., Di Marcotullio, L., and Screpanti, I. (2010). The multiple functions of Numb. *Exp. Cell Res.* **316,** 900–906.

Hatakeyama, S., Yada, M., Matsumoto, M., Ishida, N., and Nakayama, K. I. (2001). Ubox proteins as a new family of ubiquitin–protein ligases. *J. Biol. Chem.* **276,** 33111–33120.

Hooper, J. E., and Scott, M. P. (2005). Communicating with Hedgehogs. *Nat. Rev. Mol. Cell Biol.* **6,** 306–317.

Humke, E. W., Dorn, K. V., Milenkovic, L., Scott, M. P., and Rohatgi, R. (2010). The output of Hedgehog signaling is controlled by the dynamic association between Suppressor of Fused and the Gli proteins. *Genes Dev.* **24,** 670–682.

Huntzicker, E. G., Estay, I. S., Zhen, H., Lokteva, L. A., Jackson, P. K., and Oro, A. E. (2006). Dual degradation signals control Gli protein stability and tumor formation. *Genes Dev.* **20,** 276–281.

Jenuwein, T., and Allis, C. D. (2001). Translating the histone code. *Science* **293,** 1074–1080.

Jia, J., Amanai, K., Wang, G., Tang, J., Wang, B., and Jiang, J. (2002). Shaggy/GSK3 antagonizes Hedgehog signalling by regulating Cubitus interruptus. *Nature* **416,** 548–552.

Jia, J., Zhang, L., Zhang, Q., Tong, C., Wang, B., Hou, F., Amanai, K., and Jiang, J. (2005). Phosphorylation by double-time/CKIepsilon and CKIalpha targets cubitus interruptus for Slimb/beta-TRCP-mediated proteolytic processing. *Dev. Cell* **9,** 819–830.

Kelleher, F. C. (2011). Hedgehog signaling and therapeutics in pancreatic cancer. *Carcinogenesis* **32,** 445–451.

Kent, D., Bush, E. W., and Hooper, J. E. (2006). Roadkill attenuates Hedgehog esponses through degradation of Cubitus interruptus. *Development* **133,** 2001–2010.

Kimura, H., Stephen, D., Joyner, A., and Curran, T. (2005). Gli1 is important for medulloblastoma formation in Ptch1+/− mice. *Oncogene* **24,** 4026–4036.

Klein, A. L., Zilian, O., Suter, U., and Taylor, V. (2004). Murine numb regulates granule cell maturation in the cerebellum. *Dev. Biol.* **266,** 161–177.

Kogerman, P., Grimm, T., Kogerman, L., Krause, D., Unden, A. B., Sandstedt, B., Toftgard, R., and Zaphiropoulos, P. G. (1999). Mammalian suppressor-of-fused modulates nuclear-cytoplasmic shuttling of Gli-1. *Nat. Cell Biol.* **1,** 312–319.

Kuo, C. T., Mirzadeh, Z., Soriano-Navarro, M., Rasin, M., Wang, D., Shen, J., Sestan, N., Garcia-Verdugo, J., Alvarez-Buylla, A., Jan, L. Y., and Jan, Y. N. (2006). Postnatal deletion of numb/numblike reveals repair and remodeling capacity in the subventricular neurogenic niche. *Cell* **127,** 1253–1264.

Kwon, J. E., La, M., Oh, K. H., Oh, Y. M., Kim, G. R., Seol, J. H., Baek, S. H., Chiba, T., Tanaka, K., Bang, O. S., *et al.* (2006). BTB domain-containing speckle-type POZ protein (SPOP) serves as an adaptor of Daxx for ubiquitination by Cul3-based ubiquitin ligase. *J. Biol. Chem.* **281,** 12664–12672.

Lai, K., Kaspar, B. K., Gage, F. H., and Schaffer, D. V. (2003). Sonic hedgehog regulates adult neural progenitor proliferation in vitro and in vivo. *Nat. Neurosci.* **6,** 21–27.

Lee, J. D., Amanai, K., Shearn, A., and Treisman, J. E. (2002). The ubiquitin ligase Hyperplastic discs negatively regulates hedgehog and decapentaplegic expression by independent mechanisms. *Development* **129,** 5697–5706.

Lee, E. Y., Ji, H., Ouyangc, Z., Zhoud, B., Mae, W., Vokesf, S. A., McMahong, A. P., Wongd, W. H., and Scott, M. P. (2010). Hedgehog pathway-regulated gene networks in cerebellum development and tumorigenesis. *Proc. Natl. Acad. Sci.* **107,** 9736–9741.

Li, H. S., Wang, D., Shen, Q., Schonemann, M. D., Gorski, J. A., Jones, K. R., Temple, S., Jan, L. Y., and Jan, Y. N. (2003). Inactivation of Numb and Numblike in embryonic dorsal forebrain impairs neurogenesis and disrupts cortical morphogenesis. *Neuron* **40,** 1105–1118.

Lim, K. J., Bisht, S., Bar, E. E., Maitra, A., and Eberhart, C. G. (2011). A polymeric nanoparticle formulation of curcumin inhibits growth, clonogenicity and stem-like fraction in malignant brain tumors. *Cancer Biol. Ther.* **11,** 464–473.

Liu, Y. C. (2004). Ubiquitin ligases and the immune response. *Annu. Rev. Immunol.* **22,** 81–127.

Liu, S., Dontu, G., Mantle, I. D., Patel, S., Ahn, N. S., Jackson, K. W., Suri, P., and Wicha, M. S. (2006). Hedgehog signaling and Bmi-1 regulate self-renewal of normal and malignant human mammary stem cells. *Cancer Res.* **66,** 6063–6071.

Lobo, N. A., Shimono, Y., Qian, D., and Clarke, M. F. (2007). The biology of cancer stem cells. *Annu. Rev. Cell Dev. Biol.* **23,** 675–699.

Machold, R., Hayashi, S., Rutlin, M., Muzumdar, M. D., Nery, S., Corbin, J. G., Gritli-Linde, A., Dellovade, T., Porter, J. A., Rubin, L. L., Dudek, H., McMahon, A. P., *et al.* (2003). Sonic hedgehog is required for progenitor cell maintenance in telencephalic stem cell niches. *Neuron* **39,** 937–950.

Mancarelli, M. M., Zazzeroni, F., Ciccocioppo, L., Capece, D., Po, A., Murgo, S., Di Camillo, R., Rinaldi, C., Ferretti, E., Gulino, A., and Alesse, E. (2010). The tumor suppressor gene KCTD11 (REN) is regulated by Sp1 and methylation and its expression is reduced in tumors. *Mol. Cancer* **9,** 169–172.

McGill, M. A., and McGlade, C. J. (2003). Mammalian numb proteins promote Notch1 receptor ubiquitination and degradation of the Notch1 intracellular domain. *J. Biol. Chem.* **278,** 23196–23203.

Morrison, S. J., and Kimble, J. (2006). Asymmetric and symmetric stem-cell divisions in development and cancer. *Nature* **441,** 1068–1074.

Mund, T., and Pelham Hugh, R. B. (2009). Conrol of the activity of WWHECT domain E3 ubiquitin ligase by NDFIP proteins. *EMBO Rep.* **10,** 501–507.

Nusslein-Volhard, C., and Wieschaus, E. (1980). Mutations affecting segment number and polarity in Drosophila. *Nature* **287,** 795–801.

Palma, V., and Ruiz i Altaba, A. (2004). Hedgehog-GLI signaling regulates the behavior of cells with stem cell properties in the developing neocortex. *Development* **131,** 337–345.

Palma, V., Lim, D. A., Dahmane, N., Sanchez, P., Brionne, T. C., Herzberg, C. D., Gitton, Y., Carleton, A., Alvarez-Buylla, A., and Ruiz i Altaba, A. (2005). Sonic hedgehog controls stem cell behavior in the postnatal and adult brain. *Development* **132,** 335–344.

Peacock, C. D., Wang, Q., Gesell, G. S., Corcoran-Schwartz, I. M., Jones, E., Kim, J., Devereux, W. L., Rhodes, J. T., Huff, C. A., Beachy, P. A., Watkins, D. N., and Matsui, W. (2007). Hedgehog signaling maintains a tumor stem cell compartment in multiple myeloma. *Proc. Natl. Acad. Sci. USA* **104,** 4048–4053.

Pece, S., Serresi, M., Santolini, E., Capra, M., Hulleman, E., Galimberti, V., Zurrida, S., Maisonneuve, P., Viale, G., and Di Fiore, P. P. (2004). Loss of negative regulation by Numb over Notch is relevant to human breast carcinogenesis. *J. Cell Biol.* **167,** 215–221.

Petersen, P. H., Zou, K., Hwang, J. K., Jan, Y. N., and Zhong, W. (2002). Progenitor cell maintenance requires numb and numblike during mouse neurogenesis. *Nature* **419,** 929–934.

Pickart, C. M. (2001). Mechanisms underlying ubiquitination. *Annu. Rev. Biochem.* **70,** 503–533.

Po, A., Ferretti, E., Miele, E., De Smaele, E., Paganelli, A., Canettieri, G., Coni, S., Di Marcotullio, L., Biffoni, M., Massimi, L., Di Rocco, C., Screpanti, I., and Gulino, A. (2010). Hedgehog controls neural stem cells through p53-independent regulation of Nanog. *EMBO. J.* **29,** 2646–2658.

Price, M. A., and Kalderon, D. (2002). Proteolysis of the Hedgehog signaling effector Cubitus interruptus requires phosphorylation by Glycogen Synthase Kinase 3 and Casein Kinase 1. *Cell* **108,** 823–835.

Qiu, L., Joazeiro, C., Fang, N., Wang, H. Y., Elly, C., Altman, Y., Fang, D., Hunter, T., and Liu, Y. C. (2000). Recognition and ubiquitination of Notch by Itch, a hect-type E3 ubiquitin ligase. *J. Biol. Chem.* **275,** 35734–35737.

Rossi, M., De Laurenzi, V., Munarriz, E., Green, D. R., Liu, Y. C., Vousden, K. H., Cesareni, G., and Melino, G. (2005). The ubiquitin-protein ligase Itch regulates p73 stability. *EMBO J.* **24,** 836–848.

Rudin, C. M., Hann, C. L., Laterra, J., Yauch, R. L., Callahan, C. A., Fu, L., Holcomb, T., Stinson, J., Gould, S. E., Coleman, B., LoRusso, P. M., Von Hoff, D. D., *et al.* (2009). Treatment of medulloblastoma with hedgehog pathway inhibitor GDC-0449. *N. Engl. J. Med.* **361,** 1173–1178.

Ruiz i Altaba, A., Sanchez, P., and Dahmane, N. (2002). Gli and hedgehog in cancer: Tumours, embryos and stem cells. *Nat. Rev. Cancer* **2,** 361–372.

Ruppert, J. M., Vogelstein, B., and Kinzler, K. W. (1991). The zinc finger protein GLI transforms primary cells in cooperation with adenovirus E1A. *Mol. Cell. Biol.* **11,** 1724–1728.

Santolini, E., Puri, C., Salcini, A. E., Gagliani, M. C., Pelicci, P. G., Tacchetti, C., and Di Fiore, P. P. (2000). Numb is an endocytic protein. *J. Cell Biol.* **151,** 1345–1352.

Shahbazian, M. D., and Grunstein, M. (2007). Functions of sitespecific histone acetylation and deacetylation. *Annu. Rev. Biochem.* **76,** , 75–100.

Shen, Q., Zhong, W., Jan, Y. N., and Temple, S. (2002). Asymmetric Numb distribution is critical for asymmetric cell division of mouse cerebral cortical stem cells and neuroblasts. *Development* **129,** 4843–4853.

Smelkinson, M. G., and Kalderon, D. (2006). Processing of the Drosophila Hedgehog signaling effector Ci-155 to the repressor Ci-75 is mediated by direct binding to the SCF component Slimb. *Curr. Biol.* **16,** 110–116.

Song, Z., Yue, W., Wei, B., Wang, N., Li, T., Guan, L., Shi, S., Zeng, Q., Pei, X., and Chen, L. (2011). Sonic hedgehog pathway is essential for maintenance of cancer stem-like cells in human gastric cancer. *PLoS One* **6,** e17687.

Stecca, B., and Ruiz i Altaba, A. (2010). Context-dependent regulation of the GLI code in cancer by HEDGEHOG and Non-HEDGEHOG signals. *J. Mol. Cell Biol.* **2,** 84–95.

Teglund, S., and Toftgard, R. (2010). Hedgehog beyond medulloblastoma and basal cell carcinoma. *Biochim. Biophys. Acta* **1805,** 181–208.

Traweger, A., Fang, D., Liu, Y. C., Stelzhammer, W., Krizbai, I. A., Fresser, F., Bauer, H. C., and Bauer, H. (2002). The tight junction-specific protein occludin is a functional target of the E3 ubiquitin-protein ligase itch. *J. Biol. Chem.* **277,** 10201–10208.

Van Sant, C., Hagglund, R., Lopez, P., and Roizman, B. (2001). The infected cell protein 0 of herpes simplex virus 1 dynamically interacts with proteasomes, binds and activates the cdc34 E2 ubiquitin-conjugating enzyme, and possesses in vitro E3 ubiquitin ligase activity. *Proc. Natl. Acad. Sci. USA* **98,** 8815–8820.

Varnat, F., Duquet, A., Malerba, M., Zbinden, M., Mas, C., Gervaz, P., and Ruiz i Altaba, A. (2009). Human colon cancer epithelial cells harbour active HEDGEHOG-GLI signalling that is essential for tumour growth, recurrence, metastasis and stem cell survival and expansion. *EMBO Mol. Med.* **1,** 338–351.

Verdi, J. M., Bashirullah, A., Goldhawk, D. E., Kubu, C. J., Jamali, M., Meakin, S. O., and Lipshitz, H. D. (1999). Distinct human NUMB isoforms regulate differentiation vs. proliferation in the neuronal lineage. *Proc. Natl. Acad. Sci. USA* **96,** 10472–10476.

Wallace, V. A. (1999). Purkinje-cell-derived Sonic hedgehog regulates granule neuron precursor cell proliferation in the developing mouse cerebellum. *Curr. Biol.* **9,** 445–448.

Wang, Q. T., and Holmgren, R. A. (1999). The subcellular localization and activity of Drosophila cubitus interruptus are regulated at multiple levels. *Development* **126,** 5097–5106.

Wang, C., Pan, Y., and Wang, B. (2010). Suppressor of fused and Spop regulate the stability, processing and function of Gli2 and Gli3 full-length activators but not their repressors. *Development* **137,** 2001–2009.

Wechsler-Reya, R. J., and Scott, M. P. (1999). Control of neuronal precursor proliferation in the cerebellum by Sonic Hedgehog. *Neuron* **22,** 103–114.

Welchman, R. L., Gordon, C., and Mayer, R. J. (2005). Ubiquitin and ubiquitin-like proteins as multifunctional signals. *Nat. Rev. Mol. Cell Biol.* **6,** 599–609.

Yang, X. J., and Seto, E. (2007). HATs and HDACs: From structure, function and regulation to novel strategies for therapy and prevention. *Oncogene* **26,** 5310–5318.

Yauch, R. L., Dijkgraaf, G. J., Alicke, B., Januario, T., Ahn, C. P., Holcomb, T., Pujara, K., Stinson, J., Callahan, C. A., Tang, T., Bazan, J. F., Kan, Z., *et al.* (2009). Smoothened mutation confers resistance to a Hedgehog pathway inhibitor in medulloblastoma. *Science* **326,** 572–574.

Yoon, K., and Gaiano, N. (2005). Notch signaling in the mammalian central nervous system: Insights from mouse mutants. *Nat. Neurosci.* **8,** 709–715.

Zbinden, M., Duquet, A., Lorente-Trigos, A., Ngwabyt, S. N., Borges, I., and Ruiz i Altaba, A. (2010). NANOG regulates glioma stem cells and is essential in vivo acting in a cross-functional network with GLI1 and p53. *EMBO J.* **29,** 2659–2674.

Zhang, Q., Zhang, L., Wang, B., Ou, C. Y., Chien, C. T., and Jiang, J. (2006). A hedgehog-induced BTB protein modulates hedgehog signaling by degrading Ci/Gli transcription factor. *Dev. Cell* **10,** 719–729.

Zhao, C., Chen, A., Jamieson, C. H., Fereshteh, M., Abrahamsson, A., Blum, J., Kwon, H. Y., Kim, J., Chute, J. P., Rizzieri, D., Munchhof, M., VanArsdale, T., *et al.* (2009). Hedgehog signalling is essential for maintenance of cancer stem cells in myeloid leukaemia. *Nature* **458,** 776–779.

CHAPTER TEN

Palmitoylation of Hedgehog Proteins

John A. Buglino[1] *and* Marilyn D. Resh

Contents

Abstract

Hedgehog (Hh) proteins are secreted signaling proteins that contain amide-linked palmitate at the N-terminus and cholesterol at the C-terminus. Palmitoylation of Hh proteins is critical for effective long- and short-range signaling. The palmitoylation reaction occurs during transit of Hh through the secretory pathway, most likely in the lumen of the ER. Attachment of palmitate to Hh proteins is independent of cholesterol modification and autoprocessing and is catalyzed by Hhat (Hedgehog acyltransferase). Hhat is a member of the membrane bound *O*-acyltransferase (MBOAT) family, a subgroup of multipass membrane proteins that catalyze transfer of fatty acyl groups to lipids and proteins. Several classes of secreted proteins have recently been shown to be substrates for MBOAT acyltransferases, including Hh proteins and Spitz (palmitoylated by Hhat), Wg/Wnt

Cell Biology Program, Memorial Sloan-Kettering Cancer Center, New York, USA
[1] Current address: Immunology Program, Memorial Sloan-Kettering Cancer Center, New York, USA.

Vitamins and Hormones, Volume 88
ISSN 0083-6729, DOI: 10.1016/B978-0-12-394622-5.00010-9

proteins (modified with palmitate and/or palmitoleate by Porcupine) and ghrelin (octanoylated by ghrelin *O*-acyltransferase). These findings highlight protein fatty acylation as a mechanism that not only influences membrane binding of intracellular proteins but also regulates the signaling range and efficacy of secreted proteins.

I. Introduction to Protein Palmitoylation

Protein palmitoylation refers to the posttranslational modification of a protein with the 16-carbon fatty acid palmitate. In the majority of palmitoylation reactions, the palmitate donor is palmitoyl CoA (Resh, 2006; Smotrys and Linder, 2004). The term protein palmitoylation can also encompass posttranslational modification of proteins with other long chain fatty acids, as many palmitoylated proteins have been observed by mass spectrometry to incorporate heterogeneous mixtures of fatty acids at their modification site (Liang *et al.*, 2001, 2004). In nearly all cases, palmitate is attached to proteins via thioester linkage to a cysteine residue; this is known as *S*-palmitoylation. The labile nature of the thioester linkage in *S*-palmitoylated proteins allows for consecutive rounds of palmitoylation/depalmitoylation/repalmitoylation in a fashion similar to phosphorylation (Goodwin *et al.*, 2005; Loisel *et al.*, 1999; Lorentzen *et al.*, 2010; Yeh *et al.*, 1999). A small subset of palmitoylated proteins, including the *Drosophila* epidermal growth factor (EGF) like ligand Spitz (Miura *et al.*, 2006) and Hedgehog (Hh) family members (Chamoun *et al.*, 2001; Pepinsky *et al.*, 1998) are palmitoylated on a cysteine residue that is the N-terminal residue of the mature protein. *N*-Palmitoylation, in contrast to *S*-palmitoylation, is a stable modification. Palmitate attached via amide linkage is essentially as stable and long lived as the polypeptide backbone peptide bond. *N*-Palmitoylation can be readily distinguished from *S*-palmitoylation via its stability in neutral buffered hydroxylamine. The thioester linkage in *S*-palmitoylated proteins is hydrolyzed by hydroxylamine treatment while the amide linkage in *N*-palmitoylated proteins is resistant to hydrolysis.

Classes of proteins known to be modified with palmitate include cytoplasmic signaling molecules such as Src family tyrosine kinases (Alland *et al.*, 1994), the Ras family of small GTPases (Hancock *et al.*, 1989), heterotrimeric G-proteins (Kleuss and Krause, 2003), endothelial nitric oxide synthase (E-NOS) (Yeh *et al.*, 1999), and scaffolding proteins (Zhang *et al.*, 1998). Palmitoylation also occurs on several classes of membrane embedded proteins including ion channels (Gubitosi-Klug *et al.*, 2005; Qin *et al.*, 1998), transporters (Singaraja *et al.*, 2009), receptors, and tetraspanins (Resh, 2006). Finally, in addition to intracellular and membrane embedded proteins, palmitate has also been shown to be attached to secreted signaling molecules including Spitz (Miura *et al.*, 2006), members of the Wnt/wingless (Wg) family (Kurayoshi *et al.*, 2007; Willert *et al.*, 2003; Zhai *et al.*, 2004), as well as *Drosophila* and mammalian Hh family members (Buglino and Resh, 2008; Chamoun *et al.*, 2001; Pepinsky *et al.*, 1998).

Modification with palmitate has been implicated in the regulation of protein trafficking and localization, signal transduction, and enzymatic function. In general, attachment of palmitate increases the hydrophobicity of the modified protein and promotes membrane binding. Palmitoylation also enhances the partitioning of many signaling molecules into lipid rafts. Rafts are liquid ordered microdomains within cellular membranes, and raft association is implicated in mediating efficient signal transduction (Arcaro *et al.*, 2001; Fragoso *et al.*, 2003; Zhang *et al.*, 1998).

II. Palmitoylation of Hedgehog Proteins

The Hh family is the best studied example of palmitoylated, secreted signaling molecules. *Drosophila* express a single Hh protein, known simply as Hedgehog (Hh) (Nusslein-Volhard and Wieschaus, 1980). There are three Hh family members in mammals, Sonic Hedgehog (Shh), Indian Hedgehog (Ihh), and Desert Hedgehog (Dhh) (Bitgood and McMahon, 1995; Echelard *et al.*, 1993). Signaling by Hh family members plays a major role in embryonic patterning in both flies and mammals (Echelard *et al.*, 1993; Ericson *et al.*, 1995; Heemskerk and DiNardo, 1994; Ma *et al.*, 1993). All Hh family members are thought to be palmitoylated; however, direct incorporation of palmitate to hedgehog in cells has only been documented for Hh and Shh (Buglino and Resh, 2008; Pepinsky *et al.*, 1998). Palmitoylation comprises one of a series of modifications that occur on Hh proteins, as detailed below.

A. Hedgehog biosynthesis, autoprocessing, and cholesterol incorporation

All Hh family members are initially synthesized as an ~45-kDa precursor protein (Fig. 10.1). The precursor contains an amino-terminal signal sequence that directs entry into the endoplasmic reticulum (Lee *et al.*, 1992). Upon cleavage of the signal sequence, Hh is further processed in the ER by an autoproteolysis reaction catalyzed by the C-terminal half of the molecule (Chen *et al.*, 2011; Lee *et al.*, 1994). Hh autoprocessing occurs via a mechanism similar to that of intein processing in self-splicing proteins (Hall *et al.*, 1997). Autoprocessing requires an intact C-terminal domain and can be blocked by mutation of the histidine residue at position 329 or mutation of the cleavage site cysteine (C258) (Porter *et al.*, 1995, 1996a). Shh autoprocessing can also be inhibited by depletion of cellular sterols (Guy, 2000).

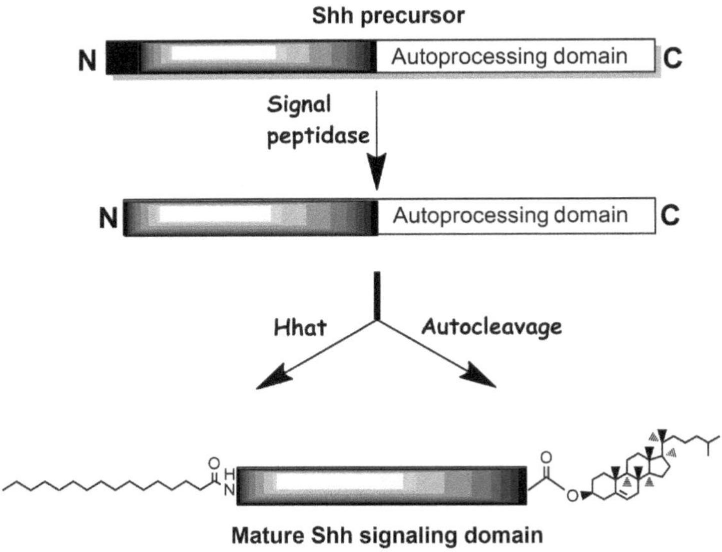

Figure 10.1 Multiple processing events generate the mature Shh signaling protein. Signal peptidase removes the N-terminal signal sequence. The autoprocessing domain mediates autocleavage and attachment of cholesterol to the C-terminus of the Shh signaling domain, while Hhat catalyzes attachment of palmitate via amide linkage to the N-terminus of the Shh signaling domain. (See Color Insert.)

The Shh cleavage reaction proceeds via a two-step mechanism. First, the sulfhydryl group of an invariant cysteine residue acts as a nucleophile to attack the carbonyl of the preceding glycine residue resulting in the formation of a thioester intermediate. This thioester linkage is then subject to nucleophilic attack by the 3β-hydroxyl group of a cholesterol molecule, resulting in the liberation of the C-terminal autoprocessing domain and the formation of an ester link between a cholesterol moiety and the carboxyl-terminal glycine residue of the 19-kDa amino-terminal fragment (Porter *et al.*, 1996a,b). Cholesterol is the only nucleophile whose incorporation into Hh has been directly observed *in vivo*. However, other related sterols can replace cholesterol in the autoprocessing reaction *in vitro* with varying degrees of efficiency (Cooper *et al.*, 1998). From *in vitro* assays, it is clear that the most important structural feature for autoprocessing is the C3 hydroxyl moiety. This group must be free of esterified adducts and there is an absolute requirement for the β orientation of the alcohol (Cooper *et al.*, 1998). The 19-kDa amino-terminal domain of Hh mediates all known signaling functions of Hh, while the C-terminal autoprocessing domain is primarily responsible for catalyzing autoprocessing and cholesterol incorporation (Fietz *et al.*, 1995; Marti *et al.*, 1995; Porter *et al.*, 1995). Following autocleavage, the C-terminal Shh fragment is rapidly degraded in the ER lumen by the ERAD (ER-associated degradation) pathway (Chen *et al.*, 2011).

B. Hedgehog palmitoylation

A molecule of palmitate is covalently attached via amide linkage to the amino-terminal cysteine residue of both Hh and Shh (Pepinsky *et al.*, 1998). The 19-kDa amino-terminal fragment of Hh/Shh containing both a C-terminal cholesterol and a N-terminal palmitate represents the mature signaling molecule and is the predominant form of Hh/Shh secreted *in vivo* (Taipale *et al.*, 2000). Shh palmitoylation is catalyzed by Hhat (Hedgehog acyltransferase), and this reaction can be recapitulated in cells and *in vitro* (Buglino and Resh, 2008, 2010).

Palmitoylation of Shh in cells occurs following cleavage of the signal sequence. Cotransfection of mammalian cells with cDNAs encoding Shh and Hhat reveals several features of the Shh palmitoylation reaction. Shh and Hhat colocalize to the endoplasmic reticulum and the Golgi (Buglino and Resh, 2008). Transit through the secretory pathway is essential, as Shh constructs that lack the N-terminal signal sequence are not palmitoylated (Buglino and Resh, 2008). These findings suggest that palmitoylation of Shh by Hhat occurs intracellularly, in the lumen of secretory organelles. Once the signal sequence is cleaved, cysteine becomes the N-terminal residue and this is the site of modification (Fig. 10.2). Palmitoylation is thought to initially occur via thioester linkage to the cysteine residue. The thioester intermediate then rearranges to an amide linkage via an intramolecular S-to-N shift, producing an amide or N-linked palmitate (Fig. 10.2A) (Mann and Beachy, 2004; Pepinsky *et al.*, 1998). However, in this scenario, the cysteine sulfhydryl group would be regenerated. Hhat could attach a second palmitate via a thioester bond, and a thioesterase would be needed to remove the second, thioester linked fatty acid. An alternative mechanism is direct attachment of the palmitate moiety to the N-terminal amide via amide linkage (Fig. 10.2B), analogous to *N*-myristoylation (Farazi *et al.*, 2001). This model is supported by the findings that N-terminally blocked Shh proteins are not substrates for Hhat and that thioester linked palmitoylated intermediates of Shh cannot be detected (Buglino and Resh, 2008). Further studies will be required to define the exact reaction mechanism.

An N-terminal cysteine is required for Shh palmitoylation. Shh mutants with substitutions of either alanine or serine have vastly reduced levels of palmitate incorporation. Signal sequence removal by signal peptidase is therefore absolutely essential as this cleavage event generates the N-terminal cysteine. Attachment of palmitate has been documented on both the 45 kDa Shh precursor protein as well as on the mature 19 kDa form of Shh. Moreover, a Shh mutant that is defective in autoprocessing and cholesterol attachment can still efficiently incorporate palmitate (Buglino and Resh, 2008). Thus, the two lipidation reactions occur independently.

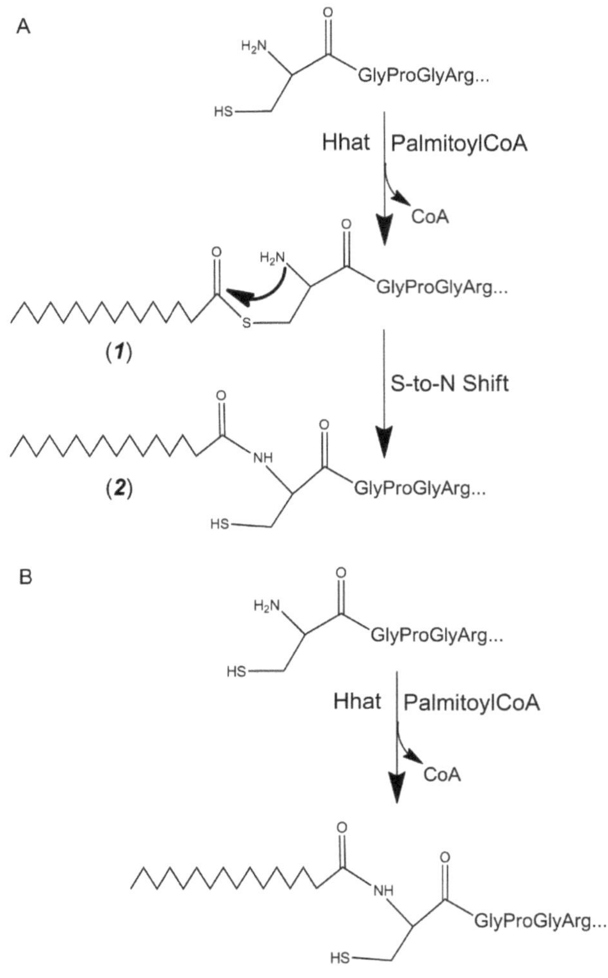

Figure 10.2 Potential mechanisms for Hhat-mediated *N*-palmitoylation of Hedgehog proteins. (A) Formation of a thioester intermediate (1) is followed by an S-to-N intramolecular shift and generation of the amide-linked, palmitoylated hedgehog protein (2). (B) Alternatively, Hhat could catalyze direct attachment of palmitate via amide bond to the amine group of the N-terminal cysteine.

C. Rasp and Hhat are Hh/Shh palmitoyl acyltransferases

Genetic studies in *Drosophila melanogaster* suggested that palmitoylation of Hh was catalyzed by a member of the membrane bound *O*-acyltransferase (MBOAT) family that was independently identified as Skinny Hedgehog, Sightless Hedgehog, Central missing, or Rasp (Amanai and Jiang, 2001; Chamoun *et al.*, 2001; Lee and Treisman, 2001; Micchelli *et al.*, 2002).

Rasp is required for induction of Hh target genes and proper patterning of both the *Drosophila* wing and eye imaginal disks, as well as for proper segmentation of *Drosophila* larva (Chamoun *et al.*, 2001; Lee and Treisman, 2001; Micchelli *et al.*, 2002). In addition, the *rasp* null phenotype closely phenocopies mutation of the palmitate acceptor site cysteine within Hh in the wing imaginal disk. Rasp activity is essential in cells of the posterior compartment that produce Hh but is not required for Hh transcription or secretion, leading to the stipulation that Rasp functions instead in the maturation of Hh signal (Chamoun *et al.*, 2001; Lee and Treisman, 2001; Micchelli *et al.*, 2002). Rasp mutant organisms display defects in Hh signaling independent of both autoprocessing and cholesterol incorporation, implying that Rasp does not play a role in these processes (Amanai and Jiang, 2001). Additional evidence that Rasp regulates Hh palmitoylation stems from the observation that Hh is significantly less hydrophobic when isolated from rasp null cells than when isolated from WT cells (Chamoun *et al.*, 2001).

The human homologue of Rasp, originally termed Skinny hedgehog, is known as Hhat (*H*edge*h*og *a*cyl*t*ransferase). Loss of Hhat function in the mouse closely phenocopies the mutant phenotype resulting from mutation of the palmitoylation site within Shh when assayed in either the limb or neural tube (Chen *et al.*, 2004). In addition, [^{3}H]-palmitate labeling of Shh is reduced when Shh is isolated from Hhat null murine embryonic fibroblasts (MEF)'s, consistent with Hhat's putative role in Shh palmitoylation (Chen *et al.*, 2004). Palmitoylation of Hh/Shh depends on the presence of a cysteine residue (position 85 in *Drosophila*, position 24 in human Shh, and position 25 in mouse Shh) immediately following the signal peptide cleavage site and on a functional form of Hhat or Rasp (Chamoun *et al.*, 2001; Pepinsky *et al.*, 1998). A Rasp construct with alanine substituted for both the aspartate at position 341 and the histidine at position 381 in tandem is unable to rescue the *rasp* mutant phenotype, implicating these residues in Rasp activity (Chamoun *et al.*, 2001).

Definitive evidence that Hhat is a palmitoyl acyltransferase for Shh was achieved when Hhat was purified to homogeneity (Buglino and Resh, 2008). Incubation of purified Hhat with purified, recombinant Shh results in near-stoichiometric incorporation of palmitate onto the N-terminal cysteine via amide linkage, features that recapitulate Shh palmitoylation in cells. The reaction occurs catalytically, uses palmitoyl CoA as the palmitate donor, and exhibits optimal activity at pH 6.5. Hhat appears to be relatively specific for Shh, as other palmitoylated proteins (H-Ras, PSD-95, Wnt proteins) are not substrates for palmitoylation by Hhat *in vitro*. A peptide containing the first 11 amino acids of the mature Shh sequence is a substrate for Hhat-mediated palmitoylation *in vitro*, thereby defining a minimal sequence for Shh palmitoylation (Buglino and Resh, 2008). In addition, the Shh substrate must contain an N-terminal cysteine with a free amino terminus, as purified Hhat cannot palmitoylate Shh peptides or proteins if the N-terminus is blocked by acetylation or by a hexa-histidine tag.

D. Role of palmitoylation in Hh signaling

The Hh-signaling cascade is utilized throughout development. Hh signaling plays a significant role in patterning of many aspects of the mammalian body plan, including the developing neural tube, limbs, axial skeleton, and induction of ventral forebrain structures (Echelard *et al.*, 1993; Ekker *et al.*, 1995; Ericson *et al.*, 1995; Lopez-Martinez *et al.*, 1995; Roelink *et al.*, 1995). Loss of Hh-signaling activity during vertebrate embryogenesis causes severe developmental disorders such as holoprosencephaly, craniofacial defects, and skeletal malformations (Cooper *et al.*, 1998; Incardona and Roelink, 2000; Lee *et al.*, 2001). In addition to its role in development, Hh has been implicated in the regulation of tissue homeostasis, stem cell biology, and the progression of many types of human malignancies, most notably mebdulloblastoma, pancreatic cancer, and other cancers of the GI tract (Feldmann *et al.*, 2007; Karhadkar *et al.*, 2004; Tian *et al.*, 2009; Yauch *et al.*, 2008).

Palmitoylation of Hh is required for proper signaling in flies. Mutation of the acceptor site cysteine produces a protein that has little to no patterning activity in *Drosophila* (Dawber *et al.*, 2005; Lee *et al.*, 2001). Expression of Hh C85S is not sufficient for rescue of the Hh phenotype of *Drosophila* larva despite the fact that it is expressed at levels similar to WT Hh and is properly autoprocessed. Misexpression of WT Hh in the posterior compartment of the wing disk leads to expansion of the anterior compartment, expanded expression of Hh target genes Ptc and Dpp, and patterning defects in the adult wing. Misexpression of Hh C85A, however, does not result in expansion of the anterior compartment or expanded expression of Patched and dpp in the wing disk, indicating that it is severely attenuated in signaling capacity (Lee *et al.*, 2001).

Mutation of the acceptor cysteine also results in a reduction of Shh signaling in mammalian tissues. However, nonpalmitoylated forms of Shh appear to be more active than their *Drosophila* counterparts. This is evidenced by the observation that ectopic overexpression of Shh C25S is still able to cause polydactyly in the mouse limb (Lee *et al.*, 2001). However, the signaling activity of nonpalmitoylated forms of Shh is clearly reduced in both the mouse limb and neural tube (Chen *et al.*, 2004). In addition, injection of retroviruses encoding WT Shh into the forebrains of E9.5 rats results in severe brain deformities, while retroviruses that encode nonpalmitoylated forms of Shh are unable to cause such defects (Kohtz *et al.*, 2001).

Studies in the mouse embryonic fibroblast line, C3H10T1/2, have revealed that palmitoylation also affects Shh potency *in vitro*. C3H10T1/2 is a mesenchymal cell line that can be induced under defined conditions to differentiate into adipocytes, chondrocytes, or bone osteoblasts (Pepinsky *et al.*, 1998). Upon differentiation into the bone lineage, there is a marked upregulation of alkaline phosphatase activity, which can be used as a marker for this process (Pepinsky *et al.*, 1998). Treatment of C3H10T1/2 cells with Shh causes a dose-dependent increase in alkaline phosphatase activity and is frequently used as a readout of Shh activity *in vitro*. Palmitoylated forms of Shh are 40–160-fold more active compared to unmodified Shh in this assay (Pepinsky *et al.*, 1998). The enhanced potency of palmitoylated forms of Shh does not correlate with enhanced receptor binding, as palmitoylated and nonpalmitoylated forms of Shh bind equally well to cells expressing Ptc (Pepinsky *et al.*, 1998). The hydrophobic nature of palmitate appears to directly contribute to its ability to enhance Shh-signaling potency. Increasing the hydrophobic character of the amino terminus of Shh, either by introducing a stretch of hydrophobic amino acids or by chemical modification, results in increased potency over WT unmodified forms. By contrast, introduction of hydrophilic residues at the N-terminus of Shh results in reduced signaling in a cell-based differentiation assay (Taylor *et al.*, 2001).

Release of Hh/Shh into the media requires the action of Dispatched, a 12-transmembrane (TM) domain containing protein (Burke *et al.*, 1999). Active Hh/Shh is released from the producing cell as a multimeric protein, and this process is facilitated by multimerization of Hh proteins on the cell surface. Dual lipidation enables Hh proteins to form nanoscale oligomers that colocalize with heparin sulfate proteoglycans (HSPGs) (Vyas *et al.*, 2008). Association with HSPGs promotes interaction of Hh proteins with ADAM17, a metalloprotease that has been implicated in Hh shedding from the cell surface (Dierker *et al.*, 2009). A recent study suggests that palmitoylation helps to position the N-terminal region of Shh for ectodomain cleavage by ADAM17 (Ohlig *et al.*, 2011). Proteolytic removal of the palmitoylated N-terminus is proposed to be required for binding of activated Shh to Patched.

After release from the cell, Hh family members act as morphogens that signal in a concentration-dependent manner (Fuccillo *et al.*, 2006; Gritli-Linde *et al.*, 2001; Heemskerk and DiNardo, 1994; Stamataki *et al.*, 2005). Hh responsive cells adopt specific cell fates, or induce different transcriptional profiles, in part depending upon the level of Hh signal received (Gritli-Linde *et al.*, 2001; Heemskerk and DiNardo, 1994). Hh protein levels and, therefore, signaling are highest at sites of Hh synthesis and decay as the distance between the source of Hh and the responding cell increases. Forming and maintaining the Hh-signaling gradient is essential for proper patterning of the cuticle in *Drosophila*, as well as the neural tube and distal limb elements during mammalian development (Gritli-Linde *et al.*, 2001; Heemskerk and DiNardo, 1994; Lee *et al.*, 2001). Posttranslational lipophilic modifications of the Hh ligand have important effects on the Hh-signaling gradient. Lipophilic modification influences partitioning of Hh proteins into lipoprotein particles, which have been implicated as playing an important role in long range Hh signaling (Callejo *et al.*, 2008; Panakova *et al.*, 2005). Modification of Hh with cholesterol and palmitate is also required for the formation of soluble multimeric forms of Hh that are freely diffusible, accumulate in a gradient, and enable the molecule to signal over long distances (Chen *et al.*, 2004; Goetz *et al.*, 2006; Zeng *et al.*, 2001).

E. Role of cholesterol modification and autoprocessing in Hh signaling

Autoprocessing and cholesterol incorporation into Hh/Shh are essential for proper tissue distribution and patterning activity. This is evidenced by the observation that several point mutations linked to holoprosencephaly occur within regions of Shh that have been implicated in autoprocessing and cholesterol incorporation (Maity *et al.*, 2005; Roessler *et al.*, 2009). However, Hh/Shh constructs that do not incorporate cholesterol still retain significant signaling activity over both short and long distances (Dawber *et al.*, 2005).

In the initial study characterizing Hh autoprocessing, ectopic overexpression of a cleavage defective form of Hh (H329A) still induced expanded Wg expression during *Drosophila* embryonic development, although to a lesser degree than when a WT construct was expressed (Lee *et al.*, 1994). This finding suggests that unprocessed Hh is able to signal but in a reduced capacity. Unprocessed forms of Hh were also less active in patterning of dorsal cuticle structures, and in patterning of the wing and eye imaginal disks (Lee *et al.*, 1994; Porter *et al.*, 1995). More recent studies have found that full-length unprocessed Hh is trafficked to the plasma membrane and can participate in direct cell-to-cell signaling, both *in vivo* and *in vitro*, but is unable to act over longer distances (Tokhunts *et al.*, 2009).

The reduction in signaling potential of unprocessed forms of Hh results from failure to liberate the amino-terminal signaling domain. Overexpression of Hh-N (the 19-kDa N-terminal fragment of Hh unmodified by cholesterol) alone is able to induce similar levels of Wg expression and causes nearly identical changes in dorsal cuticle patterning as full-length Hh (Burke *et al.*, 1999; Porter *et al.*, 1995). Hh-N is also able to rescue most Hh function in the wing imaginal disk (Burke *et al.*, 1999; Dawber *et al.*, 2005). Similarly, Shh-N still retains patterning activity in the mouse limb bud and chick neural tube (Caspary *et al.*, 2002; Yang *et al.*, 1997). In addition, purified recombinant Shh-N constructs lacking cholesterol are still able to bind to Ptc expressing cells and activate Shh signaling *in vitro* (Taylor *et al.*, 2001). When tested in the C3H10T1/2 differentiation assay, Shh-N lacking cholesterol induced alkaline phosphate upregulation but to a lesser degree than WT Shh (Taylor *et al.*, 2001). Taken together, these findings indicate that modification of Hh/Shh with cholesterol enhances signaling potency but is not strictly required for signaling activity *in vitro* or *in vivo*.

Incorporation of cholesterol into Hh/Shh regulates proper tissue distribution during development. Cholesterol incorporation into Hh/Shh restricts diffusion and promotes tethering of the molecule to membranes (Pepinsky *et al.*, 1998). Restriction of Hh/Shh diffusion is essential for forming a steep concentration gradient. Regulated overexpression of Hh-N using the GAL4:UAS system results in higher levels of Wg expression and causes more severe defects in dorsal cuticle patterning when compared to full-length Hh (Porter *et al.*, 1996a). This increased activity was correlated with a more diffuse staining pattern and reduced hydrophobic character of Hh-N, consistent with the cholesterol moiety acting to restrict the spread of Hh over long distances and promoting high levels of signaling near the site of synthesis. Similarly, reduced spread of cholesterol-modified forms of Hh has also been observed in the *Drosophila* wing disk (Dawber *et al.*, 2005).

Several lines of evidence suggest that cholesterol incorporation also restricts Shh diffusion in higher eukaryotes. The interaction of Hh with both Dispatched and HSPGs is dependent on cholesterol modification of Hh (Bellaiche *et al.*, 1998; Burke *et al.*, 1999; Kawakami *et al.*, 2002). Cholesterol incorporation has been shown to restrict the spread of Shh within the mouse limb bud (Li *et al.*, 2006). Shh-N constructs were distributed over a broader range and displayed reduced signaling potency within the limb bud when compared to full-length Shh (Li *et al.*, 2006). Similarly, when Shh-N expressed from COS-7 cells was grafted into the chick wing, it was able to diffuse further from the graft site compared to full-length Shh (Yang *et al.*, 1997). Expression of full-length human Shh in high five insect cells produces a cholesterol modified amino-terminal signaling domain which is primarily associated with the cell membrane (Pepinsky *et al.*, 1998). Mass spectrometry analysis of the minor soluble fraction of Shh remaining revealed it to be unmodified by cholesterol (Pepinsky *et al.*, 1998). Taken together, these results indicate that modification of Hh/Shh with cholesterol promotes membrane binding and favors short-range signaling. However, it is clear that cholesterol modified forms of Shh are able to directly activate signaling at a distance, and in some cases cholesterol appears to enhance long-range signaling (Briscoe *et al.*, 2001; Wang *et al.*, 2000). Thus, the question arises: how can a hydrophobic modification be required for both short- and long-range signaling?

Two possible mechanisms have been proposed to explain the requirement of cholesterol for efficient long-range Hedgehog signaling. First, cholesterol incorporation into Shh has been implicated in the formation of soluble multimeric forms of Shh (Zeng *et al.*, 2001). Second, cholesterol modification influences the partitioning of Hh into lipoprotein particles (Panakova *et al.*, 2005). Both of these structures would be able to travel longer distances and mediate long-range signaling.

III. Palmitoylation of Other Secreted Proteins

In addition to Hh family members, at least two other families of secreted signaling molecules have been shown to be palmitoylated: *Drosophila* EGF receptor (EGFR) ligands and members of the Wnt/Wg family. In both cases, palmitoylation is important for the function of the modified proteins.

A. EGF receptor ligands in flies

Spitz is a *Drosophila* EGFR ligand that is required for normal embryonic development, influencing the patterning of the eye, wing, and leg imaginal disks (Freeman, 1994; Miura *et al.*, 2006; Rutledge *et al.*, 1992). Spitz is initially synthesized as a membrane tethered precursor that is secreted following intracellular cleavage by the protease Rhomboid (Urban *et al.*, 2001). The mature secreted Spitz molecule is palmitoylated on an N-terminal cysteine residue and it is therefore likely that the palmitate attached to Spitz is amide linked (Miura *et al.*, 2006). Once secreted, palmitoylation of Spitz promotes its association with the extracellular face of the plasma membrane and acts to concentrate Spitz locally at the site of secretion (Miura *et al.*, 2006). This restriction of Spitz diffusion sets up a sharp concentration gradient at the site of synthesis. Palmitoylation is essential for proper Spitz signaling *in vivo*; nonpalmitoylated forms of Spitz are unable to rescue photoreceptor differentiation. However, palmitoylation of Spitz is not required for receptor binding or activation, as nonpalmitoylated Spitz signals in a fashion indistinguishable from palmitoylated Spitz *in vitro* (Miura *et al.*, 2006). Two other *Drosophila* EGF ligands, Keren and Gurken, also contain amino-terminal cysteine residues and are therefore also predicted to be *N*-palmitoylated (Miura *et al.*, 2006). However, direct incorporation of palmitate into either Keren or Gurken has not yet been demonstrated.

The acyltransferase responsible for Spitz palmitoylation is Rasp (Miura *et al.*, 2006). Mutation of Rasp impairs Spitz signaling in both the eye and wing imaginal disks (Miura *et al.*, 2006). Reduction of rasp mRNA levels by siRNA knockdown reduces the hydrophobic character of Spitz as judged by Triton X-114 detergent phase separation (Miura *et al.*, 2006). In addition, palmitoylation of Spitz in a heterologous system (COS-1 cells) is dependent on coexpression of a functional clone of Rasp (Miura *et al.*, 2006). Taken together, these results support the hypothesis that Rasp mediates Spitz palmitoylation.

B. Wnt and Wg proteins

Palmitoylation regulates signaling by members of the Wnt/Wg family of signaling molecules. Wnt/Wg family members are secreted glycoproteins that play important roles during development as well as in the progression of many types of human malignancies (Beachy *et al.*, 2004). Wnt/Wg proteins are heterogeneously glycosylated and palmitoylated (Tanaka *et al.*, 2002; Willert *et al.*, 2003). Palmitoylation has been directly observed in the case of Wg, Wnt3a, Wnt5a, and Wnt8 (Doubravska *et al.*, 2011; Kurayoshi *et al.*, 2007; Willert *et al.*, 2003; Zhai *et al.*, 2004). Murine Wnt3a is palmitoylated on Cysteine 77 (Willert *et al.*, 2003). This position is conserved in nearly all Wnt proteins, suggesting that other Wnt family members are also likely palmitoylated.

Modification of Wnt3a by palmitate is not essential for secretion or glycosylation of Wnt3a but is required for proper signaling. Both WT and C77A Wnt-3a (a non-palmitoylated mutant) can be efficiently recovered from conditioned media produced by transfected tissue culture cells (Willert *et al.*, 2003). However, non-palmitoylated Wnt3a is compromised in its ability to bind its receptors LRP6 and Frizzled 8 (Komekado *et al.*, 2007). In addition, purified Wnt3a C77A is unable to induce β-catenin stabilization when added to mouse L cells, indicating that palmitoylation is essential for Wnt3a signaling (Willert *et al.*, 2003). Palmitoylation serves similar functions in the case of Wnt5a. Preventing palmitoylation of Wnt5a reduces receptor binding and compromises downstream signaling but appears dispensable for secretion (Kurayoshi *et al.*, 2007). By contrast, palmitoylation of *Drosophila* Wg at position 93 is essential for Wg secretion. Treatment of *Drosophila* S2 cells with the inhibitor of protein palmitoylation, 2-bromopalmitate prevents efficient Wg secretion into the culture media (Zhai *et al.*, 2004). In addition, mutation of Cys 93 to Ala compromises secretion of Wg in the wing imaginal disk (Franch-Marro *et al.*, 2008). The C93A Wg mutant accumulates within the ER, indicating that palmitoylation is essential for intracellular trafficking and secretion of Wg (Franch-Marro *et al.*, 2008).

Wnt3a is also modified with a monounsaturated fatty acid, palmitoleic acid, on Serine 209 (Takada *et al.*, 2006). This position is well conserved within the Wnt/Wg family, suggesting that other family members might be modified in a similar fashion. Palmitoleic acid incorporation is required for efficient palmitoylation of Wnt3 at Cys 77 (Doubravska *et al.*, 2011; Takada *et al.*, 2006). Modification of Wnt3a with palmitoleic acid regulates secretion, as Wnt3a S209A protein is not recovered in the conditioned media of L cells posttransfection (Takada *et al.*, 2006). As expected given its secretion defect, S209A Wnt3a-RNA injected into *Xenopus* oocytes was unable to cause axis duplication, a hallmark of active Wnt3a signaling (Takada *et al.*, 2006).

Incorporation of palmitoleic acid has yet to be reported for any other member of the Wnt/Wg family. However, the predicted modification site, Ser 239, has been mutated in *Drosophila* Wg (Franch-Marro *et al.*, 2008). Unlike the equivalent mutation in mouse Wnt3a, Wg S239A was readily secreted (Franch-Marro *et al.*, 2008). Moreover, another *Drosophila* Wnt homologue, WntD, lacks the putative palmitoleic acid modification site but is still efficiently secreted from both S2 and L cells, indicating that palmitoleic acid modification is not essential for secretion of all Wnt/Wg family members (Ching *et al.*, 2008). Mutation of Ser 239 does compromise Wg signaling activity, suggesting that palmitoleic acid incorporation may be important for signaling by both mouse Wnt3a and *Drosophila* Wg (Franch-Marro *et al.*, 2008).

In addition to influencing secretion and signaling activity, palmitoylation of Wnt/Wg proteins also affects their extracellular transport. Enhancing palmitoylation of either Wnt1 or Wnt3a in the chick neural tube by ectopic overexpression of the putative palmitoyl transferase, Porcupine, acts to steepen the Wnt signaling gradient, presumably by preventing diffusion away from the site of secretion (Galli *et al.*, 2007). This result suggests that palmitoylation regulates extracelluar transport of Wnt/Wg in a similar fashion to Spitz. However, Wnt family members signal over longer distances. This fact, coupled with the observation that Wg copurifies with apolipophorins has led to the speculation that palmitoylation promotes Wg incorporation into lipoprotein particles (Panakova *et al.*, 2005). Once incorporated into lipoprotein particles Wg would then be able to travel and signal over longer distances.

IV. The MBOAT Family of Acyltransferases

Two groups of protein palmitoyl acyltransferases (PATs) have been identified. The first, which is responsible for palmitoylation of cytoplasmic and membrane bound proteins, is the DHHC family of PATs (Lobo *et al.*, 2002). This family is named for the presence of a conserved Asp-His-His-Cys motif that is required for activity (Lobo *et al.*, 2002). The second family of PATs, the MBOAT family, mediates attachment of palmitate and other fatty acids to lipid and protein substrates. In particular, the MBOAT family members Hhat, Porcupine, and ghrelin *O*-acyltransferase (GOAT) are responsible for fatty acylation of secreted proteins.

A. MBOAT family characteristics

The MBOAT family of acyltransferases was initially identified by bioinformatics studies. Global sequence alignment algorithms revealed a region of high homology between acyl:CoA cholesterol acyltransferases 1 and 2 (ACAT1/2), the enzymes that catalyze the formation of cholesterol esters, and Porcupine, a segment polarity gene linked to posttranslational modification of Wnt/Wg proteins (Hofmann, 2000). The defining characteristics of the MBOAT homology domain are the presence of a highly conserved His residue surrounded by a cluster of hydrophobic residues and a well-conserved Asp/Asn residue surrounded by moderately hydrophobic residues

(Hofmann, 2000). Mutation of one or both of these residues compromise enzymatic activity in all family members tested, suggesting that they may be directly involved in catalysis (Bosson *et al.*, 2006; Chamoun *et al.*, 2001; Guo *et al.*, 2005; Lin *et al.*, 2003; Shindou *et al.*, 2009; Yang *et al.*, 2008a).

All members of the MBOAT family are predicted to be multipass membrane proteins containing between 8 and 12 TM segments. Unlike the DHHC family, most MBOAT family members do not function as protein PATs, but rather catalyze transfer of long chain fatty acids to hydroxyl groups of other lipophilic molecules including cholesterol, diacylglycerol, and lysophospholipids (Cases *et al.*, 1998; Chang *et al.*, 1998; Stahl *et al.*, 2008). Members of the MBOAT family are also involved in processing glycosylphosphatidylinositol anchors and the formation of wax esters (Bosson *et al.*, 2006; Yen *et al.*, 2005). In mammalian cells, MBOAT family members with known substrates can be subdivided into three subfamilies based upon substrate preference and/or amino acid homology. These are: (1) the ACAT family, consisting of ACAT1, ACAT2, and DGAT1 (acyl:CoA diacylglycerol-acyltransferase 1); (2) the LPLAT (lysophospholipid acyltransferases) family, consisting of MBOAT1, MBOAT2, MBOAT5, MBOAT7; and (3) the PAT family consisting of Porcupine, Hhat, MBOA-4 (GOAT), and the mammalian homologue of glycerol uptake protein-1 (GUP1), also known as Hhat-like (Hhatl).

B. MBOAT family members that acylate protein substrates

Three members of the MBOAT family have been implicated in the fatty acylation of secreted proteins (Table 10.1). Hhat/Rasp are the palmitoyl acyltransferases for Hh proteins in flies and mammalian cells, and for Spitz in flies (Chamoun *et al.*, 2001). Porcupine has been implicated in Wnt/Wg palmitoylation (Zhai *et al.*, 2004). GOAT mediates incorporation of an octanoyl moiety into the appetite stimulating peptide hormone ghrelin (Gutierrez *et al.*, 2008; Yang *et al.*, 2008a). Biochemical analyses of these proteins are limited primarily due their hydrophobic nature and the difficulty in establishing *in vitro* assays to monitor fatty acid transfer to substrate proteins.

1. Hhat

Hhat is an ~50kDa multipass membrane protein that, as described in Section II.C, catalyzes attachment of palmitate to hedgehog proteins. Structure–function analysis of Hhat has begun to define regions and residues that are important for Shh palmitoylation. Transmembrane topology mapping predicts at least eight TM domains in Hhat. Truncation mutants that lack one or two TM segments from either the N- or C-terminal ends of Hhat exhibit reduced stability when expressed in Hhat-transfected cells and

Table 10.1 MBOAT family members that catalyze attachment of fatty acids to proteins

Enzyme	Protein substrate	Fatty acid substrate
Hhat	Shh, Ihh, Dhh	Palmitate (16:0)
Rasp	Hh, Spitz, Keren, Gerken	Palmitate (16:0)
Porcupine	Wg and Wnt proteins	Palmitate (16:0) Palmitoleic (16:1)
GOAT	Ghrelin	Octanoate (8:0)

are inactive in Shh palmitoylation assays (Buglino and Resh, 2010). Two large loop regions are predicted to be positioned between residues 153–235 and 311–426 in Hhat. Both of these loops contain residues that are highly conserved among MBOAT family members that acylate protein substrates. Deletions or point mutations within these loops reduce Hhat activity without compromising protein stability (Buglino and Resh, 2010). In particular, mutation of the invariant histidine (His379) that is conserved in all MBOAT proteins, reduces palmitoylation activity of Hhat toward Shh by twofold. Kinetic analysis reveals that this mutant binds and transfers palmitoyl CoA as well as the wild-type enzyme, but has an increased K_m and decreased V_{max} for Shh.

2. Porcupine

Several lines of genetic data suggest that Porcupine mediates modification of Wnt/Wg proteins with both palmitate and palmitoleic acid (Kadowaki *et al.*, 1996; Takada *et al.*, 2006; Willert *et al.*, 2003; Zhai *et al.*, 2004). The *porcupine* gene was first identified in genetic screens aimed at identifying mutations that affect segment number and polarity during early *Drosophila* development (Nusslein-Volhard and Wieschaus, 1980). Further analysis of the *porcupine* mutant phenotype indicated that loss of *porcupine* function impaired Wg signaling in *Drosophila* (Kadowaki *et al.*, 1996). Subsequent cloning and characterization of the gene responsible for the *porcupine* phenotype revealed that it encodes a multipass membrane protein that influences Wg processing during transit through the secretory pathway (Tanaka *et al.*, 2002). Loss of Porcupine function reduces the hydrophobic character of Wg as judged by reversed phase chromatography and detergent phase separation (Zhai *et al.*, 2004). In addition, the proportion of Wg present in lipid rafts is reduced when Porcupine function is impaired (Zhai *et al.*, 2004). In mammalian cells, mutation of Porcupine causes a similar reduction in the amount Wnt-3a recovered in the detergent phase after Triton X-114 extraction (Willert *et al.*, 2003). siRNA mediated knockdown of

Porcupine inhibits Wnt-3a signaling, indicating that Porcupine is required for Wnt-3a palmitoylation *in vivo* (Galli *et al.*, 2007). However, direct Porcupine mediated lipophilic modification of Wnt/Wg family member at either site has not yet been observed. Loss of Porcupine function has also been observed to result in alteration in the glycosylation status of *Drosophila* Wg, but it is not clear if these effects are direct or indirect (Kadowaki *et al.*, 1996; Tanaka *et al.*, 2002).

The C-terminal end of Porcupine has been implicated in Wg binding. C-terminal truncation mutants of Porcupine fail to Co-IP Wg to the same extent as full-length Porcupine (Tanaka *et al.*, 2002). Apart from this region, no other residues or regions of Porcupine have yet been implicated in the regulation of substrate binding or catalysis.

3. Ghrelin *O*-acyltransferase

Ghrelin is a small peptide hormone that stimulates growth hormone release from the pituitary and also acts to stimulate appetite (Cummings, 2006; Kamegai *et al.*, 2001; Kojima *et al.*, 1999). Mature ghrelin is modified by an O-linked molecule of octanoic acid on Ser-3, which is essential for all known functions (Kojima *et al.*, 1999). Loss of ghrelin by genetic knockout results in a significant reduction in obesity in mice when challenged with a high-fat diet (Kamegai *et al.*, 2001). This observation has led to increased interest in the regulation of ghrelin octanoylation as a potential drug target for prevention of obesity in humans (Cummings, 2006).

Recently, two independent groups identified the enzyme that catalyzes ghrelin octanoylation, GOAT, for ghrelin *O*-acyltransferase (Gutierrez *et al.*, 2008; Yang *et al.*, 2008a). Overexpression of GOAT stimulates ghrelin octanoylation and mice lacking GOAT are devoid of octanoylated ghrelin in circulating blood (Gutierrez *et al.*, 2008; Yang *et al.*, 2008a). These findings indicate that GOAT is the sole mediator of ghrelin octanoylation *in vivo*. GOAT activity requires the MBOAT homology Asn and His residues, as mutation of either position eliminates GOAT activity both in intact cells as well as *in vitro* (Yang *et al.*, 2008a,b). Thus far no other regions of GOAT have been implicated in catalysis or substrate binding. However, GOAT mediated octanoylation of ghrelin is influenced by the residues that surround the modification site Ser. Mutation of the Gly at position-1 or the Phe at position-4 results in almost complete inhibition of ghrelin octanoylation (Yang *et al.*, 2008b). These results indicate that the amino-terminal residues of ghrelin serve as a GOAT recognition sites (Ohgusu *et al.*, 2009; Yang *et al.*, 2008b). Further studies have shown that peptides derived from the N-terminus of ghrelin are able to inhibit the octanoylation of full-length ghrelin. In these studies, octanoylated peptides were two to four times as potent as nonoctanoylated peptides suggesting that GOAT is subject to product inhibition (Yang *et al.*, 2008b).

V. Conclusions and Future Directions

Palmitoylation has traditionally been viewed as a mechanism to enhance membrane binding of intracellular proteins. It is now clear that secreted proteins, such as Hh, Shh, and Wnt proteins, can be palmitoylated, and that palmitoylation regulates the signaling range and activity of these morphogens in the extracellular space. Hh/Shh palmitoylation is catalyzed by Hhat, is independent of autocleavage and cholesterol attachment, and occurs during passage of Hh/Shh through the secretory pathway. Hedgehog proteins enter secretory organelles by virtue of their signal sequence, but a mechanism for transporting the other substrate, palmitoyl CoA, into the lumen of the ER and/or Golgi must exist. The active site of Hhat is likely exposed to the ER lumen but details regarding Hhat TM topology remain to be determined. To date, known Hhat substrates include Hedgehog proteins and Spitz, but it is likely that other protein substrates for Hhat exist in flies and mammals. Similarities and differences in how two MBOAT proteins, Hhat and Porcupine, recognize their respective substrates, need to be delineated. We also need a better understanding of how lipidation regulates extracellular gradient formation by palmitoylated morphogens. Finally, identification of activators and inhibitors of Hhat as well as other MBOAT proteins will be useful. A more complete understanding of Hhat enzymology could potentially provide avenues for therapeutic intervention in diseases that require ligand dependent activation of the Hh-signaling cascade for growth and survival.

Acknowledgments

Research in the authors' laboratory was supported by NIH Grants GM57966, GM008539, and CA 158474.

References

Alland, L., Peseckis, S. M., Atherton, R. E., Berthiaume, L., and Resh, M. D. (1994). Dual myristylation and palmitoylation of Src family member p59fyn affects subcellular localization. *J. Biol. Chem.* **269,** 16701–16705.

Amanai, K., and Jiang, J. (2001). Distinct roles of Central missing and Dispatched in sending the Hedgehog signal. *Development* **128,** 5119–5127.

Arcaro, A., Gregoire, C., Bakker, T. R., Baldi, L., Jordan, M., Goffin, L., Boucheron, N., Wurm, F., van der Merwe, P. A., Malissen, B., and Luescher, I. F. (2001). CD8beta endows CD8 with efficient coreceptor function by coupling T cell receptor/CD3 to raft-associated CD8/p56(lck) complexes. *J. Exp. Med.* **194,** 1485–1495.

Beachy, P. A., Karhadkar, S. S., and Berman, D. M. (2004). Tissue repair and stem cell renewal in carcinogenesis. *Nature* **432,** 324–331.

Bellaiche, Y., The, I., and Perrimon, N. (1998). Tout-velu is a Drosophila homologue of the putative tumour suppressor EXT-1 and is needed for Hh diffusion. *Nature* **394,** 85–88.

Bitgood, M. J., and McMahon, A. P. (1995). Hedgehog and Bmp genes are coexpressed at many diverse sites of cell-cell interaction in the mouse embryo. *Dev. Biol.* **172,** 126–138.

Bosson, R., Jaquenoud, M., and Conzelmann, A. (2006). GUP1 of Saccharomyces cerevisiae encodes an O-acyltransferase involved in remodeling of the GPI anchor. *Mol. Biol. Cell* **17,** 2636–2645.

Briscoe, J., Chen, Y., Jessell, T. M., and Struhl, G. (2001). A hedgehog-insensitive form of patched provides evidence for direct long-range morphogen activity of sonic hedgehog in the neural tube. *Mol. Cell* **7,** 1279–1291.

Buglino, J. A., and Resh, M. D. (2008). Hhat is a palmitoylacyltransferase with specificity for N-palmitoylation of Sonic Hedgehog. *J. Biol. Chem.* **283,** 22076–22088.

Buglino, J. A., and Resh, M. D. (2010). Identification of conserved regions and residues within Hedgehog acyltransferase critical for palmitoylation of Sonic Hedgehog. *PLoS One* **5,** e11195.

Burke, R., Nellen, D., Bellotto, M., Hafen, E., Senti, K. A., Dickson, B. J., and Basler, K. (1999). Dispatched, a novel sterol-sensing domain protein dedicated to the release of cholesterol-modified hedgehog from signaling cells. *Cell* **99,** 803–815.

Callejo, A., Culi, J., and Guerrero, I. (2008). Patched, the receptor of Hedgehog, is a lipoprotein receptor. *Proc. Natl. Acad. Sci. USA* **105,** 912–917.

Cases, S., Smith, S. J., Zheng, Y. W., Myers, H. M., Lear, S. R., Sande, E., Novak, S., Collins, C., Welch, C. B., Lusis, A. J., Erickson, S. K., and Farese, R. V., Jr. (1998). Identification of a gene encoding an acyl CoA:Diacylglycerol acyltransferase, a key enzyme in triacylglycerol synthesis. *Proc. Natl. Acad. Sci. USA* **95,** 13018–13023.

Caspary, T., Garcia-Garcia, M. J., Huangfu, D., Eggenschwiler, J. T., Wyler, M. R., Rakeman, A. S., Alcorn, H. L., and Anderson, K. V. (2002). Mouse Dispatched homolog1 is required for long-range, but not juxtacrine, Hh signaling. *Curr. Biol.* **12,** 1628–1632.

Chamoun, Z., Mann, R. K., Nellen, D., von Kessler, D. P., Bellotto, M., Beachy, P. A., and Basler, K. (2001). Skinny hedgehog, an acyltransferase required for palmitoylation and activity of the hedgehog signal. *Science* **293,** 2080–2084.

Chang, C. C., Lee, C. Y., Chang, E. T., Cruz, J. C., Levesque, M. C., and Chang, T. Y. (1998). Recombinant acyl-CoA:cholesterol acyltransferase-1 (ACAT-1) purified to essential homogeneity utilizes cholesterol in mixed micelles or in vesicles in a highly cooperative manner. *J. Biol. Chem.* **273,** 35132–35141.

Chen, M. H., Li, Y. J., Kawakami, T., Xu, S. M., and Chuang, P. T. (2004). Palmitoylation is required for the production of a soluble multimeric Hedgehog protein complex and long-range signaling in vertebrates. *Genes Dev.* **18,** 641–659.

Chen, X., Tukachinsky, H., Huang, C.-H., Jao, C., Chu, Y.-R., Tang, H.-Y., Mueller, B., Schulman, S., Rapoport, T. A., and Salic, A. (2011). Processing and turnover of the Hedgehog protein in the endoplasmic reticulum. *J. Cell Biol.* **192,** 825–838.

Ching, W., Hang, H. C., and Nusse, R. (2008). Lipid-independent secretion of a Drosophila Wnt protein. *J. Biol. Chem.* **283,** 17092–17098.

Cooper, M. K., Porter, J. A., Young, K. E., and Beachy, P. A. (1998). Teratogen-mediated inhibition of target tissue response to Shh signaling. *Science* **280,** 1603–1607.

Cummings, D. E. (2006). Ghrelin and the short- and long-term regulation of appetite and body weight. *Physiol. Behav.* **89,** 71–84.

Dawber, R. J., Hebbes, S., Herpers, B., Docquier, F., and van den Heuvel, M. (2005). Differential range and activity of various forms of the Hedgehog protein. *BMC Dev. Biol.* **5,** 21.

Dierker, T., Dreier, R., Petersen, A., Bordych, C., and Grobe, K. (2009). Heparan sulfate-modulated, metalloprotease-mediated sonic hedgehog release from producing cells. *J. Biol. Chem.* **284,** 8013–8022.
Doubravska, L., Krausova, M., Gradl, D., Vojtechova, M., Tumora, L., Lukas, J., Valenta, T., Pospichalova, V., Fafilek, B., Plachy, J., Sebesta, O., and Korinek, V. (2011). Fatty acid modification of Wnt1 and Wnt3a at serine is prerequisite for lipidation at cysteine and is essential for Wnt signalling. *Cell. Signal.* **23,** 837–848.
Echelard, Y., Epstein, D. J., St-Jacques, B., Shen, L., Mohler, J., McMahon, J. A., and McMahon, A. P. (1993). Sonic hedgehog, a member of a family of putative signaling molecules, is implicated in the regulation of CNS polarity. *Cell* **75,** 1417–1430.
Ekker, S. C., McGrew, L. L., Lai, C. J., Lee, J. J., von Kessler, D. P., Moon, R. T., and Beachy, P. A. (1995). Distinct expression and shared activities of members of the hedgehog gene family of Xenopus laevis. *Development* **121,** 2337–2347.
Ericson, J., Muhr, J., Placzek, M., Lints, T., Jessell, T. M., and Edlund, T. (1995). Sonic hedgehog induces the differentiation of ventral forebrain neurons: A common signal for ventral patterning within the neural tube. *Cell* **81,** 747–756.
Farazi, T. A., Waksman, G., and Gordon, J. I. (2001). The biology and enzymology of protein N-myristoylation. *J. Biol. Chem.* **276,** 39501–39504.
Feldmann, G., Dhara, S., Fendrich, V., Bedja, D., Beaty, R., Mullendore, M., Karikari, C., Alvarez, H., Iacobuzio-Donahue, C., Jimeno, A., Gabrielson, K. L., Matsui, W., *et al.* (2007). Blockade of hedgehog signaling inhibits pancreatic cancer invasion and metastases: A new paradigm for combination therapy in solid cancers. *Cancer Res.* **67,** 2187–2196.
Fietz, M. J., Jacinto, A., Taylor, A. M., Alexandre, C., and Ingham, P. W. (1995). Secretion of the amino-terminal fragment of the hedgehog protein is necessary and sufficient for hedgehog signalling in Drosophila. *Curr. Biol.* **5,** 643–650.
Fragoso, R., Ren, D., Zhang, X., Su, M. W., Burakoff, S. J., and Jin, Y. J. (2003). Lipid raft distribution of CD4 depends on its palmitoylation and association with Lck, and evidence for CD4-induced lipid raft aggregation as an additional mechanism to enhance CD3 signaling. *J. Immunol.* **170,** 913–921.
Franch-Marro, X., Wendler, F., Griffith, J., Maurice, M. M., and Vincent, J. P. (2008). In vivo role of lipid adducts on Wingless. *J. Cell Sci.* **121,** 1587–1592.
Freeman, M. (1994). The spitz gene is required for photoreceptor determination in the Drosophila eye where it interacts with the EGF receptor. *Mech. Dev.* **48,** 25–33.
Fuccillo, M., Joyner, A. L., and Fishell, G. (2006). Morphogen to mitogen: The multiple roles of hedgehog signalling in vertebrate neural development. *Nat. Rev. Neurosci.* **7,** 772–783.
Galli, L. M., Barnes, T. L., Secrest, S. S., Kadowaki, T., and Burrus, L. W. (2007). Porcupine-mediated lipid-modification regulates the activity and distribution of Wnt proteins in the chick neural tube. *Development* **134,** 3339–3348.
Goetz, J. A., Singh, S., Suber, L. M., Kull, F. J., and Robbins, D. J. (2006). A highly conserved amino-terminal region of sonic hedgehog is required for the formation of its freely diffusible multimeric form. *J. Biol. Chem.* **281,** 4087–4093.
Goodwin, J. S., Drake, K. R., Rogers, C., Wright, L., Lippincott-Schwartz, J., Philips, M. R., and Kenworthy, A. K. (2005). Depalmitoylated Ras traffics to and from the Golgi complex via a nonvesicular pathway. *J. Cell Biol.* **170,** 261–272.
Gritli-Linde, A., Lewis, P., McMahon, A. P., and Linde, A. (2001). The whereabouts of a morphogen: Direct evidence for short- and graded long-range activity of hedgehog signaling peptides. *Dev. Biol.* **236,** 364–386.
Gubitosi-Klug, R. A., Mancuso, D. J., and Gross, R. W. (2005). The human Kv1.1 channel is palmitoylated, modulating voltage sensing: Identification of a palmitoylation consensus sequence. *Proc. Natl. Acad. Sci. USA* **102,** 5964–5968.
Guo, Z. Y., Lin, S., Heinen, J. A., Chang, C. C., and Chang, T. Y. (2005). The active site His-460 of human acyl-coenzyme A:cholesterol acyltransferase 1 resides in a hitherto undisclosed transmembrane domain. *J. Biol. Chem.* **280,** 37814–37826.

Gutierrez, J. A., Solenberg, P. J., Perkins, D. R., Willency, J. A., Knierman, M. D., Jin, Z., Witcher, D. R., Luo, S., Onyia, J. E., and Hale, J. E. (2008). Ghrelin octanoylation mediated by an orphan lipid transferase. *Proc. Natl. Acad. Sci. USA* **105,** 6320–6325.

Guy, R. K. (2000). Inhibition of sonic hedgehog autoprocessing in cultured mammalian cells by sterol deprivation. *Proc. Natl. Acad. Sci. USA* **97,** 7307–7312.

Hall, T. M., Porter, J. A., Young, K. E., Koonin, E. V., Beachy, P. A., and Leahy, D. J. (1997). Crystal structure of a Hedgehog autoprocessing domain: Homology between Hedgehog and self-splicing proteins. *Cell* **91,** 85–97.

Hancock, J. F., Magee, A. I., Childs, J. E., and Marshall, C. J. (1989). All ras proteins are polyisoprenylated but only some are palmitoylated. *Cell* **57,** 1167–1177.

Heemskerk, J., and DiNardo, S. (1994). Drosophila hedgehog acts as a morphogen in cellular patterning. *Cell* **76,** 449–460.

Hofmann, K. (2000). A superfamily of membrane-bound O-acyltransferases with implications for wnt signaling. *Trends Biochem. Sci.* **25,** 111–112.

Incardona, J. P., and Roelink, H. (2000). The role of cholesterol in Shh signaling and teratogen-induced holoprosencephaly. *Cell. Mol. Life Sci.* **57,** 1709–1719.

Kadowaki, T., Wilder, E., Klingensmith, J., Zachary, K., and Perrimon, N. (1996). The segment polarity gene porcupine encodes a putative multitransmembrane protein involved in Wingless processing. *Genes Dev.* **10,** 3116–3128.

Kamegai, J., Tamura, H., Shimizu, T., Ishii, S., Sugihara, H., and Wakabayashi, I. (2001). Chronic central infusion of ghrelin increases hypothalamic neuropeptide Y and Agouti-related protein mRNA levels and body weight in rats. *Diabetes* **50,** 2438–2443.

Karhadkar, S. S., Bova, G. S., Abdallah, N., Dhara, S., Gardner, D., Maitra, A., Isaacs, J. T., Berman, D. M., and Beachy, P. A. (2004). Hedgehog signalling in prostate regeneration, neoplasia and metastasis. *Nature* **431,** 707–712.

Kawakami, T., Kawcak, T., Li, Y. J., Zhang, W., Hu, Y., and Chuang, P. T. (2002). Mouse dispatched mutants fail to distribute hedgehog proteins and are defective in hedgehog signaling. *Development* **129,** 5753–5765.

Kleuss, C., and Krause, E. (2003). Galpha(s) is palmitoylated at the N-terminal glycine. *EMBO J.* **22,** 826–832.

Kohtz, J. D., Lee, H. Y., Gaiano, N., Segal, J., Ng, E., Larson, T., Baker, D. P., Garber, E. A., Williams, K. P., and Fishell, G. (2001). N-terminal fatty-acylation of sonic hedgehog enhances the induction of rodent ventral forebrain neurons. *Development* **128,** 2351–2363.

Kojima, M., Hosoda, H., Date, Y., Nakazato, M., Matsuo, H., and Kangawa, K. (1999). Ghrelin is a growth-hormone-releasing acylated peptide from stomach. *Nature* **402,** 656–660.

Komekado, H., Yamamoto, H., Chiba, T., and Kikuchi, A. (2007). Glycosylation and palmitoylation of Wnt-3a are coupled to produce an active form of Wnt-3a. *Genes Cells* **12,** 521–534.

Kurayoshi, M., Yamamoto, H., Izumi, S., and Kikuchi, A. (2007). Post-translational palmitoylation and glycosylation of Wnt-5a are necessary for its signalling. *Biochem. J.* **402,** 515–523.

Lee, J. D., and Treisman, J. E. (2001). Sightless has homology to transmembrane acyltransferases and is required to generate active Hedgehog protein. *Curr. Biol.* **11,** 1147–1152.

Lee, J. J., von Kessler, D. P., Parks, S., and Beachy, P. A. (1992). Secretion and localized transcription suggest a role in positional signaling for products of the segmentation gene hedgehog. *Cell* **71,** 33–50.

Lee, J. J., Ekker, S. C., von Kessler, D. P., Porter, J. A., Sun, B. I., and Beachy, P. A. (1994). Autoproteolysis in hedgehog protein biogenesis. *Science* **266,** 1528–1537.

Lee, J. D., Kraus, P., Gaiano, N., Nery, S., Kohtz, J., Fishell, G., Loomis, C. A., and Treisman, J. E. (2001). An acylatable residue of Hedgehog is differentially required in Drosophila and mouse limb development. *Dev. Biol.* **233,** 122–136.

Li, Y., Zhang, H., Litingtung, Y., and Chiang, C. (2006). Cholesterol modification restricts the spread of Shh gradient in the limb bud. *Proc. Natl. Acad. Sci. USA* **103,** 6548–6553.

Liang, X., Nazarian, A., Erdjument-Bromage, H., Bornmann, W., Tempst, P., and Resh, M. D. (2001). Heterogeneous fatty acylation of Src family kinases with polyunsaturated fatty acids regulates raft localization and signal transduction. *J. Biol. Chem.* **276,** 30987–30994.

Liang, X., Lu, Y., Wilkes, M., Neubert, T. A., and Resh, M. D. (2004). The N-terminal SH4 region of the Src family kinase Fyn is modified by methylation and heterogeneous fatty acylation: Role in membrane targeting, cell adhesion, and spreading. *J. Biol. Chem.* **279,** 8133–8139.

Lin, S., Lu, X., Chang, C. C., and Chang, T. Y. (2003). Human acyl-coenzyme A: cholesterol acyltransferase expressed in chinese hamster ovary cells: Membrane topology and active site location. *Mol. Biol. Cell* **14,** 2447–2460.

Lobo, S., Greentree, W. K., Linder, M. E., and Deschenes, R. J. (2002). Identification of a Ras palmitoyltransferase in Saccharomyces cerevisiae. *J. Biol. Chem.* **277,** 41268–41273.

Loisel, T. P., Ansanay, H., Adam, L., Marullo, S., Seifert, R., Lagace, M., and Bouvier, M. (1999). Activation of the beta(2)-adrenergic receptor-Galpha(s) complex leads to rapid depalmitoylation and inhibition of repalmitoylation of both the receptor and Galpha(s). *J. Biol. Chem.* **274,** 31014–31019.

Lopez-Martinez, A., Chang, D. T., Chiang, C., Porter, J. A., Ros, M. A., Simandl, B. K., Beachy, P. A., and Fallon, J. F. (1995). Limb-patterning activity and restricted posterior localization of the amino-terminal product of Sonic hedgehog cleavage. *Curr. Biol.* **5,** 791–796.

Lorentzen, A., Kinkhabwala, A., Rocks, O., Vartak, N., and Bastiaens, P. I. (2010). Regulation of Ras localization by acylation enables a mode of intracellular signal propagation. *Sci. Signal.* **3,** ra68.

Ma, C., Zhou, Y., Beachy, P. A., and Moses, K. (1993). The segment polarity gene hedgehog is required for progression of the morphogenetic furrow in the developing Drosophila eye. *Cell* **75,** 927–938.

Maity, T., Fuse, N., and Beachy, P. A. (2005). Molecular mechanisms of Sonic hedgehog mutant effects in holoprosencephaly. *Proc. Natl. Acad. Sci. USA* **102,** 17026–17031.

Mann, R. K., and Beachy, P. A. (2004). Novel lipid modifications of secreted protein signals. *Annu. Rev. Biochem.* **73,** 891–923.

Marti, E., Bumcrot, D. A., Takada, R., and McMahon, A. P. (1995). Requirement of 19K form of Sonic hedgehog for induction of distinct ventral cell types in CNS explants. *Nature* **375,** 322–325.

Micchelli, C. A., The, I., Selva, E., Mogila, V., and Perrimon, N. (2002). Rasp, a putative transmembrane acyltransferase, is required for Hedgehog signaling. *Development* **129,** 843–851.

Miura, G. I., Buglino, J., Alvarado, D., Lemmon, M. A., Resh, M. D., and Treisman, J. E. (2006). Palmitoylation of the EGFR ligand Spitz by Rasp increases Spitz activity by restricting its diffusion. *Dev. Cell* **10,** 167–176.

Nusslein-Volhard, C., and Wieschaus, E. (1980). Mutations affecting segment number and polarity in Drosophila. *Nature* **287,** 795–801.

Ohgusu, H., Shirouzu, K., Nakamura, Y., Nakashima, Y., Ida, T., Sato, T., and Kojima, M. (2009). Ghrelin O-acyltransferase (GOAT) has a preference for n-hexanoyl-CoA over n-octanoyl-CoA as an acyl donor. *Biochem. Biophys. Res. Commun.* **386,** 153–158.

Ohlig, S., Farshi, P., Pckhinke, U., van den Boom, J., Hoing, S., Jakuschev, S., Hoffmann, D., Dreier, R., Scholer, H. R., Dierker, T., Bordych, C., and Grobe, K. (2011). *Dev. Cell* **20,** 764–774.

Panakova, D., Sprong, H., Marois, E., Thiele, C., and Eaton, S. (2005). Lipoprotein particles are required for Hedgehog and Wingless signalling. *Nature* **435,** 58–65.

Pepinsky, R. B., Zeng, C., Wen, D., Rayhorn, P., Baker, D. P., Williams, K. P., Bixler, S. A., Ambrose, C. M., Garber, E. A., Miatkowski, K., Taylor, F. R.,

Wang, E. A., *et al.* (Check for Year). Identification of a palmitic acid-modified form of human Sonic hedgehog. *J. Biol. Chem.* **273,** 14037–14045.
Porter, J. A., von Kessler, D. P., Ekker, S. C., Young, K. E., Lee, J. J., Moses, K., and Beachy, P. A. (1995). The product of hedgehog autoproteolytic cleavage active in local and long-range signalling. *Nature* **374,** 363–366.
Porter, J. A., Ekker, S. C., Park, W. J., von Kessler, D. P., Young, K. E., Chen, C. H., Ma, Y., Woods, A. S., Cotter, R. J., Koonin, E. V., and Beachy, P. A. (1996a). Hedgehog patterning activity: Role of a lipophilic modification mediated by the carboxy-terminal autoprocessing domain. *Cell* **86,** 21–34.
Porter, J. A., Young, K. E., and Beachy, P. A. (1996b). Cholesterol modification of hedgehog signaling proteins in animal development. *Science* **274,** 255–259.
Qin, N., Platano, D., Olcese, R., Costantin, J. L., Stefani, E., and Birnbaumer, L. (1998). Unique regulatory properties of the type 2a Ca2+ channel beta subunit caused by palmitoylation. *Proc. Natl. Acad. Sci. USA* **95,** 4690–4695.
Resh, M. D. (2006). Palmitoylation of ligands, receptors, and intracellular signaling molecules. *Sci. STKE* **2006,** 14.
Roelink, H., Porter, J. A., Chiang, C., Tanabe, Y., Chang, D. T., Beachy, P. A., and Jessell, T. M. (1995). Floor plate and motor neuron induction by different concentrations of the amino-terminal cleavage product of sonic hedgehog autoproteolysis. *Cell* **81,** 445–455.
Roessler, E., El-Jaick, K. B., Dubourg, C., Velez, J. I., Solomon, B. D., Pineda-Alvarez, D. E., Lacbawan, F., Zhou, N., Ouspenskaia, M., Paulussen, A., Smeets, H. J., Hehr, U., *et al.* (2009). The mutational spectrum of holoprosencephaly-associated changes within the SHH gene in humans predicts loss-of-function through either key structural alterations of the ligand or its altered synthesis. *Hum. Mutat.* **30,** E921–E935.
Rutledge, B. J., Zhang, K., Bier, E., Jan, Y. N., and Perrimon, N. (1992). The Drosophila spitz gene encodes a putative EGF-like growth factor involved in dorsal-ventral axis formation and neurogenesis. *Genes Dev.* **6,** 1503–1517.
Shindou, H., Eto, M., Morimoto, R., and Shimizu, T. (2009). Identification of membrane O-acyltransferase family motifs. *Biochem. Biophys. Res. Commun.* **383,** 320–325.
Singaraja, R. R., Kang, M. H., Vaid, K., Sanders, S. S., Vilas, G. L., Arstikaitis, P., Coutinho, J., Drisdel, R. C., El-Husseini Ael, D., Green, W. N., Berthiaume, L., and Hayden, M. R. (2009). Palmitoylation of ATP-binding cassette transporter A1 is essential for its trafficking and function. *Circ. Res.* **105,** 138–147.
Smotrys, J. E., and Linder, M. E. (2004). Palmitoylation of intracellular signaling proteins: Regulation and function. *Annu. Rev. Biochem.* **73,** 559–587.
Stahl, U., Stalberg, K., Stymne, S., and Ronne, H. (2008). A family of eukaryotic lysophospholipid acyltransferases with broad specificity. *FEBS Lett.* **582,** 305–309.
Stamataki, D., Ulloa, F., Tsoni, S. V., Mynett, A., and Briscoe, J. (2005). A gradient of Gli activity mediates graded Sonic Hedgehog signaling in the neural tube. *Genes Dev.* **19,** 626–641.
Taipale, J., Chen, J. K., Cooper, M. K., Wang, B., Mann, R. K., Milenkovic, L., Scott, M. P., and Beachy, P. A. (2000). Effects of oncogenic mutations in Smoothened and Patched can be reversed by cyclopamine. *Nature* **406,** 1005–1009.
Takada, R., Satomi, Y., Kurata, T., Ueno, N., Norioka, S., Kondoh, H., Takao, T., and Takada, S. (2006). Monounsaturated fatty acid modification of Wnt protein: Its role in Wnt secretion. *Dev. Cell* **11,** 791–801.
Tanaka, K., Kitagawa, Y., and Kadowaki, T. (2002). Drosophila segment polarity gene product porcupine stimulates the posttranslational N-glycosylation of wingless in the endoplasmic reticulum. *J. Biol. Chem.* **277,** 12816–12823.
Taylor, F. R., Wen, D., Garber, E. A., Carmillo, A. N., Baker, D. P., Arduini, R. M., Williams, K. P., Weinreb, P. H., Rayhorn, P., Hronowski, X., Whitty, A., Day, E. S.,

et al. (1998). Enhanced potency of human Sonic hedgehog by hydrophobic modification. *Biochemistry* **40,** 4359–4371.

Tian, H., Callahan, C. A., DuPree, K. J., Darbonne, W. C., Ahn, C. P., Scales, S. J., and de Sauvage, F. J. (2009). Hedgehog signaling is restricted to the stromal compartment during pancreatic carcinogenesis. *Proc. Natl. Acad. Sci. USA* **106,** 4254–4259.

Tokhunts, R., Singh, S., Chu, T., D'Angelo, G., Baubet, V., Goetz, J. A., Huang, Z., Yuan, Z., Ascano, M., Zavros, Y., Therond, P. P., Kunes, S., *et al.* (2009). The full-length unprocessed hedgehog protein is an active signaling molecule. *J. Biol. Chem.* **285,** 2562–2568.

Urban, S., Lee, J. R., and Freeman, M. (2001). Drosophila rhomboid-1 defines a family of putative intramembrane serine proteases. *Cell* **107,** 173–182.

Vyas, N., Goswami, D., Manonmani, A., Sharma, P., Ranganath, H. A., VijayRaghaven, K., Shashidhara, L. S., Sowdhamini, R., and Mayor, S. (2008). Nano-scale organization of hedgehog is essential for long-range signaling. *Cell* **133,** 1214–1227.

Wang, B., Fallon, J. F., and Beachy, P. A. (2000). Hedgehog-regulated processing of Gli3 produces an anterior/posterior repressor gradient in the developing vertebrate limb. *Cell* **100,** 423–434.

Willert, K., Brown, J. D., Danenberg, E., Duncan, A. W., Weissman, I. L., Reya, T., Yates, J. R., 3rd, and Nusse, R. (2003). Wnt proteins are lipid-modified and can act as stem cell growth factors. *Nature* **423,** 448–452.

Yang, Y., Drossopoulou, G., Chuang, P. T., Duprez, D., Marti, E., Bumcrot, D., Vargesson, N., Clarke, J., Niswander, L., McMahon, A., and Tickle, C. (1997). Relationship between dose, distance and time in Sonic Hedgehog-mediated regulation of anteroposterior polarity in the chick limb. *Development* **124,** 4393–4404.

Yang, J., Brown, M. S., Liang, G., Grishin, N. V., and Goldstein, J. L. (2008a). Identification of the acyltransferase that octanoylates ghrelin, an appetite-stimulating peptide hormone. *Cell* **132,** 387–396.

Yang, J., Zhao, T. J., Goldstein, J. L., and Brown, M. S. (2008b). Inhibition of ghrelin O-acyltransferase (GOAT) by octanoylated pentapeptides. *Proc. Natl. Acad. Sci. USA* **105,** 10750–10755.

Yauch, R. L., Gould, S. E., Scales, S. J., Tang, T., Tian, H., Ahn, C. P., Marshall, D., Fu, L., Januario, T., Kallop, D., Nannini-Pepe, M., Kotkow, K., *et al.* (2008). A paracrine requirement for hedgehog signalling in cancer. *Nature* **455,** 406–410.

Yeh, D. C., Duncan, J. A., Yamashita, S., and Michel, T. (1999). Depalmitoylation of endothelial nitric-oxide synthase by acyl-protein thioesterase 1 is potentiated by Ca(2+)-calmodulin. *J. Biol. Chem.* **274,** 33148–33154.

Yen, C. L., Monetti, M., Burri, B. J., and Farese, R. V., Jr. (2005). The triacylglycerol synthesis enzyme DGAT1 also catalyzes the synthesis of diacylglycerols, waxes, and retinyl esters. *J. Lipid Res.* **46,** 1502–1511.

Zeng, X., Goetz, J. A., Suber, L. M., Scott, W. J., Jr., Schreiner, C. M., and Robbins, D. J. (2001). A freely diffusible form of Sonic hedgehog mediates long-range signalling. *Nature* **411,** 716–720.

Zhai, L., Chaturvedi, D., and Cumberledge, S. (2004). Drosophila wnt-1 undergoes a hydrophobic modification and is targeted to lipid rafts, a process that requires porcupine. *J. Biol. Chem.* **279,** 33220–33227.

Zhang, W., Trible, R. P., and Samelson, L. E. (1998). LAT palmitoylation: Its essential role in membrane microdomain targeting and tyrosine phosphorylation during T cell activation. *Immunity* **9,** 239–246.

CHAPTER ELEVEN

Phosphorylation Regulation of Hedgehog Signaling

Jianhang Jia

Contents

Abstract

The study of posttranslational regulation of proteins has occupied biochemists for well over a half century. Understanding balanced phosphorylation and dephosphorylation of the proteins may be the key to meeting some of the most pressing scientific challenges. A detailed examination of the phosphorylation of many components in the Hedgehog (Hh) pathway leads to a better understanding of the Hh signaling mechanisms. This chapter describes the precise phosphorylation that evolves during the phosphorylation/dephosphorylation of players in the Hh signaling cascade, including the signal transducer Smoothened and the transcription factor Ci/Gli. © 2012 Elsevier Inc.

I. Introduction

It has been known that phosphates bind to proteins since the nineteenth century. In 1954, another protein was discovered: an enzyme responsible for the transfer of a phosphate to another protein; this biological reaction is referred to as phosphorylation (Burnett and Kennedy, 1954).

Department of Molecular and Cellular Biochemistry, Markey Cancer Center, University of Kentucky College of Medicine, Lexington, Kentucky, USA

Vitamins and Hormones, Volume 88
ISSN 0083-6729, DOI: 10.1016/B978-0-12-394622-5.00011-0

The enzyme that catalyzes phosphorylation was known as a protein kinase. A year later, it was demonstrated that addition or removal of a phosphate regulates a protein involved in glycogen metabolism, suggesting that reversible phosphorylation could control the activity of a specific protein (Fischer and Krebs, 1955; Sutherland and Wosilait, 1955). The study of phosphorylation was then brought into the limelight of medical research. In 1992, Fischer and Krebs received the Nobel Prize in medicine for their pioneering work. Today, it is well known that phosphorylation/dephosphorylation is one of the most important posttranslational events in cell biology. It has been shown to be involved in many aspects of regulation such as increasing or decreasing the biological activity of a protein, facilitating the sorting of intracellular proteins, mediating protein–protein interactions, and labeling proteins for degradation. Importantly, the reception of a signal on the surface of a cell often results in a change in cellular phosphorylation patterns with various proteins being phosphorylated or dephosphorylated, which might eventually cause changes in cellular behavior.

The Hedgehog (Hh) signal transduction controls many aspects of development in both vertebrates and invertebrates and is associated with important biological phenomena such as patterning, cell growth, and morphogenesis. Abnormal activation of this pathway has been observed in several types of human cancers (Jiang and Hui, 2008); thus, it represents a major anticancer therapeutic target. Protein phosphorylation in Hh signaling has been extensively studied. Like the proteins in other signaling pathways such as Wnt and NFkB, various Hh pathway components are phosphorylated at multiple residues and most of these phosphorylation events are regulated by the upstream signal (Jia and Jiang, 2006). Dysfunction of Hh signaling contributes to numerous human disorders including birth defects and cancers such as basal cell carcinoma and medulloblastoma (Jiang and Hui, 2008); therefore, understanding how phosphorylation events in Hh signaling are normally controlled may provide new avenues for developing theraputics aimed at preventing and treating human cancers caused by aberrant Hh signaling. We begin this chapter with a discussion of the functions of protein phosphorylation in various situations of Hh signal transduction. In the second part of the chapter, we examine the balance between phosphorylation and dephosphorylation of the critical components that are involved in transducing the upstream signal. In the last part of the chapter, we discuss the idea of differential phosphorylation mediating the Hh signaling activity gradient in animal development.

Hh exerts its biological influence through a highly conserved signal transduction cascade. The mammalian Sonic Hedgehog (Shh), as well as the *Drosophila* Hh, can function as morphogens and regulate cell proliferation, embryonic patterning, and organ development (Ingham and McMahon, 2001; Ingham *et al.*, 2011). The Hh signal is transduced through a reception system that includes the twelve-span transmembrane protein,

Patched (Ptc), and the seven-span transmemberane protein, Smoothened (Smo). In the absence of Hh, Ptc inhibits Smo signaling activity by a mechanism that is still not clear, even though studies indicate that small molecules, such as lipids, could be involved (Ayers and Therond, 2010; Eaton, 2008). The presence of Hh relieves the inhibition of Smo by Ptc, activating Smo and allowing Smo to regulate the downstream signaling components. The main outcome of Hh signaling is the modulation of transcriptional responses via the cubitus interruptus (Ci)/Gli family of zinc-finger transcription factors.

The *Drosophila* system is an ideal model system with which to study Hh pathways because of the fruit-fly's well-established genetics, small genome size, minimal genetic redundancy, and ease of handling. Studies with *Drosophila* have provided invaluable information on Hh signaling. The first identified major phosphorylation event occurs at the level of the transcription factor, Ci/Gli, which promotes the proteolytic processing of these proteins. The other major phosphorylation event occurs at the level of Hh signal reception. Smo, a G-protein-coupled receptor (GPCR) family member, acts as a critical signal transducer and is phosphorylated and accumulated on the cell surface upon Hh stimulation (Denef *et al.*, 2000). Recent studies identified protein kinase A (PKA) and casein kinase 1 (CK1) as two of the kinases that phosphorylate Smo and increase its cell-surface localization and signaling activity (Apionishev *et al.*, 2005; Jia *et al.*, 2004; Zhang *et al.*, 2004). We will discuss the regulation of Smo phosphorylation in a later section. In addition to Smo and Ci/Gli, other pathway components, such as Fused (Fu), Costal2 (Cos2), and Suppressor of Fused (Sufu) are phosphorylated upon Hh stimulation (Aikin *et al.*, 2008; Lum and Beachy, 2004).

II. Phosphorylation of the Transcription Factor in Hh Signaling

Hh signaling regulates the balance between the transcriptional activator and repressor forms of Ci/Gli (Jia and Jiang, 2006; Lum and Beachy, 2004), which is mediated by phosphorylation. Ci plays dual roles that are performed by two distinct forms. In the absence of Hh, full-length Ci (Ci^{FL}) undergoes proteolytic processing to generate a truncated form (Ci^{REP}). Ci^{REP} functions as a repressor to block the expression of Hh responsive genes such as *decapentaplegic* (*dpp*) (Aza-Blanc *et al.*, 1997; Methot and Basler, 1999), which encodes a member of the TGFβ/BMP family of secreted proteins that functions as a long-range morphogen to control the growth and patterning of the wing (Lecuit *et al.*, 1996; Nellen *et al.*, 1996).

What triggers Ci proteolytic processing? The initial discovery came from a study of *Drosophila* cAMP-dependent PKA. It was reported that loss-of-function mutations in the catalytic subunit of PKA led to constitutive Hh signaling (Jiang and Struhl, 1995; Lepage *et al.*, 1995; Li *et al.*, 1995; Pan and Rubin, 1995). Subsequent work indicated that PKA phosphorylates Ci at multiple serine/threonine (Ser/Thr) residues in its C-terminal region, which is essential for processing (Chen *et al.*, 1999b; Price and Kalderon, 1999; Wang *et al.*, 1999). Later, GSK3 and CK1 were identified and characterized as two other kinases that act cooperatively with PKA to promote Ci processing (Jia *et al.*, 2002; Price and Kalderon, 2002). GSK3 and CK1 phosphorylation of Ci is primed by PKA phosphorylation at nearby Ser/Thr residues; these sites form critical phosphorylation clusters within the C-terminal region of Ci and this hyperphosphorylation promotes Ci processing (Jiang, 2002). Thus, Ci processing requires phosphorylation by multiple kinases including PKA, GSK3, and members of CK1 family.

How does phosphorylation of Ci regulate its processing? Early in 1998, an F-box/WD40 containing protein, Slimb, was identified to act downstream of PKA to promote Ci processing (Jiang and Struhl, 1998). The vertebrate homolog of Slimb, βTRCP, functions as a substrate recognition subunit of the so-called SCF (*S*kp1, *C*dc53, and *F*-box) ubiquitin ligase complex that normally targets phosphorylated substrates, such as Iκ-B and β-catenin, for ubiquitination, and subsequent proteasome-mediated proteolysis (Maniatis, 1999). The phosphorylation of Ci at multiple phosphorylation clusters creates binding sites for the SCF complex (Jia *et al.*, 2005; Smelkinson and Kalderon, 2006), leading to partial degradation of Ci, thus generating the truncated repressor form, Ci^{REP} (Hooper and Scott, 2005; Jia and Jiang, 2006).

How does Hh inhibit Ci processing? An elegant study by Zhang *et al.* describes a precise mechanism by which Hh inhibits Ci phosphorylation (Zhang *et al.*, 2005). Ci^{FL} exists in a large protein complex including Cos2 and Fu. Complex formation impedes nuclear translocation of Ci^{FL} through microtubulal-dependent and -independent mechanisms (Wang and Jiang, 2004). Interestingly, Cos2 directly associates with PKA, CK1, and GSK3 and acts as a scaffold between these kinases and Ci^{FL} to facilitate Ci^{FL} phosphorylation (Zhang *et al.*, 2005). An *in vivo* experiment performed in this study demonstrated that a Kinesin–Cos2 chimeric protein, in which the microtubule-binding domain of Cos2 is replaced by a canonical kinesin motor domain, carried Cos2-interacting proteins to the microtubule plus end. Accumulation of Kenesin-Cos2 at the plus end of microtubules allowed for a discrete intracellular localization that allowed the enrichment of PKA, CK1, and GSK3 at the same position. In addition, Cos2 was shown to be required for Ci phosphorylation by these kinases. Importantly, Hh signaling appears to inhibit Ci phosphorylation and processing by dissociating the Cos2–Ci–kinase complex (Zhang *et al.*, 2005).

In vertebrate species, three Gli proteins, Gli1, Gli2, and Gli3, are the transcriptional mediators of Hh signaling. Genetic and biochemical studies suggest that Gli2 and Gli3 are the primary mediators of Hh signaling. Gli1 is a transcriptional target of Hh signaling and provides positive feedback to reinforce the Hh signaling activity (Bai and Joyner, 2001; Bai *et al.*, 2002, 2004; Dai *et al.*, 1999; Ding *et al.*, 1998; Lee *et al.*, 1997; Matise *et al.*, 1998; Park *et al.*, 2000). Gli2 mainly functions as an activator and Gli3 as a repressor, although in some developmental contexts, the repressor activity of Gli2 and the activator activity of Gli3 have also been detected (Bai *et al.*, 2004; Buttitta *et al.*, 2003; McDermott *et al.*, 2005; Motoyama *et al.*, 2003; Tyurina *et al.*, 2005). Consistent with their ability to function as both activators and repressors, Gli2 and Gli3 are proteolytically processed to truncated forms, and there is evidence that the processing Gli2 and Gli3 are inhibited by Hh (Aza-Blanc *et al.*, 2000; Pan *et al.*, 2006; Ruiz i Altaba, 1999; Wang *et al.*, 2000a). The multiple clusters of phosphorylation sites found in Ci are also present in Gli proteins (Jia and Jiang, 2006), suggesting that Gli processing could also be regulated by multiple kinases and Slimb/β-TRCP in a similar manner to the aforementioned regulatory mechanisms for Ci processing. Indeed, studies have shown that the processing of Gli3 requires the binding of β-TRCP to phosphorylated Gli3 (Tempe *et al.*, 2006; Wang and Li, 2006), and that Gli and Ci are phosphorylated and thus processed by the same machinery (Aza-Blanc *et al.*, 2000; Marks and Kalderon, 2011).

III. Phosphorylation of Smo Receptor in Hh Signaling

The Hh signal is transduced through a reception system at the plasma membrane that includes the receptor complexes Ptc–Ihog and the signal transducer Smo (Jia and Jiang, 2006; Lum and Beachy, 2004; Zheng *et al.*, 2010). Binding of Hh to Ptc–Ihog relieves the inhibition of Smo by Ptc, which allows Smo to activate Ci/Gli transcription factors. Regulation of Smo activity remained poorly understood until recently. Smo has homology to GPCRs (Alcedo *et al.*, 1996), and a recent study revealed that Gαi, a G-protein in *Drosophila*, is essential for full activation of Hh signaling (Ogden *et al.*, 2008). In *Drosophila*, Hh induces cell-surface accumulation and phosphorylation of Smo (Denef *et al.*, 2000). Recent studies from several labs identified PKA and CK1 as two of the kinases that directly phosphorylate Smo and regulate its cell-surface expression and activity (Apionishev *et al.*, 2005; Jia *et al.*, 2004; Zhang *et al.*, 2004). Blocking PKA or CK1 activity prevents Hh-induced Smo accumulation and attenuates pathway activity, whereas increasing PKA activity promotes Smo accumulation and pathway activation (Jia *et al.*, 2004). Interestingly, similar to the phosphorylation clusters in Ci, Smo is phosphorylated by PKA and CK1 at three clusters of residues in its intracellular domain both *in vitro* and *in vivo* (Apionishev *et al.*, 2005; Jia *et al.*, 2004; Zhang *et al.*, 2004). Phosphorylation-deficient forms of Smo, in which PKA or CK1 sites are mutated to alanine (Ala), are defective in Hh signaling. Phosphorylation appears to be sufficient to activate Smo, as phosphorylation-mimicking Smo variants in which PKA and CK1 sites are converted to acidic residues, such as aspartate (Asp) and glutamate (Glu), exhibit constitutive signaling activity (Jia *et al.*, 2004; Zhang *et al.*, 2004).

It is interesting that the same set of kinases plays opposing roles and phosphorylates distinct substrates depending on signaling states. In the *Drosophila* Hh pathway, PKA and CK1 phosphorylate Ci in the absence of Hh to inhibit pathway activation. In the presence of Hh, PKA and CK1 phosphorylate Smo to stimulate its signaling activity. Similarly, in Wnt signal transduction, the absence of Wnt allows GSK3β and CK1 to phosphorylate β-catenin, thus inhibiting Wnt signaling. Upon Wnt stimulation, GSK3β and CK1 phosphorylate the Wnt coreceptor LRP5/6 to activate signaling (Davidson *et al.*, 2005; Zeng *et al.*, 2005). These observations raise the important question of how different substrates are chosen by the same kinases in different signaling states. One solution, as suggested by studies on the Wnt pathway, is that each pathway may utilize two distinct pools of kinases; one in close proximity to the transcription factor/effector and the other close to the membrane reception system (Davidson *et al.*, 2005; Zeng *et al.*, 2005).

Just like EGF receptor phosphorylation by multiple kinases at multiple residues in the C-terminal cytoplasmic-tail (C-tail) of the receptor, Smo has also been shown to be phosphorylated at 26 Ser/Thr sites in its C-tail in response to Hh stimulation (Zhang *et al.*, 2004). Several phosphorylated sites in Smo C-tail match the consensus of CK2 sites, and a recent study has found that Smo is indeed phosphorylated by CK2 at the predictable CK2 sites (Ser816, Ser817, Ser818, Ser819, and Ser843) (Jia *et al.*, 2010). Mutating these CK2 sites attenuates the activity of Smo *in vivo*. Consistently, inactivation of CK2 by RNAi attenuates Hh-induced Smo accumulation and downregulates Hh target gene expression, whereas increasing CK2 activity by overexpressing CK2 increases Smo accumulation and induces ectopic expression of Hh target gene (Jia *et al.*, 2010). Thus, one more kinase is added to the kinase pool of Smo.

Is there any other kinase also involved in Smo phosphorylation? The answer to this question is obviously positive. Several studies have suggested that GPCR kinase 2 (GRK2) promotes Hh signaling in vertebrate by regulating Smo (Chen *et al.*, 2004; Meloni *et al.*, 2006; Philipp *et al.*, 2008). Genetic studies in *Drosophila* suggest that GRK2 also modulates Hh signaling (Cheng *et al.*, 2010; Molnar *et al.*, 2007), suggesting that GRK2 plays a conserved role in regulating Hh signaling likely by modulating the activity of Smo. Researchers have then found that GRK2 is indeed a Smo kinase that phosphorylates Smo at Ser741 and Thr742 of its C-tail (Chen *et al.*, 2010). Interestingly, this phosphorylation event is regulated by

PKA and CK1 phosphorylation at adjacent Ser residues, which improve the acidic environment to felicitate the phosphorylation by GRK2. Unexpectedly, in the same study, Chen *et al.* uncovered a kinase activity-independent mechanism for GRK2 to regulate Hh signaling. They demonstrate that GRK2 forms a dimer/oligomer and binds Smo C-tail to promote the active state of Smo, which represents a novel mechanism for regulating a GPCR protein by GRK (Chen *et al.*, 2010). Above all, we have learned that multiple kinases are involved in Smo phosphorylation that is triggered by the upstream signal.

What is the consequence of Smo phosphorylation? Like many GPCRs, Smo forms homodimers and undergoes a confirmation change when it is phosphorylated by PKA and CK1 upon Hh stimulation (Zhao *et al.*, 2007a). Zhao and colleagues, using fluorescence resonance energy transfer (FRET) analysis, observed a basal level of FRET when using the donor cyan fluorescent protein (CFP) and the acceptor yellow fluorescent protein (YFP) to be fused to the C-terminal tail of Smo. They further found that this basal FRET was dramatically increased by Hh treatment. In addition, CFP- and YFP-tagged Smo N-terminal part exhibited a high level of FRET even in the absence of Hh treatment, suggesting that Smo forms a constitutive dimer/oligomer through its N-terminal region; the C-tails are within a dimer and are situated at a distance but are brought together in response to Hh through a conformational switch (Zhao *et al.*, 2007a). The dimerization model explains why the phosphorylation-deficient forms of Smo, as well as a chimeric Smo variant with its C-tail replaced with that of the Wingless (Wg) receptor Frizzled 2 (Fz2), inhibited the activity of endogenous Smo (Apionishev *et al.*, 2005; Hooper, 2003; Jia *et al.*, 2004).

How does Hh induce a conformational switch in the C-tail of Smo? The degree of conformational change in the C-tail of Smo, monitored by FRET, correlates with the level of its phosphorylation that can be manipulated by substitution of various Ser/Thr residues into either Ala or Asp (Zhao *et al.*, 2007a). Thus, a series of Smo conformational states relies on the level of Hh activity and Smo phosphorylation. More interestingly, stretches of positively charged arginine (Arg) residues residing adjacent to the phosphorylation clusters keep the C-tail of Smo in the inactive conformation. Phosphorylation of Smo at adjacent PKA and CK1 sites neutralizes the positive charges conferred by the Arg motif, leading to Smo dimerization and activation (Zhao *et al.*, 2007a). In addition to inducing a conformational change, phosphorylation also promotes Smo accumulation on the cell surface (Denef *et al.*, 2000; Jia *et al.*, 2004).

How does Hh regulate Smo phosphorylation and cell-surface accumulation? Although Smo phosphorylation promotes its cell-surface accumulation and signaling activity, the mechanisms leading to its cell-surface accumulation are still unclear. Fu and Cos2 exist in a large protein complex (Robbins *et al.*, 1997; Sisson *et al.*, 1997). Smo transduces the Hh signal by physically interacting with the Cos2–Fu complex (Hooper, 2003; Jia *et al.*, 2003; Lum *et al.*, 2003; Ogden *et al.*, 2003; Ruel *et al.*, 2003). Cell-surface recruitment of an intracellular signaling complex through accumulated Smo is thought to cause dissociation of the Cos2–Ci–Kinase complex and hence inhibition of Ci processing (Zhang *et al.*, 2005). Thus, the intracellular signaling complex containing Fu and Cos2 was thought to transduce the Hh signal downstream of Smo. Surprisingly, Liu and colleagues identified a mechanism of feedback regulation of Smo by the Fu–Cos2 protein complex (Liu *et al.*, 2007). They found that Fu is essential for Hh-induced Smo phosphorylation and cell-surface accumulation because Smo is inhibited in *fu* mutant clones or by expressing a dominant-negative (DN) form of Fu. Such inhibition is alleviated by removal of Cos2. Conversely, overexpressing Cos2 blocks Smo accumulation, which is reversed by coexpressing Fu. In addition, Fu antagonizes Cos2 by phosphorylation at Ser572, which attenuates the Cos2–Smo interaction and promotes Cos2 degradation. The authors further provided evidence that Fu and Cos2 control Smo cell-surface accumulation by regulating Smo phosphorylation (Liu *et al.*, 2007). Consistently, Claret and colleagues' observations demonstrated that Fu plays a positive role in Smo activation (Claret *et al.*, 2007). These data suggest that Hh signaling proceeds through an amplification step mediated by Smo and its downstream components. In addition to regulating the phosphorylation of Smo, Hh may regulate the dephosphorylation of Smo that is mediated by its phosphatase (see Section VI).

IV. Other Phosphorylation Events in Hh Signaling

Fu was first identified as a Ser/Thr kinase that regulates segmental polarity in *Drosophila* (Preat *et al.*, 1993; Therond *et al.*, 1993). Fu is composed of a catalytic domain and a carboxy-terminal regulatory domain; it has been shown to be essential for Hh pathway activation (Alves *et al.*, 1998). Fu is in a nonphosphorylated, inactive state in the absence of Hh and in a phosphorylated active state in the presence of Hh (Lum *et al.*, 2003; Therond *et al.*, 1996). The responsible kinase for Fu phosphorylation is unknown, although it was proposed that Fu activity is autoregulated through an intramolecular mechanism (Ascano *et al.*, 2002), suggesting that Fu could be autophosphorylated. In addition, mutational analysis of Fu identified Thr158 in its activation loop to be a critical residue that is phosphorylated upon Hh stimulation (Fukumoto *et al.*, 2001). In addition, Hh promotes Fu dimerization and activation that requires phosphorylation of its activation loop (unpublished data). The identification of the kinase that phosphorylates and thus activates Fu and the biological significance of this phosphorylation event await further investigation.

Cos2, a kinesin-like protein, exists in a large protein complex with Fu and the Ci^{FL} (Robbins *et al.*, 1997; Sisson *et al.*, 1997). Cos2 mainly acts as a negative regulator in Hh signal transduction, as loss-of-function mutations in *cos2* result in accumulation of high levels of Ci^{FL} and activation of Hh target genes (Sisson *et al.*, 1997; Wang and Holmgren, 1999). Cos2 inhibits the transcriptional activator activity of Ci^{FL} by inhibiting its nuclear translocation (Chen *et al.*, 1999a; Monnier *et al.*, 2002; Stegman *et al.*, 2000; Wang *et al.*, 2000b). Cos2 also acts as a scaffold protein to bring PKA, CK1, and glycogen synthase kinase 3 (GSK3) to Ci^{FL}, thus promoting sequential phosphorylation of Ci^{FL} by these kinases (Price, 2006; Zhang *et al.*, 2005). This leads to proteasome-mediated Ci^{FL} processing to generate its repressor form, Ci^{REP} (see below section). In addition to the role in Ci processing, Cos2 exerts a negative effect on Smo, which blocks the phosphorylation and activation of Smo, likely by masking its phosphorylation sites (Liu *et al.*, 2007). Other than its negative roles in Hh signal transduction, Cos2 has a positive role in Hh-responding cells, which correlates to its ability to form a complex with the C-terminal intracellular tail of Smo (Jia *et al.*, 2003; Lum *et al.*, 2003; Ogden *et al.*, 2003; Wang *et al.*, 2000b). In addition, Fu is diminished in cells lacking Cos2 (Lum *et al.*, 2003; Ruel *et al.*, 2003), which might explain, at least in part, why Hh signaling activity is attenuated in *cos2* mutant cells adjacent to the A/P boundary where there is Hh stimulation (Ho *et al.*, 2005). Interestingly, Cos2 is also phosphorylated in response to Hh, and this phosphorylation depends on Fu kinase activity (Lum *et al.*, 2003; Nybakken *et al.*, 2002). Phosphorylation of Cos2 at Ser572 by Fu disassociates the Cos2–Smo–Ci complex, which promotes Cos2 degradation, activates Smo, and stabilizes Ci (Liu *et al.*, 2007; Ruel *et al.*, 2007).

Sufu was first identified in *Drosophila* and its name came from the ability to suppress the wing vein defect of the *fu* mutant (Preat *et al.*, 1993). In *Drosophila* Hh signaling, Sufu forms a large protein complex with Cos2, Fu, and Ci, and is reported to regulate Ci subcellular localization and transcriptional activity in the nucleus (Methot and Basler, 2000; Wang *et al.*, 2000b). However, *Drosophila* embryos that lack Sufu display intact Hh signaling and develop into viable and fertile adults (Preat, 1992). The role of Sufu in mammalian Hh signaling is strikingly different. Loss of Sufu in mammals leads to ectopic Hh pathway activation, suggesting that Sufu is the major intracellular inhibitor of Gli activity (Jiang and Hui, 2008). It has also been speculated that Sufu might have adopted other functions during evolution (Jiang and Hui, 2008). Interestingly again, *Drosophila* Sufu is phosphorylated in response to Hh, and this phosphorylation event depends on Fu kinase activity (Dussillol-Godar *et al.*, 2006; Lum *et al.*, 2003), although there is no evidence to indicate that Fu is a direct kinase for Sufu; it is possible that other kinase(s) are involved. In addition, further studies are necessary to characterize the consequence of Sufu phosphorylation in response to Hh stimulation.

Studies of the Hh signal transduction in the past three decades have identified many components that control sending, propagating, receiving, and transducing the Hh signal (Wilson and Chuang, 2010). Phosphorylation regulation of these component proteins has brought much interest to researchers. In addition to the genetic screens using the *Drosophila* model, people are also using vertebrate systems to identify the novel kinases involved in vertebrate Hh signaling, as the functions of Fu and PKA are divergent. For example, Jacob and colleagues have carried out a genome-side RNAi screen to identify the associations between the signaling pathways and genes linked to developmental diseases. To their pleasant surprise, they identified STK11 (also known as LKB1) as a multitasking kinase likely involved in both the Hh and the Wnt signaling pathways (Jacob *et al.*, 2011). It will be very interesting to know the substrate(s) that the newly identified kinase targets. In other studies, researchers have identified some other kinases, such as DYRK2 (Varjosalo *et al.*, 2008), unc-51-like kinase 3 (ULK3) (Maloverjan *et al.*, 2010), and protein kinase C (Cai *et al.*, 2009). However, the underlying molecular mechanisms by which these newly identified kinases are involved in Hh signaling are unclear.

V. Nonconserved Phosphorylation

Vertebrate Smo proteins lack the PKA phosphorylation consensus sites found in *Drosophila* Smo, which implies that vertebrate Smo is not regulated by direct PKA phosphorylation. However, GRK2 and perhaps other kinases may regulate the conformation of Smo in vertebrates. In support of this hypothesis, in response to Hh stimulation, vertebrate Smo proteins exhibit a similar conformational switch to that utilized by *Drosophila* Smo (Zhao *et al.*, 2007a). Interestingly, a long stretch of Arg residues, which is conserved among vertebrate Smo proteins, negatively regulates mouse Smo, likely by keeping it in an inactive conformation (Zhao *et al.*, 2007a). Additionally, there is evidence that mammalian Smo is phosphorylated by GRK2 in cultured cells (Chen *et al.*, 2004). It was also tempting to speculate that Hh-induced phosphorylation of Smo through GRK2 alters the intracellular cellular localization of Smo and/or the transport of Smo into cilia. Now, Dr. Jiang and colleagues have characterized the roles of CK1α and GRK2 in regulating vertebrate Smo by direct phosphorylation. They have shown that the mammalian Smo is activated through multisite phosphorylation of its C-tail by CK1α and GRK2, and that phosphorylation of mammalian Smo induces its active conformation and simultaneously promotes its ciliary accumulation (Chen *et al.*, 2011), rather than accumulation on the cell surface like *Drosophila* Smo does.

The function of the Ser/Thr kinase Fu is also divergent among species from *Drosophila* to human. As discussed above, it is critically involved in *Drosophila* Hh signaling, but it has no effect on mouse Hh signal transduction. One possible explanation could be due to gene duplication. Understanding how Fu evolved divergent functions in distinct cellular processes is one of the topics that await further investigation.

VI. Phosphatases in Hh Signaling

Levels of cellular protein phosphorylation are often modulated by the opposing action of protein kinases and phosphatases. Phosphatases are typically classified into two main groups: Ser/Thr protein phosphatases (STPs) and protein tyrosine phosphatases (PTPs). STPs can be subdivided into the phosphoprotein phosphatase (PPP) comprising PP1, PP2A, PP2B, PP4, PP5, PP6, and PP7, and the protein phosphatase Mg^{2+}- or Mn^{2+}-dependent (PPM) family, comprised primarily of PP2C (Cohen, 1997). In the Hh signaling cascade, multiple Ser/Thr kinases are involved, including PKA, GSK3, and CK1 family members as discussed above. Are there any phosphotases involved in Hh signal transduction? In 2005, Nybakken *et al.* carried out a genome-wide RNAi screen in cultured *Drosophila* cells, in which they identified hundreds of potential new regulators of Hh signaling, including *microtubule star* (*mts*), which encodes the catalytic subunit of protein phosphatase 2A (PP2A) (Nybakken *et al.*, 2005). They showed that RNAi of Mts reduced Hh signaling. PP2A is a multimeric enzyme that consists of the catalytic subunit Mts, a regulatory A-subunit (encoded by CG33297 in Drosophila) and a B-subunit. PP2A functions are mediated largely by the regulatory B-subunit that directs the phosphatase to distinct substrates. In this study, Nybakken and colleagues found that dsRNAs targeting Widerborst (Wdb), a B-subunit, reduced Hh signaling by 50%. This indicated a likelihood that Wdb is the B-subunit that targets Mts to its substrate in the Hh signaling pathway. Consistently, it is found that B56e, the vertebrate ortholog of Wdb, regulates the Hh pathway during Xenopus eye field separation (Rorick *et al.*, 2007). In another genetic screen for modifiers of Hh signaling in *Drosophila*, Casso *et al.* found that *mts* is necessary for full activation of Hh signaling (Casso *et al.*, 2008).

Even though both of the above studies identified PP2A to be a positive regulator in Hh signaling, its relevant substrates remain undetermined. We know that regulated phosphorylation of Smo and Ci are critical events in mediating Hh signal transduction. As discussed earlier in this chapter, multiple Ser/Thr kinases regulate Hh signaling by phosphorylating Smo and Ci; however, the Smo and Ci phosphatases remained unknown until an *in vivo* RNAi screen was carried out by Jia *et al.* (2009). In this study, protein phosphatase 4 (PP4) and PP2A were identified as phosphatases that influence Hh signaling by regulating Smo and Ci, respectively. RNAi knockdown of PP4 elevates Smo phosphorylation and accumulation, leading to increased Hh signaling activity. Furthermore, the authors mapped the PP4-interacting domain in Smo and found that deletion of this domain in Smo promotes Smo phosphorylation and signaling activity. In addition, PP4 was found to regulate Hh-induced Smo cell-surface accumulation. Mechanistically, Hh was shown to downregulate the Smo–PP4 interaction that is mediated by Cos2. In the same *in vivo* RNAi screen, PP2A was identified to be a Ci phosphatase. Inactivating the regulatory subunit, Wdb, by RNAi or by a loss-of-function mutation downregulates whereas overexpressing this regulatory subunit upregulates the level and thus the signaling activity of Ci^{FL}. Furthermore, Wdb counteracts kinases to prevent Ci phosphorylation. The authors obtained evidence that Wdb attenuates Ci processing by dephosphorylating Ci. Thus, PP4 and PP2A are two phosphatases that act at different positions of the Hh signaling cascade. It is possible that other phosphatases are also involved in dephosphorylating Hh signaling components, including Smo and Ci, which would not be surprising since multiple kinases are involved.

VII. Differential Phosphorylation of Hh Signaling Components

Mammalian Shh, as well as *Drosophila* Hh, can function as a morphogen and regulate cell patterning in a concentration-dependent manner (Briscoe and Ericson, 1999, 2001; Heemskerk and DiNardo, 1994; Jia and Jiang, 2006; Roelink *et al.*, 1995; Strigini and Cohen, 1997). Hh proteins play broad roles in the development of appendages, such as wings, eyes, and legs (Ingham and McMahon, 2001; Ingham *et al.*, 2011). One of the best systems for studying Hh signaling is the *Drosophila* wing. In the developing wing (wing imaginal disc), posterior compartment (P compartment) cells express and secrete Hh proteins that act upon neighboring anterior compartment (A compartment) cells adjacent to the A/P boundary to induce the expression of *dpp*. The long-range morphogen Dpp protein then diffuses bidirectionally into both the A and P compartments and functions as a morphogen to control the growth and patterning of cells in the entire wing in a concentration-dependent manner (Campbell and Tomlinson, 1999; Lecuit *et al.*, 1996; Nellen *et al.*, 1996). Hh also specifies cell patterning at the A/P boundary by activating other genes, including the short-range morphogens *engrailed* (*en*) and *ptc* (Strigini and Cohen, 1997; Vervoort *et al.*, 1999). Low levels of Hh activity are able to induce the expression of *dpp*, whereas higher levels of Hh activity are also able to activate *ptc*.

The induction of *en* appears to require the highest doses of Hh signaling activities (Strigini and Cohen, 1997). Thus, the Hh activity levels can be monitored by expression of different responsive genes. This phenomenon is similar to the differential signal transduction of Wingless (Wg), in which different levels of Wg activity differentially regulate specific genes (Tomlinson, 2003). An important issue in developmental biology is to understand how cells perceive and transduce signal gradients and how this results in differential cell fates.

Hh family members play a role in vertebrate limb development similar to that in *Drosophila*. *Shh* is best characterized in vertebrates and is involved in an array of developmental processes (Ingham and McMahon, 2001). In the limb bud, Shh is expressed in the zone of polarizing activity (ZPA), which polarizes the digits along the A/P axis (Riddle *et al.*, 1993), which is analogous to that in the *Drosophila* wing. Ectopic expression of Shh in the anterior limb bud induces additional digits in a concentration-dependent manner. In the neural tube, Shh is produced first in the notochord and then in the floor-plate; different levels of Shh activity direct the formation of multiple neuronal subtypes (V0, V1, V2, and V3 interneurons, and motor-neurons) (Ingham and McMahon, 2001; Jessell, 2000), indicating that the protein functions to induce cell identity. Taken together, how different levels of the Hh morphogen generate distinct developmental outcomes is an important issue.

How are the phosphorylation events modulated by gradient Hh signaling activity? Although the precise mechanism(s) of how an Hh gradient is sensed and transduced inside cells is still not well understood, it is likely that the differential phosphorylation of Smo and Ci correlate to different levels of Hh stimulation. In *Drosophila* wing discs, Hh induces cell-surface accumulation and activation of Smo through phosphorylation (Jia and Jiang, 2006). Interestingly, the extent of Smo phosphorylation appears to determine both the abundance of Smo at the cell surface and its signaling potency. This raises the interesting possibility that different thresholds of Hh are transduced by differentially phosphorylated isoforms of Smo (Jia *et al.*, 2004). Low Hh only induces low levels of Smo phosphorylation and a low ratio of dimerized Smo, which suffices to activate low threshold responses such as *dpp* expression. High Hh induces high levels of Smo phosphorylation and a high ratio of Smo dimerization, which activates high threshold responses including *ptc* and *en* expression. Indeed, it has been shown that increasing the number of phosphomimetic mutations results in a gradual increase of FRET that is indicative of Smo dimerization/oligomerization (Zhao *et al.*, 2007b). It is believed that hyperphosphorylation of Smo is essential for transducing high levels of Hh signaling activity (Liu *et al.*, 2007).

How are different thresholds of Hh activity interpreted by Ci/Gli transcription factors? In *Drosophila* wing development, threshold responses to the Hh morphogen appear to be mediated by differential regulation of the two forms of Ci. Accumulation of Ci^{FL} and expression of *dpp* occur in broader domains than the activation of ptc and en, suggesting that low levels of Hh are sufficient to block Ci processing, whereas higher levels of Hh are required to stimulate the activity of Ci^{FL}. Hence, increasing the low levels of Hh activity may cause decreasing levels of Ci phosphorylation at the clusters that mediate Ci processing thus generating less Ci^{REP} repressors. On the other hand, Ci^{FL} may also need to be activated by phosphorylation/dephosphorylation. For example, blockage of Ci processing in slimb mutant clones induces *dpp* but not *ptc* (Wang *et al.*, 1999); Ci^{FL} itself is inactive, and only high levels of Hh stimulate the maturation of Ci^{FL} into a labile hyperactive form (Ohlmeyer and Kalderon, 1998). Phosphorylation of Ci by PKA not only promotes Ci processing but also inhibits the activity of Ci155 independent of Ci processing (Wang *et al.*, 1999).

A similar model has been proposed for regulating Gli proteins by graded Shh signals (Jacob and Briscoe, 2003; Stamataki *et al.*, 2005). In the neural tube, loss of Gli2, a major source of Gli activator, results in loss of progenitor cells of most ventral characters, whereas progenitor cells of ventral-lateral and lateral characters still form (Ding *et al.*, 1998; Matise *et al.*, 1998). Ventral-lateral and lateral neural progenitors lost in *Shh* mutants are partially restored in *Shh Gli3* double mutants (Litingtung and Chiang, 2000; Persson *et al.*, 2002). These results suggest that high levels of Shh specify the ventral-most progenitors via stimulating the Gli2 (and perhaps Gli3) activator, whereas low levels of Shh specify ventral-lateral and lateral progenitors through inhibiting the Gli3 repressor. Hence, a gradient of Shh could generate a gradient of Gli activator activity and a reverse gradient of Gli repressor activity.

ACKNOWLEDGMENTS

J. Jia is supported by the National Institutes of Health and the American Heart Association. The author thanks Jennifer Rogers for critically reviewing the manuscript.

REFERENCES

Aikin, R. A., Ayers, K. L., and Therond, P. P. (2008). The role of kinases in the Hedgehog signalling pathway. *EMBO Rep.* **9,** 330–336.

Alcedo, J., Ayzenzon, M., Von Ohlen, T., Noll, M., and Hooper, J. E. (1996). The Drosophila smoothened gene encodes a seven-pass membrane protein, a putative receptor for the hedgehog signal. *Cell* **86,** 221–232.

Alves, G., Limbourg-Bouchon, B., Tricoire, H., Brissard-Zahraoui, J., Lamour-Isnard, C., and Busson, D. (1998). Modulation of Hedgehog target gene expression by the Fused serine-threonine kinase in wing imaginal discs. *Mech. Dev.* **78,** 17–31.

Apionishev, S., Katanayeva, N. M., Marks, S. A., Kalderon, D., and Tomlinson, A. (2005). Drosophila Smoothened phosphorylation sites essential for Hedgehog signal transduction. *Nat. Cell Biol.* **7,** 86–92.

Ascano, M., Jr., Nybakken, K. E., Sosinski, J., Stegman, M. A., and Robbins, D. J. (2002). The carboxyl-terminal domain of the protein kinase fused can function as a dominant inhibitor of hedgehog signaling. *Mol. Cell. Biol.* **22,** 1555–1566.

Ayers, K. L., and Therond, P. P. (2010). Evaluating Smoothened as a G-protein-coupled receptor for Hedgehog signalling. *Trends Cell Biol.* **20,** 287–298.

Aza-Blanc, P., Ramirez-Weber, F. A., Laget, M. P., Schwartz, C., and Kornberg, T. B. (1997). Proteolysis that is inhibited by hedgehog targets Cubitus interruptus protein to the nucleus and converts it to a repressor. *Cell* **89,** 1043–1053.

Aza-Blanc, P., Lin, H. Y., Ruiz i Altaba, A., and Kornberg, T. B. (2000). Expression of the vertebrate Gli proteins in Drosophila reveals a distribution of activator and repressor activities. *Development* **127,** 4293–4301.

Bai, C. B., and Joyner, A. L. (2001). Gli1 can rescue the in vivo function of Gli2. *Development* **128,** 5161–5172.

Bai, C. B., Auerbach, W., Lee, J. S., Stephen, D., and Joyner, A. L. (2002). Gli2, but not Gli1, is required for initial Shh signaling and ectopic activation of the Shh pathway. *Development* **129,** 4753–4761.

Bai, C. B., Stephen, D., and Joyner, A. L. (2004). All mouse ventral spinal cord patterning by hedgehog is Gli dependent and involves an activator function of Gli3. *Dev. Cell* **6,** 103–115.

Briscoe, J., and Ericson, J. (1999). The specification of neuronal identity by graded Sonic Hedgehog signalling. *Semin. Cell Dev. Biol.* **10,** 353–362.

Briscoe, J., and Ericson, J. (2001). Specification of neuronal fates in the ventral neural tube. *Curr. Opin. Neurobiol.* **11,** 43–49.

Burnett, G., and Kennedy, E. P. (1954). The enzymatic phosphorylation of proteins. *J. Biol. Chem.* **211,** 969–980.

Buttitta, L., Mo, R., Hui, C. C., and Fan, C. M. (2003). Interplays of Gli2 and Gli3 and their requirement in mediating Shh-dependent sclerotome induction. *Development* **130,** 6233–6243.

Cai, Q., Li, J., Gao, T., Xie, J., and Evers, B. M. (2009). Protein kinase Cdelta negatively regulates hedgehog signaling by inhibition of Gli1 activity. *J. Biol. Chem.* **284,** 2150–2158.

Campbell, G., and Tomlinson, A. (1999). Transducing the Dpp morphogen gradient in the wing of Drosophila: Regulation of Dpp targets by brinker. *Cell* **96,** 553–562.

Casso, D. J., Liu, S., Iwaki, D. D., Ogden, S. K., and Kornberg, T. B. (2008). A screen for modifiers of hedgehog signaling in Drosophila melanogaster Identifies swm and mts. *Genetics* **178,** 1399–1413.

Chen, C. H., von Kessler, D. P., Park, W., Wang, B., Ma, Y., and Beachy, P. A. (1999a). Nuclear trafficking of Cubitus interruptus in the transcriptional regulation of Hedgehog target gene expression. *Cell* **98,** 305–316.

Chen, Y., Cardinaux, J. R., Goodman, R. H., and Smolik, S. M. (1999b). Mutants of cubitus interruptus that are independent of PKA regulation are independent of hedgehog signaling. *Development* **126,** 3607–3616.

Chen, W., Ren, X. R., Nelson, C. D., Barak, L. S., Chen, J. K., Beachy, P. A., de Sauvage, F., and Lefkowitz, R. J. (2004). Activity-dependent internalization of smoothened mediated by beta-arrestin 2 and GRK2. *Science* **306,** 2257–2260.

Chen, Y., Li, S., Tong, C., Zhao, Y., Wang, B., Liu, Y., Jia, J., and Jiang, J. (2010). G protein-coupled receptor kinase 2 promotes high-level Hedgehog signaling by regulating the active state of Smo through kinase-dependent and kinase-independent mechanisms in Drosophila. *Genes Dev.* **24,** 2054–2067.

Chen, Y., Sasai, N., Ma, G.,Yue, T., Jia, J., Briscoe, J., and Jiang, J. (2011). Sonic Hedgehog dependent phosphorylation by CK1α and GRK2 is required for ciliary accumulation and activation of Smoothened. *PLoS Biol.* Epub June 14.

Cheng, S., Maier, D., Neubueser, D., and Hipfner, D. R. (2010). Regulation of smoothened by Drosophila G-protein-coupled receptor kinases. *Dev. Biol.* **337,** 99–109.

Claret, S., Sanial, M., and Plessis, A. (2007). Evidence for a novel feedback loop in the hedgehog pathway involving smoothened and fused. *Curr. Biol.* **17,** 1326–1333.

Cohen, P. T. (1997). Novel protein serine/threonine phosphatases: Variety is the spice of life. *Trends Biochem. Sci.* **22,** 245–251.

Dai, P., Akimaru, H., Tanaka, Y., Maekawa, T., Nakafuku, M., and Ishii, S. (1999). Sonic Hedgehog-induced activation of the Gli1 promoter is mediated by GLI3. *J. Biol. Chem.* **274,** 8143–8152.

Davidson, G., Wu, W., Shen, J., Bilic, J., Fenger, U., Stannek, P., Glinka, A., and Niehrs, C. (2005). Casein kinase 1 gamma couples Wnt receptor activation to cytoplasmic signal transduction. *Nature* **438,** 867–872.

Denef, N., Neubuser, D., Perez, L., and Cohen, S. M. (2000). Hedgehog induces opposite changes in turnover and subcellular localization of patched and smoothened. *Cell* **102,** 521–531.

Ding, Q., Motoyama, J., Gasca, S., Mo, R., Sasaki, H., Rossant, J., and Hui, C. C. (1998). Diminished Sonic hedgehog signaling and lack of floor plate differentiation in Gli2 mutant mice. *Development* **125,** 2533–2543.

Dussillol-Godar, F., Brissard-Zahraoui, J., Limbourg-Bouchon, B., Boucher, D., Fouix, S., Lamour-Isnard, C., Plessis, A., and Busson, D. (2006). Modulation of the Suppressor of fused protein regulates the Hedgehog signaling pathway in Drosophila embryo and imaginal discs. *Dev. Biol.* **291,** 53–66.

Eaton, S. (2008). Multiple roles for lipids in the Hedgehog signalling pathway. *Nat. Rev. Mol. Cell Biol.* **9,** 437–445.

Fischer, E. H., and Krebs, E. G. (1955). Conversion of phosphorylase b to phosphorylase a in muscle extracts. *J. Biol. Chem.* **216,** 121–132.

Fukumoto, T., Watanabe-Fukunaga, R., Fujisawa, K., Nagata, S., and Fukunaga, R. (2001). The fused protein kinase regulates Hedgehog-stimulated transcriptional activation in Drosophila Schneider 2 cells. *J. Biol. Chem.* **276,** 38441–38448.

Heemskerk, J., and DiNardo, S. (1994). Drosophila hedgehog acts as a morphogen in cellular patterning. *Cell* **76,** 449–460.

Ho, K. S., Suyama, K., Fish, M., and Scott, M. P. (2005). Differential regulation of Hedgehog target gene transcription by Costal2 and Suppressor of Fused. *Development* **132,** 1401–1412.

Hooper, J. E. (2003). Smoothened translates Hedgehog levels into distinct responses. *Development* **130,** 3951–3963.

Hooper, J. E., and Scott, M. P. (2005). Communicating with Hedgehogs. *Nat. Rev. Mol. Cell Biol.* **6,** 306–317.

Ingham, P. W., and McMahon, A. P. (2001). Hedgehog signaling in animal development: Paradigms and principles. *Genes Dev.* **15,** 3059–3087.

Ingham, P. W., Nakano, Y., and Seger, C. (2011). Mechanisms and functions of Hedgehog signalling across the metazoa. *Nat. Rev. Genet.* **12,** 393–406.

Jacob, J., and Briscoe, J. (2003). Gli proteins and the control of spinal-cord patterning. *EMBO Rep.* **4,** 761–765.

Jacob, L. S., Wu, X., Dodge, M. E., Fan, C. W., Kulak, O., Chen, B., Tang, W., Wang, B., Amatruda, J. F., and Lum, L. (2011). Genome-wide RNAi screen reveals disease-associated genes that are common to Hedgehog and Wnt signaling. *Sci. Signal.* **4,** ra4.

Jessell, T. M. (2000). Neuronal specification in the spinal cord: Inductive signals and transcriptional codes. *Nat. Rev. Genet.* **1,** 20–29.

Jia, J., and Jiang, J. (2006). Decoding the Hedgehog signal in animal development. *Cell. Mol. Life Sci.* **63,** 1249–1265.
Jia, J., Amanai, K., Wang, G., Tang, J., Wang, B., and Jiang, J. (2002). Shaggy/GSK3 antagonizes Hedgehog signalling by regulating Cubitus interruptus. *Nature* **416,** 548–552.
Jia, J., Tong, C., and Jiang, J. (2003). Smoothened transduces Hedgehog signal by physically interacting with Costal2/Fused complex through its C-terminal tail. *Genes Dev.* **17,** 2709–2720.
Jia, J., Tong, C., Wang, B., Luo, L., and Jiang, J. (2004). Hedgehog signalling activity of Smoothened requires phosphorylation by protein kinase A and casein kinase I. *Nature* **432,** 1045–1050.
Jia, J., Zhang, L., Zhang, Q., Tong, C., Wang, B., Hou, F., Amanai, K., and Jiang, J. (2005). Phosphorylation by double-time/CKIepsilon and CKIalpha targets cubitus interruptus for Slimb/beta-TRCP-mediated proteolytic processing. *Dev. Cell* **9,** 819–830.
Jia, H., Liu, Y., Yan, W., and Jia, J. (2009). PP4 and PP2A regulate Hedgehog signaling by controlling Smo and Ci phosphorylation. *Development* **136,** 307–316.
Jia, H., Liu, Y., Xia, R., Tong, C., Yue, T., Jiang, J., and Jia, J. (2010). Casein kinase 2 promotes Hedgehog signaling by regulating both smoothened and Cubitus interruptus. *J. Biol. Chem.* **285,** 37218–37226.
Jiang, J. (2002). Degrading Ci: Who is Cul-pable? *Genes Dev.* **16,** 2315–2321.
Jiang, J., and Hui, C. C. (2008). Hedgehog signaling in development and cancer. *Dev. Cell* **15,** 801–812.
Jiang, J., and Struhl, G. (1995). Protein kinase A and hedgehog signaling in Drosophila limb development. *Cell* **80,** 563–572.
Jiang, J., and Struhl, G. (1998). Regulation of the Hedgehog and Wingless signalling pathways by the F-box/WD40-repeat protein Slimb. *Nature* **391,** 493–496.
Lecuit, T., Brook, W. J., Ng, M., Callega, M., Sun, H., and Cohen, S. M. (1996). Two distinct mechanisms for long-range patterning by Decapentaplegic in the Drosophila wing. *Nature* **381,** 387–393.
Lee, J., Platt, K. A., Censullo, P., and Altaba, A. R. (1997). Gli1 is a target of Sonic hedgehog that induces ventral neural tube development. *Development* **124,** 2537–2552.
Lepage, T., Cohen, S. M., Diaz-Benjumea, F. J., and Parkhurst, S. M. (1995). Signal transduction by cAMP-dependent protein kinase A in Drosophila limb patterning. *Nature* **373,** 711–715.
Li, W., Ohlmeyer, J. T., Lane, M. E., and Kalderon, D. (1995). Function of protein kinase A in hedgehog signal transduction and Drosophila imaginal disc development. *Cell* **80,** 553–562.
Litingtung, Y., and Chiang, C. (2000). Specification of ventral neuron types is mediated by an antagonistic interaction between Shh and Gli3. *Nat. Neurosci.* **3,** 979–985.
Liu, Y., Cao, X., Jiang, J., and Jia, J. (2007). Fused Costal2 protein complex regulates Hedgehog-induced Smo phosphorylation and cell-surface accumulation. *Genes Dev.* **21,** 1949–1963.
Lum, L., and Beachy, P. A. (2004). The Hedgehog response network: Sensors, switches, and routers. *Science* **304,** 1755–1759.
Lum, L., Zhang, C., Oh, S., Mann, R. K., von Kessler, D. P., Taipale, J., Weis-Garcia, F., Gong, R., Wang, B., and Beachy, P. A. (2003). Hedgehog signal transduction via Smoothened association with a cytoplasmic complex scaffolded by the atypical kinesin, Costal-2. *Mol. Cell* **12,** 1261–1274.
Maloverjan, A., Piirsoo, M., Michelson, P., Kogerman, P., and Osterlund, T. (2010). Identification of a novel serine/threonine kinase ULK3 as a positive regulator of Hedgehog pathway. *Exp. Cell Res.* **316,** 627–637.
Maniatis, T. (1999). A ubiquitin ligase complex essential for the NF-kappaB, Wnt/Wingless, and Hedgehog signaling pathways. *Genes Dev.* **13,** 505–510.
Marks, S. A., and Kalderon, D. (2011). Regulation of mammalian Gli proteins by Costal 2 and PKA in Drosophila reveals Hedgehog pathway conservation. *Development* **138,** 2533–2542.

Matise, M. P., Epstein, D. J., Park, H. L., Platt, K. A., and Joyner, A. L. (1998). Gli2 is required for induction of floor plate and adjacent cells, but not most ventral neurons in the mouse central nervous system. *Development* **125,** 2759–2770.

McDermott, A., Gustafsson, M., Elsam, T., Hui, C. C., Emerson, C. P., Jr., and Borycki, A. G. (2005). Gli2 and Gli3 have redundant and context-dependent function in skeletal muscle formation. *Development* **132,** 345–357.

Meloni, A. R., Fralish, G. B., Kelly, P., Salahpour, A., Chen, J. K., Wechsler-Reya, R. J., Lefkowitz, R. J., and Caron, M. G. (2006). Smoothened signal transduction is promoted by G protein-coupled receptor kinase 2. *Mol. Cell. Biol.* **26,** 7550–7560.

Methot, N., and Basler, K. (1999). Hedgehog controls limb development by regulating the activities of distinct transcriptional activator and repressor forms of Cubitus interruptus. *Cell* **96,** 819–831.

Methot, N., and Basler, K. (2000). Suppressor of fused opposes hedgehog signal transduction by impeding nuclear accumulation of the activator form of Cubitus interruptus. *Development* **127,** 4001–4010.

Molnar, C., Holguin, H., Mayor, F., Jr., Ruiz-Gomez, A., and de Celis, J. F. (2007). The G protein-coupled receptor regulatory kinase GPRK2 participates in Hedgehog signaling in Drosophila. *Proc. Natl. Acad. Sci. USA* **104,** 7963–7968.

Monnier, V., Ho, K. S., Sanial, M., Scott, M. P., and Plessis, A. (2002). Hedgehog signal transduction proteins: Contacts of the Fused kinase and Ci transcription factor with the kinesin-related protein Costal2. *BMC Dev. Biol.* **2,** 4.

Motoyama, J., Milenkovic, L., Iwama, M., Shikata, Y., Scott, M. P., and Hui, C. C. (2003). Differential requirement for Gli2 and Gli3 in ventral neural cell fate specification. *Dev. Biol.* **259,** 150–161.

Nellen, D., Burke, R., Struhl, G., and Basler, K. (1996). Direct and long-range action of a DPP morphogen gradient. *Cell* **85,** 357–368.

Nybakken, K. E., Turck, C. W., Robbins, D. J., and Bishop, J. M. (2002). Hedgehog-stimulated phosphorylation of the kinesin-related protein Costal2 is mediated by the serine/threonine kinase fused. *J. Biol. Chem.* **277,** 24638–24647.

Nybakken, K., Vokes, S. A., Lin, T. Y., McMahon, A. P., and Perrimon, N. (2005). A genome-wide RNA interference screen in Drosophila melanogaster cells for new components of the Hh signaling pathway. *Nat. Genet.* **37,** 1323–1332.

Ogden, S. K., Ascano, M., Jr., Stegman, M. A., Suber, L. M., Hooper, J. E., and Robbins, D. J. (2003). Identification of a functional interaction between the transmembrane protein Smoothened and the kinesin-related protein Costal2. *Curr. Biol.* **13,** 1998–2003.

Ogden, S. K., Fei, D. L., Schilling, N. S., Ahmed, Y. F., Hwa, J., and Robbins, D. J. (2008). G protein Galphai functions immediately downstream of Smoothened in Hedgehog signalling. *Nature* **456,** 967–970.

Ohlmeyer, J. T., and Kalderon, D. (1998). Hedgehog stimulates maturation of Cubitus interruptus into a labile transcriptional activator. *Nature* **396,** 749–753.

Pan, D., and Rubin, G. M. (1995). cAMP-dependent protein kinase and hedgehog act antagonistically in regulating decapentaplegic transcription in Drosophila imaginal discs. *Cell* **80,** 543–552.

Pan, Y., Bai, C. B., Joyner, A. L., and Wang, B. (2006). Sonic hedgehog signaling regulates Gli2 transcriptional activity by suppressing its processing and degradation. *Mol. Cell. Biol.* **26,** 3365–3377.

Park, H. L., Bai, C., Platt, K. A., Matise, M. P., Beeghly, A., Hui, C. C., Nakashima, M., and Joyner, A. L. (2000). Mouse Gli1 mutants are viable but have defects in SHH signaling in combination with a Gli2 mutation. *Development* **127,** 1593–1605.

Persson, M., Stamataki, D., te Welscher, P., Andersson, E., Bose, J., Ruther, U., Ericson, J., and Briscoe, J. (2002). Dorsal-ventral patterning of the spinal cord requires Gli3 transcriptional repressor activity. *Genes Dev.* **16,** 2865–2878.

Philipp, M., Fralish, G. B., Meloni, A. R., Chen, W., MacInnes, A. W., Barak, L. S., and Caron, M. G. (2008). Smoothened signaling in vertebrates is facilitated by a G protein-coupled receptor kinase. *Mol. Biol. Cell* **19,** 5478–5489.
Preat, T. (1992). Characterization of Suppressor of fused, a complete suppressor of the fused segment polarity gene of Drosophila melanogaster. *Genetics* **132,** 725–736.
Preat, T., Therond, P., Limbourg-Bouchon, B., Pham, A., Tricoire, H., Busson, D., and Lamour-Isnard, C. (1993). Segmental polarity in Drosophila melanogaster: Genetic dissection of fused in a Suppressor of fused background reveals interaction with costal-2. *Genetics* **135,** 1047–1062.
Price, M. A. (2006). CKI, there's more than one: Casein kinase I family members in Wnt and Hedgehog signaling. *Genes Dev.* **20,** 399–410.
Price, M. A., and Kalderon, D. (1999). Proteolysis of cubitus interruptus in Drosophila requires phosphorylation by protein kinase A. *Development* **126,** 4331–4339.
Price, M. A., and Kalderon, D. (2002). Proteolysis of the Hedgehog signaling effector Cubitus interruptus requires phosphorylation by Glycogen Synthase Kinase 3 and Casein Kinase 1. *Cell* **108,** 823–835.
Riddle, R. D., Johnson, R. L., Laufer, E., and Tabin, C. (1993). Sonic hedgehog mediates the polarizing activity of the ZPA. *Cell* **75,** 1401–1416.
Robbins, D. J., Nybakken, K. E., Kobayashi, R., Sisson, J. C., Bishop, J. M., and Therond, P. P. (1997). Hedgehog elicits signal transduction by means of a large complex containing the kinesin-related protein costal2. *Cell* **90,** 225–234.
Roelink, H., Porter, J. A., Chiang, C., Tanabe, Y., Chang, D. T., Beachy, P. A., and Jessell, T. M. (1995). Floor plate and motor neuron induction by different concentrations of the amino-terminal cleavage product of sonic hedgehog autoproteolysis. *Cell* **81,** 445–455.
Rorick, A. M., Mei, W., Liette, N. L., Phiel, C., El-Hodiri, H. M., and Yang, J. (2007). PP2A:B56epsilon is required for eye induction and eye field separation. *Dev. Biol.* **302,** 477–493.
Ruel, L., Rodriguez, R., Gallet, A., Lavenant-Staccini, L., and Therond, P. P. (2003). Stability and association of Smoothened, Costal2 and Fused with Cubitus interruptus are regulated by Hedgehog. *Nat. Cell Biol.* **5,** 907–913.
Ruel, L., Gallet, A., Raisin, S., Truchi, A., Staccini-Lavenant, L., Cervantes, A., and Therond, P. P. (2007). Phosphorylation of the atypical kinesin Costal2 by the kinase Fused induces the partial disassembly of the Smoothened-Fused-Costal2-Cubitus interruptus complex in Hedgehog signalling. *Development* **134,** 3677–3689.
Ruiz i Altaba, A. (1999). Gli proteins encode context-dependent positive and negative functions: Implications for development and disease. *Development* **126,** 3205–3216.
Sisson, J. C., Ho, K. S., Suyama, K., and Scott, M. P. (1997). Costal2, a novel kinesin-related protein in the Hedgehog signaling pathway. *Cell* **90,** 235–245.
Smelkinson, M. G., and Kalderon, D. (2006). Processing of the Drosophila hedgehog signaling effector Ci-155 to the repressor Ci-75 is mediated by direct binding to the SCF component slimb. *Curr. Biol.* **16,** 110–116.
Stamataki, D., Ulloa, F., Tsoni, S. V., Mynett, A., and Briscoe, J. (2005). A gradient of Gli activity mediates graded Sonic Hedgehog signaling in the neural tube. *Genes Dev.* **19,** 626–641.
Stegman, M. A., Vallance, J. E., Elangovan, G., Sosinski, J., Cheng, Y., and Robbins, D. J. (2000). Identification of a tetrameric hedgehog signaling complex. *J. Biol. Chem.* **275,** 21809–21812.
Strigini, M., and Cohen, S. M. (1997). A Hedgehog activity gradient contributes to AP axial patterning of the Drosophila wing. *Development* **124,** 4697–4705.
Sutherland, E. W., Jr., and Wosilait, W. D. (1955). Inactivation and activation of liver phosphorylase. *Nature* **175,** 169–170.
Tempe, D., Casas, M., Karaz, S., Blanchet-Tournier, M. F., and Concordet, J. P. (2006). Multisite protein kinase A and glycogen synthase kinase 3beta phosphorylation leads to Gli3 ubiquitination by SCFbetaTrCP. *Mol. Cell. Biol.* **26,** 4316–4326.

Therond, P., Busson, D., Guillemet, E., Limbourg-Bouchon, B., Preat, T., Terracol, R., Tricoire, H., and Lamour-Isnard, C. (1993). Molecular organisation and expression pattern of the segment polarity gene fused of Drosophila melanogaster. *Mechanisms Of Development* **44,** 65–80.

Therond, P. P., Knight, J. D., Kornberg, T. B., and Bishop, J. M. (1996). Phosphorylation of the fused protein kinase in response to signaling from hedgehog. *Proc. Natl. Acad. Sci. USA* **93,** 4224–4228.

Tomlinson, A. (2003). Patterning the peripheral retina of the fly: Decoding a gradient. *Dev. Cell* **5,** 799–809.

Tyurina, O. V., Guner, B., Popova, E., Feng, J., Schier, A. F., Kohtz, J. D., and Karlstrom, R. O. (2005). Zebrafish Gli3 functions as both an activator and a repressor in Hedgehog signaling. *Dev. Biol.* **277,** 537–556.

Varjosalo, M., Bjorklund, M., Cheng, F., Syvanen, H., Kivioja, T., Kilpinen, S., Sun, Z., Kallioniemi, O., Stunnenberg, H. G., He, W. W., *et al.* (2008). Application of active and kinase-deficient kinome collection for identification of kinases regulating hedgehog signaling. *Cell* **133,** 537–548.

Vervoort, M., Crozatier, M., Valle, D., and Vincent, A. (1999). The COE transcription factor Collier is a mediator of short-range Hedgehog-induced patterning of the Drosophila wing. *Curr. Biol.* **9,** 632–639.

Wang, Q. T., and Holmgren, R. A. (1999). The subcellular localization and activity of Drosophila cubitus interruptus are regulated at multiple levels. *Development* **126,** 5097–5106.

Wang, G., and Jiang, J. (2004). Multiple Cos2/Ci interactions regulate Ci subcellular localization through microtubule dependent and independent mechanisms. *Dev. Biol.* **268,** 493–505.

Wang, B., and Li, Y. (2006). Evidence for the direct involvement of {beta}TrCP in Gli3 protein processing. *Proc. Natl. Acad. Sci. USA* **103,** 33–38.

Wang, G., Wang, B., and Jiang, J. (1999). Protein kinase A antagonizes Hedgehog signaling by regulating both the activator and repressor forms of Cubitus interruptus. *Genes Dev.* **13,** 2828–2837.

Wang, B., Fallon, J. F., and Beachy, P. A. (2000a). Hedgehog-regulated processing of Gli3 produces an anterior/posterior repressor gradient in the developing vertebrate limb. *Cell* **100,** 423–434.

Wang, G., Amanai, K., Wang, B., and Jiang, J. (2000b). Interactions with Costal2 and suppressor of fused regulate nuclear translocation and activity of cubitus interruptus. *Genes Dev.* **14,** 2893–2905.

Wilson, C. W., and Chuang, P. T. (2010). Mechanism and evolution of cytosolic Hedgehog signal transduction. *Development* **137,** 2079–2094.

Zeng, X., Tamai, K., Doble, B., Li, S., Huang, H., Habas, R., Okamura, H., Woodgett, J., and He, X. (2005). A dual-kinase mechanism for Wnt co-receptor phosphorylation and activation. *Nature* **438,** 873–877.

Zhang, C., Williams, E. H., Guo, Y., Lum, L., and Beachy, P. A. (2004). Extensive phosphorylation of Smoothened in Hedgehog pathway activation. *Proc. Natl. Acad. Sci. USA* **101,** 17900–17907.

Zhang, W., Zhao, Y., Tong, C., Wang, G., Wang, B., Jia, J., and Jiang, J. (2005). Hedgehog-regulated Costal2-kinase complexes control phosphorylation and proteolytic processing of Cubitus interruptus. *Dev. Cell* **8,** 267–278.

Zhao, Y., Tong, C., and Jiang, J. (2007a). Hedgehog regulates smoothened activity by inducing a conformational switch. *Nature* **450,** 252–258.

Zhao, Y., Tong, C., and Jiang, J. (2007b). Transducing the Hedgehog signal across the plasma membrane. *Fly (Austin)* **1,** 333–336.

Zheng, X., Mann, R. K., Sever, N., and Beachy, P. A. (2010). Genetic and biochemical definition of the Hedgehog receptor. *Genes Dev.* **24,** 57–71.

CHAPTER TWELVE

Protein Kinase A Activity and Hedgehog Signaling Pathway

Tomoya Kotani

Contents

Abstract

Protein kinase A (PKA) is a well-known kinase that plays fundamental roles in a variety of biological processes. In Hedgehog-responsive cells, PKA plays key roles in proliferation and fate specification by modulating the transduction of Hedgehog signaling. In the absence of Hedgehog, a basal level of PKA activity represses the transcription of Hedgehog target genes. The main substrates of PKA in this process are the Ci/Gli family of bipotential transcription factors, which activate and repress Hedgehog target gene expression. PKA phosphorylates Ci/Gli, promoting the production of the repressor forms of Ci/Gli and thus repressing Hedgehog target gene expression. In contrast, the activation of Hedgehog signaling in response to Hedgehog increases the active forms of Ci/Gli, resulting in Hedgehog target gene expression. Because both decreased and increased levels of PKA activity cause abnormal cell proliferation and alter cell fate specification, the basal level of PKA activity in Hedgehog-responsive

Laboratory of Reproductive and Developmental Biology, Department of Biological Sciences, Faculty of Science, Hokkaido University, Sapporo, Japan

Vitamins and Hormones, Volume 88
ISSN 0083-6729, DOI: 10.1016/B978-0-12-394622-5.00012-2

cells should be precisely regulated. However, the mechanism by which PKA activity is regulated remains obscure and appears to vary between cell types, tissues, and organisms. To date, two mechanisms have been proposed. One is a classical mechanism in which PKA activity is regulated by a small second messenger, cAMP; the other is a novel mechanism in which PKA activity is regulated by a protein, Misty somites.

I. Introduction

Protein kinase A (PKA) was one of the first protein kinases to be discovered (Walsh *et al.*, 1968). Since its discovery, PKA has been found in all eukaryotes and has been demonstrated to play significant roles in a variety of biological processes, including gene transcription, cell growth, and memory. The identification of PKA as a negative regulator of Hedgehog signaling was facilitated by analyses of *Drosophila* mutants that had been isolated in genetic screens for mutations that alter the growth and patterning of adult appendages (Jiang and Struhl, 1995; Lepage *et al.*, 1995; Li *et al.*, 1995; Pan and Rubin, 1995; Strutt *et al.*, 1995). Mutations in the gene encoding the catalytic subunit of *Drosophila* PKA cause ectopic limb formation, a phenotype that resembles those caused by ectopic expression of Hedgehog. Genetic and biochemical studies have demonstrated that PKA negatively regulates the activity of the transcription factor Cubitus interruptus (Ci) in Hedgehog-responsive cells in the absence of Hedgehog signaling (Chen *et al.*, 1998; Ohlmeyer and Kalderon, 1997; Price and Kalderon, 1999; Wang *et al.*, 1999). However, in response to Hedgehog, PKA positively regulates the activity of the intracellular Hedgehog signal transducer Smoothened (Smo; Apionishev *et al.*, 2005; Jia *et al.*, 2004; Ohlmeyer and Kalderon, 1997; Zhang *et al.*, 2004).

The evolutionarily conserved role of PKA as a negative regulator of Hedgehog signaling has been demonstrated in zebrafish and mouse embryos (Concordet *et al.*, 1996; Epstein *et al.*, 1996; Fan *et al.*, 1995; Hammerschmidt *et al.*, 1996; Hynes *et al.*, 1995; Ungar and Moon, 1996). The reduction of PKA activity by ectopic expression of a dominant-negative form of PKA or a mutation in the gene encoding the catalytic subunit of PKA causes neural tube defects, small eyes, and altered patterning of the somites in vertebrate embryos (Epstein *et al.*, 1996; Hammerschmidt *et al.*, 1996; Huang *et al.*, 2002; Ungar and Moon, 1996). As in *Drosophila*, PKA negatively regulates the activity of the Gli transcription factors (Gli2 and Gli3), vertebrate homologs of the *Drosophila* Ci, in the absence of Hedgehog in vertebrates (Pan *et al.*, 2006; Wang *et al.*, 2000). To date, positive regulation of the Hedgehog signaling pathway by PKA has not been demonstrated in vertebrates.

There is evidence that alterations in PKA activity cause abnormal patterning of embryos and organs, including limbs and eyes in *Drosophila*, zebrafish, and mouse, at least in part by promoting or suppressing the expression of Hedgehog target genes irrespective of Hedgehog signaling activity (Concordet *et al.*, 1996; Epstein *et al.*, 1996; Hammerschmidt *et al.*, 1996; Jiang and Struhl, 1995; Li *et al.*, 1995; Ohlmeyer and Kalderon, 1997; Ungar and Moon, 1996). This evidence suggests that the basal PKA activity in Hedgehog-responsive cells should be precisely regulated during embryogenesis and organogenesis. Therefore, identifying the mechanism by which PKA activity is regulated is critical for understanding the mechanisms by which the growth and patterning of limbs, eyes, and other organs are regulated. PKA was originally found to function as a kinase that catalyzes cAMP-dependent protein phosphorylation in rabbit skeletal muscle (Walsh *et al.*, 1968). Since that discovery, the mechanism of PKA activation by cAMP has been well elucidated (Taylor *et al.*, 1990). In accordance with the current knowledge regarding PKA activation by cAMP, a model of cAMP-dependent activation of PKA in the regulation of PKA activity in Hedgehog-responsive cells has been proposed (Barzi *et al.*, 2010; Ogden *et al.*, 2008). Independent of this notion, the Misty somites (Mys) protein has been identified as a regulator of the PKA activity in a genetic screen of mutant zebrafish embryos (Kotani and Kawakami, 2008; Kotani *et al.*, 2010).

Although it is evident that the activation of Hedgehog signaling promotes the activation of transcription factors, Ci in *Drosophila* and Gli2 and Gli3 in vertebrates, the question of how Hedgehog signaling antagonizes the negative regulation of Ci, Gli2, and Gli3 by PKA remains largely unanswered. One recent study demonstrated that stimulation of cells with Hedgehog reduced PKA activity in Hedgehog-responsive mammalian-cultured cells, although the level of cAMP was unchanged (Barzi *et al.*, 2010). The determination of the mechanism by which PKA activity is regulated in Hedgehog-responsive cells is also important because it may lead to the elucidation of the mechanisms by which Hedgehog signaling activity is transduced through Smo.

II. Function of PKA in Hedgehog-Responsive Cells

A. Function of PKA in *Drosophila* Hedgehog signaling

Studies of *Drosophila* mutants deficient in PKA activity have revealed that PKA activity in the limb and eye imaginal discs during disc development is crucial for the growth and patterning of the wings, legs, and eyes. The basal PKA activity in imaginal discs suppresses the expression of Hedgehog target genes in the anterior region of the discs, whereas Hedgehog secreted from the posterior region of discs induces the expression of Hedgehog target genes at the anteroposterior compartment boundary. A reduction in PKA activity by mutations in the gene encoding the catalytic subunit of PKA induces ectopic expression of the Hedgehog target genes *decapentaplegic* (*dpp*) and *wingless* (*wg*) in the anterior region of the discs. The protein products of these genes act as long-range morphogens that control the growth and patterning of cells in the imaginal discs. Ectopic expression of *dpp* and *wg* in the anterior regions of limb discs causes ectopic limb formation, including anterior supernumerary wings and legs (Jiang and Struhl, 1995; Lepage *et al.*, 1995; Li *et al.*, 1995; Pan and Rubin, 1995). Similarly, ectopic expression of *dpp* in the anterior region of the eye disc causes ectopic photoreceptor differentiation in the eyes (Pan and Rubin, 1995; Strutt *et al.*, 1995).

Because sufficient amounts of PKA for *Drosophila* embryogenesis are maternally supplied in oocytes and mutations in the gene that encodes the PKA catalytic subunit cause defects in oocyte growth in adult females (Lane and Kalderon, 1993, 1994), the roles of PKA during embryogenesis cannot be analyzed with PKA-deficient mutants. Therefore, PKA activity in *Drosophila* embryos was attenuated by ectopic expression of a dominant-negative form of PKA. A reduction in PKA activity during embryogenesis revealed that PKA also plays crucial roles in the segmental patterning of embryos by modulating Hedgehog signaling (Ohlmeyer and Kalderon, 1997).

The main substrate of PKA in Hedgehog-responsive cells in *Drosophila* is Ci. PKA phosphorylates protein substrates at the consensus phosphorylation motif R-R/K-X-S/T (Taylor *et al.*, 1990). Ci contains consensus PKA phosphorylation motifs at Ser838, Ser856, and Ser892 of its carboxy-terminal region (Fig. 12.1A). In the absence of Hedgehog signaling, PKA phosphorylates Ci at these motifs, promoting Ci proteolysis from a full-length transcriptionally active form (Ci-155) to a carboxy-terminally truncated repressor form (Ci-75; Chen *et al.*, 1998; Price and Kalderon, 1999; Wang *et al.*, 1999). Ci-75 then translocates to the nucleus, binds to the promoters of the Hedgehog target genes, and inhibits their expression. Evidence from genetic studies has shown that the suppression of Hedgehog signaling pathway activity in the absence of Hedgehog depends primarily on the basal level of PKA activity. Indeed, a reduction in PKA activity causes the ectopic expression of Hedgehog target genes, even in the absence of Hedgehog ligand (Li *et al.*, 1995).

PKA also positively regulates Hedgehog signaling activity in *Drosophila*. The substrate in this action is Smo. Smo contains consensus phosphorylation motifs for PKA at Ser667, Ser687, and Ser740 in its carboxy-terminal cytoplasmic tail (Fig. 12.1A). PKA phosphorylates Smo at Ser667/687/740 in response to Hedgehog, promoting Smo accumulation at the cell surface and transduction of signaling activity through a conformational change in Smo (Apionishev *et al.*, 2005; Jia *et al.*, 2004; Zhang *et al.*, 2004; Zhao *et al.*, 2007). Mutations in the PKA phosphorylation sites in Smo eliminate the ability of wild-type Smo to rescue the *smo* mutant phenotype when a transgene

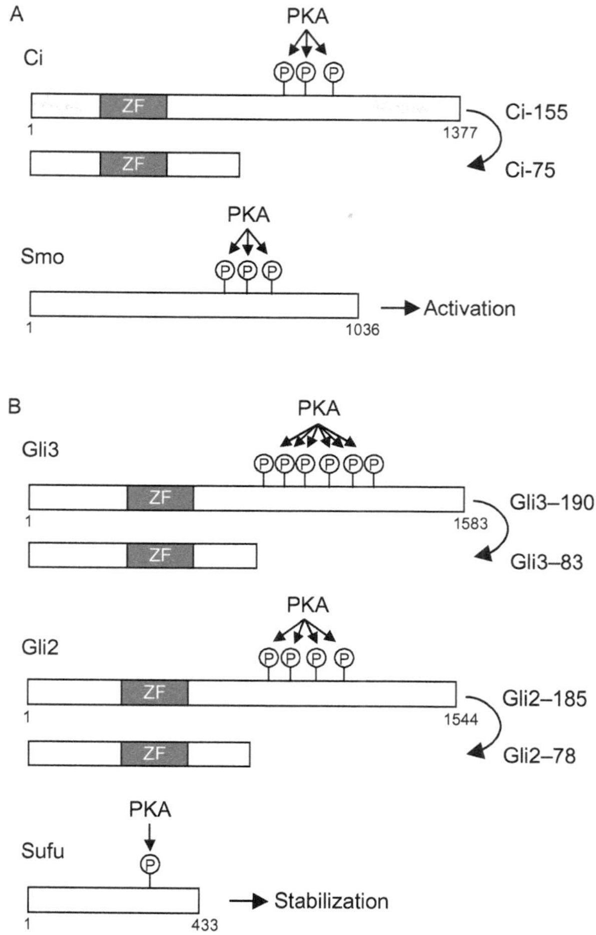

Figure 12.1 Protein substrates of PKA in *Drosophila* and vertebrate Hedgehog signaling. (A) Schematic diagrams showing *Drosophila* Cubitus interruptus (Ci) and Smoothened (Smo). In the absence of Hedgehog, PKA phosphorylates the full-length active form of Ci (Ci-155), promoting the production of the carboxy-terminally truncated repressor form Ci-75 by partial proteolysis. In response to Hedgehog, PKA phosphorylates Smo, promoting the activation of Smo. (B) Schematic diagrams showing vertebrate Gli3, Gli2, and Suppressor of fused (Sufu). In the absence of Hedgehog, PKA phosphorylates the full-length active forms of Gli3 (Gli3–190) and Gli2 (Gli2–185), promoting the production of their carboxy-terminally truncated repressor forms (Gli3–83 and Gli2–78, respectively) by partial proteolysis. In addition, PKA phosphorylates Sufu, leading to the stabilization of Sufu. ZF, zinc finger DNA-binding domain.

encoding wild-type or mutant Smo protein is introduced into *smo* mutants (Apionishev *et al.*, 2005; Jia *et al.*, 2004). Therefore, PKA-mediated activation of Smo is essential for the transduction of *Drosophila* Hedgehog signaling.

B. Function of PKA in vertebrate Hedgehog signaling

The role of PKA as a negative regulator of Hedgehog signaling is conserved in vertebrates. This conservation was first revealed by studies using mammalian embryonic explants cultured *in vitro*. In mouse presomitic mesoderm explants, purified Sonic Hedgehog (Shh) was able to induce sclerotome. Manipulating PKA activity with pharmacological agents showed that increased PKA activity antagonized Shh during the induction of sclerotome (Fan *et al.*, 1995). Similarly, in rat embryonic midbrain explants, purified Shh induced dopaminergic neurons, and increased PKA activity blocked this Shh-mediated dopaminergic neuron induction (Hynes *et al.*, 1995).

The roles of PKA in Hedgehog signaling have been further elucidated in zebrafish and mouse embryos. The ectopic expression of dominant-negative form of PKA in zebrafish embryos caused the expansion of proximal fates in the eye, ventral fates in the neural tube, and medial muscle fates in the somites, phenotypes that resemble those caused by ectopic expression of Hedgehog family members (Concordet *et al.*, 1996; Hammerschmidt *et al.*, 1996; Ungar and Moon, 1996). Similarly, in mouse embryos, the ectopic expression of dominant-negative form of PKA in the neural tube caused the ectopic induction of motor neurons in the dorsal spinal cord; these neurons are normally induced in the ventral spinal cord by Hedgehog secreted from the ventral midline (Epstein *et al.*, 1996). Several recent findings based on genetically decreased PKA activity have supported a role for PKA during vertebrate embryogenesis. The reduction of PKA activity in mouse embryos by a mutation in the gene that encodes the catalytic subunit of PKA showed that the basal level of PKA activity suppressed Hedgehog signaling in dorsal regions of the neural tube (Huang *et al.*, 2002). In addition, the reduction of PKA activity by knockdown of Mys, a regulator of PKA activity in zebrafish embryos, demonstrated that the basal level of PKA activity suppressed Hedgehog signaling in distal regions of the eye, dorsal regions of the neural tube, and medial regions of the somites (Kotani *et al.*, 2010).

The main PKA substrate in vertebrate Hedgehog-responsive cells is Gli3, a member of the Ci/Gli protein family. Gli3 contains six consensus PKA phosphorylation motifs in its carboxy-terminal region at Ser849, Ser865, Ser877, Ser907, Ser980, and Ser1006 (Fig. 12.1B). In the absence of Hedgehog signaling, PKA phosphorylates Gli3 at these six sites, promoting Gli3 proteolysis from a full-length transcriptionally active form (Gli3-190) to a carboxy-terminally truncated repressor form (Gli3-83; Wang *et al.*, 2000). Gli3-83 then translocates to the nucleus, binds to the promoters of Hedgehog target genes, and inhibits their expression. Because a large fraction of Gli3 protein exists in the Gli3-83 repressor form in embryonic limb buds, Hedgehog signaling in large regions of the embryo appears to be suppressed by PKA-promoted Gli3 proteolysis (Wang *et al.*, 2000).

The transcription factor Gli2 is an additional PKA substrate in Hedgehog-responsive vertebrate cells. Gli2 contains four consensus PKA phosphorylation motifs in its carboxy-terminal region at Ser789, Ser805, Ser817, and Ser848 (Fig. 12.1B). In the absence of Hedgehog signaling, PKA phosphorylates Gli2 at these four sites, promoting Gli2 proteolysis from a full-length active form (Gli2-185) to a carboxy-terminally truncated repressor form (Gli2-78). The PKA-catalyzed phosphorylation of Gli2 also promotes its degradation (Pan *et al.*, 2006). Because Gli2 mainly acts as a transcriptional activator of Hedgehog signaling in mouse embryos, both the proteolysis and the degradation of Gli2 attenuate Hedgehog signaling activity. One genetic study clearly demonstrated the significance of attenuating Gli2 activity by PKA-promoted Gli2 proteolysis and degradation in the Hedgehog-regulated patterning of the mouse embryo. In this study, the stabilization of Gli2–185 by mutations in the four PKA phosphorylation sites expanded the expression regions of Hedgehog target genes and caused neural tube defects, enlarged craniofacial structures, and a supernumerary digit in the anterior parts of the limbs when a gene encoding mutant Gli2 was inserted into the *Gli2* locus (Pan *et al.*, 2009).

Suppressor of fused (Sufu), a negative regulator of Hedgehog signaling, is another PKA substrate in vertebrates. Sufu sequesters the activator forms of Gli in the cytoplasm, promoting their proteolysis into truncated repressor forms (Humke *et al.*, 2010). Sufu contains four consensus PKA phosphorylation motifs in its carboxy-terminal region, of which Ser346 is critical for its negative regulation of Hedgehog signaling (Fig. 12.1B). In the absence of Hedgehog signaling, PKA phosphorylates Sufu at Ser346, stabilizing Sufu and thereby attenuating Gli transcriptional activity by promoting Gli proteolysis (Chen *et al.*, 2011; Yue *et al.*, 2009).

The role of PKA as a positive regulator of Hedgehog signaling has not been shown in vertebrates. In contrast to *Drosophila* Smo, vertebrate Smo does not contain consensus phosphorylation motifs for PKA in its carboxy-terminal cytoplasmic tail. Therefore, it is unlikely that Smo is a substrate of PKA in vertebrates. The activation of Smo through a conformational change, which is regulated by PKA-mediated phosphorylation in *Drosophila*, is promoted by G protein-coupled receptor kinase 2 (GRK2) in vertebrates (Chen *et al.*, 2004; Meloni *et al.*, 2006; Philipp *et al.*, 2008).

III. Activation of PKA

The evidence provided by mutations in the genes that encode the catalytic subunit of PKA and the ectopic expression of dominant-negative and constitutively active forms of PKA has demonstrated that the basal levels of PKA activity in *Drosophila*, zebrafish, and mouse play significant roles in regulating Hedgehog signaling to properly organize the growth and patterning of embryos and organs. Despite its importance, the mechanism by which PKA activity is regulated in Hedgehog-responsive cells remains obscure. Here, the structural features of PKA and the mechanisms of its activation as currently understood are described.

A. Conformation of PKA

PKA is a heterotetramer that consists of a regulatory subunit dimer and two catalytic subunits (Fig. 12.2A; Taylor *et al.*, 1990). The kinase core of the catalytic subunit is masked by the binding of the regulatory subunit (Kim *et al.*, 2005; Taylor *et al.*, 2008). Therefore, the PKA holoenzyme is in an inactive state. The regulatory subunit is classified as type I or type II based on its amino acid sequence (Takio *et al.*, 1984; Titani *et al.*, 1984). The biochemical property that discriminates type II from type I regulatory subunits is their potential for autophosphorylation. The type II subunits contain an autophosphorylation site, whereas the type I subunits are not autophosphorylated. In mammals, four isoforms of PKA regulatory subunits, type Iα (Prkar1a), Iβ (Prkar1b), IIα (Prkar2a), and IIβ (Prkar2b), and two isoforms of PKA catalytic subunits, type α (Prkaca) and β (Prkacb), have been isolated. In general, the transcripts encoding Prkar1a, Prkar2a, and Prkaca are expressed ubiquitously, whereas those encoding Prkar1b, Prkar2b, and Prkacb display tissue-specific expression patterns (Cadd and McKnight, 1989; Showers and Maurer, 1986; Uhler *et al.*, 1986).

The structures and conformational changes of the regulatory subunits are of crucial importance in the regulation of PKA activity. The structural features of the PKA regulatory subunits, including their various domains, are retained in all of the isoforms (Taylor *et al.*, 1990). The regulatory subunits form an asymmetric dimer via a dimerization/docking (D/D) domain located at the N-terminus (Fig. 12.2A). The D/D domain is also the docking site for A kinase-anchoring proteins (AKAPs), which target PKA to specific sites in close proximity to its protein substrates within cells. The D/D domain is adjacent to an inhibitor site, a pseudosubstrate site for type I subunits, and a substrate site for type II subunits, the latter of which is autophosphorylated by PKA. The inhibitor site is one of the regions responsible for the interaction of regulatory subunit with the catalytic subunits. Two tandem cAMP-binding domains are present in the regulatory subunits: Domain A in the middle and Domain B at the C-terminus. Domain A is also one of the regions responsible for the interaction of regulatory subunit with the catalytic subunits (Saraswat *et al.*, 1988).

Genetic and biochemical studies have shown that a large proportion of the PKA subunits in mouse and zebrafish embryos form the holoenzyme and remain in an inactive state (Amieux *et al.*, 2002; Kotani *et al.*, 2010). Only limited activation of PKA occurs in these embryos, and this activation

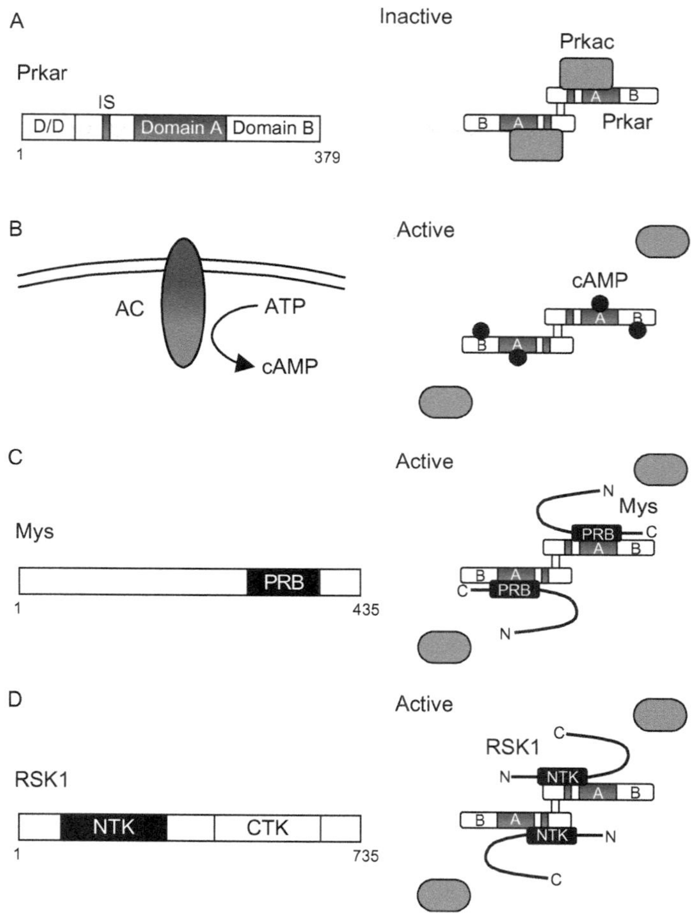

Figure 12.2 The structure and activation mechanisms of PKA. (A) The structural features of PKA. (Left) A schematic diagram showing the protein kinase A regulatory subunit (Prkar). D/D, dimerization/docking domain; IS, inhibitor site; Domain A, cAMP-binding domain A; Domain B, cAMP-binding domain B. (Right) PKA forms a heterotetramer consisting of a regulatory subunit dimer and two catalytic subunits (Prkacs). The kinase core of the catalytic subunit is masked by the binding of a regulatory subunit. (B) The mechanism of PKA activation by the small second messenger cAMP. (Left) In response to the activation of adenylate cyclase (AC) located at the membrane, cAMP is synthesized from ATP. (Right) Four molecules of cAMP bind to the regulatory subunits, promoting dissociation into a dimer of regulatory subunits and two active monomeric catalytic subunits. The monomeric catalytic subunit then phosphorylates protein substrates. (C) The mechanism of PKA activation by the protein, Misty somites (Mys). (Left) A schematic diagram showing Mys. PRB, Prkar1-binding domain. (Right) Mys binds Domain A of Prkar1 via its PRB domain, promoting the dissociation of Prkac and activating PKA. (D) The mechanism of PKA activation by the protein p90 ribosomal S6 kinase 1 (RSK1). (Left) A schematic diagram showing RSK1. NTK, N-terminal kinase domain; CTK, C-terminal kinase domain. (Right) RSK1 binds to the IS on Prkar1 via the NTK domain, promoting the dissociation of Prkac and activating PKA.

is crucial for regulating Hedgehog-mediated growth and patterning of organs during embryogenesis.

B. Activation of PKA by a second messenger, cAMP

The mechanism of PKA activation by the small second messenger cAMP has been well elucidated. After activation by a G protein-coupled receptor, the membrane-bound enzyme adenylate cyclase synthesizes cAMP from ATP (Fig. 12.2B). cAMP then binds to Domain B of each regulatory subunit of the PKA holoenzyme, causing a conformational change in PKA that makes Domain A more accessible to cAMP. An additional cAMP molecule competes with the catalytic subunit for binding to Domain A. Binding of this second cAMP to Domain A promotes dissociation of the holoenzyme into a dimer of regulatory subunits and two active monomeric catalytic subunits (Fig. 12.2B; Taylor *et al.*, 2008). The active catalytic subunits then phosphorylate protein substrates containing consensus phosphorylation motifs.

It remains uncertain whether cAMP-dependent PKA activity participates in negative regulation of Hedgehog signaling. The stimulation of *Drosophila*-cultured cells with Hedgehog reduces the level of cAMP (Ogden *et al.*, 2008). However, whether Hedgehog antagonizes PKA by the downregulation of cAMP levels remains to be determined for the following reasons. First, cAMP-insensitive PKA, which is unable to bind to the regulatory subunit and is thereby constitutively activated, can substitute for the endogenous PKA during *Drosophila* embryogenesis and disc development, implying that Hedgehog does not induce the expression of target genes by downregulating PKA activity through a reduction of cAMP levels (Jiang and Struhl, 1995; Li *et al.*, 1995; Ohlmeyer and Kalderon, 1997). Second, in Hedgehog-responsive mammalian-cultured cells, the level of cAMP is not altered by stimulation with Hedgehog (Barzi *et al.*, 2010; Murone *et al.*, 1999). However, it is possible that Hedgehog antagonizes PKA by reducing local concentrations of cAMP within Hedgehog-responding cells at levels undetectable by the assays used in these studies.

C. Activation of PKA by a protein, Misty somites

PKA activity during zebrafish embryogenesis is regulated by a protein–protein interaction between Prkar1a and Mys, a novel protein isolated in a genetic screen of mutant zebrafish embryos (Kotani and Kawakami, 2008; Kotani *et al.*, 2010). The transcripts encoding Mys and Prkar1a and their protein products are ubiquitously expressed in the zebrafish embryo (Kotani and Kawakami, 2008; Kotani *et al.*, 2010). The carboxy-terminal region of Mys (amino acids 296–389), termed the Prkar1-binding (PRB) domain, is responsible for the interaction with Prkar1 (Fig. 12.2C). The region of

Prkar1a responsible for its interaction with Mys is Domain A, which is similar to its Prkac binding region. Indeed, recombinant Mys protein competes with Prkac for binding to Domain A *in vitro*, dissociating Prkac from Prkar1a and activating PKA. Conversely, the reduction of Mys expression increases Prkac binding to Prkar1a in zebrafish embryos, decreasing PKA activity. This reduction in PKA activity causes the expansion of proximal fates in the eyes, ventral fates in the neural tube, and medial muscle fates in the somites, which collectively resembles the phenotype caused by ectopic expression of dominant-negative form of PKA (Kotani *et al.*, 2010). Therefore, PKA activity regulated by Mys is indispensable for the negative regulation of Hedgehog signaling in Hedgehog-responsive cells.

To date, the regulation of PKA activity by Mys has been observed only in zebrafish. However, the genes encoding Mys homologs are found in almost all vertebrate genomes, and their protein products are expressed in human embryonic kidney cells (HEK293 cells) and *Xenopus* embryos (Kotani and Kawakami, 2008; Kotani *et al.*, 2010). Notably, genes encoding proteins containing PRB domains (the region responsible for binding to Prkar1) are found in the following invertebrate genomes: *Ciona intestinalis* (GenBank Number: AK114792), *Apis mellifera* (XM_397072), and *Anopheles gambiae* (XM_001689177) (Fig. 12.3). Therefore, it is possible that PKA activity during embryogenesis is regulated by the interaction between the Prkar1 and Mys proteins in vertebrates and even in invertebrates.

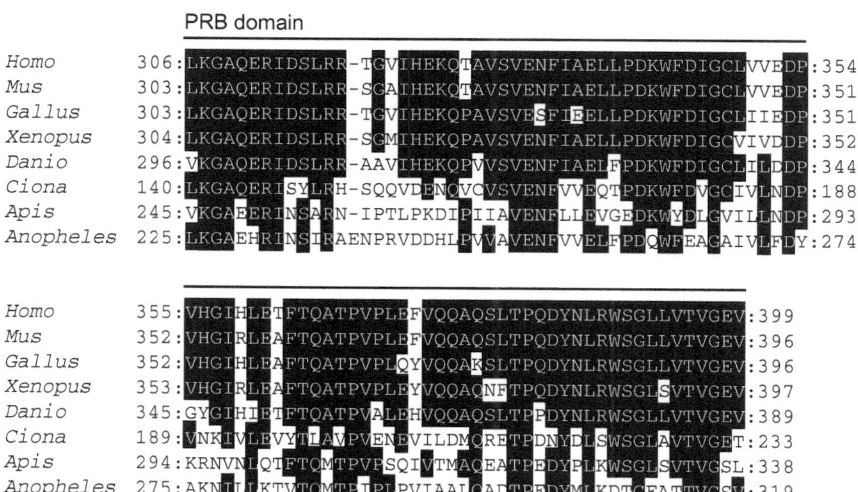

Figure 12.3 Comparison of the amino acid sequences of the PRB domain of Mys proteins in *Homo sapiens* (*Homo*), *Mus musculus* (*Mus*), *Gallus gallus* (*Gallus*), *Xenopus tropicalis* (*Xenopus*), *Danio rerio* (*Danio*), *Ciona intestinalis* (*Ciona*), *Apis mellifera* (*Apis*), and *Anopheles gambiae* (*Anopheles*) using ClustalW. Identical amino acids are highlighted in white.

D. Activation of PKA by a protein, p90 ribosomal S6 kinase 1

PKA is also activated by a protein–protein interaction between Prkar1a and p90 ribosomal S6 kinase 1 (RSK1), a kinase that plays fundamental roles in regulating a variety of biological processes, including cell proliferation and inhibition of apoptosis, downstream of the extracellular signal-regulated protein kinase 1/2 (Erk1/2; Gao *et al.*, 2010). The amino-terminal kinase (NTK) domain of RSK1 is responsible for its interaction with Prkar1a, and the pseudosubstrate region (inhibitor site) of Prkar1a, one of the regions responsible for binding to Prkac, is a binding site of RSK1 (Fig. 12.2D). Overexpression of RSK1 has been shown to dissociate Prkac from Prkar1a and increase PKA activity, whereas a reduction in RSK1 expression increased the interaction between Prkac and Prkar1a and decreased PKA activity in HEK293 cells (Gao *et al.*, 2010). Therefore, RSK1 regulates PKA activity by competing with Prkac for binding to the pseudosubstrate region of Prkar1a in a cAMP-independent manner (Fig. 12.2D). Whether RSK1 participates in regulating PKA activity in Hedgehog-responsive cells remains to be elucidated.

E. Activation of PKA by a basic lipid, sphingosine

The membrane-associated PKA that contains the regulatory type II subunit is activated by a basic lipid, sphingosine, which acts as a second messenger inducing apoptosis in various cell types (Ma *et al.*, 2005). The mechanism of PKA activation by sphingosine is different from its activation by cAMP, Mys, and RSK1. cAMP, Mys, and RSK1 activate PKA by dissociating the catalytic subunit from the regulatory subunits, whereas sphingosine activates PKA without dissociating the catalytic subunit from the regulatory subunits. Manipulation of the sphingosine levels in COS cells alters the basal level of membrane-associated PKA activity (Ma *et al.*, 2005). However, it appears unlikely that the PKA activity regulated by sphingosine participates in antagonizing Hedgehog signaling in the growth and patterning of embryos because a dominant-negative form of PKA regulatory subunit, which is unable to dissociate from the catalytic subunit because of mutations in the cAMP-binding sites, causes the ectopic expression of Hedgehog target genes.

IV. PKA Activity in Hedgehog-Responding Cells

Although it is evident that the activation of Hedgehog signaling prevents the production of the repressor forms of Ci, Gli2, and Gli3 and promotes the production of their active forms, how Hedgehog signaling antagonizes PKA remains largely unknown. In particular, the mechanism by which Smo transduces the activity of Hedgehog signaling in Hedgehog-responding cells remains to be elucidated. Because the structural features of Smo resemble those of G protein-coupled seven-transmembrane receptors, Smo is thought to transduce Hedgehog signaling activity via interactions with heterotrimeric G proteins. In accordance with this theory, exogenous Smo can activate members of the inhibitory G protein family, the Gi proteins, in frog melanophores (DeCamp *et al.*, 2000). In addition, stimulated Smo and a constitutively active form of Smo, SmoM2, can activate Gi proteins in cultured mammalian cells (Riobo *et al.*, 2006). Therefore, Smo has the potential to activate Gi proteins.

In *Drosophila* wing imaginal discs, a constitutively active form of the Gi protein induces ectopic activation of Hedgehog signaling. In contrast, mutations in the Gi protein result in reductions in Hedgehog target gene expression (Ogden *et al.*, 2008). Therefore, the *Drosophila* Gi protein participates in the transduction of Hedgehog signaling activity in Hedgehog-responding cells. In cultured wing imaginal disc-derived Cl8 cells, the basal level of cAMP is reduced in response to Hedgehog through the Gi protein, which suggests that Gi proteins regulate PKA activity by reducing cAMP levels (Ogden *et al.*, 2008). However, as discussed above, it remains to be determined whether the Gi protein modulates Hedgehog signaling activity by reducing cAMP levels or through other effectors in *Drosophila* embryos and imaginal discs because the low level of cAMP-independent PKA activity substitutes for endogenous PKA activity during *Drosophila* embryogenesis and disc development (Jiang and Hui, 2008; Jiang and Struhl, 1995; Li *et al.*, 1995; Ohlmeyer and Kalderon, 1997).

During the development of the cerebellum, Shh secreted from Purkinje cells promotes the proliferation of granule cell precursors in the external germinal layer. In mammalian-cultured granule cell precursors, stimulation with Shh promotes proliferation through Gi proteins located in the primary cilium, which functions as a signaling center for Hedgehog in vertebrates (Barzi *et al.*, 2011). In addition, stimulation with Shh induces binding of the catalytic subunit of PKA to the type II regulatory subunit at the base of the primary cilium. The accumulation of the type II regulatory subunit at the cilium base is mediated by binding of the regulatory subunit to AKAPs and is required for the Shh-induced proliferation of these cells (Barzi *et al.*, 2010). These results suggest that Shh reduces PKA activity by increasing the inactive holoenzymes containing the type II regulatory subunit at the cilium base through the activation of Gi proteins, promoting the proliferation of the granule cell precursors. However, as in *Drosophila* embryos, it remains to be determined whether Gi proteins modulate PKA activity in granule cell precursors by reducing cAMP levels or through other effectors because the level of cAMP in the cells is unchanged by stimulation with Shh.

In contrast to *Drosophila* embryos and mammalian granule cell precursors, the inhibition of Gi proteins in neural progenitor cells has little effect on Hedgehog signaling activity in chick embryos. In addition, ectopic expression of a constitutively active form of the Gi protein in the neural tube does not cause ectopic activation of Hedgehog signaling, although constitutively active SmoM2 can (Low *et al.*, 2008). These observations suggest that Gi proteins are not necessary for Hedgehog signal transduction during neural tube development in chick embryos.

In zebrafish embryos, the ectopic expression of a specific inhibitor of Gi proteins, pertussis toxin, causes phenotypes that partially resemble those generated by the ectopic expression of the constitutively active form of PKA (Hammerschmidt and McMahon, 1998). In addition, the induction of slow muscle fate by Shh *in vitro* is prevented by an increase in PKA activity, but not by the inhibition of the Gi protein by membrane-permeable pertussis toxin (Norris *et al.*, 2000). These observations suggest that Gi proteins are necessary for the transduction of Hedgehog signaling in several but not all cell types during zebrafish embryonic development.

These lines of evidence suggest the contribution of more than one signaling pathway downstream of Smo in the transduction of Hh signaling activity, which may vary between cell types, tissues, and organisms (Philipp and Caron, 2009). This hypothesis is supported by the observation that a carboxy-terminally truncated form of Smo that is able to activate Gi proteins does not promote the activation of Gli proteins in mammalian-cultured cells, implying the existence of more than one signal originating from Smo (Riobo *et al.*, 2006). The determination of the mechanisms by which PKA activity is regulated in Hedgehog-responsive cells will help resolve the pathways through which Smo transduces Hedgehog signaling activity. In this regard, the identification of Mys as a regulator of PKA activity will contribute to the elucidation of these pathways during early development in vertebrates.

V. Perspectives

Since the discovery of PKA as a kinase that negatively regulates Hedgehog signaling, genetic and biochemical studies have confirmed the importance of PKA activity for the growth and patterning of embryos and organs, from the fruit fly to vertebrates. However, the methods of analyzing the molecular mechanisms of PKA regulation at the cellular level have been limited to assays using cultured cells treated by stimulation with Hedgehog or the manipulation of PKA activity, which cannot resolve the spatial and temporal regulation of intracellular signaling in tissues and organs during normal embryogenesis and organogenesis. To precisely understand the mechanisms by which Hedgehog signaling activity is transduced, at least in part, through modulation of PKA activity, it will be necessary to visualize the level of PKA activity within individual cells of tissues and organs and the sequential changes of PKA activity in each cell during embryogenesis and organogenesis using real-time imaging (Herbst *et al.*, 2011; Zhang *et al.*, 2001, 2005). Another important question that remains unanswered is whether the local regulation of cAMP levels within Hedgehog-responsive cells participates in the transduction of Hedgehog signaling. The visualization of cAMP using real-time imaging within cells (DiPilato *et al.*, 2004; Nikolaev *et al.*, 2004; Violin *et al.*, 2008; Zaccolo *et al.*, 2000) will help answer this question. Recent reports have shown the important role of the primary cilium as a signaling center for Hedgehog in vertebrates, around which components of the Hedgehog pathway are localized. To understand the mechanisms of Hedgehog signal transduction, it will be important to perform real-time imaging of its components, including Gi proteins, regulatory subunits of PKA, catalytic subunits of PKA, adenylate cyclase, and Mys, which are distributed around the cilium area.

Although Prkar1a is the only PKA regulatory subunit isoform essential for mouse early embryonic development (Amieux *et al.*, 2002), it is possible that Hedgehog signaling in distinct tissues and organs is regulated by different isoforms of the regulatory subunits. For example, Prkar1a plays key roles during mouse and zebrafish embryogenesis (Amieux *et al.*, 2002; Kotani *et al.*, 2010), and Prkar2 plays a central role in the proliferation of mammalian granule cell precursors (Barzi *et al.*, 2010). The development of isoform-specific activators and inhibitors (Taylor *et al.*, 2008) will improve our understanding of the roles of the distinct isoforms of PKA in Hedgehog signaling in different cell types, tissues, and organs, not only during development but also in Hedgehog-related adult tissue homeostasis.

ACKNOWLEDGMENTS

I thank Dr. Masakane Yamashita (Hokkaido University) for encouragement throughout the course of this study. This work was supported by a grant from Grant-in-Aid for Young Scientists.

REFERENCES

Amieux, P. S., Howe, D. G., Knickerbocker, H., Lee, D. C., Su, T., Laszlo, G. S., Idzerda, R. L., and McKnight, G. S. (2002). Increased basal cAMP-dependent protein kinase activity inhibits the formation of mesoderm-derived structures in the developing mouse embryo. *J. Biol. Chem.* **277,** 27294–27304.

Apionishev, S., Katanayeva, N. M., Marks, S. A., Kalderon, D., and Tomlinson, A. (2005). *Drosophila* Smoothened phosphorylation sites essential for Hedgehog signal transduction. *Nat. Cell Biol.* **7,** 86–92.

Barzi, M., Berenguer, J., Menendez, A., Alvarez-Rodriguez, R., and Pons, S. (2010). Sonic-hedgehog-mediated proliferation requires the localization of PKA to the cilium base. *J. Cell Sci.* **123,** 62–69.

Barzi, M., Kostrz, D., Menendez, A., and Pons, S. (2011). Sonic Hedgehog-induced proliferation requires specific Gα inhibitory proteins. *J. Biol. Chem.* **286,** 8067–8074.

Cadd, G., and McKnight, G. S. (1989). Distinct patterns of cAMP-dependent protein kinase gene expression in mouse brain. *Neuron* **3,** 71–79.

Chen, Y., Gallaher, N., Goodman, R. H., and Smolik, S. M. (1998). Protein kinase A directly regulates the activity and proteolysis of cubitus interruptus. *Proc. Natl. Acad. Sci. USA* **95,** 2349–2354.

Chen, W., Ren, X. R., Nelson, C. D., Barak, L. S., Chen, J. K., Beachy, P. A., de Sauvage, F., and Lefkowitz, R. J. (2004). Activity-dependent internalization of smoothened mediated by β-arrestin 2 and GRK2. *Science* **306,** 2257–2260.

Chen, Y., Yue, S., Xie, L., Pu, X. H., Jin, T., and Cheng, S. Y. (2011). Dual phosphorylation of Suppressor of fused (Sufu) by PKA and GSK3β regulates its stability and localization in the primary cilium. *J. Biol. Chem.* **286,** 13502–13511.

Concordet, J. P., Lewis, K. E., Moore, J. W., Goodrich, L. V., Johnson, R. L., Scott, M. P., and Ingham, P. W. (1996). Spatial regulation of a zebrafish patched homologue reflects the roles of sonic hedgehog and protein kinase A in neural tube and somite patterning. *Development* **122,** 2835–2846.

DeCamp, D. L., Thompson, T. M., de Sauvage, F. J., and Lerner, M. R. (2000). Smoothened activates Gαi-mediated signaling in frog melanophores. *J. Biol. Chem.* **275,** 26322–26327.

DiPilato, L. M., Cheng, X., and Zhang, J. (2004). Fluorescent indicators of cAMP and Epac activation reveal differential dynamics of cAMP signaling within discrete subcellular compartments. *Proc. Natl. Acad. Sci. USA* **101,** 16513–16518.

Epstein, D. J., Marti, E., Scott, M. P., and McMahon, A. P. (1996). Antagonizing cAMP-dependent protein kinase A in the dorsal CNS activates a conserved Sonic hedgehog signaling pathway. *Development* **122,** 2885–2894.

Fan, C. M., Porter, J. A., Chiang, C., Chang, D. T., Beachy, P. A., and Tessier-Lavigne, M. (1995). Long-range sclerotome induction by sonic hedgehog: Direct role of the amino-terminal cleavage product and modulation by the cyclic AMP signaling pathway. *Cell* **81,** 457–465.

Gao, X., Chaturvedi, D., and Patel, T. B. (2010). p90 ribosomal S6 kinase 1 (RSK1) and the catalytic subunit of protein kinase A (PKA) compete for binding the pseudosubstrate region of PKAR1α: Role in the regulation of PKA and RSK1 activities. *J. Biol. Chem.* **285,** 6970–6979.

Hammerschmidt, M., and McMahon, A. P. (1998). The effect of pertussis toxin on zebrafish development: A possible role for inhibitory G-proteins in hedgehog signaling. *Dev. Biol.* **194,** 166–171.

Hammerschmidt, M., Bitgood, M. J., and McMahon, A. P. (1996). Protein kinase A is a common negative regulator of Hedgehog signaling in the vertebrate embryo. *Genes Dev.* **10,** 647–658.

Herbst, K. J., Allen, M. D., and Zhang, J. (2011). Luminescent kinase activity biosensors based on a versatile bimolecular switch. *J. Am. Chem. Soc.* **133,** 5676–5679.

Huang, Y., Roelink, H., and McKnight, G. S. (2002). Protein kinase A deficiency causes axially localized neural tube defects in mice. *J. Biol. Chem.* **277,** 19889–19896.

Humke, E. W., Dorn, K. V., Milenkovic, L., Scott, M. P., and Rohatgi, R. (2010). The output of Hedgehog signaling is controlled by the dynamic association between Suppressor of Fused and the Gli proteins. *Genes Dev.* **24,** 670–682.

Hynes, M., Porter, J. A., Chiang, C., Chang, D., Tessier-Lavigne, M., Beachy, P. A., and Rosenthal, A. (1995). Induction of midbrain dopaminergic neurons by Sonic hedgehog. *Neuron* **15,** 35–44.

Jia, J., Tong, C., Wang, B., Luo, L., and Jiang, J. (2004). Hedgehog signalling activity of Smoothened requires phosphorylation by protein kinase A and casein kinase I. *Nature* **432,** 1045–1050.

Jiang, J., and Hui, C. C. (2008). Hedgehog signaling in development and cancer. *Dev. Cell* **15,** 801–812.

Jiang, J., and Struhl, G. (1995). Protein kinase A and hedgehog signaling in *Drosophila* limb development. *Cell* **80,** 563–572.

Kim, C., Xuong, N. H., and Taylor, S. S. (2005). Crystal structure of a complex between the catalytic and regulatory (RIα) subunits of PKA. *Science* **307,** 690–696.

Kotani, T., and Kawakami, K. (2008). *misty somites*, a maternal effect gene identified by transposon-mediated insertional mutagenesis in zebrafish that is essential for the somite boundary maintenance. *Dev. Biol.* **316,** 383–396.

Kotani, T., Iemura, S., Natsume, T., Kawakami, K., and Yamashita, M. (2010). Mys protein regulates protein kinase A activity by interacting with regulatory type Iα subunit during vertebrate development. *J. Biol. Chem.* **285,** 5106–5116.

Lane, M. E., and Kalderon, D. (1993). Genetic investigation of cAMP-dependent protein kinase function in *Drosophila* development. *Genes Dev.* **7,** 1229–1243.

Lane, M. E., and Kalderon, D. (1994). RNA localization along the anteroposterior axis of the *Drosophila* oocyte requires PKA-mediated signal transduction to direct normal microtubule organization. *Genes Dev.* **8,** 2986–2995.

Lepage, T., Cohen, S. M., Diaz-Benjumea, F. J., and Parkhurst, S. M. (1995). Signal transduction by cAMP-dependent protein kinase A in *Drosophila* limb patterning. *Nature* **373,** 711–715.

Li, W., Ohlmeyer, J. T., Lane, M. E., and Kalderon, D. (1995). Function of protein kinase A in hedgehog signal transduction and *Drosophila* imaginal disc development. *Cell* **80,** 553–562.

Low, W. C., Wang, C., Pan, Y., Huang, X. Y., Chen, J. K., and Wang, B. (2008). The decoupling of Smoothened from Gαi proteins has little effect on Gli3 protein processing and Hedgehog-regulated chick neural tube patterning. *Dev. Biol.* **321,** 188–196.

Ma, Y., Pitson, S., Hercus, T., Murphy, J., Lopez, A., and Woodcock, J. (2005). Sphingosine activates protein kinase A type II by a novel cAMP-independent mechanism. *J. Biol. Chem.* **280,** 26011–26017.

Meloni, A. R., Fralish, G. B., Kelly, P., Salahpour, A., Chen, J. K., Wechsler-Reya, R. J., Lefkowitz, R. J., and Caron, M. G. (2006). Smoothened signal transduction is promoted by G protein-coupled receptor kinase 2. *Mol. Cell. Biol.* **26,** 7550–7560.

Murone, M., Rosenthal, A., and de Sauvage, F. J. (1999). Sonic hedgehog signaling by the patched-smoothened receptor complex. *Curr. Biol.* **9,** 76–84.

Nikolaev, V. O., Bunemann, M., Hein, L., Hannawacker, A., and Lohse, M. J. (2004). Novel single chain cAMP sensors for receptor-induced signal propagation. *J. Biol. Chem.* **279,** 37215–37218.

Norris, W., Neyt, C., Ingham, P. W., and Currie, P. D. (2000). Slow muscle induction by Hedgehog signalling in vitro. *J. Cell Sci.* **113**(Pt. 15), 2695–2703.

Ogden, S. K., Fei, D. L., Schilling, N. S., Ahmed, Y. F., Hwa, J., and Robbins, D. J. (2008). G protein Gαi functions immediately downstream of Smoothened in Hedgehog signalling. *Nature* **456,** 967–970.

Ohlmeyer, J. T., and Kalderon, D. (1997). Dual pathways for induction of wingless expression by protein kinase A and Hedgehog in *Drosophila* embryos. *Genes Dev.* **11,** 2250–2258.

Pan, D., and Rubin, G. M. (1995). cAMP-dependent protein kinase and hedgehog act antagonistically in regulating decapentaplegic transcription in *Drosophila* imaginal discs. *Cell* **80,** 543–552.

Pan, Y., Bai, C. B., Joyner, A. L., and Wang, B. (2006). Sonic hedgehog signaling regulates Gli2 transcriptional activity by suppressing its processing and degradation. *Mol. Cell. Biol.* **26,** 3365–3377.

Pan, Y., Wang, C., and Wang, B. (2009). Phosphorylation of Gli2 by protein kinase A is required for Gli2 processing and degradation and the Sonic Hedgehog-regulated mouse development. *Dev. Biol.* **326,** 177–189.

Philipp, M., and Caron, M. G. (2009). Hedgehog signaling: Is Smo a G protein-coupled receptor? *Curr. Biol.* **19,** R125–R127.

Philipp, M., Fralish, G. B., Meloni, A. R., Chen, W., MacInnes, A. W., Barak, L. S., and Caron, M. G. (2008). Smoothened signaling in vertebrates is facilitated by a G protein-coupled receptor kinase. *Mol. Biol. Cell* **19,** 5478–5489.

Price, M. A., and Kalderon, D. (1999). Proteolysis of cubitus interruptus in *Drosophila* requires phosphorylation by protein kinase A. *Development* **126,** 4331–4339.

Riobo, N. A., Saucy, B., Dilizio, C., and Manning, D. R. (2006). Activation of heterotrimeric G proteins by Smoothened. *Proc. Natl. Acad. Sci. USA* **103,** 12607–12612.

Saraswat, L. D., Ringheim, G. E., Bubis, J., and Taylor, S. S. (1988). Deletion mutants as probes for localizing regions of subunit interaction in cAMP-dependent protein kinase. *J. Biol. Chem.* **263,** 18241–18246.

Showers, M. O., and Maurer, R. A. (1986). A cloned bovine cDNA encodes an alternate form of the catalytic subunit of cAMP-dependent protein kinase. *J. Biol. Chem.* **261,** 16288–16291.

Strutt, D. I., Wiersdorff, V., and Mlodzik, M. (1995). Regulation of furrow progression in the *Drosophila* eye by cAMP-dependent protein kinase A. *Nature* **373,** 705–709.

Takio, K., Smith, S. B., Krebs, E. G., Walsh, K. A., and Titani, K. (1984). Amino acid sequence of the regulatory subunit of bovine type II adenosine cyclic 3′,5′-phosphate dependent protein kinase. *Biochemistry* **23,** 4200–4206.

Taylor, S. S., Buechler, J. A., and Yonemoto, W. (1990). cAMP-dependent protein kinase: Framework for a diverse family of regulatory enzymes. *Annu. Rev. Biochem.* **59,** 971–1005.

Taylor, S. S., Kim, C., Cheng, C. Y., Brown, S. H., Wu, J., and Kannan, N. (2008). Signaling through cAMP and cAMP-dependent protein kinase: Diverse strategies for drug design. *Biochim. Biophys. Acta* **1784,** 16–26.

Titani, K., Sasagawa, T., Ericsson, L. H., Kumar, S., Smith, S. B., Krebs, E. G., and Walsh, K. A. (1984). Amino acid sequence of the regulatory subunit of bovine type I adenosine cyclic 3',5'-phosphate dependent protein kinase. *Biochemistry* **23,** 4193–4199.

Uhler, M. D., Chrivia, J. C., and McKnight, G. S. (1986). Evidence for a second isoform of the catalytic subunit of cAMP-dependent protein kinase. *J. Biol. Chem.* **261,** 15360–15363.

Ungar, A. R., and Moon, R. T. (1996). Inhibition of protein kinase A phenocopies ectopic expression of hedgehog in the CNS of wild-type and cyclops mutant embryos. *Dev. Biol.* **178,** 186–191.

Violin, J. D., DiPilato, L. M., Yildirim, N., Elston, T. C., Zhang, J., and Lefkowitz, R. J. (2008). β2-adrenergic receptor signaling and desensitization elucidated by quantitative modeling of real time cAMP dynamics. *J. Biol. Chem.* **283,** 2949–2961.

Walsh, D. A., Perkins, J. P., and Krebs, E. G. (1968). An adenosine 3′,5′-monophosphate-dependant protein kinase from rabbit skeletal muscle. *J. Biol. Chem.* **243,** 3763–3765.

Wang, G., Wang, B., and Jiang, J. (1999). Protein kinase A antagonizes Hedgehog signaling by regulating both the activator and repressor forms of Cubitus interruptus. *Genes Dev.* **13,** 2828–2837.

Wang, B., Fallon, J. F., and Beachy, P. A. (2000). Hedgehog-regulated processing of Gli3 produces an anterior/posterior repressor gradient in the developing vertebrate limb. *Cell* **100,** 423–434.

Yue, S., Chen, Y., and Cheng, S. Y. (2009). Hedgehog signaling promotes the degradation of tumor suppressor Sufu through the ubiquitin-proteasome pathway. *Oncogene* **28,** 492–499.

Zaccolo, M., De Giorgi, F., Cho, C. Y., Feng, L., Knapp, T., Negulescu, P. A., Taylor, S. S., Tsien, R. Y., and Pozzan, T. (2000). A genetically encoded, fluorescent indicator for cyclic AMP in living cells. *Nat. Cell Biol.* **2,** 25–29.

Zhang, J., Ma, Y., Taylor, S. S., and Tsien, R. Y. (2001). Genetically encoded reporters of protein kinase A activity reveal impact of substrate tethering. *Proc. Natl. Acad. Sci. USA* **98,** 14997–15002.

Zhang, C., Williams, E. H., Guo, Y., Lum, L., and Beachy, P. A. (2004). Extensive phosphorylation of Smoothened in Hedgehog pathway activation. *Proc. Natl. Acad. Sci. USA* **101,** 17900–17907.

Zhang, J., Hupfeld, C. J., Taylor, S. S., Olefsky, J. M., and Tsien, R. Y. (2005). Insulin disrupts β-adrenergic signalling to protein kinase A in adipocytes. *Nature* **437,** 569–573.

Zhao, Y., Tong, C., and Jiang, J. (2007). Hedgehog regulates smoothened activity by inducing a conformational switch. *Nature* **450,** 252–258.

CHAPTER THIRTEEN

Phosphorylation of Gli by cAMP-Dependent Protein Kinase

Yoshinari Asaoka

Contents

Abstract

Hedgehog (Hh) signaling plays a central role in many developmental processes. Hh protein is a developmental morphogen that elicits a graded cellular response depending on the distance between the recipient cell and the ligand-secreting cell. Gli transcription factors are effectors that induce the expression of downstream target genes. The outline of this cascade from Hh to Gli has been elucidated, and many components have been identified. However, the interpretation of graded ligand stimulation remains to be resolved. Among the components, adenosine 3′5′-cyclic monophosphate-dependent protein kinase (PKA) functions as a negative regulator that phosphorylates a specific region of Gli, thereby inducing proteolytic cleavage to generate the repressor form. In addition, recent studies have identified different mechanisms that are followed by PKA phosphorylation of Gli. In this review, we examine Hh signaling and PKA phosphorylation and propose a possible interaction between the multiple mechanisms regulated by PKA and the gradient-dependent response.

Department of Gastroenterology, Graduate School of Medicine, The University of Tokyo, Tokyo, Japan

Vitamins and Hormones, Volume 88
ISSN 0083-6729, DOI: 10.1016/B978-0-12-394622-5.00013-4

Abbreviations

Ci	Cubitus interruptus
CKI	casein kinase I
GPCR	G-protein-coupled receptor
GSK3β	glycogen synthase kinase β
Hh	Hedgehog
PKA	cAMP-dependent protein kinase
Ptch	Patched
PTX	pertussis toxin
Smo	Smoothened
Sufu	Suppressor of Fused

I. Introduction

Gli transcription factors effect the induction of target gene expression in the Hedgehog (Hh) signaling pathway. These signaling proteins play central roles in many developmental processes in organisms from flies to humans, particularly those involving patterning of the neural tube, notochord, and prechordal mesoderm (Hooper and Scott, 2005). Deregulation of Hh signaling results in a variety of congenital malformations and cancers (Pasca di Magliano and Hebrok, 2003; Ruiz i Altaba *et al.*, 2002; Taipale and Beachy, 2001). Hh protein, the ligand that initiates the Hh signal transduction pathway, is a developmental morphogen that elicits a graded cellular response depending on the distance between the recipient cell and the ligand-secreting cell. This gradient-dependent signaling is thought to be important in cell fate determination. Given the importance of the Hh signaling pathway, there has been considerable interest in understanding the mechanism by which a target cell interprets Hh signaling to yield an appropriate response. Hh binds to its core receptor Patched (Ptch), a 12-pass transmembrane protein. In the absence of the Hh ligand, Ptch represses the activity of the seven-pass transmembrane protein Smoothened (Smo), a member of the G-protein-coupled receptor (GPCR) family, and Gli transcription factors form a complex that includes Sufu, Fused, and Costal2. The Gli proteins are then phosphorylated by adenosine 3′5′-cyclic monophosphate (cAMP)-dependent protein kinase (PKA), glycogen synthase kinase 3β (GSK3β), and casein kinase I (CKI), which target the proteins for proteasome-dependent processing and generation of the repressor forms of Gli, including Gli2, Gli3, and Cubitus interruptus (Ci). When the ligand binds and inactivates Ptch, Smo is released from Ptch and triggers an intracellular signal that inhibits the formation of the repressor form of the transcription factors (Huangfu and Anderson, 2006). However, the cascade described above is understood only as an outline of this signaling pathway, and several elements, including how to interpret the graded stimulation of the ligand, remain to be resolved.

PKA is a classic serine/threonine protein kinase and several *in vivo* studies showing that this kinase functions as a negative effector in Hh signaling were reported in 1995 (Blair, 1995; Perrimon, 1995). Since then, many studies have attempted to unveil the intracellular mechanism by which PKA functions in this signaling pathway. Most of these studies have focused on proteolytic cleavage. However, based on their amino acid sequences, they harbor different PKA phosphorylation sites from those related to cleavage, and these sites are conserved in all Gli transcription factors (Fig. 13.1). Recent reports demonstrated that different mechanisms other than proteolytic cleavage followed by PKA phosphorylation modify the activity of Gli transcription factors. These results suggest that the multiple modifications by PKA are related to the graded cellular response to Hh as a morphogen.

In this review, we examine Hh signaling and PKA phosphorylation in detail and focus on the multiple phosphorylation of Gli transcription factors by PKA. In addition, we propose a possible interaction between this mechanism and gradient-dependent signaling.

II. cAMP-Dependent Protein Kinase

cAMP-dependent protein kinase is a ubiquitous intracellular cAMP effector that regulates multiple processes. cAMP was described as the first intracellular second messenger in the late 1950s. It is present in every cell; cAMP is synthesized by adenyl cyclase from ATP and is hydrolyzed by cAMP-specific phosphodiesterases to adenosine 5′-monophosphate. The rate of cAMP production and degradation is sensitive to a wide range of extracellular and intracellular signals, such that cAMP can directly regulate a variety of cell functions, from metabolism to ion channel activation, cell growth and differentiation, gene expression, and apoptosis. Within each cell, cAMP may activate different proteins; however, its most well-known mode of action is through binding to PKA. When not bound to cAMP, PKA is an inactive tetramer of two regulatory and two catalytic subunits. When each regulatory subunit binds two molecules of cAMP, the catalytic subunits are released and in turn phosphorylate a variety of target proteins, ultimately modifying their biological activity (McKnight, 1991). PKA phosphorylates serine/threonine located in the consensus amino acid sequence K/RRXpS/T of target proteins (K represents lysine, R represents

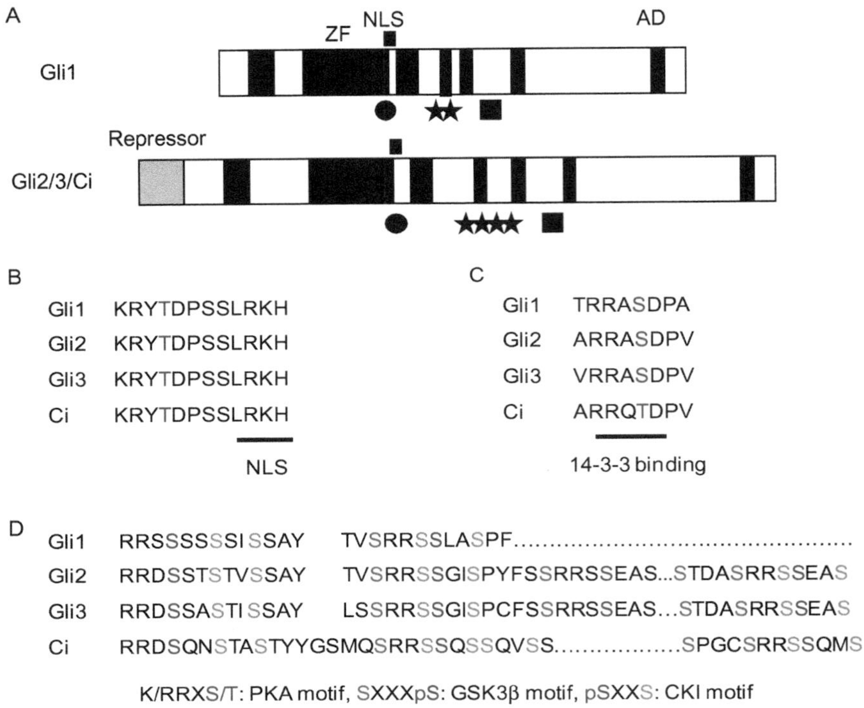

Figure 13.1 (A) Schematic drawing of Gli transcription factors. Repressor, N-terminal repressor domain; ZF, zinc finger; NLS, nuclear localization signal; AD, transactivation domain. Other conserved regions are indicated in black. (B–D) Conservation of amino acid sequences of Gli transcription factors (human Gli1, human Gli2, human Gli3, and Cubitus interruptus of *Drosophila melanogaster*). These regions are related to intracellular localization (B, black circle in (A)), 14-3-3 binding (C, black rectangle in (A)), and proteolytic cleavage (D, star in (A)). Each serine/threonine is phosphorylated by PKA, GSK3β, and CKI. The respective sites are color coded in red, blue, and green. (See Color Insert.)

arginine, X represents any amino acid, and pS/T represents serine or threonine phosphorylated by PKA; Montminy, 1997).

Cells can regulate cAMP levels induced by external signals through the activation of specific GPCRs. GPCRs are receptors linked to heterotrimeric G-proteins composed of isoforms of α, β, and γ subunits. G-protein subunits interact with and regulate the activities of various effector molecules, such as adenyl cyclases. Specificity for the activation of adenyl cyclases begins with the specific GPCR. The signal is further refined by the specific isoform composition of the receptor-associated trimeric G-protein and the specific isoform or type of the targeted adenyl cyclases. Adenyl cyclases are regulated, in part, by the balance of their exposure to specific isoforms of heteromeric G-protein α subunits; Gαs stimulates, and Gαi inhibits, the function of adenyl cyclases. External signaling through GPCRs, adenyl cyclases, and cAMP primarily regulates PKA activity (Oldham and Hamm, 2008).

III. Hedgehog Signaling and cAMP-Dependent Protein Kinase

Several genetic studies of *Drosophila* have shown that PKA is involved in the Hh signaling. Specifically, the authors described the phenotype associated with clones of imaginal disc cells homozygous for a null mutation in the catalytic subunit of PKA. In imaginal discs destined to form appendages, *Pka*$^-$ clones induced within the anterior compartment are associated with dramatic pattern duplications, while clones induced within the posterior compartment or at the AP border have no apparent effect on disc patterning. The pattern alterations extend beyond the boundaries of the *Pka*$^-$ clones and affect the growth and fate of neighboring cells. The observed pattern respecifications associated with loss of PKA activity are similar to those generated following ectopic expression of Hh (Epstein *et al.*, 1996; Heberlein *et al.*, 1995; Jiang and Struhl, 1995; Lepage *et al.*, 1995; Li *et al.*, 1995; Pan and Rubin, 1995; Strutt *et al.*, 1995). Recent studies aimed at determining the other components regulating this signal using RNAi screening or a kinome library also reconfirmed the relationship between Hh signaling and PKA (Evangelista *et al.*, 2008; Lum *et al.*, 2003; Varjosalo *et al.*, 2008).

The Hh signaling receptor Smo is an integral membrane protein containing seven membrane-spanning α helices, a long N-terminal extracellular peptide, and a C-terminal intracellular peptide (Alcedo *et al.*, 1996). Smo is most closely related to the Frizzled (Fz) family of GPCRs. Because many GPCRs function to regulate the intracellular concentration of cAMP through Gαs and Gαi as described above, the function of Smo as a GPCR was examined in several studies (Ayers and Therond, 2010; Douglas *et al.*, 2011; Kasai *et al.*, 2004; Riobo *et al.*, 2006; Waschek *et al.*, 2006). These studies showed that Smo could couple with several Gα proteins, and among them Gαi subunit is most investigated as a component of Hh signaling. Data regarding this coupling are consistent with the effects of Gαi inhibition using pertussis toxin (PTX) on Hh-induced pigment aggregation (DeCamp *et al.*, 2000), capillary morphogenesis (Kanda *et al.*, 2003), and selected aspects of eye, brain, and somite patterning (Hammerschmidt and McMahon, 1998). These data are also consistent with the effects of dsRNA-mediated knockdown of Gαi on cAMP levels and activation of Ci in *Drosophila* (Ogden *et al.*, 2008). Barzi *et al.* also showed that Hh-induced proliferation of cerebellar granular neuronal precursors is mediated by Gαi (Barzi *et al.*, 2011). However, it is unclear whether this signaling mechanism is necessary in all cell types for full Smo signaling responses. For example, although injection of PTX into zebrafish embryos causes phenotypes reminiscent of partial Shh loss (Hammerschmidt and McMahon, 1998), overexpression of a constitutively active Gαi protein did not significantly affect Hh-dependent neural cell specification in the chick neural tube (Low *et al.*, 2008). Additionally, in *Drosophila*, Smo itself is phosphorylated by PKA and upregulated to activate Hh signaling (Jia *et al.*, 2004); however, phosphorylation of Ci by PKA decreases the activity of this signaling. This opposite effect makes it difficult to investigate the relationship between Smo, Gαi, and PKA.

Many researchers have reported a negative effect of PKA on Hh signaling; however, the mechanism by which Hh-activated receptors regulate PKA function has not been fully determined.

IV. Gli Transcription Factors and cAMP-Dependent Protein Kinase

Gli transcription factors are central effectors of Hh signaling in development and tumorigenesis (Pasca di Magliano and Hebrok, 2003; Ruiz i Altaba *et al.*, 2002; Taipale and Beachy, 2001). In *Drosophila*, activation and repression of Hh target genes is mediated by a single Gli zinc finger transcription factor known as Ci (Alexandre *et al.*, 1996). In vertebrates, three Gli proteins, Gli1, Gli2, and Gli3, are involved in the transcriptional control of Hh target genes. Each Gli protein is suggested to have a unique role: Gli3 functions chiefly as a transcriptional repressor, Gli2 is a transcriptional activator, and Gli1 functions only as a transcriptional activator (Ruiz i Altaba, 1999). Although Gli1-knockout mice develop normally (Park *et al.*, 2000), Gli2 and Gli3 knockouts have serious developmental abnormalities, suggesting that Gli2 and Gli3 are likely to be intensely regulated in response to Hh ligands. In the absence of Hh ligands, Gli3 and possibly Gli2 are phosphorylated and proteolytically processed to the repressor forms (Bhatia *et al.*, 2006; Pan *et al.*, 2006; Wang *et al.*, 2000) and Gli1 is transcriptionally repressed with the consequent silencing of target genes, including Gli1 itself (Dai *et al.*, 1999). In the presence of Hh ligands, the processing of Gli2 and Gli3 is inhibited and Gli1 is induced, leading to accumulation of the activator forms. The balance of the collective activator and repressor functions of these three transcription factors seems to determine the behavior of responding cells (Stecca and Ruiz, 2010). This "GLI Code" suggests a possible interaction, with Hh functioning as a morphogen. Phosphorylation by PKA was shown by both *in vitro* and *in vivo* studies to initiate the proteolytic cleavage of Gli2 and Gli3. In addition, Gli transcription factors, including Gli1, harbor other homologous regions, which contain putative PKA phosphorylation sequences. Whereas proteolytic cleavage has been principally investigated as a suppressive mechanism followed by PKA phosphorylation, recent studies have shown that other mechanisms can regulate Gli transcription factors followed by phosphorylation of other sites. These multiple phosphorylations by PKA evoke a possible role for determining the intensity of morphogen delivery to the "GLI Code," although many aspects remain to be elucidated regarding the cascade from Hh to PKA and the regulation of Gli transcription factors by PKA. In this section, we describe three mechanisms followed by PKA phosphorylation of Gli, including proteolytic cleavage.

A. Proteolytic cleavage

Several genetic analyses of *Drosophila* have unveiled the negative effect of PKA on Hh signaling; Aza-Blanc *et al.* showed that Ci is proteolyzed, converting it to the repressor form, and this proteolysis is inhibited when Hh stimulates the receptor (Aza-Blanc *et al.*, 1997). Direct phosphorylation of Ci by PKA was shown to induce proteolytic cleavage (Chen *et al.*, 1998; Ohlmeyer and Kalderon, 1998; Price and Kalderon, 1999; Wang *et al.*, 1999). In the absence of the Hh ligand, Ptch represses the activity of Smo and Ci forms a complex including Sufu, Fused, and Costal2 (Robbins *et al.*, 1997; Sisson *et al.*, 1997; Zhang *et al.*, 2005). With the stabilization of this complex, Ci is phosphorylated by PKA, which initiates phosphorylation by GSK3β (Jia *et al.*, 2002; Price and Kalderon, 2002) and CKI (Jia *et al.*, 2005). After proteolytic processing into an N-terminal repressor form, Ci translocates to the nucleus, where it silences the expression of its target genes (Chen *et al.*, 1999). Hh signaling triggers a series of events that prevents complex formation, phosphorylation, and proteolysis of Ci in order to block generation of the repressor form and allow the full-length activated form of Ci to enter the nucleus and promote the induction of specific target genes. Several lines of evidence indicate that the proteasome is involved in the processing reaction of Ci and requires priming ubiquitination by the SCF ubiquitin ligase Slimb/β-TrCP (Jiang and Struhl, 1998). This system also participates in several other important signaling pathways, including the Wnt and NFκB pathways, where it targets phosphorylated β-catenin and IκB for ubiquitination, followed by proteasome-mediated degradation (Tian *et al.*, 2005). Although the Slimb/β-TrCP binding motifs in IκB, β-catenin, and other substrates reveal a consensus sequence (DSGXXS), this motif is absent in Ci. Jia *et al.* revealed that the multiple phosphorylated clusters in Ci harbor related sequences in each site and phosphorylation by CKI increases the affinity to Slimb/β-TrCP (Jia *et al.*, 2005). Considerable efforts by many investigators over the past 10 years have uncovered in detail how PKA represses Hh signaling.

For the Gli transcription factors, Wang *et al.* showed that Gli3 is also phosphorylated and processed in the absence of Hh; it responds to Hh by relocalizing to the nucleus without cleavage in a similar fashion to Ci (Wang *et al.*, 2000). In addition, it was confirmed that the processing of Gli3 also requires β-TrCP (Tempe *et al.*, 2006; Wang and Li, 2006). Gli2 is also regulated by this phosphorylation-dependent processing, similar to Ci and Gli3 (Bhatia *et al.*, 2006; Pan *et al.*, 2006, 2009). In contrast, it was suggested that Gli1 is not involved in this processing, as this activator does not contain the responsible phosphorylation sites; other regulation mechanisms involving degradation unrelated to phosphorylation are thought to exist (Di Marcotullio *et al.*, 2006; Huntzicker *et al.*, 2006).

B. Intracellular localization

Since several genetic analyses using *Drosophila* have demonstrated the inhibition of Hh signaling by PKA, many studies have attempted to unveil the intracellular mechanism by which PKA functions in this signaling pathway. Most of these studies have focused on proteolytic cleavage followed by phosphorylation via PKA, GSK3β, and CKI, as described above. In this process, Ci, Gli3, and Gli2 were treated as target molecules, and Gli1 was not considered. Gli1 was deduced to be a downstream target gene and to function only as a simple positive-feedback activator without modification (Dai *et al.*, 1999). However, Sheng *et al.* showed that PKA modifies the activity of Gli1 by another mechanism other than proteolytic cleavage. They reported that activation of PKA retains Gli1 in the cytoplasm, and inhibition of PKA activity promotes nuclear accumulation of Gli1 (Sheng *et al.*, 2006). In addition, they identified the major PKA site determining Gli1 protein localization. Since the phosphorylated threonine is located in the fifth zinc finger of Gli1, phosphorylation of this site was presumed to influence the affinity for the target DNA sequence. However, the authors showed that in the three-dimensional structure, this region resides adjacent to the basic residue cluster of the nuclear localization signal (NLS). Phosphorylation of this site is predicted to alter the local charge and consequently the NLS function rather than DNA binding. They then showed that the change in Gli1 localization is correlated with transcriptional activity. The sequence around the threonine is conserved in Gli2, Gli3, and Ci, thus suggesting the possibility that this mechanism is also applicable to these components.

C. Interaction with 14-3-3

Using a tandem affinity purification strategy and mass spectrometry, we found that Gli transcription factors interact with 14-3-3ε. In addition, this interaction also depends on phosphorylation by PKA. 14-3-3ε binds to Gli transcription factors and reduces their transcriptional activity. The phosphorylation sites involved in 14-3-3 binding are different from the responsible sites for proteolysis and intracellular localization described above (Asaoka *et al.*, 2010; Fig. 13.2).

The 14-3-3 proteins comprise a large family of highly conserved, small, and acidic polypeptides of 28–33 kDa that are found in all eukaryotic species (Hermeking, 2003). The first members of this protein class were identified in the late 1960s as abundant proteins in brain tissue that showed a characteristic migration pattern following two-dimensional diethylaminoethyl-cellulose chromatography and starch gel electrophoresis. This gave rise to the peculiar name 14-3-3. In every eukaryotic organism, at least two different 14-3-3 genes have been identified. In humans, seven different genes encode the highly conserved 14-3-3 isoforms. In the mid-1990s, Muslin *et al.* defined 14-3-3 proteins as phosphoserine-binding proteins that bind to the consensus motif RSXpSXP (Muslin *et al.*, 1996). Binding is dependent on phosphorylation of the central serine. 14-3-3 proteins regulate their binding partner through cytoplasmic sequestration, occupation of interaction domains and export or import sequences, prevention of degradation, activation/repression of enzymatic activity and transactivation, and facilitation of protein modifications. These effects are due to 14-3-3-mediated conformational changes or steric hindrance. For most transcription factors, binding to 14-3-3 alters their transcriptional activity by changing the subcellular localization; however, 14-3-3 binding does not change the subcellular localization of Gli transcription factors. We concluded that conformational change by 14-3-3 binding might alter their activity. The 14-3-3ε isoform, which we have identified as a Gli-interacting partner, appears to play a particularly important role in neuronal development and tumorigenesis. The 14-3-3ε gene is located on chromosome 17p13 and heterozygous deletions in this chromosome result in human neuronal disorders (Kato and Dobyns, 2003). Inappropriate activation of Hh signaling occurs in ~60% of medulloblastomas, and the most common medulloblastoma-associated genetic lesion is loss of the chromosome 17p13 region (Ferretti *et al.*, 2005).

V. Conclusions

We have discussed PKA phosphorylation of Gli transcription factors. While proteolytic cleavage primed by PKA phosphorylation has been investigated as an important regulatory system, recent studies have revealed two other mechanisms regulated by PKA, intracellular localization and 14-3-3 binding. Based on motif analysis of amino acid sequences, Gli2 and Gli3, but not Gli1, harbor the phosphorylated region priming proteolytic cleavage, but the phosphorylation sites involved in intracellular localization and 14-3-3

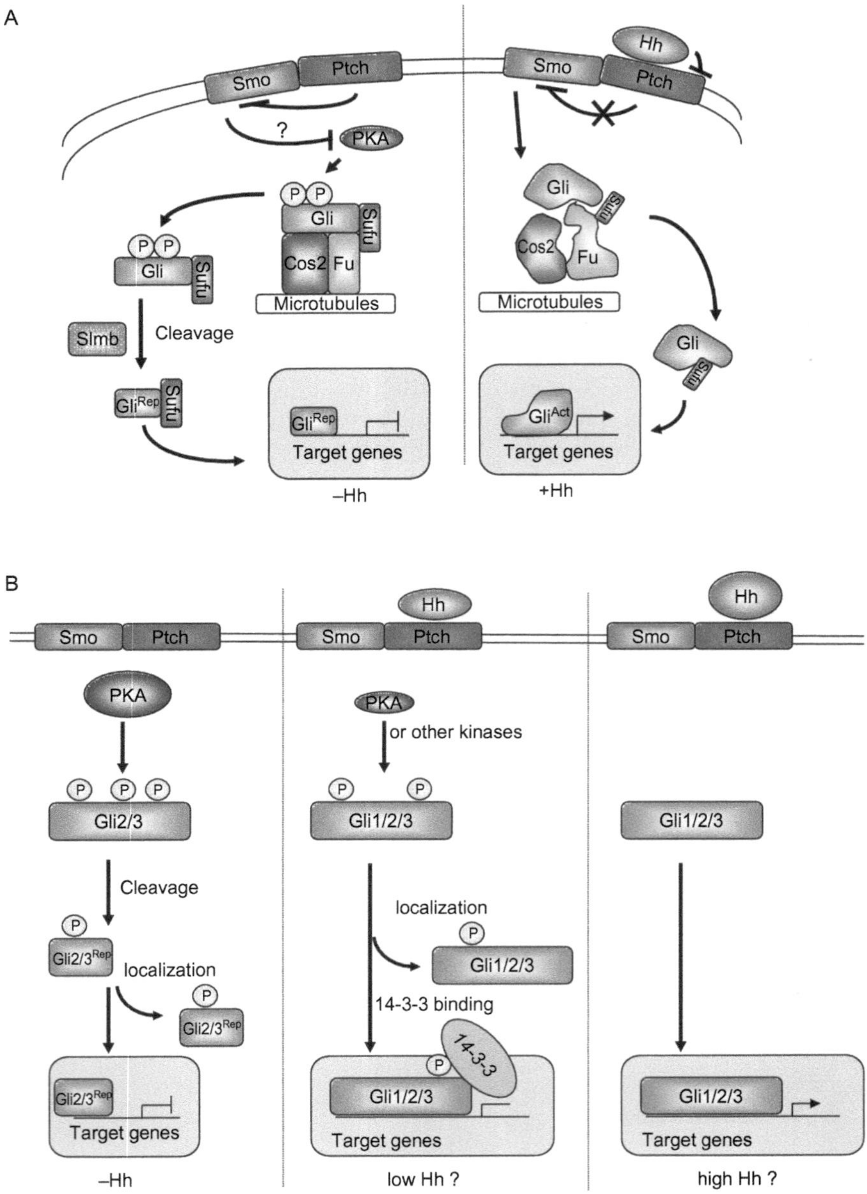

Figure 13.2 (A) The Hh pathway in the absence (left panel) or presence (right panel) of the Hh ligand. In the absence of Hh, Ptch prevents the activity of Smo, and Gli forms a complex with several components, which targets Gli for proteolytic cleavage into the repressor form. In the presence of Hh ligand, Ptch inhibition is relieved; Smo activates Gli. (B) Putative mechanism regulating Hh signaling by multiple phosphorylation of Gli by PKA. Inhibition of nuclear localization and 14-3-3 binding following PKA phosphorylation may limit the activity of Gli (middle panel). (See Color Insert.)

binding are conserved in all three Gli factors and Ci. As a morphogen, the Hh ligand is thought to stimulate the recipient cell depending on the concentration of Hh. Although signal transduction from Ptch and Smo to Gli transcription factors is under investigation by many researchers, the mechanism by which Hh concentration regulates the induction of target genes is poorly understood. The existence of three Gli transcription factors in mammals, as compared to only one factor Ci in *Drosophila*, leads us to speculate that the function of Hh as a morphogen is related to the delicate balance of Gli factors, the so-called GLI Code, in the development of a more complicated organism. In fact, Gli1 itself is a representative target gene and functions as a positive-feedback regulator to amplify this signal. The three mechanisms described in this chapter may influence each Gli protein to varying degrees, and this complicated regulation by PKA phosphorylation may be involved in mammalian development. For example, Gli3 is phosphorylated and cleaved to generate the repressor form, but if the site adjacent to the NLS site is also phosphorylated by PKA, the repressor function might be limited. When all of the Gli3 sites are phosphorylated by PKA without GSK3β, CKI, or proteins related to proteolytic cleavage, translocation to the cytoplasm and 14-3-3 binding might partially inhibit activity (Fig. 13.2). To our knowledge, no studies have attempted to shed light on the hypothesis that multiple phosphorylation mechanisms induce gradual transcriptional activity. We propose a simple hypothesis that the intensity of Hh may be correlated with the activity of PKA through its receptors, depending on the concentration of cAMP. In addition, the regulation of other components, such as GSK3β, CKI, Slimb/β-TrCP, and 14-3-3, may also be important regarding the context-dependent response of the recipient cell. JNK, which is activated in injured cells, was reported to phosphorylate some isoforms of 14-3-3, and this phosphorylation reduces the affinity of 14-3-3 to its binding partner. This mechanism is related to the activation of the apoptotic genes Bax (Tsuruta *et al.*, 2004) and c-Abl (Yoshida *et al.*, 2005), which are released from 14-3-3 and activated by extracellular stimuli. In JNK-activated cells, the Hh signal may be activated by reducing the 14-3-3 binding capacity in the context of injury and repair. Genome-wide screening to identify other genes involved in this signaling pathway may uncover other related kinases (Evangelista *et al.*, 2008; Varjosalo *et al.*, 2008). Among these kinases, we may find one that phosphorylates a limited number of the PKA sites discussed in this chapter. The candidate kinases might function as alternatives to PKA and inhibit Hh signal activity.

Declaration of interest: The author states no conflict of interest and has received no payment in preparation of this chapter.

REFERENCES

Alcedo, J., Ayzenzon, M., Von Ohlen, T., Noll, M., and Hooper, J. E. (1996). The Drosophila smoothened gene encodes a seven-pass membrane protein, a putative receptor for the hedgehog signal. *Cell* **86,** 221–232.

Alexandre, C., Jacinto, A., and Ingham, P. W. (1996). Transcriptional activation of hedgehog target genes in Drosophila is mediated directly by the cubitus interruptus protein, a member of the GLI family of zinc finger DNA-binding proteins. *Genes Dev.* **10,** 2003–2013.

Asaoka, Y., Kanai, F., Ichimura, T., Tateishi, K., Tanaka, Y., Ohta, M., Seto, M., Tada, M., Ijichi, H., Ikenoue, T., Kawabe, T., Isobe, T., *et al.* (2010). Identification of a suppressive mechanism for Hedgehog signaling through a novel interaction of Gli with 14-3-3. *J. Biol. Chem.* **285,** 4185–4194.

Ayers, K. L., and Therond, P. P. (2010). Evaluating Smoothened as a G-protein-coupled receptor for Hedgehog signalling. *Trends Cell Biol.* **20,** 287–298.

Aza-Blanc, P., Ramirez-Weber, F. A., Laget, M. P., Schwartz, C., and Kornberg, T. B. (1997). Proteolysis that is inhibited by hedgehog targets Cubitus interruptus protein to the nucleus and converts it to a repressor. *Cell* **89,** 1043–1053.

Barzi, M., Kostrz, D., Menendez, A., and Pons, S. (2011). Sonic Hedgehog-induced proliferation requires specific Galpha inhibitory proteins. *J. Biol. Chem.* **286,** 8067–8074.

Bhatia, N., Thiyagarajan, S., Elcheva, I., Saleem, M., Dlugosz, A., Mukhtar, H., and Spiegelman, V. S. (2006). Gli2 is targeted for ubiquitination and degradation by beta-TrCP ubiquitin ligase. *J. Biol. Chem.* **281,** 19320–19326.

Blair, S. S. (1995). Developmental biology. Hedgehog digs up an old friend. *Nature* **373,** 656–657.

Chen, Y., Gallaher, N., Goodman, R. H., and Smolik, S. M. (1998). Protein kinase A directly regulates the activity and proteolysis of cubitus interruptus. *Proc. Natl. Acad. Sci. USA* **95,** 2349–2354.

Chen, C. H., von Kessler, D. P., Park, W., Wang, B., Ma, Y., and Beachy, P. A. (1999). Nuclear trafficking of Cubitus interruptus in the transcriptional regulation of Hedgehog target gene expression. *Cell* **98,** 305–316.

Dai, P., Akimaru, H., Tanaka, Y., Maekawa, T., Nakafuku, M., and Ishii, S. (1999). Sonic Hedgehog-induced activation of the Gli1 promoter is mediated by GLI3. *J. Biol. Chem.* **274,** 8143–8152.

DeCamp, D. L., Thompson, T. M., de Sauvage, F. J., and Lerner, M. R. (2000). Smoothened activates Galphai-mediated signaling in frog melanophores. *J. Biol. Chem.* **275,** 26322–26327.

Di Marcotullio, L., Ferretti, E., Greco, A., De Smaele, E., Po, A., Sico, M. A., Alimandi, M., Giannini, G., Maroder, M., Screpanti, I., and Gulino, A. (2006). Numb is a suppressor of Hedgehog signalling and targets Gli1 for Itch-dependent ubiquitination. *Nat. Cell Biol.* **8,** 1415–1423.

Douglas, A. E., Heim, J. A., Shen, F., Almada, L. L., Riobo, N. A., Fernandez-Zapico, M. E., and Manning, D. R. (2011). The alpha subunit of the G protein G13 regulates activity of one or more Gli transcription factors independently of Smoothened. *J. Biol. Chem.* **286,** 30714–30722.

Epstein, D. J., Marti, E., Scott, M. P., and McMahon, A. P. (1996). Antagonizing cAMP-dependent protein kinase A in the dorsal CNS activates a conserved Sonic hedgehog signaling pathway. *Development* **122,** 2885–2894.

Evangelista, M., Lim, T. Y., Lee, J., Parker, L., Ashique, A., Peterson, A. S., Ye, W., Davis, D. P., and de Sauvage, F. J. (2008). Kinome siRNA screen identifies regulators of ciliogenesis and hedgehog signal transduction. *Sci. Signal.* **1,** ra7.

Ferretti, E., De Smaele, E., Di Marcotullio, L., Screpanti, I., and Gulino, A. (2005). Hedgehog checkpoints in medulloblastoma: The chromosome 17p deletion paradigm. *Trends Mol. Med.* **11,** 537–545.

Hammerschmidt, M., and McMahon, A. P. (1998). The effect of pertussis toxin on zebrafish development: A possible role for inhibitory G-proteins in hedgehog signaling. *Dev. Biol.* **194,** 166–171.

Heberlein, U., Singh, C. M., Luk, A. Y., and Donohoe, T. J. (1995). Growth and differentiation in the Drosophila eye coordinated by hedgehog. *Nature* **373,** 709–711.

Hermeking, H. (2003). The 14-3-3 cancer connection. *Nat. Rev. Cancer* **3,** 931–943.

Hooper, J. E., and Scott, M. P. (2005). Communicating with Hedgehogs. *Nat. Rev. Mol. Cell Biol.* **6,** 306–317.

Huangfu, D., and Anderson, K. V. (2006). Signaling from Smo to Ci/Gli: Conservation and divergence of Hedgehog pathways from Drosophila to vertebrates. *Development* **133,** 3–14.

Huntzicker, E. G., Estay, I. S., Zhen, H., Lokteva, L. A., Jackson, P. K., and Oro, A. E. (2006). Dual degradation signals control Gli protein stability and tumor formation. *Genes Dev.* **20,** 276–281.

Jia, J., Amanai, K., Wang, G., Tang, J., Wang, B., and Jiang, J. (2002). Shaggy/GSK3 antagonizes Hedgehog signalling by regulating Cubitus interruptus. *Nature* **416,** 548–552.

Jia, J., Tong, C., Wang, B., Luo, L., and Jiang, J. (2004). Hedgehog signalling activity of Smoothened requires phosphorylation by protein kinase A and casein kinase I. *Nature* **432,** 1045–1050.

Jia, J., Zhang, L., Zhang, Q., Tong, C., Wang, B., Hou, F., Amanai, K., and Jiang, J. (2005). Phosphorylation by double-time/CKIepsilon and CKIalpha targets cubitus interruptus for Slimb/beta-TRCP-mediated proteolytic processing. *Dev. Cell* **9,** 819–830.

Jiang, J., and Struhl, G. (1995). Protein kinase A and hedgehog signaling in Drosophila limb development. *Cell* **80,** 563–572.

Jiang, J., and Struhl, G. (1998). Regulation of the Hedgehog and Wingless signalling pathways by the F-box/WD40-repeat protein Slimb. *Nature* **391,** 493–496.

Kanda, S., Mochizuki, Y., Suematsu, T., Miyata, Y., Nomata, K., and Kanetake, H. (2003). Sonic hedgehog induces capillary morphogenesis by endothelial cells through phosphoinositide 3-kinase. *J. Biol. Chem.* **278,** 8244–8249.

Kasai, K., Takahashi, M., Osumi, N., Sinnarajah, S., Takeo, T., Ikeda, H., Kehrl, J. H., Itoh, G., and Arnheiter, H. (2004). The G12 family of heterotrimeric G proteins and Rho GTPase mediate Sonic hedgehog signalling. *Genes Cells* **9,** 49–58.

Kato, M., and Dobyns, W. B. (2003). Lissencephaly and the molecular basis of neuronal migration. *Hum. Mol. Genet.* **12**(Spec. No. 1), 89–96.

Lepage, T., Cohen, S. M., Diaz-Benjumea, F. J., and Parkhurst, S. M. (1995). Signal transduction by cAMP-dependent protein kinase A in Drosophila limb patterning. *Nature* **373,** 711–715.

Li, W., Ohlmeyer, J. T., Lane, M. E., and Kalderon, D. (1995). Function of protein kinase A in hedgehog signal transduction and Drosophila imaginal disc development. *Cell* **80,** 553–562.

Low, W. C., Wang, C., Pan, Y., Huang, X. Y., Chen, J. K., and Wang, B. (2008). The decoupling of Smoothened from Galphai proteins has little effect on Gli3 protein processing and Hedgehog-regulated chick neural tube patterning. *Dev. Biol.* **321,** 188–196.

Lum, L., Yao, S., Mozer, B., Rovescalli, A., Von Kessler, D., Nirenberg, M., and Beachy, P. A. (2003). Identification of Hedgehog pathway components by RNAi in Drosophila cultured cells. *Science* **299,** 2039–2045.

McKnight, G. S. (1991). Cyclic AMP second messenger systems. *Curr. Opin. Cell Biol.* **3,** 213–217.
Montminy, M. (1997). Transcriptional regulation by cyclic AMP. *Annu. Rev. Biochem.* **66,** 807–822.
Muslin, A. J., Tanner, J. W., Allen, P. M., and Shaw, A. S. (1996). Interaction of 14-3-3 with signaling proteins is mediated by the recognition of phosphoserine. *Cell* **84,** 889–897.
Ogden, S. K., Fei, D. L., Schilling, N. S., Ahmed, Y. F., Hwa, J., and Robbins, D. J. (2008). G protein Galphai functions immediately downstream of Smoothened in Hedgehog signalling. *Nature* **456,** 967–970.
Ohlmeyer, J. T., and Kalderon, D. (1998). Hedgehog stimulates maturation of Cubitus interruptus into a labile transcriptional activator. *Nature* **396,** 749–753.
Oldham, W. M., and Hamm, H. E. (2008). Heterotrimeric G protein activation by G-protein-coupled receptors. *Nat. Rev. Mol. Cell Biol.* **9,** 60–71.
Pan, D., and Rubin, G. M. (1995). cAMP-dependent protein kinase and hedgehog act antagonistically in regulating decapentaplegic transcription in Drosophila imaginal discs. *Cell* **80,** 543–552.
Pan, Y., Bai, C. B., Joyner, A. L., and Wang, B. (2006). Sonic hedgehog signaling regulates Gli2 transcriptional activity by suppressing its processing and degradation. *Mol. Cell. Biol.* **26,** 3365–3377.
Pan, Y., Wang, C., and Wang, B. (2009). Phosphorylation of Gli2 by protein kinase A is required for Gli2 processing and degradation and the Sonic Hedgehog-regulated mouse development. *Dev. Biol.* **326,** 177–189.
Park, H. L., Bai, C., Platt, K. A., Matise, M. P., Beeghly, A., Hui, C. C., Nakashima, M., and Joyner, A. L. (2000). Mouse Gli1 mutants are viable but have defects in SHH signaling in combination with a Gli2 mutation. *Development* **127,** 1593–1605.
Pasca di Magliano, M., and Hebrok, M. (2003). Hedgehog signalling in cancer formation and maintenance. *Nat. Rev. Cancer* **3,** 903–911.
Perrimon, N. (1995). Hedgehog and beyond. *Cell* **80,** 517–520.
Price, M. A., and Kalderon, D. (1999). Proteolysis of cubitus interruptus in Drosophila requires phosphorylation by protein kinase A. *Development* **126,** 4331–4339.
Price, M. A., and Kalderon, D. (2002). Proteolysis of the Hedgehog signaling effector Cubitus interruptus requires phosphorylation by Glycogen Synthase Kinase 3 and Casein Kinase 1. *Cell* **108,** 823–835.
Riobo, N. A., Saucy, B., Dilizio, C., and Manning, D. R. (2006). Activation of heterotrimeric G proteins by Smoothened. *Proc. Natl. Acad. Sci. USA* **103,** 12607–12612.
Robbins, D. J., Nybakken, K. E., Kobayashi, R., Sisson, J. C., Bishop, J. M., and Therond, P. P. (1997). Hedgehog elicits signal transduction by means of a large complex containing the kinesin-related protein costal2. *Cell* **90,** 225–234.
Ruiz i Altaba, A. (1999). Gli proteins encode context-dependent positive and negative functions: Implications for development and disease. *Development* **126,** 3205–3216.
Ruiz i Altaba, A., Sanchez, P., and Dahmane, N. (2002). Gli and hedgehog in cancer: Tumours, embryos and stem cells. *Nat. Rev. Cancer* **2,** 361–372.
Sheng, T., Chi, S., Zhang, X., and Xie, J. (2006). Regulation of Gli1 localization by the cAMP/protein kinase A signaling axis through a site near the nuclear localization signal. *J. Biol. Chem.* **281,** 9–12.
Sisson, J. C., Ho, K. S., Suyama, K., and Scott, M. P. (1997). Costal2, a novel kinesin-related protein in the Hedgehog signaling pathway. *Cell* **90,** 235–245.
Stecca, B., and Ruiz, I. A. A. (2010). Context-dependent regulation of the GLI code in cancer by HEDGEHOG and non-HEDGEHOG signals. *J. Mol. Cell Biol.* **2,** 84–95.
Strutt, D. I., Wiersdorff, V., and Mlodzik, M. (1995). Regulation of furrow progression in the Drosophila eye by cAMP-dependent protein kinase A. *Nature* **373,** 705–709.

Taipale, J., and Beachy, P. A. (2001). The Hedgehog and Wnt signalling pathways in cancer. *Nature* **411,** 349–354.
Tempe, D., Casas, M., Karaz, S., Blanchet-Tournier, M. F., and Concordet, J. P. (2006). Multisite protein kinase A and glycogen synthase kinase 3beta phosphorylation leads to Gli3 ubiquitination by SCFbetaTrCP. *Mol. Cell. Biol.* **26,** 4316–4326.
Tian, L., Holmgren, R. A., and Matouschek, A. (2005). A conserved processing mechanism regulates the activity of transcription factors Cubitus interruptus and NF-kappaB. *Nat. Struct. Mol. Biol.* **12,** 1045–1053.
Tsuruta, F., Sunayama, J., Mori, Y., Hattori, S., Shimizu, S., Tsujimoto, Y., Yoshioka, K., Masuyama, N., and Gotoh, Y. (2004). JNK promotes Bax translocation to mitochondria through phosphorylation of 14-3-3 proteins. *EMBO J.* **23,** 1889–1899.
Varjosalo, M., Bjorklund, M., Cheng, F., Syvanen, H., Kivioja, T., Kilpinen, S., Sun, Z., Kallioniemi, O., Stunnenberg, H. G., He, W. W., Ojala, P., and Taipale, J. (2008). Application of active and kinase-deficient kinome collection for identification of kinases regulating hedgehog signaling. *Cell* **133,** 537–548.
Wang, B., and Li, Y. (2006). Evidence for the direct involvement of {beta}TrCP in Gli3 protein processing. *Proc. Natl. Acad. Sci. USA* **103,** 33–38.
Wang, G., Wang, B., and Jiang, J. (1999). Protein kinase A antagonizes Hedgehog signaling by regulating both the activator and repressor forms of Cubitus interruptus. *Genes Dev.* **13,** 2828–2837.
Wang, B., Fallon, J. F., and Beachy, P. A. (2000). Hedgehog-regulated processing of Gli3 produces an anterior/posterior repressor gradient in the developing vertebrate limb. *Cell* **100,** 423–434.
Waschek, J. A., Dicicco-Bloom, E., Nicot, A., and Lelievre, V. (2006). Hedgehog signaling: New targets for GPCRs coupled to cAMP and protein kinase A. *Ann. N. Y. Acad. Sci.* **1070,** 120–128.
Yoshida, K., Yamaguchi, T., Natsume, T., Kufe, D., and Miki, Y. (2005). JNK phosphorylation of 14-3-3 proteins regulates nuclear targeting of c-Abl in the apoptotic response to DNA damage. *Nat. Cell Biol.* **7,** 278–285.
Zhang, W., Zhao, Y., Tong, C., Wang, G., Wang, B., Jia, J., and Jiang, J. (2005). Hedgehog-regulated Costal2-kinase complexes control phosphorylation and proteolytic processing of Cubitus interruptus. *Dev. Cell* **8,** 267–278.

CHAPTER FOURTEEN

ZFP932 Suppresses Cellular Hedgehog Response and Patched1 Transcription

G. Jason Huang, Zhenhua He, *and* Liang Ma

Contents

Abstract

Substantial amount of research has been devoted to the understanding of hedgehog (Hh) signal transduction in the past two decades. Although significant progresses have been made on the understanding of individual signaling components of this pathway, the transcriptional regulation of these components is not as well understood. We have recently characterized a novel Krüppel-associated box (KRAB) domain containing zinc finger transcription factor ZFP932 (formerly known as ZNF431) and showed that it transcriptionally regulates several components of the Hh pathway. ZFP932 is part of the largest KRAB-domain zinc finger transcription factor family in the mammalian genome. However, the *in vivo* functions for the majority of these transcription factors as well as their downstream targets are not fully understood. ZFP932 is a nuclear KRAB-domain zinc finger transcriptional repressor that recruits HDACs to target promoters. Using the embryonic limb mesenchymal cell line MPLB, we uncovered many potential targets for ZFP932. Among them, Patched1 (*Ptch1*) was found to be a direct transcriptional target of ZFP932. Promoter analyses using reporter assays as well as electrophoretic mobility shift assays revealed three ZFP932 binding sites in the *Ptch1* proximal promoter which are bound by

Division of Dermatology, Department of Medicine, Washington University School of Medicine, St. Louis, Missouri, USA

Vitamins and Hormones, Volume 88
ISSN 0083-6729, DOI: 10.1016/B978-0-12-394622-5.00014-6

ZFP932 *in vivo* as demonstrated by chromatin immunoprecipitation. Consistently, overexpression of ZFP932 either in MPLB cells or in *Xenopus* and mouse embryos strongly repressed *Ptch1* expression. Conversely, shRNA knockdown of *Zfp932* in MPLB cells elevated *Ptch1* expression. In addition to regulating *Ptch1* transcription, we demonstrated that ZFP932 also regulates hedgehog signaling as Hh pathway readouts are reduced in ZFP932 overexpression and elevated in ZFP932 knockdown MPLB cells. Moreover, perturbation of ZFP932 expression in C3H10T1/2 cells affected Hh-induced osteoblast differentiation. Together, these results implicate ZFP932 in Hh signaling by transcriptionally regulating Hh signal component expression.

I. Introduction

The hedgehog-signaling cascade is a well-researched pathway that plays important roles in the control of growth and development during embryogenesis. Loss of signaling leads to developmental defects, whereas inappropriate signaling leads carcinogenesis. The pleiotropic defects and cancer in mutant mice suggest the importance of homeostatic maintenance of hedgehog-signaling levels (Chiang *et al.*, 1996; Katoh and Katoh, 2006).

Three hedgehog ligands (Desert hedgehog, Indian hedgehog, and Sonic hedgehog) and two receptors (Patched1 and Patched2) exist in mammals (Lum and Beachy, 2004). Once hedgehog ligand binds to the receptor Patched, the repression on Smoothened is released which is then activated. Activated Smoothened is phosphorylated and stabilized by GRK2 (Chen *et al.*, 2010). Activated Smoothened associates with Kif3A which helps to move Smoothened to the tip of the primary cilium (Goetz *et al.*, 2009), where Rab moves Smoothened from its vesicle into the cilia membrane (Eggenschwiler *et al.*, 2006). Intraflagellar transport proteins (IFTs) and Kinesin-II proteins such as Kif7 and Kif3a are involved in the movement of molecules such as Suppressor of Fused (Sufu) and Glioma transcription factor (Gli) complex along the ciliary microtubule to the tip of the primary cilium where activated Smoothened causes the dissociation of Sufu and Gli, relieving Gli's repression (Tukachinsky *et al.*, 2010). PKA, a negative regulator of the hedgehog-signaling pathway, can block the ciliary transport system and prevent the localization and subsequent dissociation of Sufu–Gli complex despite activated Smoothened (Tukachinsky *et al.*, 2010). Once Gli is released from Sufu and activated, by Smoothened, it is most likely transported via dynamic microtubules to the nucleus (Kim *et al.*, 2009). The transcriptionally active Gli will modulate downstream target gene expression which includes *Patched* and *Gli*s (Kovacs *et al.*, 2009). Upregulated Patched then provides negative feedback to hedgehog signaling to maintain homeostasis (Katoh and Katoh, 2006; Fig. 14.1A). When Hh ligands are

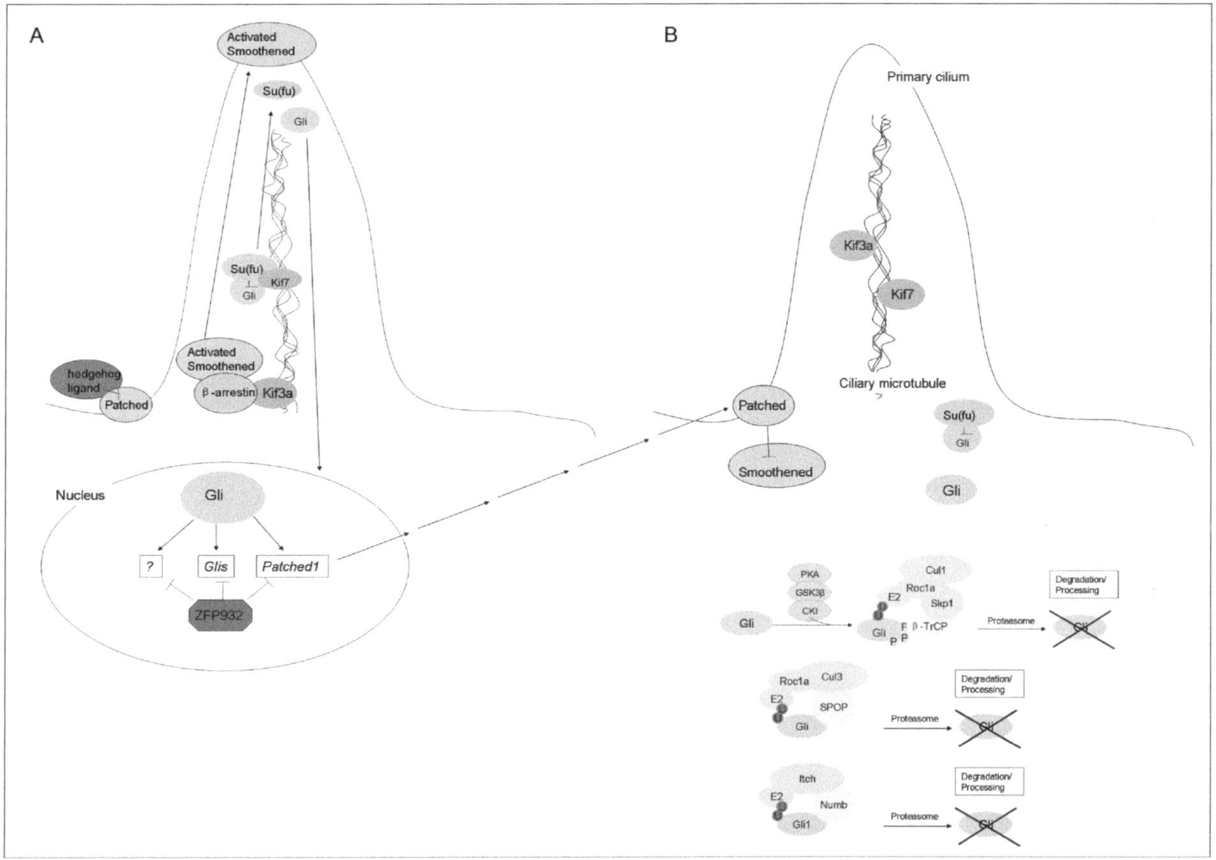

Figure 14.1 A simplified representation of the hedgehog-signaling pathway and ZFP932's role. (A) Hh pathway is activated by the Hh ligand. (B) When ligands are absent, Hh pathway is inactivated. (See Color Insert.)

absent, Patched proteins inhibit Smoothened and prevent activation downstream of the Hh-signaling pathway. With Smoothened inactivated, Sufu binds to and represses Gli. In the absence of Hh ligands, Gli transcription factors are regulated by multiple ubiquitination processes. Initial phosphorylation by PKA primes Gli2 and Gli3 for subsequent phosphorylations by GSK3 and CKI (Riobo *et al.*, 2006). This allows recognition by beta-TrCP and subsequent interaction with the ubiquitin ligase SCF complex (Skp1, Cul1, Roc1a) which ubiquitinates Glis and marks them for proteasome degradation (Bhatia *et al.*, 2006; Tempe *et al.*, 2006; Wang and Li, 2006). A possible second degradation pathway for Glis, demonstrated by study on the Drosophila Gli homolog C_i, does not require phosphorylation but instead is based on regulatory events. The BTB protein SPOP, upregulated by Hh, serves to connect C_i to Cul3 and Roc1b, which acts to ubiquitinate C_i for proteolysis (Di Marcotullio *et al.*, 2007; Jiang, 2006). A third pathway is through Numb and Itch; Numb activates ubiquitin ligase Itch and recruits Gli1 to Itch to be ubiquitinated and eventually be degraded (Di Marcotullio *et al.*, 2006; Fig. 14.1B). Ubiquitination also serves to indirectly inhibit Glis. Gli1 and Gli2 are normally acetylated and require deacetylation by HDACs to convert into active forms, while Cullin3–REN^{KCTD11} ubiquitin ligase interferes with Gli activation by ubiquitinating and subsequently degrading HDACs (Canettieri *et al.*, 2010).

Despite our extensive understanding of the Hh signal transduction, how transcriptional regulation of the signaling components regulates Hh signaling remains poorly understood. Our recent work highlighted the role of a zinc finger transcription factor, ZFP932, in regulating the transcription of multiple genes in the Hh pathway (Fig. 14.1A). ZFP932 is a nuclear transcription factor which represses transcription by binding to corepressors and recruiting histone deacetylases (HDACs; He *et al.*, 2011). This chapter will be devoted to the effect of ZFP932 on hedgehog signaling, its repression on *Ptch1*, and other potentially interesting targets and their implications.

II. *Zfp932* Characterization

The mouse *Zfp932* (per NCBI, Mouse Genome Informatics 2310001H12rik) gene contains four exons spanning over 20kb on chromosome 5. Alternative splicing generates four different mRNA isoforms. The isoform characterized (BC012405) encodes an N-terminal KRAB domain followed by 15 tandem zinc fingers at the C-terminus (Fig. 14.2A and B). All zinc fingers in ZFP932 belong to the C_2H_2 family which is the most prevalent zinc finger type in mammals containing the consensus sequence *YCX2–4CX3YX5YX2HX3,4H*, where X represents any amino acid, and Y

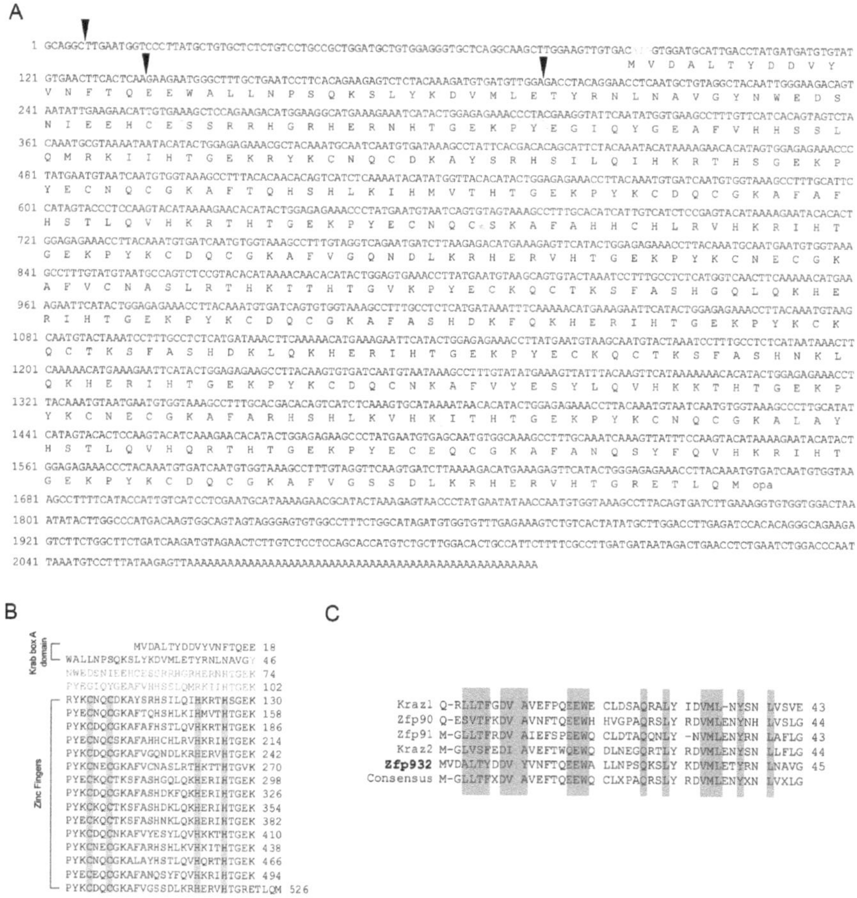

Figure 14.2 Gene and protein structure of ZFP932. (A) The cDNA and amino sequences of ZFP932. The coding sequence is from nucleotide 88 to 1668. Triangles indicate splice sites. (B) Schematic representation of the structure of ZFP932. The KRAB-A domain and the zinc finger domains are indicated with cystidines and histidines in the Zinc fingers shaded. (C) Multiple proteins and their amino acid alignment in the KRAB domains are shown. Conserved amino acids are shaded and consensus sequence is shown at the bottom. This figure was originally published in the *Journal of Biological Chemistry*. He *et al.* (2011); © The American Society for Biochemistry and Molecular Biology.

represents a hydrophobic residue (Pabo *et al.*, 2001). This sequence allows the C_2H_2 family to bind DNA in a sequence-specific manner. Sequence alignment with other KRAB-domain zinc finger proteins revealed that ZFP932 also contains a KRAB domain which can be further divided into an A and a B box (He *et al.*, 2011; Fig. 14.2B). Of the two, the KRAB-A box is more conserved (Vissing *et al.*, 1995) and participates in transcriptional repression by these zinc finger transcriptional factors (Bellefroid *et al.*, 1989; Fig. 14.2C). It has been suggested that after binding to their respective DNA sequences, these KRAB zinc finger transcription factors associate with corepressors such as Kap1, Tif1b, or Krip1 (Kim *et al.*, 1996), as well as interacting with the NuRD complexes (Schultz *et al.*, 2001), to induce heterochromatin formation at target promoters to silence gene expression (Urrutia, 2003). Thus ZFP932 belongs to a class of zinc finger proteins that comprise roughly one-third of the C_2H_2 family (Bellefroid *et al.*, 1991). These KRAB zinc finger proteins are involved in the regulation of various cell processes such as apoptosis, differentiation, proliferation, and neoplastic transformation (Hering *et al.*, 2004).

Similar to other KRAB zinc finger proteins, ZFP932 is a nuclear transcription factor and can repress transcription by recruiting HDACs. First, tagging ZFP932 with enhanced green fluorescent protein (EGFP) led to a fusion protein that is exclusively nuclear localized as compared to EGFP alone which is present both in the cytoplasm and in the nucleus. Second, in-frame fusion of ZFP932 onto the GAL4 DNA binding domain (GAL4DBD) leads to transcriptional repression of a Gal4-responsive promoter in cell culture, indicating that ZFP932 can function as a transcriptional repressor (He *et al.*, 2011). This repressive activity is mediated by the KRAB-A domain as its deletion abolished this repressive activity. On the molecular level, ZFP932 represses target gene transcription by recruiting HDACs to target promoters. Coimmunoprecipitation experiment demonstrated that endogenous as well as an HA-tagged version of ZFP932 interacted with HDAC1 and 2 in MPLB cells (He *et al.*, 2011). In fact, ZFP932 can directly interact with HDAC2 but not HDAC1 in pull-down assays using purified proteins (Fig. 14.3). This result suggests that ZFP932 first recruits HDAC2 to target promoters which then recruits HDAC1 and

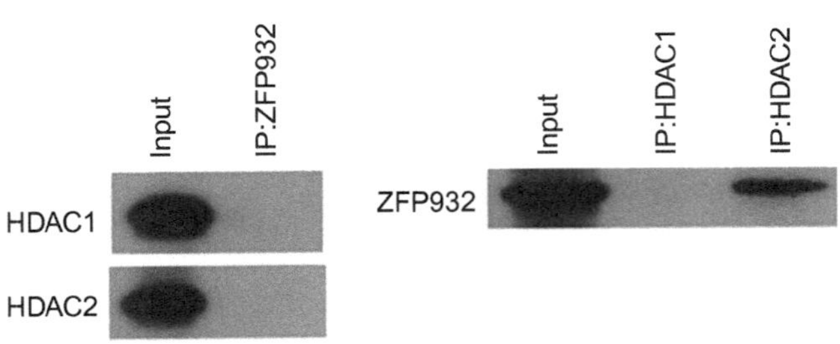

Figure 14.3 ZFP932 interaction with HDACs. *In vitro* pull-down assay showing that HDAC2 but not HDAC1 directly interacts with ZFP932. This figure was originally published in the *Journal of Biological Chemistry*. He *et al.* (2011); © The American Society for Biochemistry and Molecular Biology. (For color version of this figure, the reader is referred to the Web version of this chapter.)

maybe other corepressors to the site for chromatin modification. Consistent with the observation that the KRAB-A domain is absolutely required for transcription repression, this domain is also critically required for interaction with HDACs as mutant protein lacking this domain failed to coimmunoprecipitate with HDACs. Finally, the repressive activity of ZFP932 depends on HDAC activity as inhibition of HDAC activity by Tricostatin A partially relieved transcription repression induced by ZFP932. Taken together, these results provide molecular evidence that ZFP932 functions as a nuclear transcriptional repressor, whose activity depends on its interaction with HDACs through its KRAB-A domain and HDAC activity (He *et al.*, 2011).

III. ZFP932 Represses *Ptch1* Expression

One way to uncover the biological function of ZFP932 is to identify its transcriptional targets. For this purpose, the MPLB limb mesenchymal cell line serves as a good model system. *In situ* hybridization showed that ZFP932 is expressed in E10.5 limb mesenchyme, and thus, transcription targets identified using MPLB may have *in vivo* functional significance. Because ZFP932 is a transcriptional repressor and it is expressed at a relatively high level in MPLB cells, it should already be significantly repressing its target gene expression. And thus overexpressing more ZFP932 in MPLB cells may result in minimal changes in target gene expression and may not aid in the identification of its targets. To address this concern, ZFP932 was modified and converted from a transcriptional repressor into an activator by replacing the N-terminal KRAB-A domain with a strong viral transcription domain VP16. The resultant fusion protein will still be able to bind to ZFP932 target promoters but will now activate rather than repress transcription. Indeed, microarray results using MPLB cells transfected with this fusion construct led to the identification of approximately 200 genes whose expression was upregulated threefold or more (Table 14.1). In addition to upregulated genes, expressions of approximately 800 genes were also downregulated by this fusion protein (data not shown). These are likely secondary targets because VP15-ZFP932 should function as a transcriptional activator regardless of promoter context.

Among the many potential targets showing significant changes, *Ptch1* and Gli1 received further attention because they are components of the Hh pathway. Overexpression of ZFP932 in MPLB cells led to a 50% reduction in *Ptch1* expression, whereas *Notch1* and *Patched2* expressions were not affected. Thus, ZFP932 is not just a global nonspecific transcription repressor but rather a sequence-specific transcriptional repressor. If this is true, then reducing ZFP932 expression using shRNA knockdown in MPLB cells should relieve this repression and increase *Ptch1* expression.

Table 14.1 ZFP932 and other potential targets

GenBank	Gene symbol	Fold change	Description
NM_008079	Galc	430.41	Galactosylceramidase
NM_025408	Phca	228.63	Phytoceramidase, alkaline
XM_144310	6820424L24Rik	150.87	RIKEN cDNA 6820424L24 gene
NM_033575	Pcdhgb2	93.59	Protocadherin gamma subfamily C, 3
NM_181857	Poln	77.71	CDNA sequence BC023882
NM_024249	1810073N04Rik	72.89	RIKEN cDNA 1810073N04 gene
NM_009781	Cacna1c	72.62	Calcium channel, voltage-dependent, L-type, alpha 1C subunit
NM_009185	Sil	57.72	Tal1-interrupting locus
NM_026501	1300010M03Rik	43.75	RIKEN cDNA 1300010M03 gene
XM_485378	Tex10	43.75	Testis-expressed gene 10
NM_023813	Camk2d	40.31	Calcium/calmodulin-dependent protein kinase II, delta
NM_033077	D1Pas1	38.64	DNA segment, Chr 1, Pasteur Institute 1
BC061032	2700038I16Rik	32.76	Zinc finger protein 655
NM_134012	Mbtd1	30.12	Mbt domain-containing 1
NM_175667	Ankrd5	27.46	Ankyrin repeat domain 5
NM_031198	Tcfec	21.54	Transcription factor EC
NM_010052	Dlk1	20.97	Delta-like 1 homolog (*Drosophila*)
NM_133201	Mfn2	19.97	Mitofusin 2
NM_028617	1110012M11Rik	18.95	RIKEN cDNA 1110012M11 gene
NM_010380	H2-D1	18.79	Histocompatibility 2, D region locus 1
NM_009371	Tgfbr2	18.47	Transforming growth factor, beta receptor II
NM_008216	Has2	17.17	Hyaluronan synthase 2
NM_178927	AV344025	16.88	Expressed sequence AV344025
NM_028152	Mms19l	16.80	MMS19 (MET18 *S. cerevisiae*)-like
NM_009561	Zfp61	16.30	Zinc finger protein 61
NM_133943	Hsd3b7	16.29	Hydroxy-delta-5-steroid dehydrogenase, 3 beta-, and steroid delta isomerase 7
NM_009547	Zfp161	16.01	Zinc finger protein 161
NM_010776	Mbl2	15.78	Mannose-binding lectin, serum (C)
NM_027919	1300017K07Rik	15.76	RIKEN cDNA 1300017K07 gene

XM_145254	A630082K20Rik	14.56	RIKEN cDNA A630082K20 gene
BC063105	Cnot2	13.76	CCR4-NOT transcription complex, subunit 2
NM_009120	Sara1	13.70	SAR1a gene homolog 1 (*S. cerevisiae*)
NM_015781	Nap1l1	12.85	Nucleosome assembly protein 1-like 1
NM_026114	Eif2s1	12.35	Eukaryotic translation initiation factor 2, subunit 1 alpha
NM_139150	Carf	11.84	Calcium-response factor
BC051572	2410003A14Rik	11.48	RIKEN cDNA 2410003A14 gene
XM_132796	Fancd2	10.92	Fanconi anemia, complementation group D2
NM_177781	Anktm1	10.70	Transient receptor potential cation channel, subfamily A, member 1
NM_178681	Dgkb	9.81	Diacylglycerol kinase, beta
NM_029270	Arhgap24	9.79	Rho GTPase-activating protein 24
NM_199465	Nexn	9.72	Nexilin
AK034046	5930434B04Rik	9.50	RIKEN cDNA 5930434B04 gene
NM_199022	6230417E10Rik	9.27	RIKEN cDNA 6230417E10 gene
NM_175429	Kctd12b	9.24	Potassium channel tetramerization domain-containing 12b
NM_021463	Prps1	9.21	Phosphoribosyl pyrophosphate synthetase 1
NM_145471	E130306I01Rik	9.14	RIKEN cDNA E130306101 gene
NM_025473	1810037C20Rik	9.10	RIKEN cDNA 1810037C20 gene
NM_029963	Mrps5	8.76	Mitochondrial ribosomal protein S5
XM_143575	AK122498	8.66	Leucine-rich repeat-containing 7
NM_175126	Zcchc3	8.61	Zinc finger, CCHC domain-containing 3
NM_024412	Clcnka	8.47	Chloride channel Ka
NM_021272	Fabp7	8.28	Fatty acid-binding protein 7, brain
NM_030612	Nfkbiz	7.93	Nuclear factor of kappa light polypeptide gene enhancer in B-cells inhibitor, zeta
NM_019477	Acsl4	7.77	Acyl-CoA synthetase long-chain family member 4
NM_024174	Mrps23	7.70	Mitochondrial ribosomal protein S23
NM_011002	Olfr59	7.60	Olfactory receptor 59
NM_026323	Wfdc2	7.55	WAP four-disulfide core domain 2

(*Continued*)

Table 14.1 (*Continued*)

GenBank	Gene symbol	Fold change	Description
NM_020252	Nrxn1	7.48	Neurexin I
NM_146637	Olfr1141	7.41	Integral to membrane; G-protein-coupled receptor activity *Mus musculus* olfactory receptor; mRNA
NM_146922	Olfr376	7.40	Olfactory receptor 376
NM_028412	Ciz1	7.34	CDKN1A-interacting zinc finger protein 1
NM_013935	Ptpla	7.27	Protein tyrosine phosphatase-like (proline instead of catalytic arginine), member a
NM_007938	Epha6	7.25	Eph receptor A6
NM_010117	Rhbdf1	7.19	Rhomboid family 1 (*Drosophila*)
NM_023662	Pcm1	7.17	Pericentriolar material 1
NM_019762	Pkp3	7.10	Plakophilin 3
NM_053261	Impa2	6.95	Inositol (myo)-1 (or 4)-monophosphatase 2
NM_022417	Itm2c	6.94	Integral membrane protein 2C
NM_183125	A930006J02Rik	6.78	RIKEN cDNA A930006J02 gene
NM_019702	Hbs1l	6.64	RIKEN cDNA 4930455C13 gene
NM_178592	Bat5	6.54	HLA-B-associated transcript 5
NM_001001447	Hkr2	6.44	GLI-Kruppel family member HKR2
BC057926	Mmp16	6.34	Matrix metalloproteinase 16
XM_133912	Mki67	6.31	Antigen identified by monoclonal antibody Ki 67
BC042212	Madd	6.24	MAP kinase-activating death domain
NM_021474	Efemp2	6.08	Epidermal growth factor-containing fibulin-like extracellular matrix protein 2
NM_175087	Aqp6	5.83	Aquaporin 6
XM_355958	6430531B16Rik	5.80	RIKEN cDNA 6430531B16 gene
NM_199025	A630026F21Rik	5.79	RIKEN cDNA A630026F21 gene
NM_176968	Nt5c2l1	5.78	5′-nucleotidase, cytosolic II-like 1
NM_029763	3010019O03Rik	5.74	Polymerase (RNA) III (DNA directed) polypeptide F
NM_010678	Laf4	5.64	Hypothetical protein A730046J16
NM_008957	Ptch1	5.61	Patched homolog 1

NM_008797	Pcx	5.54	Pyruvate carboxylase
NM_009369	Tgfbi	5.53	Transforming growth factor, beta induced
NM_183017	BC055368	5.49	CDNA sequence BC055368
NM_008358	Il15ra	5.47	Interleukin 15 receptor, alpha chain
NM_177778	Armc7	5.46	Armadillo repeat-containing 7
NM_008298	Dnaja1	5.36	DnaJ (Hsp40) homolog, subfamily A, member 1
NM_182716	Nfasc	5.32	Neurofascin
NM_146985	Olfr1270	5.29	Olfactory receptor 1270
NM_008492	Ldh2	5.17	Lactate dehydrogenase 2, B chain
NM_153131	Unc5a	5.15	Unc-5 homolog A (*C. elegans*)
XM_355059	1190002H09Rik	5.13	RIKEN cDNA 1190002H09 gene
NM_022320	Gpr35	5.09	G-protein-coupled receptor 35
NM_177138	6720469N11Rik	5.07	RIKEN cDNA 6720469N11 gene
NM_144513	Gtl2	5.03	GTL2, imprinted maternally expressed untranslated mRNA
NM_172786	E230031K19	4.99	Hypothetical protein E230031K19
NM_030685	D3Ucla1	4.99	DNA segment, Chr 3, University of California at Los Angeles 1
XM_284454	Irf2bp2	4.97	Interferon regulatory factor 2-binding protein 2
NM_011428	Snap25	4.95	Synaptosomal-associated protein 25
NM_018750	Rassf5	4.75	Ras association (RalGDS/AF-6) domain family 5
NM_144892	Ncoa5	4.73	Nuclear receptor coactivator 5
NM_175141	1810030O07Rik	4.72	RIKEN cDNA 1810030O07 gene
NM_021499	Wdr8	4.70	WD repeat domain 8
NM_033648	Fxyd4	4.63	FXYD domain-containing ion transport regulator 4
NM_146687	Olfr1436	4.62	Integral to membrane; G-protein-coupled receptor activity *Mus musculus* olfactory receptor; mRNA
NM_183158	F730023N20	4.59	Hypothetical protein F730023N20
NM_009287	Stim1	4.56	Stromal interaction molecule 1
NM_011181	Pscd2	4.50	Pleckstrin homology, Sec7, and coiled-coil domains 2
NM_054078	Baz2a	4.42	Bromodomain adjacent to zinc finger domain, 2A
XM_196763	Prssl1	4.38	Protease, serine-like 1
NM_145505	AI450540	4.37	Expressed sequence AI450540
NM_007810	Cyp19a1	4.36	Cytochrome P450, family 19, subfamily a, polypeptide 1

(*Continued*)

Table 14.1 (*Continued*)

GenBank	Gene symbol	Fold change	Description
XM_125902	Xpot	4.35	Exportin, tRNA (nuclear export receptor for tRNAs)
NM_011699	Lin7c	4.30	Lin 7 homolog c (*C. elegans*)
NM_010390	H2-Q1	4.26	Histocompatibility 2, Q region locus 1
XM_132966	2810474O19Rik	4.26	RIKEN cDNA 2810474O19 gene
NM_019451	Il1f5	4.26	Interleukin 1 family, member 5 (delta)
NM_172836	9930021J03Rik	4.21	RIKEN cDNA 9930021J03 gene
NM_008400	Itgal	4.19	Integrin alpha L
NM_024182	Riok3	4.16	RIO kinase 3 (yeast)
XM_284491	Ppp2r1b	4.15	Protein phosphatase 2 (formerly 2A), regulatory subunit A (PR 65), beta isoform
XM_194204	4930539A06Rik	4.15	RIKEN cDNA 4930539A06 gene
NM_009851	Cd44	4.11	CD44 antigen
XM_356375	V2r3	4.08	Vomeronasal 2, receptor, 3
XM_149840	D330027H18Rik	4.08	RIKEN cDNA D330027H18 gene
NM_016886	Gria3	4.07	RIKEN cDNA E430013K19 gene
NM_026817	Rabl2a	4.07	RAB, member of RAS oncogene family-like 2A
NM_008337	Ifng	4.05	Interferon gamma
NM_177267	Wdr22	4.02	WD repeat domain 22
NM_007584	Ddr1	4.01	Discoidin domain receptor family, member 1
NM_175329	Ndg2	4.00	Nur77 downstream gene 2
NM_009837	Cct4	3.97	Chaperonin subunit 4 (delta)
NM_026080	Mrps24	3.96	Mitochondrial ribosomal protein S24
NM_013813	Epb4.1l3	3.92	RIKEN cDNA 2410021H03 gene
NM_133708	Gmppa	3.91	GDP-mannose pyrophosphorylase A
NM_029838	Col25a1	3.91	Procollagen, type XXV, alpha 1
NM_133832	Rdh10	3.90	Retinol dehydrogenase 10 (all-*trans*)
NM_024291	Ky	3.88	Kyphoscoliosis

XM_111357	Igh-VX24	3.85	Immunoglobulin heavy chain (X24 family)
NM_029508	0610009F02Rik	3.83	RIKEN cDNA 9530023M17 gene
NM_053266	1700012P16Rik	3.83	RIKEN cDNA 1700012P16 gene
U04227	IGHV5S17	3.82	Vh7183 family; *Mus musculus* BALB/c Ig heavy chain V region (Vh7183) gene, partial cds
NM_022980	Dscr1l2	3.80	Down syndrome critical region gene 1-like 2
NM_008087	Gas2	3.79	Growth arrest-specific 2
NM_025444	Taf13	3.77	TAF13 RNA polymerase II, TATA box-binding protein (TBP)-associated factor
NM_025286	Slc31a2	3.77	Solute carrier family 31, member 2
NM_031374	Tex15	3.76	Testis-expressed gene 15
NM_008999	Rab23	3.75	RAB23, member RAS oncogene family
NM_027427	Taf15	3.71	TAF15 RNA polymerase II, TATA box-binding protein (TBP)-associated factor
J00596	IGLJ4	3.71	Pseudo-j-lambda4; mouse germline pseudo-Ig-lambda-4 chain gene, J2-C2-J4-C4 cluster: J region
NM_133853	4732496O19Rik	3.69	RIKEN cDNA 6530407C02 gene
NM_022026	Aqp9	3.68	Aquaporin 9
NM_177095	9430076C15Rik	3.56	RIKEN cDNA 9430076C15 gene
NM_134059	Ddx41	3.56	DEAD (Asp-Glu-Ala-Asp) box polypeptide 41
XM_136506	E130106K10Rik	3.55	RIKEN cDNA 130106K10 gene
NM_007661	Cdc2l2	3.54	Cell division cycle 2-like 1
NM_008732	Slc11a2	3.53	Solute carrier family 11 (proton-coupled divalent metal ion transporters), member 2
NM_028260	1500034J20Rik	3.53	RIKEN cDNA 1500034J20 gene
NM_175425	C1qtnf7	3.50	C1q and tumor necrosis factor-related protein 7
NM_033580	Pcdhgb8	3.50	Protocadherin gamma subfamily C, 3
AK033513	6030408C04Rik	3.47	RIKEN cDNA 6030408C04 gene

(*Continued*)

Table 14.1 (*Continued*)

GenBank	Gene symbol	Fold change	Description
NM_010516	Cyr61	3.46	Cysteine-rich protein 61
NM_183390	Klhl6	3.43	Kelch-like 6 (*Drosophila*)
NM_013738	Plek2	3.39	Pleckstrin 2
NM_010357	Gsta4	3.38	Glutathione-*S*-transferase, alpha 4
NM_007717	Cmah	3.38	Cytidine monophospho-*N*-acetylneuraminic acid hydroxylase
NM_201388	Plec1	3.36	Plectin 1
BC045142	Kcnq1	3.36	Tumor-suppressing subchromosomal transferable fragment 8
AK033247	2600013E07Rik	3.34	RIKEN cDNA 2600013E07 gene
NM_009831	Ccng1	3.34	Cyclin G1
NM_010191	Fdft1	3.34	Farnesyl diphosphate farnesyl transferase 1
NM_025276	Evpl	3.32	Envoplakin
XM_355234	Kif14	3.31	Kinesin family member 14
NM_173445	E430018M08Rik	3.31	RIKEN cDNA E430018M08 gene
NM_133679	Cryzl1	3.30	Crystallin, zeta (quinone reductase)-like 1
NM_028990	8430437G11Rik	3.30	RIKEN cDNA 8430437G11 gene
NM_008524	Lum	3.29	Lumican
NM_031258	Chrdl1	3.29	Chordin-like 1
XM_132537	2410003K15Rik	3.21	RIKEN cDNA 2410003K15 gene
NM_026592	B230118H07Rik	3.21	RIKEN cDNA B230118H07 gene
NM_134189	Galnt10	3.21	UDP-*N*-acetyl-alpha-D-galactosamine:polypeptide *N*-acetylgalactosaminyltransferase 10
XM_133861	Cog7	3.21	Component of oligomeric golgi complex 7
NM_152808	1110028E10Rik	3.20	RIKEN cDNA 1110028E10 gene
NM_009835	Ccr6	3.16	Chemokine (C–C motif) receptor 6
NM_178887	Fibcd1	3.16	Fibrinogen C domain-containing 1
XM_194622	B430306D02Rik	3.16	RIKEN cDNA B430306D02 gene
NM_026410	Cdca5	3.15	Cell division cycle-associated 5

AK084926	C030045D06Rik	3.10	Hypothetical protein D430013K02
NM_030710	Slamf6	3.08	SLAM family member 6
U23025	IGHV8S10	3.04	7183 Vh gene family; *Mus musculus* C57BL/6 immunoglobulin heavy chain V region mRNA, clone CB17H-9, partial cds
XM_130347	Ndufs3	3.02	*Mus musculus*; Fe–S protein 3; mitochondrial electron transport, NADH to ubiquinone dehydrogenase; mRNA
NM_134000	AI429613	3.02	Expressed sequence AI429613
XM_110955	AI851076	3.00	Expressed sequence AI851076
NM_026267	1200016B17Rik	3.00	RIKEN cDNA 1200016B17 gene

ZFP932 was converted to activator form (VP16-ZFP932ΔN) and used to uncover potential target genes in MPLB cells. Selected genes of interest (greater than threefold), ordered from greatest fold change to least, are placed in the figure.

Indeed, three shRNA constructs that knocked *Zfp932* expression down to less than 50% all elevated *Ptch1* expression by twofold or more in MPLB cells (He *et al.*, 2011; Fig. 14.4B). The off-target effect of shRNA knockdowns was controlled by cotransfection of a rescue construct which lacks the 3′UTR where shRNA targets. Transfection of this rescue construct together with one of the shRNAs abolished shRNA-mediated *Ptch1* upregulation (Fig. 14.4B). In transient transfection assays, ZFP932 represses *Ptch1* expression within 12h after transfection suggesting a direct transcriptional regulation. It is not possible to predict the binding site for ZFP932 because it is not clear how many of its 15 zinc fingers are involved in DNA binding. Thus, to test whether ZFP932 directly regulates *Ptch1* transcription, promoter bashing was performed. *Ptch1* has two promoters, 1a and 1b. In MPLB cells, only promoter 1b is active as assayed by RT-PCR. The 1.6-kb 1b promoter fragment responded to ZFP932 transfection in luciferase reporter assays (He *et al.*, 2011). Further deletions of this 1.6-kb fragment revealed two smaller fragments that were repressed by ZFP932. Finally, three putative ZFP932 binding sites were identified and tested by EMSA. These three sites contain the GC-rich consensus sequence GCGCCC that is typical of zinc finger binding sites. Mutations in this site abolished ZFP932 binding in EMSA as well as transcriptional repression by ZFP932 in MPLB cells (He *et al.*, 2011). Based on crystallographic studies, each zinc finger binds to a three-base pair sequence. Therefore, ZFP932 uses 2 of its 15 zinc fingers when binding to the *Ptch1* promoter. However, currently, it is still not clear which two zinc fingers are involved. Binding of ZFP932 to these three sites has also been demonstrated in MPLB cells using chromatin immunoprecipitation assays. HA-tagged ZFP932 was transfected into MPLB cells and cross-linked to chromatin and pulled down by an HA antibody. PCR assays revealed that fragments containing these binding sites are enriched in antibody pull-down sample compared to controls. Together, these studies demonstrate that ZFP932 is a sequence-specific transcriptional repressor, which represses the transcription of one of its targets, *Ptch1*, through binding to three ZFP932 binding sites located in the proximal *Ptch1* 1b promoter.

Further experiments confirmed the ZFP932 repressive regulation of *Ptch1* expression in two *in vivo* overexpression systems. In *Xenopus* embryos, mouse *ZFP932* overexpression was shown to also repress the expression of *Xptc1*, the *Xenopus Ptch1* homolog. Injection of HA-ZFP932 mRNA into one of the two cells of *Xenopus laevis* embryos at the two-cell stage resulted in strongly suppressed *Xptc1* on the injected side (Fig. 14.5A) at the neurula stage. On the other hand, the uninjected side displayed strong *Xptc1* expression (Fig. 14.5A, arrows). This implies that ZFP932 response element should also exist in the *Xptc1* promoter. A search of the proximal promoter of *Xptc1* for ZFP932 response elements revealed a consensus sequence (GCGCCC) at

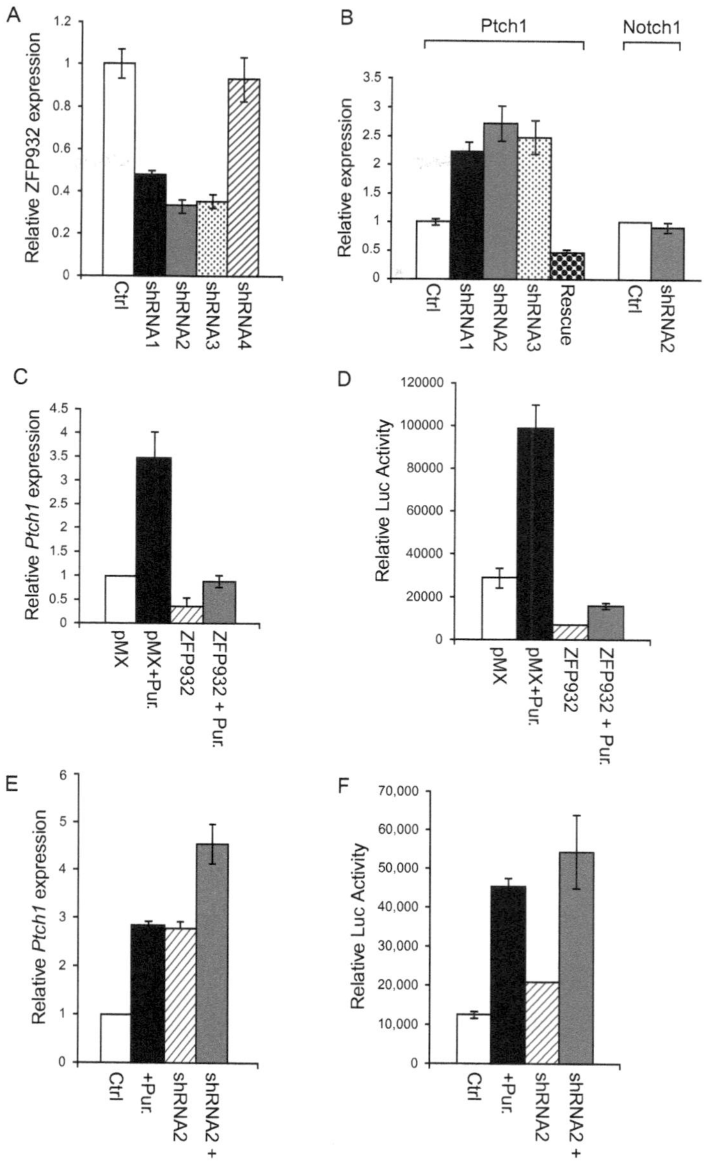

Figure 14.4 ZFP932 represses Ptch1 expression and Hh response. (A) Three of four shRNAs against Zfp932 knocked ZFP932 expression down to less than 50%. (B) ZFP932 knockdown does not affect Notch1 expression but elevates Ptch1 expression in MPLB cells, which is reversed when co-transfected with a Zfp932-rescue construct-insensitive shRNA knockdown. (C–F) Cellular responsiveness to Hh signaling is affected by differing ZFP932 levels. Treatment of purmorphamine (Pur.) activates Hh pathway. (C and D) ZFP932 reduced both basal expression levels of Ptch1 and Gli-Luc and their inductions by Hh. (E–F) Knocking down ZFP932 elevated both Ptch1 and Gli-Luc basal expression as well as induction by purmorphamine. This figure was originally published in the *Journal of Biological Chemistry*. He *et al*. (2011); © the American Society for Biochemistry and Molecular Biology.

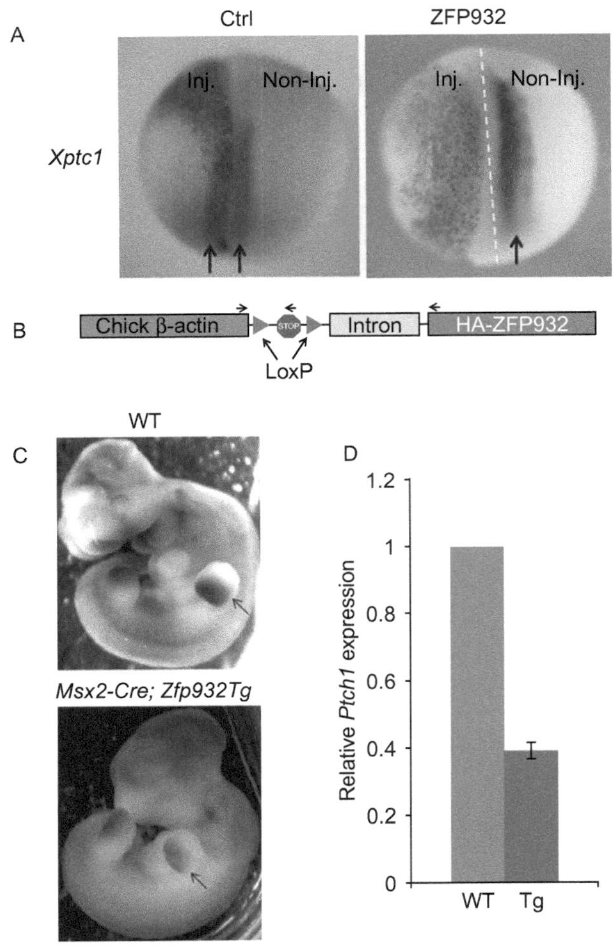

Figure 14.5 ZFP932 overexpression repressed *Ptch1* expression in *Xenopus* and transgenic mouse embryos. (A) *Xenopus* embryos showing the injected (Inj.) halves as indicated by X-gal staining (blue) and Non-injected (Non-Inj.) halves. Xptc1 staining by *in situ* hybridization is shown in purple. (B) HA-tagged ZFP932 was driven by the chicken beta-actin promoter preceded by a LoxP-flanked transcriptional stop cassette and a beta-globin intron. This construct was used to create transgenic mouse lines. Relative positions of genotyping primers are indicated by arrows. (C) Wild-type embryos (upper); HA-ZFP932 transgenic embryos (lower). *Ptch1* staining in the limb bud is indicated by arrows. (D) qPCR quantification of *Ptch1* expression in wild-type or HA-ZFP932 transgenic embryos is shown. The results are shown as relative *Ptch1* expression. This figure was originally published in the *Journal of Biological Chemistry*. He *et al.* (2011); © the American Society for Biochemistry and Molecular Biology. (See Color Insert.)

−315 relative to the transcription start site. This suggests that the regulatory relationship between *Zfp932* and *Ptch1* is preserved in *Xenopus*. ZFP932 homolog identification in *Xenopus* was complicated by the number of putative homologous proteins as well as the fact that only 2 of the 15 zinc fingers may be responsible for DNA binding (He *et al.*, 2011).

In transgenic mouse, global overexpression of ZFP932 led to a 60% reduction in *Ptch1* expression compared to littermate controls without the transgene (Fig. 14.5D). *In situ* hybridization showed global *Ptch1* expression reduction, especially obvious in somites and the developing limb bud (Fig. 14.5C). The major isoform expressed in all tissues was the *Ptch1* 1b. *ZFP932* overexpression specifically downregulated the 1b but not the 1a isoform (He *et al.*, 2011).

IV. ZFP932 Represses Cellular Hedgehog-Signaling Response

The microarray results reveal that both *Ptch1* and *Gli1* are targets of ZFP932. Because expressions of both genes are regulated by the Hh pathway, it raises the possibility that ZFP932 modulates cellular Hh response. First, *Zfp932* itself could be regulated by Hh signaling, but that turns out not to be the case as Hh activation by purmorphamine treatment (Sinha and Chen, 2006) had minimal effect on *Zfp932* expression in MPLB cells (He *et al.*, 2011). However, ZFP932 level appears crucial for the magnitude of hedgehog-signaling response. Overexpression of *Zfp932* in MPLB cells dramatically reduced *Ptch1* and Gli1-luc induction by purmorphamine which activates the Hh pathway. On the other hand, knocking down *ZFP932* in MPLB cells elevated *Ptch1* and Gli1-luc basal- and Hh-induced expression. These studies appear at odds with the fact that ZFP932 directly represses *Ptch1* expression. As Patched1 is a negative regulator of Hh signaling and functions by inactivating Smoothened, suppressing Patched1 expression should elevate Hh signaling. Nonetheless, the data suggest that ZFP932 affects more than just *Ptch1* in the Hh pathway. RT-PCR survey of Hh-signaling component expression revealed that five genes, *Smo*, *Kif7*, *Gli1*, *Gli2*, and *Gli3*, are also regulated by ZFP932 (Fig. 14.6). Their expression was suppressed by ZFP932 overexpression, induced by VP16-ZFP932 and elevated upon *Zfp932* knockdown. Therefore, ZFP932 is able to dampen Hh signal response by repressing the expression of several Hh pathway components which may be unrelated to its repressive activity on the *Ptch1* promoter. Consistently, overexpression of *Zfp932* in MPLB cells also abolished Hh-induced gene expression changes in these five genes further supporting a suppressive role for ZFP932 in Hh signaling. Altogether, these studies implicate ZFP932 in dampening cellular Hh signal

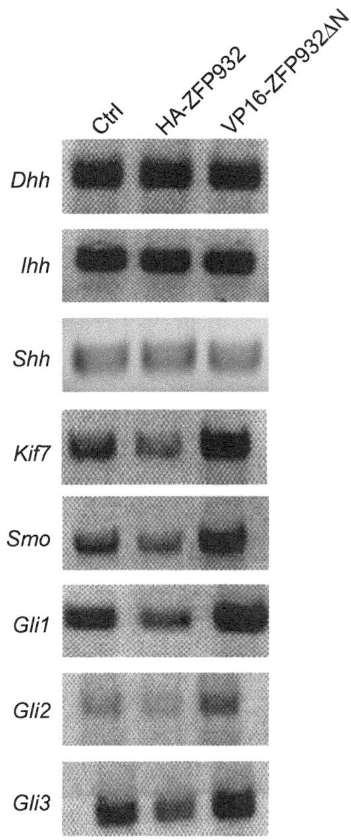

Figure 14.6 Hh components and ZFP932. *Dhh*, *Ihh*, and *Shh* expressions are not affected by ZFP932. On the other hand, expressions of *Kif7*, *Smo*, *Gli*, *2*, *3* are reduced by ZFP932 overexpression but elevated by the activator form of ZFP932, VP16-ZFP932ΔN which instead increases gene expression. This figure was originally published in the *Journal of Biological Chemistry*. He *et al*. (2011); © the American Society for Biochemistry and Molecular Biology. (For color version of this figure, the reader is referred to the Web version of this chapter.)

response by modulating the expression of several components of the signaling pathway (He *et al.*, 2011).

V. ZFP932 Regulates Osteoblast Differentiation

The demonstration that ZFP932 functions to dampen Hh signaling begs the question whether this finding has any physiological relevance. To address this question, we utilized the C3H10T1/2 cell system, a multipotent mesenchymal progenitor cell capable of differentiating into osteoblast upon growth factor stimulation (Wu *et al.*, 2002, 2004). In response to Shh signaling, C3H10T1/2 cells differentiate into osteoblasts and strongly express alkaline phosphatases (ALP). We treated C3H10T1/2 cells with purmorphamine to induce Hh signaling for 6 days and assayed for ALP activity. A sevenfold induction of ALP activity was observed between cells treated with purmorphamine and those treated with DMSO (control; Fig. 14.7A and B). Next, we tested the effect of *Zfp932* overexpression in this process by transfecting these C3H10T1/2 cells on the fourth day of treatment with either pMX-HA-ZFP932 or the control vector and measured ALP activity 48 h later. Overexpression of ZFP932 significantly attenuated purmorphamine-induced ALP activity in C3H10T1/2 cells. Similarly, transfection of VP16-ZFP932 which is an activator form of ZFP932 mimicked Hh activation and led to an increase in ALP activity in

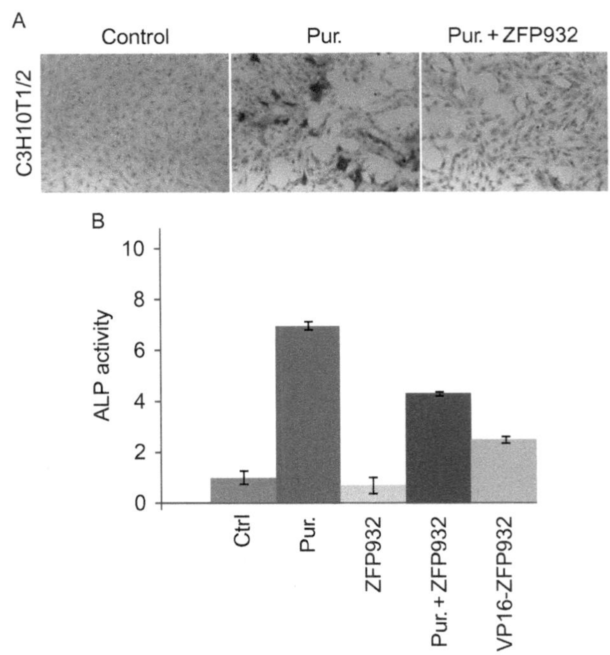

Figure 14.7 ZFP932 interferes with Hh-induced osteoblast differentiation. (A) Purmorphamine (Pur.) treatment induced osteoblast differentiation in C3H10T1/2 cells shown by ALP staining. Overexpression of ZFP932 partially blocked Hh induction shown by reduced ALP staining. (B) The level of osteoblast differentiation was quantitatively measured by the ALP activity. Purmorphamine induced a sevenfold increase in ALP activity which was much reduced when ZFP932 was overexpressed. The activator form (VP16-ZFP932ΔN) induced ALP activity in the absence of purmorphamine, mimicking Hh activation. (See Color Insert.)

the absence of purmorphamine treatment (Fig. 14.7B). These studies implicate important function for ZFP932 in inhibiting osteoblast differentiation.

VI. Conclusion

ZFP932 is a nuclear transcriptional repressor which represses gene transcription by recruiting HDACs and other corepressors to target promoters. The recruited HDAC1 and HDAC2 are known components of two well-studied repressive complexes, NuRD and Sin3A (Ahringer, 2000). While the exact repressive complex responsible for ZFP932 interaction is currently unknown, the N-terminal KRAB-A domain of ZFP932 is absolutely required for HDAC recruitment and subsequent transcriptional repression.

From a large list of target genes determined by microarray analyses, *Ptch1* was chosen for further study because of its involvement in Hh signaling. We found that ZFP932 represses *Ptch1* by direct binding to three ZFP932 binding sites in the mouse *Ptch1* promoter 1b as demonstrated by EMSA and chromatin immunoprecipitation. Mutations in all three sites were necessary to eliminate ZFP932-mediated transcriptional repression. The binding sites, moreover, did not overlap with the Gli binding sites in the *Ptch1* promoter (Agren *et al.*, 2004), suggesting independent regulation of *Ptch1* expression by ZFP932. This repressive regulation on *Ptch1* expression appears to be relevant *in vivo* as endogenous *Ptch1* expression in *Xenopus* and mouse embryos was downregulated upon ZFP932 overexpression. These results suggest that the function of ZFP932 in *Ptch1* regulation may be evolutionarily conserved.

In addition to repressing *Ptch1* expression, ZFP932 also represses the expression of several additional players in the Hh pathway including *Kif7*, *Smo*, *Gli1*, *Gli2*, and *Gli3*. Consequently, overexpression of ZFP932 in cultured cells dampened Hh response evidenced by reporter assays. Moreover, genetic knockdown of *Zfp932* sensitized these cells to Hh signaling implicating a functional role of ZFP932 in Hh signal transduction. Finally, in a physiologically relevant osteoblast differentiation model, ZFP932 expression attenuated Hh-induced osteoblast differentiation. Together, these data strongly implicate ZFP932 in regulating Hh signaling by affecting the expression of several key components of the Hh pathway through transcriptional repression.

ACKNOWLEDGMENTS

This work was supported by NIH Grant ES016597. The authors would like to thank Qian Cao and Daniel Scott Hayes for critical reading of the chapter.

REFERENCES

Agren, M., Kogerman, P., Kleman, M. I., Wessling, M., and Toftgard, R. (2004). Expression of the PTCH1 tumor suppressor gene is regulated by alternative promoters and a single functional Gli-binding site. *Gene* **330,** 101–114.

Ahringer, J. (2000). NuRD and SIN3 histone deacetylase complexes in development. *Trends Genet.* **16,** 351–356.

Bellefroid, E. J., Lecocq, P. J., Benhida, A., Poncelet, D. A., Belayew, A., and Martial, J. A. (1989). The human genome contains hundreds of genes coding for finger proteins of the Kruppel type. *DNA* **8,** 377–387.

Bellefroid, E. J., Poncelet, D. A., Lecocq, P. J., Revelant, O., and Martial, J. A. (1991). The evolutionarily conserved Kruppel-associated box domain defines a subfamily of eukaryotic multifingered proteins. *Proc. Natl. Acad. Sci. USA* **88,** 3608–3612.

Bhatia, N., Thiyagarajan, S., Elcheva, I., Saleem, M., Dlugosz, A., Mukhtar, H., and Spiegelman, V. S. (2006). Gli2 is targeted for ubiquitination and degradation by beta-TrCP ubiquitin ligase. *J. Biol. Chem.* **281,** 19320–19326.

Canettieri, G., Di Marcotullio, L., Greco, A., Coni, S., Antonucci, L., Infante, P., Pietrosanti, L., De Smaele, E., Ferretti, E., Miele, E., Pelloni, M., De Simone, G., *et al.* (2010). Histone deacetylase and Cullin3-REN(KCTD11) ubiquitin ligase interplay regulates Hedgehog signalling through Gli acetylation. *Nat. Cell Biol.* **12,** 132–142.

Chen, Y., Li, S., Tong, C., Zhao, Y., Wang, B., Liu, Y., Jia, J., and Jiang, J. (2010). G protein-coupled receptor kinase 2 promotes high-level Hedgehog signaling by regulating the active state of Smo through kinase-dependent and kinase-independent mechanisms in Drosophila. *Genes Dev.* **24,** 2054–2067.

Chiang, C., Litingtung, Y., Lee, E., Young, K. E., Corden, J. L., Westphal, H., and Beachy, P. A. (1996). Cyclopia and defective axial patterning in mice lacking Sonic hedgehog gene function. *Nature* **383,** 407–413.

Di Marcotullio, L., Ferretti, E., Greco, A., De Smaele, E., Po, A., Sico, M. A., Alimandi, M., Giannini, G., Maroder, M., Screpanti, I., and Gulino, A. (2006). Numb is a suppressor of Hedgehog signalling and targets Gli1 for Itch-dependent ubiquitination. *Nat. Cell Biol.* **8,** 1415–1423.

Di Marcotullio, L., Ferretti, E., Greco, A., De Smaele, E., Screpanti, I., and Gulino, A. (2007). Multiple ubiquitin-dependent processing pathways regulate hedgehog/gli signaling: Implications for cell development and tumorigenesis. *Cell Cycle* **6,** 390–393.

Eggenschwiler, J. T., Bulgakov, O. V., Qin, J., Li, T., and Anderson, K. V. (2006). Mouse Rab23 regulates hedgehog signaling from smoothened to Gli proteins. *Dev. Biol.* **290,** 1–12.

Goetz, S. C., Ocbina, P. J., and Anderson, K. V. (2009). The primary cilium as a Hedgehog signal transduction machine. *Methods Cell Biol.* **94,** 199–222.

He, Z., Cai, J., Lim, J. W., Kroll, K., and Ma, L. (2011). A novel KRAB domain-containing zinc finger transcription factor ZNF431 directly represses Patched1 transcription. *J. Biol. Chem.* **286,** 7279–7289.

Hering, T. M., Kazmi, N. H., Huynh, T. D., Kollar, J., Xu, L., Hunyady, A. B., and Johnstone, B. (2004). Characterization and chondrocyte differentiation stage-specific expression of KRAB zinc-finger protein gene ZNF470. *Exp. Cell Res.* **299,** 137–147.

Jiang, J. (2006). Regulation of Hh/Gli signaling by dual ubiquitin pathways. *Cell Cycle* **5,** 2457–2463.

Katoh, Y., and Katoh, M. (2006). Hedgehog signaling pathway and gastrointestinal stem cell signaling network (review). *Int. J. Mol. Med.* **18,** 1019–1023.

Kim, S. S., Chen, Y. M., O'Leary, E., Witzgall, R., Vidal, M., and Bonventre, J. V. (1996). A novel member of the RING finger family, KRIP-1, associates with the KRAB-A transcriptional repressor domain of zinc finger proteins. *Proc. Natl. Acad. Sci. USA* **93,** 15299–15304.

Kim, J., Kato, M., and Beachy, P. A. (2009). Gli2 trafficking links Hedgehog-dependent activation of Smoothened in the primary cilium to transcriptional activation in the nucleus. *Proc. Natl. Acad. Sci. USA* **106,** 21666–21671.

Kovacs, J. J., Hara, M. R., Davenport, C. L., Kim, J., and Lefkowitz, R. J. (2009). Arrestin development: Emerging roles for beta-arrestins in developmental signaling pathways. *Dev. Cell* **17,** 443–458.

Lum, L., and Beachy, P. A. (2004). The Hedgehog response network: Sensors, switches, and routers. *Science* **304,** 1755–1759.

Pabo, C. O., Peisach, E., and Grant, R. A. (2001). Design and selection of novel Cys2His2 zinc finger proteins. *Annu. Rev. Biochem.* **70,** 313–340.

Riobo, N. A., Lu, K., Ai, X., Haines, G. M., and Emerson, C. P., Jr. (2006). Phosphoinositide 3-kinase and Akt are essential for Sonic Hedgehog signaling. *Proc. Natl. Acad. Sci. USA* **103,** 4505–4510.

Schultz, D. C., Friedman, J. R., and Rauscher, F. J., 3rd. (2001). Targeting histone deacetylase complexes via KRAB-zinc finger proteins: The PHD and bromodomains of KAP-1 form a cooperative unit that recruits a novel isoform of the Mi-2alpha subunit of NuRD. *Genes Dev.* **15,** 428–443.

Sinha, S., and Chen, J. K. (2006). Purmorphamine activates the Hedgehog pathway by targeting Smoothened. *Nat. Chem. Biol.* **2,** 29–30.

Tempe, D., Casas, M., Karaz, S., Blanchet-Tournier, M. F., and Concordet, J. P. (2006). Multisite protein kinase A and glycogen synthase kinase 3beta phosphorylation leads to Gli3 ubiquitination by SCFbetaTrCP. *Mol. Cell. Biol.* **26,** 4316–4326.

Tukachinsky, H., Lopez, L. V., and Salic, A. (2010). A mechanism for vertebrate Hedgehog signaling: Recruitment to cilia and dissociation of SuFu-Gli protein complexes. *J. Cell Biol.* **191,** 415–428.

Urrutia, R. (2003). KRAB-containing zinc-finger repressor proteins. *Genome Biol.* **4,** 231.

Vissing, H., Meyer, W. K., Aagaard, L., Tommerup, N., and Thiesen, H. J. (1995). Repression of transcriptional activity by heterologous KRAB domains present in zinc finger proteins. *FEBS Lett.* **369,** 153–157.

Wang, B., and Li, Y. (2006). Evidence for the direct involvement of {beta}TrCP in Gli3 protein processing. *Proc. Natl. Acad. Sci. USA* **103,** 33–38.

Wu, X., Ding, S., Ding, Q., Gray, N. S., and Schultz, P. G. (2002). A small molecule with osteogenesis-inducing activity in multipotent mesenchymal progenitor cells. *J. Am. Chem. Soc.* **124,** 14520–14521.

Wu, X., Walker, J., Zhang, J., Ding, S., and Schultz, P. G. (2004). Purmorphamine induces osteogenesis by activation of the hedgehog signaling pathway. *Chem. Biol.* **11,** 1229–1238.

CHAPTER FIFTEEN

A New Era for an Ancient Drug: Arsenic Trioxide and Hedgehog Signaling

Elspeth M. Beauchamp *and* Aykut Üren

Contents

Abstract

Arsenic has been used for ages as a therapeutic agent. Currently, it is an FDA approved drug to treat acute promyelocytic leukemia where it leads to degradation of the PML–RAR fusion protein. It has been shown to have various other targets in cells such as JNK, NFκB, thioredoxin reductase, and MAPK pathways. Most of its effects in cells have been through arsenic's ability to bind to thiol groups in cysteine residues. Recent evidence has shown that arsenic can inhibit the Hedgehog pathway by inhibiting GLI proteins. The proposed mechanism of action is through direct binding. Potential binding sites include the critical cysteine residues in GLI zinc finger domains. The role of the Hedgehog pathway has been implicated in many cancers such as basal cell carcinoma, medulloblastoma, Ewing sarcoma, and rhabdoid tumors. Current Hedgehog pathway inhibitors have been fraught with resistance issues and so arsenic trioxide may provide an alternative therapy when combined with these other inhibitors or after acquired resistance.

Lombardi Comprehensive Cancer Center, Georgetown University Medical Center, Washington, District of Columbia, USA

Vitamins and Hormones, Volume 88
ISSN 0083-6729, DOI: 10.1016/B978-0-12-394622-5.00015-8

I. Historic Use of Arsenic in Medicine

Arsenic compounds have been used for more than 2000 years as a therapeutic for many diseases such as psoriasis, syphilis, and trypanosomiasis. Arsenic sulfides were first used by Hippocrates in approximately 400 BC to treat skin ulcers by putting a paste on them (Hippocrates and Adams, 1886). In approximately 70–90 AD, Dioscorides used arsenic as an antiseptic and depilatory. In his fifth book of De Materia Medica, he recommends it as an "antiseptic, astringent, and scab forming with a burning, strong, biting strength, and it is one of those medicines that repress abnormal growths and make the hair fall out" (Dioscorides, 2000). In the 1400s, arsenical compounds were used against the plague by Angelus Salva. Jean de Gorris was a French physician and used arsenic as a sudorific or agent to cause sweating in the 1500s as a way to relieve fever. In the 1600s, arsenic was used as an antimalarial agent (Haller, 1994). Starting in the seventeenth century, arsenic was the primary therapy for treating leukemia, specifically chronic myelogenous leukemia (CML) (Cutler and Bradford, 1879). In 1786, Thomas Fowler invented Fowler's solution, which was 1% arsenic trioxide (ATO) in potassium bicarbonate. This became the mainstay of nineteenth century medicine, where it was used to treat a variety of ailments including asthma, chorea, eczema, pemphigus, psoriasis, Hodgkin's lymphoma, and leukemia (Waxman and Anderson, 2001).

Paul Ehrlich is the father of the idea of targeted therapy. He theorized an agent could be fashioned as a "magic bullet" to specifically target a disease-causing organism. In 1910, Paul Ehrlich invented an arsenical compound called Salvarsan, which became a cure for both syphilis and trypanosomiasis and was his "magic bullet." However, what he thought his "magic bullet" was targeting was not in fact the real target. Salvarsan was used until the 1940s when the newly discovered penicillin supplanted it (Dilda and Hogg, 2007). Arsenic in the form of melarsoprol is still used today to treat trypanosomiasis (Dilda and Hogg, 2007; Pepin and Milord, 1994).

Arsenic's medicinal use in oncology declined upon the introduction of radiotherapy and chemotherapy (Waxman and Anderson, 2001). Much of its decline in use was due to concerns over toxicity and potential carcinogenicity. During the height of its use in the nineteenth century, some physicians recommended caution in using arsenic and that it should be used in small doses, whereas others recommended using it in heroic doses and that it should be used until "the toxic effects show themselves" (Haller, 1994). Often higher doses of arsenic than needed were probably used which lead to subsequent poisonings and toxicities. In the 1930s, CML patients were reported to develop chronic arsenic poisoning (Kandel and LeRoy, 1937). In the 1970s and 1980s, retrospective studies of patients who used Fowler's solution in the nineteenth and early twentieth centuries showed many developed dermatological diseases such as keratoses, hyperpigmentation, and carcinomas of the skin due to arsenic (Cuzick *et al.*, 1982; Jackson and Grainge, 1975). Some patients also developed other malignancies. One of the studies showed that toxicity correlated with greater age and higher doses of arsenic (Cuzick *et al.*, 1982). Also at this time, the International Agency for Research on Cancer (IARC) classified arsenic as group I "carcinogenic to humans" based on epidemiological evidence in humans and limited evidence of carcinogenicity in animal models (IARC Monographs, 1987a,b). In Bangladesh, tube wells were installed to access deeper water sources and prevent gastrointestinal diseases from groundwater. However, these water sources are contaminated with arsenic, and it is estimated that out of the 125+ million people in Bangladesh, 35–77 million are exposed to high levels of arsenic in their drinking water (Smith *et al.*, 2000). Several studies in places such as Bangladesh; West Bengal, India; Argentina; and Taiwan have linked the presence of arsenic in drinking water to the development of skin lesions and malignancies (Chen *et al.*, 1992; Guha Mazumder *et al.*, 1998; Hopenhayn-Rich *et al.*, 1998; Tondel *et al.*, 1999).

Arsenic compounds have been used in China as part of traditional Chinese medicine for many hundreds of years. In the past and still today, it is used in some forms to devitalize the pulp of diseased teeth, and as a treatment for psoriasis, rheumatic diseases, and syphilis (Miller *et al.*, 2002; Waxman and Anderson, 2001). The end of the use of arsenic in western medicine would have occurred if not for Chinese studies in the 1970s and 1980s that showed that ATO was particularly effective at treating acute promyelocytic leukemia (APL) (Douer and Tallman, 2005). A study that summarized the results of 242 APL patients treated with ATO in China showed a complete response rate of approximately 75% (Zhang, 1999). Because of the success of these studies, it leads to studies conducted in the USA. Results from a pilot study and the subsequent multicenter trial showed a combined complete response rate of 87% (Soignet *et al.*, 1998, 2001). In 2000, the US Food and Drug Administration (FDA) approved ATO for the treatment of APL in patients who do not respond or relapse on all-*trans*-retinoic acid (ATRA) therapy. Currently, there are multiple clinical trials throughout the USA to investigate the potential of ATO alone or combined with other agents for the treatment of other leukemia types, myelodysplastic syndromes, multiple myeloma, and various solid tumor types (Dilda and Hogg, 2007; Douer and Tallman, 2005; Table 15.1).

II. Molecular Mechanisms of Arsenic in Cancer Therapy

ATO is a FDA approved treatment for APL. APL is a type of acute myeloid leukemia (AML) that is characterized by a specific translocation occurring between chromosomes 15 and 17, which results in the fusion

Table 15.1 Currently active US clinical trials using ATO that are registered with NIH (from www.clinicaltrials.gov)

Agents used	Phase	Cancer type
ATO, decitabine, and ascorbic acid	I	MDS, AML
ATO, temozolomide, and radiation therapy	I and II	Brain and central nervous system tumors
Disulfiram and ATO	I	Metastatic melanoma
ATO	II	Endometrial carcinoma
Oral ATO and ascorbic acid	I	Myelofibrosis
ATO	II	CLL/SLL
Tamibarotene and ATO	I	APL
ATO, autologous transplantation, allogenic transplantation, and ATRA	IV	APL
ATO, Ara-C	0	Myelofibrosis
ATRA, ATO, and idarubicin	II	APL
ATO, Ara-C, daunorubicin, mercaptopurine, methotrexate, and ATRA	II	APL
ATO, Ara-C, idarubicin, mercaptopurine, methotrexate, mitoxantrone, and ATRA	III	APL
ATO, gemtuzumab, mercaptopurine, methotrexate, and ATRA	II	APL
ATO and vitamin D	II	MDS

Trial start dates span from March 2005 to July 2010.

protein promyelocytic leukemia (PML)–retinoic acid receptor (RAR). In APL patients, the first line of therapy is to treat with ATRA, which causes the leukemia cells to differentiate into normal hematopoietic cells. In APL cells, the PML–RAR protein acts like a dominant negative RAR in the absence of ATRA. Therefore, when cells are treated with ATRA they differentiate due to activation of RAR target genes (Kakizuka *et al.*, 1991). ATO's proposed mechanism of action in APL is via degradation of the PML–RAR fusion protein (Nasr *et al.*, 2008). Retinoic acid targets the RAR portion of the fusion protein, whereas ATO appears to be targeting the PML part of the protein. In APL, the fusion protein alters the localization of PML as well as other proteins from nuclear bodies, which contributes to aberrant cell growth (Dyck *et al.*, 1994; Weis *et al.*, 1994). Arsenic targets both PML and PML–RAR to nuclear bodies in APL cells and leads to subsequent degradation (Fig. 15.1) (Zhu *et al.*, 1997). Targeting PML

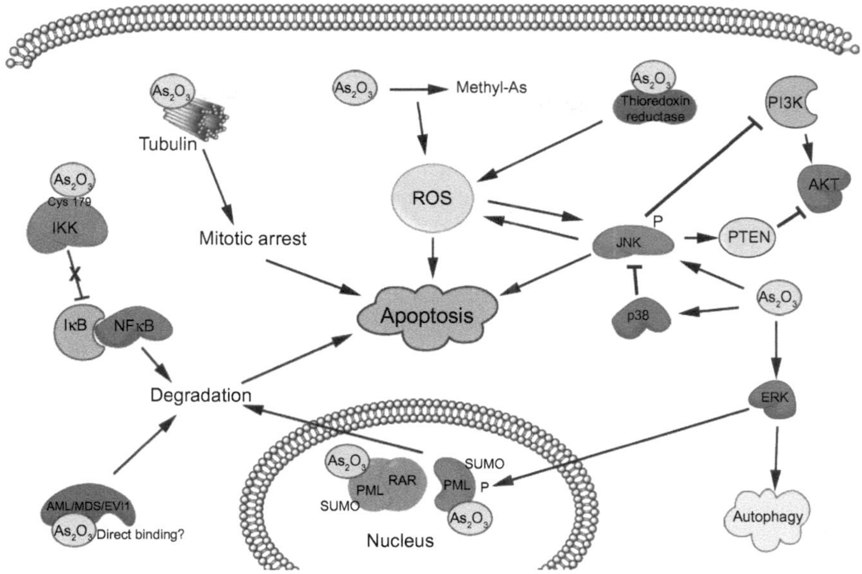

Figure 15.1 Pathways in cells involved in ATO cytotoxicity.

degradation with arsenic has been shown in CML as well (Ito *et al.*, 2008). A recent publication has shown that arsenic specifically binds to the PML zinc finger domain at cysteine residues displacing the zinc and causing a shift in secondary structure as well as aggregation that leads to increased sumoylation and subsequent degradation (Jeanne *et al.*, 2010; Zhang *et al.*, 2010). Trivalent arsenic, which is the form found in ATO or other medicinal arsenic compounds, readily binds to thiol groups in cysteines and thus alters proper protein folding, structure, or function (Dilda and Hogg, 2007; Jiang *et al.*, 2003; Ramadan *et al.*, 2009).

In another subtype of AML, arsenic has been shown to degrade another fusion protein AML1/MDS1/EVI1 (Fig. 15.1; Shackelford *et al.*, 2006). It was shown that arsenic specifically targets the MDS1 and EVI1 portion of the protein. The EVI1 portion contains two zinc finger DNA-binding domains. Zinc finger DNA-binding domains require cysteine residues for coordination of zinc metal ions. Since arsenic has a propensity to bind to thiol groups, and given the evidence with PML that the arsenic binds to the zinc finger domain, then it is possible that arsenic is also binding to the cysteine residues. This is therefore causing the protein to become degraded.

The NFκB pathway has also shown to be inhibited by ATO in cancer cells (Fig. 15.1). When the NFκB pathway is not active, IκB binds to NFκB and this leads to the degradation of NFκB. In cancer, NFκB can be activated by cytokines or TNF-α, which leads to evasion of apoptosis. Activation of this pathway either through TRAF2 in the case of TNF-α or PI3K/AKT leads to activation of IKK. IKK is responsible for phosphorylation of the negative regulator IκB. IκB no longer binds to NFκB, where it is now able to translocate to the nucleus and activate genes required for the evasion of apoptosis and cell proliferation. In multiple myeloma cells, arsenic was shown to prevent NFκB activation by TNF-α (Hayashi *et al.*, 2002). IKK is inhibited by arsenic directly binding to it. Interestingly, arsenic specifically binds to cysteine residue 179 in the activation loop of the catalytic subunit of IKK and inhibits its activity (Kapahi *et al.*, 2000).

Much of trivalent arsenic's effects in cells are due to its ability to bind to thiol groups. Arsenic can bind to cysteines in microtubules and other cytoskeletal proteins to effect polymerization and mitosis (Fig. 15.1; Li and Broome, 1999; Ramirez *et al.*, 1997). Arsenic can also increase production of reactive oxygen species (ROS), which is in part due to its ability to bind to enzymes involved in regulating the cellular redox state. Arsenic directly binds to thioredoxin reductase via its sulfur selenium group and inhibits its activity (Fig. 15.1; Lu *et al.*, 2007). Arsenic metabolism also leads to the production of ROS when it is methylated (Fig. 15.1; Kojima *et al.*, 2009; Tezuka *et al.*, 1993).

Arsenic has also shown to affect various branches of the MAPK pathway in part through production of ROS, which includes p38, JNK, and ERK (Fig. 15.1). ATO increases activation of p38 MAPK in several leukemia types (Giafis *et al.*, 2006). However, inhibition of p38 or its downstream effectors attenuated the cytotoxic effects of ATO and increased JNK activation (Dolniak *et al.*, 2008; Kannan-Thulasiraman *et al.*, 2006; Verma *et al.*, 2002). This indicates that p38 is activated in a negative feedback loop in cancer cells. JNK activation appears to be an important event in ATO-induced apoptosis in leukemia cells (Davison *et al.*, 2004; Huang *et al.*, 1999; Mann *et al.*, 2006). ATO-resistant APL cell lines showed little activation of JNK due to upregulation of glutathione (GSH) (Davison *et al.*, 2004). Treating cell with compounds that deplete GSH in cells enhances ATO's cytotoxic effects (Chen *et al.*, 2006; Davison *et al.*, 2004). Increased GSH has been correlated with a lack of sensitivity to arsenic (Dai *et al.*, 1999; Davison *et al.*, 2003; Jing *et al.*, 1999). This effect could be through GSH directly binding arsenic as part of its metabolism and excretion or through decreasing the amount of ROS in the cell. In chronic lymphocytic leukemia (CLL), JNK activation was an early event leading to upregulation of PTEN, which results in PI3K, AKT, and NFκB inhibition (Redondo-Munoz *et al.*, 2010). ROS production in this model was secondary to JNK activation. Nuclear receptor function has been shown to be inhibited by ATO though JNK activation and a subsequent decrease of phosphorylation by the retinoid X receptor (RXR) (Mann *et al.*, 2005). Autophagy is shown to be activated by arsenic in leukemia cells and that the ERK pathway is required for this activation (Goussetis *et al.*, 2010). ATO-dependent ERK2 phosphorylation of PML has also been shown to lead to increased sumoylation/degradation of the PML protein and subsequently lead to apoptosis (Hayakawa and Privalsky, 2004).

III. Role of Hedgehog Signaling in Cancer

While long recognized as critical in developmental biology, Hedgehog (Hh) signaling has become a pathway of intense study in oncology. The Hh pathway is activated in several cancers such as basal cell carcinoma (BCC) and medulloblastoma (Berman *et al.*, 2002; Di Marcotullio *et al.*, 2004; Hahn *et al.*, 1996; Hatton *et al.*, 2008; Kimura *et al.*, 2005; Pietsch *et al.*, 1997; Regl *et al.*, 2002; Reifenberger *et al.*, 1998; Taylor *et al.*, 2002). The pathway is composed of three Hh ligands (Sonic, Indian, and Desert Hh) that bind to the Patched receptor. In the absence of Hh ligand, Patched inhibits another transmembrane protein, Smoothened (SMO) (Fig. 15.2). When Patched is engaged by the Hh ligand, SMO is activated due to diminished inhibitory signal from Patched. The signal is then transduced to the important downstream effectors GLI1, GLI2, and GLI3, which act as transcription factors (Fig. 15.3).

While many Hh-driven human cancers contain either inactivated Patched1 (PTCH) or activation of SMO, some tumors activate the pathway by increasing the activity of GLI. Loss of inhibitory molecules SUFU/REN has been described in medulloblastomas (Di Marcotullio *et al.*, 2004; Taylor *et al.*, 2002). There are also cases of *GLI* gene amplifications in

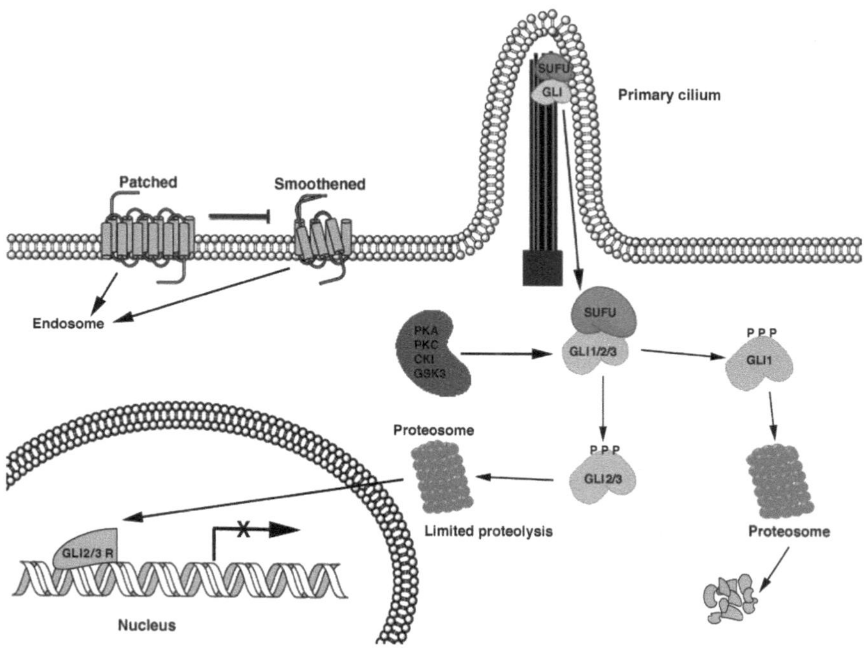

Figure 15.2 Hedgehog pathway without ligand activation.

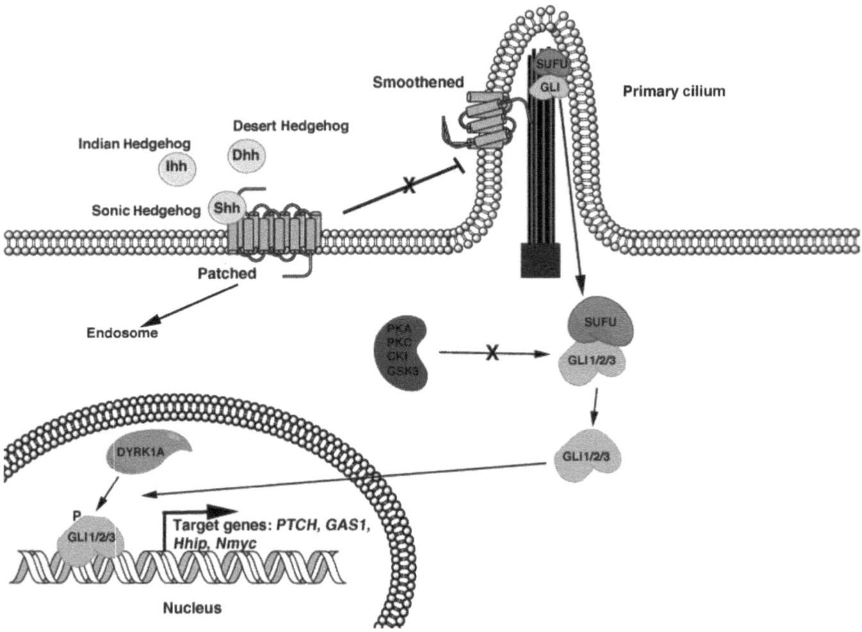

Figure 15.3 Hedgehog pathway with ligand activation.

rhabdomyosarcoma, osteosarcoma, glioma, and breast cancer (Kinzler *et al.*, 1987; Nessling *et al.*, 2005; Roberts *et al.*, 1989); chromosomal translocations involving *GLI* in pericytoma (Dahlen *et al.*, 2004); and increased GLI protein stability in prostate cancer (Bhatia *et al.*, 2006) that results in increased GLI function. Recent evidence has also shown that Hh-independent activation of GLI1 by the RAS or transforming growth factor-β (TGF-β) signaling pathways is important for tumor formation of many cancer types, including pancreas, lung, and colon (Dennler *et al.*, 2007; Ji *et al.*, 2007; Nolan-Stevaux *et al.*, 2009; Pasca di Magliano *et al.*, 2006; Stecca *et al.*, 2007). Even so, the effect of RAS and TGF–β signaling on GLI1 activation has not been shown to be a direct mechanism. In Ewing sarcoma, there is a novel mechanism of Hh-independent GLI1 activation whereby EWS-FLI1 directly activates GLI1 by binding to the *GLI1* promoter, leading to an increase in abundance of the GLI1 protein. Most Ewing sarcoma cell lines showed no detectable Hh ligand expression or very low levels compared with the control. Consistent with this finding, cyclopamine did not inhibit GLI1 transcriptional activity in Ewing sarcoma cells (Beauchamp *et al.*, 2009).

Recent evidence has shown that primary cilia play an important role for Hh pathway signal transduction (Corbit *et al.*, 2005; Han *et al.*, 2008; Huangfu *et al.*, 2003; Liu *et al.*, 2005; May *et al.*, 2005; Rohatgi *et al.*, 2007;

Spassky *et al.*, 2008). Primary cilia are immotile, antenna-like structures made up of microtubules found in any cell of the body that function to receive both mechanical and chemical signals from other cells and the environment, and to further transmit these signals. Genetic mutations that disrupt the function of primary cilia result in many disorders, including polycystic kidney disease, hepatic and pancreatic abnormalities, skeletal malformations, obesity, and severe developmental defects (Michaud and Yoder, 2006). This cilia–Hh connection was first discovered when mice carrying mutations for intraflagellar transport proteins required for cilia formation showed developmental defects similar to that seen in Hh signaling mutants (Huangfu *et al.*, 2003; Liu *et al.*, 2005). SMO is localized to primary cilia during pathway activation and in the absence of ligand is recycled via the endosome (Figs. 15.2 and 15.3). In medulloblastoma and BCC, it has been shown that primary cilia are required for tumorigenesis in cases of upstream pathway activation such as constitutively active SMO. However, if a tumor is activated at the level of GLI, then the presence of primary cilia is inhibitory (Wong *et al.*, 2009; Yauch *et al.*, 2009). Primary cilia have also been shown to be important for PDGF signaling transduction in addition to Hh signaling (Schneider *et al.*, 2005).

In cases of epithelial tumors, as found in prostate and pancreatic cancers, there is debate about whether the expression of ligand by these cancers acts in an autocrine manner or in a paracrine manner on the stroma. A recent study by Yauch *et al.* argues that epithelial tumors do not respond to Hh ligand but the stroma does (Yauch *et al.*, 2008). However, this does not exclude the small population of what is considered epithelial cancer stem cells that appear to have active Hh signaling and the ability to be affected with Hh pathways drugs (Feldmann *et al.*, 2008; Stecca and Ruiz i Altaba, 2009). Interestingly, the more stem-like pancreatic cancer cells exhibit a greater mesenchymal phenotype (Rasheed *et al.*, 2010). In the prostate, active Hh signaling is found only in advanced cancers (Sheng *et al.*, 2004). The Yauch *et al.* study also does not account for the fact that some epithelial tumors may have downstream pathway activations at the level of GLI, and inhibiting Hh ligand or SMO will not affect these tumors.

Posttranslational modification of GLI is important for GLI activity. Phosphorylation by several kinases by GLI1 has an inhibitory effect in the absence of active upstream pathway (Cai *et al.*, 2009; Riobo *et al.*, 2006). However, it has been determined that phosphorylation of GLI1 in the nucleus by Dryk1 kinase contributes to the retention of GLI1 in the nucleus and enhances its transcriptional activity (Mao *et al.*, 2002). A recent paper reveals the importance of acetylation on GLI1 and GLI2 activity (Canettieri *et al.*, 2010). In the absence of upstream pathway activation, GLI1 is acetylated, a process which inhibits its activity. The Hh pathway inhibitory member REN links Cul3 E3 ligase to HDAC1, which stimulates its ubiquitination and degradation via the proteosome. During pathway activation, REN is no longer able to inhibit HDAC1; HDAC1 then deacetylates GLI1 and increases its transcriptional activity. In addition, REN has also been shown to antagonize Dryk1-associated GLI1 translocation to the nucleus (Di Marcotullio *et al.*, 2004).

IV. Effect of Arsenic on Hh Signaling

Recent work has shown that arsenic can inhibit Hh signaling at the level of GLI. The inhibition was shown to be at the level of GLI because ATO was able to inhibit activated Hh signaling in $SUFU^{-/-}$ MEFS, as well as when GLI1 or GLI2 was overexpressed. SMO or other upstream pathway inhibitors cannot inhibit Hh signaling in this context (Beauchamp *et al.*, 2011; Kim *et al.*, 2010). Also ATO is able to inhibit the growth of both upstream activated medulloblastoma cell lines as well as Ewing sarcoma cells lines which have activation of GLI1 irrespective of SMO. One study examined the effects of ATO specifically on GLI2. It was shown that ATO inhibits GLI2 accumulation in cilia and long-term treatment lead to GLI2 reduced protein stability (Kim *et al.*, 2010). However, GLI1 activity was inhibited before any protein degradation was observed, which does not corroborate an effect on GLI protein stability. Further, ATO was able to inhibit GLI1 function independent of primary cilia and even with GLI1 trapped in nucleus (Beauchamp *et al.*, 2011). Therefore, preventing GLI1 localization to cilia is unlikely to be the primary mechanism of action. One has to acknowledge the differences may be due to cell line selection or experimental methods. Also functional differences between GLI1 and GLI2 and that they are processed differently in the cell may also account for these differences. Even though there were some apparent paradoxical observations related to mechanisms of action, both studies conclude that ATO inhibits Hh signaling at the level of GLI transcription factors.

Earlier work showed that ATO is cytotoxic in ESFT cells at similar concentrations but did not show an effect on GLI proteins. Jung *et al.* showed ATO-induced apoptosis via an increase in the phosphorylation status of JNK (Jung *et al.*, 2006). In the ESFT cell line TC71, ATO did activate JNK, but if a pharmacological inhibitor of JNK was used in combination with ATO, it inhibited JNK activation but not ATO cytotoxicity or its ability to inhibit GLI1 activity. Matheiu *et al.* provided evidence that ATO caused apoptosis of ESFT in a p53-independent manner and proposed that ATO inhibition of NFκB was the possible mechanism (Mathieu and Besancon, 2006). However, NFκB knockdown in ESFT cells has no effect on viability under normal growth conditions (White and Burchill, 2008). ATO did not inhibit NFκB in ESFT cell line TC71, and activation of NFκB did not alter cell viability in the presence or absence of ATO. Therefore, it is unlikely that inhibition of NFκB by ATO is the molecular mechanism for ATO-mediated ESFT cytotoxicity.

Prior investigations have established GLI1 as an important transcription factor for ESFT tumorigenesis (Beauchamp *et al.*, 2009; Joo *et al.*, 2009; Zwerner *et al.*, 2007). GLI activation as a downstream target of the Hh pathway has been actively studied in tumors with constitutively activated SMO (due to *PTCH1* mutation or loss) such as BCC and medulloblastoma. ATO is cytotoxic in ESFT and medulloblastoma cells, which is most likely at least in part through the inhibition of GLI proteins. The exact mechanism of this inhibition still needs further investigation. As mentioned previously, arsenic has been shown throughout the literature that it can bind to thiol groups in cysteines. Arsenic can bind to PML–RAR by directing binding to cysteine residues in the zinc finger domain of the PML portion of the fusion protein (Zhang *et al.*, 2010). The authors showed that this interaction occurs in the nucleus and results in a conformational change that allows increased sumoylation and subsequent ubiquitilation leading to degradation. GLI1 is a cysteine-rich protein (28 total residues) and has five zinc finger binding domains that contain cysteine residues. Given the evidence, it is highly plausible ATO binds to the zinc finger domains in GLI1 or other cysteines in the protein. How this direct binding inhibits GLI is still unknown and necessitates further investigation. The effect could occur in the nucleus by preventing protein–protein interactions of transcriptional coactivators that bind to GLI1. Another possible mechanism may involve increased acetylation of GLI1 by ATO, thus inhibiting its activity.

A recent study showed that arsenic activated Hh signaling in primary lung cells and NIH3T3 cells and tried to link this discovery to arsenic-induced bladder cancer (Fei *et al.*, 2010). The authors showed that arsenic activated GLI1/Hh signaling in these cells by inhibiting the GLI3 repressor. However, in this study sodium arsenite was used, which is a different compound than ATO. Perhaps sodium arsenite has preferential binding to GLI3 over GLI1 and GLI2 and thus activates signaling instead of repressing it. Also in specific cell types that do not have GLI1 or GLI2 activation perhaps inhibiting GLI3, will activate GLI1/2. However, it would be possible that long-term treatment of these cells would eventually lead to inhibition of all GLI proteins expressed. The difference in activation versus repression could also be species specific, given that a majority of the data in the study used mouse cells instead of human cells.

In APL where arsenic is an FDA approved treatment, the role of GLI/Hh signaling has not been clearly established. There has been only one report that investigated the expression of Hh pathway components in APL patients (Bai *et al.*, 2008). In this study, 100% of APL cases expressed GLI1. However, their cohort included only three APL cases and the antibody used for immunohistochemistry has not been properly validated for specific detection of GLI1. A study needs to be done with a larger sample size and GLI1 expression confirmed by other methods such as *in situ* hybridization, Western blot, or qPCR. If the results are confirmatory, it is plausible that ATO's effect on APL cells could be the result of inhibiting both PML–RAR and GLI.

Many studies have demonstrated that ATO is cytotoxic in other leukemia types besides APL. Historically, arsenic was used to treat CML up until the 1930s. The presence of the fusion protein BCR–ABL in CML was shown to correlate with ATO sensitivity (Puccetti *et al.*, 2000). Other studies demonstrated that the combination of imatinib and arsenic sulfide was better than either agent alone in a mouse model of CML (Zhang *et al.*, 2009). A recent publication indicates that expansion of the BCR–ABL leukemic stem cells was dependent on SMO and active Hh signaling. They propose SMO as a potential drug target for patients who relapse after imatinib treatment (Dierks *et al.*, 2008). However, in CLL cells, it was shown that inhibition of GLI1 pharmacologically or by siRNA but not SMO induced apoptosis (Desch *et al.*, 2010). Another recent study suggests an effect of GLI1, independent of SMO, on the regulation of proliferation and differentiation of hematopoietic stem cells and myeloid progenitors (Merchant *et al.*, 2010). These studies also indicated that using SMO inhibitors may not be sufficient in treating all leukemia and that GLI inhibitors may be more effective. ATO was shown to be cytotoxic in CLL and the main proposed mechanism was through PI3 kinase inhibition and JNK activation (Redondo-Munoz *et al.*, 2010). However, since GLI1 is important in CLL, one could propose that it may cytotoxic through inhibiting both pathways. Due to increasing evidence for the role of Hh signaling in leukemia, the cytotoxicity of ATO may be in part due to its inhibition of GLI in the leukemia stem cell compartment.

Recent evidence has shown that GLI activation, independent of upstream pathway components, in Snf5 mutant rhabdoid tumors is important for growth of these tumor cells (Jagani *et al.*, 2010). Therefore, inhibition of GLI by ATO in these cells could be a potential therapy.

In BCCs, mutations in Patched or SMO that lead to activation of the Hh pathway have been shown to play an important role in cancer formation (Hahn *et al.*, 1996; Regl *et al.*, 2002; Reifenberger *et al.*, 1998). A Dutch population study examined the correlation between cancer risk and the presence of low doses of arsenic concentrations in drinking water. In contrast to the studies done in Bangladesh and other parts of the world with higher doses of arsenic, they found that arsenic did not increase the risk of multiple cancer types. In fact, increasing concentration of arsenics in the water correlated with a decreased risk of non-melanoma skin cancers (Baastrup *et al.*, 2008). This reduction of non-melanoma skin cancer (which includes BCC) may be due to genetic differences as these two populations are ethnically different. However, the risk may be partially due to the effects of arsenic on Hh signaling.

A recent phase I trial of an Hh pathway inhibitor, GDC-0449, that targets SMO, showed promising results in BCC (Von Hoff *et al.*, 2009). As a result, ATO potentially could be a new therapy for BCC as well. The Hh pathway inhibitor GDC-0449 used in the BCC trial was also administered to one medulloblastoma patient with metastatic disease who was refractory to multiple treatments. The patient first showed a remarkable regression of their multifocal metastases after 2 months of treatment but eventually relapsed (Rudin *et al.*, 2009). The cause of the relapse was discovered to be a mutation in SMO that prevented the binding of the drug (Fig. 15.4). The same resistance to the drug was also recapitulated in a medulloblastoma mouse model (Yauch *et al.*, 2009). ATO was able to inhibit cells with expression of this specific SMO mutant *in vitro* (Kim *et al.*, 2010). Two other subsequent SMO inhibitors have been evaluated in mouse models. In this case, the tumors did not develop resistance because the drug did not bind to SMO but either had GLI2 amplification or activation of the PI3 kinase pathway (Fig. 15.4; Buonamici *et al.*, 2010; Dijkgraaf *et al.*, 2011). Since ATO is a GLI inhibitor and inhibits the pathway downstream of SMO, it thus may be an effective therapy in medulloblastoma patients that develop resistance to SMO inhibitors (Fig. 15.4). It is possible that combining the two agents up front could also prevent resistance in the first place. For those that develop resistance due to PI3 kinase pathway activation, ATO may be effective in these tumors as it has been shown to downregulate PI3 kinase signaling in CLL and therefore with ATO you may be able to target both pathways.

V. Conclusions

In summary, there is strong evidence that ATO inhibits GLI protein function. ATO-mediated inhibition of GLI is independent of upstream Hh signaling. ATO exhibited marked cytotoxicity on ESFT and medulloblastoma cell lines *in vitro* and showed promising effects *in vivo* on ESFT and medulloblastoma xenografts, as well as a transgenic medulloblastoma mouse model. The pharmacokinetics from a published pediatric phase I trial indicates that the peak ATO levels at the maximum tolerated dose are in the 0.5-μM range, with prolonged exposure to levels of about 0.1 μM (Fox *et al.*, 2008). The average IC50 range for ESFT and medulloblastoma cell lines is right within this range at 0.7 μM. Currently, there are at least 44 Hh pathway inhibitors in development. Of these, 29 are SMO inhibitors and only 4 are classified as GLI inhibitors (Tremblay *et al.*, 2009). There is growing evidence that current SMO inhibitors will not suffice as single agent therapies due to acquired resistance as well their inability to inhibit the growth of cancers with activation at the level of GLI. Continued preclinical

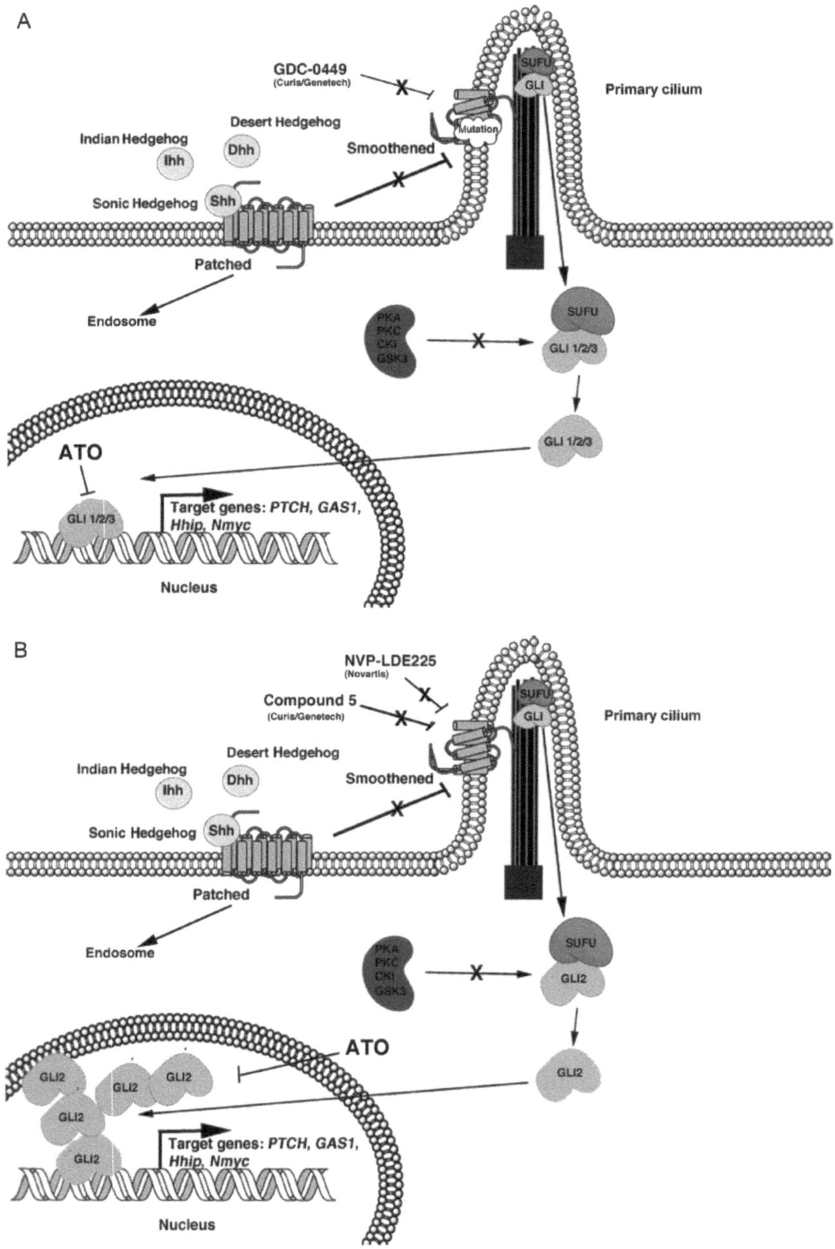

Figure 15.4 Mechanisms of resistance of Hedgehog pathway inhibitors. (A) Resistance to Smoothened inhibitor GDC-0449. (B) Resistance to Smoothened inhibitors compound 5 and NVP-LDE225.

evaluation of ATO in tumors that either have upstream pathway activation or transcriptional activation of GLI proteins is warranted. Because of the significant clinical experience with ATO, further insight into the mechanism of action of ATO on GLI should lead to extended clinical applications of ATO as a Hh pathway inhibitor.

REFERENCES

Genetic and related effects: An updating of selected IARC monographs from volumes 1 to 42. (1987a). *IARC Monogr. Eval. Carcinog. Risks Hum. Suppl.* **6,** 1–729.

Overall evaluations of carcinogenicity: An updating of IARC monographs volumes 1 to 42. (1987b). *IARC Monogr. Eval. Carcinog. Risks Hum. Suppl.* **7,** 1–440.

Baastrup, R., Sorensen, M., Balstrom, T., Frederiksen, K., Larsen, C. L., Tjonneland, A., Overvad, K., and Raaschou-Nielsen, O. (2008). Arsenic in drinking-water and risk for cancer in Denmark. *Environ. Health Perspect.* **116,** 231–237.

Bai, L. Y., Chiu, C. F., Lin, C. W., Hsu, N. Y., Lin, C. L., Lo, W. J., and Kao, M. C. (2008). Differential expression of Sonic hedgehog and Gli1 in hematological malignancies. *Leukemia* **22,** 226–228.

Beauchamp, E., Bulut, G., Abaan, O., Chen, K., Merchant, A., Matsui, W., Endo, Y., Rubin, J. S., Toretsky, J., and Uren, A. (2009). GLI1 is a direct transcriptional target of EWS-FLI1 oncoprotein. *J. Biol. Chem.* **284,** 9074.

Beauchamp, E. M., Ringer, L., Bulut, G., Sajwan, K. P., Hall, M. D., Lee, Y. C., Peaceman, D., Ozdemirli, M., Rodriguez, O., Macdonald, T. J., Albanese, C., Toretsky, J. A., *et al.* (2011). Arsenic trioxide inhibits human cancer cell growth and tumor development in mice by blocking Hedgehog/GLI pathway. *J. Clin. Invest.* **121,** 148–160.

Berman, D. M., Karhadkar, S. S., Hallahan, A. R., Pritchard, J. I., Eberhart, C. G., Watkins, D. N., Chen, J. K., Cooper, M. K., Taipale, J., Olson, J. M., and Beachy, P. A. (2002). Medulloblastoma growth inhibition by hedgehog pathway blockade. *Science* **297,** 1559–1561.

Bhatia, N., Thiyagarajan, S., Elcheva, I., Saleem, M., Dlugosz, A., Mukhtar, H., and Spiegelman, V. S. (2006). Gli2 is targeted for ubiquitination and degradation by beta-TrCP ubiquitin ligase. *J. Biol. Chem.* **281,** 19320–19326.

Buonamici, S., Williams, J., Morrissey, M., Wang, A., Guo, R., Vattay, A., Hsiao, K., Yuan, J., Green, J., Ospina, B., Yu, Q., Ostrom, L., *et al.* (2010). Interfering with resistance to smoothened antagonists by inhibition of the PI3K pathway in medulloblastoma. *Sci. Transl. Med.* **2,** 51ra70.

Cai, Q., Li, J., Gao, T., Xie, J., and Evers, B. M. (2009). Protein kinase Cdelta negatively regulates hedgehog signaling by inhibition of Gli1 activity. *J. Biol. Chem.* **284,** 2150–2158.

Canettieri, G., Di Marcotullio, L., Greco, A., Coni, S., Antonucci, L., Infante, P., Pietrosanti, L., De Smaele, E., Ferretti, E., Miele, E., Pelloni, M., De Simone, G., *et al.* (2010). Histone deacetylase and Cullin3-REN(KCTD11) ubiquitin ligase interplay regulates Hedgehog signalling through Gli acetylation. *Nat. Cell Biol.* **12,** 132–142.

Chen, C. J., Chen, C. W., Wu, M. M., and Kuo, T. L. (1992). Cancer potential in liver, lung, bladder and kidney due to ingested inorganic arsenic in drinking water. *Br. J. Cancer* **66,** 888–892.

Chen, D., Chan, R., Waxman, S., and Jing, Y. (2006). Buthionine sulfoximine enhancement of arsenic trioxide-induced apoptosis in leukemia and lymphoma cells is mediated via activation of c-Jun NH2-terminal kinase and up-regulation of death receptors. *Cancer Res.* **66,** 11416–11423.

Corbit, K. C., Aanstad, P., Singla, V., Norman, A. R., Stainier, D. Y., and Reiter, J. F. (2005). Vertebrate Smoothened functions at the primary cilium. *Nature* **437,** 1018–1021.

Cutler, E. G., and Bradford, E. H. (1879). Changes of the globular richness of human blood. *J. Physiol.* **1,** 427–433.

Cuzick, J., Evans, S., Gillman, M., and Price Evans, D. A. (1982). Medicinal arsenic and internal malignancies. *Br. J. Cancer* **45,** 904–911.

Dahlen, A., Fletcher, C. D., Mertens, F., Fletcher, J. A., Perez-Atayde, A. R., Hicks, M. J., Debiec-Rychter, M., Sciot, R., Wejde, J., Wedin, R., Mandahl, N., and Panagopoulos, I. (2004). Activation of the GLI oncogene through fusion with the beta-actin gene (ACTB) in a group of distinctive pericytic neoplasms: Pericytoma with t(7;12). *Am. J. Pathol.* **164,** 1645–1653.

Dai, J., Weinberg, R. S., Waxman, S., and Jing, Y. (1999). Malignant cells can be sensitized to undergo growth inhibition and apoptosis by arsenic trioxide through modulation of the glutathione redox system. *Blood* **93,** 268–277.

Davison, K., Cote, S., Mader, S., and Miller, W. H. (2003). Glutathione depletion overcomes resistance to arsenic trioxide in arsenic-resistant cell lines. *Leukemia* **17,** 931–940.

Davison, K., Mann, K. K., Waxman, S., and Miller, W. H., Jr. (2004). JNK activation is a mediator of arsenic trioxide-induced apoptosis in acute promyelocytic leukemia cells. *Blood* **103,** 3496–3502.

Dennler, S., Andre, J., Alexaki, I., Li, A., Magnaldo, T., ten Dijke, P., Wang, X. J., Verrecchia, F., and Mauviel, A. (2007). Induction of sonic hedgehog mediators by transforming growth factor-beta: Smad3-dependent activation of Gli2 and Gli1 expression in vitro and in vivo. *Cancer Res.* **67,** 6981–6986.

Desch, P., Asslaber, D., Kern, D., Schnidar, H., Mangelberger, D., Alinger, B., Stoecher, M., Hofbauer, S. W., Neureiter, D., Tinhofer, I., Aberger, F., Hartmann, T. N., *et al.* (2010). Inhibition of GLI, but not Smoothened, induces apoptosis in chronic lymphocytic leukemia cells. *Oncogene* **29,** 4885–4895.

Di Marcotullio, L., Ferretti, E., De Smaele, E., Argenti, B., Mincione, C., Zazzeroni, F., Gallo, R., Masuelli, L., Napolitano, M., Maroder, M., Modesti, A., Giangaspero, F., *et al.* (2004). REN(KCTD11) is a suppressor of Hedgehog signaling and is deleted in human medulloblastoma. *Proc. Natl. Acad. Sci. USA* **101,** 10833–10838.

Dierks, C., Beigi, R., Guo, G. R., Zirlik, K., Stegert, M. R., Manley, P., Trussell, C., Schmitt-Graeff, A., Landwerlin, K., Veelken, H., and Warmuth, M. (2008). Expansion of Bcr-Abl-positive leukemic stem cells is dependent on Hedgehog pathway activation. *Cancer Cell* **14,** 238–249.

Dijkgraaf, G. J., Alicke, B., Weinmann, L., Januario, T., West, K., Modrusan, Z., Burdick, D., Goldsmith, R., Robarge, K., Sutherlin, D., Scales, S. J., Gould, S. E., *et al.* (2011). Small molecule inhibition of GDC-0449 refractory smoothened mutants and downstream mechanisms of drug resistance. *Cancer Res.* **71,** 435–444.

Dilda, P. J., and Hogg, P. J. (2007). Arsenical-based cancer drugs. *Cancer Treat. Rev.* **33,** 542–564.

Dioscorides, P. (2000). De Materia Medica. IBIDIS Press, Johannesburg, South Africa.

Dolniak, B., Katsoulidis, E., Carayol, N., Altman, J. K., Redig, A. J., Tallman, M. S., Ueda, T., Watanabe-Fukunaga, R., Fukunaga, R., and Platanias, L. C. (2008). Regulation of arsenic trioxide-induced cellular responses by Mnk1 and Mnk2. *J. Biol. Chem.* **283,** 12034–12042.

Douer, D., and Tallman, M. S. (2005). Arsenic trioxide: New clinical experience with an old medication in hematologic malignancies. *J. Clin. Oncol.* **23,** 2396–2410.

Dyck, J. A., Maul, G. G., Miller, W. H., Jr., Chen, J. D., Kakizuka, A., and Evans, R. M. (1994). A novel macromolecular structure is a target of the promyelocyte-retinoic acid receptor oncoprotein. *Cell* **76,** 333–343.

Fei, D. L., Li, H., Kozul, C. D., Black, K. E., Singh, S., Gosse, J. A., DiRenzo, J., Martin, K. A., Wang, B., Hamilton, J. W., Karagas, M. R., and Robbins, D. J. (2010). Activation of Hedgehog signaling by the environmental toxicant arsenic may contribute to the etiology of arsenic-induced tumors. *Cancer Res.* **70,** 1981–1988.

Feldmann, G., Fendrich, V., McGovern, K., Bedja, D., Bisht, S., Alvarez, H., Koorstra, J. B., Habbe, N., Karikari, C., Mullendore, M., Gabrielson, K. L., Sharma, R., *et al.* (2008). An orally bioavailable small-molecule inhibitor of Hedgehog signaling inhibits tumor initiation and metastasis in pancreatic cancer. *Mol. Cancer Ther.* **7,** 2725–2735.

Fox, E., Razzouk, B. I., Widemann, B. C., Xiao, S., O'Brien, M., Goodspeed, W., Reaman, G. H., Blaney, S. M., Murgo, A. J., Balis, F. M., and Adamson, P. C. (2008). Phase 1 trial and pharmacokinetic study of arsenic trioxide in children and adolescents with refractory or relapsed acute leukemia, including acute promyelocytic leukemia or lymphoma. *Blood* **111,** 566–573.

Giafis, N., Katsoulidis, E., Sassano, A., Tallman, M. S., Higgins, L. S., Nebreda, A. R., Davis, R. J., and Platanias, L. C. (2006). Role of the p38 mitogen-activated protein kinase pathway in the generation of arsenic trioxide-dependent cellular responses. *Cancer Res.* **66,** 6763–6771.

Goussetis, D. J., Altman, J. K., Glaser, H., McNeer, J. L., Tallman, M. S., and Platanias, L. C. (2010). Autophagy is a critical mechanism for the induction of the antileukemic effects of arsenic trioxide. *J. Biol. Chem.* **285,** 29989–29997.

Guha Mazumder, D. N., Haque, R., Ghosh, N., De, B. K., Santra, A., Chakraborty, D., and Smith, A. H. (1998). Arsenic levels in drinking water and the prevalence of skin lesions in West Bengal, India. *Int. J. Epidemiol.* **27,** 871–877.

Hahn, H., Wicking, C., Zaphiropoulous, P. G., Gailani, M. R., Shanley, S., Chidambaram, A., Vorechovsky, I., Holmberg, E., Unden, A. B., Gillies, S., Negus, K., Smyth, I., *et al.* (1996). Mutations of the human homolog of Drosophila patched in the nevoid basal cell carcinoma syndrome. *Cell* **85,** 841–851.

Haller, J. S. (1994). Medical Protestants: The Eclectics in American Medicine, 1825–1939. Southern Illinois University Press, Carbondale.

Han, Y. G., Spassky, N., Romaguera-Ros, M., Garcia-Verdugo, J. M., Aguilar, A., Schneider-Maunoury, S., and Alvarez-Buylla, A. (2008). Hedgehog signaling and primary cilia are required for the formation of adult neural stem cells. *Nat. Neurosci.* **11,** 277–284.

Hatton, B. A., Villavicencio, E. H., Tsuchiya, K. D., Pritchard, J. I., Ditzler, S., Pullar, B., Hansen, S., Knoblaugh, S. E., Lee, D., Eberhart, C. G., Hallahan, A. R., and Olson, J. M. (2008). The Smo/Smo model: Hedgehog-induced medulloblastoma with 90% incidence and leptomeningeal spread. *Cancer Res.* **68,** 1768–1776.

Hayakawa, F., and Privalsky, M. L. (2004). Phosphorylation of PML by mitogen-activated protein kinases plays a key role in arsenic trioxide-mediated apoptosis. *Cancer Cell* **5,** 389–401.

Hayashi, T., Hideshima, T., Akiyama, M., Richardson, P., Schlossman, R. L., Chauhan, D., Munshi, N. C., Waxman, S., and Anderson, K. C. (2002). Arsenic trioxide inhibits growth of human multiple myeloma cells in the bone marrow microenvironment. *Mol. Cancer Ther.* **1,** 851–860.

Hippocrates Adams, F. (1886). The Genuine Works of Hippocrates. W. Wood and Company, New York.

Hopenhayn-Rich, C., Biggs, M. L., and Smith, A. H. (1998). Lung and kidney cancer mortality associated with arsenic in drinking water in Cordoba, Argentina. *Int. J. Epidemiol.* **27,** 561–569.

Huang, C., Ma, W. Y., Li, J., and Dong, Z. (1999). Arsenic induces apoptosis through a c-Jun NH2-terminal kinase-dependent, p53-independent pathway. *Cancer Res.* **59,** 3053–3058.

Huangfu, D., Liu, A., Rakeman, A. S., Murcia, N. S., Niswander, L., and Anderson, K. V. (2003). Hedgehog signalling in the mouse requires intraflagellar transport proteins. *Nature* **426,** 83–87.

Ito, K., Bernardi, R., Morotti, A., Matsuoka, S., Saglio, G., Ikeda, Y., Rosenblatt, J., Avigan, D. E., Teruya-Feldstein, J., and Pandolfi, P. P. (2008). PML targeting eradicates quiescent leukaemia-initiating cells. *Nature* **453,** 1072–1078.

Jackson, R., and Grainge, J. W. (1975). Arsenic and cancer. *Can. Med. Assoc. J.* **113,** 396–401.

Jagani, Z., Mora-Blanco, E. L., Sansam, C. G., McKenna, E. S., Wilson, B., Chen, D., Klekota, J., Tamayo, P., Nguyen, P. T., Tolstorukov, M., Park, P. J., Cho, Y. J., *et al.* (2010). Loss of the tumor suppressor Snf5 leads to aberrant activation of the Hedgehog-Gli pathway. *Nat. Med.* **16,** 1429–1433.

Jeanne, M., Lallemand-Breitenbach, V., Ferhi, O., Koken, M., Le Bras, M., Duffort, S., Peres, L., Berthier, C., Soilihi, H., Raught, B., and de The, H. (2010). PML/RARA oxidation and arsenic binding initiate the antileukemia response of As_2O_3. *Cancer Cell* **18,** 88–98.

Ji, Z., Mei, F. C., Xie, J., and Cheng, X. (2007). Oncogenic KRAS activates hedgehog signaling pathway in pancreatic cancer cells. *J. Biol. Chem.* **282,** 14048–14055.

Jiang, G., Gong, Z., Li, X. F., Cullen, W. R., and Le, X. C. (2003). Interaction of trivalent arsenicals with metallothionein. *Chem. Res. Toxicol.* **16,** 873–880.

Jing, Y., Dai, J., Chalmers-Redman, R. M., Tatton, W. G., and Waxman, S. (1999). Arsenic trioxide selectively induces acute promyelocytic leukemia cell apoptosis via a hydrogen peroxide-dependent pathway. *Blood* **94,** 2102–2111.

Joo, J., Christensen, L., Warner, K., States, L., Kang, H. G., Vo, K., Lawlor, E. R., and May, W. A. (2009). GLI1 is a central mediator of EWS/FLI1 signaling in Ewing tumors. *PLoS One* **4,** e7608.

Jung, H. S., Kim, H. S., Lee, M. J., Shin, H. Y., Ahn, H. S., Ryu, K. H., Seoh, J. Y., Kim, C. J., and Jang, J. J. (2006). Arsenic trioxide concentration determines the fate of Ewing's sarcoma family tumors and neuroblastoma cells in vitro. *FEBS Lett.* **580,** 4969–4975.

Kakizuka, A., Miller, W. H., Jr., Umesono, K., Warrell, R. P., Jr., Frankel, S. R., Murty, V. V., Dmitrovsky, E., and Evans, R. M. (1991). Chromosomal translocation t (15;17) in human acute promyelocytic leukemia fuses RAR alpha with a novel putative transcription factor, PML. *Cell* **66,** 663–674.

Kandel, E., and LeRoy, G. (1937). Chronic arsenical poisoning during the treatment of chronic myeloid leukemia. *Arch. Intern. Med.* **60,** 846–866.

Kannan-Thulasiraman, P., Katsoulidis, E., Tallman, M. S., Arthur, J. S., and Platanias, L. C. (2006). Activation of the mitogen- and stress-activated kinase 1 by arsenic trioxide. *J. Biol. Chem.* **281,** 22446–22452.

Kapahi, P., Takahashi, T., Natoli, G., Adams, S. R., Chen, Y., Tsien, R. Y., and Karin, M. (2000). Inhibition of NF-kappa B activation by arsenite through reaction with a critical cysteine in the activation loop of Ikappa B kinase. *J. Biol. Chem.* **275,** 36062–36066.

Kim, J., Lee, J. J., Gardner, D., and Beachy, P. A. (2010). Arsenic antagonizes the Hedgehog pathway by preventing ciliary accumulation and reducing stability of the Gli2 transcriptional effector. *Proc. Natl. Acad. Sci. USA* **107,** 13432–13437.

Kimura, H., Stephen, D., Joyner, A., and Curran, T. (2005). Gli1 is important for medulloblastoma formation in Ptc1+/− mice. *Oncogene* **24,** 4026–4036.

Kinzler, K. W., Bigner, S. H., Bigner, D. D., Trent, J. M., Law, M. L., O'Brien, S. J., Wong, A. J., and Vogelstein, B. (1987). Identification of an amplified, highly expressed gene in a human glioma. *Science* **236,** 70–73.

Kojima, C., Ramirez, D. C., Tokar, E. J., Himeno, S., Drobna, Z., Styblo, M., Mason, R. P., and Waalkes, M. P. (2009). Requirement of arsenic biomethylation for oxidative DNA damage. *J. Natl. Cancer Inst.* **101,** 1670–1681.

Li, Y. M., and Broome, J. D. (1999). Arsenic targets tubulins to induce apoptosis in myeloid leukemia cells. *Cancer Res.* **59,** 776–780.

Liu, A., Wang, B., and Niswander, L. A. (2005). Mouse intraflagellar transport proteins regulate both the activator and repressor functions of Gli transcription factors. *Development* **132,** 3103–3111.

Lu, J., Chew, E. H., and Holmgren, A. (2007). Targeting thioredoxin reductase is a basis for cancer therapy by arsenic trioxide. *Proc. Natl. Acad. Sci. USA* **104,** 12288–12293.

Mann, K. K., Padovani, A. M., Guo, Q., Colosimo, A. L., Lee, H. Y., Kurie, J. M., and Miller, W. H., Jr. (2005). Arsenic trioxide inhibits nuclear receptor function via SEK1/JNK-mediated RXRalpha phosphorylation. *J. Clin. Invest.* **115,** 2924–2933.

Mann, K. K., Davison, K., Colombo, M., Colosimo, A. L., Diaz, Z., Padovani, A. M., Guo, Q., Scrivens, P. J., Gao, W., Mader, S., and Miller, W. H., Jr. (2006). Antimony trioxide-induced apoptosis is dependent on SEK1/JNK signaling. *Toxicol. Lett.* **160,** 158–170.

Mao, J., Maye, P., Kogerman, P., Tejedor, F. J., Toftgard, R., Xie, W., Wu, G., and Wu, D. (2002). Regulation of Gli1 transcriptional activity in the nucleus by Dyrk1. *J. Biol. Chem.* **277,** 35156–35161.

Mathieu, J., and Besancon, F. (2006). Clinically tolerable concentrations of arsenic trioxide induce p53-independent cell death and repress NF-kappa B activation in Ewing sarcoma cells. *Int. J. Cancer* **119,** 1723–1727.

May, S. R., Ashique, A. M., Karlen, M., Wang, B., Shen, Y., Zarbalis, K., Reiter, J., Ericson, J., and Peterson, A. S. (2005). Loss of the retrograde motor for IFT disrupts localization of Smo to cilia and prevents the expression of both activator and repressor functions of Gli. *Dev. Biol.* **287,** 378–389.

Merchant, A., Joseph, G., Wang, Q., Brennan, S., and Matsui, W. (2010). Gli1 regulates the proliferation and differentiation of HSCs and myeloid progenitors. *Blood* **115,** 2391–2396.

Michaud, E. J., and Yoder, B. K. (2006). The primary cilium in cell signaling and cancer. *Cancer Res.* **66,** 6463–6467.

Miller, W. H., Jr., Schipper, H. M., Lee, J. S., Singer, J., and Waxman, S. (2002). Mechanisms of action of arsenic trioxide. *Cancer Res.* **62,** 3893–3903.

Nasr, R., Guillemin, M. C., Ferhi, O., Soilihi, H., Peres, L., Berthier, C., Rousselot, P., Robledo-Sarmiento, M., Lallemand-Breitenbach, V., Gourmel, B., Vitoux, D., Pandolfi, P. P., *et al.* (2008). Eradication of acute promyelocytic leukemia-initiating cells through PML-RARA degradation. *Nat. Med.* **14,** 1333–1342.

Nessling, M., Richter, K., Schwaenen, C., Roerig, P., Wrobel, G., Wessendorf, S., Fritz, B., Bentz, M., Sinn, H. P., Radlwimmer, B., and Lichter, P. (2005). Candidate genes in breast cancer revealed by microarray-based comparative genomic hybridization of archived tissue. *Cancer Res.* **65,** 439–447.

Nolan-Stevaux, O., Lau, J., Truitt, M. L., Chu, G. C., Hebrok, M., Fernandez-Zapico, M. E., and Hanahan, D. (2009). GLI1 is regulated through Smoothened-independent mechanisms in neoplastic pancreatic ducts and mediates PDAC cell survival and transformation. *Genes Dev.* **23,** 24–36.

Pasca di Magliano, M., Sekine, S., Ermilov, A., Ferris, J., Dlugosz, A. A., and Hebrok, M. (2006). Hedgehog/Ras interactions regulate early stages of pancreatic cancer. *Genes Dev.* **20,** 3161–3173.

Pepin, J., and Milord, F. (1994). The treatment of human African trypanosomiasis. *Adv. Parasitol.* **33,** 1–47.

Pietsch, T., Waha, A., Koch, A., Kraus, J., Albrecht, S., Tonn, J., Sorensen, N., Berthold, F., Henk, B., Schmandt, N., Wolf, H. K., von Deimling, A., *et al.* (1997). Medulloblastomas of the desmoplastic variant carry mutations of the human homologue of Drosophila patched. *Cancer Res.* **57,** 2085–2088.

Puccetti, E., Guller, S., Orleth, A., Bruggenolte, N., Hoelzer, D., Ottmann, O. G., and Ruthardt, M. (2000). BCR-ABL mediates arsenic trioxide-induced apoptosis independently of its aberrant kinase activity. *Cancer Res.* **60,** 3409–3413.

Ramadan, D., Rancy, P. C., Nagarkar, R. P., Schneider, J. P., and Thorpe, C. (2009). Arsenic(III) species inhibit oxidative protein folding in vitro. *Biochemistry* **48,** 424–432.

Ramirez, P., Eastmond, D. A., Laclette, J. P., and Ostrosky-Wegman, P. (1997). Disruption of microtubule assembly and spindle formation as a mechanism for the induction of aneuploid cells by sodium arsenite and vanadium pentoxide. *Mutat. Res.* **386,** 291–298.

Rasheed, Z. A., Yang, J., Wang, Q., Kowalski, J., Freed, I., Murter, C., Hong, S. M., Koorstra, J. B., Rajeshkumar, N. V., He, X., Goggins, M., Iacobuzio-Donahue, C., *et al.* (2010). Prognostic significance of tumorigenic cells with mesenchymal features in pancreatic adenocarcinoma. *J. Natl. Cancer Inst.* **102,** 340–351.

Redondo-Munoz, J., Escobar-Diaz, E., Hernandez Del Cerro, M., Pandiella, A., Terol, M. J., Garcia-Marco, J. A., and Garcia-Pardo, A. (2010). Induction of B-chronic lymphocytic leukemia cell apoptosis by arsenic trioxide involves suppression of the phosphoinositide 3-kinase/Akt survival pathway via c-jun-NH2 terminal kinase activation and PTEN upregulation. *Clin. Cancer Res.* **16,** 4382–4391.

Regl, G., Neill, G. W., Eichberger, T., Kasper, M., Ikram, M. S., Koller, J., Hintner, H., Quinn, A. G., Frischauf, A. M., and Aberger, F. (2002). Human GLI2 and GLI1 are part of a positive feedback mechanism in basal cell carcinoma. *Oncogene* **21,** 5529–5539.

Reifenberger, J., Wolter, M., Weber, R. G., Megahed, M., Ruzicka, T., Lichter, P., and Reifenberger, G. (1998). Missense mutations in SMOH in sporadic basal cell carcinomas of the skin and primitive neuroectodermal tumors of the central nervous system. *Cancer Res.* **58,** 1798–1803.

Riobo, N. A., Lu, K., Ai, X., Haines, G. M., and Emerson, C. P., Jr. (2006). Phosphoinositide 3-kinase and Akt are essential for Sonic Hedgehog signaling. *Proc. Natl. Acad. Sci. USA* **103,** 4505–4510.

Roberts, W. M., Douglass, E. C., Peiper, S. C., Houghton, P. J., and Look, A. T. (1989). Amplification of the gli gene in childhood sarcomas. *Cancer Res.* **49,** 5407–5413.

Rohatgi, R., Milenkovic, L., and Scott, M. P. (2007). Patched1 regulates hedgehog signaling at the primary cilium. *Science* **317,** 372–376.

Rudin, C. M., Hann, C. L., Laterra, J., Yauch, R. L., Callahan, C. A., Fu, L., Holcomb, T., Stinson, J., Gould, S. E., Coleman, B., LoRusso, P. M., Von Hoff, D. D., *et al.* (2009). Treatment of medulloblastoma with hedgehog pathway inhibitor GDC-0449. *N. Engl. J. Med.* **361,** 1173–1178.

Schneider, L., Clement, C. A., Teilmann, S. C., Pazour, G. J., Hoffmann, E. K., Satir, P., and Christensen, S. T. (2005). PDGFRalphaalpha signaling is regulated through the primary cilium in fibroblasts. *Curr. Biol.* **15,** 1861–1866.

Shackelford, D., Kenific, C., Blusztajn, A., Waxman, S., and Ren, R. (2006). Targeted degradation of the AML1/MDS1/EVI1 oncoprotein by arsenic trioxide. *Cancer Res.* **66,** 11360–11369.

Sheng, T., Li, C., Zhang, X., Chi, S., He, N., Chen, K., McCormick, F., Gatalica, Z., and Xie, J. (2004). Activation of the hedgehog pathway in advanced prostate cancer. *Mol. Cancer* **3,** 29.

Smith, A. H., Lingas, E. O., and Rahman, M. (2000). Contamination of drinking-water by arsenic in Bangladesh: A public health emergency. *Bull. World Health Organ.* **78,** 1093–1103.

Soignet, S. L., Maslak, P., Wang, Z. G., Jhanwar, S., Calleja, E., Dardashti, L. J., Corso, D., DeBlasio, A., Gabrilove, J., Scheinberg, D. A., Pandolfi, P. P., and Warrell, R. P., Jr. (1998). Complete remission after treatment of acute promyelocytic leukemia with arsenic trioxide. *N. Engl. J. Med.* **339,** 1341–1348.

Soignet, S. L., Frankel, S. R., Douer, D., Tallman, M. S., Kantarjian, H., Calleja, E., Stone, R. M., Kalaycio, M., Scheinberg, D. A., Steinherz, P., Sievers, E. L., Coutre, S., *et al.* (2001). United States multicenter study of arsenic trioxide in relapsed acute promyelocytic leukemia. *J. Clin. Oncol.* **19,** 3852–3860.

Spassky, N., Han, Y. G., Aguilar, A., Strehl, L., Besse, L., Laclef, C., Ros, M. R., Garcia-Verdugo, J. M., and Alvarez-Buylla, A. (2008). Primary cilia are required for cerebellar development and Shh-dependent expansion of progenitor pool. *Dev. Biol.* **317,** 246–259.

Stecca, B., and Ruiz i Altaba, A. (2009). A GLI1-p53 inhibitory loop controls neural stem cell and tumour cell numbers. *EMBO J.* **28,** 663–676.

Stecca, B., Mas, C., Clement, V., Zbinden, M., Correa, R., Piguet, V., Beermann, F., and Ruiz, I. A. A. (2007). Melanomas require HEDGEHOG-GLI signaling regulated by interactions between GLI1 and the RAS-MEK/AKT pathways. *Proc. Natl. Acad. Sci. USA* **104,** 5895–5900.

Taylor, M. D., Liu, L., Raffel, C., Hui, C. C., Mainprize, T. G., Zhang, X., Agatep, R., Chiappa, S., Gao, L., Lowrance, A., Hao, A., Goldstein, A. M., *et al.* (2002). Mutations in SUFU predispose to medulloblastoma. *Nat. Genet.* **31,** 306–310.

Tezuka, M., Hanioka, K., Yamanaka, K., and Okada, S. (1993). Gene damage induced in human alveolar type II (L-132) cells by exposure to dimethylarsinic acid. *Biochem. Biophys. Res. Commun.* **191,** 1178–1183.

Tondel, M., Rahman, M., Magnuson, A., Chowdhury, I. A., Faruquee, M. H., and Ahmad, S. A. (1999). The relationship of arsenic levels in drinking water and the prevalence rate of skin lesions in Bangladesh. *Environ. Health Perspect.* **107,** 727–729.

Tremblay, M. R., Nesler, M., Weatherhead, R., and Castro, A. C. (2009). Recent patents for Hedgehog pathway inhibitors for the treatment of malignancy. *Expert Opin. Ther. Pat.* **19,** 1039–1056.

Verma, A., Mohindru, M., Deb, D. K., Sassano, A., Kambhampati, S., Ravandi, F., Minucci, S., Kalvakolanu, D. V., and Platanias, L. C. (2002). Activation of Rac1 and the p38 mitogen-activated protein kinase pathway in response to arsenic trioxide. *J. Biol. Chem.* **277,** 44988–44995.

Von Hoff, D. D., LoRusso, P. M., Rudin, C. M., Reddy, J. C., Yauch, R. L., Tibes, R., Weiss, G. J., Borad, M. J., Hann, C. L., Brahmer, J. R., Mackey, H. M., Lum, B. L., *et al.* (2009). Inhibition of the hedgehog pathway in advanced basal-cell carcinoma. *N. Engl. J. Med.* **361,** 1164–1172.

Waxman, S., and Anderson, K. C. (2001). History of the development of arsenic derivatives in cancer therapy. *Oncologist* **6**(Suppl. 2), 3–10.

Weis, K., Rambaud, S., Lavau, C., Jansen, J., Carvalho, T., Carmo-Fonseca, M., Lamond, A., and Dejean, A. (1994). Retinoic acid regulates aberrant nuclear localization of PML-RAR alpha in acute promyelocytic leukemia cells. *Cell* **76,** 345–356.

White, D. E., and Burchill, S. A. (2008). BAY 11–7082 induces cell death through NF-kappaB-independent mechanisms in the Ewing's sarcoma family of tumours. *Cancer Lett.* **268,** 212–224.

Wong, S. Y., Seol, A. D., So, P. L., Ermilov, A. N., Bichakjian, C. K., Epstein, E. H., Jr., Dlugosz, A. A., and Reiter, J. F. (2009). Primary cilia can both mediate and suppress Hedgehog pathway-dependent tumorigenesis. *Nat. Med.* **15,** 1055–1061.

Yauch, R. L., Gould, S. E., Scales, S. J., Tang, T., Tian, H., Ahn, C. P., Marshall, D., Fu, L., Januario, T., Kallop, D., Nannini-Pepe, M., Kotkow, K., *et al.* (2008). A paracrine requirement for hedgehog signalling in cancer. *Nature* **455,** 406–410.

Yauch, R. L., Dijkgraaf, G. J., Alicke, B., Januario, T., Ahn, C. P., Holcomb, T., Pujara, K., Stinson, J., Callahan, C. A., Tang, T., Bazan, J. F., Kan, Z., *et al.* (2009). Smoothened mutation confers resistance to a Hedgehog pathway inhibitor in medulloblastoma. *Science* **326,** 572–574.

Zhang, P. (1999). The use of arsenic trioxide (As_2O_3) in the treatment of acute promyelocytic leukemia. *J. Biol. Regul. Homeost. Agents* **13,** 195–200.

Zhang, Q. Y., Mao, J. H., Liu, P., Huang, Q. H., Lu, J., Xie, Y. Y., Weng, L., Zhang, Y., Chen, Q., Chen, S. J., and Chen, Z. (2009). A systems biology understanding of the synergistic effects of arsenic sulfide and Imatinib in BCR/ABL-associated leukemia. *Proc. Natl. Acad. Sci. USA* **106,** 3378–3383.

Zhang, X. W., Yan, X. J., Zhou, Z. R., Yang, F. F., Wu, Z. Y., Sun, H. B., Liang, W. X., Song, A. X., Lallemand-Breitenbach, V., Jeanne, M., Zhang, Q. Y., Yang, H. Y., *et al.* (2010). Arsenic trioxide controls the fate of the PML-RARalpha oncoprotein by directly binding PML. *Science* **328,** 240–243.

Zhu, J., Koken, M. H., Quignon, F., Chelbi-Alix, M. K., Degos, L., Wang, Z. Y., Chen, Z., and de The, H. (1997). Arsenic-induced PML targeting onto nuclear bodies: Implications for the treatment of acute promyelocytic leukemia. *Proc. Natl. Acad. Sci. USA* **94,** 3978–3983.

Zwerner, J. P., Joo, J., Warner, K. L., Christensen, L., Hu-Lieskovan, S., Triche, T. J., and May, W. A. (2007). The EWS/FLI1 oncogenic transcription factor deregulates GLI1. *Oncogene* **27,** 3282–3291.

CHAPTER SIXTEEN

Aberrations and Therapeutics Involving the Developmental Pathway Hedgehog in Pancreatic Cancer

Fergal C. Kelleher*,† *and* Raymond McDermott‡

Contents

Abstract

To conduct a systematic review of the role that the hedgehog signaling pathway has in pancreatic cancer tumorigenesis. A PubMed search from 2000 to 2010 and literature-based references were sourced. It was found that in 2009 a genetic analysis of pancreatic cancers discovered that a core set of 12 cellular signaling pathways including hedgehog were genetically altered in 67–100% of cases. Second, *in vitro* and *in vivo* studies of

* Department of Medical Oncology, St. Vincent's University Hospital, Dublin, Ireland
† Department of Medical Oncology, Peter Mac Callum Cancer Centre, Melbourne, Victoria, Australia
‡ St. Vincent's University Hospital, Dublin, Ireland and Adelaide and Meath Hospital, Dublin, Ireland

Vitamins and Hormones, Volume 88
ISSN 0083-6729, DOI: 10.1016/B978-0-12-394622-5.00016-X

treatment with cyclopamine (a naturally occurring antagonist of the hedgehog signaling pathway component; Smoothened) have shown that inhibition of hedgehog can abrogate pancreatic cancer metastasis. Third, experimental evidence has demonstrated that sonic hedgehog (Shh) is correlated with desmoplasia in pancreatic cancer. This is important because targeting the Shh pathway potentially may facilitate chemotherapeutic drug delivery as pancreatic cancers tend to have a dense fibrotic stroma that extrinsically compressed the tumor vasculature leading to a hypoperfusing intratumoral circulation. It is probable that patients with locally advanced pancreatic cancer will derive the greatest benefit from treatment with Smoothened antagonists. Fourth, it has been found that ligand-dependent activation by hedgehog occurs in the tumor stromal microenvironment in pancreatic cancer, a paracrine effect on tumorigenesis. Finally, in pancreatic cancer, cells with the CD44+CD24+ESA+ immunophenotype select a population enriched for cancer initiating stem cells. Shh is increased 46-fold in CD44+CD24+ESA+ cells compared with normal pancreatic epithelial cells. Medications that destruct pancreatic cancer initiating stem cells are a potentially novel strategy in cancer treatment.

In conclusion, aberrant hedgehog signaling occurs in pancreatic cancer tumorigenesis and therapeutics that target the transmembrane receptor Smoothened abrogate hedgehog signaling and may improve the outcomes of patients with pancreatic cancer.

I. Introduction

The Nobel Prize in Medicine was awarded in 1995 for discoveries concerning "the genetic control of early embryonic development." Two of the German prize recipients in Heidelberg identified genes that control the segmentation pattern of *Drosophila melanogaster* (fruit fly) embryos. Their genetic screens found that loss of a gene function later to be called hedgehog caused a mutant embryonic phenotype such that the embryos were covered with projections similar in appearance to a hedgehog. There are three hedgehog genes designated by the prefixes Sonic, Desert, and Indian. Sonic hedgehog (*Shh*) is a developmental morphogen in humans such that one function of Shh expression is that it forms a gradient that is primarily responsible for setting up the anterior–posterior axis in the developing limb (Genetics in medicine, Thompson and Thompson). More recently, aberrant hedgehog signaling has been implicated in tumorigenesis, and hedgehog directed therapeutics have a potential role in diseases other than cancer such as neurodegeneration and stroke.

II. The Hedgehog Pathway

A. Structure, cancer, and therapeutics

The hedgehog pathway is activated by secreted hedgehog molecules. These hedgehog ligands bind to a transmembrane protein called Patched (Ptch). This ligand receptor binding inhibits the repression of Smoothened another transmembrane receptor that is tonically repressed by the Patched receptor in the absence of hedgehog ligand. Therefore, Smoothened functions only when Hh is present. Smoothened transduces the signal intracellular via Gli zinc finger transcription factors (Molecular biology intelligence unit. Ariel Ruiz i Altaba).

The hedgehog pathway is most frequently pharmacologically inhibited by targeting the Smoothened receptor. The classic inhibitor is cyclopamine, a naturally occurring teratogenic alkaloid, and more recently, small molecule inhibitors have been discovered, such as GDC-0449, IPI-926, and XL139. The latter medication XL139 (BMS-833923) has shown phase I tumor efficacy against basal cell carcinomas (BCCs;) (Siu *et al.*, 2009). GDC-0449 also has been shown to be efficacy in BCCs. Additionally, a phase I study of IPI-926 in patients with advanced or metastatic solid tumors found partial responses in patients with BCC, and three non-BCC patients had stable disease for at least 6months (adenoid cystic carcinoma of the nasopharynx, a neuroendocrine tumor, and a case of chondrosarcoma; ESMO meeting, Milan Italy, October 2010, C. M. Rudin *et al.*).

Another theoretical mechanism to inhibit hedgehog signaling is to target Gli-mediated gene transcription using candidate medicinal compounds such as the GANT inhibitors, GANT 58 and GANT 61 (see Fig. 16.1). Finally, an important intracellular compound is suppressor of fused (SUFU). It is not usually targeted by medicinal compounds but is clinically important as it is a tumor suppressor gene, and mutations in SUFU can lead to the formation of multiple tumors.

In cancer treatment, much of the evidence for the benefit of inhibiting the hedgehog pathway has been accrued from preclinical model systems. The effect that inhibition of the hedgehog pathway has on medulloblastoma has been evaluated by investigators at St. Jude Children's Research Hospital in Memphis Tennessee. They demonstrated that in a Ptc1(+/−) p53(−/−) mouse model 100% of which are destined to develop medulloblastoma, treatment with a hedgehog antagonist resulted in complete tumor eradication (Romer *et al.*, 2004). The Hedgehog pathway may be unregulated in tumor cells because of intrinsic mutations in the hedgehog pathway such as is found in medulloblastomas leading to "oncogene addiction." The hedgehog pathway may also be upregulated because of increased hedgehog ligand. This occurs in stomach and pancreatic cancer

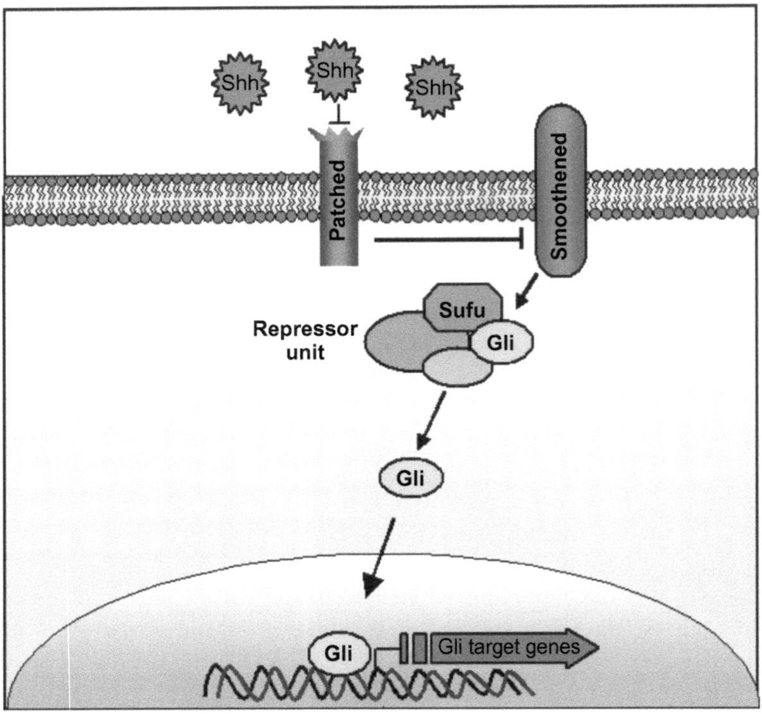

Figure 16.1 The hedgehog pathway. Diagram of the sonic hedgehog pathway. Sonic hedgehog binds to the Ptch receptor; this relieves inhibition of Smoothened. Downstream this causes upregulation of Gli zinc finger transcription factors. SUFU is an intracellular molecule that forms a repressor unit that inhibits Gli activities.

(Berman *et al.*, 2003). The exact mechanism how Hh acts in this later group of tumors is uncertain but may be due to autocrine signaling, selective activation of cancer initiating stem cells, or paracrine signaling. Further in pancreatic cancer, mutant KRAS activates Gli1 (Evangelista *et al.*, 2006; Ji *et al.*, 2007; Merchant and Matsui, 2010), and Gli activity in pancreatic cancer may be modulated by altering the activity of pathways that interact with hedgehog signaling via intercellular cross talk such as the RAF/MEK/MAPK pathway.

It may also be desirable to target different aspects of hedgehog signaling or to be more selective by undertaking localized tissue delivery of hedgehog antagonists to obviate deleterious side effects. For instance, it has been found that transient inhibition of hedgehog signaling in young mice cause permanent structural bone deficits. This is of importance if inhibitors of hedgehog signaling are to be used to treat medulloblastomas, a tumor that affects children and young adults (Kimura *et al.*, 2008).

Last, PF-04449913 is an inhibitor of Hedgehog signaling. A phase I/II study involving this drug in myeloid malignancies exemplified the risk of contaminating studies with clinically similar tumours that have differing patterns of Hedgehog pathway activity. This is usually detrimental to study findings. In that study, it was notable that Hedgehog signaling including intact Smoothened expression is required for over 95% of cases of chronic myeloid leukemia that have the BCR-ABL chimeric fusion gene but not leukemogenesis due to the mixed lineage leukemia MLL:AF9 chimeric fusion gene (Dierks *et al.*, 2008). This is an ever present danger in contemporary trial design.

B. Inferences from tumors derived from ectodermal tissue

Gorlin's syndrome a disorder arising from inherited germline mutations in the Patched gene is a unifying phenotypic expression of the role Hedgehog signaling has in endodermal and ectodermally derived tumors. It is associated with an increased predisposition to the development of multiple BCC, medulloblastoma, and rhabdomyosarcoma as well as prostate, breast, and pancreatic cancer. Two of the tumor that this disorder confers an increased risk of basal cell carcinoma and medulloblastoma were the subject of landmark papers involving GDC-0449 in September 2009 (Rudin *et al.*, 2009; von Hoff *et al.*, 2009). Loss-of-function mutations occur in the gene encoding Patched 1 in BCC and 30% of medulloblastomas particularly the desmoplastic variant (Dellovade *et al.*, 2006; García de Marcos *et al.*, 2009; Rubin and de Sauvage, 2006). First, in patients with metastatic or locally advanced basal cell carcinoma (n=33), 55% of patients had an objective response (2 complete response, 16 partial response) (Von Hoff *et al.*, 2009). Of the other 15 patients, 11 had stable disease for up to 10.8months and 4 experienced progressive diseases. Hedgehog also has a critical role in cerebellar development in directing the spatial and temporal migration of granule cell neural precursors from the external to the internal granule cell layer. In the second study which was a clinical case report, a patient with treatment refractory metastatic medulloblastoma received GDC-0449 and experienced a rapid but transient tumor response (Rudin *et al.*, 2009). In this case, the patient's medulloblastoma had loss of heterozygosity and a somatic mutation of the Patched 1 gene. Prior to treatment with GDC-0449, molecular profiling revealed a somatic mutation in Patched 1 (PTCH1-W844C) and upregulation of hedgehog target genes. The PTCH1-W844C mutant was not capable of suppression of Smoothened activity in an Hh responsive, Gli-luciferase reporter cell line (Yauch *et al.*, 2009). The loss of a functional PTCH transmembrane receptor causes a loss of the normal tonic repression of the Smoothened receptor by the PTCH receptor with consequent tumor growth attributable to hedgehog pathway activation.

C. Management of acquired molecular resistance to hedgehog directed therapeutics

In the mentioned case of metastatic medulloblastoma, the transient improvement with GDC-0449 in the tumor extent did not persist, and a biopsy or recrudescent tumor confirmed the previously detected PTCH-W844C mutation, accompanied by loss of heterozygosity of PTCH leading to constitutive activation of the hedgehog pathway due to a lack of functional PTCH.

In general, mutations in Smoothened account for acquired molecular resistance to GDC-0449 in medulloblastoma. In an evaluating of the mechanism of acquired resistance to GDC-0449 of the aforementioned patient's medulloblastoma, an amino acid substitution of a conserved aspartic acid residue disrupted GDC-0499 binding to the serpentine Smoothened receptor (Yauch *et al.*, 2009). The investigators identified a heterozygous guanine to cytosine (G to C) misses mutation at position 1697 in the Smoothened locus which was predicted to change codon 473 from Asp to His (D473H) in the GDC-499 resistant progressive medulloblastoma, a change that was not detected in the initial tumor tissue. This mutant allele was found in the recurrent tumor tissue and not in the patient's initial primary/metastatic tumor or skin biopsy. An analogous well-established scenario is that of chronic myelogenous leukemia (CML) that is refractory or develops resistance to imatinib but is responsive to dasatinib. This type of CML is resistant *in vitro* and *in vivo* to dasatinib when it acquires the T315I mutation (Talpaz *et al.*, 2006).

In a mouse model interrogating the mechanism for GDC-0449 resistance in medulloblastoma (a drug-resistant subcutaneous allograft of medulloblastoma in Ptch1 +/−; p53−/− mice selected for GDC-0449 resistance), three separate GDC-0449 resistant tumor lines were established. In one arising drug-resistant tumor cell line model, Smoothened was sequenced and a heterozygous missense mutation at 1944 causing an aspartic acid-477 to glycine (D477G) change was found that was not in the parental model. In fascinating evidence, the corresponding residue in human Smoothened is aspartic acid at position 473, the affected locus that was found in the case of the aforementioned relapsed GDC-449-resistant medulloblastoma (Yauch *et al.*, 2009). Predictive modeling concluded that the Asp-473 residue is at the C-terminal end of the sixth transmembrane segment, near the extracellular lip of the heptahelical transmembrane Smoothened receptor.

A correlative example of acquired molecular resistance to targeted therapeutics in cellular components other than those of the hedgehog pathway was described recently in a patient with EML4 (echinoderm microtubule-associated protein-like 4)–ALK (anaplastic lymphoma kinase) subtype, non-small cell lung cancer (Choi *et al.*, 2010). Two secondary mutations arose within the kinase domain of the protein product arising from the EML4–ALK fusion gene in tumor cells isolated from a patient during the relapse phase after 5 months of treatment with the ALK inhibitor crizotinib. These alterations were G→A and C→A changes at positions corresponding to nucleotides 4374 and 4493 of wild-type *ALK* cDNA. These corresponded to C1156Y and L1196M mutants; the cysteine at position 1156 resides close to the upper edge of the ATP-binding pocket and leucine at position 1196 resided within the kinase domain. Interestingly, L1196 of ALK corresponds to the threonine at position 315 in ABL and at position 790 in EGFR, each of which is the site of the most frequently acquired mutations that confer resistance to tyrosine kinase inhibitors in these kinases. Though these are fascinating examples from other molecularly targeted tumor components, it is to be recalled that the evidence from other targeted pathways is predominated by altered kinase function whereas Smoothened is a transmembrane receptor proto-oncogene which can incur molecular mutations but is not a tyrosine kinase or serine threonine kinase.

Ultimately, it may be inferred from preclinical experiments and correlative evidence from other targeted therapeutic model systems that resistance to Smoothened antagonists may be due to molecular mutations in the Smoothened receptor. This resistance is more likely to emerge in tumor subclones rather than homogenously within tumor masses and the practical implication of this is that withdrawal of a Smoothened antagonist in the absence of an alternative medication could theoretically lead to a rebound tumor flare in the repressed hedgehog-reliant residual component of the tumor.

The scenario of aberrant activation of the hedgehog pathway intracellular to Smoothened where Smoothened antagonists are ineffective has been described in cases of medulloblastoma, glioma pericytomas, and prostate cancer (Lauth *et al.*, 2007). Smoothened independent mechanisms for hedgehog pathway activation include loss of heterozygosity of the tumor suppressor gene Suppressor of Fused (SUFU), amplification or chromosomal translocation of the Gli genes, or stabilization of the translated Gli protein. In the particular case of pancreatic cancer, resistance to Smoothened antagonists may arise by subversion of the pathway by cross talk from the RAS/Raf/MEK pathway. Cyclopamine-resistant Gli1(+) cell lines derived from adenocarcinoma of the pancreas have previously been reported (Thayer *et al.*, 2003). Candidate screened small molecule inhibitors to tackle the difficultly of Hedgehog activation independent of Smoothened and downstream of SUFU have given rise to the GANT compounds 58 and 61. They both act in the nucleus to block Gli-mediated transcription and GANT61 also blocks Gli1 DNA binding (Lauth *et al.*, 2007).

III. Hedgehog Therapeutics

A. Preclinical studies of hedgehog and pancreatic cancer

In considering the clinical implications of experimental evidence for hedgehog pathway inhibition in pancreatic cancer, it is important to appreciate if the evidence is from pancreatic cell lines, tumor xenograft, or mouse models that more truly replicate ductal adenocarcinoma of the pancreas. The latter is demonstrated in KPC mouse models that conditionally express mutant K-ras and p53 in pancreatic cells and have the fibro-inflammatory tumor investing tissue that occurs in human pancreatic cancer. The dense desmoplastic stromal reaction of pancreatic cancer is an established impediment to delivery of chemotherapy, and pancreatic cancers have a hypoperfusing vasculature. Experimental evidence has demonstrated that Shh is correlated with desmoplasia in pancreatic cancer (Bailey *et al.*, 2008). Therefore, targeting the *Shh* pathway potentially could facilitate the delivery of chemotherapeutic drugs by diminishing the desmoplastic stromal reaction. This concept was investigated by a multicentre group using KPC mice. KPC mice demonstrated resistance to gemcitabine compared to transplantation mouse models, and Olive *et al.* in a series of experiments showed that impaired drug delivery was the probable cause. IPI-926 a semisynthetic derivative of cyclopamine that inhibits *Smo* was studied with gemcitabine. Combination treatment with IPI-926/gemcitabine extended median survival of KPC mice from 11 to 25 days (p=0.001; log rank test), most tumors decreased in size, and IPI-926/gemcitabine treatment significantly decreased metastasis to the liver (p=0.015; Fischer's exact test). Treatment benefits were transient, however. Theoretically, combination treatment with a hedgehog antagonist and gemcitabine may increase the tumors bioavailability of the cytotoxic drug (Olive *et al.*, 2009). Further in a separate study, Feldmann *et al.* (2008) in a genetically engineered mouse model of pancreatic cancer, found that hedgehog inhibition with cyclopamine prolonged survival and abrogates systemic metastasis. That study from Johns Hopkins found that the Hedgehog pathway is activated in Pdx1-Cre, LsL-Kras (G12D), Ink4a/Arf (Iox/Iox) transgenic mice models of pancreatic cancer. Hedgehog inhibition with cyclopamine significantly prolonged median survival from 61 to 67 days, p=0.02. *In vitro* data indicates that Hedgehog activation is at least partly attributable to activation of KRAS signaling similar to medulloblastoma studies. Finally, hedgehog antagonists may be therapeutically effective in pancreatic cancer because pancreatic cancer has increased activation of the hedgehog pathway. A study of digestive tract tumors found that tumors originating from the pancreas have increased Hh pathway activity suppressible by an Hh antagonist (Berman *et al.*, 2003).

A second mechanism by which hedgehog antagonism may be therapeutically efficacious in pancreatic cancer is through paracrine effects on the tumor stromal cells. Yauch and colleagues have demonstrated that ligand-dependent activation by hedgehog occurs in the tumor stromal environment. Exploiting interspecies differences, it was shown that inhibition of hedgehog signaling using an anti-Hh antibody or genetic deletion of *Smo* in mouse stroma resulted in tumor growth inhibition in xenograft tumor models (Yauch *et al.*, 2009). Further when Hedgehog inhibitors were tested in 125 cell lines, no correlation was seen between the extent of basal Hh pathway activity and cellular sensitivity to Hh Antag. Mechanistically, it appears that hedgehog secreted by tumors acts on Smo in the stromal cells leading to changes in the extracellular matrix, structural support to tumor growth, and stromal cytokine release such as Wnt and insulin-like growth factor (IGF) that leads to tumor growth.

B. Pancreatic cancer stem cells and hedgehog

Cancer stem cells have the properties of self-renewal and multilineage differentiation. They are at the apex of tumor biology architecture. Fluorescent-activated cell sorting and tumor xenograft models in immunocompromised mice have been used to identify cancer stem cells. In pancreatic cancer, cells with the CD44+CD24+ESA+ immunophenotype select a population of cells enriched for cancer stem cells (Li *et al.*, 2007). Reverse transcriptase polymerase chain reaction has found that Sonic hedgehog is increased 46-fold in CD44+CD24+ESA+ cells compared with normal pancreatic epithelial cells. Therefore, Sonic hedgehog along with other developmental pathway genes is important in stem cell biology, and targeting Shh may be of particular benefit in eradication of pancreatic cancer stems cells.

In 2007, investigators at Ludwig—Maximilians—University in Munich, Germany identified a subpopulation of migrating CD133(+) CXCR4(+) cancer stem cells that are essential for tumor metastasis (Hermann *et al.*, 2007). It was found that pancreatic cancer tissue contains these cells that are tumorigenic and highly resistant to standard chemotherapy. CD133(+) CXCR4(+) cells occupy the invasive front at the tumor host interface of pancreatic tumors and determine the individual tumors metastatic phenotype. Depletion of the cancer stem cell pool for these migrating cancer stem cells virtually abrogated the metastatic phenotype of pancreatic tumors without affecting their tumorigenic potential. A majority of patients with pancreatic cancer present with locally advanced or metastatic disease, and the life expectancies for non-metastatic disease are 6–10months and 3–6 months for metastatic disease, respectively (Hawes *et al.*, 2000). This emphasizes the importance of effectively targeting the subpopulation of cells in pancreatic cancer responsible for distant disease. If sonic hedgehog signaling was found to be an important intracellular pathway for the stem cell subcategory defined by CD133(+) CXCR(+) immunophenotypic positivity that would be a potent theoretical advance in targeted therapeutics and pancreatic cancer. Experimental evidence remains lacking in this specific regard; however, it is likely that upregulation of the hedgehog pathway occurs in CD133(+) CXCR4(+) pancreatic cancer stem cells as it is increased in pancreatic cancer stem cells defined by the CD44+CD24+ESA+ immunophenotype.

In general, therapeutics directed against pancreatic cancer stems cells offer new strategies to treat the disease. As stem cell utilize self-renewal pathways including PTEN, WNT, BMI-1, and Notch in addition to hedgehog, therapeutics directed against these pathways are particularly selective for stem cells. As stem cells frequently evade cytoreductive chemotherapy because they are relatively quiescent and divide infrequently, therapeutics directed against stem cells may decrease tumor repopulation posttreatment. Also as CD133(+) CXCR4(+) determines the metastatic phenotype incorporation of medications that antagonize the hedgehog pathway into adjuvant treatment, paradigms may prove to be particularly efficacious. However, preclinical work remains to be done, as it has not been definitively shown that CD133(+) CXCR4(+) cells have increased levels of Sonic hedgehog pathway activity compared to other pancreatic cancer cells.

C. Hedgehog and canonical Wnt signaling in pancreatic adenocarcinoma

WNTs are cytosine-rich glycoproteins that function as locally active molecules that bind to cellular receptors. They are expressed in a spatially restricted and dynamic pattern in embryos and adults, and there are 19 different WNTs in mammals. If the local concentration of Wnts exceeds the buffering capacity of the inhibitors, they can interact with Frizzled receptors and LRP coreceptors. This interaction can activate the canonical β-catenin/TCF pathway and noncanonical pathways (Kelleher *et al.*, 2006).

In the canonical/β-catenin pathway, in the absence of Wnt ligands, β-catenin is recruited into a destruction complex. This contains the APC gene and AXIN, which facilitate the phosphorylation of β-catenin by casein kinase 1 and then by glycogen synthase kinase 3. This leads to the ubiquitylation and proteosomal degradation of β-catenin. Conversely, when β-catenin levels are increased, it is translocated to the nucleus where it binds to T-cell factor (TCF)/lymphoid enhancer factor (LEF) transcription elements. Therefore, Wnt induces destabilization of the β-catenin destruction complex and stabilization of β-catenin. This potentially affects the expression of downstream targets of wnt signaling (Fig. 16.2).

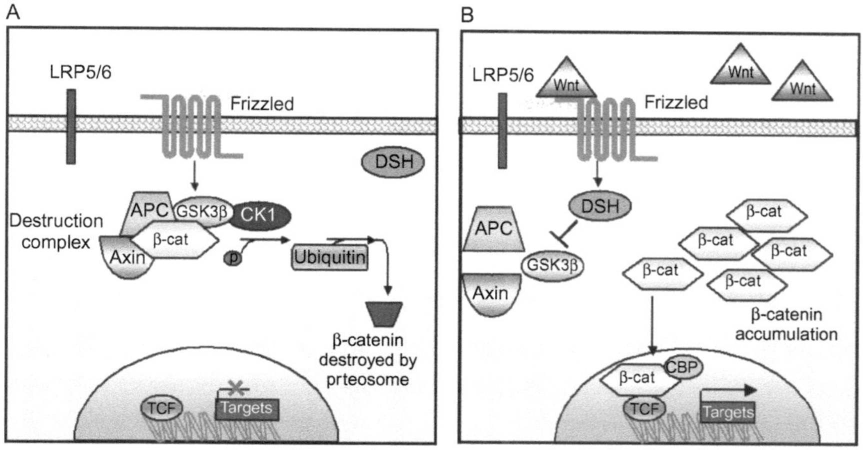

Figure 16.2 Wnt signalling in neoplasia has been best characterised in colon cancer and this diagram is a generic example of the role the Wnt signalling pathway can have in cancer inferred from this disease. Diagram of the Wnt signaling pathway. (A) In the absence of Wnt ligand, β-catenin undergoes intracellular destruction. (B) In the presence of Wnt ligand, β-catenin accumulates in the cytoplasm and translocates to the nucleus, where it binds to TCF/LEF transcription elements. This effects the transcription of downstream targets.

Developmental pathways like Wnt, Notch, and Hedgehog require interaction for appropriate embryonic development and dysregulated Hedgehog may be influenced by Wnt. Wnt signaling had previously been shown to be activated to high levels in early pancreatic development (Heiser *et al.*, 2006; Heller *et al.*, 2002). University of California San Francisco (UCSF) investigators found that inhibition of Wnt signaling reduces cell proliferation and increases apoptosis in pancreatic adenocarcinoma cells. Further Hedgehog signaling was found to regulate Wnt signaling in untransformed pancreatic ductal epithelial cells and in pancreatic cancer. Therefore, Hedgehog activity may regulate the effects of β-catenin signaling in pancreatic cancer. Temporally induction of β-catenin signaling occurs during PanIN formation and hedgehog signaling has been implicated in both the early and late stages of pancreatic cancer tumorigenesis.

Finally, it was shown by the UCSF investigators that Hedgehog signaling regulates expression of TCF4 in both normal and cancerous pancreatic ductal cells (Pasca di Magliano *et al.*, 2007). TCF4 is expressed in the gastrointestinal tract and in gastrointestinal neoplasms. Tcf4mRNA was expressed at high level in all pancreatic cancer cell lines tested. Only low Tcf4 protein was found in a small proportion of pancreatic ductal cells, but GLI2, the hedgehog effector, leads to marked Tcf4 expression in pancreatic ductal cells. The cell line studies found that Hedgehog signaling regulates the Wnt pathway at least partly by affecting the concentration of nuclear

TCF4 in pancreatic cancer cells. Xenograft mice models also showed that epithelial-specific activation of Hedgehog signaling is sufficient to induce Wnt/β-catenin signaling in pancreatic cells. The investigators concluded that their experiments showed that Wnt/β-catenin signaling is frequently activated in preneoplastic lesions and pancreatic ductal adenocarcinoma, and activity of Wnt is at least partly regulated by Hedgehog signaling.

D. Hedgehog and human adenocarcinoma of the pancreas

Pancreatic adenocarcinoma arises from precursor lesions, pancreatic intraepithelial neoplasia (PanIN), that progress from PanIN-1–3. PanIN-1 lesions have K-ras mutations and overexpression of Her2. In PanIN-2, p16 mutations occur. In PanIN-3, DPC4, P53 and BRCA2 mutations predominate. Shh expression is both a feature of PanIN and pancreatic adenocarcinoma implicating hedgehog as an early and late mediator of pancreatic cancer tumorigenesis. Recurrently, altered signaling pathways in pancreatic cancers were determined by Jones *et al.* by a comprehensive genetic analysis of 24 pancreatic cancers. They sequenced pancreatic cancer transcripts and searched for homozygous deletions and amplifications by microarrays containing probes for approximately 1 million SNPs. It was found that pancreatic cancers contain an average of 63 genetic alterations, the majority of which are point mutations. A core set of 12 cellular signaling pathways one of which was the Hedgehog pathway were identified and they were each genetically altered in 67–100% of the tumors. In the Hedgehog pathway, 19 altered genes were identified and 100% of tumors had alterations in at least one of the hedgehog pathway genes. Representative genes in the hedgehog pathway included SOX3, LRP2, TBX 5, GLI1, GLI3, BOC, BMPR2, and CREBBP. Of the other altered pathways, they were involved in apoptosis, DNA damage control, regulation in G1/S phase transition, homophilic cell adhesion, integrin signaling, cJun N-terminal kinase signaling, K-ras signaling, regulation of invasion, small GTPase dependent signaling, TGF-beta signaling, and Wnt/Notch signaling (Jones *et al.*, 2008) (Table 16.1).

E. Therapeutics in pancreatic cancer exploiting the hedgehog pathway

Pancreatic cancer can be divided into localized, locally advanced, and metastatic disease. Each subcategory has differing management algorithm and prognosis. Integration of inhibitors or hedgehog signaling also confers different management and outcome possibilities in each subcategory.

First, considering localized and therefore resectable disease. There is existing phase III evidence supporting adjuvant treatment with gemcitabine of pancreatic cancer in the CONKO-001 trial (Oettle *et al.*, 2007). *In vitro* and *in vivo* studies of treatment with cyclopamine have shown that

Table 16.1 Studies implicating sonic hedgehog in pancreatic cancer

Principle author	Study title	Study design	Study findings
Thayer *et al.* (2003) Massachusetts General Hospital *Nature*	Hedgehog is an early and late mediator of pancreatic cancer tumorigenesis	Transgenic mice models Screening human adenocarcinoma cell lines for expression of Hh components Gli-luciferase reporter construct, cell line transfection	Hedgehog may have an early and critical role in pancreatic carcinogenesis aberrant proliferation and tumorigenesis
Berman *et al.* (2003) Howard Hughes Medical Institute, Johns Hopkins *Nature*	Widespread requirement for hedgehog ligand stimulation in growth of digestive tract tumors	Cell line Shh, Ihh mRNA Hh-inducible Gli-luciferase reporter Pancreatic carcinoma xenografts Cyclopamine, 5E1 monoclonal Ab studies	Digestive tract tumors: esophagus, stomach, billiard tract, and pancreas but not colon display increased Hh pathway activity
Jones *et al.*, 2008 Johns Hopkins *Science*	Core signaling pathways in human pancreatic carcinomas revealed by global genomic analysis	Genetic analysis of 24 pancreatic cancers. Determined sequences of 23,219 transcripts representing 20,661 protein-coding genes. Searched for homozygous deletions and amplifications	Core set of 12 signaling pathways altered in 67–100% of tumors. Included Hedgehog pathway
Bailey *et al.*, 2008 Eppley Institute, University of Nebraska *Clin. Cancer Res.*	Sonic hedgehog promotes desmoplasia in pancreatic cancer	Expressed Shh in transformed cell line from human pancreas Inhibited Shh activity *in vivo* in mice	Shh contributes to the formation of desmoplasia

(Continued)

Table 16.1 (*Continued*)

Principle author	Study title	Study design	Study findings
Morton *et al.* (2007) University of Massachusetts *PNAS*	Sonic hedgehog acts at multiple stages during pancreatic tumorigenesis	Transgenic mice Tumor cell lines	Enhances proliferation of pancreatic ductal epithelial cells and protects against apoptosis. Shh cooperates with activated K-ras and reduces dependence on sustained activation of MAPK and PI 3-Kinase /Akt/ mTOR
Bailey *et al.*, 2009 Eppley Institute *Oncogene*	Sonic hedgehog paracrine signaling regulates metastasis and lymphangiogenesis in pancreatic cancer	Orthoptic transplantation of cell lines expressing Shh Treatment with 5E1	Supported paracrine signaling evidence for Shh
Tian *et al.*, 2009 Genentech *PNAS*	Hedgehog signaling is restricted to the stromal compartment during pancreatic carcinogenesis	An oncogenic allele of Smoothened *SmoM2* was expressed to autonomously activate Hh in mouse pancreas	Hedgehog pathway was not activated by expression of *SmoM2* in epithelial cells but was activated by expression in mesenchymal cells

inhibition of hedgehog can abrogate pancreatic cancer metastasis. Treatment of pancreatic cancer cell lines inhibited epithelial to mesenchymal transition by upregulating E-cadherin and downregulating snail (a transcription regulator that downregulates the expression of E-cadherin). Also overexpression of Gli1 *in vitro* cell line studies lead to marked tumor invasion. Also in murine orthoptic xenograft models, cyclopamine inhibited metastatic spread and synergised with gemcitabine such that metastasis were completely eliminated (Feldmann *et al.*, 2007). The accruing evidence suggests that there may be a future trial of a hedgehog inhibitor incorporated into established treatment paradigms for the adjuvant treatment of pancreatic cancer (refer Table 16.2).

What of locally advanced pancreatic cancer (LAPC)? In the United States, 40% of patients diagnosed with pancreatic cancer have locally advanced disease (Willett *et al.*, 2005). This disease is deemed unresectable because of local invasion into retroperitoneal vessels without detectable metastasis. In a retrospective cohort study of data from the SEER program conducted by the Dana Farber Cancer Institute, only 24% of patients with LAPC received what was considered the standard of care, radiotherapy with concomitant 5-fluorouracil (Krzyanowska *et al.*, 2003). As Smoothened antagonists decrease the desmoplastic stroma of pancreatic cancer, LAPC represents a treatment group with the promise of important therapeutic improvements if hedgehog inhibitors are incorporated onto the relevant treatment algorithms.

A number of clinical trials have been designed to target the desmoplastic stroma that invests pancreatic cancer cells. This is of importance as depletion of the fibroblastic stroma that contributes to the hypoperfusing tumor vasculature may improve the bioavailability of administered chemotherapeutics. Additionally, diminished desmoplasia may remove involvement or encasement of the superior mesenteric artery or superior mesenteric vein portal venous confluence during traditional contraindications to pancreaticoduodenectomy. A caution to note is that a study of 593 patients found that the need for vascular resection did not adversely affect postoperative morbidity, mortality, or overall survival post-pancreaticoduodenectomy (Martin *et al.*, 2009).

Targeting the fibroblastic tumor stroma of pancreatic cancer is not the exclusive domain of Smoothened antagonists. Nab-paclitaxel (Abraxane later called Abraxix) is a chemotherapeutic agent that was developed to obviate Cremophor from paclitaxel preparations and consists of albumin bound nanoparticles. The albumin in nab-paclitaxel binds to blood vessels in the tumor microenvironment and to a protein entitled SPARC (secreted protein acid rich in cysteine), that is, highly expressed in pancreatic cancer. This has the effect of concentrating the drug within the tumor. A phase I/II study of 63 patients by Von Hoff *et al.* evaluating the treatment efficacy and toxicity of the combination of gemcitabine and nab-paclitaxel in metastatic

Table 16.2 National Cancer Institute www.cancer.gov, search terms Hedgehog, pancreatic cancer. Search performed November 2010

Study title	Phase, protocol ID's	Lead sponsor	Trial design	Primary end point
Hedgehog inhibition for pancreatic ductal adenocarcinoma in the preoperative setting	Phase II OCRD 201014, NCT01096732	Addenbrooke's Hospital Roche Genetech	Operable localized pancreatic cancer treated with GDC-0449 for approximately 2 weeks from diagnosis to surgery. Initial biopsy and surgical samples	To detect a change in Hedgehog signaling in peritumoral pancreatic tissue
Hedgehog inhibitors for metastatic adenocarcinoma of the pancreas	Phase II J1013, NA 00036883, NCT01088815	Sidney Kimmel Comprehensive Cancer Center at Johns Hopkins Stand Up To Cancer Genentech, Abraxis Oncology	Metastatic adenocarcinoma of the pancreas treated with GDC-0449 in combination with chemotherapy (gemcitabine and nab-Paclitaxel). Open label single arm, multicentre trial.	Progression free survival
Pilot study of the effect of Hedgehog antagonist GDC-0449 and gemcitabine hydrochloride on cancer stem cells in patients with advanced pancreatic cancer	No phase specified UMCC-2010-003 2010-003, 8417, NCT01143415	NCI	To evaluate the number and percentage of pancreatic cancer stem cells before and after treatment with GDC-0449 in patients with advanced pancreatic cancer. To assess the effect of treatment with GDC-0449 alone and in combination with gemcitabine hydrochloride on total circulating tumor cells	Proportion of pancreatic cancer stem cells before and after treatment with GDC-0449. Number of circulating tumor cells at 0, 15, and 43 days.

Phase II randomized study of gemcitabine hydrochloride with versus without Hedgehog antagonist GDC-0449 in patients with recurrent or metastatic pancreatic cancer	Phase II UCCRC-8418 8418, NCT01064622	NCI	Patients with recurrent or metastatic pancreatic cancer treated with gemcitabine hydrochloride with versus without hedgehog antagonist GDC-0449	Compare PFS of patients with recurrent or metastatic pancreatic cancer treated with gemcitabine hydrochloride with versus without hedgehog antagonist GDC-0449.
Phase I study of Hedgehog antagonist GDC-0449 and erlotinib hydrochloride with or without gemcitabine hydrochloride in patients with metastatic pancreatic cancer or unresectable solid tumors	Phase I MAYO-MC0811 MC0811, 8231, NCT00878163	NCI	GDC-0449 and erlotinib hydrochloride with or without gemcitabine hydrochloride in treating patients with metastatic pancreatic cancer or solid tumors that cannot be removed by surgery	To determine the maximum tolerated dose of erlotinib hydrochloride and GDC-0449 with or without gemcitabine hydrochloride in patients with unresectable solid tumors

(*Continued*)

Table 16.2 (*Continued*)

Study title	Phase, protocol ID's	Lead sponsor	Trial design	Primary end point
A study evaluating IPI-926 in combination with gemcitabine in patients with metastatic pancreatic cancer	Phase II, phase I IPI-926-03 NCT01130142	Infinity Pharmaceuticals	IPI-926-03 is a Phase Ib/2 clinical trial to evaluate IPI 926 in combination with gemcitabine in patients with previously untreated metastatic pancreatic cancer. Phase II is designed as a randomized, double-blind (investigator/patient), placebo-controlled study.	Evaluation of IPI-926 in combination with gemcitabine.

pancreatic cancer was reported at the American Society of Clinical Oncology annual meeting in 2009. In that study, serial PET scans of 53 pts showed a 23% complete responses rate, 55% partial response rate, and 8% stable disease. Of 49 patients evaluated by RECIST criteria, 1 (2%) had a complete response, 12 (24%) had a partial response, and 20 (41%) had stable disease. The median survival was 9months at time of assessment. Patients that were SPARC positive (8/27) were more likely to be responders (75%) than patients who were SPARC negative (26%), $p=0.03$. It also appears that a positive SPARC status in these patients treated with nab-paclitaxel and gemcitabine was associated with a longer progression free survival (Abstract, 2009). Of relevance, a phase II trial NCT01088815 being led by The Sidney Kimmel Comprehensive Cancer Centre at Johns Hopkins, Baltimore, Stand up to Cancer and industry sponsors is listed by the NCI as involving the treatment of metastatic adenocarcinoma of the pancreas with GDC-0449 in combination with chemotherapy (gemcitabine and nab-paclitaxel).

The remarkable efficacy of the nab-paclitaxel gemcitabine combination in metastatic pancreatic cancer has been shown to be due to increased bioavailability of the chemotherapeutic medications in a mouse xenograft tumor model due to stromal collapse by Maitra and Hidago (Garber, 2010). Further experimental data has been accrued on tumor cell:stromal interactions in pancreatic cancer. A recent study found that Smoothened is upregulated in pancreatic cancer associated fibroblasts (CAF) relative to control fibroblasts and that CAF transduce the sonic hedgehog ligand downstream intracellular signal to activate expression of Gli1. In microarray analysis of primary pancreatic cancers, intraductal papillary mucinous neoplasms (IPMN), and chronic pancreatitis tissue samples, stromal fibroblasts in chronic pancreatitis were weakly positive for Smoothened expression in 73.0% of cases, Smoothened expression in IPMN was 80.0% through weaker than in pancreatic cancers. Smo was heterogeneously expressed in stromal fibroblasts in 96.2% of pancreatic cancers and siRNA knockdown of SMO blocked the ability of CAF to induce Gli1 expression (Walter *et al.*, 2010).

Gemcitabine is the currently accepted standard of care in advanced pancreatic cancer. Efforts to improve the efficacy of gemcitabine have been mainly unsuccessful. A study by the National Cancer Institute of Canada Clinical Trials Group found a survival benefit for the combination of gemcitabine and erlotinib over gemcitabine alone for 191 days compared with 177days (HR for death 0.82; $p=0.02$;) (Moore *et al.*, 2007). Though statistically significant, the incremental gain in overall survival was not of appreciable clinical importance; however, the paradigm of small molecule targeted therapeutics having efficacy in this disease was established. Presently, there are six ongoing studies of Smoothened antagonists (GDC-0449 and IPI-926) incorporation onto treatment paradigms for pancreatic cancer being sponsored by the US National Cancer Institute (see Table 16.2).

F. Inhibition of Hedgehog signaling: A therapeutic advance

Hedgehog-directed therapeutics offer new hope in the treatment of pancreatic cancer. It potentially may cause tumor regression by intrinsic inhibition of the hedgehog pathway, decrease the cancer stem population, have a paracrine effect on the tumor stromal cells, and improve the therapeutic efficacy of cytotoxic drugs by affecting the desmoplastic fibrotic response and improving chemotherapeutic drug bioavailability. The NCI sponsored trial of GDC-0449 and erlotinib hydrochloride in metastatic pancreatic cancer or unresectable solid tumor is one potentially important advance. Trials with Smoothened antagonists are most likely to have positive findings when the study design involves a Smoothened antagonist given as part of a therapeutic combination. A study in prostate cancer showed that cytotoxicity was induced when cyclopamine and gefitinib a small molecule inhibitor of EGFR were combined (Mimeault *et al.*, 2006; Zhou *et al.*, 2009). Studies have also shown that the phosphoinositide 3-kinase/AKT pathway is essential for Sonic Hedgehog signaling (Riobó *et al.*, 2006). Combination treatment with PI-3 kinase or mTOR inhibitors and a Smoothened antagonist is a potentially important combination in treatment of malignant diseases in which hedgehog is implicated. However, it should be noted with regard to the specific case of pancreatic cancer KRAS has been shown to activate hedgehog in pancreatic ductal adenocarcinoma cells by the RAF/MEK/MAPK pathway and not the PI3K/AKT pathway. A clinical trial with combined treatment with a MEK inhibitor and Smoothened antagonist may arise in the future (Ji *et al.*, 2007; Riobó *et al.*, 2006). It is also notable that an analysis of RAS and Hedgehog signaling found that oncogenic RAS induces Sonic hedgehog ligand expression and inhibits the canonical (HH-Ptch-Smo-initiated) Hedgehog pathway. It further modulates the (TGF-β-initiated) Hedgehog pathway; all these molecular consequences are done by activating DYRK1B. The implication of these findings is that mutant KRAS induces hedgehog signaling in neighboring cells and simultaneously blocks hedgehog signaling in Sonic hedgehog producing tumor cells (Lauth *et al.*, 2010).

Yauch *et al.* (2008) in their experiments establishing a paracrine requirement for hedgehog signaling in tumorigenesis found that inhibition of Hh signaling in the stroma can also modulate factors, such as insulin-like growth factor and Wnt signaling pathway components. Strands of evidence exist implicating both wnt and insulin-like growth factor in pancreatic cancer, reemphasising the potential therapeutic importance of inhibition of the hedgehog pathway. In a study of human pancreatic ductal adenocarcinoma cell lines and their xenografts, the following was discovered: (i) hedgehog promotes neovascularisation in pancreatic cancers by regulating Ang-1 and IGF-1 expression in bone-marrow derived proangiogenic cells and (ii) host-derived Ang-1 and IGF-1 mRNA levels were downregulated by cyclopamine in tumor xenografts (Nakamura *et al.*, 2010). Regarding wnt, *in vitro* and *in vivo* studies in human pancreatic tumor cell lines and ectopic mouse model attempted to resensitise gemcitabine refractory pancreatic cells with masitinib (a multitargeted tyrosine kinase inhibitor). Masitinib was found not to have single agent activity but it sensitized Mia Paca2 cells to gemcitabine with a 400-fold reduction in the IC_{50}. Interestingly in transcriptional analysis, the Wnt/β-catenin signaling pathway was downregulated in the cell lines resensitized by the masitinib/gemcitabine combination.

Greater treatment efficacy in pancreatic cancer is likely to emerge in clinical trials of chemotherapeutics when Smoothened antagonists which target both tumor and stromal cells are used. Pancreatic cancer tumor stromal cells are responsible for hypovascular tumor perfusion due to desmoplasia. Locally advanced and localized pancreatic cancers offer the most favorable theoretical opportunity in this respect. Further, the benefit of selective targeting of pancreatic cancer initiating stem cells, tumor cell proliferation, and growth offers additional potential therapeutic benefit.

IV. Conclusion

The 5-year survival rate for pancreatic cancer is 4% compared with the 5-year survival of an American patient diagnosed with breast cancer of 86% and 97% for prostate cancer (US, National Cancer Institute, Cancer Statistics). Preclinical and clinical evidence exists for the important role hedgehog signaling has in pancreatic cancer tumorigenesis with over expression of Hedgehog ligand occurring in 70% of cases. LAPC representing 30–40% of pancreatic cancers is the most likely subset to derive benefit from inhibition of hedgehog signaling by targeting Smoothened. A National Cancer Institute phase II study of GDC-0449 and erlotinib in pancreatic cancer and solid tumors is one potentially important therapeutic advance, and further NCI trials involving Smoothened antagonists are accruing or planned. A randomize phase II trial of gemcitabine +/− GDC-0449 for metastatic pancreatic cancer is also currently ongoing. Hedgehog inhibition incorporated onto treatment paradigms hopefully will offer improved outcomes in this comparatively chemo- and radio-resistant malignancy.

REFERENCES

Bailey, J. M., Swanson, B. J., Hamada, T., Eggers, J. P., Singh, P. K., Caffery, T., Ouellette, M. M., and Hollingsworth, M. A. (2008). Sonic hedgehog promotes desmoplasia in pancreatic cancer. *Clin. Cancer Res.* **14**(19), 5995–6600.

Bailey, J. M., Mohr, A. M., and Hollingsworth, M. A. (2009). Sonic hedgehog paracrine signalling regulates metastasis and lymphangiogenesis in pancreatic cancer. *Oncogene* **28**(40), 3513–3525.

Berman, D. M., Karhadkar, S. S., Maitra, A., Montes De Oca, R., Gerstenblith, M. R., Briggs, K., Parker, A. R., Shimada, Y., Eshleman, J. R., Watkins, D. N., and Beachy, P. A. (2003). Widespread requirement for Hedgehog ligand stimulation in growth of digestive tract tumours. *Nature* **425**(6960), 846–851.

Choi, Y. L., Soda, M., Yamashita, Y., Ueno, T., Takashima, J., Nakajima, T., Takeuchi, K., Hamada, T., Haruta, H., Ishikawa, Y., Kimura, H., Mitsudomi, T., *et al.* (2010). EML4-ALK mutations in lung cancer that confer resistance to ALK inhibitors. *N. Engl. J. Med.* **363,** 1734–1739.

Dellovade, T., Romer, J. T., Curran, T., and Rubin, L. L. (2006). The hedgehog pathway and neurological disorders. *Annu. Rev. Neurosci.* **29,** 539–563.

Developmental genetics and birth defects. Chapter 14, P441. Genetics in medicine. Thompson and Thompson, ISBN 9781416030805.

Dierks, C., Beigi, R., Guo, G. R., Zirlik, K., Stegert, M. R., Manley, P., Trussell, C., Schmitt-Graeff, A., Landwerlin, K., Veelken, H., and Warmuth, M. (2008). Expansion of Bcr-Abl-positive leukemic stem cells is dependent on Hedgehog pathway activation. *Cancer Cell* **14**(3), 238–249.

Evangelista, M., Tian, H., and de Sauvage, F. J. (2006). The hedgehog signaling pathway in cancer. *Clin. Cancer Res.* **12**(20), 5924–5928.

Feldmann, G., Dhara, S., Fendrich, V., Bedja, D., Beaty, R., Mullendore, M., Karikari, C., Alvarez, H., Iacobuzio-Donahue, C., Jimeno, A., Gabrielson, K. L., Matsui, W., *et al.* (2007). Blockade of hedgehog signalling inhibits pancreatic cancer invasion and metastases: A new paradigm for combination therapy in solid cancers. *Cancer Res.* **67**(5), 2187–2196.

Feldmann, G., Habbe, N., Dhara, S., Bisht, S., Alvarez, H., and Fendrich, V. (2008). Hedgehog inhibition prolongs survival in a genetically engineered mouse model of pancreatic cancer. *Gut* **57**(10), 1420–1430.

Garber, K. (2010). Stromal depletion goes on trial in pancreatic cancer. *J. Natl. Cancer Inst.* **102**(7), 448–450. Cited abstract (2009) *J. Clin. Oncol.* **27**,15s (suppl; abstr 4525).

García de Marcos, J. A., Dean-Ferrer, A., Arroyo Rodríguez, S., Calderón-Polanco, J., Alamillos Granados, F. J., and Poblet, E. (2009). Basal cell nevus syndrome: Clinical and genetic diagnosis. *Oral Maxillofac. Surg.* **13**(4), 225–230.

Hawes, R. H., Xiong, Q., Waxman, I., Chang, K. J., Evans, D. B., and Abbruzzese, J. L. (2000). A multispeciality approach to approach to the diagnosis and management of pancreatic cancer. *Am. J. Gastroenterol.* **95,** 17–31.

Hedgehog signalling in human disease. Molecular biology intelligence unit. Ariel Ruiz i Altaba. Chapter 1, p2.

Heiser, P. W., Lau, J., Taketo, M. M., Herrera, P. L., and Hebrok, M. (2006). Stabilization of beta-catenin impacts pancreas growth. *Development* **133**(10), 2023–2032.

Heller, R. S., Dichmann, D. S., Jensen, J., Miller, C., Wong, G., Madsen, O. D., and Serup, P. (2002). Expression patterns of Wnts, Frizzleds, sFRPs, and misexpression in transgenic mice suggesting a role for Wnts in pancreas and foregut pattern formation. *Dev. Dyn.* **225**(3), 260–270.

Hermann, P. C., Huber, S. L., Herrler, T., Aicher, A., Ellwart, J. W., Guba, M., Bruns, C. J., and Heeschen, C. (2007). Distinct populations of cancer stem cells determine tumor growth and metastatic activity in human pancreatic cancer. *Cell Stem Cell* **1**(3), 313–323.

Ji, Z., Mei, F. C., Xie, J., and Cheng, X. (2007). Oncogenic KRAS activates hedgehog signaling pathway in pancreatic cancer cells. *J. Biol. Chem.* **282**(19), 14048–14055.

Jones, S., Zhang, X., Parsons, D. W., Lin, J. C., Leary, R. J., Angenendt, P., *et al.* (2008). Core signalling pathways in human pancreatic cancers revealed by global genomic analyses. *Science* **321**(5897), 1801–1806.

Kelleher, F. C., *et al.* (2006). Common critical pathways in embryogenesis and cancer. *Acta Oncol.* **45**(4), 375–388.

Kimura, H., Ng, J. M., and Curran, T. (2008). Transient inhibition of the Hedgehog pathway in young mice causes permanent defects in bone structure. *Cancer Cell* **13**(3), 249–260.

Krzyanowska, M. K., Weeks, J. C., and Earle, C. C. (2003). Treatment of locally advanced pancreatic cancer in the real world: Population-based practices and effectiveness. *J. Clin. Oncol.* **21**(18), 3409–3414.

Lauth, M., Bergström, A., Shimokawa, T., and Toftgård, R. (2007). Inhibition of GLI-mediated transcription and tumor cell growth by small-molecule antagonists. *Proc. Natl. Acad. Sci. USA* **104**(20), 8455–8460.

Lauth, M., Bergström, A., Shimokawa, T., Tostar, U., Jin, Q., Fendrich, V., Guerra, C., Barbacid, M., and Toftgård, R. (2010). DYRK1B-dependent autocrine-to-paracrine shift of Hedgehog signaling by mutant RAS. *Nat. Struct. Mol. Biol.* **17**(6), 718–725.

Li, C., Heidt, D. G., Dalerba, P., Burant, C. F., Zhang, L., Adsay, V., Wicha, M., Clarke, M. F., and Simeone, D. M. (2007). Identification of pancreatic cancer stem cells. *Cancer Res.* **67,** 1030–1037.

Martin, R. C., 2nd, Scoggins, C. R., Egnatashvili, V., Staley, C. A., McMasters, K. M., and Kooby, D. A. (2009). Arterial and venous resection for pancreatic adenocarcinoma: Operative and long-term outcomes. *Arch. Surg.* **144**(2), 154–159.

Merchant, A. A., and Matsui, W. (2010). Targeting hedgehog—A cancer stem cell pathway. *Clin. Cancer Res.* **16**(12), 3130–3150.

Mimeault, M., Moore, E., Moniaux, N., Hénichart, J. P., Depreux, P., Lin, M. F., and Batra, S. K. (2006). Cytotoxic effects induced by a combination of cyclopamine and gefitinib, the selective hedgehog and epidermal growth factor receptor signalling inhibitors, in prostate cancer cells. *Int. J. Cancer* **118**(4), 1022–1031.

Moore, M. J., Goldstein, D., Hamm, J., Figer, A., Hecht, J. R., Gallinger, S., Au, H. J., Murawa, P., Walde, D., Wolff, R. A., Campos, D., Lim, R., *et al.* (2007). Erlotinib plus gemcitabine compared with gemcitabine alone in patients with advanced pancreatic cancer: A phase III trial of the National Cancer Institute of Canada Clinical Trials Group. *J. Clin. Oncol.* **25,** 1960.

Morton, J. P., Mongeau, M. E., Klimstra, D. S., Morris, J. P., Lee, Y. C., Kawaguchi, Y., *et al.* (2007). Sonic hedgehog acts at multiple stages during pancreatic tumorigenesis. *Proc. Natl. Acad. Sci. USA* **104**(12), 5103–5108.

Nakamura, K., Sasajima, J., Mizukami, Y., Sugiyama, Y., Yamazaki, M., Fujii, R., Kawamoto, T., Koizumi, K., Sato, K., Fujiya, M., Sasaki, K., Tanno, S., *et al.* (2010). Hedgehog promotes neovascularisation in pancreatic cancers by regulating Ang-1 and IGF-1 expression in bone-marrow derived pro-angiogenic cells. *PLoS One* **5**(1), e8824.

Oettle, H., Post, S., Neuhaus, P., Gellert, K., Langrehr, J., Ridwelski, K., Schramm, H., Fahlke, J., Zuelke, C., Burkart, C., Gutberlet, K., Kettner, E., *et al.* (2007). Adjuvant chemotherapy with gemcitabine vs observation in patients undergoing curative-intent resection of pancreatic cancer: A randomized controlled trial. *JAMA* **297**(3), 267–277.

Olive, K. P., Jacobetz, M. A., Davidson, C. J., Gopinathan, A., McIntyre, D., Honess, D., Madhu, B., Goldgraben, M. A., Caldwell, M. E., Allard, D., Frese, K. K., Denicola, G., *et al.* (2009). Inhibition of Hedgehog signalling enhances delivery of chemotherapy in a mouse model of pancreatic cancer. *Science* **324**(5933), 1457–1461.

Pasca di Magliano, M., Biankin, A. V., Heiser, P. W., Cano, D. A., Gutierrez, P. J., Deramaudt, T., Segara, D., Dawson, A. C., Kench, J. G., Henshall, S. M., Sutherland, R. L., Dlugosz, A., *et al.* (2007). Common activation of canonical Wnt signaling in pancreatic adenocarcinoma. *PLoS One* **2**(11), e1155.

Riobó, N. A., Lu, K., Ai, X., Haines, G. M., and Emerson, C. P., Jr. (2006). Phosphoinositide 3-kinase and Akt are essential for Sonic Hedgehog signalling. *Proc. Natl. Acad. Sci. USA* **103**(12), 4505–4510.

Romer, J. T., Kimura, H., Magdaleno, S., Sasai, K., Fuller, C., Baines, H., Connelly, M., Stewart, C. F., Gould, S., Rubin, L. L., and Curran, T. (2004). Suppression of the Shh pathway using a small molecule inhibitor eliminates medulloblastoma in Ptc1(+/−)p53 (−/−) mice. *Cancer Cell* **6**(3), 229–240.

Rubin, L. L., and de Sauvage, F. J. (2006). Targeting the Hedgehog pathway in cancer. *Nat. Rev. Drug Discov.* **5**(12), 1026–1033.

Rudin, C. M., Hann, C. L., Laterra, J., Yauch, R. L., Callahan, C. A., Fu, L., Holcomb, T., Stinson, J., Gould, S. E., Coleman, B., LoRusso, P. M., Von Hoff, D. D., *et al.* (2009). Treatment of medulloblastoma with hedgehog pathway inhibitor GDC-0449. *N. Engl. J. Med.* **361**(12), 1173–1178.

Siu, L. L., Papadopoulos, K. P., Alberts, S. R., Taylor, S., Patnaik, A., Chen, E. X., *et al.* (2009). A first-in-human, phase I study of an oral hedgehog pathway antagonist, BMS-833923 (XL139), in subjects with advanced or metastatic solid tumors. *Mol. Cancer Ther.* **8,** A5510.1158/1535-7163.TARG-09-A55. Abstract.

Talpaz, M., Shah, N. P., Kantarjian, H., Donato, N., Nicoll, J., Paquette, R., Cortes, J., O'Brien, S., Nicaise, C., Bleickardt, E., Blackwood-Chirchir, M. A., Iyer, V., *et al.* (2006). Dasatinib in imatinib-resistant Philadelphia chromosome-positive leukemias. *N. Eng. J. Med.* **354,** 2531–2541.

Thayer, S. P., di Magliano, M. P., Heiser, P. W., Nielsen, C. M., Roberts, D. J., Lauwers, G. Y., Qi, Y. P., Gysin, S., Fernández-del Castillo, C., Yajnik, V., Antoniu, B., McMahon, M., *et al.* (2003a). Hedgehog is an early and late mediator of pancreatic cancer tumorigenesis. *Nature* **425**(6960), 851–856.

Tian, H., Callahan, C. A., DuPree, K. J., Darbonne, W. C., Ahn, C. P., Scales, S. J., and de Sauvage, F. J. (2009). Hedgehog signalling is restricted to the stromal compartment during pancreatic carcinogenesis. *Proc. Natl. Acad. Sci. USA* **106**(11), 4254–4259.

Von Hoff, D. D., LoRusso, P. M., Rudin, C. M., Reddy, J. C., Yauch, R. L., Tibes, R., Weiss, G. J., Borad, M. J., Hann, C. L., Brahmer, J. R., Mackey, H. M., Lum, B. L., *et al.* (2009). Inhibition of the hedgehog pathway in advanced basal-cell carcinoma. *N. Engl. J. Med.* **361**(12), 1164–1172.

Walter, K., Omura, N., Hong, S. M., Griffith, M., Vincent, A., Borges, M., and Goggins, M. (2010). Overexpression of smoothened activates the sonic hedgehog signaling pathway in pancreatic cancer-associated fibroblasts. *Clin. Cancer Res.* **16**(6), 1781–1789.

Willett, C. G., Czito, B. G., Bendell, J. C., and Ryan, D. P. (2005). *J. Clin. Oncol.* **23**(20), 4538–4544.

Yauch, R. L., Gould, S. E., Scales, S. J., Tang, T., Tian, H., Ahn, C. P., Marshall, D., Fu, L., Januario, T., Kallop, D., Nannini-Pepe, M., Kotkow, K., *et al.* (2008). A paracrine requirement for hedgehog signalling in cancer. *Nature* **455**(7211), 406–410.

Yauch, R. L., Dijkgraaf, G. J., Alicke, B., Januario, T., Ahn, C. P., Holcomb, T., Pujara, K., Stinson, J., Callahan, C. A., Tang, T., Bazan, J. F., Kan, Z., *et al.* (2009). Smoothened mutation confers resistance to a Hedgehog pathway inhibitor in medulloblastoma. *Science* **326**(5952), 572–574.

Zhou, B. B., Zhang, H., Damelin, M., Geles, K. G., Grindley, J. C., and Dirks, P. B. (2009). Tumour-initiating cells: Challenges and opportunities for anticancer drug discovery. *Nat. Rev. Drug Discov.* **8**(10), 806–823.

CHAPTER SEVENTEEN

Sonic Hedgehog Signaling and Potential Therapeutic Indications

Nicholas C. Bambakidis *and* Kaine Onwuzulike

Contents

Abstract

Sonic hedgehog (Shh) is a glycoprotein molecule that is expressed throughout the central nervous system (CNS). It is important during neurodevelopment, particularly through its induction of endogenous neural precursor cells and neural stem cells. The signaling system by which Shh has its diffuse effects on multiple end organs depends on a downstream signaling cascade initiated by binding to the receptor Patched (PTC) on Shh-specific target cells. The downstream signaling cascade has widespread effects on multiple organ systems. Research into tumor suppressive effects and germline mutations has uncovered important implications in neuro-oncology. Modulating the neuroproliferative effects of the Shh signaling cascade implicated a potential role in ameliorating the effects of CNS injury, particularly in animal models of spinal cord injury and cerebral cortical ischemia. This chapter provides a summary of the Shh signaling mechanism and a review of its impact in these neurological disorders.

Department of Neurological Surgery, University Hospitals Case Medical Center, Cleveland, Ohio, USA

Vitamins and Hormones, Volume 88
ISSN 0083-6729, DOI: 10.1016/B978-0-12-394622-5.00017-1

I. Sonic Hedgehog Discovered (Shh)

Two sentinel scientific events herald the discovery of the Shh gene, the first of which occurred around the mid-1950s when a series of experiments on chicken embryos led to the discovery of what is known as the Zone of Polarizing Activity (ZPA) found in the limb bud of developing embryos. Researchers John Saunders and Edgar Zwilling demonstrated that moving tissue within the limb buds of embryos altered the development of limbs. Another breakthrough came in the 1970s, with the work of Eric Wieschaus and Christiane Nusslein-Volhard in *Drosophila melanogaster* (house fly) unzipping a gene that caused asymmetry in the body segments. This mutation gives the Drosophila embryo a dramatic "spiny" phenotype reminiscent of the ubiquitous old-world mammal, the hedgehog (Hh), hence the term was coined. Soon thereafter, correlations were made between the ZPA and an Hh gene leading to many speculations regarding its developmental role in the appendages of mammals and humans (Shubin, 2008). However, the vertebrate limb exhibits a mode of development dissimilar from that seen in the Drosophila species. In vertebrates, limbs develop from contributions of both the mesoderm and ectoderm.

Shh is a glycoprotein molecule encoded by a single gene and expressed widely throughout the central nervous system (CNS) that has been shown to be associated with proliferative capacity of both endogenous neural precursor cells during embryonic development and neural stem cells in the adult subventricular zone (SVZ). It has a diffuse effect on multiple end organs during development and has been implicated as a modulator of neural progenitors in neoplastic processes such as prostate cancer and medulloblastoma. The myriad downstream effects of Shh are widespread yet not completely understood. It has been shown to exert influences in the development of neural tissue embryonic notochord whereby it modulates the induction of motor neurons and oligodendrocyte precursor cells. The second messenger response to Shh signaling is mediated by receptors, cytoplasmic factors, and transcription factors, which act positively and negatively to modulate Shh activity in the patterning of diverse cell types in the ventral neural tube (Capdevila and Johnson, 2000; Fig. 17.1).

II. Shh Signaling in Vertebrates

During embryonic vertebrate development, the spatial and temporally coordinated development of an extraordinary array of highly specialized neural tissues occurs. This process of neural induction results in the acquisition of specialized neural properties of the neural plate, which is a sheet of

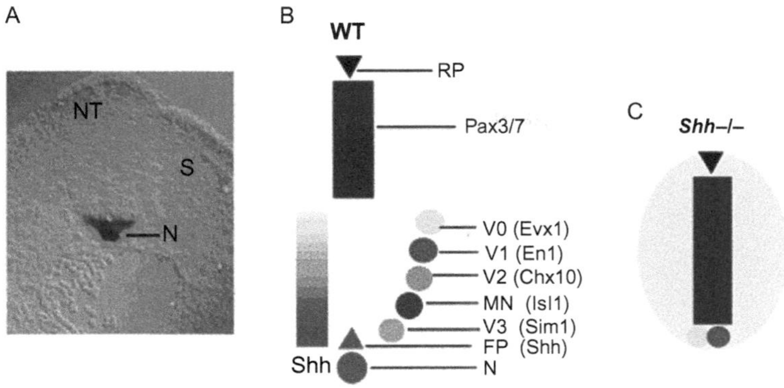

Figure 17.1 Neural patterning in spinal cords. Reprinted with permission from Litingtung and Chiang (2000). (See Color Insert.)

epithelial columnar cells that overlies the axial mesoderm cells. The foremost sign of neurulation occurs as the neural plate involutes at the ventral midline through a process involving neuroepithelial signaling. Eventually, the dorsal edges of the neural tube converge at midline forming the neural tube, which heralds the establishment of dorsal versus ventral orientation as evidenced by regional specific gene expression. At the dorsal and ventral midline, specialized cells collectively known as the roof and floor plate differentiate forming a neuropore (Capdevila and Johnson, 2000; Ingham and Placzek, 2000; Ribes *et al.*, 2010). During this gastrulation and somitogenesis, the neuropore is dependent upon a gradient concentration of Shh. For example, higher Shh concentrations results in the formation of ventral cell types in close proximity to the notochord, whereas lower Shh concentration induces cells further from the notochord. Specific evidence regarding the directionality in response to Shh signaling can be found on the cellular and embryonic levels. Transplantation experiments in chick embryos confirmed the notochord and floor plate to be dynamic regulators in embryonic directionality, releasing ventralizing signals, while concomitantly suppressing markers of dorsal formation. Conversely, removal of the notochord *in vivo* inhibited floor plate and motor neuron differentiation and ultimately the expression of dorsal regional markers within ventral aspects of the neural tube (Ribes *et al.*, 2010; Fig. 17.2). On a cellular level, strong support can be summoned from experimentation showing the development of specific cell types within neural explants when neural tissue is cultured with an Shh signaling source (Dierks, 2010; Farzan *et al.*, 2008; Ingham and Placzek, 2000; Patten and Placzek, 2000). Moreover, induction of Shh precursors is specifically inhibited by a monoclonal antibody that binds to Shh and prevents its downstream signaling and induction. Further, reinforcement of the signaling role of Shh comes via well-designed gain-of-function genetic studies demonstrating ectopic differentiation of distinct

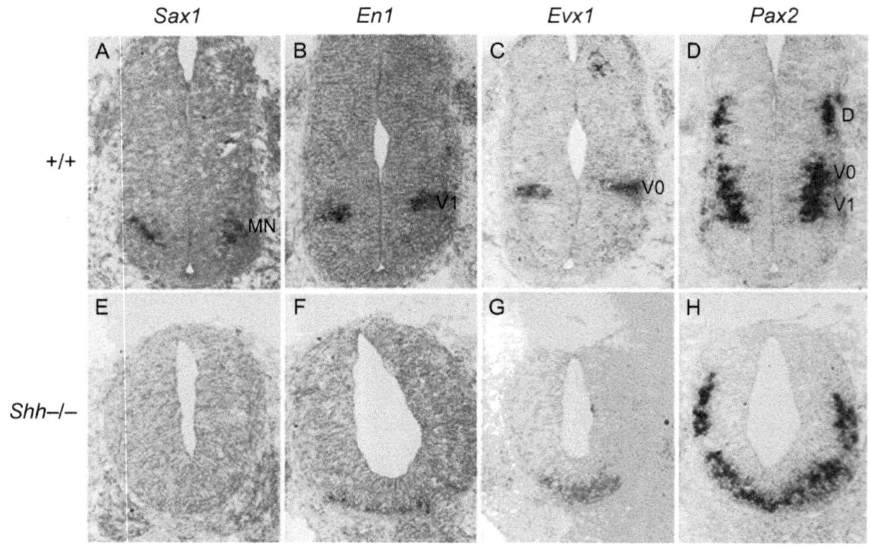

Figure 17.2 Absence of motor neurons and disrupted dorsal-ventral patterning. Reprinted with permission from Litingtung and Chiang (2000). (See Color Insert.)

ventral cell types from ectopic Shh transposition *in vivo* (Ribes *et al.*, 2010). In complementary gain-of-function studies, a constitutively activated Shh signaling pathway results in the ectopic differentiation of floor plate and motor neurons. Conversely, loss-of-function paradigms in Shh-null mice results in a multitude of abnormalities including cyclopia, poor neural patterning, lack of limb growth, and abnormal development of organs and foregut. These studies lend credence to the necessity of Shh in the formation and development of ventral cells and motor neurons along the dorsal-ventral axis (Patten and Placzek, 2000).

In addition to ventral midline cells, a variety of diverse cell types are induced by the signaling action of Shh. These include the hindbrain, spinal interneurons, and oligodendrocytes, which together function in regulating motor and sensory function. In the brainstem, Shh is serotonergic and influences dopaminergic cell proliferation, which plays a key role in human biopsychological function, coordinated movement and disease processes (such as Parkinson disease), and numerous psychiatric disorders (Fig. 17.3).

One of the most intriguing entities in embryologic development is the manner in which cells that become tissues, which form organs that cooperate to maintain homeostasis, obtain their individual identities. The details governing Shh diversification has been extensively investigated. These studies collectively endorse the dose–response nature of neural explants.

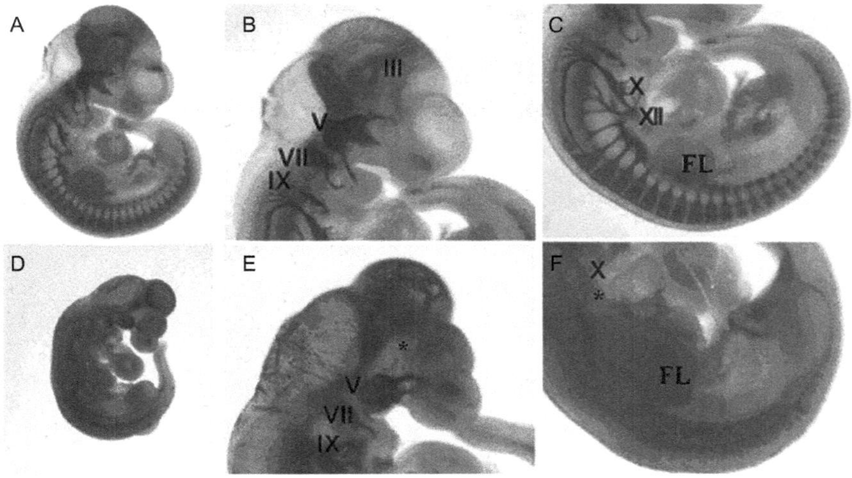

Figure 17.3 Lack of cranial motor axons in Shh. Reprinted with permission from Litingtung and Chiang (2000). (See Color Insert.)

III. Shh Second Messenger Systems

The Shh signaling pathway modulates tissue formation and reconstruction via Shh-dependent autocrine and paracrine mechanisms, which influence the diversification of Shh-responsive cell populations. The innate second messenger system required to enact the downstream effects of Shh ligand binding is daunting and complex, with many subsets and pathways, each with its distinct effects. Ligand-producing cells manufacture and release the Shh ligand. Soon thereafter, the signal sequence is cleaved to yield a 45-kDa precursor protein, which is further cleaved into two discrete peptides that yield an amino terminal domain and a carboxy terminal domain, which then interface and bind with the Shh receptor Patched (PTC) on Shh-specific target cells, initiating a signaling cascade (Patten and Placzek, 2000; Fig. 17.4). This initial ligand binding dissociates the coreceptor Smoothened (Smo) from inhibition by PTC and facilitates Shh-initiated downstream signaling, with the end result of nuclear aggregation and binding of glial family transcription factors to Shh-target genes. Shh pathway activation facilitates the growth and viability of Shh-responsive cells; conversely, attenuating or blocking Shh signaling can result in apoptosis. Together, this dynamic PTC, which binds the Shh proteins, and Smo, which transduces Shh signals intracellularly, is responsible for the unique cellular composition of various neural tissues. In the absence of Shh, PTC represses Shh-target genes, and binding of Shh to PTC relieves

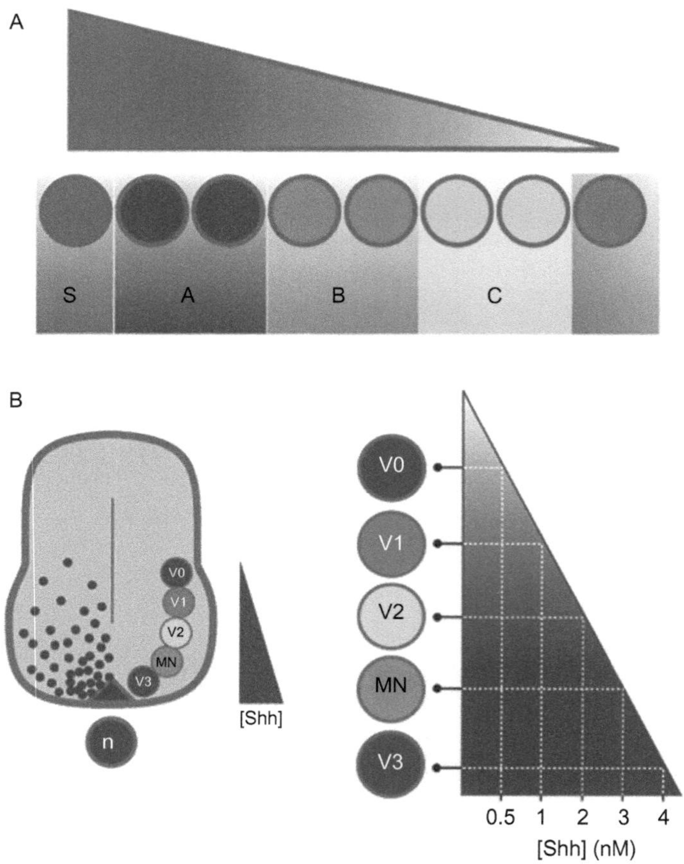

Figure 17.4 Shh acts in a graded manner to pattern the ventral neural tube. Reprinted with permission from Briscoe (2009). (See Color Insert.)

this inhibition. Once Shh is released, its activity and propagation is governed in a temporally and spatially precise manner (Fig. 17.5). A variety of accessory proteins and coreceptors are required to enact this regulation of Shh activity. Heparan sulfate proteoglycans are large extracellular molecules of heparan sulfate glycosaminoglycan polymers, which are part of the milieu regulating Shh signaling via two mechanisms: indirectly via the facilitation of movement of N-palmitoylated Hh and directly via signal reception by the Shh receiving cell (Toftgard, 2000). A very unique feature of the Shh family is the discovery that the active signaling form of the protein is biologically coupled to cholesterol, which seems to influence the range of its signaling activity.

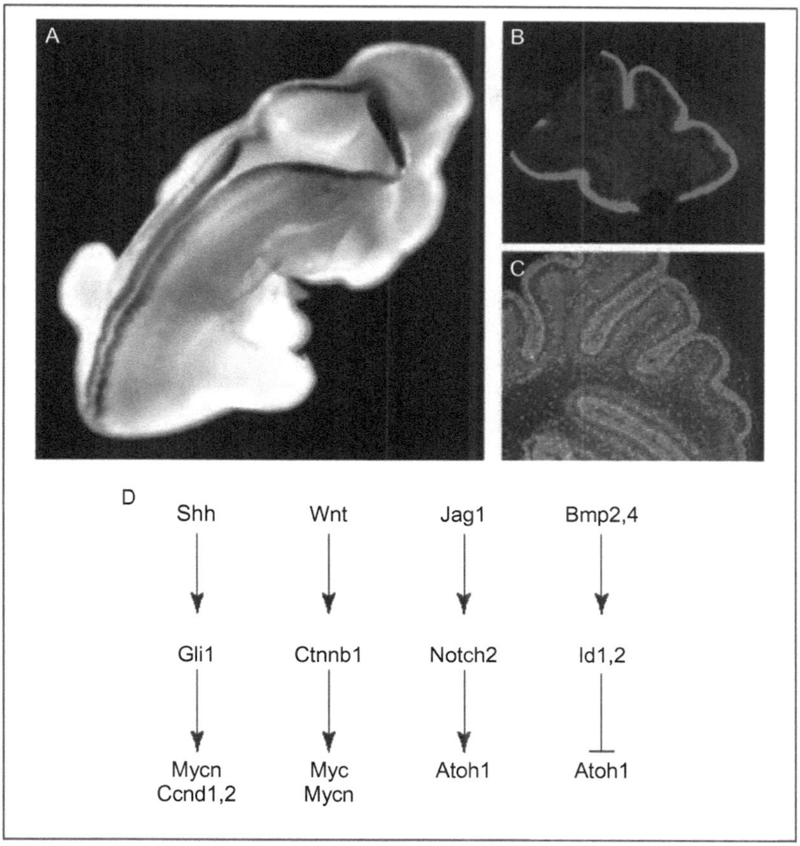

Figure 17.5 Changing patterns of gene expression and signaling pathways in cerebellar granule cell progenitor expansion. Reprinted with permission from Hatten and Roussel (2011). (See Color Insert.)

IV. Shh in Stem Cell Proliferation

Stem cell therapy in the setting of CNS injury is a major area of exciting and promising research, adding a powerful autologous mechanism to the scant armamentarium for the restoration of function after CNS injury. In theory, the pluripotent nature of these precursor cell populations makes it possible for them to develop along any neural lineage. As such, their progeny potentially can aid in the process of neural circuitry restoration, such as neuronal regeneration, remyelination, prevention of demyelination, and amelioration of scar formation (Bambakidis *et al.*, 2008b).

Since their discovery, it was widely believed that stem cells were present briefly during mammalian embryonic development; however, we now know that these cells maintain highly proliferative and migratory characteristics that concomitantly contribute to their potential contribution in the restoration of neurologic function after injury. Additional investigation has revealed the discrete anatomical zones in which these cells primarily reside, which include but are not limited to the SVZ of the lateral ventricle, the hippocampus, the central canal of the spinal cord, and ependymal layer of the fourth ventricle, hypothalamus, optic nerve, and substantia nigra. Although normally quiescent, it is now believed that these cells remain available for replication and maturation throughout life. Injury to any of these key areas results in a unique proliferative response whereby most of the progenitor cells mature along an astroglial lineage. In rats, overexpression of Shh in the developing brain results in increased stem cell proliferation in the dorsal columns of the spinal cord. In the same accord, it has been experimentally observed that administration of complementary DNA encoding Shh expression in the adult rat brain results in a threefold increase in the production of endogenous stem cell DNA, which develops into mature neural glial cells (Bambakidis *et al.*, 2005, 2008a,b). Moreover, the oral administration of Shh agonists to stimulate the Shh pathway increases the proliferation of stem cells in the mature rodent cerebrum (Sousa and Fishell, 2010).

The biological factors that influence the development of neural progenitor cells along a particular lineage are likewise beginning to come to light. For example, erythrocyte growth factor (EGF), insulin-like growth factor-I, and fibroblast growth factor-2 (FGF-2) have been shown to have direct effects on the development of adult neural stem cells. Administration of FGF-2 and EGF results in a 50-fold increase in the degree of neurogenesis after ischemic injury. The molecular interaction of these cytokines is an integral part in the signaling cascade that influences the development of stem cells down a particular lineage (Liebner and Plate, 2010).

In the ongoing research to apply stem cell therapy to neurological diseases, the capacity to harness the pluripotent ability of endogenous stem cell populations is evident. These cell populations proliferate in response to a variety of injury states in the CNS, but in the absence of a supportive microenvironment, they contribute little if any significant functional recovery. Novel research on the regenerative potential of these stem cells in various disease states has demonstrated that proliferative potential of endogenous stem cells can be greatly upregulated in response to appropriate growth signals. Further investigation, with respect to the potential of such multipotent cells and the mechanisms that underlie the creation of a favorable microenvironment to this end, will hopefully lead to additional therapeutic alternatives in the setting of neurological diseases. Because endogenous source of pluripotent stem cells would require no exogenous stem cell transplantation, there is instant alleviation of the major ethical and moral considerations regarding their use. These characteristics play a promising role in the setting of trauma in which a combination of disruption of neuronal networks, demyelination, and astrocytic scar formation independently and collectively contribute to end organ injury. Models of traumatic brain injury are promising for the promotion of functional neurologic recovery.

V. Shh in Oncogenesis

Tumors of embryonal origin constitute the largest group of malignant brain tumors in children and are unfortunately associated with high morbidity and mortality. Recently, numerous collective studies have demonstrated that the signaling and molecular milieu underlying development, when inappropriately regulated, become the driving force for the formation of cancer. Individuals with germline mutations in the Shh pathway have a significantly increased incidence of medulloblastoma, the most common malignant brain tumor of infancy (Figs. 17.6–17.8). The annual incidence of medulloblastoma ranges from 0.48 in females to 0.75 in males per 100,000 children. The peak age at presentation is 7 years (de Bont *et al.*, 2008; Dierks, 2010). It occurs in the cerebellum during embryogenesis and is recognized by proliferation of cells from the external granular layer. This process occurs in response to growth factor secretion from Purkinje neurons, such as Shh. A particularly ominous characteristic of medulloblastoma is its ability to metastasize throughout the neuraxis. Numerous signaling pathways have been associated with medulloblastoma occurrence and proliferations including Shh. Currently, in clinical trials, Shh pathway inhibitors are strategically positioned to become the first approved targeted therapies for the treatment of medulloblastoma (Hatten and Roussel, 2011; Ridgway *et al.*, 2011).

With the recent discovery of the Shh receptor PTC as a tumor suppressor gene in which mutations underlie the human nevoid basal cell carcinoma syndrome (NBCCS), mutations of Shh signaling pathway have become indelibly linked to cancer. In addition to its cancerous predisposition, NBCCS is also associated with a number of developmental defects, including macrocephaly, calcification of falx cerebri, and polydactyly (Pfister *et al.*, 2010). Constitutive activation of the Shh pathway via mutations in PTC1 or coreceptor Smo are linked with the formation of numerous tumor types, such as rhabdomyosarcoma. Fetal rhabdomyomas and rhabdomyosarcomas are observed in a small subset of NBCCS patients, with a strong correlation to defects in Shh signaling supported by the presence of similar neoplasms in PTC1+/− murine lineages. Currently, point mutations in PTC1 have been reported in trichoepithelioma bladder and esophageal squamous cell

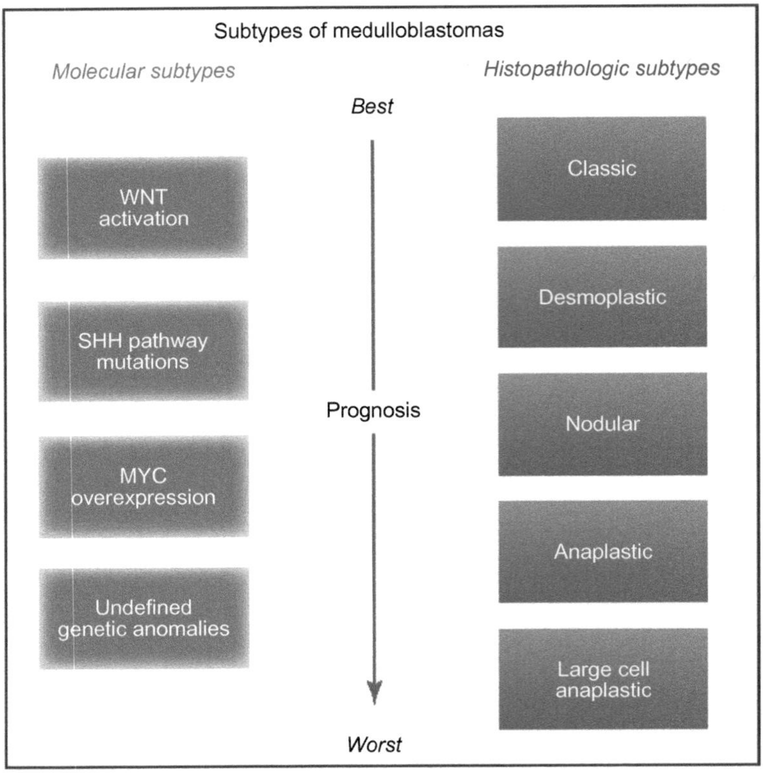

Figure 17.6 Subtypes of medulloblastomas. Reprinted with permission from Hatten and Roussel (2011). (See Color Insert.)

carcinoma. PTC mutations occur at a high rate (>50%) in both familial and sporadic NBCCS. These mutations result in functional inactivation and loss of function, actions consistent with PTC function as a tumor suppressor gene (de Bont *et al.*, 2008; Dierks, 2010; Hatten and Roussel, 2011; Omenetti and Diehl, 2008; Parkin and Ingham, 2008; Pfister *et al.*, 2010).

Shh pathway inhibitors have been shown to block the growth of hepatoblastoma cells in culture, raising the possibility that inappropriate activation of Shh pathway signaling is involved in hepatocarcinogenesis. To this end, dysregulation of the Shh pathway has also been implicated in the development of cholangiocarcinoma, hepatoblastoma specific subgroups, and hepatocellular carcinoma (Omenetti and Diehl, 2008). Basal cell carcinoma (BCC) is the most prevalent tumor in the Western world and has an estimated yearly incidence of greater than one million cases in the United States alone. BCCs are keratinocyte-derived regionally invasive tumors that rarely metastasize. Mutations in PTC1 occur at a fairly high rate in both familial and sporadic phenotypes. Loss of heterozygosity at the PTC1 locus in BCC

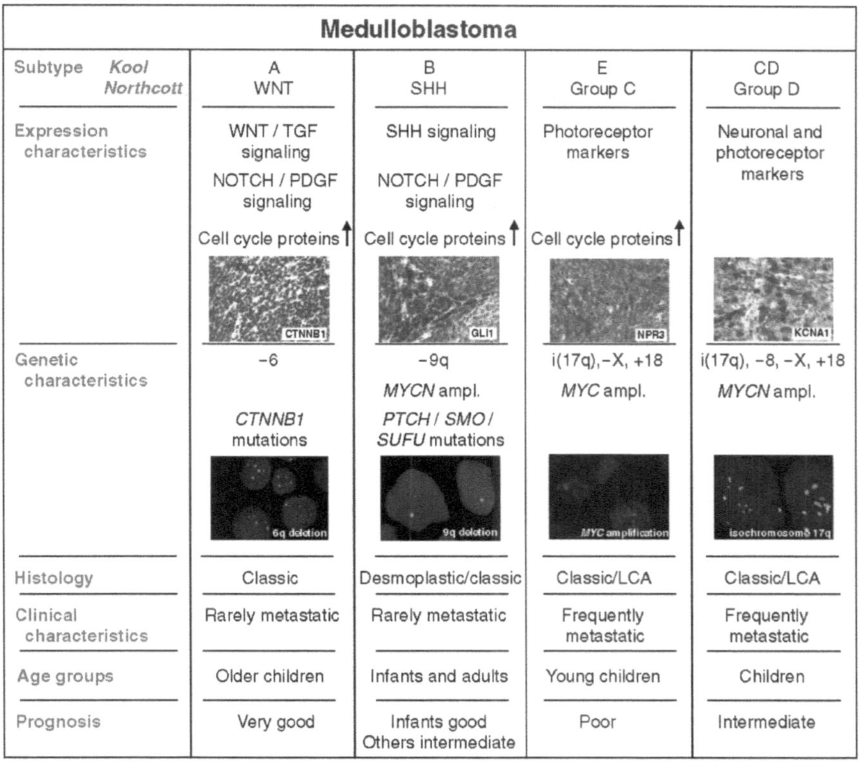

Figure 17.7 Medulloblastoma diagnostics. Reprinted with permission from Pfister *et al.* (2010). (See Color Insert.)

occurs with a frequency of 50–70%, suggesting that the vast majority of BCCs contain mutations associated with activation of the Shh signaling pathway.

Implications of Shh signaling defects with respect to breast cancer are sparse; however, two somatic missense mutations in PTC1 have been reported to be associated in the development of breast carcinoma (Dierks, 2010). PTC1 expression occurs during murine postnatal mammary gland development, and PTC1+/− heterozygous mice develop a number of breast tissue abnormalities, including ductal hyperplasias and dysplasias.

VI. Shh in Injury of the CNS

Animal models of CNS injury are active areas of intense scientific investigation. Common to all such models that include traumatic brain and spinal injury as well as some models of ischemic injury is a progressive

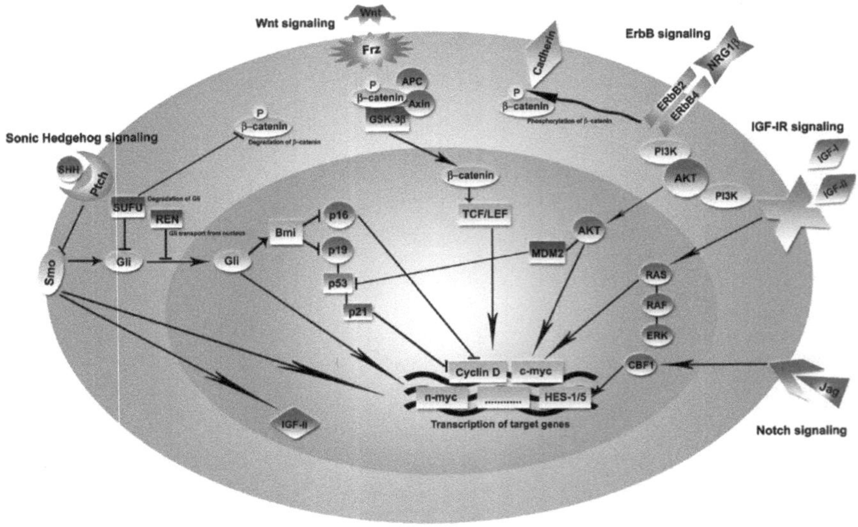

Figure 17.8 Signaling pathways involved in the development of the brain and pathogenesis of medulloblastoma and ependymoma. Reprinted with permission from de Bont *et al.* (2008). (See Color Insert.)

disruption of axonal transmission and subsequent neuronal and axonal degeneration (Figs. 17.9–17.11). These processes are preludes to the formation of astrocytic scar tissue. Any treatment algorithm must include the amelioration of the initial effects of ischemia and demyelination, followed by the promotion of neural regeneration. In contradistinction to low-impact or slowly progressive injuries, such as those caused by tumors, the first step in treatment of acute CNS injury must be removal of the offending agent. Given the chronic ischemic state of injured tissue in this setting, adequate tissue perfusion must occur. In the long term, regenerative therapies, including techniques to promote remyelination or prevent dysmyelination, can then be employed.

Potential therapeutic effects of modulation of the Shh signaling pathway have shown promise in the treatment of animal models of CNS injury. For example, administration of Shh agonist in a model of midthoracic spinal cord contusion in rodents has resulted in behavioral improvement over time. Research regarding spinal cord injury (SCI) has been particularly challenging, given the heterogeneity of clinical scenarios of SCI and the limited ability of the spinal cord to regenerate in response to contusion or crush injury (Bambakidis *et al.*, 2003, 2010).

The first scientific support of the regenerative ability of the injured spinal cord came from studies conducted by Richardson *et al.* (1980). This finding helped to promote a shift toward the development of research strategies

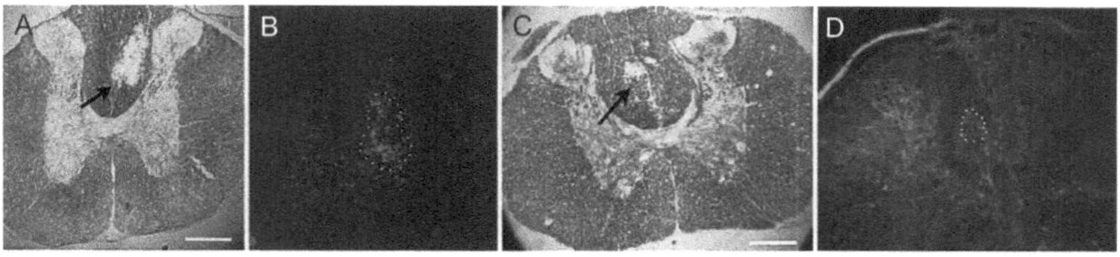

Figure 17.9 Photomicrographs showing midthoracic spinal cord cross-sections of a rat receiving spinal cord injury without Shh treatment (A and B) and following Shh injection (C and D). Reprinted with permission from Bambakidis *et al.* (2003). (See Color Insert.)

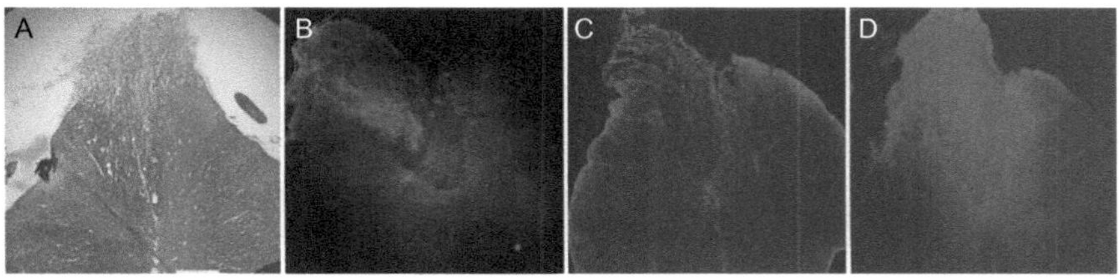

Figure 17.10 Photomicrographs showing posterior outgrowth characteristically observed in animals treated with a lesion and Shh. Reprinted with permission from Bambakidis *et al.* (2003). (See Color Insert.)

aimed at restoring neurologic function. Although the results in animal models have provided insight and been promising, models of treatment transferable to human patients are sparse. SCI represents a complex injury cascade. Therefore, it is reasonable to assert that no single "magic bullet" will sufficiently lead to functional recovery. Instead, a multimodality approach is needed to make appreciable gains in clinical treatment, which can be divided into those that focus on ameliorating the secondary injury cascade (neuroprotection) and those that target remyelination and axonal and neuronal regeneration (neuroregeneration). To date, most work on the potential benefits of stem cell therapy in treating SCI has concentrated on neuroregeneration. To this end, numerous cell types have emerged as candidates in the repertoire of cellular-based support and repair mechanism in SCI, based on their potential to aid in the formation of myelin, drive axonal growth, and potentiate the secretion of trophic factors that have neuroprotective effects and promote plasticity in the injured spinal cord. The activation and promotion of endogenous stem cells are of particular interest because 500,000–2,000,000 new cells are produced at the site of the injury within 1 month. These endogenous cells are upregulated after a contusion injury to the spinal cord, with the highest level of cellular induction occurring 3–7 days after injury. In their native state, these cells develop primarily along astrocytic lineages, thus contributing to scar formation and thereby inhibiting a

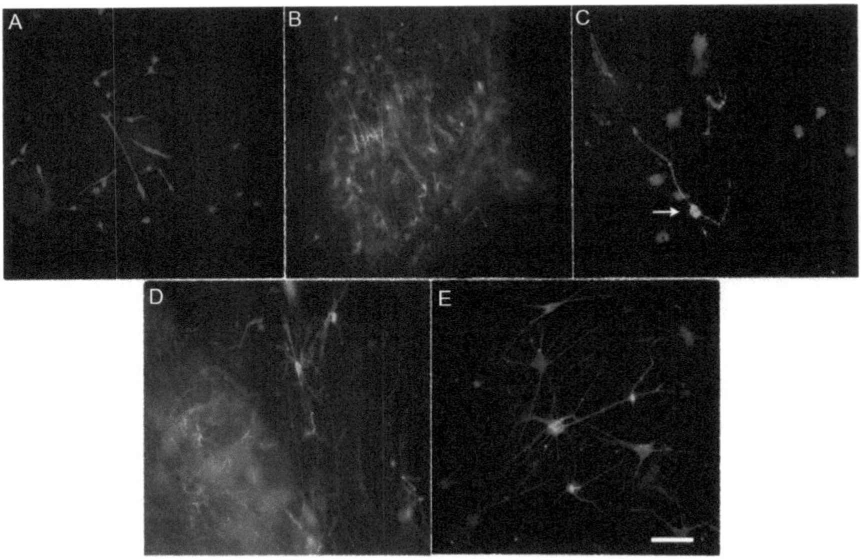

Figure 17.11 Photomicrographs showing that Shh treatment increases the number of neural precursors in posterior spinal cord cultures. Reprinted with permission from Bambakidis *et al.* (2003). (See Color Insert.)

restorative response. Current research efforts focus on the modification of this counterproductive inhibitory response.

The Shh protein, which is critical in oligodendroglial development and neurogenesis, dramatically increases the number of neuronal progenitor cells in the spinal cord. Interestingly, a similar phenomenon occurs in rats with contusion injury to the spinal cord. After oligodendrocytic precursors have been transplanted, significantly reduced demyelination and increased conduction velocity are observed across lesions. The availability of small-molecule analogs to the Shh pathway, which can be administered intravenously, allow replication of these experimental findings, obviating the need for direct injection into the injured spinal cord (Bambakidis *et al.*, 2003, 2005, 2008a,b, 2010). If significant strides in the treatment and recovery of individuals with SCI are to be made, a multimodal approach is required in a sequential manner. In the acute period, aggressive clinical treatment should be optimized. During the subacute period, neuroprotective treatment should be employed. Finally, during the delayed period, neuroregenerative therapy should ensue. With more research in stem cell biology, in particular, human stem cells, significant advances in the treatment of SCI are possible.

Similar work has shown that administration of Shh analog intrathecally in a rodent model of middle cerebral artery occlusion results in behavioral improvement, reductions in infarct volume, and upregulation of endogenous stem cell proliferation. Further work is ongoing to elucidate the exact mechanisms by which modulation of the Shh pathway results in such observations. Nevertheless, these findings, as well as the observation that intraperitoneally administered Shh agonists can upregulate the Shh pathway in the CNS (Bambakidis *et al.*, 2008a,b), may lead to novel therapeutic possibilities in a large group of CNS disorders in which current treatments are inadequate.

VII. Conclusion

Shh shows a spatially oriented pattern of expression in ventrally located neural cells and governs the differentiation of diverse cell types throughout the development of the neural tube along the dorsal-ventral axis. The Shh signaling pathway governs tissue development and reconstruction. This process is enacted via Shh-dependent autocrine/paracrine mechanisms that modulate the development of Shh-responsive cell populations. It has been repeatedly demonstrated that expression of Shh is integral for the specification of neuronal identity. Its activity results in the sequential and temporal expression of the transcriptional determinants fundamental in the patterning of the ventral spinal cord. After ventral patterning occurs, Shh signaling remains necessary for various embryologic processes. These include the regulation of mitosis, the maintenance of neurogenesis, and the differentiation of various neuronal and glial cellular components. In addition to their role as morphogens, Shh has been shown to exhibit mitogenic, cell survival, and axon guidance properties. In accordance with the intimate role Shh plays in human embryonic development and tissue regeneration, faulty Shh signaling is implicated in numerous pathological conditions, such as holoproscencephaly, Gorlin syndrome, and cancer. Given its far-reaching role in development and numerous disease states, understanding the regulation of Shh function and activity is essential. Many reports have substantiated that stem cell transplantation in the setting of CNS injury results in functional neurologic recovery in animal models. The study of endogenous stem cell populations in the adult mammalian CNS continues to demonstrate potential for the development of specific treatment paradigms in various neurological diseases entities.

REFERENCES

Bambakidis, N. C., Wang, R., Franic, L., and Miller, R. H. (2003). Sonic hedgehog-induced neural precursor proliferation after adult rodent spinal cord injury. *J. Neurosurg. Spine* **99,** 70–75.

Bambakidis, N. C., Theodore, N., Nakaji, P., Harvey, A., Sontag, V., Preul, M., and Miller, R. (2005). Endogenous stem cell proliferation after central nervous system injury: Alternative therapeutic options. *Neurosurg. Focus* **19,** E1.

Bambakidis, N., Butler, H., Horn, E., Wang, X., Preul, M., Theodore, N., Spetzler, R., and Sontag, V. (2008a). Stem cell biology and its therapeutic applications in the setting of spinal cord injury. *Neurosurg. Focus* **24,** E19.

Bambakidis, N. C., Horn, E. M., Nakaji, P., Theodore, N., Bless, E., Ma, C., Preul, M. C., Coons, S. W., Spetzler, R. F., and Sonntag, V. K. H. (2008b). Endogenous stem cell proliferation is induced by intravenous hedgehog agonist administration after contusion injury in the adult rat spinal cord. *J. Neurosurg.* **10,** 171–176.

Bambakidis, N. C., Wang, X., Lucas, R., Spetzler, R., Sonntag, V., and Pruel, M. (2010). Intravenous hedgehog agonist induces proliferation of neural and oligodendrocyte precursors in rodent spinal cord injury. *Neurosurgery* **67,** 1709–1715.

Briscoe, J. (2009). Making a grade: Sonic hedgehog signalling and the control of neural cell fate. *EMBO J.* **28,** 457–465.

Capdevila, J., and Johnson, R. (2000). Hedgehog signaling in vertebrate and invertebrate limb patterning. *Cell. Mol. Life Sci.* **57,** 1682–1694.

de Bont, J. M., Packer, R. J., Michiels, E. M., den Boer, M. L., and Pieters, R. (2008). Biological background of pediatric medulloblastoma and ependymoma: A review from a translational research perspective. *Neuro Oncol.* **10,** 1040–1060.

Dierks, C. (2010). GDC-0449-Targeting the hedgehog signaling pathway. *In* "Small Molecules in Oncology, Recent Results in Cancer Research," (Uwe M. Martens, Ed.). 10.1007/978-3-642-01222-8_17. Springer Verlag, Berlin, Heidelberg.

Farzan, S., Singh, S., Schilling, N., and Robbins, D. (2008). The adventures of sonic hedgehog in development and repair: Hedgehog processing and biological activity. *Am. J. Physiol. Gastrointest. Liver Physiol.* **294,** G844–G849.

Hatten, M. E., and Roussel, M. F. (2011). Development and cancer of the cerebellum. *Trends Neurosci.* **34,** 134–142.

Ingham, P. W., and Placzek, M. (2000). Introduction: Preface to the Hedgehog family of proteins (review volume). *Cell. Mol. Life Sci.* **57,** 1671.

Liebner, S., and Plate, K. (2010). Differentiation of the brain vasculature: The answer came blowing in the wind (review). *J. Angiogenes. Res.* **2,** 1–10.

Litingtung, Y., and Chiang, C. (2000). Control of Shh activity and signaling in the neural tube. *Dev. Dyn.* **219,** 143–154.

Omenetti, A., and Diehl, A. M. (2008). The adventures of sonic hedgehog in development and repair: Sonic hedgehog and liver development, inflammation and cancer. *Am. J. Physiol. Gastrointest. Liver Physiol.* **294,** G595–G598.

Parkin, C., and Ingham, P. (2008). The adventures of sonic hedgehog in development and repair: Hedgehog signaling in gastrointestinal development and disease. *Am. J. Physiol. Gastrointest. Liver Physiol.* **294,** G363–G367.

Patten, I., and Placzek, M. (2000). The role of Sonic hedgehog in neural tube patterning. *Cell. Mol. Life Sci.* **57,** 1695–1708.

Pfister, S. M., Korshunov, A., Kool, M., Hasselblattm, M., Eberhart, C., and Taylor, M. D. (2010). Molecular diagnostics of CNS embryonal tumors. *Acta Neuropathol.* **120,** 553–566.

Ribes, V., Balaskas, N., Sasai, N., Cruz, C., Dessaud, E., Cayuso, S., Yang, L., Novitch, B., Marti, E., and Briscoe, J. (2010). Distinct sonic hedgehog signaling dynamics specify floor plate and ventral neuronal progenitors in the vertebrate neural tube. *Genes Dev.* **24,** 1186–1200.

Richardson, P., McGuinness, U., and Aguayo, A. (1980). Axons from CNS neurones regenerate into PNS grafts. *Nature* **284,** 264–265.

Ridgway, L., Wetzel, M., and Marchetti, D. (2011). Heparanase modulates shh and wnt3a signaling in human medulloblastoma cells. *Exp. Ther. Med.* **2,** 229–238.

Shubin, N. (2008). Your Inner Fish: A Journey into the 3.5-Billion-Year History of the Human Body. Random House, New York.

Sousa, V., and Fishell, G. (2010). Sonic Hedgehog functions through dynamic changes in temporal competence in the developing ventral telencephalon. *Curr. Opin. Genet. Dev.* **20,** 391–399.

Toftgard, R. (2000). Hedgehog signalling in cancer. *Cell. Mol. Life Sci.* **57,** 1720–1731.

CHAPTER EIGHTEEN

Sonic Hedgehog on Microparticles and Neovascularization

Raffaella Soleti *and* Maria Carmen Martinez

Contents

Abstract

Neovascularization represents a pivotal process consisting in the development of vascular network during embryogenesis and adult life. Postnatally, it arises mainly through angiogenesis, which has physiological and pathological roles in health and disease. Blood vessel formation results as tightly regulated multi-step process which needs coordination and precise regulation of the balance of proangiogenic and antiangiogenic factors.

Sonic Hedgehog (SHH), a morphogen belonging to Hedgehog (HH) family proteins, is implicated in a remarkably wide variety of process, including vessel development. Recent evidence demonstrate that, in addition to the classic factors, microvesicles (MVs), both microparticles (MPs) and exosomes,

INSERM U1063, Université d'Angers, Angers, France

Vitamins and Hormones, Volume 88
ISSN 0083-6729, DOI: 10.1016/B978-0-12-394622-5.00018-3

small vesicles released distinct cellular compartments, are involved in modulation of neovascularization. MPs generated from T lymphocytes undergoing both activation and apoptosis harbor at their surface SHH and play a crucial role in modulation of neovascularization. They are able to modulate the different steps implicated in angiogenesis process *in vitro* and to enhance postischemic neovascularization *in vivo*. As the consequence, we suggest that the MPs carrying SHH contribute to generation of a vascular network and may represent a new therapeutic approach to treat pathologies associated with failed angiogenesis.

I. Introduction

Impairment in new vasculature formation is strictly linked to different pathologies like tissue damage after reperfusion of ischemic tissue, cardiac failure, or diabetes. Hence, to improve these disease conditions, is necessary in the restoration of correct supply of oxygen and nutrients as well as removal of waste products by formation of new collateral vessels.

During recent years, several lines of evidence have highlight the key role of microparticles (MPs), small fragments generated from plasma membrane, as efficient biological effectors able to transfer long-range information between cells. In this context, it has been shown that MPs deliver messages related to modulation of different stages of neovascularization. Moreover, the several angiogenic responses elicited by MPs are connected to stimulus at their origins, at the cells which they stem from and at their concentration.

At light of these evidence, we have shown that human T lymphocytes undergoing both activation and apoptosis generate MPs bearing Sonic Hedgehog (SHH, MPs^{SHH+}) (Martínez *et al.*, 2006). SHH is a crucial morphogen implicated not only during embryonic development, but also in adult life and orchestrates diverse processes, including new vessel formation. Activation of SHH cascade induces formation of capillary-like structures *in vitro* (Kanda *et al.*, 2003) and new vessel generation *in vivo* (Pola *et al.*, 2001). Furthermore, the effect promoted by SHH affects modulation of vascular endothelial growth factor (VEGF) and endothelial nitric oxide (NO) synthase (eNOS) activities (Podlasek *et al.*, 2005) and implicates PI3K/Akt pathway, which also belongs to NO release.

In addition, in endothelial cells (ECs), MPs^{SHH+} trigger changes in expression and activation of enzymes related to NO pathway, by a mechanism directly mediated by SHH (Agouni *et al.*, 2007). Very recently we have demonstrated that these MPs have an essential role in modulation of vessel formation. *In vitro*, they by act on steps linking to ECs ability to form capillary-like structure, adhesion, and expression of proangiogenic factors (Soleti *et al.*, 2009). *In vivo*, they modulate neovascularization during hindlimb ischemia, favoring vessel formation, recovery of blood flow, NO production, and acting on expression of factors implicated in regulation of angiogenic process.

Furthermore, in both studies, some of the effects elicited by these MPs are reversed by inhibition of SHH signaling, suggesting that a direct stimulation of vascular cells by SHH associated with MPs is mandatory to favor formation of neovessels.

Comprehension of mechanism triggered by MPs is essential for the development of new therapeutic strategies for diseases characterized by altered angiogenesis in order to redirect angiogenic switch.

II. Neovascularization

A. General aspects

Although from the earliest stage, the embryo develops in absence of vascularization, receiving its nutrition by diffusion (Noden, 1989), in an ordered and sequential manner, the embryo rapidly evolves an elaborate network of capillary plexuses and blood vessels. Then, tissue development is dependent on new vessel formation for the supply of oxygen and nutrients as well as removal of waste products (Griffioen and Molema, 2000).

Blood vessels develop early during embryogenesis and are derived from mesodermal cells called angioblasts, which share a common precursor, hemangioblasts, with hematopoietic cell precursors. Angioblasts divide and form dense syncytial masses which become networks of solid tubes that eventually canalize to form blood vessels (Carmeliet, 2004). In this process, termed vasculogenesis, angioblasts differentiate into ECs forming a vascular network (Murohara, 2001). This primordial plexus is refined into a functional network by a process termed angiogenesis where vessels undergo extensive elongation, inosculation, intussusception, and sprouting (Nguyen and D'Amore, 2001).

Angiogenesis process is based on endothelial sprouting or intussusceptive (nonsprouting) microvascular growth (Ausprunk and Folkman, 1977; Risau, 1997). Intussusceptive angiogenesis is a mode of blood vessel formation and remodeling, which occurs by internal division of the preexisting capillary plexus without sprouting. Intussusception is a central and probably widespread process, which plays a role not only in capillary network growth and expansion (intussusceptive microvascular growth), but also in vascular plexus remodeling and tree formation (intussusceptive arborization). In contrast to sprouting, intussusceptive angiogenesis does not require intense local EC proliferation; it is implemented primarily by rearrangement and attenuation of the EC plates (Djonov *et al.*, 2000). Finally, the functional modifications of largest arteries, such addiction of a thick muscular coat concomitant with acquisition of viscoelastic and vasomotor properties, are referred to as arteriogenesis.

Postnatal neovascularization arise mainly through angiogenesis, although vasculogenesis have physiological and pathological roles in health and disease in adults. Even if, as a general principle, establishment of the vasculature of most organs occurs by angiogenesis, development of the vascular network of certain endodermal organs, including the liver, lungs, pancreas, stomach/intestine, and spleen, occurs by vasculogenesis (Pardanaud and Dieterlen-Lièvre, 1999). The existence of a postnatal vasculogenesis is sustained by the demonstration that both ECs and endothelial progenitor cells (EPCs) coexist in the circulation. Moreover, EPCs are also recruited to sites of neovascularization in mature mammals from a circulating, marrow-derived population of progenitor cells (Asahara *et al.*, 1997). The distinction between vasculogenesis and angiogenesis is not categorical, in fact, they possess common characteristics. Both need EC proliferation, migration, three-dimensional reorganization of newly formed structures and utilize analogous extracellular matrix (ECM) adhesive mechanisms (Drake *et al.*, 1995). Moreover, vasculogenesis and angiogenesis are not reciprocally exclusive, since angioblasts can be integrated into expanding preexisting blood vessels (Auerbach and Auerbach, 1997).

Sprouting angiogenesis results as tightly regulated multistep process. Stimuli able to initiate the formation of new vessels include hypoxia, inflammation, and mechanical factors such as shear stress and stretch. They either directly or indirectly activate ECs, by initiating autocrine or paracrine production and release of growth factors or cytokines (Milkiewicz *et al.*, 2006), which, in turn, cause increased EC permeability through dissolution of adherent junctions (Pepper, 2001). EC proliferation occurs early in angiogenesis and continues as the new capillary sprout elongates. In order to promote EC invasion into the surrounding interstitial matrix, proteolysis of basement membrane components is necessary. Thus, cellular migration is triggered and the sprouting tip of the ECs proceeds into the interstitium. Lumen formation occurs as the sprout forms a multicellular structure. The new capillary channel forms an anastamosis with a preexisting capillary, creating a new patent capillary. The final stage requires stabilization of the capillary through the construction of basement membrane, adherent junctions, and cessation of EC activation (Milkiewicz *et al.*, 2006).

Angiogenesis is a complex process which needs coordination and precise regulation of the balance of proangiogenic and antiangiogenic factors. The stability of angiogenic switch determines the time of initiation of the subsequent angiogenic process. When angiogenic stimulators prevail on angiogenic inhibitors, as in the case of normal wound healing and female endometrial repair, the angiogenic switch will be turned on and angiogenesis will proceed. Furthermore, during the process, each step is strictly mediated by the balance of different types of angiogenic stimulators or inhibitors. However, in some pathological cases, as in tumor growth, the angiogenic switch remains in the "ON" mode which leads to "non-stop" formation of new blood vessels (Yue *et al.*, 2007). This condition of excessive vascularization is characteristic of many diseases, including cancer (both solid and hematologic tumors), cardiovascular pathologies (atherosclerosis), chronic inflammation (rheumatoid arthritis, Crohn's disease), diabetic retinopathy, psoriasis, and endometriosis (Griffioen and Molema, 2000). In contrast, other pathologies are associated with impairment of new vasculature formation. Hence, tissue damage after reperfusion of ischemic tissue, cardiac failure or diabetes needs formation of new collateral vessels to improve disease conditions (Carmeliet *et al.*, 1999; Ferrara and Alitalo, 1999).

Several elements play critical function within the complex angiogenic cascade, under both physiological and pathological conditions. Different phases included in sprouting form of angiogenesis are triggered by production and activation of growth factors, their receptors and intracellular signaling mediators. A description of some these components will precede analysis of the phases of angiogenic sprouting.

B. Angiogenic mediators

Under a large variety of conditions, several mediators generated from different cells control angiogenesis. These mediators are soluble/ECM, membrane-bound growth factors or components of ECM themselves. Among the angiogenic players the best known are the families of VEGF and fibroblast growth factor (FGF) (Jacobs, 2007; Tunyogi-Csapo *et al.*, 2007), nitric oxide (NO) (Lau and Ma, 1996), hepatocyte growth factor (HGF) (Tong *et al.*, 2006), and angiopoietins (Ang1 and 2) (Asahara *et al.*, 1998).

1. Vascular endothelial growth factor

VEGF is a potent EC-specific mitogen and angiogenic factor that has been shown to play a central role in neovascular responses that accompany a number of physiological and pathological processes (Benjamin and Keshet, 1997; Ferrara, 1996; Hashimoto *et al.*, 1995; Plate *et al.*, 1992). VEGF is produced by a variety of cell types, including keratinocytes (Brown *et al.*, 1992), macrophages (Berse *et al.*, 1992), mast cells (Grützkau *et al.*, 1998), smooth muscle cells (SMC) (Ferrara *et al.*, 1991), and ECs (Gerber *et al.*, 1998).

The VEGF family includes at least seven members. The most abundant and potently mitogenic is the VEGF-A. The two VEGF A-specific tyrosine kinase receptors, VEGFR-1 (Flt-1) and VEGFR-2 (KDR in human, Flk-1 in mice), are expressed on vascular endothelium. Activation of VEGFR-2 by VEGF interaction is a critical requirement to induce the full spectrum of VEGF responses. Thus, in ECs, intracellular signal transduction pathway through dimerization leads to permeability enhancement, cellular proliferation and migration (Griffioen and Molema, 2000).

VEGF transcription is stimulated greatly by hypoxia, as a result of hypoxia inducible factor (HIF-α) binding to a hypoxia response element within the VEGF promoter (Semenza, 2001). VEGF production is also increased by inflammatory mediators, such as interleukin (IL)-1α and IL-1β, transforming growth factor β (TGFβ), prostaglandin E2 (PGE2), or cyclooxygenase 2 (COX-2) activation (McColl *et al.*, 2004), as well as mechanical forces of shear stress and cell stretch (Li *et al.*, 1997; Milkiewicz *et al.*, 2001).

VEGF promotes EC survival through activation of PI3K/Akt pathway and through association with $\alpha_v\beta_3$ integrin and activation of focal adhesion kinase (FAK) (Zachary, 2003). VEGF induces EC proliferation and migration through numerous pathways, including activation of the MAPK, extracellular signal-regulated kinase (ERK), p38 and c-jun N-terminal kinase (JNK), and Rho-GTPase family members (Zachary, 2003).

The angiogenic property of VEGF has been attributed to several distinct functions associated with this cytokine. VEGF is an EC mitogen and permeability-enhancing factor that influences the egress of plasma proteins and cells that both directly and indirectly stimulate angiogenesis (Dvorak *et al.*, 1995; Leung *et al.*, 1989; Mandriota and Pepper, 1997).

2. Fibroblast growth factor

Members of the FGF family are also potent inducers of angiogenesis. Cellular responses mediated by FGFs include cell migration, proliferation, and differentiation (Kanda *et al.*, 1997). The FGF family consists of at least 22 known members (McKeehan *et al.*, 1998), of which prototypical are FGF-1 (acid-FGF) and FGF-2 (basic-FGF). FGFs stimulate a variety of cellular functions by binding to cell surface tyrosine kinase receptors (FGFR) (Klein *et al.*, 1996) in the presence of heparin proteoglycans. Binding of FGF-2 to its receptor initiates PKC-dependent signaling and/or GRB2/SOS-mediated activation of MAPK pathway. FGFR1 signaling stimulates migration and differentiation whereas FGFR2 only migration (Kanda *et al.*, 2004). Importantly, FGF-2 regulates EC expression of VEGF, which is a major autocrine mediator of FGF-2-induced angiogenesis (Seghezzi *et al.*, 1998). Receptor activation gives rise to a signal transduction cascade that leads to gene activation and diverse biological responses, like cell differentiation, proliferation, matrix dissolution, thus initiating a process of mitogenic activity particularly for ECs, fibroblasts, and smooth muscle cells.

3. Hepatocyte growth factor

HGF was first discovered to strongly promote liver cell growth (Trusolino and Comoglio, 2002). Subsequently, HGF was found to play a significant biologic role in health and disease (Grierson *et al.*, 2000; Yoshida *et al.*, 2003).

HGF is a potent stimulator of new vessel formation and an important angiogenic factor in vascular retinopathies (Boulton, 1999; Colombo *et al.*, 2007; Lashkari *et al.*, 1999, 2000; Ramos-Lopez *et al.*, 2003). HGF is secreted by mesenchyme-derived cells as an inactive precursor that is activated by proteolytic cleavage by urokinase or tissue plasminogen activator (uPA or tPA) (Zhang and Vande Woude, 2003). HGF binds to a tyrosine kinase receptor, c-met, found predominantly on epithelial and ECs. HGF binding activates c-met via intracellular phosphorylation initiating RAS-RAF-ERK, and PI3K-AKT-mTOR signaling as well as several other pathways (Seiwert *et al.*, 2009). HGF/c-met signaling leads to increased cell growth, cell motility, angiogenesis, wound healing, and tissue regeneration, in part because it induces production of VEGF by the endothelium (Reisinger *et al.*, 2003; Zhang and Vande Woude, 2003). Also, HGF can induce angiogenesis independently of the VEGF pathway (Sengupta *et al.*, 2003). HGF-dependent activation of the transcription factor Ets-1 generates a positive feedback loop leading to further production and release of HGF, and thus a continuous activation of EC (Hashiya *et al.*, 2004). Another effect of HGF signaling is the negative regulation of thrombospondin 1 (TSP1) in tumor cells (Zhang *et al.*, 2003). TSP1 is an ECM protein associated with antiangiogenesis signaling, including activation of EC apoptosis.

4. Nitric oxide (NO)

NO is an important mediator of angiogenesis, in addition to its well-recognized vasodilatory properties. Capillary EC growth and differentiation is linked to cyclic GMP (cGMP) generation, suggesting the existence of an autocrine/paracrine loop in NO effect (Ziche and Morbidelli, 2000). The regulation of capillary growth by NO is complex because both angiogenic and angiostatic effects of NO have been demonstrated (Lau and Ma, 1996; Murohara *et al.*, 1998; RayChaudhury *et al.*, 1996; Ziche *et al.*, 1994). NO generate upon constitutive eNOS activation can stimulate soluble guanylyl cyclase (sGC) in an autocrine manner in ECs or diffuse outside. Following the stimulation of sGC, cGMP is formed and activates kinase cascades including the protein kinase G (PKG) and the mitogen-activated protein kinase MEKK and ERK1/2, leading the membrane-linked signal to the nuclear level. The transcription of specific genes, as well as the growth factor FGF-2, leads to the cellular events associated with angiogenesis as proliferation, migration, and matrix degradation or cell survival (Donnini and Ziche, 2002). Interestingly, while VEGF-induced angiogenesis is mediated by NO, the capillary growth stimulated by FGF-2 can be both NO independent and inhibited by NO (RayChaudhury *et al.*, 1996; Ziche *et al.*, 1997). On the other hand, it has been shown that high levels of NO produced by the inducible NOS (iNOS) activity can affect different intracellular signaling pathways. NO can act on gene transcription via S-nitrosylation of redox-sensitive transcription factors like AP-l and NF-κB, leading to the transcription of angiogenic factor. Moreover, iNOS activity can promote migration and proliferation of ECs by inhibiting PKCδ activity or the EC survival by blocking the caspase signaling pathway (for review see Donnini and Ziche, 2002).

Thus, NO, depending on the nature, the timing, and the degree of the enzyme activity releasing it, may display opposite effects on ECs, resulting in physiological and/or pathological angiogenesis.

5. Angiopoietins

Two isoforms of angiopoietins (Ang), Ang1 and Ang2, and their tyrosine kinase receptors (Tie1, Tie2) regulate sprouting angiogenesis, vascular remodeling and the transition from the quiescent to the activated EC phenotype. Ang1 is associated with vessels developing and remodeling, leading to formation of structures with a wide diameter; its absence provokes defects in vascular remodeling (Thurston, 2003). Ang2 is abundantly expressed by angiogenic ECs (Stratmann *et al.*, 1998) and is found in tissues such as ovary, uterus, and placenta that undergo transient or periodic growth and vascularization, followed by regression (Maisonpierre *et al.*, 1997).

Quiescent ECs are covered by perivascular support cells including pericytes and smooth muscle cells, which secrete Ang1. Ang1 induces Tie2 clustering at interendothelial cell–cell junctions, Tie2 activation and EC survival and stabilization. It also seems to promote perivascular support cell recruitment and stabilization of interactions between endothelial and perivascular support cells. However, in activated endothelium, ECs liberate from Weibel–Palade bodies stored Ang2, which counteracts Ang1 by binding to Tie2 and inducing its clustering at cell-ECM contacts. This may facilitate endothelial sprouting induced by VEGF, for instance (Saharinen *et al.*, 2010).

6. Other angiogenic mediators

Other angiogenic players implicate in the switch are HIF-1, the transforming growth factor (TGF)-β and the platelet-derived growth factor (PDGF)-B. HIF-1 is a master regulator of hypoxic/ischemic vascular responses, driving transcriptional activation of hundreds of genes involved in vascular reactivity, angiogenesis, arteriogenesis, and the mobilization and homing of bone marrow-derived angiogenic cells (Rey and Semenza, 2010). TGF-β can act as an angiostatic (higher doses) or angiogenic (lower doses) molecule (Distler *et al.*, 2003). It has the ability to influence proliferation, migration, apoptosis, and accumulation of ECM. Moreover, TGF-β has been reported to direct neural crest cells and mesenchymal cells toward a vascular SMC lineage (Hirschi *et al.*, 1998; Shah *et al.*, 1996). PDGF-B is secreted by ECs, presumably in response to VEGF and is required for the recruitment and differentiation of pericytes (Lindahl *et al.*, 1997).

It is important to note that the activity of an angiogenesis-regulating cytokine depends on the presence and concentration of other factors or cytokines in the environment of the responding endothelium (Pepper *et al.*, 1998). As example, exogenous factors such as hormones can affect condition leading to angiogenesis (Schiffenbauer *et al.*, 1997). Moreover it is well appreciated that immune system cells such as monocytes/macrophages, lymphocytes, and mast cells can affect pro- and antiangiogenic balance (Blair *et al.*, 1997; Sunderkötter *et al.*, 1996).

Furthermore, evidence has accumulated that, in addition to the classic factors, many other endogenous peptides (erythropoietin, angiotensin II, endothelins, proadrenomedullin-derived peptides, urotensin II, adipokines, neuropeptide-Y, vasoactive intestinal peptide, pituitary adenylate cyclase-activating polypeptide and substance P) play an important regulatory role in angiogenesis, especially under pathological conditions (Ribatti *et al.*, 2007).

Moreover, microRNAs, short noncoding RNAs, are a relatively recent discovery as important regulators of gene expression. They have been identified in ECs, and their role in the regulation of different aspects of the angiogenic process has been recently investigated (Suárez and Sessa, 2009).

Finally, current evidences illustrate that microvesicles (MVs), both MPs and exosomes, small vesicles released from plasma membrane and multi-vesicular compartments, respectively, are involved in modulation of neo-vascularization. MPs and exosomes from platelets (Brill *et al.*, 2005; Kim *et al.*, 2004), from ECs (Taraboletti *et al.*, 2002), MPs derived from tumor cells (Kim *et al.*, 2002); as well as lymphocytic MPs generated in vitro (Benameur *et al.*, 2010a; Soleti *et al.*, 2009; Yang *et al.*, 2008) contribute to angiogenesis. Major details on MPs and their involvement in angiogenic process are discussed below.

C. Angiogenesis process

1. Vessel dilatation, endothelial permeability increasing and ECM degradation

Vasodilatation of existing vessels is one of the earliest steps in angiogenesis. VEGF is a major player in neovessels initiation; based on its ability to induce vasodilatation via endothelial NO production and its EC permeability-increasing effect (Ziche *et al.*, 1997). The observation that VEGF production is under control of HIF-1 reinforce the suggestion of an early involvement of VEGF in the angiogenic response. Moreover, VEGFR expression is upregulated under hypoxic or ischemic conditions as well (Forsythe *et al.*, 1996). These effects are due to redistribution of adhesion molecules including platelet/endothelial cell adhesion molecule (PECAM)-1 and vascular endothelial (VE)-cadherin and to activation of different kinases leading to modification of membrane properties (Eliceiri *et al.*, 1999). Then, in order to create a provisional network support, the extravasion of plasma proteins follows (Dvorak *et al.*, 1995), allowing subsequent migration of activated ECs. Vascular wall permeability directs by VEGF is controlled by Ang1–Tie2 system, a natural antipermeability factor, which provides protection and balance against excessive plasma leakage (Thurston, 2003). However, at angiogenic sites, Ang2 favors detachment of SMC facilitating EC migration (Gale and Yancopoulos, 1999; Maisonpierre *et al.*, 1997).

Degradation of ECM requires a variety of proteinases which act, not only to create a migration area, but also to release growth factors, including bFGF, VEGF, and insulin-like growth factor-1 (IGF-1), which otherwise remain sequestered within the matrix. Over twenty matrix metalloproteinases (MMPs) have been described and implicated in angiogenesis, tumorigenesis, and cell proliferation (Nelson *et al.*, 2000). Inhibitors of MMPs include circulating protease inhibitors, such as tissue-localized inhibitors of metalloproteinases (TIMPs) (Brew *et al.*, 2000). MMP activation is coordinated and participates to degradation of all ECM elements. As example, basal membrane is the first obstacle encounter by ECs, therefore MMP2 is recruited to degrade one of its constituent, collagene. Then, activation of MMP9, which degrades elastic fibers and proteoglycans occurs. It is, at least partly, through the secretion of MMP2, MMP3, and MMP9, and suppression of TIMP2 that Ang1 induces sprouting (Kim *et al.*, 2000). Interaction with other proteins, like antiangiogenic TSP1, prevent MMP2 and MMP9 activation (Bein and Simons, 2000), but also plasmin, another proteinase, enables ECs migration, by interfering to in matrix degradation (Pepper, 2001).

2. EC proliferation and migration

Once physical barriers are dissolved, proliferating ECs are free to sprout to distant sites under influence of different factors. This stage involves interplay between the various forms of VEGF, Ang, FGF, and their receptors, all of which are responsible in mediating angiogenesis, although additional factors have also been implicated. Angiogenic factor mechanisms are redundant, therefore is very difficult to attribute an exclusive role to each of them. Thus, a part of triggering function, VEGF also affects EC proliferation, by activating MAPK family through cGMP (Yu and Sato, 1999). Ang1, via phosphorylation of Tie2, induces sprouting and stimulates the interaction between ECs and peri-ECs (Gale *et al.*, 2002; Suri *et al.*, 1996). FGF promotes EC growth and, by recruitment of mesenchymal and/or inflammatory cells, allows further production of angiogenic factors (Carmeliet, 2000). Several chemokines, including monocyte chemotactic protein-1 (MCP-1), are able to induce endothelial growth (Belperio *et al.*, 2000). Additionally, during proliferation and migration of ECs, they also contact with other cells, in part by signaling through integrins $\alpha_v\beta_3$ and $\alpha_5\beta_1$ (Eliceiri and Cheresh, 1999), PECAM-1 (Ilan *et al.*, 1999) and Eph/ephrin receptor–ligand pairs (Huynh-Do *et al.*, 1999; Shima and Mailhos, 2000; Wilkinson, 2000). Finally, EC junctions are established with gap proteins such as VE-cadherin and members of the connexin family (Corada *et al.*, 1999; Knudsen *et al.*, 1998).

3. EC assemblage, lumen formation and stabilization

Once ECs migrate into the ECM far from origin vessel, they assemble in solid cords. Sprouting of one or two cells may form a lumen by intracellular canalization, which occurs through fusion of cytoplasmic vesicles, or by the alternative process in which a lumen is created by the membrane apposition of two different cells (Egginton and Gerritsen, 2003). Lumen diameter is tightly regulated by interactions between various VEGF isoforms, Ang, and their receptors as well as different integrins ($\alpha_v\beta_3$ and $\alpha_5\beta_1$) (Bayless *et al.*, 2000; Suri *et al.*, 1998).

The interaction between ECs, ECM, and mesenchymal cells is essential to form a stable vasculature. Therefore, newly endothelial tubular structures need to surround vessel layer composed of mural cells, as pericytes in small vessels and SMCs in large vessels. In this stage, communication between different cells is sustained by Ang/Tie system (Maisonpierre *et al.*, 1997). ECs, through synthesis and secretion of PDGF, a mitogen chemoattractant for a variety of mesenchymal cells, contribute to this phase. Subsequent differentiation of mural precursor cells into pericytes and SMC is probably due to a process dependent to cell contact (Griffioen and Molema, 2000). On EC–mural cell contact, a latent form of TGF-β, producing by both types of cells, is activated in a plasmin-mediated process and induce changes in myofibroblasts and pericytes. These mechanisms contribute to the formation of a quiescent vessel, ECM production and maintenance of growth control. With the maturation of vascular network, ECs acquire highly specialized characteristics to assure functional needs within specific tissues and organs.

III. Sonic Hedgehog

A. General aspects

SHH is a secreted morphogen belonging to Hedgehog (HH) family proteins. The central components of this family being first identified in *Drosophila melanogaster* larvae (Nüsslein-Volhard and Wieschaus, 1980). Larvae are covered by denticles, this phenotype gave them the aspect of a HH, which inspired the name.

During vertebrate evolution, genomic duplication originated several Hh genes. As a result, mammals genome includes three *HH* genes named *Sonic hedgehog* (*SHH*), after a popular video game character, *Desert hedgehog*

(*DHH*), after an Egyptian species of HH (*Hemiechinus auritus*), and *Indian hedgehog* (*IHH*), a HH species endemic in Pakistan (*Hemiechinus micropus*) (Katoh and Katoh, 2005a,b; Marigo *et al.*, 1995; van den Brink, 2007). These genes are greatly conserved between mouse and human, although *SHH* is the most widely expressed and the most actively explored.

Like other signals regulating embryonic development, SHH protein is involved in a remarkably wide variety of process, ranging from control of central nervous systems development, lateral asymmetry and anterior–posterior limb axis (Chiang *et al.*, 1996), as well as lung (Bellusci *et al.*, 1997), gut (van den Brink, 2007) and vasculature (Pola *et al.*, 2003) development. Furthermore, in the adult, SHH pathway controls the formation or the persistence of certain stem- and precursor cell populations (Adolphe *et al.*, 2004; Machold *et al.*, 2003; Watkins *et al.*, 2003), maintenance of tissue homeostasis and tissue repair during chronic persistent inflammation (Bijlsma *et al.*, 2004; Hooper and Scott, 2005; Lum and Beachy, 2004). In addition, deficits in one or more components of SHH signaling translate into developmental defects (Dellovade *et al.*, 2006; Ruiz i Altaba, 1999), and unrepressed signaling underlies several cancers (Briscoe and Therond, 2005; Kasper *et al.*, 2006; Katoh and Katoh, 2005c; Pasca di Magliano *et al.*, 2003; Taipale and Beachy, 2001).

B. SHH biogenesis and pathway

Even though several characteristic of the molecular processes by which SHH binding to cells induces cellular fate have been elucidated, it is important to note that SHH cascade is one of the least deciphered systems in mammalian cells, and a variety of aspects of this pathway remain unclear. It is evident that SHH signaling is of extraordinary complexity and utilizes unconventional mechanisms not detect in other biological systems.

Typically, SHH is synthesized as precursor which undergo to different reactions during posttranscriptional processing, leading to its activation and explosion on cells.

SHH genes encode precursor polypeptides of approximately 45 kDa that undergo both amino-terminal signal sequence trimming as well as internal proteolysis at a conserved Gly-Cys-Phe tripeptide. Endoproteolytic cleavage between Gly-Cys residues of tripeptide is catalyzed by the processing activity associated with the carboxy-terminal domain of SHH precursor which has no known additional function (Mann and Beachy, 2000). The reaction produces a 19-kDa amino-terminal segment (SHH-N) with which all known signaling activities are associated, and a 26-kDa cardoxy-terminal product (SHH-C) which functions as a cholesterol transferase. The reaction proceeds by two sequential nucleophilic displacements, the first a rearrangement to replace the main chain peptide linkage between Gly and Cys with a thioester involving the Cys side chain (Porter *et al.*, 1996a,b). The second step of autoprocessing reaction involves attack upon the same carbonyl by a second nucleophile, displacing the sulfur and severing the link between SHH-N and SHH-C.

A second lipophilic modification of the SHH signaling protein occurs on a proportion of the amino-terminal signaling domain of human SHH (Pepinsky *et al.*, 1998). This additional modifying adduct is a fatty acid, usually palmitate, in an amide linkage with the amine of the amino-terminal cysteine that is exposed by signal sequence cleavage. This Cys residue is the first of a pentapeptide, Cys-Gly-Pro-Gly-Arg, that is widely conserved among species, suggesting the possibility that these conserved residues and others nearby may constitute an important determinant for the palmitoylation reaction. The fatty acylation is proposed to occur via a thioester intermediate involving the side chain of the amino-terminal cysteine, followed by a rearrangement to form the amide (Mann and Beachy, 2000). Consequently, these lipid modifications have great and paradoxal influence on SHH protein properties: on one hand, they favorite retention of SHH on cell surface, on the other hand, they are necessary in establishing range of Shh action.

An important outcome of cholesterol modification is that it allows SHH retention on producing cell membranes (Roelink *et al.*, 1995), and as a result, it is initially considerate responsible for short-range signaling of SHH. Nevertheless, further studies have been shown that cholesterol adduct is indispensable for protein activity, and acts to assist the spreading of SHH over the long-range (Lewis *et al.*, 2001). Besides, SHH palmitoylation is essential for its activity as well as the generation of a protein gradient (Chen *et al.*, 2004).

Due to its great hydrophobic character, SHH diffusion capacity is strongly limited in medium surrounds cells. Since this evidence represents a problem for long-range signaling, different mechanisms of SHH distribution have been proposed: diffusion that generates an extracellular gradient, serial transfers between neighboring cells that generates intracellular gradient, direct transfer at points of cytoneme contact and secondary signals (Chuang and Kornberg, 2000). Moreover, it has been reported existence of soluble multimeric protein complex, which requires lipid modification for its organization and constitutes the major active component for SHH signaling soluble. Therefore, SHH multimeric complex travels in the morphogenetic field to activate SHH signaling in distant responsive cells (Chen *et al.*, 2004). Another mechanism consists of transport of SHH by low-density lipoprotein (or other particles), the mammalian equivalent of *Drosophila* lipophorin, which have inspired this hypothesis (Bijlsma *et al.*, 2006a). Concomitantly, one more transport system it has been depicted, based on SHH ability to be retain on cell membrane by its lipid portions. SHH long-range signaling is generated when morphogen is carried by MPs (Martínez *et al.*, 2006) (see below for details).

In addition to these important effects, lipid modifications are also involved in secretion and diffusion of SHH. Therefore, a specific protein, Dispatched (DISP), shifts the morphogen to its target in SHH-expressing cells (Burke *et al.*, 1999). DISP functions in SHH-secreting cells to promote bioavailability of lipid-modified SHH, but does not affect SHH that lacks lipid modifications (Burke *et al.*, 1999). Thus, DISP is not needed for membrane targeting of the SHH signal, nor for its secretion, but is necessary for its movement through tissues (Hooper and Scott, 2005).

SHH signaling is received in target cells by two membrane proteins: patched (PTCH) and smoothened (SMO). Two PTCH homologs (PTCH 1 and PTCH 2) are found in humans. PTCH 1, PTCH 2, and DISP are 12-pass transmembrane proteins. All three have amino- and carboxytermini intracellularly located. PTCH 1 and PTCH 2 have different amino- and carboxy-terminal domains, including the absence of 150 amino acid residues from the C-terminal end of PTCH 2. Two large extracellular loops on all three proteins determine where SHH binding occurs (Cohen, 2010). Moreover, these proteins contain a sterol-sensing domain, consisting of approximately 180 amino acids that make up five membrane-spanning domains, indicating that it may required for SHH secretion or binding (Ma *et al.*, 2008). SMO has seven hydrophobic membrane-spanning domains, an extracellular amino-terminal region, and an intracellular carboxy-terminal region (Cohen, 2010).

Most membrane-bound receptors activate downstream signaling upon ligand binding. In contrast, SHH receptor PTCH is repressed by its ligand, freeing SMO for downstream signaling.

Four models for the regulation of SMO by PTCH and SHH have been proposed. The classical interpretation of the relationship between PTCH and SMO transmembrane proteins indicates that PTCH suppresses SMO from downstream signaling. SHH binding to PTCH alleviates PTCH-mediated repression of SMO by a conformational change that frees SMO for downstream signaling (Alcedo *et al.*, 2000; Chen and Struhl, 1996; Cohen, 2010; Murone *et al.*, 1999; Stone *et al.*, 1996). A second model postulates that SHH activates SMO by causing dissociation of the PTCH/SMO complex. In a third model, SHH binding to PTCH alters the activity of a diffusible intermediate that mediates the inhibition between PTCH and SMO, allowing SMO to become activated. In a fourth model, PTCH acts catalytically via a small molecule to suppress SMO, and on SHH binding to PTCH, SMO is activated by becoming dissociated from PTCH and the small molecule (Cohen, 2010).

SMO activity can be modulated by many synthetic small molecules (Chen *et al.*, 2002a; Frank-Kamenetsky *et al.*, 2002) and natural products, including the steroidal alkaloids cyclopamine and jervine (Chen *et al.*, 2002b). The structural similarity between cyclopamine and sterols (Cooper *et al.*, 1998) suggests that endogenous sterols might regulate

SMO activity. Oxysterols (Corcoran and Scott, 2006) and vitamin D3 derivatives (Bijlsma *et al.*, 2006b) have been suggested to be the endogenous metabolites that modulate SMO activity. Of these, vitamin D3 appears to bind to SMO (Bijlsma *et al.*, 2006b).

SMO downstream signaling leads to the activation and nuclear translocation of GLI proteins, the SHH-specific zinc-finger transcription factors. Three GLI proteins (GLI 1, GLI 2, and GLI 3) have been found in mammals (Jacob and Briscoe, 2003; Ruiz i Altaba *et al.*, 2007). GLI 1 and GLI 2 are responsible for most activator functions and have similar activities at protein level (Bai and Joyner, 2001). GLI 1 expression is induced by SHH, and its function appears to be primarily to provide positive feedback and to prolong cellular responses to SHH. GLI 3, in turn, functions primarily as a repressor (Litingtung *et al.*, 2002; Wang *et al.*, 2000). In the absence of SHH, GLI proteins are retained in the cytoplasm by KIF7/27, a kinesin family members, and Suppressor of Fused (SUFU) to limit transcriptional activation (Katoh and Katoh, 2005c). Moreover, GLI 3 is phosphorylated, recognized by the ubiquitin ligase complex β-TrCP, and proteolytically processed to a truncated repressor form (Pan *et al.*, 2006; Wang and Li, 2006). Whether similar processing of GLI 2 results in complete degradation or generation of a truncated repressor form is unclear (Pan *et al.*, 2006; Wang and Li, 2006). Addition of SHH leads to inhibition of processing and accumulation of full-length forms of both GLI 2 and GLI 3 (Varjosalo and Taipale, 2008).

Thus, in the absence of SHH signaling, full-length GLI family members are accumulated in the nucleus to activate the transcription of SHH target genes, while N-terminal fragment of GLI family members are accumulated in the nucleus to repress the transcription of SHH target genes. SMO activation by SHH signaling leads to STK36-dependent phosphorylation of SUFU, which leads to nuclear accumulation of full-length GLI family members by preventing nuclear export of GLI family members (Katoh and Katoh, 2005c).

Several additional cell surface and ECM proteins regulated SHH signaling, in addition to PTCH 1. HIP (hedgehog-interacting protein), CDO (cell-adhesion-molecule related/downregulated by oncogenes; also known as CDON), and BOC (bioregional CDO related/downregulated by oncogenes (Cdon) binding protein) have been identified as SHH pathway components. HIP is a secreted glycoprotein interacting with SHH, with an affinity comparable to that of PTCH 1. It functions as HH inhibitor, probably mediating endocytotic uptake and degradation of SHH (Katoh and Katoh, 2005c). CDO and BOC are negative targets of SHH signaling in the mouse in several contexts, but act to positively regulate this cascade in vertebrate by directly binding SHH protein (Tenzen *et al.*, 2006; Yao *et al.*, 2006; Zhang *et al.*, 2006). CDO and BOC bind SHH via conserved fibronectin repeats (Wang *et al.*, 2007) and enhance signaling in what appears to be a synergistic interaction with PTCH 1 (Yao *et al.*, 2006).

Other regulating proteins involved in SHH signaling are growth arrest-specific gene 1 (GAS 1) and megalin (also known as gp330 or low-density lipoprotein receptor-related protein (LRP)-2) (McCarthy *et al.*, 2002). Given that SHH appears to have at least a partially unique surface for GAS 1 binding, it leaves open the possibility that a ternary complex may form between GAS 1, SHH, and PTCH 1 and that GAS 1 may function as a coreceptor for PTCH 1 (Martinelli and Fan, 2009). Moreover, SHH is known to bind to megalin with high affinity and this may be aided by interaction with heparan sulfate proteoglycans (HSPGs). Thus, three possible models have been supposed for megalin–SHH interactions. In the first, once bound, the SHH–megalin complex may result in intracellular signaling via a megalin pathway. An alternative hypothesis is that the SHH–megalin complex interacts with PTCH on the cell surface and then signaling ensues. In the second, SHH binds to megalin on the cell surface and undergoes endocytosis, possibly delivering SHH to vesicular pools of PTCH. The process requires interaction with receptor-associated protein (RAP). In the third, SHH undergoes transcytosis whereas megalin internalizes PTCH and SMO. SHH may act as a ligand bridge for a complex that includes megalin, SHH, PTCH, and SMO, and it may be that RAP and/or HSPGs are involved. PTCH and SMO are internalized whereas megalin delivers SHH to the extracellular surface of the cell, and then is recycled back to the cell membrane (Fisher and Howie, 2006).

Besides, glypicans (GPCs) are glycosylphosphatidylinositol-linked HSPGs implicate in regulation of SHH signaling. In mammalian systems, GPC 3 is a negative regulator of this pathway. Thus, PTCH 1 and GPC 3 compete for binding to SHH. Binding of SHH to PTCH 1 leads to internalization of the complex and activation of SMO. When SHH binds to GPC3, it is also internalized but it is targeted to a degradative compartment, thus reducing the amount of SHH available for signaling (Beckett *et al.*, 2008).

Actually, only few SHH gene target have been identified. *PTCH 1* and *HIP* genes are transcriptional targets of SHH signaling pathway. Because PTCH 1 and HIP are SHH signaling inhibitors, they are implicated in the negative feedback mechanism of this cascade as partially mentioned above. Cyclin D1 and D2, c-Myc, N-and L-MYC and Forkhead-box transcription factors are reported to be upregulated by SHH signaling (Katoh and Katoh, 2005c). Cathepsin B (Hwang *et al.*, 2009) and insulin-like growth factor II (IGF2) (Ingram *et al.*, 2002) are also a downstream targets of this cascade.

Thus, the canonical pathway concerns the binding of SHH to PTCH 1, which relieves the repression of SMO. In turn, this prevents the processing of GLI-transcription factors into their repressor forms allowing the accumulation of transcriptional activators which drive the expression of SHH-responsive target genes. However, several evidences support the existence of noncanonical SHH signaling, which may be defined as deviation from classical paradigm. Three different mechanisms have been envisaged: (1) signaling that involves SHH pathway components but which is independent of GLI-mediated transcription; (2) direct interaction of SHH signaling components with components of other molecular pathways; and (3) "noncontiguous" or "atypical" interaction of core SHH pathway components with another elements (Jenkins, 2009). Hence, the complexity of SHH network consists of coexistence of these two transduction forms, which, in turn, can be subjected to regulation by other signaling cascades or can be cross talk with their components.

C. SHH functions

Several roles have been illustrated for SHH during development, acting as a morphogen, mitogen, or differentiation factor. In vertebrate, one of the first function of SHH to emerge was in the zone of polarizing activity (ZPA), a region at the posterior margin of the limb bud responsible for normal anteroposterior patterning. SHH is expressed in this region and its ectopic localization can alter limb patterning in way similar to ZPA grafting (Riddle *et al.*, 1993). Moreover, SHH acts as a mitogen for neuron precursors in the developing cerebellum, as a differentiation factor for Bergmann glia and as a survival factor for different cell types, including motor and dopaminergic neurons, tooth and neural crest cells (Mullor *et al.*, 2002).

Concerning cardiovascular development, it is interesting to note that SHH is involved in the morphogenesis of the heart and blood vessel. Evidence in this implication is the capacity of SHH to induce VEGF and Ang in human fibroblasts (Pola *et al.*, 2001). Regarding heart development, it has been shown that *SHH* mutant mice present many difference from wild type, for instance, a common atrium, an absent pulmonary valve, reduced right ventricle, extended left ventricle (Kim *et al.*, 2001).

In addition to its role during development, SHH protein and the component of its cascade are expressed in postnatal and adult tissue. SHH signaling is recapitulated, for example, in the adult skeletal muscle and myocardium in response to ischemia (Kusano *et al.*, 2005; Pola *et al.*, 2003). In addition, the importance of proper SHH cascade is established by knowledge that defects in its signalization affect both embryonic life and adulthood. Loss of activity of the SHH pathway can cause a variety of human birth defects, including holoprosencephaly, polydactyly, craniofacial defects and skeletal malformations (McMahon *et al.*, 2003). Inappropriate activation of SHH signaling is responsible for nearly all basal cell carcinomas, some medulloblastomas and rhabdomyosarcomas, and has been implicated in other tumors (Pasca di Magliano and Hebrok, 2003). Smith-Lemli-Opitz syndrome is known to affect SHH signaling either by improper sterolation of SHH protein (Kelley and Hennekam, 2000) or by reduced responsiveness of cells to SHH (Cooper *et al.*, 2003), moreover, in these patients are common congenital heart disorders (Kelley and Hennekam, 2000). In addition, recent findings about the relationship between primary cilia and the mouse SHH pathway (Huangfu *et al.*, 2003) suggest that SHH cascade may be affected in human syndromes caused by defects in cilia, including Bardet Biedl syndrome, Kartagener syndrome, polycystic kidney disease and retinal degeneration (Pan *et al.*, 2005).

SHH signaling is also an important regulator of T-cell differentiation, at the transition from the double-negative (DN, $CD4^{-}CD8^{-}$) 1 to DN 2 stage, at the transition from DN 3 to double-positive (DP, $CD4^{+}CD8^{+}$) stage, in T-cell receptor-repertoire selection at the transition from DP to single-positive ($CD4^{+}$ or $CD8^{+}$) stage and in peripheral T-cell activation (Crompton *et al.*, 2007).

Many more examples of SHH functions have been described, but especially interesting to the scope of this review is the involvement of SHH in the formation of blood vessels.

D. SHH and neovascularization

It is known the implication of SHH pathway in morphogenesis of blood vessel. Primarily, SHH protein is pivotal in vascular remodeling in yolk sac (Byrd and Grabel, 2004). In mesoderm, blood islands, which represent the first hallmarks of vascular structures in yolk sac, are formed in response to FGF, although, *in vitro*, a requirement for SMO was found. Blood islands fuse to create vascular plexus and angioblasts differentiate in ECs. In a model of *SMO* mutant mice, development arrests in this phase, without further modifications (Byrd *et al.*, 2002). Moreover, in SHH-deficient mice, capillary plexus is remodeling to early vessels, nevertheless these result smaller and less organized than in wild-type vasculature (Bijlsma *et al.*, 2006a). However, mice that overexpress SHH in dorsal neural tube show an hypervascularization of the neuroectoderm (Nagase *et al.*, 2005). Furthermore, SHH proteins are needed in dorsal aorta formation (Vokes *et al.*, 2004) and are involved in the fusion of bilateral dorsal aortas at the midline to form the dorsal aorta (Garriock *et al.*, 2010).

A part of its function in *de novo* vascularization during embryonic life, SHH cascade plays an active role in postnatal physiology being central in inducing vessel formation. Cells in the adult cardiac and vascular tissues express PTCH 1 and can respond to exogenous administration of SHH by upregulating PTCH 1. Also, in corneal and ischemic hindlimb models of angiogenesis, SHH induce robust neovascularization. This process is characterized by formation of functional vascular network composed of several venous structures with arterovenus shunts. The angiogenic activity of SHH appears to be indirect because it might be triggered through SHH/PTCH 1 signaling, specifically in mesenchymal cells, where induce an array of angiogenic growth factors, including three VEGF-1 isoforms as well as Ang1 and Ang2 (Pola *et al.*, 2001).

Following these observations, it has been displayed that SHH is postnatally recapitulated in response to skeletal muscle ischemia (Pola *et al.*, 2003). In muscle regeneration after ischemia, angiogenesis plays a pivotal role (Couffinhal *et al.*, 1998). The coexpression of SHH and PTCH 1 in the interstitial mesenchymal fibroblasts indicates the presence of an autocrine mechanism regulating SHH signaling in ischemic muscles. Although the same cells do not represent the majority of VEGF-producing cells during ischemia, the activation of the SHH signaling is central for the overall production of VEGF and the related angiogenic response (Pola *et al.*, 2003).

Based on these findings it has been found that SHH gene therapy increases blood flow, capillary, and arteriole densities upon peripheral limb ischemia. This treatment also enhances vasculogenesis, by increasing the number of circulating endothelial precursors and improving the contribution of these cells to the process of neovascularization. Finally, SHH gene therapy induces upregulation of prototypical angiogenic, arteriogenic, and vasculogenic factors, such as VEGF, Ang 1 and stromal cell-derived factor-1 (SDF-1α), a trafficking chemokine for hematopoietic stem cells (Palladino *et al.*, 2011).

There were no direct action of SHH on cellular responses by cultured ECs, such as proliferation, migration (Kanda *et al.*, 2003; Pola *et al.*, 2003) and serum-deprived survival (Pola *et al.*, 2003). However, even if SHH increases both PI3-K activity and transcriptional regulation pathways in ECs, followed by induction of capillary morphogenesis, PI3-K cascade is not involved in SHH-induced GLI 1 nuclear translocation (Kanda *et al.*, 2003).

Recently, it has been shown that SHH triggers proangiogenic responses in ECs via noncanonical pathways. Thus, SHH activates the small GTPase RhoA in ECs and stimulates tubulogenesis by a SMO- and Gi protein-dependent but GLI-independent pathway. In addition, SHH reduces caspase activation and promotes survival by inactivation of the PTCH 1 proapoptotic activity, independently of SMO modulation (Chinchilla *et al.*, 2010). Like this, SHH induces EC migration and capillary morphogenesis by activation of Rho/ROCK pathway, which, in turn, enhances the expression of downstream target such as MMP9 and osteopontin (Renault *et al.*, 2010).

Besides, intramyocardial transfer of naked DNA encoding human SHH triggers a beneficial effect on recovery from myocardial ischemia in adult animals. This gene transfer upregulates SHH pathway in fibroblast and cardiomyocytes. Myocardial tissue preservation and repair occur not only by promoting neovascularization, in contribution with EPC recruitment, but also by preserving left ventricular function and by reducing fibrosis and cardiac apoptosis (Kusano *et al.*, 2005).

Consistent with these results is that activation of SHH signaling is both critical for coronary development and sufficient to promote formation of new coronary vessels in the embryonic and adult heart. SHH signaling orchestrates coronary development by controlling the expression of multiple proangiogenic factors, including VEGF isoforms and Ang2. Given that SHH signaling regulates coronary vascular formation, promotes neovascularization in the adult heart, and induces expression of numerous signaling molecules, it has been suggested that SHH pathway constitutes an essential regulator of coronary vessel growth (Lavine *et al.*, 2006).

Moreover, activation of SHH pathway by topical gene therapy promotes a beneficial effect on microvascular remodeling during cutaneous wound healing and thereby accelerates wound closure, in diabetic mouse model. Indeed, it has been observed that during would healing PTCH 1 is upregulated. Moreover, dermal fibroblasts are activated by exogenous SHH stimulation, resulting in the expression of several angiogenic cytokines and induction of proliferative activity. The effects of SHH on wound healing are mediated to a large extent by the augmentation of neovascularization via VEGF upregulation and are not the result of stimulating reepithelialization alone. SHH induces microvasculature spouting toward the central part of the wound. This morphogenic effect of SHH combined with its capacity to induce expression of angiogenic cytokines from fibroblasts may contribute to the formation of mature vessels. In addition, SHH exerts several direct effects on EPCs, including induction of proliferation, enhanced adhesion, increased migration and tube formation. SHH also increases the expression of SDF-1α, providing an explanation for the enhanced recruitment of progenitor or stem cells (Asai *et al.*, 2006).

In fact, SHH protein promotes an angiogenic phenotype, including EPC proliferation and migration and VEGF production. All of these processes are requisite steps in new vessel formation. The mechanisms underlying these effects include activation of PI3-K/Akt pathway and are not completely dependent on VEGF (Fu *et al.*, 2006).

Thus, vessel formation may depend on the ability of SHH to induce VEGF. Though the exact mechanism remains elusive, it may very well involve IGF, which is a downstream target of the SHH cascade (Ingram *et al.*, 2002) and has a demonstrated ability to synergize with SHH (Hahn *et al.*, 2000; Rao *et al.*, 2004). Moreover, IGF2 has also been shown to induce VEGF (Kim and Kim, 2005; Kim *et al.*, 1998; Kwon *et al.*, 2004). Thus, IGF2 may mediate the angiogenic effects of SHH, and provide the critical link between SHH and VEGF (Chao and D'Amore, 2008).

All these studies demonstrate that activation of SHH signaling in embryonic and adult life leads to increased neovascularization. This ability of SHH has been evaluating by observing expression of endogenous proteins, using recombinant proteins, gene transfer or genetically modified mice. In addition other studies showed SHH involvement in new vessel formation using MPs as vectors.

IV. Microparticles

A. General aspects

Multicellular organisms have developed complex communication strategies to orchestrate and coordinate cell activities, in order to allow an accurate exchange of information during their functioning. A large amount of this transfer take place at cell interface and is controlled by sophisticate structural machinery which requires also secreted extracellular bioactive molecules. In the last years, this classical and well-characterized network is been enlarged to new vectors of cellular communication: MPs. They represent carriers able to exchange biological message between different type of cells, not only in the proximity area of their generation, but also far from cells they stem from. This ability situates MPs among the important players of transfer information, indeed, MP action results a new, efficient, and complex phenomenon, however, not completely elucidated, and therefore biological significance of MPs can be not fully appreciate.

MPs are mobile and small vesicles (0.05–1 μm) surrounded by the phospholipid bilayer and released by blebbing from the plasma membrane (Fig. 18.1). They are generated from several cell types upon activation by agonists, physical or chemical stress, or apoptosis. They frequently expose at their surface phosphatidylserine (PS) and express antigenic profile characteristic of the cell they originate. MPs have been widely detected in various biological fluids including peripheral blood, urine, and ascitic fluids.

Heterogeneous composition and variable concentration of MPs can be considered an explanation of their beneficial or deleterious effects observed during physiological or pathological conditions (Freyssinet, 2003). Despite being previously considered inert dust without specific function, MPs actively orchestrate important physiological functions and pathophysiological processes in vascular diseases. MPs have been implicated in hemostasis and thrombosis, diabetes, inflammation, atherosclerosis, angiogenesis, tumor progression, apoptosis, vascular cell proliferation, and outgrowth of transplanted hematopoietic stem cells (Azevedo *et al.*, 2007; Baj-Krzyworzeka *et al.*, 2002; Janowska-Wieczorek *et al.*, 2001).

In vitro, the release of MPs has been shown from ECs, SMCs, platelets, leukocytes, lymphocytes, erythrocytes, fibroblasts, and tumor cells. The most abundant circulating MPs are that of platelet origin (PMPs); however, blood plasma contains MPs from lymphocyte (LMPs), monocyte, and endothelial (EMPs) origins. MPs are physiological constituents of blood plasma (detected in healthy individuals), but their amount results modified during pathological conditions. There are obvious alteration in number, cellular origins and composition of some of MP populations in various disease states. However, the real impact of these changes on their *in vivo* effect is still not fully understood.

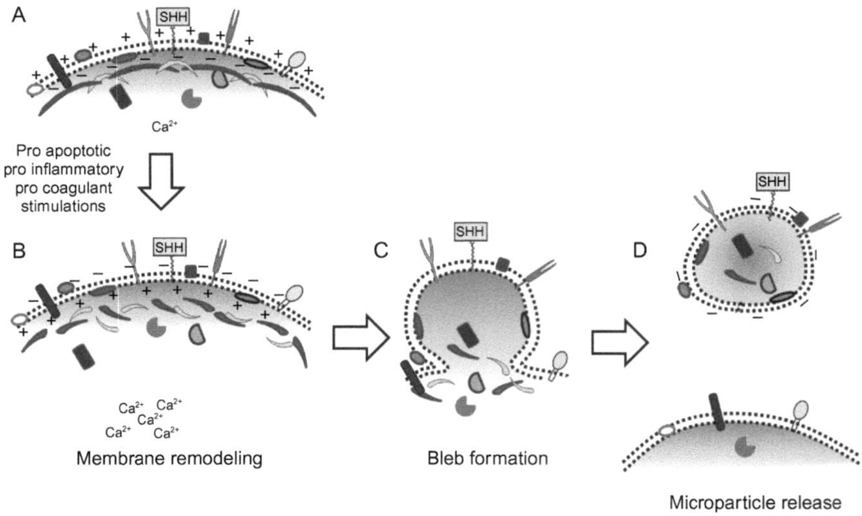

Figure 18.1 Membrane remodeling and microparticles (MPs) formation. (A) In eukaryotic cells, phospholipids are asymmetrically distributed across plasma membrane; in this condition, the amine-containing glycerophospholipids, negatively charged, such as phosphatidylserine (PS), are located preferentially on the cytoplasmic leaflet. The maintenance of transbilayer asymmetry is ensured by combined action of specific transporters. (B) Follow different stimulations leading to cell activation and/or apoptosis, a sustained increase of Ca^{2+} directs various membrane modifications. This perturbation causes an alteration of transporters activities and favors loss of transmembrane phospholipid asymmetry, with consequent exposure of PS in the external monolayer which became negatively charged. (C) These modifications are followed by kinase activation, phosphatase inhibition, cytoskeleton degradation by Ca^{2+}-dependent proteolysis and, then, an increase in bleb formation takes place. (D) Hence, released MPs are highly variable in size (0.05–1 μm), in cargo (proteins, phospholipids, nucleic acids), depending from both cell types and process triggering their formation, which, in turn determines their different functions.

The ability of MPs to circulate freely throughout the vasculature favors and encourages the exchange and the transmission of biological messages. Additionally, the mechanisms by which MPs mediate intercellular communication are various (Tual-Chalot *et al.*, 2011). They can bear single or combinations of ligands that would simultaneously engage different cell surface receptors. Thus, MPs may provide interaction between cells without the need for direct cell contact. They can transfer antigens by binding target cell membrane, which acquires new surface antigens and then new biologic properties and activities. Alternatively, MPs can interact with target cells by fusion or internalization mechanisms which allow the transfer bioactive molecules.

All these evidences sustain the idea of MPs as vectors able to carry transcellular messages and thus, allow to consider them as authentic players in the regulation of physiological and pathological process.

B. Generation of MPs

MPs are a heterogeneous population of small membrane-coated vesicles virtually generated from any cell types during activation or apoptosis. Their release is a well-regulated process, although these vesicles are highly variable in size, composition, and function.

In eukaryotic cells, distribution of lipids across biological membranes is asymmetric (Bretscher, 1972). The choline-containing lipids, phosphatidylcholine (PC) and sphingomyelin (SM), are present primarily on the external layer of the plasma membrane. In contrast, the amine-containing glycerophospholipids, phosphatidylethanolamine (PE) and PS, are located preferentially on the cytoplasmic surface. Loss of transmembrane phospholipid asymmetry, with consequent exposure of PS in the external monolayer, occurs in both normal and pathologic conditions. PS externalization occurs during apoptotic process (Fadok *et al.*, 1998) and cell activation (Bevers *et al.*, 1982). This phenomenon causes alteration of cell surface properties, like acquisition of a procoagulant state (Lubin *et al.*, 1981), enhanced adhesiveness (Schlegel *et al.*, 1985), aggregation (Wali *et al.*, 1987), and recognition by phagocytic cells (Fadok *et al.*, 1998). In normal cell development and homeostasis, these processes are indispensable for normal functioning; however, unregulated loss of PS asymmetry contribute to heart disease and stroke and are associated with pathologies that have high cardiovascular risk (Wali *et al.*, 1988; Wilson *et al.*, 1993).

During cell activation by different agents (thrombin, collagen, ADP, calcium ionophores), the asymmetrical distribution of phospholipids is lost. The intracellular mechanisms underlying the release of MPs seem to be associated to sustained increase of Ca^{2+} that positively and negatively regulates scramblase and translocase activities, respectively. This enhancement provokes the exposure of PS at the external leaflet of the plasma membrane and, probably, represents the major feature of the collapse of transbilayer asymmetry in mammalian cells. These modifications are followed by kinase activation, phosphatase inhibition, cytoskeleton degradation by Ca^{2+}-dependent proteolysis, facilitating membrane blebbing and release of MPs. However, the fact that no all MPs bear PS suggest that other mechanisms must be implicated.

All these events represent a general model at basis of MP formation; however, the mechanisms involved during apoptosis or activation are different. As example, MP formation during apoptosis results from Rho-associated kinase (ROCK I) activity, due to caspase 3 activation. ROCK I promotes increased actin–myosin force generation, couples actin–myosin filaments to the plasma membrane and, as consequence, leads to disruption of membrane skeleton structure and formation of membrane blebs (Coleman *et al.*, 2001). Conversely, other authors have shown that MP shedding induced by thrombin from ECs involves ROCK II activation via caspase 2 pathway, despite an absence of cell death (Sapet *et al.*, 2006). These examples illustrate the complexity of pathways that lead to the formation of MPs.

C. Composition of MPs

Despite their small size, MPs are enriched in surface bioactive molecules (lipids and proteins) and contain cytoplasmic proteins and nucleic acids. Their composition differs depending on cellular origin and process triggering their formation.

MP membrane bilayer consists mainly of phospholipids and proteins and results negatively charged due to the presence of PS. As examples, MPs from healthy humans present phospholipids composition consisting mainly of PC (60%) (Weerheim *et al.*, 2002); whereas MPs from synovial fluid of inflamed joints of arthritis patients contain PC, PE, SM, lysophospholipids (all 20–25%) and small amounts of PS (Fourcade *et al.*, 1995). Also, MPs from ECs exposed to an oxidative stress present oxidized phospholipids, whereas there are absent in MPs derived from the same cells undergoing stimulation with calcium ionophore (Huber *et al.*, 2002).

Furthermore, MPs carry on their surface antigens typical of the parental cell and their state, as well as other membrane and cytoplasmic constituents. This characteristic allows the determination of their cell origin using antibodies directed against specific antigens. PMPs expose glycoproteins (GP)Ib (CD42b), platelet-endothelium adhesion molecule-1 (PECAM 1; CD31) and the fibrinogen receptor, the integrin αIIbβ3 (GPIIb–IIIa). In addition, they can expose markers which consent to evaluate status of cells they originate, for instance, MPs exposing P-selectin (CD62P),derive from activated platelets (Diamant *et al.*, 2004).

Protein composition of MPs is also related to stimulus at their origin. The comparison of protein composition obtained from the CEM T-cell line MPs, either in mitogenic (phytohemagglutinin, PHA) and apoptotic (actinomycin D, ActD) conditions, shows several differences. In total, 390 proteins were identified in MPs, among which 34% were described to be associated or localized in the plasma membrane. Only very few nuclear, mitochondrial, Golgi, or endoplasmic reticulum proteins were detected (less than 10% in total). Histone proteins were only identified in apoptotic conditions, as expected following cell death induction. Half of the detected proteins are intracellular proteins. They can be grouped into cytoskeleton or cytoskeleton-associated proteins, heat shock proteins, translation-associated proteins, and metabolism enzymes. These cytoplasmic proteins are mostly the same in MPs generated from apoptosis and mitogenic conditioned cells, except for ribosomal proteins which were dramatically increased under apoptotic stimulation. Concerning plasma membrane proteins, 17 hematopoietic clusters of differentiation are present in MPs derived from both stimulation conditions, except for CD99, CD98, and CD81 only present under apoptotic conditions. Numerous adhesion, transport, signal transduction, and cytoskeleton-anchoring proteins are also identified (Miguet *et al.*, 2006). Finally, MPs generated from activated (PHA and phorbol-myristate-acetate, PMA) and apoptotic (ActD) CEM T lymphocytes or lymphocytes from diabetic patients expose on their surface the morphogen SHH. Whereas treatment of same cells with PHA alone, PMA alone, and ActD alone generates MPs lacking in SHH, as well as diabetic patient MPs elicited under apoptotic conditions (Martínez *et al.*, 2006).

Furthermore, because MPs engulf some cytoplasm, they shuttle specific subset of mRNA or microRNA that are then delivered to target cells. In this transfer of mRNA, MPs act as a kind of "naturally-engineered liposomes" (Ratajczak, 2011).

D. Vessel formation and MVs

MVs are involved in different process, including inflammation, coagulation, and also changes in endothelial functions. In fact, it has been shown that different stages of vessel formation are mediated by MVs. The effects triggered by MVs depend on the mechanics of stimulation and the activation status of the cell from which they originate and also on their concentration. Different responses trigger by MVs can be acute, by releasing several factors, or delayed, implying changes in expression of genes involved in structural and functional regulation of vascular wall.

ECs shed protease-containing vesicles that stimulate endothelial proteolitic activity during cell migration and three-dimensional morphological organization during angiogenesis. Endothelial MVs (EMPs) contain MMP2 and MMP9 as well as membrane type 1-matrix metalloproteinase (MT1-MMP) in both active and proenzyme forms, and are therefore ready to promote matrix degradation. TIMP-1 and TIMP-2, endogenous inhibitors of MMPs, are present in the same MVs and could exert a regulatory effect on the activity of the protease. Low concentrations stimulated cord formation, higher ones being inhibitory. Small amounts of MMPs are essential for the onset of morphogenetic program, whereas an excess of proteolytic activity prevents tubulogenesis and cause the reabsorptions of the formed vasculature. In this respect, the need for a delicate balance between proteases and their inhibitors, in each phase of the angiogenic process, could provide an additional functional explanation for the presence of TIMPs in vesicles shed by ECs (Taraboletti *et al.*, 2002). By contrast, another study shown that EMVs are able to impairs angiogenesis *in vitro* by affecting all parameters of the capillary network formation by uniformly decreasing total capillary length and numbers of meshes and branching points and increasing mesh area. One of the possible mechanisms of this phenomenon may be the increase in oxidative stress in cells treated with

EMVs. This hypothesis is supported by the finding that treatment with a superoxide dismutase (SOD) mimetic restores all parameters of angiogenesis affected by EMVs. Moreover, these MVs increase apoptotic rate in a dose–response manner which suggest the role of this mechanism in the altered angiogenic patter (Mezentsev *et al.*, 2005). The explanation of different effects observed in these studies may be related to concentration of MVs used. Physiological concentration of EMVs, such as those found in the circulation of healthy subjects, do not affect any of the parameters of angiogenesis *in vivo*. At low concentration, MVs do not affect the endothelium (Brodsky *et al.*, 2004; Taraboletti *et al.*, 2002). However, when the number of circulating EMPs exceeds a certain threshold, the EMPs became important factors in pathophysiology, directly affecting the endothelium and other circulating cells (Horstman *et al.*, 2004; Jy *et al.*, 2004). Moreover, EMPs provide a catalytic surface for the conversion of plasminogen into plasmin by expressing uPA and its receptor (uPAR). They are able to modulate angiogenic responses of EPCs *in vitro*. EMPs affect EPCs angiogenesis in a concentration-dependent manner. Thus, while low amounts of EMPs increase tube formation, higher concentrations inhibit it (Lacroix *et al.*, 2007).

Furthermore, MPs derived from EPCs are able to trigger an angiogenic program in mature quiescent ECs. In fact, they are incorporated in ECs by interaction with integrins expressed on MP surface. *In vitro*, these MPs promoted EC survival, proliferation, apoptosis resistance, and organization in capillary-like structures. *In vivo*, MP-stimulated human ECs organize in patent vessels. Moreover, when incubated with RNase MPs failed to induce *in vitro* and *in vivo* effects, suggesting that they are able to transfer angiogenic program also through mRNA. In fact, analysis of mRNA extract indicate that these MPs are shuttling a specific subset of cellular mRNA, such as mRNA associated with PI3K/Akt signaling pathway and with eNOS, known to be involved in angiogenesis (Deregibus *et al.*, 2007).

PMPs may trigger various biological responses in target cells by stimulating them via their surface proteins, by exposing them their bioactive lipids and by transferring different surface proteins. This kind of MPs is able to affect many of functions of human hematopoietic cells by stimulating their proliferation, survival, adhesion, and chemotaxis (Baj-Krzyworzeka *et al.*, 2002).

Moreover, PMPs stimulate proliferation, chemotaxis, and tube formation of ECs *via* the Pertussis toxin-sensitive G protein, ERK and the PI3K pathway. All these effects are mediated by their lipid component, with a minor contribution of protein component (Kim *et al.*, 2004).

Another study has shown that platelet MVs (PMVs) induce an angiogenic response, both *in vitro* and *in vivo*. In rat aortic ring model, this effect is mediated by cytokines, such as VEGF, bFGF, and PDGF. PMVs exert their effect via PI3K, Src kinase and ERK. Moreover PMVs induce invasion of

ECs through Matrigel®, and this effect is also mediated by VEGF, heparanase, and PDGF. Furthermore, these authors have shown that in a rat model of *in vivo* chronic myocardial ischemia, injection of PMVs into the myocardium induces angiogenesis indicating their strong implication on development of blood vessel *de novo* (Brill *et al.*, 2005).

MPs derived from activated platelets interact with lung cancer cells and enhance their metastatic and angiogenic potential. PMPs transferred the platelet-derived integrin CD41 to different lung cancer cells and stimulated the phosphorylation of mitogen-activated protein kinase (MAPK) p42/44 and serine/threonine kinase AKT, as well as the expression of MT1-MMP. PMPs are able to induce chemotaxis, proliferation, and invasion of lung cancer cells. Furthermore, in these cells, PMPs upregulate cyclin D2 expression, stimulate mRNA expression for angiogenic factors such as MMP9, VEGF, IL-8, and HGF, as well as adhesion to fibrinogen and ECs. Based on these findings, it has been suggested that PMPs play an important role in tumor progression/metastasis and angiogenesis in lung cancer (Janowska-Wieczorek *et al.*, 2005).

Shed-membrane MPs isolated from human atherosclerotic lesion stimulate EC proliferation and *in vivo* neovessel formation after CD40 ligation. The effect on endothelial proliferation is specific for MPs isolated from human plaque, because an equivalent concentration of MPs isolated from the blood of the same patients did not affect this phenomenon. Also, a crucial role in proliferation is attributed to CD40L bound to macrophage-derived MPs isolated from human atherosclerotic plaques. Furthermore, *in vivo* neovessel formation induced by plaque MPs requires the presence of functional CD40 and involves VEGF and PI3K/Akt pathways, suggesting that these MPs act as potential endogenous triggers of neovascularization and growth of atherosclerotic plaque (Leroyer *et al.*, 2008).

Vitreous MPs isolated from patients with diabetic retinopathy derive from platelet, endothelial, and retinal cells (photoreceptor and microglial cells) and stimulate EC proliferation and formation of new vessels, suggesting that they could contribute to retinal angiogenesis and disease progression (Chahed *et al.*, 2010).

MPs can also bear proteins such as peroxisome proliferator-activated receptor-alpha (PPARα), which results necessary for EPC differentiation and angiogenesis. Indeed, MPs harboring PPARα increase EPC differentiation and EC marker expression, enhance EPC differentiation that is mediated, at least in part, by PPARα-NF-kB and Akt pathways; modulate *in vitro* angiogenic-related properties of ECs; and stimulate *in vivo* bone marrow-derived cell-associated neovascularization. Altogether, these findings provide a novel mechanism underlying EPC differentiation and describe the proangiogenic property of these MPs (Benameur *et al.*, 2010b).

On the other hand, it has been shown that MPs shed by apoptotic lymphocytes, inhibit *in vitro* and *in vivo* angiogenesis, by suppressing vascular cell survival, proliferation, and migration. They induce ROS production by enhancing both NOX activity and expression. Increased ROS levels occur upstream of induction of CD36 (TSP1 receptor) with subsequent suppression of VEGF/VEGFR2 signaling pathway. Moreover, in a model of *in vivo* neovascularization, these MPs are able to antagonize the proangiogenic effects induced by VEGF, and then, inhibit cell migration (Yang *et al.*, 2008). Furthermore, it has been shown that MPs from apoptotic lymphocytes decrease NO production via PI3K pathway. Decreased NO generation is associated with enhanced phosphorylation of eNOS on its inhibitory site and overexpression of Cav-1 (Mostefai *et al.*, 2008b). Indeed, antiangiogenic effects of LMPs are linked to oxidative stress and reduced NO release from ECs (Mostefai *et al.*, 2008a).

Furthermore, it has been shown that these LMPs inhibit lung tumor growth and microvessel density and limited local production of VEGF-A. Decreased lung carcinoma cell viability and proliferation is associated with an increased apoptosis. These MPs inhibit tumor activity in a VEGF-A-dependent manner. Indeed, they limit VEGF-A expression and reduce VEGF-induced cell migration. These data support the notion that LMPs suppress tumor growth *in vivo* by targeting tumor angiogenesis and cell growth by interfering with the VEGF-A pathway. In addition, the antitumor effects of LMPs in lung carcinoma cells are, at least in part, dependent on low-density lipoprotein receptor (LDLR) activity. This finding allow to hypothesize that LMPs transfer into tumor cells may suppress proangiogenic and progrowth response under pathophysiological conditions (Yang *et al.*, 2010).

Taking in consideration that stimulation at basis of MPs formation is pivotal to determining their functional effects, in the section below we will discuss the evidence that MPs generated from T lymphocytes undergoing both activation and apoptosis have a crucial role in modulation of neovascularization.

E. MPs harboring SHH and neovascularization

As described below, composition of MPs depends on type cell from they have originated and the type of stimulation. We have shown that activation of T lymphocytes with phytohemagglutin and phorbol ester and further induction of apoptosis with ActD generates MPs expressing the morphogen SHH (MPs^{SHH+}) on their surface, whereas MPs released from T lymphocytes treated with ActD alone do not express SHH (MPs^{SHH-}) (Martínez *et al.*, 2006). Interestingly, the different composition of these MPs, that is, the presence or the absence of SHH, may explain the observed differences on target cells (Soleti *et al.*, 2009). Indeed, *in vitro* treatment of human umbilical vein endothelial cells (HUVECs) with MPs^{SHH+} induced proangiogenic changes of these cells, mainly the increase of cell proliferation and endothelial adhesion through the increase of intercellular adhesion molecule (ICAM)-1 expression. These effects involve Rho kinase pathway and are inhibited by silencing of SHH cascade, suggesting the interaction between SHH harbored by MPs and SHH pathway-associated receptors. In addition, MPs^{SHH+} induced an increase in mRNA and protein expression of several proangiogenic factors (HGF, VEGF, and IL-1β) indicating that SHH carried by MPs acts on a large number of target genes that regulate angiogenesis at different phases (Fig. 18.2). It should be noted that although the concentration of SHH carried by MPs has not been determined, it must be largely lower than 10 μg/ml, since this was the total concentration of MP proteins.

Moreover, MPs^{SHH+} were able to favor postischemic neovascularization. Notably, *in vivo* treatment of hindlimb ischemic mice with MPs^{SHH+} improves blood flow in the ischemic leg as well as the number of capillaries when compared with the non ischemic leg (Benameur *et al.*, 2010a). These effects were due to a dual action of MPs^{SHH+}. On the one hand, the interaction between ligand (SHH) and its receptors (PTCH/SMO) was evidenced by blocking of the canonical pathway of SHH. On the other hand, MPs^{SHH+}-stimulated endogenous SHH expression on mouse ischemic muscle. In addition, an increase in NO production, as a consequence of the activation of endothelial NO-synthase, and an enhanced expression of

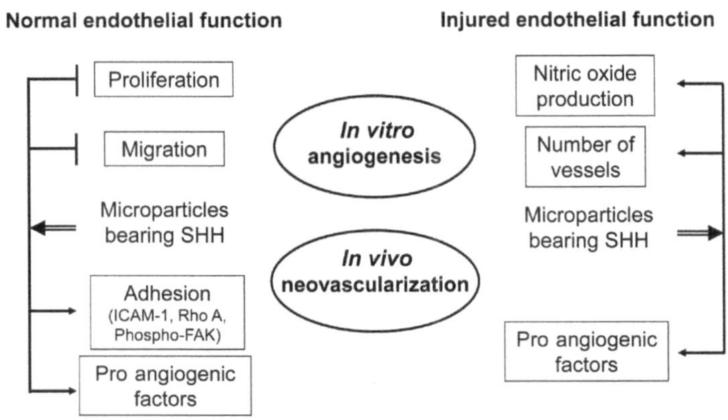

Figure 18.2 Effects evoked by microparticles bearing Sonic Hedgehog (SHH) on angiogenesis depending on normal endothelial function and dysfunction. Microparticles decrease both endothelial cell proliferation and migration, but they enhance mRNA levels of proangiogenic mediators, regulate endothelial cell adhesion by augmenting expression of the intracellular adhesion molecule-1 (ICAM-1), Rho A and activity of focal adhesion kinase (FAK). Altogether, microparticles contribute to generation of *in vitro* angiogenesis. Under physiopathological conditions, microparticles increase nitric oxide production and proangiogenic mediator expressions leading to recovery of blood flow and neovascularization.

several proangiogenic factors, including FGF5, FGF2, and VEGF, were described in the muscle from the ischemic leg after MPs^{SHH+} treatment. Altogether, these results suggest that MPs^{SHH+} are potent regulators of angiogenesis during muscle regeneration after ischemia in mice and they might be potential tools in diseases associated with failed angiogenesis (Fig. 18.2).

However, in other models such as chronic liver diseases, angiogenesis is a pathological hallmark and the signal carried by MPs^{SHH+} could be deleterious. In particular, during cirrhosis, angiogenesis contributes to vascular remodeling and new approaches indicate that antiangiogenic therapy related to reduce vascularization for blocking or slowing down the disease progression seem to be promising. In this respect, it has been shown that, in cirrhotic livers, myofibroblastic hepatic stellate cells (MF-HSC) and cholangiocytes released exosome-enriched MPs containing biologically active HH ligands that induced changes in gene expression of hepatic sinusoidal ECs (Witek *et al.*, 2009). Among the changes reported in hepatic sinusoidal ECs, an increase of iNOS and CD31 expressions that were associated with GLI2 expression confirming that the changes that occur in these cells after MPs^{SHH+} treatment result from the activation of HH pathway. These data suggest a novel mechanism for cirrhotic vasculopathy implicating MPs^{SHH+} released from HH-producing liver cells that accumulate in fibrotic livers and promote remodeling of the hepatic sinusoids during cirrhosis.

In addition, MPs that are generated from apoptotic T lymphocytes and which do not express SHH (Martínez *et al.*, 2006) inhibit angiogenesis as demonstrated in two models, *in vitro* aortic rings and *in vivo* cornea neovascularization in mice (Yang *et al.*, 2008). Yang *et al.* have demonstrated that MPs^{SHH-} suppress angiogenesis and inhibit both EC survival and proliferation through the increase of reactive oxygen species generation probably associated with an increase of NADPH activity. The effects of MPs^{SHH-} were associated with the downregulation of VEGFR2 and ERK pathways. More recently, these authors proposed MPs^{SHH-} as promising antiangiogenic agents for the treatment of lung carcinomas. In fact, intratumoral injection of these MPs decreased tumor size and vascularization, and VEGF levels in mice (Yang *et al.*, 2010).

V. Conclusion

Taken together, we have shown dual effects of MPs harboring SHH. On the one hand, MPs can therefore be considered true vectors in the transfer of morphogen-borne biologic information to remote responsive cells, and thereby contribute to the maintenance of homeostasis. On the other hand, the release of SHH associated to membrane vesicles is able to modulate important cellular functions, notably, the formation of new vessels. Taken advantage of these interesting properties, MPs could be used as therapeutic tools in pathologies linked to failed angiogenesis.

REFERENCES

Adolphe, C., Narang, M., Ellis, T., Wicking, C., Kaur, P., and Wainwright, B. (2004). An *in vivo* comparative study of sonic, desert and Indian hedgehog reveals that hedgehog pathway activity regulates epidermal stem cell homeostasis. *Development* **131,** 5009–5019.

Agouni, A., Mostefai, H. A., Porro, C., Carusio, N., Favre, J., Richard, V., Henrion, D., Martínez, M. C., and Andriantsitohaina, R. (2007). Sonic hedgehog carried by microparticles corrects endothelial injury through nitric oxide release. *FASEB J.* **21,** 2735–2741.

Alcedo, J., Zou, Y., and Noll, M. (2000). Posttranscriptional regulation of smoothened is part of a self-correcting mechanism in the Hedgehog signalling system. *Mol. Cell* **6,** 457–465.

Asahara, T., Murohara, T., Sullivan, A., Silver, M., Zee, R. V. D., Li, T., Witzenbichler, B., Schattemen, G., and Isner, J. M. (1997). Isolation of putative progenitor endothelial cells for angiogenesis. *Science* **275,** 964–967.

Asahara, T., Chen, D., Takahashi, T., Fujikawa, K., Kearney, M., Magner, M., Yancopoulos, G. D., and Isner, J. M. (1998). Tie2 receptor ligands, angiopoietin-1 and angiopoietin-2, modulate VEGF-induced postnatal neovascularization. *Circ. Res.* **83,** 233–240.

Asai, J., Takenaka, H., Kusano, K. F., Ii, M., Luedemann, C., Curry, C., Eaton, E., Iwakura, A., Tsutsumi, Y., Hamada, H., Kishimoto, S., Thorne, T., *et al.* (2006). Topical sonic hedgehog gene therapy accelerates wound healing in diabetes by enhancing endothelial progenitor cell-mediated microvascular remodeling. *Circulation* **113,** 2413–2424.

Auerbach, H. R., and Auerbach, W. (1997). Profond effects on vascular development caused by perturbations of during organogenesis. *Am. J. Pathol.* **151,** 1183–1186.

Ausprunk, D. H., and Folkman, J. (1977). Migration and proliferation of endothelial cells in preformed and newly formed blood vessels during angiogenesis. *Microvasc. Res.* **14,** 53–65.

Azevedo, L. C., Pedro, M. A., and Laurindo, F. R. (2007). Circulating microparticles as therapeutic targets in cardiovascular diseases. *Recent Pat. Cardiovasc. Drug Discov.* **2,** 41–51.

Bai, C. B., and Joyner, A. L. (2001). Gli1 can rescue the *in vivo* function of Gli2. *Development* **128,** 5161–5172.

Baj-Krzyworzeka, M., Majka, M., Pratico, D., Ratajczak, J., Vilaire, G., Kijowski, J., Reca, R., Janowska-Wieczorek, A., and Ratajczak, M. Z. (2002). Platelet-derived microparticles stimulate proliferation, survival, adhesion, and chemotaxis of hematopoietic cells. *Exp. Hematol.* **30,** 450–459.

Bayless, K. J., Salazar, R., and Davis, G. E. (2000). RGD-dependent vacuolation and lumen formation observed during endothelial cell morphogenesis in three-dimensional fibrin matrices involves the alpha(v)beta(3) and alpha(5)beta(1) integrins. *Am. J. Pathol.* **156,** 1673–1683.

Beckett, K., Franch-Marro, X., and Vincent, J. P. (2008). Glypican-mediated endocytosis of Hedgehog has opposite effects in flies and mice. *Trends Cell Biol.* **18,** 360–363.

Bein, K., and Simons, M. (2000). Thrombospondin type 1 repeats interact with matrix metalloproteinase 2. Regulation of metalloproteinase activity. *J. Biol. Chem.* **275,** 32167–32173.

Bellusci, S., Furuta, Y., Rush, M. G., Henderson, R., Winnier, G., and Hogan, B. L. (1997). Involvement of Sonic hedgehog (Shh) in mouse embryonic lung growth and morphogenesis. *Development* **124,** 53–63.

Belperio, J. A., Keane, M. P., Arenberg, D. A., Addison, C. L., Ehlert, J. E., Burdick, M. D., and Strieter, R. M. (2000). CXC chemokines in angiogenesis. *J. Leukoc. Biol.* **68,** 1–8.

Benameur, T., Soleti, R., Porro, C., Andriantsitohaina, R., and Martínez, M. C. (2010a). Microparticles carrying Sonic hedgehog favor neovascularization through the activation of nitric oxide pathway in mice. *PLoS One* **5,** e12688.

Benameur, T., Tual-Chalot, S., Andriantsitohaina, R., and Martínez, M. C. (2010b). PPARalpha is essential for microparticle-induced differentiation of mouse bone marrow-derived endothelial progenitor cells and angiogenesis. *PLoS One* **5,** e12392.

Benjamin, L. E., and Keshet, E. (1997). Conditional switching of vascular endothelial growth factor (VEGF) expression in tumors: Induction of endothelial cell shedding and regression of hemangioblastoma-like vessels by VEGF withdrawal. *Proc. Natl. Acad. Sci. USA* **94,** 8761–8766.

Berse, B., Brown, L. F., Van de Water, L., Dvorak, H. F., and Senger, D. R. (1992). Vascular permeability factor (vascular endothelial growth factor) gene is expressed differentially in normal tissues, macrophages, and tumors. *Mol. Biol. Cell* **3,** 211–220.

Bevers, E. M., Comfurius, P., van Rijn, J. L., Hemker, H. C., and Zwaal, R. F. (1982). Generation of prothrombin-converting activity and the exposure of phosphatidylserine at the outer surface of platelets. *Eur. J. Biochem.* **122,** 429–436.

Bijlsma, M. F., Spek, C. A., and Peppelenbosch, M. P. (2004). Hedgehog: An unusual signal transducer. *Bioessays* **26,** 387–394.

Bijlsma, M. F., Peppelenbosch, M. P., and Spek, C. A. (2006a). Hedgehog morphogen in cardiovascular disease. *Circulation* **114,** 1985–1991.

Bijlsma, M. F., Spek, C. A., Zivkovic, D., van de Water, S., Rezaee, F., and Peppelenbosch, M. P. (2006b). Repression of smoothened by patched-dependent (pro-)vitamin D3 secretion. *PLoS Biol.* **4,** e232.

Blair, R. J., Meng, H., Marchese, M. J., Ren, S., Schwartz, L. B., Tonnesen, M. G., and Gruber, B. L. (1997). Human mast cells stimulate vascular tube formation. Tryptase is a novel, potent angiogenic factor. *J. Clin. Invest.* **99,** 2691–2700.

Boulton, M. (1999). A role for hepatocyte growth factor in diabetic retinopathy. *Br. J. Ophthalmol.* **83,** 763–764.

Bretscher, M. S. (1972). Asymmetric lipid bilayer structure for biological membranes. *Nat. New Biol.* **236,** 11–12.

Brew, K., Dinakarpandian, D., and Nagase, H. (2000). Tissue inhibitors of metalloproteinases: Evolution, structure and function. *Biochim. Biophys. Acta* **1477,** 267–283.

Brill, A., Dashevsky, O., Rivo, J., Gozal, Y., and Varon, D. (2005). Platelet-derived microparticles induce angiogenesis and stimulate post-ischemic revascularization. *Cardiovasc. Res.* **67,** 30–38.

Briscoe, J., and Therond, P. (2005). Hedgehog signaling: From the Drosophila cuticle to anti-cancer drugs. *Dev. Cell* **8,** 143–151.

Brodsky, S. V., Zhang, F., Nasjletti, A., and Goligorsky, M. S. (2004). Endothelium-derived microparticles impair endothelial function *in vitro*. *Am. J. Physiol. Heart Circ. Physiol.* **286,** 1910–1915.

Brown, L. F., Yeo, K. Y., Berse, B., Yeo, T. K., Senger, D. R., Dvorak, H. F., and Van de Water, L. (1992). Expression of vascular permeability factor (vascular endothelial growth factor) by epidermal keratinocytes during wound healing. *J. Exp. Med.* **176,** 1375–1379.

Burke, R., Nellen, D., Bellotto, M., Hafen, E., Senti, K. A., Dickson, B. J., and Basler, K. (1999). Dispatched, a novel sterol-sensing domain protein dedicated to the release of cholesterol-modified hedgehog from signaling cells. *Cell* **99,** 803–815.

Byrd, N., and Grabel, L. (2004). Hedgehog signaling in murine vasculogenesis and angiogenesis. *Trends Cardiovasc. Med.* **14,** 308–313.
Byrd, N., Becker, S., Maye, P., Narasimhaiah, R., St-Jacques, B., Zhang, X., McMahon, J., McMahon, A., and Grabel, L. (2002). Hedgehog is required for murine yolk sac angiogenesis. *Development* **129,** 361–372.
Carmeliet, P. (2000). Fibroblast growth factor-1 stimulates branching and survival of myocardial arteries: A goal for therapeutic angiogenesis? *Circ. Res.* **87,** 176–178.
Carmeliet, P. (2004). Manipulating angiogenesis in medicine. *J. Intern. Med.* **255,** 538–561.
Carmeliet, P., Ng, Y. S., Nuyens, D., Theilmeier, G., Brusselmans, K., Cornelissen, I., Ehler, E., Kakkar, V. V., Stalmans, I., Mattot, V., Perriard, J. C., Dewerchin, M., *et al.* (1999). Impaired myocardial angiogenesis and ischemic cardiomyopathy in mice lacking the vascular endothelial growth factor isoforms VEGF164 and VEGF188. *Nat. Med.* **5,** 495–502.
Chahed, S., Leroyer, A. S., Benzerroug, M., Gaucher, D., Georgescu, A., Picaud, S., Silvestre, J. S., Gaudric, A., Tedgui, A., Massin, P., and Boulanger, C. M. (2010). Increased vitreous shedding of microparticles in proliferative diabetic retinopathy stimulates endothelial proliferation. *Diabetes* **59,** 694–701.
Chao, W., and D'Amore, P. A. (2008). IGF2: Epigenetic regulation and role in development and disease. *Cytokine Growth Factor Rev.* **19,** 111–120.
Chen, Y., and Struhl, G. (1996). Dual roles for Patched in sequestering and transducing Hedgehog. *Cell* **87,** 553–563.
Chen, J. K., Taipale, J., Young, K. E., Maiti, T., and Beachy, P. A. (2002a). Small molecule modulation of Smoothened activity. *Proc. Natl. Acad. Sci. USA* **99,** 14071–14076.
Chen, J. K., Taipale, J., Cooper, M. K., and Beachy, P. A. (2002b). Inhibition of Hedgehog signaling by direct binding of cyclopamine to Smoothened. *Genes Dev.* **16,** 2743–2748.
Chen, M. H., Li, Y. J., Kawakami, T., Xu, S. M., and Chuang, P. T. (2004). Palmitoylation is required for the production of a soluble multimeric Hedgehog protein complex and long-range signaling in vertebrates. *Genes Dev.* **18,** 641–659.
Chiang, C., Litingtung, Y., Lee, E., Young, K. E., Corden, J. L., Wsetphal, H., and Beachy, P. A. (1996). Cyclopia and defective axial patterning in mice lacking Sonic hedgehog gene function. *Nature* **383,** 407–413.
Chinchilla, P., Xiao, L., Kazanietz, M. G., and Riobo, N. A. (2010). Hedgehog proteins activate pro-angiogenic responses in endothelial cells through non-canonical signaling pathways. *Cell Cycle* **9,** 570–579.
Chuang, P. T., and Kornberg, T. B. (2000). On the range of hedgehog signaling. *Curr. Opin. Genet. Dev.* **10,** 515–522.
Cohen, M. M., Jr. (2010). Hedgehog signaling update. *Am. J. Med. Genet. A.* **152A,** 1875–1914.
Coleman, M. L., Sahai, E. A., Yeo, M., Bosch, M., Dewar, A., and Olson, M. F. (2001). Membrane blebbing during apoptosis results from caspase-mediated activation of ROCK I. *Nat. Cell Biol.* **3,** 339–345.
Colombo, E. S., Menicucci, G., McGuire, P. G., and Das, A. (2007). Hepatocyte growth factor/scatter factor promotes retinal angiogenesis through increased urokinase expression. *Invest. Ophthalmol. Vis. Sci.* **48,** 1793–1800.
Cooper, M. K., Porter, J. A., Young, K. E., and Beachy, P. A. (1998). Teratogen-mediated inhibition of target tissue response to Shh signaling. *Science* **280,** 1603–1607.
Cooper, M. K., Wassif, C. A., Krakowiak, P. A., Taipale, J., Gong, R., Kelley, R. I., Porter, F. D., and Beachy, P. A. (2003). A defective response to Hedgehog signaling in disorders of cholesterol biosynthesis. *Nat. Genet.* **33,** 508–513.
Corada, M., Mariotti, M., Thurston, G., Smith, K., Kunkel, R., Brockhaus, M., Lampugnani, M. G., Martin-Padura, I., Stoppacciaro, A., Ruco, L., McDonald, D. M., Ward, P. A., *et al.* (1999). Vascular endothelial-cadherin is an

important determinant of microvascular integrity in vivo. *Proc. Natl. Acad. Sci. USA* **96,** 9815–9820.

Corcoran, R. B., and Scott, M. P. (2006). Oxysterols stimulate Sonic hedgehog signal transduction and proliferation of medulloblastoma cells. *Proc. Natl. Acad. Sci. USA* **103,** 8408–8413.

Couffinhal, T., Silver, M., Zheng, L. P., Kearney, M., Witzenbichler, B., and Isner, J. M. (1998). Mouse model of angiogenesis. *Am. J. Pathol.* **152,** 1667–1679.

Crompton, T., Outram, S. V., and Hager-Theodorides, A. L. (2007). Sonic hedgehog signalling in T-cell development and activation. *Nat. Rev. Immunol.* **7,** 726–735.

Dellovade, T., Romer, J. T., Curran, T., and Rubin, L. L. (2006). The Hedgehog pathway and neurological disorders. *Annu. Rev. Neurosci.* **29,** 539–563.

Deregibus, M. C., Cantaluppi, V., Calogero, R., Lo Iacono, M., Tetta, C., Biancone, L., Bruno, S., Bussolati, B., and Camussi, G. (2007). Endothelial progenitor cell derived microvesicles activate an angiogenic program in endothelial cells by a horizontal transfer of mRNA. *Blood* **110,** 2440–2448.

Diamant, M., Tushuizen, M. E., Sturk, A., and Nieuwland, R. (2004). Cellular microparticles: New players in the field of vascular disease? *Eur. J. Clin. Invest.* **34,** 392–401.

Distler, J. H., Hirth, A., Kurowska-Stolarska, M., Gay, R. E., Gay, S., and Distler, O. (2003). Angiogenic and angiostatic factors in the molecular control of angiogenesis. *Q. J. Nucl. Med.* **47,** 149–161.

Djonov, V., Schmid, M., Tschanz, S. A., and Burri, P. H. (2000). Intussusceptive angiogenesis: Its role in embryonic vascular network formation. *Circ. Res.* **86,** 286–292.

Donnini, S., and Ziche, M. (2002). Constitutive and inducible nitric oxide synthase: Role in angiogenesis. *Antioxid. Redox Signal.* **4,** 817–823.

Drake, C. J., Cheresh, D. A., and Little, C. D. (1995). An antagonist of integrin $a_v\beta_3$ prevents maturation of blood vessels during embryonic neovascularization. *J. Cell Sci.* **108,** 2655–2661.

Dvorak, H. F., Brown, L. F., Detmar, M., and Dvorak, A. M. (1995). Vascular permeability factor/vascular endothelial growth factor, vascular hyperpermeability, and angiogenesis. *Am. J. Pathol.* **146,** 1029–1039.

Egginton, S., and Gerritsen, M. (2003). Lumen formation: *In vivo* versus *in vitro* observations. *Microcirculation* **10,** 45–61.

Eliceiri, B. P., and Cheresh, D. A. (1999). The role of alphav integrins during angiogenesis: Insights into potential mechanisms of action and clinical development. *J. Clin. Invest.* **103,** 1227–1230.

Eliceiri, B. P., Paul, R., Schwartzberg, P. L., Hood, J. D., Leng, J., and Cheresh, D. A. (1999). Selective requirement for Src kinases during VEGF-induced angiogenesis and vascular permeability. *Mol. Cell* **4,** 915–924.

Fadok, V. A., Bratton, D. L., Frasch, S. C., Warner, M. L., and Henson, P. M. (1998). The role of phosphatidylserine in recognition of apoptotic cells by phagocytes. *Cell Death Differ.* **5,** 551–562.

Ferrara, N. (1996). Vascular endothelial growth factor. *Eur. J. Cancer* **32A,** 2413–2422.

Ferrara, N., and Alitalo, K. (1999). Clinical applications of angiogenic growth factors and their inhibitors. *Nat. Med.* **5,** 1359–1364.

Ferrara, N., Winer, J., and Burton, T. (1991). Aortic smooth muscle cells express and secrete vascular endothelial growth factor. *Growth Factors* **5,** 141–148.

Fisher, C. E., and Howie, S. E. (2006). The role of megalin (LRP-2/Gp330) during development. *Dev. Biol.* **296,** 279–297.

Forsythe, J. A., Jiang, B. H., Iyer, N. V., Agani, F., Leung, S. W., Koos, R. D., and Semenza, G. L. (1996). Activation of vascular endothelial growth factor gene transcription by hypoxia-inducible factor 1. *Mol. Cell. Biol.* **16,** 4604–4613.

Fourcade, O., Simon, M. F., Viodé, C., Rugani, N., Leballe, F., Ragab, A., Fournié, B., Sarda, L., and Chap, H. (1995). Secretory phospholipase A2 generates the novel lipid mediator lysophosphatidic acid in membrane microvesicles shed from activated cells. *Cell* **80,** 919–927.

Frank-Kamenetsky, M., Zhang, X. M., Bottega, S., Guicherit, O., Wichterle, H., Dudek, H., Bumcrot, D., Wang, F. Y., Jones, S., Shulok, J., Rubin, L. L., and Porter, J. A. (2002). Small-molecule modulators of Hedgehog signaling: Identification and characterization of Smoothened agonists and antagonists. *J. Biol.* **1,** 10.

Freyssinet, J. M. (2003). Cellular microparticles: What are they bad or good for? *J. Thromb. Haemost.* **1,** 1655–1662.

Fu, J. R., Liu, W. L., Zhou, J. F., Sun, H. Y., Xu, H. Z., Luo, L., Zhang, H., and Zhou, Y. F. (2006). Sonic hedgehog protein promotes bone marrow-derived endothelial progenitor cell proliferation, migration and VEGF production via PI 3-kinase/Akt signaling pathways. *Acta Pharmacol. Sin.* **27,** 685–693.

Gale, N. W., and Yancopoulos, G. D. (1999). Growth factors acting via endothelial cell-specific receptor tyrosine kinases: VEGFs, angiopoietins, and ephrins in vascular development. *Genes Dev.* **13,** 1055–1066.

Gale, N. W., Thurston, G., Hackett, S. F., Renard, R., Wang, Q., McClain, J., Martin, C., Witte, C., Witte, M. H., Jackson, D., Suri, C., Campochiaro, P. A., *et al.* (2002). Angiopoietin-2 is required for postnatal angiogenesis and lymphatic patterning, and only the latter role is rescued by Angiopoietin-1. *Dev. Cell* **3,** 411–423.

Garriock, R. J., Czeisler, C., Ishii, Y., Navetta, A. M., and Mikawa, T. (2010). An anteroposterior wave of vascular inhibitor downregulation signals aortae fusion along the embryonic midline axis. *Development* **137,** 3697–3706.

Gerber, H. P., McMurtrey, A., Kowalski, J., Yan, M., Keyt, B. A., Dixit, V., and Ferrara, N. (1998). Vascular endothelial growth factor regulates endothelial cell survival through the phosphatidylinositol 3′-kinase/Akt signal transduction pathway. Requirement for Flk-1/KDR activation. *J. Biol. Chem.* **273,** 30336–30343.

Grierson, I., Heathcote, L., Hiscott, P., Hogg, P., Briggs, M., and Hagan, S. (2000). Hepatocyte growth factor/scatter factor in the eye. *Prog. Retin. Eye Res.* **19,** 779–802.

Griffioen, A. W., and Molema, G. (2000). Angiogenesis: Potentials for pharmacologic intervention in the treatment of cancer, cardiovascular diseases, and chronic inflammation. *Pharmacol. Rev.* **52,** 237–268.

Grützkau, A., Krüger-Krasagakes, S., Baumeister, H., Schwarz, C., Kögel, H., Welker, P., Lippert, U., Henz, B. M., and Möller, A. (1998). Synthesis, storage, and release of vascular endothelial growth factor/vascular permeability factor (VEGF/VPF) by human mast cells: Implications for the biological significance of VEGF206. *Mol. Biol. Cell* **9,** 875–884.

Hahn, H., Wojnowski, L., Specht, K., Kappler, R., Calzada-Wack, J., Potter, D., Zimmer, A., Müller, U., Samson, E., Quintanilla-Martinez, L., and Zimmer, A. (2000). Patched target Igf2 is indispensable for the formation of medulloblastoma and rhabdomyosarcoma. *J. Biol. Chem.* **275,** 28341–28344.

Hashimoto, M., Ohsawa, M., Ohnishi, A., Naka, N., Hirota, S., Kitamura, Y., and Aosaza, K. (1995). Expression of vascular endothelial growth factor and its receptor mRNA in angiosarcoma. *Lab. Invest.* **73,** 859–863.

Hashiya, N., Jo, N., Aoki, M., Matsumoto, K., Nakamura, T., Sato, Y., Ogata, N., Ogihara, T., Kaneda, Y., and Morishita, R. (2004). *In vivo* evidence of angiogenesis induced by transcription factor Ets-1: Ets-1 is located upstream of angiogenesis cascade. *Circulation* **109,** 3035–3041.

Hirschi, K. K., Rohovsky, S. A., and D'Amore, P. A. (1998). PDGF, TGF-beta, and heterotypic cell-cell interactions mediate endothelial cell-induced recruitment of 10T1/2 cells and their differentiation to a smooth muscle fate. *J. Cell Biol.* **141,** 805–814.

Hooper, J. F., and Scott, M. P. (2005). Communicating with Hedgehogs. *Nat. Rev. Mol. Cell Biol.* **6,** 306–317.

Horstman, L. L., Jy, W., Jimenez, J. J., and Ahn, Y. S. (2004). Endothelial microparticles as markers of endothelial dysfunction. *Front. Biosci.* **9,** 1118–1135.
Huangfu, D., Liu, A., Rakeman, A. S., Murcia, N. S., Niswander, L., and Anderson, K. V. (2003). Hedgehog signalling in the mouse requires intraflagellar transport proteins. *Nature* **426,** 83–87.
Huber, J., Vales, A., Mitulovic, G., Blumer, M., Schmid, R., Witztum, J. L., Binder, B. R., and Leitinger, N. (2002). Oxidized membrane vesicles and blebs from apoptotic cells contain biologically active oxidized phospholipids that induce monocyte-endothelial interactions. *Arterioscler. Thromb. Vasc. Biol.* **22,** 101–107.
Huynh-Do, U., Stein, E., Lane, A. A., Liu, H., Cerretti, D. P., and Daniel, T. O. (1999). Surface densities of ephrin-B1 determine EphB1-coupled activation of cell attachment through alphavbeta3 and alpha5beta1 integrins. *EMBO J.* **18,** 2165–2173.
Hwang, J. H., Lee, S. H., Lee, K. H., Lee, K. Y., Kim, H., Ryu, J. K., Yoon, Y. B., and Kim, Y. T. (2009). Cathepsin B is a target of Hedgehog signaling in pancreatic cancer. *Cancer Lett.* **273,** 266–272.
Ilan, N., Mahooti, S., Rimm, D. L., and Madri, J. A. (1999). PECAM-1 (CD31) functions as a reservoir for and a modulator of tyrosine-phosphorylated beta-catenin. *J. Cell Sci.* **112,** 3005–3014.
Ingram, W. J., Wicking, C. A., Grimmond, S. M., Forrest, A. R., and Wainwright, B. J. (2002). Novel genes regulated by Sonic Hedgehog in pluripotent mesenchymal cells. *Oncogene* **21,** 8196–8205.
Jacob, J., and Briscoe, J. (2003). Gli proteins and the control of spinal-cord patterning. *EMBO Rep.* **4,** 761–765.
Jacobs, J. (2007). Combating cardiovascular disease with angiogenic therapy. *Drug Discov. Today* **12,** 1040–1045.
Janowska-Wieczorek, A., Majka, M., Kijowski, J., Baj-Krzyworzeka, M., Reca, R., Turner, A. R., Ratajczak, J., Emerson, S. G., Kowalska, M. A., and Ratajczak, M. Z. (2001). Platelet-derived microparticles bind to hematopoietic stem/progenitor cells and enhance their engraftment. *Blood* **98,** 3143–3149.
Janowska-Wieczorek, A., Wysoczynski, M., Kijowski, J., Marquez-Curtis, L., Machalinski, B., Ratajczak, J., and Ratajczak, M. Z. (2005). Microvesicles derived from activated platelets induce metastasis and angiogenesis in lung cancer. *Int. J. Cancer* **113,** 752–760.
Jenkins, D. (2009). Hedgehog signalling: Emerging evidence for non-canonical pathways. *Cell. Signal.* **21,** 1023–1034.
Jy, W., Minagar, A., Jimenez, J. J., Sheremata, W. A., Mauro, L. M., Horstman, L. L., Bidot, C., and Ahn, Y. S. (2004). Endothelial microparticles (EMP) bind and activate monocytes: Elevated EMP-monocyte conjugates in multiple sclerosis. *Front. Biosci.* **9,** 3137–3144.
Kanda, S., Hodgkin, M. N., Woodfield, R. J., Wakelam, M. J., Thomas, G., and Claesson-Welsh, L. (1997). Phosphatidylinositol 3'-kinase-independent p70 S6 kinase activation by fibroblast growth factor receptor-1 is important for proliferation but not differentiation of endothelial cells. *J. Biol. Chem.* **272,** 23347–23353.
Kanda, S., Mochizuki, Y., Suematsu, T., Miyata, Y., Nomata, K., and Kanetake, H. (2003). Sonic hedgehog induces capillary morphogenesis by endothelial cells through phosphoinositide 3-kinase. *J. Biol. Chem.* **278,** 8244–8249.
Kanda, S., Miyata, Y., and Kanetake, H. (2004). Role of focal adhesion formation in migration and morphogenesis of endothelial cells. *Cell. Signal.* **16,** 1273–1281.
Kasper, M., Regl, G., Frischaf, A. M., and Aberger, F. (2006). Gli transcription factors: Mediators of oncogenic Hedgehog signalling. *Eur. J. Cancer* **42,** 437–445.
Katoh, Y., and Katoh, M. (2005a). Identification and characterization of rat Desert hedgehog and Indian hedgehog genes in silico. *Int. J. Oncol.* **26,** 545–549.

Katoh, Y., and Katoh, M. (2005b). Comparative genomics on Sonic hedgehog orthologs. *Oncol. Rep.* **14,** 1087–1090.

Katoh, Y., and Katoh, M. (2005c). Hedgehog signaling in gastric cancer. *Cancer Biol. Ther.* **4,** 1050–1054.

Kelley, R. I., and Hennekam, R. C. (2000). The Smith-Lemli-Opitz syndrome. *J. Med. Genet.* **37,** 321–335.

Kim, H. J., and Kim, T. Y. (2005). Regulation of vascular endothelial growth factor expression by insulin-like growth factor-II in human keratinocytes, differential involvement of mitogen-activated protein kinases and feedback inhibition of protein kinase C. *Br. J. Dermatol.* **152,** 418–425.

Kim, K. W., Bae, S. K., Lee, O. H., Bae, M. H., Lee, M. J., and Park, B. C. (1998). Insulin-like growth factor II induced by hypoxia may contribute to angiogenesis of human hepatocellular carcinoma. *Cancer Res.* **58,** 348–351.

Kim, I., Kim, H. G., Moon, S. O., Chae, S. W., So, J. N., Koh, K. N., Ahn, B. C., and Koh, G. Y. (2000). Angiopoietin-1 induces endothelial cell sprouting through the activation of focal adhesion kinase and plasmin secretion. *Circ. Res.* **86,** 952–959.

Kim, P. C., Mo, R., and Hui, Cc.C. (2001). Murine models of VACTERL syndrome: Role of sonic hedgehog signaling pathway. *J. Pediatr. Surg.* **36,** 381–384.

Kim, C. W., Lee, H. M., Lee, T. H., Kang, C., Kleinman, H. K., and Gho, Y. S. (2002). Extracellular membrane vesicles from tumor cells promote angiogenesis via sphingomyelin. *Cancer Res.* **62,** 6312–6317.

Kim, H. K., Song, K. S., Chung, J. H., Lee, K. R., and Lee, S. N. (2004). Platelet microparticles induce angiogenesis *in vitro*. *Br. J. Haematol.* **124,** 376–384.

Klein, S., Bikfalvi, A., Birkenmeier, T. M., Giancotti, F. G., and Rifkin, D. B. (1996). Integrin regulation by endogenous expression of 18-kDa fibroblast growth factor-2. *J. Biol. Chem.* **271,** 22583–22590.

Knudsen, K. A., Frankowski, C., Johnson, K. R., and Wheelock, M. J. (1998). A role for cadherins in cellular signaling and differentiation. *J. Cell. Biochem. Suppl.* **30–31,** 168–176.

Kusano, K. F., Pola, R., Murayama, T., Curry, C., Kawamoto, A., Iwakura, A., Shintani, S., Ii, M., Asai, J., Tkebuchava, T., Thorne, T., Takenaka, H., *et al.* (2005). Sonic hedgehog myocardial gene therapy: Tissue repair through transient reconstitution of embryonic signaling. *Nat. Med.* **11,** 1197–1204.

Kwon, Y. W., Kwon, K. S., Moon, H. E., Park, J. A., Choi, K. S., Kim, Y. S., Jang, H. S., Oh, C. K., Lee, Y. M., Kwon, Y. G., Lee, Y. S., and Kim, K. W. (2004). Insulin-like growth factor-II regulates the expression of vascular endothelial growth factor by the human keratinocyte cell line HaCaT. *J. Invest. Dermatol.* **123,** 152–158.

Lacroix, R., Sabatier, F., Mialhe, A., Basire, A., Pannell, R., Borghi, H., Robert, S., Lamy, E., Plawinski, L., Camoin-Jau, L., Gurewich, V., Angles-Cano, E., *et al.* (2007). Activation of plasminogen into plasmin at the surface of endothelial microparticles: A mechanism that modulates angiogenic properties of endothelial progenitor cells *in vitro*. *Blood* **110,** 2432–2439.

Lashkari, K., Rahimi, N., Hirose, T., McMeel, J. W., and Kazlauskas, A. (1999). Hepatocyte growth factor & vascular endothelial growth factor are elevated in subretinal fluid from advanced retinopathy of prematurity. *Invest. Ophthalmol. Vis. Sci.* **40**(S9–9).

Lashkari, K., Hirose, T., Yazdany, J., McMeel, J. W., Kazlauskas, A., and Rahimi, N. (2000). Vascular endothelial growth factor and hepatocyte growth factor levels are differentially elevated in patients with advanced retinopathy of prematurity. *Am. J. Pathol.* **156,** 1337–1344.

Lau, Y. T., and Ma, W. C. (1996). Nitric oxide inhibits migration of cultured endothelial cells. *Biochem. Biophys. Res. Commun.* **221,** 670–674.

Lavine, K. J., White, A. C., Park, C., Smith, C. S., Choi, K., Long, F., Hui, C. C., and Ornitz, D. M. (2006). Fibroblast growth factor signals regulate a wave of Hedgehog activation that is essential for coronary vascular development. *Genes Dev.* **20,** 1651–1666.

Leroyer, A. S., Rautou, P. E., Silvestre, J. S., Castier, Y., Lesèche, G., Devue, C., Duriez, M., Brandes, R. P., Lutgens, E., Tedgui, A., and Boulanger, C. M. (2008). CD40 ligand+ microparticles from human atherosclerotic plaques stimulate endothelial proliferation and angiogenesis a potential mechanism for intraplaque neovascularization. *J. Am. Coll. Cardiol.* **52,** 1302–1311.

Leung, D. W., Cachianes, G., Kuang, W. J., Goeddel, D. V., and Ferrara, N. (1989). Vascular endothelial growth factor is a secreted angiogenic mitogen. *Science* **246,** 1306–1309.

Lewis, P. M., Dunn, M. P., McMahon, J. A., Logan, M., Martin, J. F., St-Jacques, B., and McMahon, A. P. (2001). Cholesterol modification of sonic hedgehog is required for long-range signaling activity and effective modulation of signaling by Ptc1. *Cell* **105,** 599–612.

Li, J., Hampton, T., Morgan, J. P., and Simons, M. (1997). Stretch-induced VEGF expression in the heart. *J. Clin. Invest.* **100,** 18–24.

Lindahl, P., Johansson, B. R., Levéen, P., and Betsholtz, C. (1997). Pericyte loss and microaneurysm formation in PDGF-B-deficient mice. *Science* **277,** 24224–24225.

Litingtung, Y., Dahn, R. D., Li, Y., Fallon, J. F., and Chiang, C. (2002). Shh and Gli3 are dispensable for limb skeleton formation but regulate digit number and identity. *Nature* **418,** 979–983.

Lubin, B., Chiu, D., Bastacky, J., Roelofsen, B., and Van Deenen, L. L. (1981). Abnormalities in membrane phospholipid organization in sickled erythrocytes. *J. Clin. Invest.* **67,** 1643–1649.

Lum, L., and Beachy, P. A. (2004). The Hedgehog response network: Sensors, switches, and routers. *Science* **304,** 1755–1759.

Ma, G., Xiao, Y., and He, L. (2008). Recent progress in the study of Hedgehog signaling. *J. Genet. Genomics* **35,** 129–137.

Machold, R., Hayashi, S., Rutlin, M., Muzumdar, M. D., Nery, S., Corbin, J. G., Gritli-Linde, A., Dellovade, T., Porter, J. A., and Rubin, L. L. (2003). Sonic hedgehog is required for progenitor cell maintenance in telencephalic stem cell niches. *Neuron* **39,** 937–950.

Maisonpierre, P. C., Suri, C., Jones, P. F., Bartunkova, S., Wiegand, S. J., Radziejewski, C., Compton, D., McClain, J., Aldrich, T. H., Papadopoulos, N., Daly, T. J., Davis, S., *et al.* (1997). Angiopoietin-2, a natural antagonist for Tie2 that disrupts *in vivo* angiogenesis. *Science* **277,** 55–60.

Mandriota, S. J., and Pepper, M. S. (1997). Vascular endothelial growth factor-induced *in vitro* angiogenesis and plasminogen activator expression are dependent on endogenous basic fibroblast growth factor. *J. Cell Sci.* **110,** 2293–2302.

Mann, R. K., and Beachy, P. A. (2000). Cholesterol modification of proteins. *Biochim. Biophys. Acta* **1529,** 188–202.

Marigo, V., Roberts, D. J., Lee, S. M., Tsukurov, O., Levi, T., Gastier, J. M., Epstein, D. J., Gilbert, D. J., Copeland, N. G., Seidman, C. E., Jenkins, N. A., Seidman, J. G., *et al.* (1995). Cloning, expression, and chromosomal location of SHH and IHH: Two human homologues of the Drosophila segment polarity gene hedgehog. *Genomics* **28,** 44–51.

Martinelli, D. C., and Fan, C. M. (2009). A sonic hedgehog missense mutation associated with holoprosencephaly causes defective binding to GAS1. *J. Biol. Chem.* **284,** 19169–19172.

Martínez, M. C., Larbret, F., Zobairi, F., Coulombe, J., Debili, N., Vainchenker, W., Ruat, M., and Freyssinet, J. M. (2006). Transfer of differentiation signal by membrane microvesicles harboring hedgehog morphogens. *Blood* **108,** 3012–3020.

McCarthy, R. A., Barth, J. L., Chintalapudi, M. R., Knaak, C., and Argraves, W. S. (2002). Megalin functions as an endocytic sonic hedgehog receptor. *J. Biol. Chem.* **277,** 25660–25667.

McColl, B. K., Stacker, S. A., and Achen, M. G. (2004). Molecular regulation of the VEGF family—Inducers of angiogenesis and lymphangiogenesis. *APMIS* **112,** 463–480.

McKeehan, W. L., Wang, F., and Kan, M. (1998). The heparan sulfate-fibroblast growth factor family: Diversity of structure and function. *Prog. Nucleic Acid Res. Mol. Biol.* **59,** 135–176.

McMahon, A. P., Ingham, P. W., and Tabin, C. J. (2003). Developmental roles and clinical significance of hedgehog signaling. *Curr. Top. Dev. Biol.* **53,** 1–114.

Mezentsev, A., Merks, R. M., O'Riordan, E., Chen, J., Mendelev, N., Goligorsky, M. S., and Brodsky, S. V. (2005). Endothelial microparticles affect angiogenesis *in vitro*: Role of oxidative stress. *Am. J. Physiol. Heart Circ. Physiol.* **289,** 1106–1114.

Miguet, L., Pacaud, K., Felden, C., Hugel, B., Martinez, M. C., Freyssinet, J. M., Herbrecht, R., Potier, N., van Dorsselaer, A., and Mauvieux, L. (2006). Proteomic analysis of malignant lymphocyte membrane microparticles using double ionization coverage optimization. *Proteomics* **6,** 153–171.

Milkiewicz, M., Brown, M. D., Egginton, S., and Hudlicka, O. (2001). Association between shear stress, angiogenesis, and VEGF in skeletal muscles *in vivo*. *Microcirculation* **8,** 229–241.

Milkiewicz, M., Ispanovic, E., Doyle, J. L., and Haas, T. L. (2006). Regulators of angiogenesis and strategies for their therapeutic manipulation. *Int. J. Biochem. Cell Biol.* **38,** 333–357.

Mostefai, H. A., Andriantsitohaina, R., and Martínez, M. C. (2008a). Plasma membrane microparticles in angiogenesis: Role in ischemic diseases and in cancer. *Physiol. Res.* **57,** 311–320.

Mostefai, H. A., Agouni, A., Carusio, N., Mastronardi, M. L., Heymes, C., Henrion, D., Andriantsitohaina, R., and Martinez, M. C. (2008b). Phosphatidylinositol 3-kinase and xanthine oxidase regulate nitric oxide and reactive oxygen species productions by apoptotic lymphocyte microparticles in endothelial cells. *J. Immunol.* **180,** 5028–5035.

Mullor, J. L., Sánchez, P., and Ruiz i Altaba, A. (2002). Pathways and consequences: Hedgehog signaling in human disease. *Trends Cell Biol.* **12,** 562–569.

Murohara, T. (2001). Therapeutic vasculogenesis using human cord blood-derived endothelial progenitors. *Trends Cardiovasc. Med.* **11,** 303–307.

Murohara, T., Asahara, T., Silver, M., Bauters, C., Masuda, H., Kalka, C., Kearney, M., Chen, D., Symes, J. F., Fishman, M. C., Huang, P. L., and Isner, J. M. (1998). Nitric oxide synthase modulates angiogenesis in response to tissue ischemia. *J. Clin. Invest.* **101,** 2567–2578.

Murone, M., Rosenthal, A., and de Sauvage, F. J. (1999). Sonic hedgehog signaling by the Patched-Smoothened receptor complex. *Curr. Biol.* **12,** 562–569.

Nagase, T., Nagase, M., Yoshimura, K., Fujita, T., and Koshima, I. (2005). Angiogenesis within the developing mouse neural tube is dependent on sonic hedgehog signaling: Possible roles of motor neurons. *Genes Cells* **10,** 595–604.

Nelson, A. R., Fingleton, B., Rothenberg, M. L., and Matrisian, L. M. (2000). Matrix metalloproteinases: Biologic activity and clinical implications. *J. Clin. Oncol.* **18,** 1135–1149.

Nguyen, L. L., and D'Amore, P. A. (2001). Cellular interactions in vascular growth and differentiation. *Int. Rev. Cytol.* **204,** 1–48.

Noden, D. M. (1989). Embryonic origins and assembly of blood vessels. *Am. Rev. Respir. Dis.* **140,** 1097–1103.

Nüsslein-Volhard, C., and Wieschaus, E. (1980). Mutations affecting segment number and polarity in Drosophila. *Nature* **287,** 795–801.

Palladino, M., Gatto, I., Neri, V., Straino, S., Silver, M., Tritarelli, A., Piccioni, A., Smith, R. C., Gaetani, E., Losordo, D. W., Crea, F., Capogrossi, M., *et al.* (2011). Pleiotropic beneficial effects of sonic hedgehog gene therapy in an experimental model of peripheral limb ischemia. *Mol. Ther.* **19,** 658–666.

Pan, J., Wang, Q., and Snell, W. J. (2005). Cilium-generated signaling and cilia-related disorders. *Lab. Invest.* **85,** 452–463.

Pan, Y., Bai, C. B., Joyner, A. L., and Wang, B. (2006). Sonic hedgehog signaling regulates Gli2 transcriptional activity by suppressing its processing and degradation. *Mol. Cell. Biol.* **26,** 3365–3377.

Pardanaud, L., and Dieterlen-Lièvre, F. (1999). Manipulation of the angiopoietic/hemangiopoietic commitment in the avian embryo. *Development* **126,** 617–627.

Pasca di Magliano, M., and Hebrok, M. (2003). Hedgehog signalling in cancer formation and maintenance. *Nat. Rev. Cancer* **3,** 903–911.

Pepinsky, R. B., Zeng, C., Wen, D., Rayhorn, P., Baker, D. P., Williams, K. P., Bixler, S. A., Ambrose, C. M., Garber, E. A., Miatkowski, K., Taylor, F. R., Wang, E. A., *et al.* (1998). Identification of a palmitic acid-modified form of human Sonic hedgehog. *J. Biol. Chem.* **273,** 14037–14045.

Pepper, M. S. (2001). Role of the matrix metalloproteinase and plasminogen activator-plasmin systems in angiogenesis. *Arterioscler. Thromb. Vasc. Biol.* **21,** 1104–1117.

Pepper, M. S., Mandriota, S. J., Jeltsch, M., Kumar, V., and Alitalo, K. (1998). Vascular endothelial growth factor (VEGF)-C synergizes with basic fibroblast growth factor and VEGF in the induction of angiogenesis *in vitro* and alters endothelial cell extracellular proteolytic activity. *J. Cell. Physiol.* **177,** 439–452.

Plate, K. H., Breiser, G., Weich, H. A., and Risau, W. (1992). Vascular endothelial growth factor is a potential tumor angiogenesis factor *in vivo*. *Nature* **359,** 845–848.

Podlasek, C. A., Meroz, C. L., Korolis, H., Tang, Y., McKenna, K. E., and McVary, K. T. (2005). Sonic hedgehog, the penis and erectile dysfunction: A review of sonic hedgehog signaling in the penis. *Curr. Pharm. Des.* **11,** 4011–4027.

Pola, R., Ling, L. E., Silver, M., Corbley, M. J., Kearney, M., Blake Pepinsky, R., Shapiro, R., Taylor, F. R., Baker, D. P., Asahara, T., and Isner, J. M. (2001). The morphogen Sonic hedgehog is an indirect angiogenic agent upregulating two families of angiogenic growth factors. *Nat. Med.* **7,** 706–711.

Pola, R., Ling, L. E., Aprahamian, T. R., Barban, E., Bosch-Marce, M., Curry, C., Corbley, M., Kearney, M., Isner, J. M., and Losordo, D. W. (2003). Postnatal recapitulation of embryonic hedgehog pathway in response to skeletal muscle ischemia. *Circulation* **108,** 479–485.

Porter, J., Young, K., and Beachy, P. (1996a). Cholesterol modification of hedgehog signaling proteins in animal development. *Science* **274,** 255–259.

Porter, J. A., Ekker, S. C., Park, W. J., von Kessler, D. P., Young, K. E., Chen, C. H., Ma, Y., Woods, A. S., Cotter, R. J., Koonin, E. V., and Beachy, P. A. (1996b). Hedgehog patterning activity-role of a lipophilic modification mediated by the carboxy-terminal autoprocessing domain. *Cell* **86,** 21–34.

Ramos-Lopez, M., Hirose, T., and Lashkari, K. (2003). Hepatocyte growth factor receptor is selectively activated in the vascular compartment of stage 5 retinopathy of prematurity. *Invest. Ophthalmol. Vis. Sci.* **44,** U112–U.

Rao, G., Pedone, C. A., Valle, L. D., Reiss, K., Holland, E. C., and Fults, D. W. (2004). Sonic hedgehog and insulin-like growth factor signaling synergize to induce medulloblastoma formation from nestin-expressing neural progenitors in mice. *Oncogene* **23,** 6156–6162.

Ratajczak, M. Z. (2011). The emerging role of microvesicles in cellular therapies for organ/tissue regeneration. *Nephrol. Dial. Transplant.* **26,** 1453–1456.

RayChaudhury, A., Frischer, H., and Malik, A. B. (1996). Inhibition of endothelial cell proliferation and bFGF-induced phenotypic modulation by nitric oxide. *J. Cell. Biochem.* **63,** 125–134.

Reisinger, K., Kaufmann, R., and Gille, J. (2003). Increased Sp1 phosphorylation as a mechanism of hepatocyte growth factor (HGF/SF)-induced vascular endothelial growth factor (VEGF/VPF) transcription. *J. Cell Sci.* **116,** 225–238.

Renault, M. A., Roncalli, J., Tongers, J., Thorne, T., Klyachko, E., Misener, S., Volpert, O. V., Mehta, S., Burg, A., Luedemann, C., Qin, G., Kishore, R., *et al.* (2010). Sonic hedgehog induces angiogenesis via Rho kinase-dependent signaling in endothelial cells. *J. Mol. Cell. Cardiol.* **49,** 490–498.

Rey, S., and Semenza, G. L. (2010). Hypoxia-inducible factor-1-dependent mechanisms of vascularization and vascular remodelling. *Cardiovasc. Res.* **86,** 236–242.

Ribatti, D., Conconi, M. T., and Nussdorfer, G. G. (2007). Nonclassic endogenous novel regulators of angiogenesis. *Pharmacol. Rev.* **59,** 185–205.

Riddle, R. D., Johnson, R. L., Laufer, E., and Tabin, C. (1993). Sonic hedgehog mediates the polarizing activity of the ZPA. *Cell* **75,** 1401–1416.

Risau, W. (1997). Mechanisms of angiogenesis. *Nature* **386,** 671–674.

Roelink, H., Porter, J. A., Chiang, C., Tanabe, Y., Chang, D. T., Beachy, P. A., and Tessell, T. M. (1995). Floor plate and motor neuron induction by different concentrations of the amino terminal cleavage product of Sonic hedgehog autoproteolysis. *Cell* **81,** 445–455.

Ruiz i Altaba, A. (1999). Gli proteins and Hedgehog signaling: Development and cancer. *Trends Genet.* **15,** 418–425.

Ruiz i Altaba, A., Mas, C., and Stecca, B. (2007). The Gli code: An information nexus regulating cell fate, stemness and cancer. *Trends Cell Biol.* **17,** 438–447.

Saharinen, P., Bry, M., and Alitalo, K. (2010). How do angiopoietins Tie in with vascular endothelial growth factors? *Curr. Opin. Hematol.* **17,** 198–205.

Sapet, C., Simoncini, S., Loriod, B., Puthier, D., Sampol, J., Nguyen, C., Dignat-George, F., and Anfosso, F. (2006). Thrombin-induced endothelial microparticle generation: Identification of a novel pathway involving ROCK-II activation by caspase-2. *Blood* **108,** 1868–1876.

Schiffenbauer, Y. S., Abramovitch, R., Meir, G., Nevo, N., Holzinger, M., Itin, A., Keshet, E., and Neeman, M. (1997). Loss of ovarian function promotes angiogenesis in human ovarian carcinoma. *Proc. Natl. Acad. Sci. USA* **94,** 13203–13208.

Schlegel, R. A., McEvoy, L., and Williamson, P. (1985). Membrane phospholipid asymmetry and the adherence of loaded red blood cells. *Bibl. Haematol.* **51,** 150–156.

Seghezzi, G., Patel, S., Ren, C. J., Gualandris, A., Pintucci, G., Robbins, E. S., Shapiro, R. L., Galloway, A. C., Rifkin, D. B., and Mignatti, P. (1998). Fibroblast growth factor-2 (FGF-2) induces vascular endothelial growth factor (VEGF) expression in the endothelial cells of forming capillaries: An autocrine mechanism contributing to angiogenesis. *J. Cell Biol.* **141,** 1659–1673.

Seiwert, T. Y., Jagadeeswaran, R., Faoro, L., Janamanchi, V., Nallasura, V., El Dinali, M., Yala, S., Kanteti, R., Cohen, E. E., Lingen, M. W., Martin, L., Krishnaswamy, S., *et al.* (2009). The MET receptor tyrosine kinase is a potential novel therapeutic target for head and neck squamous cell carcinoma. *Cancer Res.* **69,** 3021–3031.

Semenza, G. L. (2001). Regulation of hypoxia-induced angiogenesis: A chaperone escorts VEGF to the dance. *J. Clin. Invest.* **108,** 39–40.

Sengupta, S., Gherardi, E., Sellers, L. A., Wood, J. M., Sasisekharan, R., and Fan, T. P. (2003). Hepatocyte growth factor/scatter factor can induce angiogenesis independently of vascular endothelial growth factor. *Arterioscler. Thromb. Vasc. Biol.* **23,** 69–75.

Shah, N. M., Groves, A. K., and Anderson, D. J. (1996). Alternative neural crest cell fates are instructively promoted by TGFbeta superfamily members. *Cell* **85,** 331–343.

Shima, D. T., and Mailhos, C. (2000). Vascular developmental biology: Getting nervous. *Curr. Opin. Genet. Dev.* **10,** 536–542.

Soleti, R., Benameur, T., Porro, C., Panaro, M. A., Andriantsitohaina, R., and Martínez, M. C. (2009). Microparticles harboring Sonic Hedgehog promote angiogenesis through the upregulation of adhesion proteins and proangiogenic factors. *Carcinogenesis* **30,** 580–588.

Stone, D. M., Hynes, M., Armanini, M., Swanson, T. A., Gu, Q., Johnson, R. L., Scott, M. P., Pennica, D., Goddard, A., Phillips, H., Noll, M., Hooper, J. E., *et al.* (1996). The tumour-suppressor gene patched encodes a candidate receptor for Sonic hedgehog. *Nature* **384,** 129–134.

Stratmann, A., Risau, W., and Plate, K. H. (1998). Cell type-specific expression of angiopoietin-1 and angiopoietin-2 suggests a role in glioblastoma angiogenesis. *Am. J. Pathol.* **153,** 1459–1466.

Suárez, Y., and Sessa, W. C. (2009). MicroRNAs as novel regulators of angiogenesis. *Circ. Res.* **104,** 442–454.

Sunderkötter, C., Steinbrink, K., Henseleit, U., Bosse, R., Schwarz, A., Vestweber, D., and Sorg, C. (1996). Activated T cells induce expression of E-selectin *in vitro* and in an antigen-dependent manner *in vivo*. *Eur. J. Immunol.* **26,** 1571–1579.

Suri, C., Jones, P. F., Patan, S., Bartunkova, S., Maisonpierre, P. C., Davis, S., Sato, T. N., and Yancopoulos, G. D. (1996). Requisite role of angiopoietin-1, a ligand for the TIE2 receptor, during embryonic angiogenesis. *Cell* **87,** 1171–1180.

Suri, C., McClain, J., Thurston, G., McDonald, D. M., Zhou, H., Oldmixon, E. H., Sato, T. N., and Yancopoulos, G. D. (1998). Increased vascularization in mice overexpressing angiopoietin-1. *Science* **282,** 468–471.

Taipale, J., and Beachy, P. (2001). The Hegehog and Wnt signalling pathways in cancer. *Nature* **411,** 349–354.

Taraboletti, G., D'Ascenzo, S., Borsotti, P., Giavazzi, R., Pavan, A., and Dolo, V. (2002). Shedding of the matrix metalloproteinases MMP-2, MMP-9, and MT1-MMP as membrane vesicle-associated components by endothelial cells. *Am. J. Pathol.* **160,** 673–680.

Tenzen, T., Allen, B. L., Cole, F., Kang, J. S., Krauss, R. S., and McMahon, A. P. (2006). The cell surface membrane proteins Cdo and Boc are components and targets of the Hedgehog signaling pathway and feedback network in mice. *Dev. Cell* **10,** 647–656.

Thurston, G. (2003). Role of Angiopoietins and Tie receptor tyrosine kinases in angiogenesis and lymphangiogenesis. *Cell Tissue Res.* **314,** 61–68.

Tong, Q., Zheng, L., Li, B., Wang, D., Huang, C., Matuschak, G. M., and Li, D. (2006). Hypoxia-induced mitogenic factor enhances angiogenesis by promoting proliferation and migration of endothelial cells. *Exp. Cell Res.* **312,** 3559–3569.

Trusolino, L., and Comoglio, P. M. (2002). Scatter-factor and semaphorin receptors: Cell signalling for invasive growth. *Nat. Rev. Cancer* **2,** 289–300.

Tual-Chalot, S., Leonetti, D., Andriantsitohaina, R., and Martínez, M. C. (2011). Microvesicles: Intercellular vectors of biological messages. *Mol. Interv.* **11,** 88–94.

Tunyogi-Csapo, M., Koreny, T., Vermes, C., Galante, J. O., Jacobs, J. J., and Glant, T. T. (2007). Role of fibroblasts and fibroblast-derived growth factors in periprosthetic angiogenesis. *J. Orthop. Res.* **25,** 1378–1388.

van den Brink, G. R. (2007). Hedgehog signaling in development and homeostasis of the gastrointestinal tract. *Physiol. Rev.* **87,** 1343–1375.

Varjosalo, M., and Taipale, J. (2008). Hedgehog: functions and mechanisms. *Genes Dev.* **22,** 2454–2472.

Vokes, S. A., Yatskievych, T. A., Heimark, R. L., McMahon, J., McMahon, A. P., Antin, P. B., and Krieg, P. A. (2004). Hedgehog signaling is essential for endothelial tube formation during vasculogenesis. *Development* **131,** 4371–4380.

Wali, R. K., Jaffe, S., Kumar, D., Sorgente, N., and Kalra, V. K. (1987). Increased adherence of oxidant treated human and bovine erythrocytes to cultured endothelial cells. *J. Cell. Physiol.* **133,** 25–36.

Wali, R. K., Jaffe, S., Kumar, D., and Kalra, V. K. (1988). Alterations in organization of phospholipids in erythrocytes as a factor in adherence to endothelial cells in diabetes mellitus. *Diabetes* **37,** 104–111.
Wang, B., and Li, Y. (2006). Evidence for the direct involvement of {β}TrCP in Gli3 protein processing. *Proc. Natl. Acad. Sci.* **103,** 33–38.
Wang, B., Fallon, J. F., and Beachy, P. A. (2000). Hedgehog-regulated processing of Gli3 produces an anterior/posterior repressor gradient in the developing vertebrate limb. *Cell* **100,** 423–434.
Wang, Y., McMahon, A. P., and Allen, B. L. (2007). Shifting paradigms in Hedgehog signaling. *Curr. Opin. Cell Biol.* **19,** 159–165.
Watkins, D. N., Berman, D. M., Burkholder, S. G., Wang, B., Beachy, P. A., and Baylin, S. B. (2003). Hedgehog signalling within airway epithelial progenitors and in small-cell lung cancer. *Nature* **422,** 313–317.
Weerheim, A. M., Kolb, A. M., Sturk, A., and Nieuwland, R. (2002). Phospholipid composition of cell-derived microparticles determined by one-dimensional high-performance thin-layer chromatography. *Anal. Biochem.* **302,** 191–198.
Wilkinson, D. G. (2000). Eph receptors and ephrins: Regulators of guidance and assembly. *Int. Rev. Cytol.* **196,** 177–244.
Wilson, M. J., Richter-Lowney, K., and Daleke, D. L. (1993). Hyperglycemia induces a loss of phospholipid asymmetry in human erythrocytes. *Biochemistry* **32,** 11302–11310.
Witek, R. P., Yang, L., Liu, R., Jung, Y., Omenetti, A., Syn, W. K., Choi, S. S., Cheong, Y., Fearing, C. M., Agboola, K. M., Chen, W., and Diehl, A. M. (2009). Liver cell-derived microparticles activate hedgehog signaling and alter gene expression in hepatic endothelial cells. *Gastroenterology* **136,** 320–330.
Yang, C., Mwaikambo, B. R., Zhu, T., Gagnon, C., Lafleur, J., Seshadri, S., Lachapelle, P., Lavoie, J. C., Chemtob, S., and Hardy, P. (2008). Lymphocytic microparticles inhibit angiogenesis by stimulating oxidative stress and negatively regulating VEGF-induced pathways. *Am. J. Physiol. Regul. Integr. Comp. Physiol.* **294,** 467–476.
Yang, C., Gagnon, C., Hou, X., and Hardy, P. (2010). Low density lipoprotein receptor mediates anti-VEGF effect of lymphocyte T-derived microparticles in Lewis lung carcinoma cells. *Cancer Biol. Ther.* **10,** 448–456.
Yao, S., Lum, L., and Beachy, P. (2006). The ihog cell-surface proteins bind Hedgehog and mediate pathway activation. *Cell* **125,** 343–357.
Yoshida, S., Yamaguchi, Y., Itami, S., Yoshikawa, K., Tabata, Y., Matsumoto, K., and Nakamura, T. (2003). Neutralization of hepatocyte growth factor leads to retarded cutaneous wound healing associated with decreased neovascularization and granulation tissue formation. *J. Invest. Dermatol.* **120,** 335–343.
Yu, Y., and Sato, J. D. (1999). MAP kinases, phosphatidylinositol 3-kinase, and p70 S6 kinase mediate the mitogenic response of human endothelial cells to vascular endothelial growth factor. *J. Cell. Physiol.* **178,** 235–246.
Yue, P. Y. K., Mak, N. K., Cheng, Y. K., Leung, K. W., Ng, T. B., Fan, D. T. B., Yeung, H. W., and Wong, R. N. S. (2007). Pharmacogenomics and the Yin/Yang actions of ginseng: Anti-tumor, angiomodulating and steroid-like activities of ginsenosides. *Chin. Med.* **2,** 6.
Zachary, I. (2003). VEGF signalling: Integration and multi-tasking in endothelial cell biology. *Biochem. Soc. Trans.* **31,** 1171–1177.
Zhang, Y. W., and Vande Woude, G. F. (2003). HGF/SF-met signaling in the control of branching morphogenesis and invasion. *J. Cell. Biochem.* **88,** 408–417.
Zhang, Y. W., Su, Y., Volpert, O. V., and Vande Woude, G. F. (2003). Hepatocyte growth factor/scatter factor mediates angiogenesis through positive VEGF and negative thrombospondin 1 regulation. *Proc. Natl. Acad. Sci. USA* **100,** 12718–12723.

Zhang, W., Kang, J. S., Cole, F., Yi, M. J., and Krauss, R. S. (2006). Cdo functions at multiple points in the Sonic Hedgehog pathway, and Cdo-deficient mice accurately model human holoprosencephaly. *Dev. Cell* **10,** 657–665.

Ziche, M., and Morbidelli, L. (2000). Nitric oxide and angiogenesis. *J. Neurooncol.* **50,** 139–148.

Ziche, M., Morbidelli, L., Masini, E., Amerini, S., Granger, H. J., Maggi, C. A., Geppetti, P., and Ledda, F. (1994). Nitric oxide mediates angiogenesis *in vivo* and endothelial cell growth and migration *in vitro* promoted by substance P. *J. Clin. Invest.* **94,** 2036–2044.

Ziche, M., Morbidelli, L., Choudhuri, R., Zhang, H. T., Donnini, S., Granger, H. J., and Bicknell, R. (1997). Nitric oxide synthase lies downstream from vascular endothelial growth factor-induced but not basic fibroblast growth factor-induced angiogenesis. *J. Clin. Invest.* **99,** 2625–2634.

CHAPTER NINETEEN

"PATCH"-ING UP THE NEURONS: REVIVAL OR ENERVATION?

Sayantani Ghosh, Arunabha Chakrabarti, *and* Debashis Mukhopadhyay

Contents

Abstract

Since its identification Patched1 (Ptch1) has gained importance for playing a cardinal role in developmental patterning through Hedgehog (Hh) pathway, acting as a transmembrane receptor. Involvement of this protein in diverse aspects of the neuronal system, from development to regeneration and protection, including uncontrolled proliferation in oncogenic perspectives, makes it an intriguing candidate for investigation in neurobiology. Stem cell population of adult nervous system is also found to be regulated by Ptch1. Though not elaborately studied, research in this field for the past one decade has suggested a new spectrum of Ptch1 function through an alternative route independent of Hh. In this chapter, the available knowledge about Ptch1 in neuronal system is critically reviewed and further functional insights about this protein are evaluated.

Structural Genomics Division, Saha Institute of Nuclear Physics, Kolkata, West Bengal, India

Vitamins and Hormones, Volume 88
ISSN 0083-6729, DOI: 10.1016/B978-0-12-394622-5.00019-5

I. Introduction

Basic understanding of the molecular mechanisms of cell fate determination, tissue organization, and maintenance of homeostasis requires the appreciation of the complex interplay of developmental pathways that exert combinatorial control over all these processes. This realization dramatically augmented the research interest in molecular studies of the signaling networks in developmental and regenerative processes.

Among the multitude of developmental signaling cascades, Hedgehog perhaps remains the most attractive one. Exhaustive research in this field has paved a way toward the discovery of exciting molecular players of this pathway that interacts in a coordinated fashion with a host of other molecules and controls a series of biochemical processes pivotal to regulation of growth, patterning, and morphogenesis of multicellular organisms. In both flies and mammals, Hedgehog (Hh) cascade is initiated through the binding of the protein Hh to its receptor Patched (Ptch), thereby activating succession of signaling events which ultimately culminates in the induction of zinc finger-domain-bearing protein Ci (cubitus interruptus in drosophila) or Gli (in mammals) resulting in the expression of an array of different target genes (Ingham and McMahon, 2001).

The downstream activation of gene expression requires the binding of the secreted Hh complex to a trimeric, 12 pass transmembrane receptor protein Ptch on the responding cell. Internalization occurs by a dynamine-mediated endocytosis procedure, assisted by a few other members like BOC, CDO, and glycophosphatidylinositol (GPI)-anchored Gas1 (Tenzen *et al.*, 2006; Yavari *et al.*, 2010). In mammals, the ligand binding to Ptch1 faces competition from another transmembrane receptor GPC3, a GPI-linked glypican subfamily of HSPGs (heparan sulfate proteoglycans), and HHiP (Hedgehog interacting proteins), both of which act as antagonistic modulators of Hh pathway (The *et al.*, 1999). In Hh nonresponsive cells, Ptch1 represses Smo (Smoothened) accumulation from intracellular endosomes to cell surface (in flies) or primary cilium (in mammals). Association of Hh with Ptch1 relieves the repression on Smo by some nonphysical interaction possibly mediated through another protein phospholipid phosphatidyl-inositol-4-phosphate (PI4P) (currently demonstrated in drosophila) (Yavari *et al.*, 2010).

The signal transduction event carried out by Hh pathway is adapted principally by changing the modification pattern and nuclear transport of Ci in flies and its mammalian counterpart Gli protein, both of which have a bipartite role in the whole regulation. In absence of Hh, PKA, GSK3, and Caesin kinase1 (CK1) phosphorylate Ci/Gli protein at specific conserved sites targeting them for a proteasome-mediated degradation. Full length forms of Ci/Gli function as transcriptional activator, while truncated N-terminal forms repress transcription. Presence of Hh activates Smo, a canonical GPCR protein, which, through downregulation of cAMP, deactivates PKA, and repressor Gli/Ci cannot be formed. In vertebrates, Gli1 lacks the N-terminal region and functions only as an activator, while Gli2 and 3 both can have the dual role. Activated Smo carries out sequential phosphorylation events which disassemble the Hh signaling complex and allows full length Ci/Gli to translocate to nucleus subsequently to induce the target gene expression. In uninduced cells where Smo is inactive, the Hh signaling complex constituted by Kif7, SUFU, Cos2, PKA, CK1, and GSK3 among others harbors the Gli/Ci repressor. In the responding cells, the internalized Ptch1–Hh complex is degraded later (Fig. 19.1; reviewed by Milenkovic and Scott, 2010; Riobo and Manning, 2007; Varjosalo and

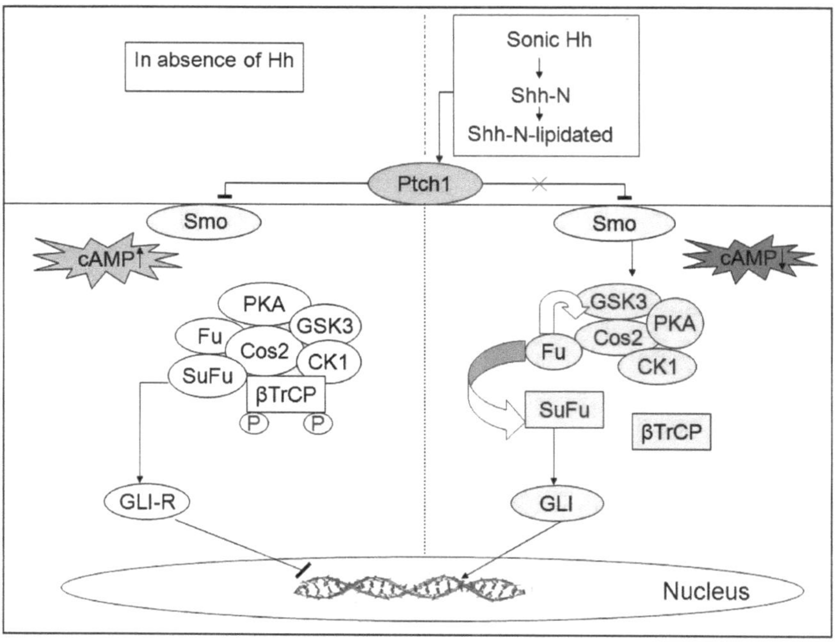

Figure 19.1 Regulatory role of Ptch1 in the Hh pathway. The left side of the panel shows how in absence of Hh, Ptch1 inhibits Smo, resulting is an upsurge of cAMP, which, in turn, helps PKA to gain activity, finally leading to the formation of Gli-repressor and subsequent repression of gene expression. On the right side, the contrasting event happens. In the presence of Hh, Ptch1 is deactivated and Smo activation occurs by phosphorylation by PKA, GSK3, and CK1. Subsequent phosphorylation of Cos2 and SUFU by Fused occurs, releasing Gli full length form that translocates to the nucleus and activates transcription of target genes. (For color version of this figure, the reader is referred to the Web version of this chapter.)

Taipale, 2008; Wilson and Chuang, 2010). At the interface of Hh-mediated signal transduction, therefore, Ptch1 acts as a "gate keeper," profoundly influencing the downstream fate. Several recent observations emphasize the emerging significance of the receptor in the whole process.

II. The "Patched" Transmission

Ptch, one of the least characterized proteins of Hh signaling network, acts as a repressor. Till date, two homologs of Ptch have been identified in vertebrates—Ptch1 and Ptch2. Among them, Ptch1 appears to be the major role player in embryogenesis and maintenance of homeostasis of central nervous system (CNS) in adults (Carpenter *et al.*, 1998; Goodrich *et al.*, 1996; Marigo and Tabin, 1996; Marigo *et al.*, 1996a,b; Motoyama *et al.*, 1998; Pearse *et al.*, 2001).

Ptch1 binds to its immediate upstream player Hh with a nanomolar affinity (Ho and Scott, 2002). One important strategy in proper transduction of Hh signal is a negative feedback loop assumed by Ptch1 which attenuates the signal with subsequent building up of inhibitor molecules, in this case the receptor itself. A protein gradient, thus formed, is also essential for the appropriate outcome of the event. Presence of Hh abrogates Ptch1-mediated repression of Smo and activates transcription of different target genes. Ptch1 itself is one of the major targets of Hh signaling and is expressed in high abundance in Hh responsive cells. The auxiliary function of ligand binding to Ptch1 is the sequestration of Hh in its zone of secretion which otherwise would manifest a global paracrine activity in the system. Mutation of Ptch1 in drosophila cells allows Hh to work in a diffused manner resulting in an incongruous developmental outcome which proves the importance of Ptch1 in restricting Hh movement (Briscoe *et al.*, 2001). Transgenic expression study demonstrates that though a basal level of Ptch1 is capable of antagonizing the downstream activity of the pathway in absence of Hh, ligand-dependent feedback cycle is a prerequisite for redundant Ptch1 expression for controlling the range of ligand diffusion. Thus, Hh-induced Ptch1 expression poses a diffusion barrier and, in turn, sharpens the morphogen gradient in responding cellular environment (Chen and Struhl, 1996).

Though there exists relatively low homology in the sequence of Ptch1 in drosophila and mammals (only 36% similarity in the amino acid sequence), the activity of Ptch1 is fairly conserved. In fact, mutated Dptch1 (drosophila) can be functionally rescued by introduction of Hptc1 (human) into drosophila (Varjosalo *et al.*, 2006). However, these proteins share conserved sequence motifs in their extracellular sterol-binding domains. Some other conserved motifs are also present inside the transmembrane loops in a fashion similar to that of most of the ATP-binding transporter cassettes (Varjosalo *et al.*, 2006). Ptch1 also displays sequence similarity with a bacterial proton-driven hopanoid (sterol homolog in bacteria) transporter, resistance-nodulation-division family protein, indicating to its functional ancestry (Hausmann *et al.*, 2009; Tseng *et al.*, 1999). Mutation in these loops impairs downstream Smo repression keeping the Hh binding and dynamin-mediated endocytic internalization phenomenon unaltered, suggesting that the signaling and sequestration activities of Ptch1 are probably separable (Strutt *et al.*, 2001; Varjosalo *et al.*, 2006).

Ptch1 shares homology domains with NPC1 that has role in normal trafficking of organelles like late endosomes along with fatty acids. One possibility includes the regulation of vesicular trafficking of its ligand Hh by Ptch1 in a manner markedly similar with that of NPC1. Mutation in the second sterol-binding extracellular domain (SSD) causes ectopic accumulation of mutant protein in the late endosomes implying that SSD is essential for intracellular movements of Ptch1. Moreover, the observation that in cultured Cos1 cells Ptch1 and NPC1 colocalize also convincingly indicates to a functional convergence between the two intracellular trafficking pathways taken by both of these proteins. (Incardona *et al.*, 2000; Ko *et al.*, 2001; Martin *et al.*, 2001).

These evolutionarily conserved domains with broader functional significance point to novel roles of Ptch1 as a transmembrane receptor beyond that of Hh-mediated signaling.

III. Ptch1 Expression in Neural Tissues

The CNS is endowed with the responsibility of controlling the functional coordination and maintaining the behavioral repertoire of the whole body. In vertebrates, developmental program of CNS proceeds through a series of episodes, starting from the induction of neuroectoderm to the formation of neural tube followed by cephalic folding, proliferation, differentiation, migration, synaptogenesis, and myelination. All of these events acquire extremely high precision owing to the prodigious coordination among multiple signaling networks. Rigorous experimental efforts over the past few decades have unequivocally identified Hh as one of the major role players in processing high-fidelity neurogenesis (Marti and Bovolenta, 2002). Though in vertebrates three homologs of Hh viz. Sonic Hh (Shh), Indian Hh (Ihh), and Desert Hh (Dhh) are present, Shh is the one to be predominantly present in the course of neural embryogenesis and in adult CNS. Apart from specifying multiple neural cell fates from a multipotent population of progenitor cells in discrete zones of CNS and perpetuating neuronal equilibrium in matured organisms, Hh signaling also has been diagnosed to play a massive role in retrieving neuronal plasticity under several neurodegenerative conditions (Gulino *et al.*, 2007).

A. Expression in developing CNS

Comprehensive investigation about the distribution of Ptch1 during CNS development has unveiled the presence of this protein in a temporal manner at the two main signaling centers responsible for generating both the dorsoventral and rostrocaudal patterning of CNS (Marti *et al.*, 1995). As the immediate downstream effector of Hh protein, Ptch1 is responsible to initiate myriad of outcomes with respect to morphogenetic patterning during different developmental stages in a tissue-specific manner. Location of this protein at the advent of the whole cascade makes it probably the most crucial player in the regulation of the entire signaling system.

In case of vertebrate neural development, signaling stems out from the midline and guides the dorsoventral polarity of the cells in neural tube and somites. The signal is processed from ventral to dorsal sides (Brand-Saberi *et al.*, 1993; Johnson and Tabin, 1995; Pourquie *et al.*, 1993). During early neuroaxis determination, the main ventral cell types which are induced include floor plate cells, motor neurons, and interneuron. The latter two, being more dorsally located, are induced later. In response to a concentration gradient of Hh, the neuronal precursor cells are destined to specific fates. In a developing neural tube, Shh is initially secreted from the notochord to induce the formation of floor plate cells along the ventral midline, followed by its secretion from both the notochord and floor plate to initiate formation of motor neurons and interneurons (Marti *et al.*, 1995). A higher dose is responsible for the generation of floor plate cells along the ventral midline, whereas lower concentration selectively persuades dorsal cell types. Graded expression of purified Shh in the *in vitro* naive explants culture has reported to induce Ptch1 expression in neural tube and paraxial mesoderm (Patten and Placzek, 2000). Ptch1 constitutively represses the downstream transcriptional targets of the Hh pathway including its own transcription. High abundance of Ptch1 is observed in tissues actively responding to Shh (reviewed by Fuccillo *et al.*, 2006; Ho and Scott, 2002; Jessell, 2000; Marti and Bovolenta, 2002; Miao *et al.*, 1997; Patten and Placzek, 2000)

Cryosectioning of whole mount chicken embryo followed by *in situ* hybridization has elegantly demonstrated the spatiotemporal expression pattern of Ptch1. Developmentally older sections of rostral regions have revealed strong Ptch1 expression compared to those of caudal segments. Expression was detected strongly in a more anterior section of ventral neural tube, floor plate, and notochord, whereas only low level of Ptch1 was found in a posterior section of the same developmental stage. In the later stages, the expression becomes robust in notochord, ventral somites, and splanchnic mesoderm with the strongest being in the cells of neural tube adjacent to the floor plate. As an axial symmetry is established, the expression of Ptch1 subdues caudally. This is followed by a uniform appearance of ptch1 transcript along the entire anterior–posterior axis of neural tube, most strongly at the luminal surface (Marigo and Tabin, 1996).The spatiotemporal distribution of Ptch1 during development is shown in Table 19.1. Similar distribution is expected in case of higher eukaryotes as well.

B. Expression in adult CNS

Apart from its role in generating inductive signal for fate determination of spinal motor, basal forebrain cholinergic, midbrain dopaminergic, and also serotonergic neurons in developing embryos (Ericson *et al.*, 1995; Hynes *et al.*, 1995; McMahon *et al.*, 2003; Roelink *et al.*, 1994; Ye *et al.*, 1998), Shh participates in enhancing neuroprotective and neurodegenerative activity in matured organisms. Detailed mapping of Ptch1 transcript has pinpointed its discrete localization in adult CNS (Fig. 19.2) (Traiffort *et al.*, 1999). In contrast to the predominant presence of Ptch1 mRNA in thalamic and hypothalamic regions of adult rats, spinal cord contains almost inconceivable amount of transcript. Along with a few areas of basal forebrain, the coincidence of bountiful transcription of both Shh and Ptch1 in Purkinje cell layers and facial nucleus indicates their active functional association to maintain a local signaling circuitry in these regions of adult CNS, possibly instrumental in maintaining the neuronal homeostasis. In other brain areas like median eminence, granule cell layers, medial vestibular nucleus, and solitary tract nucleus which are devoid of Shh, unrivaled presence of Ptch1 transcript reflects continued mode of Smo repression in absence of ligand-based stimulation. Or it might also justify a role of Ptch1 in signal transduction in a yet unidentified Shh-independent manner either through an alternate pathway or through binding to a substrate similar to Shh (Traiffort *et al.*, 1998, 1999).

Overexpression of Ptch1 in developing mouse embryo has reported to inhibit induction of Shh target genes including Ptch1 itself and also caused abnormal dorsalization of developing neural tube. Hydrocephaly and spinal bifida are two major consequences of erroneous Ptch1 expression in mammals. Further analysis of neuronal phenotype in cells with aberrant Ptch1 expression has demonstrated that a legitimate balance between Shh and Ptch1 in establishment of accurate neuronal patterning is of utmost importance (Goodrich *et al.*, 1999).

IV. Ptch1 in Disease, Development, and Injury

Considering the significant role played by Ptch1 at different stages of development, it would be prudent to review its role in proliferative, regenerative, and degenerative events.

Table 19.1 Distribution of Ptch1 (protein and mRNA) in developing chicken CNS according to different embryonic stages (shown as levels)

Developmental stages							
Names of tissue	12	13	14	15	16	17	18
Notochord	Present/*present*			High/*low*			*Basal*
Floor plate	Present			High/*low*			*Basal*
Neural tube		Present		High/*low*			
Somites	Present	Present					
Splanchnic mesoderm				Present			
Stomodeum					Present		
Hind brain				Present		Present	
Posterior branchial arches					Present		
CNS					Present	Present	
Ventral neural tube	Feeble in caudal and robust in rostral section			Present			
Sclerotome							*Present*

Qualitative pattern says that from stage 12 level, the expression starts to rise and reaches a maximum at stage 15. Afterwards, it gradually decreases to achieve a basal level by stage 18. Italicized fonts signify mRNA. Blank spaces represent lack of information. The data are adopted from Marigo and Tabin (1996).

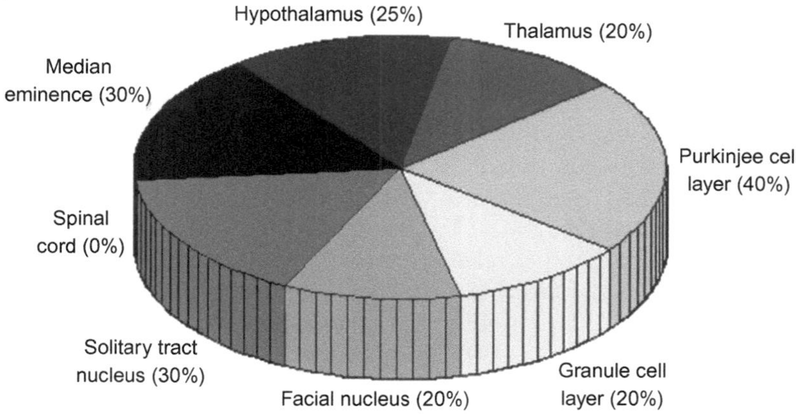

Figure 19.2 Relative distribution of Ptch1 mRNA in adult rat CNS shown as percentage expression taking the data for spinal cord (%) since the baseline as matured spinal cord is known to contain almost no Ptch1. (For color version of this figure, the reader is referred to the Web version of this chapter.)

A. Ptch1 and medulloblastoma

Several reports indicated a putative connection between Nevoid basal cell carcinoma (NBCC) with germ line mutation in the human homolog of *Ptch1* (Pan, 2010). NBCC or more broadly Gorlin–Goltz syndrome is characterized primarily by increased hypersensitivity toward radiation, developmental abnormalities, skin tumors, and predisposition to medulloblastoma (MB), an invasive pediatric neuroectodermal brain tumor. Normal cerebellar development requires a plethora of consecutive events guided by the proper activity of developmental signaling cascades. After originating from the rhombic lips near the ventricular zone (VZ), granular neural progenitors (GNP) proliferate rapidly to the external germinal layer (EGL) of the cerebellum. In response to the Shh secretion from the underlying Purkinje cell layer, GNPs move from the EGL through the Purkinje layer below to form an internal granular layer. Enroute suitable programmed modulation of the Hh system instructs the cells to stop proliferating and undergo differentiation to develop postmitotic mature granular neurons (Dahmane and Ruiz i Altaba, 1999; Wallace, 1999; Wechsler-Reya and Scott, 1999, 2001). Experimental animal model has helped to demonstrate that anomaly in the processing of the GNPs due to aberrant activity of Hh pathway can account for the progress of MB. In human subjects with this disease, thorough investigation of the chromosomal abnormality has helped to associate loss or mutation in the 9q22.3–q32 region, harboring the *Ptch1* allele (Ellison, 2002; Gilbertson, 2002; Goussia *et al.*, 2000). Homozygous *Ptch1−/−* mice cannot survive over 9.5 days of gestation owing to acute systemic defects in nervous and cardiovascular development. In single mutant mouse (*Ptch1+/−*) brain, though heterozygosity of *Ptch1* locus is sufficient to initiate the disease, preneoplastic cells, having characteristics of GNPs and destined to the fate of MB cells, show gradual decay in expression of the remaining wild-type *Ptch1* allele. The disappearance of Ptch1 expression as quantified by real-time PCR analysis suggests that loss of the remaining wild-type *Ptch1* is essential for the growth of tumor in heterozygous condition (Berman *et al.*, 2002; Oliver *et al.*, 2005). Previously, a few studies reported a continued basal level expression of wild-type *Ptch1* by RT-PCR, Northern blot analysis of total RNA, sequencing and *in situ* hybridization of cDNA isolated from these primary tumors (Romer *et al.*, 2004; Wetmore *et al.*, 2000; Zurawel *et al.*, 2000). Another study refers that DNA methylation induced silencing of the wild-type copy of *Ptch1* may be responsible for the proliferation of the metastatic tumor (Berman *et al.*, 2002). Dearth in the activation of the ATR–Chk pathway in these mutant mice probably accounts for the ionizing radiation-induced genomic instability in MB (Leonard *et al.*, 2008). MB is a disease with variable expressivity and incomplete penetrance. Detailed genetic study has revealed that a combination of different genes including SUFU, P53, and Gli in connection to Ptch1 can activate or repress the outcome of MB. Elevation in the production of growth factors as Igf2 is also a key to tumor production in *Ptch1* heterozygous animals (Hahn *et al.*, 2000). Homozygous deletion of *Igf2* in *Ptch1+/−* mouse has shown similar occurrence of tumor with that of *Ptch1+/−* only, but they cannot develop to matured tumors (Corcoran *et al.*, 2008). This observation suggests toward the fact that Igf2, a target of Hh pathway, is essential for tumor development but not initiaton in *Ptch1* mutant. Mutation in *Ptch2* allele does not show any significant increase in the perpetuation of disease and there is no disposition to this disease found in case of *Ptch2−/−* mice already having *Ptch1+/−* mutation (Lee *et al.*, 2006; Nieuwenhuis *et al.*, 2006).

Additionally, deregulation of miRNA networks can cause fatal outcome with rapid progress of MB. Retroviral expression of miR17–92 cluster in purified GNP from *Ptch1+/−* and *Ink4c−/−* and orthotopicaly injecting them back in *Ptch1+/−* mice significantly enhanced the MB incidence compared to nontransduced cells (Uziel *et al.*, 2009). The perpetual reduction of wild-type *Ptch1* allele in these cells points toward a strong correlation of the regulations of Hh pathway by this cluster. In this regard, it would be interesting to find out if Ptch1 contains binding site for any of the members of this cluster.

Administration of cyclopamine, a potent inhibitor of Hh pathway, has demonstrated reduced expression of Ptch1 in tumor cell as well as depletion of cell number. But while being treated with another Hh antagonist HhAntag-691, although the tumors were completely eradicated from the animals, there was not much effect on the level of Ptch1 in the cells (Romer *et al.*, 2004). This indicates that although the initiation of tumorigenesis is activated by aberrant function of Ptch1, the gene takes a rear seat in the development scenario.

B. Function of Ptch1 in maintaining stem cell niche in adult vertebrates

In the developmental phase, expansion and differentiation of neuronal progenitor cells are important for the proper maturation of appropriate postmitotic neurons. In adults, the knowledge of the maintenance of this stem cell pool is slowly emerging. Previously, it was thought that neural progenitor cells are transiently present during neural development. However, in adult mammalian CNS, significant amount of stem cell population is present in a few particular anatomical locations like (1) subventricular zone (SVZ) of the lateral ventricle and in hippocampus in case of brain and in the ependymal layer of the fourth ventricle and around the central canal in case of spinal cord and (2) Optic nerve, hypothalamus, substantia nigra (Gage, 2000; Horner *et al.*, 2000; Lie *et al.*, 2002, 2004). In response to signaling molecules like Hh, under certain conditions, they start differentiating into new lineages especially oligodendrocytes and neurons. Under normal condition in postnatal and adult CNS, the Shh signaling system helps in a renewed maintenance of these cells and regulates the electrophysiological features of matured neurons. In adulthood, induced expression of Shh in the hippocampus modulates the proliferation of adult neural progenitors in the subgranular zones of the dentate gyrus (Lai *et al.*, 2003). *In vivo* administration of Shh-N and Smo also activated the Shh signaling in the adult SVZ (Charytoniuk *et al.*, 2002; Machold *et al.*, 2003; McMahon *et al.*, 2003). Direct injection of Shh in the left ventricular SVZ has shown an upregulaiton of Ptch1 and Gli1 in many areas of brain including the corpus callosum and cerebral cortex. The Ptch1 expression coincides with that of olig1+ cells (oligodendroglial progenitors) confirming the activation of the Hh pathway in the stem cells (Loulier *et al.*, 2006).

C. Ptch1 in spinal cord injury

Despite the advances in therapeutic intervention to fix an injured spinal cord, the regeneration of axon following an atrocious injury still remains a challenge. The postmitotic cells of adult CNS cannot recuperate the massive neuronal degeneration which results in persistent functional loss. However, in particular, the inhospitable environment surrounding the severed CNS axons inhibits their ability to regrow. The obstacles include extrinsic axonal growth inhibiting factors secreted from immunoreactive cells, decreased intrinsic regenerative potential of CNS axons, absence of remyelination, and development of a cystic tissue defect. Abortive sprouting of damaged axons and addition of new neurons at specific areas of adult CNS still continuously occur. These phenomena need induction of neuronal stem cells. Remedial roles exhibited by stem cell therapeutics can assume any of the following strategies: (a) induction of fresh myelination and prevention of demyelination. (b) Inhibition of immunologic response generation and prevention of cystic cavitation formation (Bambakidis *et al.*, 2005).

The function of Hh signaling is not much delineated in adult vertebrate organisms. The principal cause of severe secondary disability of spinal cord injury is the loss of myelinaiton that results from the death of oligodendrocyte (Rosenbluth *et al.*, 1997). If involvement of Hh plays a potent role in repair and regeneration following injury, administration of small molecule modulators of the pathway can be a putative therapeutic approach. With this notion, several studies have been conducted to investigate the effect of escalating doses of Shh to recover demyelinating lesion in spinal cord. Diffused proliferative responses in the focal lesion exposed to Shh possibly are generated from the latent stem cell population specific for CNS precursors. This proliferation and subsequent differentiation culminate in the production of functional oligodendrocytes and motor neurons (Bambakidis *et al.*, 2003). Similar result was further substantiated by administration of Hh agonist (small molecules potentiating the function of Hh) in mouse suffering from laminectomy. Direct intravenous application of Shh has reported to promote expression of another neural proliferative marker Gli1. In addition, induction of CNS precursors along the rim of the central cystic cavity formed as a response of injury is also reported. Injection through other routes has evoked lesser response possibly because of different pharmacokinetics associated with its release in the system (Bambakidis *et al.*, 2003, 2010).

Migration and proliferation of oligodendrocyte precursor cells have been shown to be extremely sensible to the presence and dosage of Hh in its immediate vicinity. Therefore, stimulating OPC by Hh can considerably modify its fate and instigate it to regenerate myelin along neighboring axon. Specifically, stem cells that become oligodendrocytes have been reported to maximally stimulate axonal revival in terms of conduction velocity (Gould *et al.*, 1999; Lai *et al.*, 2003; Nakatomi *et al.*, 2002; Rowitch *et al.*, 1999). Shh can bring about induction of nestin, musashi-1, and O4 positive endogenous neural stem cell population *in vivo* as well as in explants culture *in vitro*. Subjecting the host cells to raise a renewed response is always easy and appreciable than going through the hazardous procedure of exogenous precursor transplantation. Following an external trauma, environmental cues like FGF2, IGF5, and EGF from the surrounding tissues lead to the neural precursor cells maturation mostly along reactive astroglial lineage. But postinjury Shh delivery directs a surge of stem cell proliferation and channelizes their fates toward functional oligodendrocytes and motor neurons (Bambakidis and Miller, 2004). Apart from CNS, crush injury to peripheral nervous system is also highly responsive to Shh. Study of expression in terms of mRNA following sciatic nerve insult has hinted toward an elevated level of Hh in Schwann cells around the site of injury. This is accompanied by an increase in BDNF, NT3, Gli1, NT4/5, GDNF (all neurotrophic factors) affirming the role of Hh employing these molecules in neuroregeneration. Presence of Ptch1 and Smo in cultured Schwann cells perhaps suggests its responsiveness to Hh as well (Hashimoto *et al.*, 2008).

The rudimentary knowledge about involvement of Hh signaling in reversal of spinal trauma, cumulatively derived from all these studies by Bambakidis *et al.*, can now help speculate a role of Ptch1 in these events. Since all the previous reports suggest that upregulation of Shh results in the consequent expression of Ptch1, a strong correlation between Ptch1 and stimulation of a regenerative response in the aforementioned scenarios can be anticipated from further downstream studies.

In recent past detailed experimental analysis has thrown some light on the understanding of the Ptch1-mediated pattern of regenerative response in cord contused zebrafish. Basal level expression of all the major players of the pathway in unlesioned spinal cord turned out to be robust, a couple of weeks postinjury. The expression exactly reiterates the developmental pattern. Shh gives a potent signal in the VZ and Ptch1 in its adjacent location. Ptch1 shows a tailed off expression signature toward the dorsal spinal cord. The region of intense Ptch1 expression is seen to be overlapped with cells having markers of motor neuron precursors, again indicating its role in the transcriptional activation of autogenous stem cell population (Reimer *et al.*, 2009). Strong upregulation of Ptch1, in this context, possibly indicates lesion-induced activity of Hh signaling in the precursor cells. This assumption is further affirmed by strong to weak expression gradient of Ptch1 and Smo along ventral to dorsal zone of the spinal cord with the strongest being in the ependymoradial glial cells (pMN like) (Reimer *et al.*, 2009). The identical embryogenic developmental paradigm, assumed by the proteins in a regenerating system, corroborates that the primitive embryonic developmental cascades can be somewhat recapitulated in neuroregenerative scenarios during adulthood.

D. Ptch1 in neurodegenerative diseases and aneuploidy

Having discussed the roles of Ptch1 in neuroregenerative scenarios, let us shift our focus toward common neurodegenerative disorders. There are reports that establish the role of Hh signaling pathway in the functional recovery in neurodegenerative disorders ranging from Parkinson's disease to peripheral neuropathy, and experiments have been done very convincingly to support this neuroprotective and regeneration activities simultaneously to the activation of this pathway (Dellovade *et al.*, 2006). The Hh signaling is able to recruit endogenous stem cells for the process of repairing the nervous system and regeneration of it after neurodegeneration (Gulino *et al.*, 2007). Therefore, the Hh pathway is involved in the responsive mechanisms of the brain after a neurodegeneration had taken place. Small molecule agonists of this pathway, in this purpose, are being isolated and characterized to exploit these molecules as therapeutics for nervous system disorders.

Another neurodegenerative disorder, the Niemann-Pick-C1 (NPC) disease, is caused due to abnormal accumulation of cholesterol in neuronal cells (Gazi *et al.*, 2009). It is shown experimentally that a drosophila model for Ptch mutant develop this kind of neurodegeneration containing membranous and lamellar inclusions in their brain. The number of synaptic terminals is also reduced in case of mutant adult flies. Feeding cholesterol to the wild-type flies generates this kind of inclusions in the brain but does not develop the disease since Ptch is functional in the wild-type flies. This indicates that Ptch plays a suppressive role for this neurodegenerative disease and loss of function of Ptch in the mutant turns on the neurodegeneration process. Repression of the disease occurs through indirect effect of Ptch where it functions via regulation of Smo in the brain (Gazi *et al.*, 2009). The regulation of Smo activity by Ptch1 plays very crucial role in respect to Shh signaling. Overexpression of Ptch1 inhibits Shh signaling, playing an inhibitory role in the regulation of cell proliferation (Goodrich *et al.*, 1999; Taipale *et al.*, 2002).

This activity of Ptch1 is well established in case of Down syndrome (DS) (Trazzi *et al.*, 2011). It has been shown recently that the reduced proliferation of cerebellar granule cell precursors from Ts65Dn mice (DS mouse model) is related to an attenuated response to Shh pathway (Roper *et al.*, 2006). The wide expression of Shh pathway in different brain neurogenic zones (Arsic *et al.*, 2007; Ishibashi *et al.*, 2005) indicates that this pathway exerts critical role in regulating neural precursor proliferation, and attenuation of the response to Shh (probably through the activity of Ptch1) may result in widespread neurogenic impairments, characteristic of trisomic brain. A very recent quantitative RT-PCR study from the NPCs of the SVZ and hippocampus of Ts65Dn mice (Trazzi *et al.*, 2011) showed an increased expression of Ptch1, maintaining Shh pathway in a repressed state. The expression of Ptch1 in NPCs is significantly high in the protein level too. This indicates that increased expression of Ptch1 in trisomic NPCs is a major determining factor for the derangement of Shh pathway and cell proliferation. In the same study with Ts65Dn mice, partial silencing of Ptch1 expression restored cell proliferation (thus Shh signaling) indicating that proliferation impairment of trisomic NPCs was due to Ptch1 overexpression.

A report from our group suggests that the intracellular C-terminal domain of amyloid precursor protein (AICD) upregulates Ptch1 expression in neuroblastoma cells (Raychaudhuri and Mukhopadhyay, 2011). Ptch1 is also overexpressed in neuronal precursors of human fetuses with DS (Trazzi

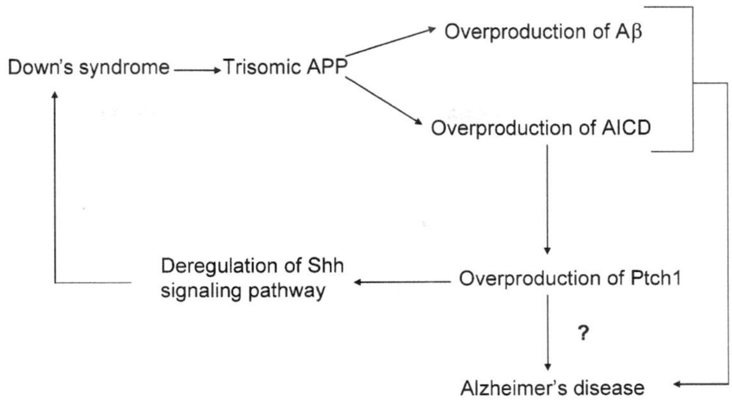

Figure 19.3 Schematic representation showing possible interrelation of Alzheimer's disease and Down's syndrome. AICD and Ptch1 might act as important bridges between these correlated disorders.

et al., 2011). Impairment of Shh signaling pathway is one of the characteristics of DS, and the deregulation of this pathway in DS is due to the overexpression of Ptch1. On the other hand, since the overexpression of Ptch1 is in concurrence with the overexpression of the trisomic APP gene, which, in turn, promotes transactivation of Ptch1 through AICD, a strong relationship between DS and AD is apparent (Trazzi *et al.*, 2011). In addition, AICD transgenic mice also exhibit impaired neurogenesis similar to trisomic mice Ts65Dn (Ghosal *et al.*, 2010). In summary, the neurogenesis defects in DS are due to the repression of Shh pathway by Ptch1 overexpression which is actually mediated by the protein APP (mainly through AICD). Therefore, AICD and Ptch1, acting as bridges between AD and DS, may be resulting in a consequence that most individuals with DS should naturally develop AD after the third or fourth decade of life (see a possible schema in Fig. 19.3).

V. Conclusion

Despite efforts to identify the role of Ptch1 in neural cells, intensive study of the molecular intricacies about how Ptch1 regulates different aspect of development, maintenance, and regeneration in neural cells is still an enigma. Apart from its major roles in anterior–posterior patterning of neural tube development, regulation of oligodendrogenesis, maintenance of stem cell niche, continuous renewal of neurons in mature CNS, and axonal pathfinding, Ptch1 has another significant influence in amelioration of neurodegenerative condition. Delving deeper into the activity of Ptch1 in adult neuronal circuitry, it seems likely that a huge amount of information will be forthcoming regarding its multifunctional roles both in developmental and protective perspectives. It would be interesting to further elucidate the roles of Ptch1 in the two fundamental physiological aspects: In depth understanding of basic biology of the signaling and the therapeutic roles of Ptch1.

Overexpression of Ptch1 in AICD-transfected cells may point toward its role in regulating a neurodegenerative pathway in cells. On the contrary, elevated expression of Ptch1 and subsequent stimulation of neural progenitor cells following CNS injury speak for its protective role in the extreme degenerative condition. Administration of Shh/agonists has shown to elicit regenerative response through an increased Ptch1 level. Considering the multifaceted regulatory direction of Ptch1 ranging from a morphogen in embryogenesis to mitogen in tumor progression, it seems likely that Ptch1 can prove itself to be a very interesting target of research in neurobiology. The interaction of other signaling cascades with the induction and expression of Ptch1 in signal propagation throughout the huge repertoire of development to regenerative platform in adults presents an extremely enticing field of study. Through conserved motifs like sterol-binding domains, Ptch1 is likely to elicit its role in neural survival. Another fascinating approach can be the finding of putative miRNA targets of Ptch1. Thus, manipulating them might be a novel therapeutic proposition in conjunction with the already available strategies. A better understanding of the cellular and molecular mechanisms underlying the neuroplasticity might be helpful for adaptation of neurorehabiliation strategies to promote functional recovery following such degenerative disorders and neural trauma. In near future, neuroscience research concerning Ptch1 is going to address a broader dimension beyond Hh signaling.

ACKNOWLEDGMENTS

A. C. and S. G. acknowledge Council of Scientific and Industrial Research (Govt. of India). The work is supported by SPGHGD project.

REFERENCES

Arsic, D., Beasley, S. W., and Sullivan, M. J. (2007). Switched-on Sonic hedgehog: A gene whose activity extends beyond fetal development–to oncogenesis. *J. Paediatr. Child Health* **43,** 421–423.

Bambakidis, N. C., and Miller, R. H. (2004). Transplantation of oligodendrocyte precursors and sonic hedgehog results in improved function and white matter sparing in the spinal cords of adult rats after contusion. *Spine J.* **4,** 16–26.

Bambakidis, N. C., Wang, R. Z., Franic, L., and Miller, R. H. (2003). Sonic hedgehog-induced neural precursor proliferation after adult rodent spinal cord injury. *J. Neurosurg.* **99,** 70–75.

Bambakidis, N. C., Theodore, N., Nakaji, P., Harvey, A., Sonntag, V. K., Preul, M. C., and Miller, R. H. (2005). Endogenous stem cell proliferation after central nervous system injury: Alternative therapeutic options. *Neurosurg. Focus* **19,** E1.

Bambakidis, N. C., Wang, X., Lukas, R. J., Spetzler, R. F., Sonntag, V. K., and Preul, M. C. (2010). Intravenous hedgehog agonist induces proliferation of neural and oligodendrocyte precursors in rodent spinal cord injury. *Neurosurgery* **67,** 1709–1715, discussion 1715.

Berman, D. M., Karhadkar, S. S., Hallahan, A. R., Pritchard, J. I., Eberhart, C. G., Watkins, D. N., Chen, J. K., Cooper, M. K., Taipale, J., Olson, J. M., and Beachy, P. A. (2002). Medulloblastoma growth inhibition by hedgehog pathway blockade. *Science* **297,** 1559–1561.

Brand-Saberi, B., Ebensperger, C., Wilting, J., Balling, R., and Christ, B. (1993). The ventralizing effect of the notochord on somite differentiation in chick embryos. *Anat Embryol (Berl).* **188,** 239–245.

Briscoe, J., Chen, Y., Jessell, T. M., and Struhl, G. (2001). A hedgehog-insensitive form of patched provides evidence for direct long-range morphogen activity of sonic hedgehog in the neural tube. *Mol. Cell* **7,** 1279–1291.

Carpenter, D., Stone, D. M., Brush, J., Ryan, A., Armanini, M., Frantz, G., Rosenthal, A., and de Sauvage, F. J. (1998). Characterization of two patched receptors for the vertebrate hedgehog protein family. *Proc. Natl. Acad. Sci. USA* **95,** 13630–13634.

Charytoniuk, D., Traiffort, E., Hantraye, P., Hermel, J. M., Galdes, A., and Ruat, M. (2002). Intrastriatal sonic hedgehog injection increases Patched transcript levels in the adult rat subventricular zone. *Eur. J. Neurosci.* **16,** 2351–2357.

Chen, Y., and Struhl, G. (1996). Dual roles for patched in sequestering and transducing Hedgehog. *Cell* **87,** 553–563.

Corcoran, R. B., Bachar Raveh, T., Barakat, M. T., Lee, E. Y., and Scott, M. P. (2008). Insulin-like growth factor 2 is required for progression to advanced medulloblastoma in patched1 heterozygous mice. *Cancer Res.* **68,** 8788–8795.

Dahmane, N., and Ruiz i Altaba, A. (1999). Sonic hedgehog regulates the growth and patterning of the cerebellum. *Development* **126,** 3089–3100.

Dellovade, T., Romer, J. T., Curran, T., and Rubin, L. L. (2006). The hedgehog pathway and neurological disorders. *Annu. Rev. Neurosci.* **29,** 539–563.

Ellison, D. (2002). Classifying the medulloblastoma: Insights from morphology and molecular genetics. *Neuropathol. Appl. Neurobiol.* **28,** 257–282.

Ericson, J., Muhr, J., Placzek, M., Lints, T., Jessell, T. M., and Edlund, T. (1995). Sonic hedgehog induces the differentiation of ventral forebrain neurons: A common signal for ventral patterning within the neural tube. *Cell* **81,** 747–756.

Fuccillo, M., Joyner, A. L., and Fishell, G. (2006). Morphogen to mitogen: The multiple roles of hedgehog signalling in vertebrate neural development. *Nat. Rev. Neurosci.* **7,** 772–783.

Gage, F. H. (2000). Mammalian neural stem cells. *Science* **287,** 1433–1438.

Gazi, M., Shyamala, B. V., and Bhat, K. M. (2009). A neurodegenerative disease affecting synaptic connections in Drosophila mutant for the tumor suppressor morphogen Patched. *Dev. Biol.* **334,** 311–323.

Ghosal, K., Stathopoulos, A., and Pimplikar, S. W. (2010). APP intracellular domain impairs adult neurogenesis in transgenic mice by inducing neuroinflammation. *PLoS One* **5,** e11866.

Gilbertson, R. (2002). Paediatric embryonic brain tumours. Biological and clinical relevance of molecular genetic abnormalities. *Eur. J. Cancer* **38,** 675–685.

Goodrich, L. V., Johnson, R. L., Milenkovic, L., McMahon, J. A., and Scott, M. P. (1996). Conservation of the hedgehog/patched signaling pathway from flies to mice: Induction of a mouse patched gene by Hedgehog. *Genes Dev.* **10,** 301–312.

Goodrich, L. V., Jung, D., Higgins, K. M., and Scott, M. P. (1999). Overexpression of ptc1 inhibits induction of Shh target genes and prevents normal patterning in the neural tube. *Dev. Biol.* **211,** 323–334.

Gould, E., Reeves, A. J., Graziano, M. S., and Gross, C. G. (1999). Neurogenesis in the neocortex of adult primates. *Science* **286,** 548–552.

Goussia, A. C., Bruner, J. M., Kyritsis, A. P., Agnantis, N. J., and Fuller, G. N. (2000). Cytogenetic and molecular genetic abnormalities in primitive neuroectodermal tumors of the central nervous system. *Anticancer Res.* **20,** 65–73.

Gulino, A., Di Marcotullio, L., Ferretti, E., De Smaele, E., and Screpanti, I. (2007). Hedgehog signaling pathway in neural development and disease. *Psychoneuroendocrinology* **32**(Suppl. 1), S52–S56.

Hahn, H., Wojnowski, L., Specht, K., Kappler, R., Calzada-Wack, J., Potter, D., Zimmer, A., Muller, U., Samson, E., and Quintanilla-Martinez, L. (2000). Patched target Igf2 is indispensable for the formation of medulloblastoma and rhabdomyosarcoma. *J. Biol. Chem.* **275,** 28341–28344.

Hashimoto, M., Ishii, K., Nakamura, Y., Watabe, K., Kohsaka, S., and Akazawa, C. (2008). Neuroprotective effect of sonic hedgehog up-regulated in Schwann cells following sciatic nerve injury. *J. Neurochem.* **107,** 918–927.

Hausmann, G., von Mering, C., and Basler, K. (2009). The hedgehog signaling pathway: Where did it come from? *PLoS Biol.* **7,** e1000146.

Ho, K. S., and Scott, M. P. (2002). Sonic hedgehog in the nervous system: Functions, modifications and mechanisms. *Curr. Opin. Neurobiol.* **12,** 57–63.

Horner, P. J., Power, A. E., Kempermann, G., Kuhn, H. G., Palmer, T. D., Winkler, J., Thal, L. J., and Gage, F. H. (2000). Proliferation and differentiation of progenitor cells throughout the intact adult rat spinal cord. *J. Neurosci.* **20,** 2218–2228.

Hynes, M., Porter, J. A., Chiang, C., Chang, D., Tessier-Lavigne, M., Beachy, P. A., and Rosenthal, A. (1995). Induction of midbrain dopaminergic neurons by Sonic hedgehog. *Neuron* **15,** 35–44.

Incardona, J. P., Gaffield, W., Lange, Y., Cooney, A., Pentchev, P. G., Liu, S., Watson, J. A., Kapur, R. P., and Roelink, H. (2000). Cyclopamine inhibition of Sonic hedgehog signal transduction is not mediated through effects on cholesterol transport. *Dev. Biol.* **224,** 440–452.

Ingham, P. W., and McMahon, A. P. (2001). Hedgehog signaling in animal development: Paradigms and principles. *Genes Dev.* **15,** 3059–3087.

Ishibashi, M., Saitsu, H., Komada, M., and Shiota, K. (2005). Signaling cascade coordinating growth of dorsal and ventral tissues of the vertebrate brain, with special reference to the involvement of Sonic Hedgehog signaling. *Anat. Sci. Int.* **80,** 30–36.

Jessell, T. M. (2000). Neuronal specification in the spinal cord: Inductive signals and transcriptional codes. *Nat. Rev. Genet.* **1,** 20–29.

Johnson, R. L., and Tabin, C. (1995). The long and short of hedgehog signaling. *Cell* **81,** 313–316.

Ko, D. C., Gordon, M. D., Jin, J. Y., and Scott, M. P. (2001). Dynamic movements of organelles containing Niemann-Pick C1 protein: NPC1 involvement in late endocytic events. *Mol. Biol. Cell* **12,** 601–614.

Lai, K., Kaspar, B. K., Gage, F. H., and Schaffer, D. V. (2003). Sonic hedgehog regulates adult neural progenitor proliferation *in vitro* and *in vivo*. *Nat. Neurosci.* **6,** 21–27.

Lee, Y., Miller, H. L., Russell, H. R., Boyd, K., Curran, T., and McKinnon, P. J. (2006). Patched2 modulates tumorigenesis in patched1 heterozygous mice. *Cancer Res.* **66,** 6964–6971.

Leonard, J. M., Ye, H., Wetmore, C., and Karnitz, L. M. (2008). Sonic Hedgehog signaling impairs ionizing radiation-induced checkpoint activation and induces genomic instability. *J. Cell Biol.* **183,** 385–391.

Lie, D. C., Dziewczapolski, G., Willhoite, A. R., Kaspar, B. K., Shults, C. W., and Gage, F. H. (2002). The adult substantia nigra contains progenitor cells with neurogenic potential. *J. Neurosci.* **22,** 6639–6649.

Lie, D. C., Song, H., Colamarino, S. A., Ming, G. L., and Gage, F. H. (2004). Neurogenesis in the adult brain: New strategies for central nervous system diseases. *Annu. Rev. Pharmacol. Toxicol.* **44,** 399–421.

Loulier, K., Ruat, M., and Traiffort, E. (2006). Increase of proliferating oligodendroglial progenitors in the adult mouse brain upon Sonic hedgehog delivery in the lateral ventricle. *J. Neurochem.* **98,** 530–542.

Machold, R., Hayashi, S., Rutlin, M., Muzumdar, M. D., Nery, S., Corbin, J. G., Gritli-Linde, A., Dellovade, T., Porter, J. A., Rubin, L. L., Dudek, H., McMahon, A. P., *et al.* (2003). Sonic hedgehog is required for progenitor cell maintenance in telencephalic stem cell niches. *Neuron* **39,** 937–950.

Marigo, V., and Tabin, C. J. (1996). Regulation of patched by sonic hedgehog in the developing neural tube. *Proc. Natl. Acad. Sci. USA* **93,** 9346–9351.

Marigo, V., Davey, R. A., Zuo, Y., Cunningham, J. M., and Tabin, C. J. (1996a). Biochemical evidence that patched is the Hedgehog receptor. *Nature* **384,** 176–179.

Marigo, V., Scott, M. P., Johnson, R. L., Goodrich, L. V., and Tabin, C. J. (1996b). Conservation in hedgehog signaling: Induction of a chicken patched homolog by Sonic hedgehog in the developing limb. *Development* **122,** 1225–1233.

Marti, E., and Bovolenta, P. (2002). Sonic hedgehog in CNS development: One signal, multiple outputs. *Trends Neurosci.* **25,** 89–96.

Marti, E., Takada, R., Bumcrot, D. A., Sasaki, H., and McMahon, A. P. (1995). Distribution of Sonic hedgehog peptides in the developing chick and mouse embryo. *Development* **121,** 2537–2547.

Martin, V., Carrillo, G., Torroja, C., and Guerrero, I. (2001). The sterol-sensing domain of Patched protein seems to control Smoothened activity through Patched vesicular trafficking. *Curr. Biol.* **11,** 601–607.

McMahon, A. P., Ingham, P. W., and Tabin, C. J. (2003). Developmental roles and clinical significance of hedgehog signaling. *Curr. Top. Dev. Biol.* **53,** 1–114.

Miao, N., Wang, M., Ott, J. A., D'Alessandro, J. S., Woolf, T. M., Bumcrot, D. A., Mahanthappa, N. K., and Pang, K. (1997). Sonic hedgehog promotes the survival of specific CNS neuron populations and protects these cells from toxic insult *in vitro*. *J. Neurosci.* **17,** 5891–5899.

Milenkovic, L., and Scott, M. P. (2010). Not lost in space: Trafficking in the hedgehog signaling pathway. *Sci. Signal.* **3,** pe14.

Motoyama, J., Takabatake, T., Takeshima, K., and Hui, C. (1998). Ptch2, a second mouse Patched gene is co-expressed with Sonic hedgehog. *Nat. Genet.* **18,** 104–106.

Nakatomi, H., Kuriu, T., Okabe, S., Yamamoto, S., Hatano, O., Kawahara, N., Tamura, A., Kirino, T., and Nakafuku, M. (2002). Regeneration of hippocampal pyramidal neurons after ischemic brain injury by recruitment of endogenous neural progenitors. *Cell* **110,** 429–441.

Nieuwenhuis, E., Motoyama, J., Barnfield, P. C., Yoshikawa, Y., Zhang, X., Mo, R., Crackower, M. A., and Hui, C. C. (2006). Mice with a targeted mutation of patched2 are viable but develop alopecia and epidermal hyperplasia. *Mol. Cell. Biol.* **26,** 6609–6622.

Oliver, T. G., Read, T. A., Kessler, J. D., Mehmeti, A., Wells, J. F., Huynh, T. T., Lin, S. M., and Wechsler-Reya, R. J. (2005). Loss of patched and disruption of granule cell development in a pre-neoplastic stage of medulloblastoma. *Development* **132,** 2425–2439.

Pan, S., Dong, Q., Sun, L., and Li, T. (2010). Mechanisms of Inactivation of PTCH1 Gene in Nevoid Basal Cell Carcinoma Syndrome: Modification of the Two-Hit Hypothesis. *Clin. Cancer Res.* **16**(2), 442–450.

Patten, I., and Placzek, M. (2000). The role of Sonic hedgehog in neural tube patterning. *Cell. Mol. Life Sci.* **57,** 1695–1708.

Pearse, R. V., 2nd, Vogan, K. J., and Tabin, C. J. (2001). Ptc1 and Ptc2 transcripts provide distinct readouts of Hedgehog signaling activity during chick embryogenesis. *Dev. Biol.* **239,** 15–29.

Pourquie, O., Coltey, M., Teillet, M. A., Ordahl, C., and Le Douarin, N. M. (1993). Control of dorsoventral patterning of somitic derivatives by notochord and floor plate. *Proc. Natl. Acad. Sci. USA* **90,** 5242–5246.

Raychaudhuri, M., and Mukhopadhyay, D. (2011). AICD overexpression in neuro 2A cells regulates expression of PTCH1 and TRPC5. *Int J Alzheimer's Dis.* **2011,** 1–5.

Reimer, M. M., Kuscha, V., Wyatt, C., Sorensen, I., Frank, R. E., Knuwer, M., Becker, T., and Becker, C. G. (2009). Sonic hedgehog is a polarized signal for motor neuron regeneration in adult zebrafish. *J. Neurosci.* **29,** 15073–15082.

Riobo, N. A., and Manning, D. R. (2007). Pathways of signal transduction employed by vertebrate Hedgehogs. *Biochem. J.* **403,** 369–379.

Roelink, H., Augsburger, A., Heemskerk, J., Korzh, V., Norlin, S., Ruiz i Altaba, A., Tanabe, Y., Placzek, M., Edlund, T., Jessell, T. M., *et al.* (1994). Floor plate and motor neuron induction by vhh-1, a vertebrate homolog of hedgehog expressed by the notochord. *Cell* **76,** 761–775.

Romer, J. T., Kimura, H., Magdaleno, S., Sasai, K., Fuller, C., Baines, H., Connelly, M., Stewart, C. F., Gould, S., Rubin, L. L., and Curran, T. (2004). Suppression of the Shh pathway using a small molecule inhibitor eliminates medulloblastoma in Ptc1(+/−)p53 (−/−) mice. *Cancer Cell* **6,** 229–240.

Roper, R. J., Baxter, L. L., Saran, N. G., Klinedinst, D. K., Beachy, P. A., and Reeves, R. H. (2006). Defective cerebellar response to mitogenic Hedgehog signaling in Down [corrected] syndrome mice. *Proc. Natl. Acad. Sci. USA* **103,** 1452–1456.

Rosenbluth, J., Schiff, R., Liang, W. L., Menna, G., and Young, W. (1997). Xenotransplantation of transgenic oligodendrocyte-lineage cells into spinal cord-injured adult rats. *Exp. Neurol.* **147,** 172–182.

Rowitch, D. H., Benoit, St.-Jacques1, Lee, S. M., Flax, J. D., Snyder, E. Y., and McMahon, A. P. (1999). Sonic hedgehog regulates proliferation and inhibits differentiation of CNS precursor cells. *J. Neurosci.* **19,** 8954–8965.

Strutt, H., Thomas, C., Nakano, Y., Stark, D., Neave, B., Taylor, A. M., and Ingham, P. W. (2001). Mutations in the sterol-sensing domain of Patched suggest a role for vesicular trafficking in Smoothened regulation. *Curr. Biol.* **11,** 608–613.

Taipale, J., Cooper, M. K., Maiti, T., and Beachy, P. A. (2002). Patched acts catalytically to suppress the activity of Smoothened. *Nature* **418,** 892–897.

Tenzen, T., Allen, B. L., Cole, F., Kang, J. S., Krauss, R. S., and McMahon, A. P. (2006). The cell surface membrane proteins Cdo and Boc are components and targets of the Hedgehog signaling pathway and feedback network in mice. *Dev. Cell* **10,** 647–656.

The, I., Bellaiche, Y., and Perrimon, N. (1999). Hedgehog movement is regulated through tout velu-dependent synthesis of a heparan sulfate proteoglycan. *Mol. Cell* **4,** 633–639.

Traiffort, E., Charytoniuk, D. A., Faure, H., and Ruat, M. (1998). Regional distribution of Sonic Hedgehog, patched, and smoothened mRNA in the adult rat brain. *J. Neurochem.* **70,** 1327–1330.

Traiffort, E., Charytoniuk, D., Watroba, L., Faure, H., Sales, N., and Ruat, M. (1999). Discrete localizations of hedgehog signalling components in the developing and adult rat nervous system. *Eur. J. Neurosci.* **11,** 3199–3214.

Trazzi, S., Mitrugno, V. M., Valli, E., Fuchs, C., Rizzi, S., Guidi, S., Perini, G., Bartesaghi, R., and Ciani, E. (2011). APP-dependent up-regulation of Ptch1 underlies proliferation impairment of neural precursors in Down syndrome. *Hum. Mol. Genet.* **20,** 1560–1573.

Tseng, T. T., Gratwick, K. S., Kollman, J., Park, D., Nies, D. H., Goffeau, A., and Saier, M. H., Jr. (1999). The RND permease superfamily: An ancient, ubiquitous and diverse family that includes human disease and development proteins. *J. Mol. Microbiol. Biotechnol.* **1,** 107–125.

Uziel, T., Karginov, F. V., Xie, S., Parker, J. S., Wang, Y. D., Gajjar, A., He, L., Ellison, D., Gilbertson, R. J., Hannon, G., and Roussel, M. F. (2009). The miR-17~92 cluster collaborates with the Sonic Hedgehog pathway in medulloblastoma. *Proc. Natl. Acad. Sci. USA* **106,** 2812–2817.

Varjosalo, M., and Taipale, J. (2008). Hedgehog: Functions and mechanisms. *Genes Dev.* **22,** 2454–2472.

Varjosalo, M., Li, S. P., and Taipale, J. (2006). Divergence of hedgehog signal transduction mechanism between Drosophila and mammals. *Dev. Cell* **10,** 177–186.

Wallace, V. A. (1999). Purkinje-cell-derived Sonic hedgehog regulates granule neuron precursor cell proliferation in the developing mouse cerebellum. *Curr. Biol.* **9,** 445–448.

Wechsler-Reya, R. J., and Scott, M. P. (1999). Control of neuronal precursor proliferation in the cerebellum by Sonic Hedgehog. *Neuron* **22,** 103–114.

Wechsler-Reya, R., and Scott, M. P. (2001). The developmental biology of brain tumors. *Annu. Rev. Neurosci.* **24,** 385–428.

Wetmore, C., Eberhart, D. E., and Curran, T. (2000). The normal patched allele is expressed in medulloblastomas from mice with heterozygous germ-line mutation of patched. *Cancer Res.* **60,** 2239–2246.

Wilson, C. W., and Chuang, P. T. (2010). Mechanism and evolution of cytosolic Hedgehog signal transduction. *Development* **137,** 2079–2094.

Yavari, A., Nagaraj, R., Owusu-Ansah, E., Folick, A., Ngo, K., Hillman, T., Call, G., Rohatgi, R., Scott, M. P., and Banerjee, U. (2010). Role of lipid metabolism in smoothened derepression in hedgehog signaling. *Dev. Cell* **19,** 54–65.

Ye, W., Shimamura, K., Rubenstein, J. L., Hynes, M. A., and Rosenthal, A. (1998). FGF and Shh signals control dopaminergic and serotonergic cell fate in the anterior neural plate. *Cell* **93,** 755–766.

Zurawel, R. H., Allen, C., Wechsler-Reya, R., Scott, M. P., and Raffel, C. (2000). Evidence that haploinsufficiency of Ptch leads to medulloblastoma in mice. *Genes Chromosomes Cancer* **28,** 77–81.

CHAPTER TWENTY

Activation of Hedgehog Pathway in Gastrointestinal Cancers

Ling Yang,* Xiulan Su,* *and* Jingwu Xie†

Contents

Abstract

The hedgehog (Hh) pathway is a major regulator for cell differentiation, tissue polarity, and cell proliferation in embryonic development and homeostasis in adult tissue. Studies from many laboratories reveal activation of this pathway in a variety of human cancer, including basal cell carcinomas (BCCs), medulloblastomas, leukemia, gastrointestinal, lung, ovarian, breast, and prostate cancers. It is thus believed that targeted inhibition of Hh signaling may be effective in treatment and prevention of human cancer. Even more exciting is the discovery and synthesis of specific signaling antagonists for the Hh pathway, which have significant clinical implications in novel cancer therapeutics. In this review, we summarize major advances in the past 2 years in our understanding of Hh signaling activation in human gastrointestinal cancer and their potential in clinical treatment with Hh pathway inhibitors. © 2012 Elsevier Inc.

I. Introduction

The *hedgehog* (Hh) gene was identified in 1980 through genetic analysis of segmentation of fruit fly *Drosophila* (Nusslein-Volhard and Wieschaus, 1980). In the early 1990s, three homologues of the *Hh* gene

* Clinical Research Center of the Affiliated Hospital, Inner Mongolia Medical College, Hohhot, Inner Mongolia, China
† Wells Center for Pediatric Research, Department of Pediatrics and The Simon Cancer Center, Indiana University, Indianapolis, Indiana, USA

Vitamins and Hormones, Volume 88
ISSN 0083-6729, DOI: 10.1016/B978-0-12-394622-5.00020-1

were identified in vertebrates (Chang *et al.*, 1994; Echelard *et al.*, 1993; Krauss *et al.*, 1993; Riddle *et al.*, 1993; Roelink *et al.*, 1994). As an essential developmental signaling pathway, the Hh pathway is critical for maintaining tissue polarity and stem cell population. Inactivation of this pathway causes developmental defects such as holoprosencephaly (Bale, 2002), whereas hyperactivation of this pathway is found in most basal cell carcinomas (BCCs) and many extracutaneous cancers (Xie, 2005, 2008a,b). The emerging role of Hh signaling in human cancer further emphasizes the relevance of studying this pathway to human health.

Overall, the general signaling mechanisms of the Hh pathway are conserved from fly to the humans (Ingham and Placzek, 2006). The seven transmembrane domain containing protein smoothened (SMO) serves as the key player for signal transduction of this pathway, whose function is inhibited by another transmembrane protein Patched (PTC) in the absence of Hh ligands. In the presence of active Hh ligands, binding of Hh to its receptor PTC releases this inhibition, allowing SMO to signal downstream, eventually to Gli transcription factors. As transcription factors, Gli molecules can regulate target gene expression by direct association with a specific consensus sequence located in the promoter region of the target genes (Kinzler and Vogelstein, 1990; Sasaki *et al.*, 1997). Figure 20.1 shows the simplified diagram of Hh signaling in the presence or absence of Hh.

Six Hh pathway inhibitors (e.g., GDC-0449, BMS-833923, IPI-926, and PF-04449913) have been used in clinical trials for treatment of BCC, medulloblastoma, ovarian cancer, and lung cancer (Sheridan, 2009). GDC-0449, which was developed by Curis and Genentech, had great effects in therapy of advanced BCC and medulloblastoma (Dlugosz and Talpaz, 2009; Von Hoff *et al.*, 2009). However, to our disappointment, clinical trials of GDC-0449 treatment in ovarian cancer and colorectal cancer have been suspended due to limited benefits for the patients. The remarkable differences in effectiveness of Hh pathway inhibitor in some types of cancer indicated that the Hh pathway plays different roles in different types of cancer. Therefore, in future, thorough study is required before Hh pathway inhibitors are advanced into clinical trials. In this review, we summarize major advances in Hh signaling activation in human gastrointestinal cancer to provide in-depth information for evaluation of clinical treatment with Hh pathway inhibitors.

II. Activation of the Hh Pathway in Esophageal Cancer

Esophageal cancer is the sixth cause of cancer-related death worldwide and the seventh cause of cancer-related death in American men (Jemal *et al.*, 2008). The two types of esophageal cancer, squamous cell carcinoma

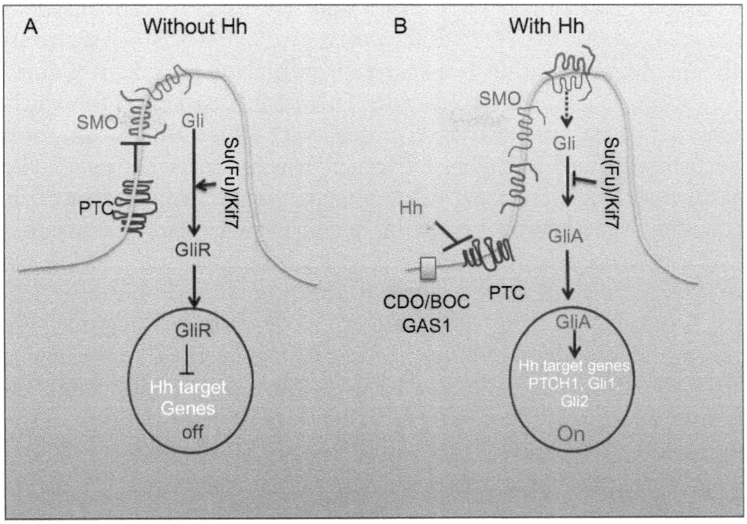

Figure 20.1 A simplified diagram of Hh signaling in vertebrates. A. In the absence of Hh ligands, ciliary PTC somehow inhibits SMO signaling, which results in formation of repressor forms of Gli transcriptional factors Gli2 and Gli3. B. In the presence of Hh ligands, ciliary SMO, no longer affected by PTC, is signaling to downstream effectors, leading to formation of active forms of Gli transcriptional factors and ultimately activation of the target genes. This process is negatively regulated by Su(Fu) and Kif7. Coreceptors of Hh ligands include CDO/BOC and GAS1. This figure was from Liu *et al.* (2011) with permission.

(ESCC) and adenocarcinoma of esophagus (ACE), have different incidence in different geographic regions: ESCC is still the predominant type worldwide, whereas in the United States, the incidence of adenocarcinomas approaches the incidence of ESCC (Brown *et al.*, 2008). China is one of the countries with the highest incidence of esophageal cancer (He *et al.*, 2005). It is known that the etiology of ACE includes long-standing acid/bile reflux esophagitis and development of Barrett's esophagus, an intestinal type metaplasia of the normal squamous epithelium which further progresses to dysplasia and carcinoma. Similarly, ESCC is considered to arise from multiple steps through progression of precancerous lesion, squamous dysplasia to invasive ESCC. Most esophageal cancers are diagnosed in the advanced stage, resulting in a high mortality. Therefore, identifying sensitive and specific biomarkers for esophageal cancer will help early diagnosis and may help design novel strategies for more effective targeted therapy for patients with advanced diseases.

Hh signaling activation has been reported in esophageal cancer. Previously, others and we found activation of Hh signaling in esophageal cancer (Ma *et al.*, 2006; Thayer *et al.*, 2003; Xie and Abbruzzese, 2003). Elevated expression of *sonic hedgehog* (Shh) and its target genes has been found in several esophageal cancer cell lines and esophageal cancer specimens (Berman *et al.*, 2003; Ma *et al.*, 2005; Mori *et al.*, 2006). Studies also suggest that Hh signaling activation is associated with poor prognosis (Mori *et al.*, 2006) of esophageal cancer. Cross talks among Hh signaling, the epithelial–mesenchymal transition pathway (Isohata *et al.*, 2009), the phosphinositide-3 kinase pathway, and mitogen-activated protein kinase pathway (Wei and Xu, 2010) have been reported in esophageal cancer. Activation of the Hh pathway after chemoradiotherapy in esophageal cancer promotes tumor reoccurrence (Sims-Mourtada *et al.*, 2006; Yoshikawa *et al.*, 2008), indicating that combination of Hh pathway inhibitor and chemoradiotherapy may improve prognosis of esophageal cancer. Recent study finds that Gli1 is a novel target of ursodeoxycholic acid–aspirin-combined treatment (Rizvi *et al.*, 2010). Compared with study of Hh pathway in esophageal cancer, little is known about Hh pathway activation in precancerous lesions of esophageal cancer. Recent study indicates that Shh is induced in Barrett's esophagus to mediate paracrine Hh signaling (Wang *et al.*, 2010). We also found that expression of PTCH1 is associated with occurrence of severe squamous dysplasia/carcinoma *in situ* and Barrett's esophagus (our unpublished data). Consistent with the human specimen data, we found a high percentage of Hh signaling activation in precancerous lesions in rat models (our unpublished data). These data indicate that Hh pathway activation is an early molecular event in the development of esophageal cancer and may be a target for early diagnosis and novel therapy for esophageal cancer. Activation of Hh pathway in ACE is much higher than in ESCC (our unpublished data), indicating that Hh pathway in esophageal cancer may be context dependent and affected by tumor heterogeneity. Without knowing the status of Hh signaling, it will be difficult to predict the outcomes of Hh pathway inhibitors in cancer treatment. Several animal models have been used in the study of esophageal cancer, including surgical bilious reflux injury-induced rat/mouse esophageal cancer model (Sui *et al.*, 2006; Wang *et al.*, 2010), NMBA induction of rat ESCC model (Chen *et al.*, 2004) and esophagogastroduodenal anastomosis-induced rat ACE model (Su *et al.*, 2004). These are useful tools in evaluation of effects of Hh pathway inhibitors for esophageal cancer development and progression.

III. Activation of the Hh Pathway in Gastric Cancer

Recent studies have shown that the Hh pathway is involved in gastrointestinal development and gastric regeneration (Dimmler *et al.*, 2003; El-Zaatari *et al.*, 2009; Fukaya *et al.*, 2006; Kang *et al.*, 2009; Kim *et al.*, 2005; van den Brink, 2007). Molecules such as the transcriptional factors GATA-4, GATA-6

(Haveri *et al.*, 2008), FoxF1, FoxL1 (Madison *et al.*, 2009), COUP-TFII (Takamoto *et al.*, 2005), E-cadherin (Xiao *et al.*, 2010), ERK (Osawa *et al.*, 2006), and epithelial–mesenchymal transition pathways (Ohta *et al.*, 2009) are reported to be associated with Hh signaling in this process. Paracrine Hh signaling is found in gastric patterning throughout murine embryonic and adult life (Kolterud *et al.*, 2009). Gastric acid induces *Shh* gene expression (Dimmler *et al.*, 2003; El-Zaatari *et al.*, 2010; Minegishi *et al.*, 2007). Aberrant expression of Shh is involved in the pathogenesis of gastric diseases such as gastric atrophy (Faller and Kirchner, 2005; Waghray *et al.*, 2010). All these studies will help us design better strategy to treat gastric diseases.

Increasing evidence shows that Hh signaling plays a role in gastric cancer. It is estimated that about 90% of gastric cancers are adenocarcinomas. According to Lauren classification, there are two major types of gastric adenocarcinoma: intestinal and diffuse type. Several studies about activation of the Hh pathway in these two types show inconsistent results. Hh signaling is found to be more frequently and highly activated in the diffuse form, not than the intestinal type (Fukaya *et al.*, 2006). Another study reports that expression of Shh is stronger in the intestinal type than in the diffuse type (Lee *et al.*, 2007). Consistent with the role of Hh signaling in gastric cancer, loss of Gli3, an inhibitory component of the Shh pathway, results in intestinal transformation of gastric epithelium (Kim *et al.*, 2005). Expression of *Shh*, *Ihh*, pathway components, and target genes increase in gastric cancer (Berman *et al.*, 2003; Ma *et al.*, 2005), and gastric lesions are associated with the methylation status of the *Shh* promoter (Wang *et al.*, 2006). Nuclear translocation of Gli1 was found to be higher in undifferentiated-type tumors and to be positively correlated with lymph node metastasis in gastric carcinoma (Yanai *et al.*, 2007). Hh signaling was found to promote gastric cancer cells proliferation (Ohta *et al.*, 2005), epithelial–mesenchymal transition (Ohta *et al.*, 2009), mobility and invasiveness (Yoo *et al.*, 2008), and inhibit apoptosis (Han *et al.*, 2009; Ma *et al.*, 2005). *Helicobacter pylori* infection suppresses expression of Shh in gastric mucosa prior to neoplastic transformation (Shiotani *et al.*, 2005) and eradication of the infection restore Shh expression (Nishizawa *et al.*, 2007, 2009; Shiotani *et al.*, 2007). *H. pylori* infection is also implicated in activation of Hh pathway in gastric cancer (Kim *et al.*, 2010a; Lee *et al.*, 2010). These studies indicate that *H. pylori* have different roles in Shh pathway in temporal and spacial dependent manners during gastric carcinogenesis. Mitogen-activated protein kinase cascade promotes activation of the Hh pathway in gastric cancer cells (Seto *et al.*, 2009). Activation of the Wnt pathway is reversely correlated with activation of the Hh pathway in gastric cancer (Kim *et al.*, 2010b; Yanai *et al.*, 2008), probably through inhibition of the Wnt pathway by Hh target gene sFRP1 (He *et al.*, 2006; Kim *et al.*, 2010b). A recent study provides evidence that the estrogen receptor-alpha pathway promotes gastric cancer cells proliferation by activating the Hh pathway (Kameda *et al.*, 2010). Expression of Shh signaling components is also found in precancerous lesions of gastric cancer (Lee *et al.*, 2007), indicating that the Shh pathway is involved early during gastric carcinogenesis. Cyclopamine, a SMO antagonist, inhibits proliferation of gastric cancer cells (Ma *et al.*, 2005); however, in acid environment, reduced efficiency of cyclopamine is found (Wilson *et al.*, 2010). Therefore, the design of acid-proof cyclopamine analogues, such as IPI-926, may help improve treatment in gastric cancer.

IV. Activation of the Hh Pathway in Colorectal Cancer

Loss of function of adenomatous polyposis coli, a component in the Wnt pathway, is responsible for familial adenomatous polyposis coli and 90% sporadic colorectal cancer (van den Brink *et al.*, 2004). Several lines of evidence indicate that activation of the Hh pathway is also implicated in colorectal cancer and cancer stem cell. However, cross talks between Wnt and Hh signaling are still unclear. Activation of these two pathways rarely coexist in colorectal cancer (Akiyoshi *et al.*, 2006; Varnat *et al.*, 2010). Studies indicate that several Hh signaling components negatively regulate Wnt signaling. Gli1 inhibits the proliferation of colon cancer cells with constitutive activation of the Wnt pathway (Akiyoshi *et al.*, 2006). SMO knockdown reduces proliferation of colon cancer cells *in vitro* and *in vivo* by downregulation of β-catenin expression, not via Gli1-dependent transcription (Arimura *et al.*, 2009). Suppressor of fused also negatively regulates Wnt signaling (Meng *et al.*, 2001). Expression of Indian Hh induces differentiation of colorectal epithelial cells and cancer cells (Fu *et al.*, 2009; van den Brink *et al.*, 2004).

Expression of *Shh*, target genes, and pathway components has been reported in colorectal adenocarcinomas (Bian *et al.*, 2007; Douard *et al.*, 2006; Oniscu *et al.*, 2004), and expression of PTCH1 and SMO is associated with tumor progression (Yoshikawa *et al.*, 2009). Activation of Hh signaling has not been detected in seven colon cancer cell lines (Chatel *et al.*, 2007). The difference in Hh signaling activation in colorectal cancer specimens and cell lines may be partly explained by different methylation status of Hh ligand (Fu *et al.*, 2010). Some reports indicate that Hh signaling drives colon cancer growth, recurrence, metastasis, and stem cell expansion (Mazumdar *et al.*, 2010; Varnat *et al.*, 2009). Colorectal cancer cells treated with cyclopamine as well as Gli1 inhibitor GANT61 undergo apoptosis (Mazumdar *et al.*, 2010; Qualtrough *et al.*, 2004). Additionally, cyclopamine increases the cytotoxicity of paclitaxel and ionizing radiation in pancreatic cancer cells with activation of Hh signaling (Shafaee *et al.*, 2006). Therefore, Hh signaling inhibitors such as cyclopamine may be used as an adjuvant agent in combining with radiation therapy for colorectal cancer treatment.

V. Summary

In summary, increasing evidence indicate that Hh signaling is involved in gastrointestinal cancers. Understanding of Hh pathway provides great opportunities for developing novel therapeutic strategies for human gastrointestinal cancers with activated Hh signaling.

REFERENCES

Akiyoshi, T., Nakamura, M., Koga, K., Nakashima, H., Yao, T., Tsuneyoshi, M., Tanaka, M., and Katano, M. (2006). Gli1, downregulated in colorectal cancers, inhibits proliferation of colon cancer cells involving Wnt signalling activation. *Gut* **55,** 991–999.

Arimura, S., Matsunaga, A., Kitamura, T., Aoki, K., Aoki, M., and Taketo, M. M. (2009). Reduced level of smoothened suppresses intestinal tumorigenesis by down-regulation of Wnt signaling. *Gastroenterology* **137,** 629–638.

Bale, A. E. (2002). Hedgehog signaling and human disease. *Annu. Rev. Genomics Hum. Genet.* **3,** 47–65.

Berman, D. M., Karhadkar, S. S., Maitra, A., Montes De Oca, R., Gerstenblith, M. R., Briggs, K., Parker, A. R., Shimada, Y., Eshleman, J. R., Watkins, D. N., and Beachy, P. A. (2003). Widespread requirement for Hedgehog ligand stimulation in growth of digestive tract tumours. *Nature* **425,** 846–851.

Bian, Y. H., Huang, S. H., Yang, L., Ma, X. L., Xie, J. W., and Zhang, H. W. (2007). Sonic hedgehog-Gli1 pathway in colorectal adenocarcinomas. *World J. Gastroenterol.* **13,** 1659–1665.

Brown, L. M., Devesa, S. S., and Chow, W. H. (2008). Incidence of adenocarcinoma of the esophagus among white Americans by sex, stage, and age. *J. Natl. Cancer Inst.* **100,** 1184–1187.

Chang, D. T., Lopez, A., von Kessler, D. P., Chiang, C., Simandl, B. K., Zhao, R., Seldin, M. F., Fallon, J. F., and Beachy, P. A. (1994). Products, genetic linkage and limb patterning activity of a murine hedgehog gene. *Development* **120,** 3339–3353.

Chatel, G., Ganeff, C., Boussif, N., Delacroix, L., Briquet, A., Nolens, G., and Winkler, R. (2007). Hedgehog signaling pathway is inactive in colorectal cancer cell lines. *Int. J. Cancer* **121,** 2622–2627.

Chen, W., Ren, X. R., Nelson, C. D., Barak, L. S., Chen, J. K., Beachy, P. A., de Sauvage, F., and Lefkowitz, R. J. (2004). Activity-dependent internalization of smoothened mediated by beta-arrestin 2 and GRK2. *Science* **306,** 2257–2260.

Dimmler, A., Brabletz, T., Hlubek, F., Hafner, M., Rau, T., Kirchner, T., and Faller, G. (2003). Transcription of sonic hedgehog, a potential factor for gastric morphogenesis and gastric mucosa maintenance, is up-regulated in acidic conditions. *Lab. Invest.* **83,** 1829–1837.

Dlugosz, A. A., and Talpaz, M. (2009). Following the hedgehog to new cancer therapies. *N. Engl. J. Med.* **361,** 1202–1205.

Douard, R., Moutereau, S., Pernet, P., Chimingqi, M., Allory, Y., Manivet, P., Conti, M., Vaubourdolle, M., Cugnenc, P. H., and Loric, S. (2006). Sonic Hedgehog-dependent proliferation in a series of patients with colorectal cancer. *Surgery* **139,** 665–670.

Echelard, Y., Epstein, D. J., St-Jacques, B., Shen, L., Mohler, J., McMahon, J. A., and McMahon, A. P. (1993). Sonic hedgehog, a member of a family of putative signaling molecules, is implicated in the regulation of CNS polarity. *Cell* **75,** 1417–1430.

El-Zaatari, M., Saqui-Salces, M., Waghray, M., Todisco, A., and Merchant, J. L. (2009). Sonic hedgehog in gastric physiology and neoplastic transformation: Friend or foe? *Curr. Opin. Endocrinol. Diabetes Obes.* **16,** 60–65.

El-Zaatari, M., Zavros, Y., Tessier, A., Waghray, M., Lentz, S., Gumucio, D., Todisco, A., and Merchant, J. L. (2010). Intracellular calcium release and protein kinase C activation stimulate sonic hedgehog gene expression during gastric acid secretion. *Gastroenterology* **139**2061.e2–2071.e2.

Faller, G., and Kirchner, T. (2005). Immunological and morphogenic basis of gastric mucosa atrophy and metaplasia. *Virchows Arch.* **446,** 1–9.

Fu, X., Yang, X., and Zhao, L. (2009). Indian hedgehog, a neglected member of hedgehog pathway, may offer a novel avenue for colorectal cancer therapy. *Cancer Biother. Radiopharm.* **24,** 733–735.

Fu, X., Deng, H., Zhao, L., Li, J., Zhou, Y., and Zhang, Y. (2010). Distinct expression patterns of hedgehog ligands between cultured and primary colorectal cancers are associated with aberrant methylation of their promoters. *Mol. Cell. Biochem.* **337,** 185–192.

Fukaya, M., Isohata, N., Ohta, H., Aoyagi, K., Ochiya, T., Saeki, N., Yanagihara, K., Nakanishi, Y., Taniguchi, H., Sakamoto, H., Shimoda, T., Nimura, Y., *et al.* (2006). Hedgehog signal activation in gastric pit cell and in diffuse-type gastric cancer. *Gastroenterology* **131,** 14–29.

Han, M. E., Lee, Y. S., Baek, S. Y., Kim, B. S., Kim, J. B., and Oh, S. O. (2009). Hedgehog signaling regulates the survival of gastric cancer cells by regulating the expression of Bcl-2. *Int. J. Mol. Sci.* **10,** 3033–3043.

Haveri, H., Westerholm-Ormio, M., Lindfors, K., Maki, M., Savilahti, E., Andersson, L. C., and Heikinheimo, M. (2008). Transcription factors GATA-4 and GATA-6 in normal and neoplastic human gastrointestinal mucosa. *BMC Gastroenterol.* **8,** 9.

He, Y. T., Hou, J., Chen, Z. F., Qiao, C. Y., Song, G. H., Meng, F. S., Jin, H. X., and Chen, C. (2005). Decrease in the esophageal cancer incidence rate in mountainous but not level parts of Cixian County, China, over 29 years. *Asian Pac. J. Cancer Prev.* **6,** 510–514.

He, J., Sheng, T., Stelter, A. A., Li, C., Zhang, X., Sinha, M., Luxon, B. A., and Xie, J. (2006). Suppressing Wnt signaling by the hedgehog pathway through sFRP-1. *J. Biol. Chem.* **281,** 35598–35602.

Ingham, P. W., and Placzek, M. (2006). Orchestrating ontogenesis: Variations on a theme by sonic hedgehog. *Nat. Rev. Genet.* **7,** 841–850.

Isohata, N., Aoyagi, K., Mabuchi, T., Daiko, H., Fukaya, M., Ohta, H., Ogawa, K., Yoshida, T., and Sasaki, H. (2009). Hedgehog and epithelial-mesenchymal transition signaling in normal and malignant epithelial cells of the esophagus. *Int. J. Cancer* **125,** 1212–1221.

Jemal, A., Siegel, R., Ward, E., Hao, Y., Xu, J., Murray, T., and Thun, M. J. (2008). Cancer statistics, 2008. *CA Cancer J. Clin.* **58,** 71–96.

Kameda, C., Nakamura, M., Tanaka, H., Yamasaki, A., Kubo, M., Tanaka, M., Onishi, H., and Katano, M. (2010). Oestrogen receptor-alpha contributes to the regulation of the hedgehog signalling pathway in ERalpha-positive gastric cancer. *Br. J. Cancer* **102,** 738–747.

Kang, D. H., Han, M. E., Song, M. H., Lee, Y. S., Kim, E. H., Kim, H. J., Kim, G. H., Kim, D. H., Yoon, S., Baek, S. Y., Kim, B. S., Kim, J. B., *et al.* (2009). The role of hedgehog signaling during gastric regeneration. *J. Gastroenterol.* **44,** 372–379.

Kim, J. H., Huang, Z., and Mo, R. (2005). Gli3 null mice display glandular overgrowth of the developing stomach. *Dev. Dyn.* **234,** 984–991.

Kim, J. H., Choi, Y. J., Lee, S. H., Shin, H. S., Lee, I. O., Kim, Y. J., Kim, H., Yang, W. I., and Lee, Y. C. (2010a). Effect of Helicobacter pylori infection on the sonic hedgehog signaling pathway in gastric cancer cells. *Oncol. Rep.* **23,** 1523–1528.

Kim, J. H., Shin, H. S., Lee, S. H., Lee, I., Lee, Y. S., Park, J. C., Kim, Y. J., Chung, J. B., and Lee, Y. C. (2010b). Contrasting activity of Hedgehog and Wnt pathways according to gastric cancer cell differentiation: Relevance of crosstalk mechanisms. *Cancer Sci.* **101,** 328–335.

Kinzler, K. W., and Vogelstein, B. (1990). The GLI gene encodes a nuclear protein which binds specific sequences in the human genome. *Mol. Cell. Biol.* **10,** 634–642.

Kolterud, A., Grosse, A. S., Zacharias, W. J., Walton, K. D., Kretovich, K. E., Madison, B. B., Waghray, M., Ferris, J. E., Hu, C., Merchant, J. L., Dlugosz, A. A., Kottmann, A. H., *et al.* (2009). Paracrine Hedgehog signaling in stomach and intestine: New roles for hedgehog in gastrointestinal patterning. *Gastroenterology* **137,** 618–628.

Krauss, S., Concordet, J. P., and Ingham, P. W. (1993). A functionally conserved homolog of the Drosophila segment polarity gene hh is expressed in tissues with polarizing activity in zebrafish embryos. *Cell* **75,** 1431–1444.

Lee, S. Y., Han, H. S., Lee, K. Y., Hwang, T. S., Kim, J. H., Sung, I. K., Park, H. S., Jin, C. J., and Choi, K. W. (2007). Sonic hedgehog expression in gastric cancer and gastric adenoma. *Oncol. Rep.* **17,** 1051–1055.

Lee, K. M., Lee, J. S., Jung, H. S., Park, D. K., Park, H. S., and Hahm, K. B. (2010). Late reactivation of sonic hedgehog by Helicobacter pylori results in population of gastric epithelial cells that are resistant to apoptosis: Implication for gastric carcinogenesis. *Cancer Lett.* **287,** 44–53.

Liu, H., Gu, D., and Xie, J. (2011). Clinical implications of hedgehog signaling pathway inhibitors. *Chin. J. Cancer* **30,** 13–26.

Ma, X., Chen, K., Huang, S., Zhang, X., Adegboyega, P. A., Evers, B. M., Zhang, H., and Xie, J. (2005). Frequent activation of the hedgehog pathway in advanced gastric adenocarcinomas. *Carcinogenesis* **26,** 1698–1705.

Ma, X., Sheng, T., Zhang, Y., Zhang, X., He, J., Huang, S., Chen, K., Sultz, J., Adegboyega, P. A., Zhang, H., and Xie, J. (2006). Hedgehog signaling is activated in subsets of esophageal cancers. *Int. J. Cancer* **118,** 139–148.

Madison, B. B., McKenna, L. B., Dolson, D., Epstein, D. J., and Kaestner, K. H. (2009). FoxF1 and FoxL1 link hedgehog signaling and the control of epithelial proliferation in the developing stomach and intestine. *J. Biol. Chem.* **284,** 5936–5944.

Mazumdar, T., Devecchio, J., Ting, S., Jones, J., Agyeman, A., and Houghton, J. A. (2010). Hedgehog signaling drives cellular survival in human colon carcinoma cells. *Cancer Res.* **71,** 1092–1102.

Meng, X., Poon, R., Zhang, X., Cheah, A., Ding, Q., Hui, C. C., and Alman, B. (2001). Suppressor of fused negatively regulates beta-catenin signaling. *J. Biol. Chem.* **276,** 40113–40119.

Minegishi, Y., Suzuki, H., Arakawa, M., Fukushima, Y., Masaoka, T., Ishikawa, T., Wright, N. A., and Hibi, T. (2007). Reduced Shh expression in TFF2-overexpressing lesions of the gastric fundus under hypochlorhydric conditions. *J. Pathol.* **213,** 161–169.

Mori, Y., Okumura, T., Tsunoda, S., Sakai, Y., and Shimada, Y. (2006). Gli-1 expression is associated with lymph node metastasis and tumor progression in esophageal squamous cell carcinoma. *Oncology* **70,** 378–389.

Nishizawa, T., Suzuki, H., Masaoka, T., Minegishi, Y., Iwasahi, E., and Hibi, T. (2007). Helicobacter pylori eradication restored sonic hedgehog expression in the stomach. *Hepatogastroenterology* **54,** 697–700.

Nishizawa, T., Suzuki, H., Nakagawa, I., Minegishi, Y., Masaoka, T., Iwasaki, E., and Hibi, T. (2009). Early Helicobacter pylori eradication restores sonic hedgehog expression in the gastric mucosa of Mongolian gerbils. *Digestion* **79,** 99–108.

Nusslein-Volhard, C., and Wieschaus, E. (1980). Mutations affecting segment number and polarity in Drosophila. *Nature* **287,** 795–801.

Ohta, M., Tateishi, K., Kanai, F., Watabe, H., Kondo, S., Guleng, B., Tanaka, Y., Asaoka, Y., Jazag, A., Imamura, J., Ijichi, H., Ikenoue, T., *et al.* (2005). p53-independent negative regulation of p21/cyclin-dependent kinase-interacting protein 1 by the sonic hedgehog-glioma-associated oncogene 1 pathway in gastric carcinoma cells. *Cancer Res.* **65,** 10822–10829.

Ohta, H., Aoyagi, K., Fukaya, M., Danjoh, I., Ohta, A., Isohata, N., Saeki, N., Taniguchi, H., Sakamoto, H., Shimoda, T., Tani, T., Yoshida, T., *et al.* (2009). Cross talk between hedgehog and epithelial-mesenchymal transition pathways in gastric pit cells and in diffuse-type gastric cancers. *Br. J. Cancer* **100,** 389–398.

Oniscu, A., James, R. M., Morris, R. G., Bader, S., Malcomson, R. D., and Harrison, D. J. (2004). Expression of Sonic hedgehog pathway genes is altered in colonic neoplasia. *J. Pathol.* **203,** 909–917.

Osawa, H., Ohnishi, H., Takano, K., Noguti, T., Mashima, H., Hoshino, H., Kita, H., Sato, K., Matsui, H., and Sugano, K. (2006). Sonic hedgehog stimulates the proliferation of rat gastric mucosal cells through ERK activation by elevating intracellular calcium concentration. *Biochem. Biophys. Res. Commun.* **344,** 680–687.

Qualtrough, D., Buda, A., Gaffield, W., Williams, A. C., and Paraskeva, C. (2004). Hedgehog signalling in colorectal tumour cells: Induction of apoptosis with cyclopamine treatment. *Int. J. Cancer* **110,** 831–837.

Riddle, R. D., Johnson, R. L., Laufer, E., and Tabin, C. (1993). Sonic hedgehog mediates the polarizing activity of the ZPA. *Cell* **75,** 1401–1416.

Rizvi, S., Demars, C. J., Comba, A., Gainullin, V. G., Rizvi, Z., Almada, L. L., Wang, K., Lomberk, G., Fernandez-Zapico, M. E., and Buttar, N. S. (2010). Combinatorial chemoprevention reveals a novel smoothened-independent role of GLI1 in esophageal carcinogenesis. *Cancer Res.* **70,** 6787–6796.

Roelink, H., Augsburger, A., Heemskerk, J., Korzh, V., Norlin, S., Ruiz i Altaba, A., Tanabe, Y., Placzek, M., Edlund, T., Jessell, T. M., *et al.* (1994). Floor plate and motor neuron induction by vhh-1, a vertebrate homolog of hedgehog expressed by the notochord. *Cell* **76,** 761–775.

Sasaki, H., Hui, C., Nakafuku, M., and Kondoh, H. (1997). A binding site for Gli proteins is essential for HNF-3beta floor plate enhancer activity in transgenics and can respond to Shh in vitro. *Development* **124,** 1313–1322.

Seto, M., Ohta, M., Asaoka, Y., Ikenoue, T., Tada, M., Miyabayashi, K., Mohri, D., Tanaka, Y., Ijichi, H., Tateishi, K., Kanai, F., Kawabe, T., *et al.* (2009). Regulation of the hedgehog signaling by the mitogen-activated protein kinase cascade in gastric cancer. *Mol. Carcinog.* **48,** 703–712.

Shafaee, Z., Schmidt, H., Du, W., Posner, M., and Weichselbaum, R. (2006). Cyclopamine increases the cytotoxic effects of paclitaxel and radiation but not cisplatin and gemcitabine in Hedgehog expressing pancreatic cancer cells. *Cancer Chemother. Pharmacol.* **58,** 765–770.

Sheridan, C. (2009). Genentech obtains proof of concept for hedgehog inhibition. *Nat. Biotechnol.* **27,** 968–969.

Shiotani, A., Iishi, H., Uedo, N., Ishiguro, S., Tatsuta, M., Nakae, Y., Kumamoto, M., and Merchant, J. L. (2005). Evidence that loss of sonic hedgehog is an indicator of Helicobacter pylori-induced atrophic gastritis progressing to gastric cancer. *Am. J. Gastroenterol.* **100,** 581–587.

Shiotani, A., Uedo, N., Iishi, H., Tatsuta, M., Ishiguro, S., Nakae, Y., Kamada, T., Haruma, K., and Merchant, J. L. (2007). Re-expression of sonic hedgehog and reduction of CDX2 after Helicobacter pylori eradication prior to incomplete intestinal metaplasia. *Int. J. Cancer* **121,** 1182–1189.

Sims-Mourtada, J., Izzo, J. G., Apisarnthanarax, S., Wu, T. T., Malhotra, U., Luthra, R., Liao, Z., Komaki, R., van der Kogel, A., Ajani, J., and Chao, K. S. (2006). Hedgehog: An attribute to tumor regrowth after chemoradiotherapy and a target to improve radiation response. *Clin. Cancer Res.* **12,** 6565–6572.

Su, Y., Chen, X., Klein, M., Fang, M., Wang, S., Yang, C. S., and Goyal, R. K. (2004). Phenotype of columnar-lined esophagus in rats with esophagogastroduodenal anastomosis: Similarity to human Barrett's esophagus. *Lab. Invest.* **84,** 753–765.

Sui, G., Bonde, P., Dhara, S., Broor, A., Wang, J., Marti, G., Feldmann, G., Duncan, M., Montgomery, E., Maitra, A., and Harmon, J. W. (2006). Epidermal growth factor receptor and hedgehog signaling pathways are active in esophageal cancer cells from rat reflux model. *J. Surg. Res.* **134,** 1–9.

Takamoto, N., You, L. R., Moses, K., Chiang, C., Zimmer, W. E., Schwartz, R. J., DeMayo, F. J., Tsai, M. J., and Tsai, S. Y. (2005). COUP-TFII is essential for radial and anteroposterior patterning of the stomach. *Development* **132,** 2179–2189.

Thayer, S. P., di Magliano, M. P., Heiser, P. W., Nielsen, C. M., Roberts, D. J., Lauwers, G. Y., Qi, Y. P., Gysin, S., Fernandez-del Castillo, C., Yajnik, V., Antoniu, B., McMahon, M., *et al.* (2003). Hedgehog is an early and late mediator of pancreatic cancer tumorigenesis. *Nature* **425,** 851–856.

van den Brink, G. R. (2007). Hedgehog signaling in development and homeostasis of the gastrointestinal tract. *Physiol. Rev.* **87,** 1343–1375.

van den Brink, G. R., Bleuming, S. A., Hardwick, J. C., Schepman, B. L., Offerhaus, G. J., Keller, J. J., Nielsen, C., Gaffield, W., van Deventer, S. J., Roberts, D. J., and Peppelenbosch, M. P. (2004). Indian Hedgehog is an antagonist of Wnt signaling in colonic epithelial cell differentiation. *Nat. Genet.* **36,** 277–282.

Varnat, F., Duquet, A., Malerba, M., Zbinden, M., Mas, C., Gervaz, P., and Ruiz i Altaba, A. (2009). Human colon cancer epithelial cells harbour active HEDGEHOG-GLI signalling that is essential for tumour growth, recurrence, metastasis and stem cell survival and expansion. *EMBO Mol. Med.* **1,** 338–351.

Varnat, F., Siegl-Cachedenier, I., Malerba, M., Gervaz, P., and Ruiz i Altaba, A. (2010). Loss of WNT-TCF addiction and enhancement of HH-GLI1 signalling define the metastatic transition of human colon carcinomas. *EMBO Mol. Med.* **2,** 440–457.

Von Hoff, D. D., LoRusso, P. M., Rudin, C. M., Reddy, J. C., Yauch, R. L., Tibes, R., Weiss, G. J., Borad, M. J., Hann, C. L., Brahmer, J. R., Mackey, H. M., Lum, B. L., *et al.* (2009). Inhibition of the hedgehog pathway in advanced basal-cell carcinoma. *N. Engl. J. Med.* **361,** 1164–1172.

Waghray, M., Zavros, Y., Saqui-Salces, M., El-Zaatari, M., Alamelumangapuram, C. B., Todisco, A., Eaton, K. A., and Merchant, J. L. (2010). Interleukin-1beta promotes gastric atrophy through suppression of Sonic Hedgehog. *Gastroenterology* **138,** 562–572, 572e1–e2.

Wang, L. H., Choi, Y. L., Hua, X. Y., Shin, Y. K., Song, Y. J., Youn, S. J., Yun, H. Y., Park, S. M., Kim, W. J., Kim, H. J., Choi, J. S., and Kim, S. H. (2006). Increased expression of sonic hedgehog and altered methylation of its promoter region in gastric cancer and its related lesions. *Mod. Pathol.* **19,** 675–683.

Wang, D. H., Clemons, N. J., Miyashita, T., Dupuy, A. J., Zhang, W., Szczepny, A., Corcoran-Schwartz, I. M., Wilburn, D. L., Montgomery, E. A., Wang, J. S., Jenkins, N. A., Copeland, N. A., *et al.* (2010). Aberrant epithelial-mesenchymal Hedgehog signaling characterizes Barrett's metaplasia. *Gastroenterology* **138,** 1810–1822.

Wei, L., and Xu, Z. (2010). Cross-signaling among phosphinositide-3 kinase, mitogen-activated protein kinase and sonic hedgehog pathways exists in esophageal cancer. *Int. J. Cancer* **129,** 275–284.

Wilson, S. R., Strand, M. F., Krapp, A., Rise, F., Petersen, D., and Krauss, S. (2010). Hedgehog antagonist cyclopamine isomerizes to less potent forms when acidified. *J. Pharm. Biomed. Anal.* **52,** 707–713.

Xiao, C., Ogle, S. A., Schumacher, M. A., Schilling, N., Tokhunts, R. A., Orr-Asman, M. A., Miller, M. L., Robbins, D. J., Hollande, F., and Zavros, Y. (2010). Hedgehog signaling regulates E-cadherin expression for the maintenance of the actin cytoskeleton and tight junctions. *Am. J. Physiol. Gastrointest. Liver Physiol.* **299,** G1252–G1265.

Xie, J. (2005). Hedgehog signaling in prostate cancer. *Future Oncol.* **1,** 331–338.
Xie, J. (2008a). Hedgehog signaling pathway: Development of antagonists for cancer therapy. *Curr. Oncol. Rep.* **10,** 107–113.
Xie, J. (2008b). Molecular biology of basal and squamous cell carcinomas. *Adv. Exp. Med. Biol.* **624,** 241–251.
Xie, K., and Abbruzzese, J. L. (2003). Developmental biology informs cancer: The emerging role of the hedgehog signaling pathway in upper gastrointestinal cancers. *Cancer Cell* **4,** 245–247.
Yanai, K., Nagai, S., Wada, J., Yamanaka, N., Nakamura, M., Torata, N., Noshiro, H., Tsuneyoshi, M., Tanaka, M., and Katano, M. (2007). Hedgehog signaling pathway is a possible therapeutic target for gastric cancer. *J. Surg. Oncol.* **95,** 55–62.
Yanai, K., Nakamura, M., Akiyoshi, T., Nagai, S., Wada, J., Koga, K., Noshiro, H., Nagai, E., Tsuneyoshi, M., Tanaka, M., and Katano, M. (2008). Crosstalk of hedgehog and Wnt pathways in gastric cancer. *Cancer Lett.* **263,** 145–156.
Yoo, Y. A., Kang, M. H., Kim, J. S., and Oh, S. C. (2008). Sonic hedgehog signaling promotes motility and invasiveness of gastric cancer cells through TGF-beta-mediated activation of the ALK5-Smad 3 pathway. *Carcinogenesis* **29,** 480–490.
Yoshikawa, R., Nakano, Y., Tao, L., Koishi, K., Matsumoto, T., Sasako, M., Tsujimura, T., Hashimoto-Tamaoki, T., and Fujiwara, Y. (2008). Hedgehog signal activation in oesophageal cancer patients undergoing neoadjuvant chemoradiotherapy. *Br. J. Cancer* **98,** 1670–1674.
Yoshikawa, K., Shimada, M., Miyamoto, H., Higashijima, J., Miyatani, T., Nishioka, M., Kurita, N., Iwata, T., and Uehara, H. (2009). Sonic hedgehog relates to colorectal carcinogenesis. *J. Gastroenterol.* **44,** 1113–1117.

CHAPTER TWENTY-ONE

The Role of Sonic Hedgehog as a Regulator of Gastric Function and Differentiation

Rui Feng,[1] Chang Xiao,[1] *and* Yana Zavros

Contents

Abstract

The Hedgehog (Hh) genes play a key role in the regulation of embryonic development and govern processes such as cell differentiation, cell proliferation, and tissue patterning. In vertebrate embryos, Hh gene expression regulates correct formation of limbs, skeleton, muscles, and organs including stomach. In the adult, the Hh pathway functions in tissue repair and regeneration, along with maintenance of stem cells. Sonic Hedgehog (Shh) signaling has been extensively studied for its role in developmental and cancer biology. Recent advances in the field of gastroenterology show that in the stomach, Shh is responsible for proper differentiation of the gastric glands. The aberrant activity of the Shh signaling pathway leads to an altered gastric differentiation program and loss of gastric acid secretion that is the predominant function of the stomach. In this chapter, we review the most recent findings that reveal the role of Shh as a regulator of gastric function and differentiation and how this signaling is dysregulated during the development of gastric cancer in response bacterial infection.

Department of Molecular and Cellular Physiology, University of Cincinnati College of Medicine, Cincinnati, Ohio, USA
[1] Authors contributed equally.

Vitamins and Hormones, Volume 88
ISSN 0083-6729, DOI: 10.1016/B978-0-12-394622-5.00021-3

I. Introduction

In 1980, using a mutagenesis screen in *Drosophila*, Nüsslein-Volhard and Wieschaus first discovered the Hedgehog (Hh) gene (Nüsslein-Volhard and Wieschaus, 1980). In these experiments, it was demonstrated that one of these mutants alters the patterning within the segment in the larva. In the wild-type *Drosophila*, each segment contained an anterior half covered with the presence of a band of denticles and a posterior half that was smooth. In the mutant, each segment was depleted of the smooth posterior half with only the anterior half present, causing the entire larva to be covered by denticles that resembled a "Hedgehog," and thus the origin of this funny name (Nüsslein-Volhard and Wieschaus, 1980). Later, the corresponding Hh gene was described in *Drosophila*, followed by the identification of three Hh homologs that are highly conserved from the fruit fly to humans. The three homologs were named Sonic Hedgehog (Shh), named after a Sega game character, Desert Hedgehog (Dhh), based on an Egyptian species of Hh and Indian Hedgehog (Ihh), a species of Hh that resides in Pakistan. Although little is known about how the transcription of these genes is controlled, a quite comprehensive recognition about the protein products has been achieved by intensive studies, such as the protein secretion, the signaling induction, and the physiological function. In the stomach, Hh is known to control cellular functions including proliferation, differentiation, and acid secretion. Therefore, identifying the role of Shh as a regulator of adult stomach physiological function and differentiation is important for understanding the pathogenesis of gastric disease.

II. Hh in the Adult Stomach

A. Shh expression and processing

Shh is synthesized as a 45kDa precursor protein, containing a 19kDa amino-terminal fragment and a 26kDa carboxy-terminal fragment. The 26kDa fragment is the known enzymatic component that cleaves the precursor protein into two fragments and catalyzes the transfer of a cholesterol moiety to the carboxy terminus of the 19kDa fragment. The cholesterol-modified 19kDa fragment contains all signaling activities but requires further modification by a palmitate group at its amino-terminal end (Martí *et al.*, 1995; Lee *et al.*, 1994; Porter *et al.*, 1995). This palmitoylation is catalyzed by a membrane-bound *O*-acyltransferase that is commonly known as Skinny Hedgehog (Ski) and is essential to the biological activity of the processed Hh proteins (Chen *et al.*, 2004).

Although most of the initial studies of Hh processing were performed in invertebrate systems, our understanding of Hh processing in the mammalian system has been advanced by Zavros *et al.* (2007). By using primary parietal cell cultures and tissue extracts from mouse and human stomachs, Zavros *et al.* (2007) demonstrated that Shh is process by an acid- and protease-dependent mechanism. Pepsin A processing is acid dependent where acid facilitates the conversion from the inactive pre-enzyme pepsinogen A to the active enzyme pepsin. The investigators explain that the absence of processed Shh in human gastric cancer tissue may be a consequence of hypochlorhydria present due to the parietal cell atrophy (Zavros *et al.*, 2007). Interestingly, recently, it is revealed that the unprocessed full-length Shh precursor protein is also biologically active (Tokhunts *et al.*, 2010). It is shown that full-length Shh also binds to Patched (Ptch) and induces the similar target gene expression in the developing chick neural tube and rescues the eye developmental defect in *Drosophila* Hh mutants (Tokhunts *et al.*, 2010). Whether the full-length and processed Shh ligands play the same role in the adult tissues is unknown but may have significant implications in the role of Shh in the proliferation of cancer stem cells.

Shh processing is also shown to be regulated by gastrin (El-Zaatari *et al.*, 2010; Zavros *et al.*, 2007). Studies using a mouse model of gastrin deficiency show that there is a significant decrease in Shh protein expression, and the reinfusion of gastrin successfully restored Shh expression, indicating the important regulatory role of gastrin (Zavros *et al.*, 2007). In addition, recent studies by Xiao *et al.* using a mouse model expressing a parietal cell-specific deletion of Shh (HKCre/Shh^{KO}) demonstrated that these mice developed severe hypergastrinemia reflective of increased circulating gastrin concentration (Xiao *et al.*, 2010a,b). Xiao *et al.* also showed that with loss of Shh and accompanied hypergastrinemia, mice expressed increased Ihh within the surface mucous pit epithelium. Somatostatin is a known physiological inhibitor of gastrin and when HKCre/Shh^{KO} mice were treated with the somatostatin analogue octreotide, hypergastrinemia was normalized and Ihh expression decreased to levels similar to controls groups (Xiao *et al.*, 2010a,b). Such data suggested that gastrin not only regulates Shh processing but also has the potential to also regulate Ihh.

Using isolated canine parietal cell and mouse organ cultures to identify the precise mechanism that gastrin regulates Shh, El-Zaatari *et al.* recently demonstrated that intracellular calcium release and protein kinase C (PKC) activation stimulate Shh gene expression (El-Zaatari *et al.*, 2010). Based on previous reports that intracellular calcium release stimulates acid secretion in parietal cells, gadolinium-, thapsigargin-, and carbachol-mediated release of intracellular calcium induced Shh expression (El-Zaatari *et al.*, 2010).

The role of the gastric hormone gastrin in acid secretion of the stomach was first recognized by Edkins (1905). The gastrin gene located on chromosomes 17q21 is expressed by a variety of cell in the body. Among them, gastrin is predominantly manufactured by the antral mucosa of the stomach, where G-cells secrete this peptide hormone for the purpose of stimulating gastric acid secretion (McGuigan, 1968; Taylor *et al.*, 1982). The actions of gastrin are mediated by G-protein-coupled receptors, cholecystokinin (CCK)/gastrin receptors. Two types of CCK receptors, CCK_1-R and CCK_2-R (formerly CCKAR and CCKBR), have been identified. CCK_1-R has high affinity for CCK (reviewed in Noble and Roques, 1999), another gastrointestinal hormone regulating digestion. The CCK_2-R, originally defined in central nervous system, has a high affinity for both CCK and gastrin (Dufresne *et al.*, 2006a,b).

In stomach, CCK_2-R is mainly located on parietal cells (Kopin *et al.*, 1992) and enterochromaffin-like (ECL) cells within gastric mucosa, where it mediates action of gastrin in acid secretion (reviewed in Dockray *et al.*, 2005). The precise cellular localization of the CCK_2-R by *in situ* hybridization and immunohistochemistry demonstrates that expression is also detected in gastric precursor cells (Kulaksiz *et al.*, 2000). Presence of CCK_2-R on precursor cells supports a number of previously unrecognized roles for gastrin as a regulator of cellular proliferation, migration, and differentiation.

Like typical G-protein-coupled receptor, gastrin binding with CCK_2-R activates classical signaling pathway, such as phospholipase C (PLC)-β/diacylglycerol (DAG)/Ca^{2+}/PKC cascade or the adenyl cyclase (AC) pathway (reviewed in Dufresne *et al.*, 2006a,b). Binding of gastrin activates Gq-coupled receptor, which subsequently activates the PLC family (principally PLC-β isoforms). PLC then hydrolyzed plasma membrane-bound phosphatidyl inositol 4,5-bisphosphate (PIP_2) into the second messengers inositol trisphophate (IP_3) and DAG. IP_3 binds to a ligand-gated Ca^{2+} channels in ER and release Ca^{2+} from intracellular stores, while DAG activates PKC (Roche *et al.*, 1990; Yassin and Murthy, 1991). PKC and/or Ca^{2+} mobilization, in turn, activate a cascade of mitogen-activated protein kinases (MAPKs), including extracellular signal-related kinase (ERK) and c-jun NH_2-terminal kinase (JUK) in several studied cell models (Hocker, 2004; Höcker *et al.*, 1997). Activation of ERK-dependent transcription factors has been shown to be involved in cell growth, differentiation, survival, and apoptosis (Daulhac *et al.*, 1999; Dockray *et al.*, 2005). CCK_2-R also induces several other signaling pathways that are activated by tyrosine kinase receptors (Dabrowski *et al.*, 1997; Stepan *et al.*, 1999). In addition to the signaling pathways described above that mediate acid secretion, cellular proliferation, and migration, the transcription factor NF-κB activated by CCK_2-R leads to the expression of interleukin-8 and other proinflammatory gene products (Hiraoka *et al.*, 2001). Through signaling, gastrin increases Ca^{2+} release in target cells (Tsunoda *et al.*, 1988). Thus, the increase of intracellular Ca^{2+} by gastrin may also be a plausible mechanism regulating Ihh gene expression but requires further investigation (Fig. 21.1).

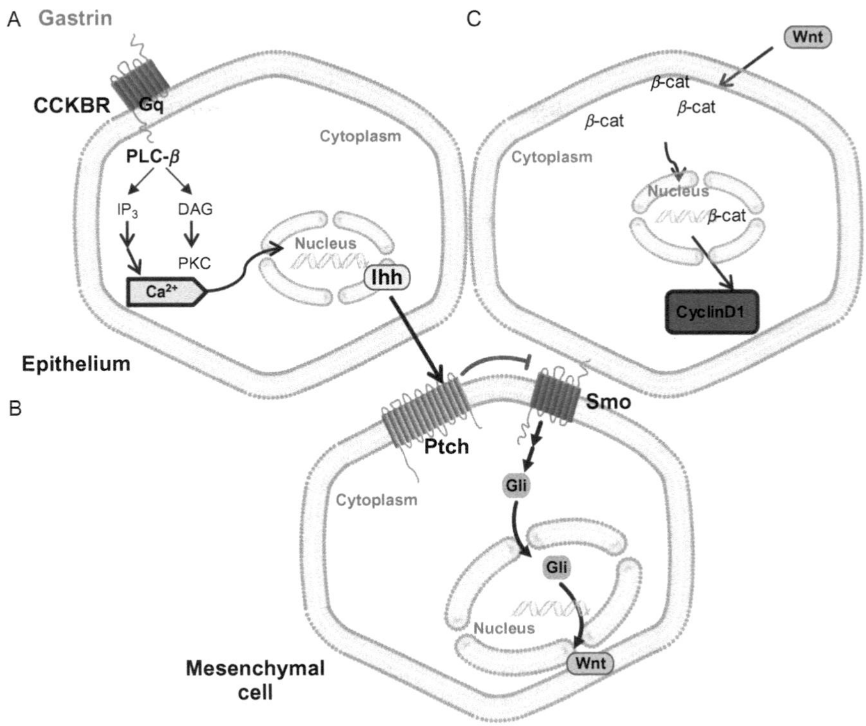

Figure 21.1 Proposed mechanism for the role of Ihh as a regulator of gastrin-induced proliferation. (A) Gastrin binding with Gq-associated CCKBR activates phospholipase C (PLC)-β which may regulate calcium mobilization by producing inositol trisphosphate (IP_3) and diacylglycerol (DAG). Ihh may be induced by gastrin in PKC/Ca^{2+}-dependent mechanism. (B) Ihh secreted from the surface pit epithelium targets Gli positive mesenchymal cells, inducing activation of Wnt pathway. (C) Activation of the Wnt pathway causes translocation of β-catenin from cytoplasm to nucleus and production of CyclinD1. Wnt-mediated CyclinD1 is a key regulator of cell proliferation. (For color version of this figure, the reader is referred to the Web version of this chapter.)

B. Hh signaling in the adult stomach

In vertebrates, the Hh signal is triggered by Hh proteins and transmitted by Smoothened (Smo) that is a seven-span transmembrane receptor. However, Hh proteins do not activate Smo through a direct interaction. Instead, they regulate the activity of Smo via a Ptch-mediated mechanism. Ptch is a 12-span transmembrane protein receptor for Hh. In the absence of Hh, Ptch inhibits Smo (Bijlsma *et al.*, 2006). Binding of Hh to Ptch terminates the inhibitory restraint on Smo and enables Smo to relay the signaling cascade. Activated Smo prevents the processing of Hh transcription factor—Glioblastoma (Gli) family proteins into their repressor forms (Gli-R), causing the accumulation of the transcriptional activators (Gli-A). Gli-A then induces the transcription of

Hh-responsive target genes including some Hh pathway components themselves, such as Ptch and Gli1 (reviewed in van den Brink, 2007).

Recent studies have provided emerging evidence for the existence of a "noncanonical" pathway that may be Gli- and Smo independent. Studies of nervous system development in chick embryos showed that Shh expression is crucial for normal neural cell migration while Smo antagonist cyclopamine or interference RNA targeting Smo did not alter this process, providing evidence for the involvement of Shh in a noncanonical and Smo-independent pathway (Bourikas *et al.*, 2005; Jarov *et al.*, 2003; Testaz *et al.*, 2001). In a Gli1 null mutant mouse model, injection of Shh-expressing retroviral vectors into the embryonic central nervous system induces medulloblastoma, suggesting Gli1 is not required for Shh to initiate tumorigenesis in the brain (Weiner *et al.*, 2002). Further support for the existence of such a noncanonical Hh signaling pathway comes from a recent *in vitro* study using C3H10T1/2 mouse fibroblast cells treated with soluble Shh protein. Shh-treated mouse fibroblast cells exhibited actin cytoskeletal reorganization within 10min, a time frame much shorter than required to induce the expression of Hh-responsive target genes after treatment (Bijlsma *et al.*, 2007). Another *in vitro* study using 293T cells identifies a physiological interaction between Ptch1 and CyclinB1, a critical cell division regulator whose phosphorylation and nuclear entry promote G2/M phase cell cycle progression. In this study, the Ptch1–CyclinB1 complex exhibits an antiproliferative effect on 293T cells. Exposure of these cells to Shh dissociates this complex, relocalizes CyclinB1 into the nucleus, and reverses the cell proliferation, suggesting Shh and Ptch1 participate in G2/M phase checkpoint through regulation of the cellular distribution of CyclinB1 (Barnes *et al.*, 2001). The interaction between Ptch and CyclinB1 has been further confirmed in Madin–Darby canine kidney (MDCK) cells where Ptch1 is colocalized with endogenous CyclinB1 at the apical surface. Treatment of MDCK cells with Shh causes degradation of Ptch1 accompanied by cytoplasmic and nuclear relocalization of CyclinB1 (Jenkins, 2009). In an *in vivo*, Ptch1−/− mouse model, ablation of Ptch1 results in increased number in basal cells expressing nuclear CyclinB1 and formation of skin tumors arising from Ptch1-ablated cells, further supporting the role of Ptch in G2/M phase checkpoint (Adolphe *et al.*, 2006). Collectively, these data suggest the existence of a noncanonical Hh signaling pathway that regulates cell proliferation in a Smo- and Gli-independent and CyclinB1-mediated mechanism.

C. Hh as a regulator of gastric epithelial cell function and differentiation

Several studies indicate that Shh functions as a morphogen in the adult stomach to guide gastric epithelial cell differentiation and function. The first of these studies treated mice with the Hh inhibitor cyclopamine demonstrating enhanced gastric epithelial cell proliferation and diminished protein expression of BMP4, islet-1, and Hnf3β, all of which are transcription targets of Shh signaling participating in tissue-specific differentiation (van den Brink *et al.*, 2001). Subsequent studies further indicated that downregulating Shh expression is a mechanism by which *Helicobacter pylori* infection triggers gastric cancer, since the consequence of *H. pylori* infection is parietal cell atrophy and gastric hypochlorhydria. Shh positive cells are dramatically reduced in the intestinal metaplastic areas of *H. pylori*-positive human stomach (Shiotani *et al.*, 2005a,b). Other studies suggest that *H. pylori* infection may be the direct cause of downregulation of Shh in human gastric mucosa and that the early eradication restores the expression of Shh and the differentiation of the stomach (Shiotani *et al.*, 2006, 2007). The causative relationship between *H. pylori* infection and decreased Shh expression has been further confirmed by a study in Mongolian gerbils (Suzuki *et al.*, 2005), an established animal model of *H. pylori* infection recapitulating the histopathological changes observed in humans (Hirayama *et al.*, 1996; Watanabe *et al.*, 1998). Recent work by Waghray *et al.* (2010) using a mouse model of *H. pylori* infection, it was demonstrated that Shh gene expression in parietal cells is suppressed and precedes parietal cell atrophy. In the study, the inhibitory effect of *H. pylori* on Shh expression was mediated by the proinflammatory cytokine interleukin-1β via a reduction in acid secretion and intracellular calcium release (Waghray *et al.*, 2010). Together, these studies support the hypothesis of the *H. pylori* infection results in loss of Shh that subsequently contributes to preneoplastic transformation and eventually the development of gastric cancer. Nevertheless, all above evidence are built on the observations or studies that are associated with *H. pylori* infection. *H. pylori* infection induces not only repressed Shh expression but also complicated immune and inflammatory response, thus it is hard to verify the weight of loss of Shh *per se* in contributing to the development of gastric cancer.

The development of a mouse model expressing a parietal cell-specific deletion of Shh (HKCre/Shh^{KO} mice) has allowed us to assay changes in gastric epithelial cell differentiation and function independent of factors such as inflammation induced by *H. pylori* infection (Xiao *et al.*, 2010a,b). Compared to the normal gastric pathology of the control animals in these studies, every HKCre/Shh^{KO} mouse demonstrated an age-dependent increase in the number of surface pit mucous cells that was reminiscent of foveolar hyperplasia. Interestingly, the phenotype observed in the HKCre/Shh^{KO} mouse was similar to that observed in the overexpressing TGFα transgenic mice (Bockman *et al.*, 1995; Goldenring *et al.*, 1996; Nomura *et al.*, 2005) and in patients with Menetrier's disease (Larsen *et al.*, 1987; Wolfsen *et al.*, 1993). However, unlike both Menetrier's disease and *H. pylori*-infected patients, the HKCre/Shh^{KO} mouse model did not develop atrophy. Given that HKCre/Shh^{KO} mice lacked inflammation, additional factors are required, such as inflammatory cytokines, for parietal cell atrophy to occur (Xiao *et al.*, 2010a,b).

Emerging evidence also suggests that regulating the formation of cell junctions may be one of the mechanisms by which Hh signaling controls the homeostasis of gastric epithelium. The maintenance of functional adherens- and tight junctions is essential for epithelial morphology, physiological function, and differentiation. A study in rat kidney epithelial cells (RK3E) shows that Snail, a suppressor of E-cadherin, is a transcriptional target of Hh signaling (Li *et al.*, 2006), suggesting Hh signaling is involved in the regulation of E-cadherin expression. E-cadherin is a transmembrane protein with its intracellular segment directly binding to β-catenin, which, in turn, binds to actin filament binding protein α-catenin, to form the adherens junction complex. The adherens junction cadherin/β-catenin/α-catenin protein complex is morphologically associated with actin filaments. Loss of E-cadherin expression has been involved in epithelial-to-mesenchymal transition (EMT). EMT is characterized by loss of connections between cells, expression of mesenchymal markers within epithelial cells, and increased cell motility. This molecular event is necessary for embryonic organogenesis, inflammatory response, and normal wound healing; however, under certain pathological condition such as cancer, EMT contributes to tumor invasion and metastasis. In immortalized human pancreatic ductal epithelial cells, Gli1 overexpression causes almost a complete loss of E-cadherin and an invasive phenotype of transformation. In contrast, in invasive pancreatic cancer cell lines, treatment of cyclopamine represses Snail and restores E-cadherin expression, thus inhibiting EMT and invasive capacity *in vitro* (Feldmann *et al.*, 2007). Loss of membrane-bound E-cadherin and abnormal β-catenin localization is also observed in gastric cancer with a correlation with the risk of early submucosal invasion and lymphoid node metastasis (Cheng *et al.*, 2005; Tanaka *et al.*, 2002).

The integrity of adherens junctions is a crucial factor determining epithelial morphology, physiological function, and differentiation (Braga, 2002; Braga *et al.*, 1997; Hopkins *et al.*, 2003; Samarin *et al.*, 2007). The HKCre/Shh^{KO} mice were later used to demonstrate that Shh signaling regulates cell–cell adhesion (Xiao *et al.*, 2010a,b). Recent studies from our laboratory using the HKCre/Shh^{KO} demonstrated that the loss of Shh triggers a number of molecular changes (Xiao *et al.*, 2010a,b) that were consistent with EMT of gastric epithelial cells (Ikenouchi *et al.*, 2003; Li *et al.*, 2006). Such molecular changes included loss of E-cadherin expression and translocation of β-catenin to the nucleus (Xiao *et al.*, 2010a,b). Nuclear translocation of β-catenin binds to the DNA-binding proteins Tcf/Lef1 that subsequently regulates target genes including CyclinD1 (Nelson and Nusse, 2004) important in proliferation (Luo *et al.*, 2008). Consistent with our published data, the HKCre/Shh^{KO} mice also exhibited hyperproliferation of the surface mucous cells (Xiao *et al.*, 2010a,b). Hyperproliferation of surface mucous cells often disrupts the differentiation of other cell lineages such as the zymogen cells (Bredemeyer *et al.*, 2009). The HKCre/Shh^{KO} mice have a delay in the differentiation of zymogen cells from mucous neck cells, whereby zymogen cells located in the base of the gland frequently coexpressed mucous neck cell markers (Xiao *et al.*, 2010a,b). Intriguingly, disruption of the Hh signaling pathway in the transduced IMGE-5^{ski} cells resulted in the loss of zymogen cell marker pepsinogen and expression of mucous neck cell marker MUC6 when compared to the IMGE-5^{scram} controls. Expression of MUC6 in IMGE-5^{ski} cells was accompanied by an increase in cell proliferation and dissociation of the E-cadherin/β-catenin complex (Xiao *et al.*, 2010a,b). In support of our findings, Bredemeyer *et al.* (2009) discovered that in a model knocking out the cytoskeleton-regulating gene *Cd2ap* in *Mist1*$^{-/-}$ mice, zymogen cells reorganize their cytoskeleton as they differentiate to form a polarized epithelium. Our data support findings that show that aberrant E-cadherin expression causes changes in zymogenic cell morphology and differentiation (Bredemeyer *et al.*, 2009).

Our recent studies demonstrate that adherens junction protein E-cadherin may be a downstream target of the Hh signaling pathway. Shh regulates critical pathways in epithelial cell differentiation but we are only beginning to understand the mechanism (Shiotani *et al.*, 2005a,b; Stepan *et al.*, 2005; Suzuki *et al.*, 2005; van den Brink *et al.*, 2002; Xiao *et al.*, 2010a). The integrity of adherens junctions is a crucial factor determining not only epithelial morphology, physiological function, and differentiation (Braga, 2002; Braga *et al.*, 1997; Hopkins *et al.*, 2003; Samarin *et al.*, 2007) but also tumor growth, invasiveness, and metastasis (reviewed in Birchmeier *et al.*, 1993; Yilmaz and Christofori, 2009). Given the crucial role of Shh during gastric homeostasis and disease, understanding the role of Hh signaling as a regulator of the adherens junctions advances our knowledge by which Shh may regulate gastric epithelial cell physiology and pathophysiology.

III. The Potential Role of Sonic Hh as a Gastric Hormone

The predominant gastric hormone gastrin has two major roles in the gastrointestinal tract, the first of which is the major role in stimulation of gastric acid secretion in the stomach. Gastrin released from G-cells in the gastric antrum acts upon ECL cells, stimulating the release of histamine, again dependent on increased intracellular Ca^{2+} both from stores and from activation of calcium channels (Prinz *et al.*, 1993; Zeng *et al.*, 1996). Histamine then interacts with parietal cells, stimulating the upregulation of apical H^+/K^+ proton pumps and thus the secretion of acid. Gastrin and Ach also directly bind with the receptor on parietal cells, increasing acid secretion. Therefore, the role of gastrin in acid secretion is either direct or indirect by increasing histamine release. As acid increases, D endocrine cells release somatostatin, which acts as an inhibitory regulator on parietal cells to inhibit acid secretion and on ECL cells to inhibit histamine secretion (Allen *et al.*, 1986). As a result, acid secretion is regulated by this negative feedback control mechanism.

Gastrin also stimulates and maintains the proliferation of the gastric epithelium (reviewed in Dockray, 1999). Hypertrophic gastric mucosa is observed in patients with gastrin-producing tumors and also in rats infused continuously with gastrin (Ryberg *et al.*, 1990). In contrast, hypotrophic or atrophic mucosa results from fasting in rats whose serum gastrin levels become low (Solomon, 1986). Gastrin does this by upregulating its own receptor expressing on ECL cells and on mucous neck cells in the gastric mucosa, which can further differentiate into other cells types, including parietal cells and surface pit cells. The acceptance of gastrin as a growth factor for the gastric mucosa is also proven by trophic effect of gastrin and its receptor CCK_2-R during gastric regeneration. CCK_2 receptors are expressed and localized to the regenerative mucosa in the ulcer margin. Binding with the receptors, gastrin shows significant trophic effects on ulcer margin, enhancing wound healing (Ashurst *et al.*, 2008; Schmassmann and Reubi, 2002).

Another significant characteristic of the HKCre/Shh^{KO} mice was the lack of acid secretion in response to histamine stimulation. Ultrastructural analysis showed altered secretory canalicular membranes that were distorted and disorganized in the HKCre/Shh^{KO} parietal cells. Similar changes have recently been reported in the Huntington-interacting protein 1-related 1 (Hip1r)-deficient mouse model (Jain *et al.*, 2008). The Hip1r protein has been shown to be crucial for the normal dynamics of the secretory canalicular membrane of the parietal cells (Jain *et al.*, 2008). In addition, similar parietal cell morphological changes were observed between the HKCre/Shh^{KO} mouse and the basolateral Ae2 Cl^-/HCO_3^- exchanger (Slc4a2)-deficient, Atp4a(−/−) and NHE4 mice (Recalde *et al.*, 2006). Collectively, such findings suggest that Shh may be involved in the development of the secretory membrane dynamics that is involved in parietal cell function.

The HKCre/Shh^{KO} mice lacked the ability to secrete acid that was accompanied by severe hypergastrinemia and decreased somatostatin expression (Xiao *et al.*, 2010a,b). Hypergastrinemia has been reported previously in a pharmacological model of Hh signaling inhibition, using cyclopamine (El-Zaatari *et al.*, 2008). In fact, treatment of HKCre/Shh^{KO} mice, with the somatostatin analogue octreotide, significantly suppressed hypergastrinemia and subsequently hyperproliferation. Hypergastrinemia was associated with increased Ihh consistent with observations made in human stomach where Ihh was predominantly expressed in the pit cells where it induces pit cell differentiation in primary mouse gastric cells, while Shh in the gland/neck region (Fukaya *et al.*, 2006). Octreotide treatment of the HKCre/Shh^{KO}

mice resulted in decreased Ihh expression similar to levels measured in the control group (Xiao *et al.*, 2010a,b). We propose that the increased Ihh gene expression may be a consequence of hypergastrinemia. Further studies using the gastrin-deficient mouse model and the current HKCre/Shh^{KO} mice will be necessary to test the role of gastrin as a direct regulator of Ihh and hyperproliferation. Thus, the phenotype observed with loss of Shh may be attributed to increases in circulating gastrin concentrations due to loss of somatostatin. Our current understanding of hypergastrinemia is now extended because, besides the proposed role as a morphogen for the gastric epithelium, Shh may also be a fundamental regulator of the gastrin–gastric acid negative feedback mechanism (Fig. 21.2).

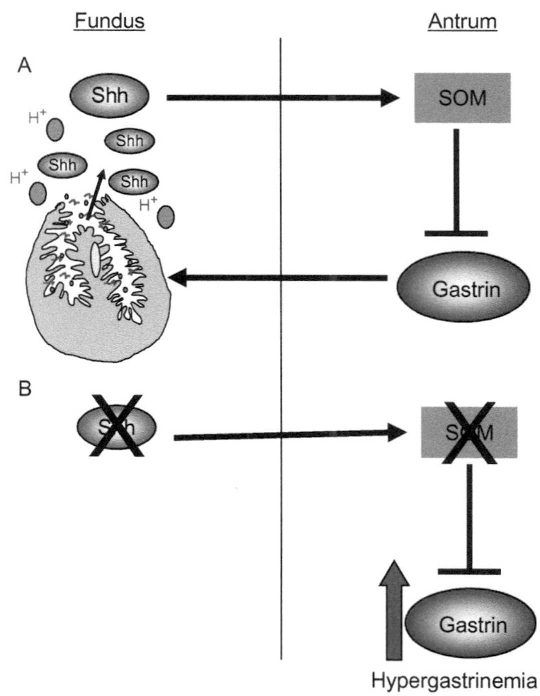

Figure 21.2 Shh as a regulator of the gastrin–gastric acid negative feedback. (A) During normal gastric function, gastrin induces acid secretion that is accompanied by Shh processing and release from the parietal cell. As a regulator of gastric acid secretion, Shh may induce somatostatin (SOM) secretion, which, in turn, inhibits the gastrin production from G-cells and acid secretion from the parietal cells. (B) Loss of gastric Shh results in diminished production of SOM and the subsequent increase in serum gastrin concentrations (hypergastrinemia) and thus an overall disruption in the gastrin–gastric acid negative feedback mechanism. (For color version of this figure, the reader is referred to the Web version of this chapter.)

The potential role of Shh to act as a gastric hormone would suggest that the protein would be secreted from the parietal cell via the basolateral membrane. The first suggestion for the secretion of Shh from the basolateral membrane came from our earlier studies using isolated rabbit gastric glands demonstrating that Shh expressed not only in parietal cells and associated with the H^+,K^+-ATPase-containing tubulovesicles and secretory membranes but also colocalized with the gamma-actin at the basolateral membrane (Zavros *et al.*, 2008). Moreover, in a polarized Shh- and H^+,K^+-ATPase-expressing human gastric cancer cell line (23132/87) (Dimmler *et al.*, 2003), histamine treatment induces secretion of the processed 19kDa Shh protein from both the apical and basolateral membranes (Zavros *et al.*, 2008). Interestingly, recent investigations during development also show the trafficking of Hh to the basolateral membrane to form a morphogenetic gradient in the *Drosophila* wing imaginal disk epithelium (Callejo *et al.*, 2011). Functional analysis of the 12-transmembrane protein Dispatched, the glypican Dally-like protein (Dlp), and the Ig-like and FNNIII domains of protein, Interference Hh (Ihog) revealed that Dispatched may be involved in the regulation of vesicular trafficking for basolateral release of Hh, Dlp, and Ihog (Callejo *et al.*, 2011). This important study also revealed that during development, Dlp is needed in the Hh-producing cells to allow for protein release. Ihog, a Hh coreceptor, then anchors Hh to the basolateral part of the disk epithelium (Callejo *et al.*, 2011). Whether a similar mechanism regulates the release of basolateral Shh from the gastric parietal cell remains to be determined. However, collectively these findings support the potential role of Shh as gastric hormone in particular as a regulator of acid secretion.

IV. Conclusions and Future Directions

Shh is a parietal cell-secreted protein regulating gastric cell proliferation, differentiation, and function. Recent studies reveal that Shh may be secreted basolaterally into the circulation not only during development (Callejo *et al.*, 2011) but also during gastric acid secretion (Zavros *et al.*, 2008). These exciting findings advance our understanding of the role of Shh as a potential hormone and chemoattractant. Dunaeva *et al.* shows that Shh activates classical intracellular signal transduction pathways related to human monocytes recruitment *in vitro* (Dunaeva *et al.*, 2010). An intriguing question that remains is whether Shh acts as a chemokine for monocyte recruitment during *H. pylori* infection. Another question that remains unanswered is the identity of the major organ producing Shh. What is the role of the stomach as a Shh-producing organ? Given the emerging role of Shh as a regulator of the immune response, adult tissue function and homeostasis and gastric disease, this appears to be a promising area of research and will certainly broaden our understanding of this developmental morphogen in adult stomach.

ACKNOWLEDGMENT

This work was supported by NIH 1R01DK083402-01 Grant (Y. Zavros).

REFERENCES

Adolphe, C., Hetherington, R., Ellis, T., and Wainwright, B. (2006). Patched1 functions as a gatekeeper by promoting cell cycle progression. *Cancer Res.* **66**(4), 2081–2088.

Allen, J. M., Bishop, A. E., Daly, M. J., Larsson, H., Carlsson, E., Polak, J. M., and Bloom, S. R. (1986). Effect of inhibition of acid secretion on the regulatory peptides in the rat stomach. *Gastroenterology* **90**(4), 970–977.

Ashurst, H. L., Varro, A., and Dimaline, R. (2008). Regulation of mammalian gastrin/CCK receptor (CCK2R) expression in vitro and in vivo. *Exp. Physiol.* **93**(2), 223–236.

Barnes, E. A., Kong, M., Ollendorff, V., and Donoghue, D. J. (2001). Patched1 interacts with cyclin B1 to regulate cell cycle progression. *EMBO J.* **20**(9), 2214–2223.

Bijlsma, M. F., Spek, C. A., Zivkovic, D., van de Water, S., Rezaee, F., and Peppelenbosch, M. P. (2006). Repression of smoothened by patched-dependent (pro-) vitamin D3 secretion. *PLoS Biol.* **4**(8), e232.

Bijlsma, M. F., Borensztajn, K. S., Roelink, H., Peppelenbosch, M. P., and Spek, C. A. (2007). Sonic hedgehog induces transcription-independent cytoskeletal rearrangement and migration regulated by arachidonate metabolites. *Cell. Signal.* **19**(12), 2596–2604.

Birchmeier, W., Weidner, K. M., Hülsken, J., and Behrens, J. (1993). Molecular mechanisms leading to cell junction (cadherin) deficiency in invasive carcinomas. *Semin. Cancer Biol.* **4**(4), 231–239.

Bockman, D. E., Sharp, R., and Merlino, G. (1995). Regulation of terminal differentiation of zymogenic cells by transforming growth factor alpha in transgenic mice. *Gastroenterology* **108**(2), 447–454.

Bourikas, D., Pekarik, V., Baeriswyl, T., Grunditz, A., Sadhu, R., Nardó, M., and Stoeckli, E. T. (2005). Sonic hedgehog guides commissural axons along the longitudinal axis of the spinal cord. *Nat. Neurosci.* **8**(3), 297–304.

Braga, V. M. (2002). Cell-cell adhesion and signalling. *Curr. Opin. Cell Biol.* **14**(5), 546–556.

Braga, V. M., Machesky, L. M., Hall, A., and Hotchin, N. A. (1997). The small GTPases Rho and Rac are required for the establishment of cadherin-dependent cell-cell contacts. *J. Cell Biol.* **137**(6), 1421–1431.

Bredemeyer, A. J., Geahlen, J. H., Weis, V. G., Huh, W. J., Zinselmeyer, B. H., Srivatsan, S., Miller, M. J., Shaw, A. S., and Mills, J. C. (2009). The gastric epithelial progenitor cell niche and differentiation of the zymogenic (chief) cell lineage. *Dev. Biol.* **325**(1), 211–224.

Callejo, A., Bilioni, A., Mollica, E., Gorfinkiel, N., Andrés, G., Ibáñez, C., Torroja, C., Doglio, L., Sierra, J., and Guerrero, I. (2011). Dispatched mediates Hedgehog basolateral release to form the long-range morphogenetic gradient in the Drosophila wing disk epithelium. *Proc. Natl. Acad. Sci. USA* **108**(31), 12591–12598.

Chen, M. H., Li, Y. J., Kawakami, T., Xu, S. M., and Chuang, P. T. (2004). Palmitoylation is required for the production of a soluble multimeric Hedgehog protein complex and long-range signaling in vertebrates. *Genes Dev.* **18,** 641–659.

Cheng, X. X., Wang, Z. C., Chen, X. Y., Sun, Y., Kong, Q. Y., Liu, J., Gao, X., Guan, H. W., and Li, H. (2005). Frequent loss of membranous E-cadherin in gastric cancers: A cross-talk with Wnt in determining the fate of beta-catenin. *Clin. Exp. Metastasis* **22**(1), 85–93.

Dabrowski, A., Detjen, K. M., Logsdon, C. D., and Williams, J. A. (1997). Stimulation of both CCK-A and CCK-B receptors activates MAP kinases in AR42J and receptor-transfected CHO cells. *Digestion* **58**(4), 361–367.

Daulhac, L., Kowalski-Chauvel, A., Pradayrol, L., Vaysse, N., and Seva, C. (1999). Gastrin stimulates the formation of a p60Src/p125FAK complex upstream of the phosphatidylinositol 3-kinase signaling pathway. *FEBS Lett.* **445**(2–3), 251–255.

Dimmler, A., Brabletz, T., Hlubek, F., Häfner, M., Rau, T., Kirchner, T., and Faller, G. (2003). Transcription of sonic hedgehog, a potential factor for gastric morphogenesis and gastric mucosa maintenance, is up-regulated in acidic conditions. *Lab. Invest.* **83**(12), 1829–1837.

Dockray, G. J. (1999). Topical review. Gastrin and gastric epithelial physiology. *J. Physiol.* **518**(Pt. 2), 315–324.

Dockray, G., Dimaline, R., and Varro, A. (2005). Gastrin: Old hormone, new functions. *Pflugers Arch.* **449**(4), 344–355.

Dufresne, M., Escrieut, C., Clerc, P., Le Huerou-Luron, I., Prats, H., Bertrand, V., Le Meuth, V., Guilloteau, P., Vaysse, N., and Fourmy, D. (2006a). Molecular cloning, developmental expression and pharmacological characterization of the CCKB/gastrin receptor in the calf pancreas. *Eur. J. Pharmacol.* **297**(1–2), 165–179.

Dufresne, M., Seva, C., and Fourmy, D. (2006b). Cholecystokinin and gastrin receptors. *Physiol. Rev.* **86**(3), 805–847.

Dunaeva, M., Voo, S., van Oosterhoud, C., and Waltenberger, J. (2010). Sonic hedgehog is a potent chemoattractant for human monocytes: Diabetes mellitus inhibits Sonic hedgehog-induced monocyte chemotaxis. *Basic Res. Cardiol.* **105**(1), 61–71.

Edkins, J. S. (1905). On the chemical mechanism of gastric secretion. *Proc. R. Soc. Lond. B Biol. Sci.* **76,** 376.

El-Zaatari, M., Grabowska, A. M., McKenzie, A. J., Powe, D. G., Scotting, P. J., and Watson, S. A. (2008). Cyclopamine inhibition of the sonic hedgehog pathway in the stomach requires concomitant acid inhibition. *Regul. Pept.* **146**(1–3), 131–139.

El-Zaatari, M., Zavros, Y., Tessier, A., Waghray, M., Lentz, S., Gumucio, D., Todisco, A., and Merchant, J. L. (2010). Intracellular calcium release and protein kinase C activation stimulate sonic hedgehog gene expression during gastric acid secretion. *Gastroenterology* **139**(6), 2061–2071.e2.

Feldmann, G., Dhara, S., Fendrich, V., Bedja, D., Beaty, R., Mullendore, M., Karikari, C., Alvarez, H., Iacobuzio-Donahue, C., Jimeno, A., Gabrielson, K. L., Matsui, W., *et al.* (2007). Blockade of hedgehog signaling inhibits pancreatic cancer invasion and metastases: A new paradigm for combination therapy in solid cancers. *Cancer Res.* **67**(5), 2187–2196.

Fukaya, M., Isohata, N., Ohta, H., Aoyagi, K., Ochiya, T., Saeki, N., Yanagihara, K., Nakanishi, Y., Taniguchi, H., Sakamoto, H., Shimoda, T., Nimura, Y., *et al.* (2006). Hedgehog signal activation in gastric pit cell and in diffuse-type gastric cancer. *Gastroenterology* **131**(1), 14–29.

Goldenring, J. R., Ray, G. S., Soroka, C. J., Smith, J., Modlin, I. M., Meise, K. S., and Coffey, R. J., Jr. (1996). Overexpression of transforming growth factor-alpha alters differentiation of gastric cell lineages. *Dig. Dis. Sci.* **41**(4), 773–784.

Hiraoka, S., Miyazaki, Y., Kitamura, S., Toyota, M., Kiyohara, T., Shinomura, Y., Mukaida, N., and Matsuzawa, Y. (2001). Gastrin induces CXC chemokine expression in gastric epithelial cells through activation of NF-kappaB. *Am. J. Physiol. Gastrointest. Liver Physiol.* **281**(3), G735–G742.

Hirayama, F., Takagi, S., Yokoyama, Y., Iwao, E., and Ikeda, Y. (1996). Establishment of gastric Helicobacter pylori infection in Mongolian gerbils. *J. Gastroenterol.* **31**(Suppl. 9), 24–28.

Hocker, M. (2004). Molecular mechanisms of gastrin-dependent gene regulation. *Ann. N. Y. Acad. Sci.* **1014,** 97–109.

Höcker, M., Henihan, R. J., Rosewicz, S., Riecken, E. O., Zhang, Z., Koh, T. J., and Wang, T. C. (1997). Gastrin and phorbol 12-myristate 13-acetate regulate the human histidine decarboxylase promoter through Raf-dependent activation of extracellular signal-regulated kinase-related signaling pathways in gastric cancer cells. *J. Biol. Chem.* **272**(43), 27015–27024.

Hopkins, A. M., Walsh, S. V., Verkade, P., Boquet, P., and Nusrat, A. (2003). Constitutive activation of Rho proteins by CNF-1 influences tight junction structure and epithelial barrier function. *J. Cell Sci.* **116**(Pt. 4), 725–742.

Ikenouchi, J., Matsuda, M., Furuse, M., and Tsukita, S. (2003). Regulation of tight junctions during the epithelium-mesenchyme transition: Direct repression of the gene expression of claudins/occludin by Snail. *J. Cell Sci.* **116**(Pt. 10), 1959–1967.

Jain, R. N., Al-Menhali, A. A., Keeley, T. M., Ren, J., El-Zaatari, M., Chen, X., Merchant, J. L., Ross, T. S., Chew, C. S., and Samuelson, L. C. (2008). Hip1r is expressed in gastric parietal cells and is required for tubulovesicle formation and cell survival in mice. *J. Clin. Invest.* **118**(7), 2459–2470.

Jarov, A., Williams, K. P., Ling, L. E., Koteliansky, V. E., Duband, J. L., and Fournier-Thibault, C. (2003). A dual role for Sonic hedgehog in regulating adhesion and differentiation of neuroepithelial cells. *Dev. Biol.* **261**(2), 520–536.

Jenkins, D. (2009). Hedgehog signalling: Emerging evidence for non-canonical pathways. *Cell. Signal.* **21**(7), 1023–1034.

Kopin, A. S., Lee, Y. M., McBride, E. W., Miller, L. J., Lu, M., Lin, H. Y., Kolakowski, L. F., Jr., and Beinborn, M. (1992). Expression cloning and characterization of the canine parietal cell gastrin receptor. *Proc. Natl. Acad. Sci. USA* **89**(8), 3605–3609.

Kulaksiz, H., Arnold, R., Göke, B., Maronde, E., Meyer, M., Fahrenholz, F., Forssmann, W. G., and Eissele, R. (2000). Expression and cell-specific localization of the cholecystokinin B/gastrin receptor in the human stomach. *Cell Tissue Res.* **299**(2), 289–298.

Larsen, B., Tarp, U., and Kristensen, E. (1987). Familial giant hypertrophic gastritis (Ménétrier's disease). *Gut* **28**(11), 1517–1521.

Lee, J. J., Ekker, S. C., von Kessler, D. P., Porter, J. A., Sun, B. I., and Beachy, P. A. (1994). Autoproteolysis in hedgehog protein biogenesis. *Science* **266**(5190), 1492–1493.

Li, X., Deng, W., Nail, C. D., Bailey, S. K., Kraus, M. H., Ruppert, J. M., and Lobo-Ruppert, S. M. (2006). Snail induction is an early response to Gli1 that determines the efficiency of epithelial transformation. *Oncogene* **25**(4), 609–621.

Luo, G. Q., Li, J. H., Wen, J. F., Zhou, Y. H., Hu, Y. B., and Zhou, J. H. (2008). Effect and mechanism of the Twist gene on invasion and metastasis of gastric carcinoma cells. *World J. Gastroenterol.* **14**(16), 2487–2493.

Martí, E., Bumcrot, D. A., Takada, R., and McMahon, A. P. (1995). Requirement of 19K form of Sonic hedgehog for induction of distinct ventral cell types in CNS explants. *Nature* **375**(6529), 322–325.

McGuigan, J. I. (1968). Gastrin mucosal intracellular localization of gastrin by immunofluorescence. *Gastroenterology* **55,** 315–327.

Nelson, W. J., and Nusse, R. (2004). Convergence of Wnt, beta-catenin, and cadherin pathways. *Science* **303**(5663), 1483–1487.

Noble, F., and Roques, B. P. (1999). CCK-B receptor: Chemistry, molecular biology, biochemistry and pharmacology. *Prog. Neurobiol.* **58**(4), 349–379.

Nomura, S., Settle, S. H., Leys, C. M., Means, A. L., Peek, R. M., Jr., Leach, S. D., Wright, C. V., Coffey, R. J., and Goldenring, J. R. (2005). Evidence for repatterning of the gastric fundic epithelium associated with Ménétrier's disease and TGFalpha overexpression. *Gastroenterology* **128**(5), 1292–1305.

Nüsslein-Volhard, C., and Wieschaus, E. (1980). Mutations affecting segment number and polarity in Drosophila. *Nature* **287**(5785), 795–801.

Porter, J. A., von Kessler, D. P., Ekker, S. C., Young, K. E., Lee, J. J., Moses, K., and Beachy, P. A. (1995). The product of *hedgehog* autoproteolytic cleavage active in local and long-range signaling. *Nature* **374**(6520), 363–366.

Prinz, C., Kajimura, M., Scott, D. R., Mercier, F., Helander, H. F., and Sachs, G. (1993). Histamine secretion from rat enterochromaffinlike cells. *Gastroenterology* **105**(2), 449–461.

Recalde, S., Muruzábal, F., Looije, N., Kunne, C., Burrell, M. A., Sáez, E., Martínez-Ansó, E., Salas, J. T., Mardones, P., Prieto, J., Medina, J. F., and Elferink, R. P. (2006). Inefficient chronic activation of parietal cells in Ae2a, β(−/−) mice. *Am. J. Pathol.* **169**(1), 165–176.

Roche, S., Bali, J. P., and Magous, R. (1990). Involvement of a pertussis toxin-sensitive G protein in the action of gastrin on gastric parietal cells. *Biochim. Biophys. Acta* **1055**(3), 287–294.

Ryberg, B., Axelson, J., Håkanson, R., Sundler, F., and Mattsson, H. (1990). Trophic effects of continuous infusion of [Leu15]-gastrin-17 in the rat. *Gastroenterology* **98**(1), 33–38.

Samarin, S. N., Ivanov, A. I., Flatau, G., Parkos, C. A., and Nusrat, A. (2007). Rho/Rho-associated kinase-II signaling mediates disassembly of epithelial apical junctions. *Mol. Biol. Cell* **18**(9), 3429–3439.

Schmassmann, A., and Reubi, J. C. (2002). Cholecystokinin-B/gastrin receptors enhance wound healing in the rat gastric mucosa. *J. Clin. Invest.* **106**(8), 1021–1029.

Shiotani, A., Iishi, H., Uedo, N., Ishiguro, S., Tatsuta, M., Nakae, Y., Kumamoto, M., and Merchant, J. L. (2005). Evidence that loss of sonic hedgehog is an indicator of Helicobacter pylori-induced atrophic gastritis progressing to gastric cancer. *Am. J. Gastroenterol.* **100**(3), 581–587.

Shiotani, A., Iishi, H., Uedo, N., Ishihara, R., Ishiguro, S., Tatsuta, M., Nakae, Y., Kumamoto, M., Hinoi, T., and Merchant, J. L. (2006). Helicobacter pylori-induced atrophic gastritis progressing to gastric cancer exhibits sonic hedgehog loss and aberrant CDX2 expression. *Aliment. Pharmacol. Ther.* **24**(Suppl. 4), 71–80.

Shiotani, A., Uedo, N., Iishi, H., Tatsuta, M., Ishiguro, S., Nakae, Y., Kamada, T., Haruma, K., and Merchant, J. L. (2007). Re-expression of sonic hedgehog and reduction of CDX2 after Helicobacter pylori eradication prior to incomplete intestinal metaplasia. *Int. J. Cancer* **121**(6), 1182–1189.

Solomon, T. E. (1986). Trophic effects of pentagastrin on gastrointestinal tract in fed and fasted rats. *Gastroenterology* **91**(1), 108–116.

Stepan, V. M., Dickinson, C. J., del Valle, J., Matsushima, M., and Todisco, A. (1999). Cell type-specific requirement of the MAPK pathway for the growth factor action of gastrin. *Am. J. Physiol.* **276**(6 Pt. 1), G1363–G1372.

Stepan, V., Ramamoorthy, S., Nitsche, H., Zavros, Y., Merchant, J. L., and Todisco, A. (2005). Regulation and function of the sonic hedgehog signal transduction pathway in isolated gastric parietal cells. *J. Biol. Chem.* **280**(16), 15700–15708.

Suzuki, H., Minegishi, Y., Nomoto, Y., Ota, T., Masaoka, T., van den Brink, G. R., and Hibi, T. (2005). Down-regulation of a morphogen (sonic hedgehog) gradient in the gastric epithelium of Helicobacter pylori-infected Mongolian gerbils. *J. Pathol.* **206**(2), 186–197.

Tanaka, M., Kitajima, Y., Edakuni, G., Sato, S., and Miyazaki, K. (2002). Abnormal expression of E-cadherin and beta-catenin may be a molecular marker of submucosal invasion and lymph node metastasis in early gastric cancer. *Br. J. Surg.* **89**(2), 236–244.

Taylor, I. L., Byrne, W. J., Christie, D. L., Ament, M. E., and Walsh, J. H. (1982). Effect of individual l-amino acids on gastric acid secretion and serum gastrin and pancreatic polypeptide release in humans. *Gastroenterology* **83**(1 Pt. 2), 273–278.

Testaz, S., Jarov, A., Williams, K. P., Ling, L. E., Koteliansky, V. E., Fournier-Thibault, C., and Duband, J. L. (2001). Sonic hedgehog restricts adhesion and migration of neural crest cells independently of the Patched- Smoothened-Gli signaling pathway. *Proc. Natl. Acad. Sci. USA* **98**(22), 12521–12526.

Tokhunts, R., Singh, S., Chu, T., D'Angelo, G., Baubet, V., Goetz, J. A., Huang, Z., Yuan, Z., Ascano, M., Zavros, Y., Thérond, P. P., Kunes, S., *et al.* (2010). The full-length unprocessed hedgehog protein is an active signaling molecule. *J. Biol. Chem.* **285** (4), 2562–2568.

Tsunoda, Y., Takeda, H., Otaki, T., Asaka, M., Nakagaki, I., and Sasaki, S. (1988). A role for Ca^{2+} in mediating hormone-induced biphasic pepsinogen secretion from the chief cell determined by luminescent and fluorescent probes and X-ray microprobe. *Biochim. Biophys. Acta* **941**(1), 83–101.

van den Brink, G. R. (2007). Hedgehog signaling in development and homeostasis of the gastrointestinal tract. *Physiol. Rev.* **87**(4), 1343–1375.

van den Brink, G. R., Hardwick, J. C., Tytgat, G. N., Brink, M. A., Ten Kate, F. J., Van Deventer, S. J., and Peppelenbosch, M. P. (2001). Sonic hedgehog regulates gastric gland morphogenesis in man and mouse. *Gastroenterology* **121**(2), 317–328.

van den Brink, G. R., Hardwick, J. C., Nielsen, C., Xu, C., ten Kate, F. J., Glickman, J., van Deventer, S. J., Roberts, D. J., and Peppelenbosch, M. P. (2002). Sonic hedgehog expression correlates with fundic gland differentiation in the adult gastrointestinal tract. *Gut* **51**(5), 628–633.

Waghray, M., Zavros, Y., Saqui-Salces, M., El-Zaatari, M., Alamelumangapuram, C. B., Todisco, A., Eaton, K. A., and Merchant, J. L. (2010). Interleukin-1beta promotes gastric atrophy through suppression of Sonic Hedgehog. *Gastroenterology* **138**(2), 562–572,572. e1–2.

Watanabe, T., Tada, M., Nagai, H., Sasaki, S., and Nakao, M. (1998). Helicobacter pylori infection induces gastric cancer in Mongolian gerbils. *Gastroenterology* **115**(3), 642–648.

Weiner, H. L., Bakst, R., Hurlbert, M. S., Ruggiero, J., Ahn, E., Lee, W. S., Stephen, D., Zagzag, D., Joyner, A. L., and Turnbull, D. H. (2002). Induction of medulloblastomas in mice by sonic hedgehog, independent of Gli1. *Cancer Res.* **62**(22), 6385–6389.

Wolfsen, H. C., Carpenter, H. A., and Talley, N. J. (1993). Menetrier's disease: A form of hypertrophic gastropathy or gastritis? *Gastroenterology* **104**(5), 1310–1319.

Xiao, C., Ogle, S. A., Schumacher, M. A., Orr-Asman, M. A., Miller, M. L., Lertkowit, N., Varro, A., Hollande, F., and Zavros, Y. (2010a). Loss of parietal cell expression of Sonic hedgehog induces hypergastrinemia and hyperproliferation of surface mucous cells. *Gastroenterology* **138**(2), 550–561561.e1–8.

Xiao, C., Ogle, S. A., Schumacher, M. A., Schilling, N., Tokhunts, R. A., Orr-Asman, M. A., Miller, M. L., Robbins, D. J., Hollande, F., and Zavros, Y. (2010b). Hedgehog signaling regulates E-cadherin expression for the maintenance of the actin cytoskeleton and tight junctions. *Am. J. Physiol. Gastrointest. Liver Physiol.* **299**(6), G1252–G1265.

Yassin, R. R., and Murthy, S. N. (1991). Possible involvement of protein kinase C in mediating gastrin-induced response in rat colonic epithelium. *Peptides* **12**(5), 925–927.

Yilmaz, M., and Christofori, G. (2009). EMT, the cytoskeleton, and cancer cell invasion. *Cancer Metastasis Rev.* **28**(1–2), 15–33.

Zavros, Y., Waghray, M., Tessier, A., Bai, L., Todisco, A. L., Gumucio, D., Samuelson, L. C., Dlugosz, A., and Merchant, J. L. (2007). Reduced pepsin A processing of sonic hedgehog in parietal cells precedes gastric atrophy and transformation. *J. Biol. Chem.* **282**(46), 33265–33274.

Zavros, Y., Orr, M. A., Xiao, C., and Malinowska, D. H. (2008). Sonic hedgehog is associated with H+−K+−ATPase-containing membranes in gastric parietal cells and secreted with histamine stimulation. *Am. J. Physiol. Gastrointest. Liver Physiol.* **295**(1), G99–G111.

Zeng, N., Walsh, J. H., Kang, T., Helander, K. G., Helander, H. F., and Sachs, G. (1996). Selective ligand-induced intracellular calcium changes in a population of rat isolated gastric endocrine cells. *Gastroenterology* **110**(6), 1835–1846.

CHAPTER TWENTY-TWO

Sonic Hedgehog-Mediated Synergistic Effects Guiding Angiogenesis and Osteogenesis

Sabine Fuchs, Eva Dohle, *and* Charles James Kirkpatrick

Contents

Abstract

Sonic hedgehog (Shh) is a morphogen controlling the skeletal and vascular development in the embryo but is also reactivated during adult repair processes. Thus, this molecule holds great therapeutic potential for biotechnological and biomedical approaches aiming to enhance tissue regeneration or to replace damaged tissues. According to present knowledge, Shh signaling controls the expression of several families of growth factors involved in neovascularization and vessel maturation and acts upstream of the most prominent angiogenic growth factor, vascular endothelial growth factor. In this context, a very interesting feature of Shh is that it controls both angiogenic activity and vessel stabilization by mural cells. In parallel, Shh seems to have a direct effect on endothelial cell tube formation and seems to trigger the differentiation process of mesenchymal stem cells toward the osteogenic lineage. In this chapter, we will therefore shortly summarize the multifaceted potential of Shh

Institute of Pathology, Langenbeckstr. 1, Johannes Gutenberg University, Mainz, Germany

Vitamins and Hormones, Volume 88
ISSN 0083-6729, DOI: 10.1016/B978-0-12-394622-5.00022-5

for bone repair and vascularization according to the current literature. In addition, we will show how coculture models based on outgrowth endothelial cells and primary osteoblasts can be used to reveal some of the relevant mechanisms by which Shh drives and connects bone regeneration and vascularization.

I. General Introduction

Bone regeneration is a complex process including the interaction of various cell types such as mesenchymal stem cells, chondrocytes, osteoblasts, osteoclasts, and endothelial cells.

In order to orchestrate the individual phases and mechanisms in bone repair, some of the key control elements which regulate bone formation during embryonic development seem to be reactivated. The molecules guiding these developmental or repair processes are often multitasking molecules affecting multiple differentiation and repair processes, as well as cells from several lineages. Thus, these molecules might serve as potential target molecules to improve the repair process as a whole, including the individual subevents such as callus formation, new bone formation, and vascularization. One group of molecules currently discussed as control elements during embryonic development as well as mediators of bone repair processes in adults are the morphogens indian hedgehog (Ihh) and sonic hedgehog (Shh). In this chapter, we give an overview on the role of these molecules during bone repair and discuss specifically Shh as a potent agent driving and coordinating osteogenic differentiation and vascularization.

II. General Aspects of Shh Signaling

Hedgehog (Hh) proteins are morphogens playing an important role during embryonic development, during which they control the pattern of different tissues and organs (Chiang *et al.*, 1996; Johnson *et al.*, 1994b; Nagase *et al.*, 2007; Ruiz i Altaba, 1994). Hh, the founding protein of the Hh family, was first described in *Drosophila melanogaster* in 1980 (Nusslein-Volhard and Wieschaus, 1980). Hh was found to be expressed in the developing fly larva and was functionally identified as a segment polarity gene in the fly embryo. Today, three vertebrate homologues of Drosophila Hh gene have been described, Shh, desert Hh, and Ihh (Bitgood *et al.*, 1996; Echelard *et al.*, 1993; Fietz *et al.*, 1994). Shh is the most studied member of the vertebrate Hh family and is known to regulate the development of the neural tube and limb development during embryogenesis (Ekker *et al.*, 1995; Fietz *et al.*, 1994; Johnson *et al.*, 1994a). Shh protein is translated as an ~45 kDa precursor and undergoes posttranslational autocatalytic processing to produce an ~20 kDa N-terminal signaling domain and an ~25 kDa C-terminal domain with no known signaling role (Bumcrot *et al.*, 1995). Shh signaling acts through the interaction with the receptor patched1 and the coreceptor smoothened (Cohen, 2003; Stone *et al.*, 1996). When Shh ligand is absent, patched1 inhibits the coreceptor smoothened which leads to conversion of the transcription factor Gli to its repressor form. Binding of Shh ligand to patched1 represses patched1 and abolishes smoothened inhibition which finally leads to the activation of Gli family transcription factors (Marigo *et al.*, 1996; Stone *et al.*, 1996). Gli accumulates in the nucleus and controls the transcription of Shh target genes, including patched-1 and Gli themselves.

III. The Role of Hh Morphogens in Bone Development and Repair

In endochondral and intramembraneous bone formation, the morphogens Ihh and Shh seem to fulfill individual functions, although details in the mechanisms are far from being completely elucidated. During endochondral bone formation mesenchymal stem cells condense and differentiate into chondrocytes forming cartilage. In this context, Ihh is considered as one of the key elements controlling endochondral bone formation as described in the following section.

Ihh is produced by prehyperthrophic and hypertrophic chondrocytes (Iwasaki *et al.*, 1997; Vortkamp *et al.*, 1998) in endochondral ossification where it induces the upregulation of PTHrP and results in hypertrophic cartilage differentiation. In contrast, intramembranous bone formation seems to be not severely affected in Ihh mutants (St-Jacques *et al.*, 1999). Besides the function of Hh molecules as control elements during development, they also seem to play an active role in bone repair in adult organisms. Ihh, for instance, is upregulated in the bone marrow (Ito *et al.*, 1999) during the early phases of bone repair processes. Studies investigating the localization and the role of Ihh and Shh during fracture repair have shown that Shh, Gli, and BMP-4 mRNA are found in callus-forming cells shortly after the fracture and are specifically prominent in the proliferating periosteum, soft tissues, and medullatory cavity at day 2 postfracture (Miyaji *et al.*, 2003). Accordingly, Hh molecules might also serve as potential targets for therapeutic intervention to enhance bone repair processes. In this context, the potential benefit of Hh molecules on the repair process is investigated in several studies but needs to be critically discussed in terms of potential side effects. It has been shown that intramuscular transplantation of Shh-expressing fibroblasts leads to ectopic bone formation in a mouse model

(Kinto *et al.*, 1997). Nevertheless, other studies reported that systemic administration of Shh results in a primary increase in osteoblasts and their precursors but is also associated with altered trabecular morphology, decreased bone volume, and decreased compressive strength (Kiuru *et al.*, 2009b). Further, Shh seems to affect the hematopoietic niche in the bone and in particular lymphoid differentiation, resulting in a decrease in lymphoid progenitor engraftment and differentiation (Kiuru *et al.*, 2009a). These findings highlight the potential risks and underline the need to understand the role of action of Hh molecules during bone repair. In addition they highlight the multifaceted potential of hedgehog molecules.

IV. Shh: Potential Candidate Linking Angiogenesis and Osteogenesis

In endochondral bone fracture healing, chrondrocyte apoptosis and cartilage degradation lead to the definition of a structural pattern and also result in the ingrowth of blood vessels from existing vessels in the periosteum. This process is mediated by two major groups of angiogenic growth factors, such as vascular endothelial growth factor (VEGF) and angiopoietins. Both classes of angiogenic molecules can be triggered by the morphogen Shh (Pola *et al.*, 2001), which is also induced in the early phases of bone repair as described above. Taking into account, these observations Shh might be one of the key molecules linking osteogenesis and angiogenesis during bone repair and thus is also considered as an attractive therapeutic tool to enhance simultaneously angiogenesis and osteogenesis. The individual effects of Shh on neovascularization and osteogenesis will be discussed in more detail in the following sections and are schematically highlighted in Fig. 22.1.

V. Effects of Shh on Angiogenesis, Vessel Maturation, and Osteogenesis

The role of Shh in the development of functional vasculature in embryonal stages has been observed for various tissues (Nagase *et al.*, 2005, 2006b; Pepicelli *et al.*, 1998). The scientific interest in Shh as one of the key elements guiding angiogenesis and vasculogenesis derives from the observation that Shh acts upstream of most of the angiogenic growth factors studied so far, including the most prominent one, VEGF. In addition, Shh contributes to the neovascularization process by attracting endothelial progenitor cells to ischemic regions (Asai *et al.*, 2006; Yamazaki *et al.*, 2008). Shh induces the expression of at least two major groups of angiogenic growth factors, namely VEGF and the angiopoietins, in interstitial

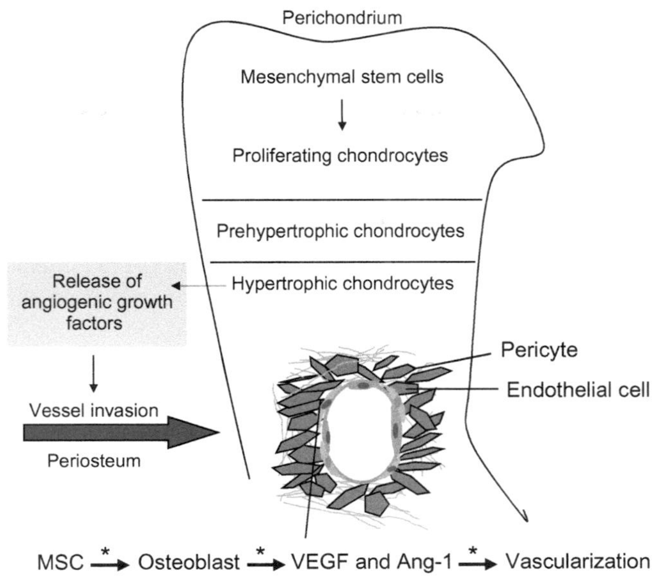

Figure 22.1 Sonic hedgehog mediated processes linking angiogenesis and osteogenesis. (For color version of this figure, the reader is referred to the Web version of this chapter.)

mesenchymal cells, thus stimulating angiogenesis through paracrine mechanisms between endothelial cells (Pola *et al.*, 2001) and cells of mesenchymal origin. During vascular rearrangement and turnover, the angiopoietins-1 and -2 have opposing effects which are controlled by their competitive binding to the Tie-2 receptor, thus creating a delicate balance which determines vessel formation, stabilization, or regression (Carmeliet, 2003). As an upstream control element of angiogenesis, Shh seems to be involved in regulating this balance of angiopoietin-1 and -2 in collaboration with FGF-2 (Fujii and Kuwano, 2010). Nevertheless, besides these indirect pathways controlling angiogenic activity, Shh also acts as a direct mediator for endothelial activation and tube formation based on Gli-independent pathways as discussed in the following section.

According to its impact on factors involved in vascular stabilization such as angiopoietin-1, Shh contributes to the recruitment of mural cells and thus offers additional beneficial effects on vascular stabilization (Nagase *et al.*, 2006a). It has been shown that Shh improves the proliferation of smooth muscle cells (SMCs) mediated by the G1 cyclin-retinoblastoma axis (Li *et al.*, 2010). In addition, combinatorial approaches based on Shh gene therapy and AMD3100 to induce progenitor cell mobilization, facilitated the development of capillaries and SMC-covered vessels in a myocardial infarction model (Roncalli *et al.*, 2011). Further, Shh promotes the upregulation of PDGFRβ guiding the wrapping of SMCs around micro-vessels (Frontini *et al.*, 2011) and is involved in the maintenance of vascular smooth muscle progenitors in the arterial adventitia (Passman *et al.*, 2008). As a consequence, Shh seems to coordinate several crucial events during vessel formation and stabilization and thus possesses a significant therapeutic potential for vascular regeneration or engineering.

The potential of Shh to support osteogenic differentiation (Nakamura *et al.*, 1997; van der Horst *et al.*, 2003) has also promoted interest in Shh as a therapeutic agent in bone tissue engineering, although the mechanisms of action are not completely understood. Shh is considered as an early inductive signal in skeletal development (Ingham and McMahon, 2001). Some studies suggest that the effects of Shh on osteogenic differentiation are associated with an increase in bone morphogenetic proteins, including BMP-2 (Yuasa *et al.*, 2002) and BMP-4 (Miyaji *et al.*, 2003) or with parathyroid hormone-related peptide (PTHrP) (Jemtland *et al.*, 2003). In this context, several studies focused on the therapeutic potential in bone regeneration and tissue engineering. Approaches using genetically modified Shh-expressing cells embedded in alginate collagen constructs demonstrated the potential of Shh in bone regeneration in a calvarial defect model (Edwards *et al.*, 2004). According to several studies Shh signaling drives the osteogenic differentiation of mesenchymal stem cells originating from different sources in tissue engineering or regenerative medicine-based approaches (James *et al.*, 2010; van der Horst *et al.*, 2003; Wang *et al.*, 2010). This might include direct cell signaling mechanisms in the differentiation as well as paracrine effects, as recently shown in a coculture approach (Levi *et al.*, 2011) using adipose tissue-derived stem cells and calvarial osteoblasts, thus highlighting the need for appropriate cell culture models to understand the role of Hh molecules in tissue repair.

VI. Coculture Models as *In Vitro* Models to Investigate Effects of Shh on Osteogenesis and Angiogenesis

In order to investigate the distinct role and the mechanisms by which Shh contributes to repair processes both animal models and *in vitro* models have been used, as described above by several groups in this field. To analyze the role of Shh-mediated signaling processes in the direct and indirect interaction of individual cell types involved in bone repair and neovascularization, advanced *in vitro* systems such as coculture systems (Dohle *et al.*, 2010, 2011; Levi *et al.*, 2011) might be a valuable scientific tool. Such cell culture systems permit on the one hand study of complex cellular and molecular relationships in well-defined models and under controlled conditions. On the other hand, they are able to mimic a kind of regenerative niche, resulting in a close approximation to the repair process within a tissue. In this context, we have developed coculture systems-based outgrowth endothelial cells (OECs) and primary osteoblasts (Fuchs *et al.*, 2007, 2009b). Coculture approaches are currently discussed as a potential therapeutic approach to improve vascularization of bone tissues and can also be used as advanced model systems to assess the mechanisms underlying the cellular interaction during bone repair and vascularization (Grellier *et al.*, 2009; Rivron *et al.*, 2008).

OECs are a unique cell population appearing as distinct colonies with endothelial characteristics in culture of mononuclear cells from the peripheral blood (Fuchs *et al.*, 2006, 2010). These cells are discussed as a potential autologous cell source for vascularization strategies in tissue engineering. In coculture with osteoblasts or other cell types, these cells develop a primitive vascular network which might also facilitate a fast connection of tissue engineered constructs to the vascular supply after implantation (Au *et al.*, 2008; Fuchs *et al.*, 2009a; Melero-Martin *et al.*, 2007). These effects seem to be associated with the natural supply of angiogenic growth factors and proangiogenic matrix components provided by the osteoblasts, which positively enhance the angiogenic activity of OECs in the system. On the other hand, also osteogenic differentiation of osteoblasts in such systems was observed, suggesting that these cocultures are mimicking a kind of bone repair niche in a simplified but effective way.

VII. Effects of Shh on Angiogenesis in a Coculture Model

In previous studies, we have shown that Shh might also guide the angiogenic activities of OEC in the coculture system in which both direct effects of Shh on the angiogenic activity and indirect effects mediated by the paracrine factors downstream of Shh play a significant role. Direct effects of Shh on the angiogenic activity of OEC can be observed for instance in simple angiogenesis experiments by seeding OEC on a Matrigel® matrix. Treatment of OEC in this experimental setting leads to the formation of a primitive vascular network (Fig. 22.2), whereas in the presence of the Shh inhibitor cyclopamine the proangiogenic effect of Shh is blocked. Nevertheless, little is known about the mechanisms behind these direct proangiogenic effects of Shh on endothelial cells. These phenomena might be at least partly mediated by a pathway independent of the classical Shh signaling cascade through patched and Gli such as rho kinase-dependent signaling pathways guiding tubulogenesis in endothelial cells, as recently shown by studies from two different groups (Chinchilla *et al.*, 2010; Renault *et al.*, 2010).

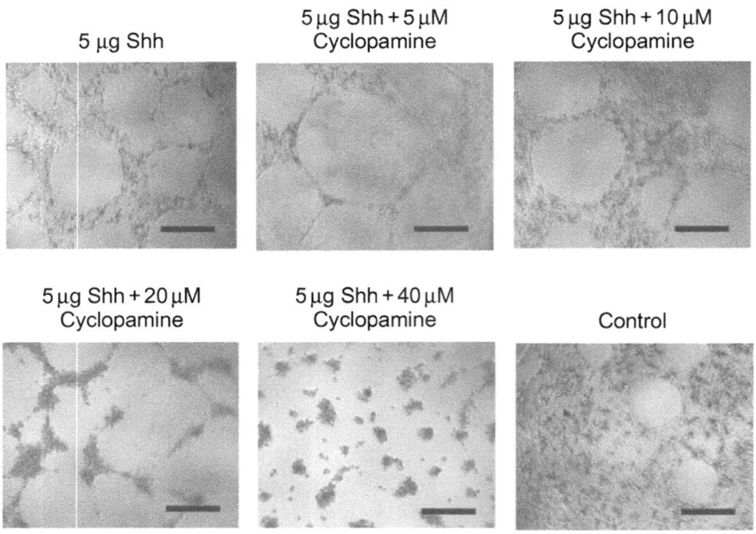

Figure 22.2 OEC seeded on Matrigel® Basement Membrane Matrix: Simultaneous treatment of OEC monocultures with recombinant human sonic hedgehog and different concentrations of the Shh inhibitor cyclopamine (5–40 µM) for 24 h. $n=3$. Scale bar: 200 µm. (See Color Insert.)

Such direct effects of Shh might as well play a role in the coculture, as indicated by similar blocking experiments based on cyclopamine. On the other hand, the proangiogenic effect of Shh is also mediated by indirect effects through the upregulation of proangiogenic factors in non-endothelial cell types, such as fibroblasts, mesenchymal interstitial cells, or cardiomyocytes (Kusano *et al.*, 2005; Palladino *et al.*, 2011; Pola *et al.*, 2003). In the coculture of primary osteoblasts and OEC, the proangiogenic effect of Shh is partly mediated by the stimulation of angiopoietin-1, as well as VEGF produced by the osteoblasts in the coculture (Dohle *et al.*, 2010). At the same time, Shh also has a positive effect on angiopoietin-2 expression by the endothelial cell population in the coculture system. This interaction seems to favor a balance of factors guiding angiogenic activation and the formation of vascular structures mediated by VEGF and angiopoietin-2 on one hand and vessel stabilization through angiopoietin-1 on the other. Shh might offer several additional advantages in comparison to VEGF, which is commonly used as a pro-angiogenic agent in the field of bone tissue engineering (Elcin *et al.*, 2001; Sun *et al.*, 2010). Shh seems to have the potential to promote both elemental processes in neovascularization, new vessel formation, and stabilization via mural cells as previously shown in a comparative study (Dohle *et al.*, 2011) (compare also fig. 22.3). In this study, we have shown that, in contrast to VEGF and in accordance with its multifunctional

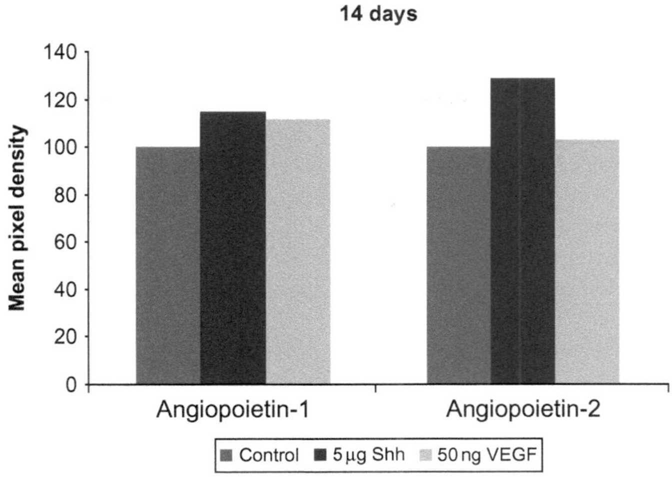

Figure 22.3 Determination of the intracellular protein level of angiopoietin-1 and angiopoietin-2 in cocultures in response to Shh or VEGF stimulation. The protein amount in the cell lysates of cocultures was evaluated using a Proteome Profiler™ Human Angiogenesis Array (n=3). Results are referred to a positive control and depicted as mean pixel densities, which were finally converted into percentage values. All individual values were referred to control coculture set at 100%.

potential, Shh induces several factors involved in vessel maturation and stabilization. The missing vascular stabilization is still a major problem and a limiting factor of VEGF-based approaches to enhance the vascularization process. Shh treatment resulted in effects on several groups of molecules involved in vessel stabilization including angiopoietin-1, PDGF, and TGF-β, as well as in an increase of smooth muscle actin at both the protein and gene expression level in coculture systems. Further, in the presence of the Shh pathway inhibitor, cyclopamine, this functional relationship between Shh and factors involved in vascular stabilization was abolished.

VIII. Effects of Shh on Osteogenesis in a Coculture Model

In addition to the effects of Shh on the angiogenic potential of endothelial cells and on factors involved in vessel maturation in the coculture, Shh further enhanced multiple key features involved in osteogenic differentiation, suggesting a synergistic effect of Shh on both processes during bone formation. In the cocultures, Shh treatment resulted in an increased osteogenic differentiation as detected by increased levels of marker expression associated with osteogenic differentiation, alkaline phosphatase (ALP) activity, and higher rates of calcification. We have already shown that Shh treatment of cocultures consisting of primary osteoblasts and OECs resulted in an upregulation of osteogenic markers, such as osteocalcin, osteonectin, osteopontin, and alkaline phosphatase at the mRNA level after 24h of stimulation (Dohle *et al.*, 2010). In accordance, the concentration of alizarin red, used as marker for mineralization processes in the coculture system in response to Shh or VEGF treatment, was clearly higher in Shh-treated cocultures after 14 days of stimulation, compared to the cultures treated with VEGF and control cultures (Fig. 22.4A).

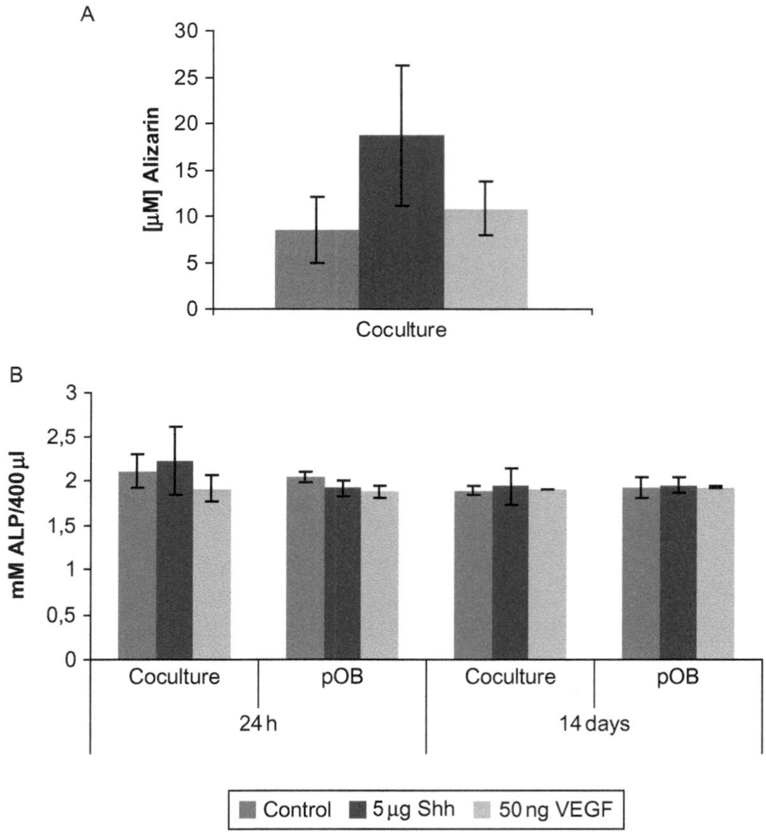

Figure 22.4 Effect of Shh and VEGF on osteoblastic differentiation in the coculture system consisting of pOB and OEC. (A) Quantification of calcification (alizarin red) in Shh-treated cocultures (dark bars) after 14 days (n=3) of Shh and VEGF stimulation compared to unstimulated cultures. Values represent averages of at least three independent experiments in μM alizarin red/mg protein. (B) ALP activity within the supernatants of Shh and VEGF treated cocultures and pOB monocultures after 14 days of stimulation compared with unstimulated cultures. n=3.

In addition, the activity of ALP was determined in cocultures in response to the different treatments, as depicted in Fig. 22.4B. The highest ALP activity could be found in Shh-stimulated cocultures after 14 days.

In summary, Shh treatment simultaneously enhanced angiogenesis, vascular maturation, and osteogenic differentiation, again highlighting the multifaceted potential of this molecule. In addition, the coculture model based on endothelial cells and osteoblasts can be considered as a valuable tool to understand the role of Shh in cellular interaction and might be applied as an advanced *in vitro* model. In this context, these models also have the potential to complement as well as to reduce animal studies in the search for therapeutically relevant molecules.

IX. Future Therapeutical Options and Delivery Strategies

The beneficial effects of Shh on bone regeneration and vascularization might also be used for therapeutic applications. A recent *in vivo* study based on sequential delivery of basic fibroblast growth factor and Shh by cotransduction of mesenchymal stem cells resulted in an increase of osteogenic markers and bone area, as well as in higher vascular densities than in control groups (Song *et al.*, 2011). Also in terms of vascularization, gene therapy approaches based on Shh result in pleiotropic effects, including the improvement of angiogenesis and vasculogenesis as well as in the upregulation of angiogenic growth factors in a peripheral limb ischemia model (Palladino *et al.*, 2011). These kind of synergistic effects highlight the therapeutic potential of Shh. Nevertheless, considering Shh as a possible therapeutic agent, potential side effects have to be thoroughly investigated. Such side effects might include effects on tissue patterning which is strictly controlled during embryonic development by the concentration profiles of Shh. Further, effects in the hematopoietic niche after systematic application of Shh, resulting in a decrease in hematopoietic stem cells need to be considered in systemic delivery approaches. Moreover, altered Shh signaling is associated with malformations in various tissues and even tumor development. It is certainly a delicate task to develop delivery strategies for Shh which cope with the strict control of Shh profiles found in the tissue. Spatial Hh profiles in the tissue are regulated by binding to cholesterol in the cellular plasma membrane, which then enables the creation of distinct gradients within cellular assemblies (Ingham, 2008). It is well known that Shh signaling is gradient dependent with tissue- or site-specific threshold levels (Ingham, 2008; Young *et al.*, 2010). A lipid-modified, soluble form fraction is released from Shh-secreting cells as micelle-like complexes or as lipoprotein particles, thus allowing a more rapid distribution of Shh away from the source. These two mechanisms guiding short- and long-distance activity of Shh highlight the complexity in regulation of Hh distribution during tissue development or repair. Thus, choosing a suitable application method from several options, such as systematic delivery, locally applied dosage forms, recombinant viral delivery strategies, DNA vectors, or *in vitro* pretreatment of tissue constructs, might be one of the key issues to ensure the therapeutic success and to minimize the undesired and potentially harmful side effects of such potent molecules.

REFERENCES

Asai, J., Takenaka, H., Kusano, K. F., Ii, M., Luedemann, C., Curry, C., Eaton, E., Iwakura, A., Tsutsumi, Y., Hamada, H., Kishimoto, S., Thorne, T., *et al.* (2006). Topical sonic hedgehog gene therapy accelerates wound healing in diabetes by enhancing endothelial progenitor cell-mediated microvascular remodeling. *Circulation* **113,** 2413–2424.

Au, P., Daheron, L. M., Duda, D. G., Cohen, K. S., Tyrrell, J. A., Lanning, R. M., Fukumura, D., Scadden, D. T., and Jain, R. K. (2008). Differential in vivo potential of endothelial progenitor cells from human umbilical cord blood and adult peripheral blood to form functional long-lasting vessels. *Blood* **111,** 1302–1305.

Bitgood, M. J., Shen, L., and McMahon, A. P. (1996). Sertoli cell signaling by Desert hedgehog regulates the male germline. *Curr. Biol.* **6,** 298–304.

Bumcrot, D. A., Takada, R., and McMahon, A. P. (1995). Proteolytic processing yields two secreted forms of sonic hedgehog. *Mol. Cell. Biol.* **15,** 2294–2303.

Carmeliet, P. (2003). Angiogenesis in health and disease. *Nat. Med.* **9,** 653–660.

Chiang, C., Litingtung, Y., Lee, E., Young, K. E., Corden, J. L., Westphal, H., and Beachy, P. A. (1996). Cyclopia and defective axial patterning in mice lacking Sonic hedgehog gene function. *Nature* **383,** 407–413.

Chinchilla, P., Xiao, L., Kazanietz, M. G., and Riobo, N. A. (2010). Hedgehog proteins activate pro-angiogenic responses in endothelial cells through non-canonical signaling pathways. *Cell Cycle* **9,** 570–579.

Cohen, M. M., Jr. (2003). The hedgehog signaling network. *Am. J. Med. Genet. A* **123A,** 5–28.

Dohle, E., Fuchs, S., Kolbe, M., Hofmann, A., Schmidt, H., and Kirkpatrick, C. J. (2010). Sonic hedgehog promotes angiogenesis and osteogenesis in a coculture system consisting of primary osteoblasts and outgrowth endothelial cells. *Tissue Eng. Part A* **16,** 1235–1237.

Dohle, E., Fuchs, S., Kolbe, M., Hofmann, A., Schmidt, H., and Kirkpatrick, C. J. (2011). Comparative study assessing effects of sonic hedgehog and VEGF in a human co-culture model for bone vascularisation strategies. *Eur. Cell. Mater.* **21,** 144–156.

Echelard, Y., Epstein, D. J., St-Jacques, B., Shen, L., Mohler, J., McMahon, J. A., and McMahon, A. P. (1993). Sonic hedgehog, a member of a family of putative signaling molecules, is implicated in the regulation of CNS polarity. *Cell* **75,** 1417–1430.

Edwards, P. C., Ruggiero, S., Fantasia, J., Burakoff, R., Moorji, S. M., Paric, E., Razzano, P., Grande, D. A., and Mason, J. M. (2004). Sonic hedgehog gene-enhanced tissue engineering for bone regeneration. *Gene Ther.* **12,** 75–86.

Ekker, S. C., Ungar, A. R., Greenstein, P., von Kessler, D. P., Porter, J. A., Moon, R. T., and Beachy, P. A. (1995). Patterning activities of vertebrate hedgehog proteins in the developing eye and brain. *Curr. Biol.* **5,** 944–955.

Elcin, Y. M., Dixit, V., and Gitnick, G. (2001). Extensive in vivo angiogenesis following controlled release of human vascular endothelial cell growth factor: Implications for tissue engineering and wound healing. *Artif. Organs* **25,** 558–565.

Fietz, M. J., Concordet, J. P., Barbosa, R., Johnson, R., Krauss, S., McMahon, A. P., Tabin, C., and Ingham, P. W. (1994). The hedgehog gene family in Drosophila and vertebrate development. *Dev. Supplement S* 43–51.

Frontini, M. J., Nong, Z., Gros, R., Drangova, M., O'Neil, C., Rahman, M. N., Akawi, O., Yin, H., Ellis, C. G., and Pickering, J. G. (2011). Fibroblast growth factor 9 delivery during angiogenesis produces durable, vasoresponsive microvessels wrapped by smooth muscle cells. *Nat. Biotech.* **29,** 421–427.

Fuchs, S., Hermanns, M. I., and Kirkpatrick, C. J. (2006). Retention of a differentiated endothelial phenotype by outgrowth endothelial cells isolated from human peripheral blood and expanded in long-term cultures. *Cell Tissue Res.* **326,** 79–92.

Fuchs, S., Hofmann, A., and Kirkpatrick, C. J. (2007). Microvessel-like structures from outgrowth endothelial cells from human peripheral blood in 2-dimensional and 3-dimensional co-cultures with osteoblastic lineage cells. *Tissue Eng.* **13,** 2577–2588.

Fuchs, S., Ghanaati, S., Orth, C., Barbeck, M., Kolbe, M., Hofmann, A., Eblenkamp, M., Gomes, M., Reis, R. L., and Kirkpatrick, C. J. (2009a). Contribution of outgrowth endothelial cells from human peripheral blood on in vivo vascularization of bone tissue engineered constructs based on starch polycaprolactone scaffolds. *Biomaterials* **30,** 526–534.

Fuchs, S., Jiang, X., Schmidt, H., Dohle, E., Ghanaati, S., Orth, C., Hofmann, A., Motta, A., Migliaresi, C., and Kirkpatrick, C. J. (2009b). Dynamic processes involved in the pre-vascularization of silk fibroin constructs for bone regeneration using outgrowth endothelial cells. *Biomaterials* **30,** 1329–1338.

Fuchs, S., Dohle, E., Kolbe, M., and Kirkpatrick, C. J. (2010). Outgrowth endothelial cells: Sources, characteristics and potential applications in tissue engineering and regenerative medicine. *Adv. Biochem. Eng. Biotechnol.* **123,** 201–217.

Fujii, T., and Kuwano, H. (2010). Regulation of the expression balance of angiopoietin-1 and angiopoietin-2 by Shh and FGF-2. *In Vitro Cell. Dev. Biol. Anim.* **46,** 487–491.

Grellier, M., Bordenave, L., and Amedee, J. (2009). Cell-to-cell communication between osteogenic and endothelial lineages: Implications for tissue engineering. *Trends Biotechnol.* **27,** 562–571.

Ingham, P. W. (2008). Hedgehog signalling. *Curr. Biol.* **18,** R238–R241.

Ingham, P. W., and McMahon, A. P. (2001). Hedgehog signaling in animal development: Paradigms and principles. *Genes Dev.* **15,** 3059–3087.

Ito, H., Akiyama, H., Shigeno, C., Iyama, K.-i., Matsuoka, H., and Nakamura, T. (1999). Hedgehog signaling molecules in bone marrow cells at the initial stage of fracture repair. *Biochem. Biophys. Res. Commun.* **262,** 443–451.

Iwasaki, M., Le, A. X., and Helms, J. A. (1997). Expression of indian hedgehog, bone morphogenetic protein 6 and gli during skeletal morphogenesis. *Mech. Dev.* **69,** 197–202.

James, A. W., Leucht, P., Levi, B., Carre, A. L., Xu, Y., Helms, J. A., and Longaker, M. T. (2010). Sonic hedgehog influences the balance of osteogenesis and adipogenesis in mouse adipose-derived stromal cells. *Tissue Eng. Part A* **16,** 2605–2616.

Jemtland, R., Divieti, P., Lee, K., and Segre, G. V. (2003). Hedgehog promotes primary osteoblast differentiation and increases PTHrP mRNA expression and iPTHrP secretion. *Bone* **32,** 611–620.

Johnson, R. L., Laufer, E., Riddle, R. D., and Tabin, C. (1994a). Ectopic expression of Sonic hedgehog alters dorsal-ventral patterning of somites. *Cell* **79,** 1165–1173.

Johnson, R. L., Riddle, R. D., Laufer, E., and Tabin, C. (1994b). Sonic hedgehog: A key mediator of anterior-posterior patterning of the limb and dorso-ventral patterning of axial embryonic structures. *Biochem. Soc. Trans.* **22,** 569–574.

Kinto, N., Iwamoto, M., Enomoto-Iwamoto, M., Noji, S., Ohuchi, H., Yoshioka, H., Kataoka, H., Wada, Y., Yuhao, G., Takahashi, H. E., Yoshiki, S., and Yamaguchi, A. (1997). Fibroblasts expressing Sonic hedgehog induce osteoblast differentiation and ectopic bone formation. *FEBS Lett.* **404,** 319–323.

Kiuru, M., Hidaka, C., Hubner, R. H., Solomon, J., Krause, A., Leopold, P. L., and Crystal, R. G. (2009a). Sonic hedgehog expands diaphyseal trabecular bone altering bone marrow niche and lymphocyte compartment. *Mol. Ther.* **17,** 1442–1452.

Kiuru, M., Solomon, J., Ghali, B., van der Meulen, M., Crystal, R. G., and Hidaka, C. (2009b). Transient overexpression of sonic hedgehog alters the architecture and mechanical properties of trabecular bone. *J. Bone Miner. Res.* **24,** 1598–1607.

Kusano, K. F., Pola, R., Murayama, T., Curry, C., Kawamoto, A., Iwakura, A., Shintani, S., Ii, M., Asai, J., Tkebuchava, T., Thorne, T., Takenaka, H., *et al.* (2005). Sonic hedgehog myocardial gene therapy: Tissue repair through transient reconstitution of embryonic signaling. *Nat. Med.* **11,** 1197–1204.

Levi, B., James, A. W., Nelson, E. R., Li, S., Peng, M., Commons, G. W., Lee, M., Wu, B., and Longaker, M. T. (2011). Human adipose-derived stromal cells stimulate autogenous skeletal repair via paracrine hedgehog signaling with calvarial osteoblasts. *Stem Cells Dev.* **20,** 243–257.

Li, F., Duman-Scheel, M., Yang, D., Du, W., Zhang, J., Zhao, C., Qin, L., and Xin, S. (2010). Sonic hedgehog signaling induces vascular smooth muscle cell proliferation via induction of the G1 cyclin-retinoblastoma axis. *Arterioscler. Thromb. Vasc. Biol.* **30,** 1787–1794.

Marigo, V., Scott, M. P., Johnson, R. L., Goodrich, L. V., and Tabin, C. J. (1996). Conservation in hedgehog signaling: Induction of a chicken patched homolog by Sonic hedgehog in the developing limb. *Development* **122,** 1225–1233.

Melero-Martin, J. M., Khan, Z. A., Picard, A., Wu, X., Paruchuri, S., and Bischoff, J. (2007). In vivo vasculogenic potential of human blood-derived endothelial progenitor cells. *Blood* **109,** 4761–4768.

Miyaji, T., Nakase, T., Iwasaki, M., Kuriyama, K., Tamai, N., Higuchi, C., Myoui, A., Tomita, T., and Yoshikawa, H. (2003). Expression and distribution of transcripts for sonic hedgehog in the early phase of fracture repair. *Histochem. Cell Biol.* **119,** 233–237.

Nagase, T., Nagase, M., Yoshimura, K., Fujita, T., and Koshima, I. (2005). Angiogenesis within the developing mouse neural tube is dependent on sonic hedgehog signaling: Possible roles of motor neurons. *Genes Cells* **10,** 595–604.

Nagase, M., Nagase, T., Koshima, I., and Fujita, T. (2006a). Critical time window of hedgehog-dependent angiogenesis in murine yolk sac. *Microvasc. Res.* **71,** 85–90.

Nagase, T., Nagase, M., Yoshimura, K., Machida, M., and Yamagishi, M. (2006b). Defects in aortic fusion and craniofacial vasculature in the holoprosencephalic mouse embryo under inhibition of sonic hedgehog signaling. *J. Craniofac. Surg.* **17,** 736–744.

Nagase, T., Nagase, M., Machida, M., and Yamagishi, M. (2007). Hedgehog signaling: A biophysical or biomechanical modulator in embryonic development? *Ann. N. Y. Acad. Sci.* **1101,** 412–438.

Nakamura, T., Aikawa, T., Iwamoto-Enomoto, M., Iwamoto, M., Higuchi, Y., Pacifici, M., Kinto, N., Yamaguchi, A., Noji, S., Kurisu, K., and Matsuya, T. (1997). Induction of osteogenic differentiation by hedgehog proteins. *Biochem. Biophys. Res. Commun.* **237,** 465–469.

Nusslein-Volhard, C., and Wieschaus, E. (1980). Mutations affecting segment number and polarity in Drosophila. *Nature* **287,** 795–801.

Palladino, M., Gatto, I., Neri, V., Straino, S., Silver, M., Tritarelli, A., Piccioni, A., Smith, R. C., Gaetani, E., Losordo, D. W., Crea, F., Capogrossi, M., *et al.* (2011). Pleiotropic beneficial effects of sonic hedgehog gene therapy in an experimental model of peripheral limb ischemia. *Mol. Ther.* **19,** 658–666.

Passman, J. N., Dong, X. R., Wu, S.-P., Maguire, C. T., Hogan, K. A., Bautch, V. L., and Majesky, M. W. (2008). A sonic hedgehog signaling domain in the arterial adventitia supports resident Sca1+ smooth muscle progenitor cells. *Proc. Natl. Acad. Sci. USA* **105,** 9349–9354.

Pepicelli, C. V., Lewis, P. M., and McMahon, A. P. (1998). Sonic hedgehog regulates branching morphogenesis in the mammalian lung. *Curr. Biol.* **8,** 1083–1086.

Pola, R., Ling, L. E., Silver, M., Corbley, M. J., Kearney, M., Blake Pepinsky, R., Shapiro, R., Taylor, F. R., Baker, D. P., Asahara, T., and Isner, J. M. (2001). The morphogen Sonic hedgehog is an indirect angiogenic agent upregulating two families of angiogenic growth factors. *Nat. Med.* **7,** 706–711.

Pola, R., Ling, L. E., Aprahamian, T. R., Barban, E., Bosch-Marce, M., Curry, C., Corbley, M., Kearney, M., Isner, J. M., and Losordo, D. W. (2003). Postnatal recapitulation of embryonic hedgehog pathway in response to skeletal muscle ischemia. *Circulation* **108,** 479–485.

Renault, M. A., Roncalli, J., Tongers, J., Thorne, T., Klyachko, E., Misener, S., Volpert, O. V., Mehta, S., Burg, A., Luedemann, C., Qin, G., Kishore, R., *et al.* (2010). Sonic hedgehog induces angiogenesis via Rho kinase-dependent signaling in endothelial cells. *J. Mol. Cell. Cardiol.* **49,** 490–498.

Rivron, N. C., Liu, J. J., Rouwkema, J., de Boer, J., and van Blitterswijk, C. A. (2008). Engineering vascularised tissues in vitro. *Eur. Cell. Mater.* **15,** 27–40.

Roncalli, J., Renault, M.-A., Tongers, J., Misener, S., Thorne, T., Kamide, C., Jujo, K., Tanaka, T., Ii, M., Klyachko, E., and Losordo, D. W. (2011). Sonic Hedgehog-induced functional recovery after myocardial infarction is enhanced by AMD3100-mediated progenitor-cell mobilization. *J. Am. Coll. Cardiol.* **57,** 2444–2452.

Ruiz i Altaba, A. (1994). Pattern formation in the vertebrate neural plate. *Trends Neurosci.* **17,** 233–243.

Song, K., Rao, N. J., Chen, M. L., Huang, Z. J., and Cao, Y. G. (2011). Enhanced bone regeneration with sequential delivery of basic fibroblast growth factor and sonic hedgehog. *Injury*.

St-Jacques, B., Hammerschmidt, M., and McMahon, A. P. (1999). Indian hedgehog signaling regulates proliferation and differentiation of chondrocytes and is essential for bone formation. *Genes Dev.* **13,** 2072–2086.

Stone, D. M., Hynes, M., Armanini, M., Swanson, T. A., Gu, Q., Johnson, R. L., Scott, M. P., Pennica, D., Goddard, A., Phillips, H., Noll, M., Hooper, J. E., *et al.* (1996). The tumour-suppressor gene patched encodes a candidate receptor for Sonic hedgehog. *Nature* **384,** 129–134.

Sun, Q., Silva, E. A., Wang, A., Fritton, J. C., Mooney, D. J., Schaffler, M. B., Grossman, P. M., and Rajagopalan, S. (2010). Sustained release of multiple growth factors from injectable polymeric system as a novel therapeutic approach towards angiogenesis. *Pharm. Res.* **27,** 264–271.

van der Horst, G., Farih-Sips, H., Lowik, C. W., and Karperien, M. (2003). Hedgehog stimulates only osteoblastic differentiation of undifferentiated KS483 cells. *Bone* **33,** 899–910.

Vortkamp, A., Pathi, S., Peretti, G. M., Caruso, E. M., Zaleske, D. J., and Tabin, C. J. (1998). Recapitulation of signals regulating embryonic bone formation during postnatal growth and in fracture repair. *Mech. Dev.* **71,** 65–76.

Wang, Q., Huang, C., Zeng, F., Xue, M., and Zhang, X. (2010). Activation of the Hh pathway in periosteum-derived mesenchymal stem cells induces bone formation in vivo: Implication for postnatal bone repair. *Am. J. Pathol.* **177,** 3100–3111.

Yamazaki, M., Nakamura, K., Mizukami, Y., Ii, M., Sasajima, J., Sugiyama, Y., Nishikawa, T., Nakano, Y., Yanagawa, N., Sato, K., Maemoto, A., Tanno, S., *et al.* (2008). Sonic hedgehog derived from human pancreatic cancer cells augments angiogenic function of endothelial progenitor cells. *Cancer Sci.* **99,** 1131–1138.

Young, N. M., Chong, H. J., Hu, D., Hallgrimsson, B., and Marcucio, R. S. (2010). Quantitative analyses link modulation of sonic hedgehog signaling to continuous variation in facial growth and shape. *Development* **137,** 3405–3409.

Yuasa, T., Kataoka, H., Kinto, N., Iwamoto, M., Enomoto-Iwamoto, M., Iemura, S., Ueno, N., Shibata, Y., Kurosawa, H., and Yamaguchi, A. (2002). Sonic hedgehog is involved in osteoblast differentiation by cooperating with BMP-2. *J. Cell. Physiol.* **193,** 225–232.

CHAPTER TWENTY-THREE

Hedgehog Inhibition as an Anti-Cancer Strategy

G. Praveen Raju *and* Diane Pham

Contents

Abstract

Dysregulated Hedgehog (Hh) signaling has been implicated in a growing number of human cancers. Although first identified as an important developmental signaling pathway crucial for cellular proliferation, differentiation, and migration during organogenesis in invertebrates, these fundamental processes have been co-opted in human cancers. Initial evidence for the Hh pathway in tumor biology comes from mutations of signaling pathway components in a hereditary cancer syndrome that typically results in basal-cell carcinoma and medulloblastoma. Subsequent analysis revealed that Hh pathway mutations are found in sporadic tumors as well as activated Hh signaling in several epithelial cancers independent of Hh pathway mutation status. Further, recent evidence has demonstrated paracrine Hh signaling within stromal cells of the tumor microenvironment with implications for drug delivery. Several Hh antagonists targeting the Hh receptor, Smoothened (SMO), have been developed and show efficacy in preclinical studies and early-stage clinical trials in humans. However, major issues with these small molecule compounds include rapid

Department of Pediatrics, Division of Child Neurology, Weill Cornell Medical College, New York, USA

Vitamins and Hormones, Volume 88
ISSN 0083-6729, DOI: 10.1016/B978-0-12-394622-5.00023-7

acquired resistance, potential developmental toxicities secondary to use in children, and limited efficacy in cancers driven by Hh signaling downstream of the SMO receptor.

I. Introduction

The secreted signaling molecule, Hedgehog (Hh), and several components of its signal transduction pathway were initially identified secondary to their roles in early embryonic patterning through classical genetic mutagenesis screens in Drosophila (Nusslein-Volhard and Wieschaus, 1980). Subsequent analysis has revealed that most of the components of the pathway are conserved in higher organisms, including humans. Although the Hh signal transduction pathway has been described in detail in other chapters of this volume and in recent reviews (Ingham *et al.*, 2011; Ng and Curran, 2011), the pathway in vertebrates will be summarized here as an introduction to this chapter that describes the dysregulation of Hh signaling in various cancers as well as within stromal cells of the tumor microenvironment.

In brief, Hh ligands bind to a 12-pass transmembrane receptor, Patched 1 (PTCH1), a negative regulator of the pathway, that results in its internalization and subsequent degradation. This results in release of the negative influence of PTCH1 on the 16-pass transmembrane protein, Smoothened (SMO), which then can enter the primary cilia through a poorly understood mechanism. Activated SMO then promotes dissociation of a Suppressor of fused (SUFU)-glioma associated oncogene homolog (GLI) protein complex that results in alterations in proteolytic processing of GLI proteins. This results in nuclear translocation and activation of the GLI2 transcription factor and degradation of the repressor, GLI3. The activated GLI2 protein stimulates the transcription of Hh pathway target genes, many of which are critical for important processes including cell proliferation, differentiation, and migration. Normally, the precise timing and duration of Hh signaling results in proper embryonic patterning as well as tissue differentiation during organogenesis (McMahon *et al.*, 2003). However, alterations of the Hh pathway, primarily through mutations of pathway components or through mechanisms that result in upregulated GLI transcriptional activity, often result in dysregulated cell proliferation thus resulting in cancer.

II. Hh Signaling in Normal Development

The multiple roles for Hh signaling in normal developmental processes must be emphasized to better understand the mechanisms by which Hh pathway alterations contribute to oncogenesis as well as the potential consequences of Hh pathway inhibition for cancer therapy. In mammals, three related proteins, Desert Hedgehog (Dhh), Indian Hedgehog (Ihh), and Sonic Hedgehog (Shh), function as ligands for PTCH1. All mammalian Hh proteins are thought to have similar physiological effects on proliferation and/or differentiation; however, the differences in their roles in development likely result from the diverse and unique expression patterns of the ligands in various tissues (McMahon *et al.*, 2003).

Shh is the most broadly expressed mammalian Hh ligand. The initial evidence for the importance of Hh signaling in mammalian development was provided by observations that mutations in Shh cause holoprosencephaly (Belloni *et al.*, 1996; Chiang *et al.*, 1996; Roessler *et al.*, 1996), a developmental disorder that affects midline morphogenesis of the face and nervous system resulting in cyclopia and holoprosencephaly, respectively. Interestingly, a similar malformation phenotype had long been associated with the exposure of developing lambs to naturally occurring teratogenic alkaloids such as cyclopamine (Binns *et al.*, 1962), which binds the SMO protein (Chen *et al.*, 2002b; Cooper *et al.*, 1998; Incardona *et al.*, 1998; Scales and de Sauvage, 2009) and has served as the basis for the use of several Hh pathway inhibitors currently in clinical development.

Although most studies of Hh signaling focus on the Shh pathway, the other two ligands, Dhh and Ihh, share many of the same downstream signaling components, thus suggesting that Hh pathway inhibition will likely have toxicities related to inhibiting their normal functions. The Dhh ligand appears to be more abundant in the reproductive organs of both males and females, including Sertoli cells of testis and granulosa cells of ovaries (Bitgood *et al.*, 1996; Varjosalo and Taipale, 2008). Consistent with this, male mice lacking Dhh are infertile due to the complete absence of mature sperm (Bitgood *et al.*, 1996). Ihh plays an important role in skeletal development, primarily in cortical bone and long bone formation. The most severe manifestation of Ihh deficiency can be seen in children with acrocapitofemoral dysplasia, who have short stature and bone defects (Hellemans *et al.*, 2003).

III. Hh Pathway Alterations in Cancer

A. Ligand-independent mechanisms

Genetic studies in Drosophila had identified the Hh receptor, PTCH1, as a negative regulator of Hh signal transduction. Binding of Hh ligand to PTCH1 relieves this inhibitory effect on the pathway, thus resulting in pathway activation. The initial evidence for the role of Hh signaling in cancer centered on identified mutations within the PTCH1 gene in patients with hereditary cancer syndromes. Specifically, patients with the hereditary

Gorlin's syndrome, or basal-cell nevus syndrome, develop basal-cell carcinoma skin tumors, and less frequently the cerebellar brain tumor, medulloblastoma. Through genetic linkage studies of families with basal-cell nevus syndrome, the locus carrying the mutant gene was mapped to chromosome 9q22 and subsequently to the PTCH1 gene (Hahn *et al.*, 1996; Johnson *et al.*, 1996; Lench *et al.*, 1997). Subsequent tumor analyses in patients with sporadic basal-cell carcinoma and medulloblastoma similarly found point mutations in the PTCH1 gene (Raffel *et al.*, 1997; Xie *et al.*, 1997a,b), firmly establishing a clear contribution of increased Hh signaling to cancer. These genetic alterations typically resulted in either chromosomal loss of the region containing PTCH1 or point mutations resulting in truncations and thus inactive PTCH1 protein. Both of the mechanisms would presumably result in constitutive, ligand-independent activation of Hh signaling.

Subsequent mutational analyses of other Hh signaling pathway component have identified additional genetic alterations in basal-cell carcinoma and medulloblastoma independent of the PTCH1 gene. In basal-cell carcinoma, approximately 10% of tumors containing activating point mutations in the downstream SMO gene, which presumably alters SMO to be resistant to PTCH1 inhibition, therefore, similarly resulting in ligand-independent constitutive pathway activation (Reifenberger *et al.*, 2005; Xie *et al.*, 1997a,b). In medulloblastoma patients, activating SMO mutations in addition to inactivating mutations of the negative SUFU gene have been similarly identified (Taylor *et al.*, 2002).

Finally, molecular lesions that result in increased transcriptional activity of the downstream GLI1 and GLI2 transcription factors have been described in several other types of cancer. GLI1 amplifications have been observed in adult patients with the brain tumor, glioblastoma, as well as in childhood sarcoma (Kinzler *et al.*, 1987; Roberts *et al.*, 1989). However, a direct role for GLI gene amplification has been difficult to elicit in these cases as the amplicons have been found to include other putative oncogenes in the adjacent chromosomal regions. Additionally, translocations of GLI1 and ACTB, encoding the β-actin protein, have been identified and define a new subtype of pericytoma (Dahlen *et al.*, 2004).

B. Ligand-dependent mechanisms

After the identification of cyclopamine as a Hh antagonist, several additional tumor types without evidence of Hh pathway mutations were shown to be sensitive to SMO inhibition. Although the initial interpretation of these *in vitro* and *in vivo* studies suggested an autocrine feedback loop in tumors that produced Hh ligands, it was subsequently shown in several xenograft studies in mice that the Hh target genes were actually upregulated in stromal cells of the microenvironment or the inhibition of tumor cell growth may be secondary to off-target effects of high dose cyclopamine independent of

Hh inhibition (Yauch *et al.*, 2008). However, the paracrine role of Hh signaling has been implicated in a growing list of epithelial tumors including small-cell lung cancer, pancreatic, and other gastrointestinal tract malignancies, as well as prostate cancer (Berman *et al.*, 2003; Theunissen and de Sauvage, 2009; Watkins *et al.*, 2003). Finally, a reverse mode of paracrine signaling from Hh ligand produced in stromal cells of the microenvironment to support hematological malignancies including some B-cell lymphomas, multiple myeloma, and leukemias has been described (Yang *et al.*, 2010).

In summary, several Hh signaling mechanisms encompassing ligand-independent and ligand-dependent mechanisms have been described in several cancers (see Fig. 23.1). Importantly, in addition to targeting tumor

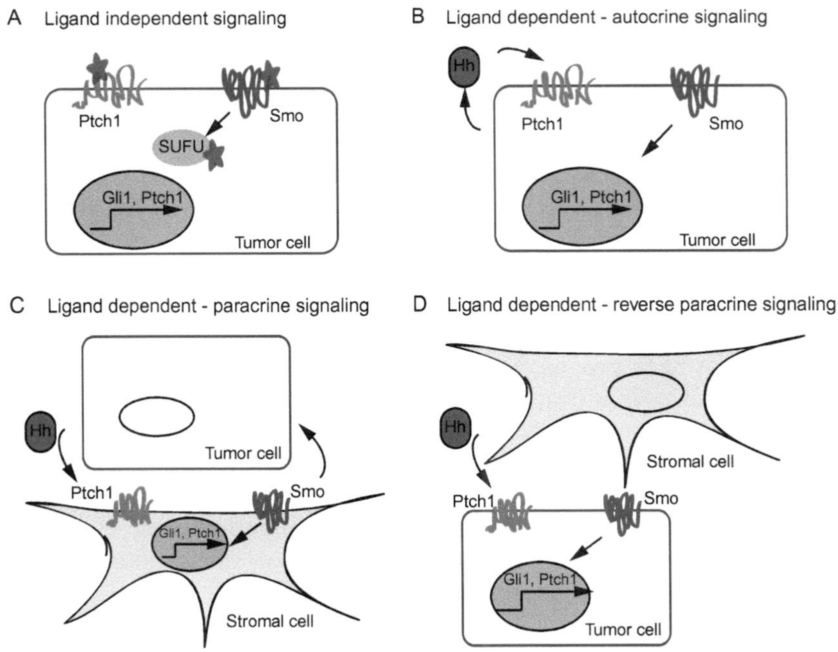

Figure 23.1 Mechanisms of Hedgehog pathway activation in cancer. (A) Ligand-independent signaling typically occurs as a consequence of mutations (denoted by star) in pathway components resulting in cell autonomous constitutive Hh signaling. (B) Ligand-dependent autocrine signaling occurs secondary to overproduction of Hh ligands that result in cell autonomous Hh signaling. (C) Ligand-dependent paracrine signaling typically found in epithelial tumors that produce Hh ligands that signal to neighboring stromal cells that in turn provide reciprocal trophic support to the tumor cell. (D) Ligand-dependent reverse paracrine signaling where neighboring stromal cells of the tumor microenvironment produce Hh ligands that signal to tumor cells for growth and survival. Activation of hedgehog (Hh) target gene (Gli1 and PTCH1) expression is depicted in nuclei of either tumor or stromal cells depending on the mode of Hh signaling. (See Color Insert.)

cells directly, the inhibition of the Hh pathway in neighboring stromal cells may have benefit in the treatment of Hh dependent cancers. Although much of the data on Hh signaling and the initial use of SMO antagonists such as cyclopamine has been obtained on basal-cell carcinoma and medulloblastoma tumors, it is currently unclear which of the other tumors with evidence of Hh ligand expression or Hh pathway activation as evidenced by target gene expression will actually benefit from targeted treatment strategies against Hh signaling.

IV. Targeted Therapies for Hedgehog Dependent Cancers

A. SMO antagonists

Based on preclinical mouse modeling through tissue-specific Hh activation that recapitulates the biology of the human tumors, several Hh antagonists have been developed through *in vitro* screening of chemical libraries and proven to be efficacious in preclinical studies both *in vitro* and *in vivo* (Tremblay *et al.*, 2010). Over 50 compounds have been identified as having an inhibitory effect on the Hh pathway. To date, most of these Hh antagonists have targeted the Hh receptor, SMO.

SMO small molecule antagonists consist of natural and synthetic products. Cyclopamine, a plant-derived steroidal alkaloid, binds directly to SMO and has been proposed to hinder a conformational change of SMO required for pathway activation (Chen *et al.*, 2002a). As a therapeutic agent, oral treatment with cyclopamine showed a marked decrease in UV-induced basal-cell carcinoma and medulloblastoma growth in Ptch1+/− mice (Athar *et al.*, 2004; Sanchez *et al.*, 2005). However, there are several issues with cyclopamine as an anti-Hh therapeutic compound. One concern is that although cyclopamine has some specificity to Hh signaling at lower concentrations (typically $<10\,\mu M$), it can cause cell death at higher concentrations independent of Hh signaling inhibition depending on the cell type (Yauch *et al.*, 2008). Further, cyclopamine's insoluble features, low potency, rapid clearance, varied bioavailability depending on type of administration, and nonspecific toxicity precluded its further clinical development as a drug (Lipinski *et al.*, 2008).

Novel inhibitors of SMO were identified through cell-based screening approaches using GLI-dependent transcriptional reporter assays (Mas and Ruiz i Altaba, 2010; Tremblay *et al.*, 2010). Interestingly, many of the compounds appear to be distinct from cyclopamine, yet appear to bind the same site on SMO to inhibit Hh signaling. Further, several of the compounds that have been advanced for clinical development demonstrated a 100-fold increase in potency of Hh inhibition *in vitro* (Robarge *et al.*, 2009;

Tremblay *et al.*, 2009; Mahindroo *et al.*, 2009). Several synthetic products that target SMO have been generated through medicinal chemistry and currently there are seven Hh pathway inhibitors that are in clinical development encompassing 44 independent clinical trials (Ng and Curran, 2011). These include GDC-0449 (Genentech), LDE225 (Novartis), XL139 (BMS/Exelixis), IPI-926 (Infinity Pharm), PF-04449913 (Pfizer), LEQ506 (Novartis), TAK-441 (Millennium). These compounds have shown varying levels of preclinical efficacy in several types of cancer cells *in vitro* and *in vivo* xenografts likely secondary to issues of dose and type of administration (Lipinski *et al.*, 2008; Yauch *et al.*, 2008). However, use of an orally deliverable benzimidazole, HhAntag (Curis, Genentech), in animal models of basal-cell carcinoma and medulloblastoma led to dramatic eradication of tumors at their endogenous locations providing the strongest preclinical evidence for the potential efficacy of SMO antagonists in treating Hh dependent cancers (Romer *et al.*, 2004; Williams *et al.*, 2003). The xenograft studies, however, did demonstrate inhibition of tumor growth in several studies in mice either through effects on tumor cells directly or via the inhibition of stromal support suggesting a dual benefit to Hh inhibition in the cancer setting (Yauch *et al.*, 2008). Interestingly, in a genetically engineered model of pancreatic cancer, use of IPI-926 in conjunction with gemcitabine, showed a reduction in the dense fibrotic reaction associated with this tumor model and led to an increase in survival (Heretsch *et al.*, 2010; Olive *et al.*, 2009). Mechanistically, IPI-926 appeared to increase tumor neovascularization, thus improving gemcitabine chemotherapy delivery to previously inaccessible regions of the tumor, demonstrating another potential role for Hh inhibition in improving drug delivery through the stromal compartment of some tumors (Olive *et al.*, 2009).

B. Issues with Hedgehog inhibitors

1. Acquired resistance mechanisms

The most widely used SMO antagonist in current clinical trials for various cancers is GDC-0449. The results of the first Phase I trial using GDC-0449 reported that nearly 60% of patients with basal-cell carcinoma had either a partial or complete response, and an adult patient with advanced metastatic medulloblastoma similarly had a dramatic transient response to this therapy (Rudin *et al.*, 2009; Von Hoff *et al.*, 2009). As with the use of standard chemotherapeutic agents and/or radiation treatment for many cancers, unfortunately, a rapid relapse occurred in the adult medulloblastoma patient after 3 months posttreatment with GDC-0449 (Yauch *et al.*, 2009). Interestingly, molecular analysis of a biopsy specimen from this patient after relapse revealed a new heterozygous G-to-C missense mutation at position 1697, that is predicted to change codon 473 from Asp to His (D473H). This SMO mutation confirms drug resistance by interfering with the ability of

GDC-0449 to bind the SMO protein. Interestingly, the same drug resistance mutation was identified in xenograft studies in mice treated with GDC-0449 (Yauch *et al.*, 2009).

In preclinical studies in mice using SMO inhibitors, it has been shown that alternative resistance mechanisms occur including new SMO mutations that prevent drug binding or downstream mutations including either Gli2 amplification or CyclinD1 gene amplification (Buonamici *et al.*, 2010; Dijkgraaf *et al.*, 2011). Based on preclinical data and the recent clinical trials using SMO antagonists as well as other targeted therapeutics in other cancer settings, it is likely that drug resistance will likely be an issue. The identification of new compounds that can overcome SMO resistance are currently under preclinical development, including inhibition of alternate signaling pathways such as PI3K activity (Buonamici *et al.*, 2010; Dijkgraaf *et al.*, 2011). However, resistance mechanisms that occur downstream of SMO, or alternate pathways such as PI3K, are likely to be overcome with this approach. Similarly, Hh activation downstream of SMO, such as SUFU mutations or GLI amplifications, will also not likely benefit from initial treatment with SMO inhibitors as will be described later.

Importantly, other than these two clinical trials using SMO antagonists for basal-cell carcinoma or medulloblastoma patients, there has been no major clinical response in Phase I studies of GDC-0449 in other cancer settings (LoRusso *et al.*, 2011). An important consideration in these studies is that patients are not routinely preselected as having tumors with evidence of Hh activation, thus obscuring the potential positive effects of Hh targeted therapy. In addition to molecular stratification of patients, it will be necessary to validate the treatment effects directly on tumor tissue at early and late timepoints during the course of therapy in order to better assess efficacy of Hh antagonists.

2. Potential developmental toxicities

Although SMO inhibitors have remarkably few side effects in adult patients (LoRusso *et al.*, 2011), the consequences of Hh inhibition in children particularly during crucial periods of development may be significant. Preclinical studies in juvenile mice using the SMO inhibitor, HhAntag, showed dramatic and permanent defects in bone growth even when the drug was administered for a brief 2–4-day period (Kimura *et al.*, 2008). Importantly, this bone phenotype which mimics the conditional knockout studies of the Ihh gene (Maeda *et al.*, 2007) cannot be rescued by the administration of parathyroid hormone-related protein, or PTHrP, which functions downstream of the Ihh to maintain chondrocytes in a proliferative state (Brechbiel *et al.*, 2011). Similarly, the role of Shh signaling in the postnatal development of the cerebellum, particularly in the proliferation of granule cell precursors, raises additional concerns with Hh pathway inhibition in this age group. Finally, as several developmental signaling pathways, including Wnt and bone morphogenetic proteins, are coordinated with Hh signaling in the proper patterning and morphogenesis of several tissues, dysregulation of this balance through Hh inhibition in young patients may have other unsatisfactory side effects on normal development.

C. Alternative Hh inhibition strategies

1. Inhibition of Hh ligand

Although the identification of small molecule compounds that inhibit the Hh pathway have focused on binding and inhibition of SMO, proteins upstream or downstream of SMO may serve as additional therapeutic targets. As paracrine Hh signaling from tumor cells to stroma and vice versa have been implicated in several cancers, recent studies have identified the first small molecule inhibitor of the Hh ligand (Stanton *et al.*, 2009). This molecule, robotnikinin, is a Hh pathway antagonist that binds the amino terminal region of Shh (ShhN) and inhibits pathway activation in a dose-dependent manner without substantial cytotoxicity in cell viability assays. Importantly, the activity of this Hh ligand antagonist has only been assessed *in vitro* in keratinocytes and in cell line reporter systems, and it is unknown its efficacy and toxicity *in vivo*. Further, the specificity of robotnikinin toward the three mammalian Hh ligands is not known. A Hh ligand specificity may be useful in minimizing the bone defects associated with Ihh inhibition if there is a selectivity associated with this compound.

2. Inhibition of Hh components downstream of SMO

As previously mentioned, the usefulness of SMO inhibitors for cancers with Hh activation downstream of SMO is unclear. Amplification of GLI1 or GLI2 has been implicated in some gliomas and childhood sarcomas, inactivation of SUFU in medulloblastomas, and the development of resistance to SMO inhibitors has sparked recent interest in developing molecular therapeutics against proteins that function downstream of SMO such as the GLI proteins. There are currently several strategies to inhibiting GLI protein function. One is by altering the posttranslational modifications that occur after Hh ligand stimulation that proteolytically convert GLI1 and GLI2 to activator forms. A second strategy would be to limit or interfere with GLI transcriptional activity by either interfering with GLI DNA binding or inhibiting the transcriptional activation domains of the GLI proteins (Merchant and Matsui, 2010).

A cellular based screen for compounds that inhibit GLI-mediated transcription initially identified Gli-antagonist (GANT)-58 and GANT-61 (Lauth *et al.*, 2007). These two compounds inhibit tumor growth *in vitro* and *in vivo* in a human prostate cancer xenograft model; however, their exact mechanism of action is unclear, although GANT-61 does appear to somewhat limit DNA binding by GLI proteins. Four additional compounds that inhibit GLI1 transcriptional activity were identified in another screen and termed Hh pathway inhibitors, or HPI-1 through HPI-4 (Hyman *et al.*, 2009). Similar to the GANT compounds, the exact mechanism of action of the HPI compounds is also poorly understood. However, it appears that HPI-1 inhibits both GLI1 and GLI2 possibly through their posttranslational modification and HPI-2 and HPI-3 may prevent GLI2 processing to an activator form.

Recent studies have identified arsenic trioxide (ATO) as having inhibitory effects on GLI1 and/or GLI2 (Raju, 2011). One recent study suggested that ATO directly binds GLI1 and GLI2 and alters the transcriptional activity without altering DNA binding or subcellular localization of these proteins (Beauchamp *et al.*, 2011). ATO inhibited growth of Ewing sarcoma and medulloblastoma cell lines *in vitro* and in xenograft and transgenic medulloblastoma tumor models *in vivo*. Meanwhile, another study identified ATO's mechanism of action in antagonizing Hh signaling primarily through interference with GLI2 accumulation in the primary cilium and through inhibition of GLI2 stability (Kim *et al.*, 2010). The differences in ATO mechanisms might be related to technical differences or the different cell types that were used in these studies and will need to be sorted out in subsequent studies. Regardless, as ATO is already an FDA approved drug for the treatment of acute promyelocytic leukemia, this raises the possibility of use of this molecule in Hh dependent cancers or in the setting of SMO inhibitor resistance.

V. Conclusions and Future Directions

A combination of intense investigation in both developmental biology and cancer research has led to the identification of Hh signaling as a critical component in both disciplines. The basic science understanding of this pathway and its effects on proliferation and differentiation of both normal cells and transformed cells has resulted in significant breakthroughs that are beginning to be translated into the clinical setting. The contributions of the medicinal chemistry field and biotechnology companies to the rapid development of small molecule antagonists that interfere with Hh signaling at multiple levels of the pathway (see Fig. 23.2) has further spurned advances to bring these compounds to the clinical setting. Table 23.1 lists small molecule Hh antagonists that are currently in preclinical or clinical trials.

There are still several challenges that need to be overcome before success in treating Hh dependent cancers can be achieved. Although success with the use of lead compounds such as GDC-0449 have led to transient clinical responses in patients with basal-cell carcinoma and medulloblastoma, it is still unclear the true Hh dependence of other cancers. Additional preclinical

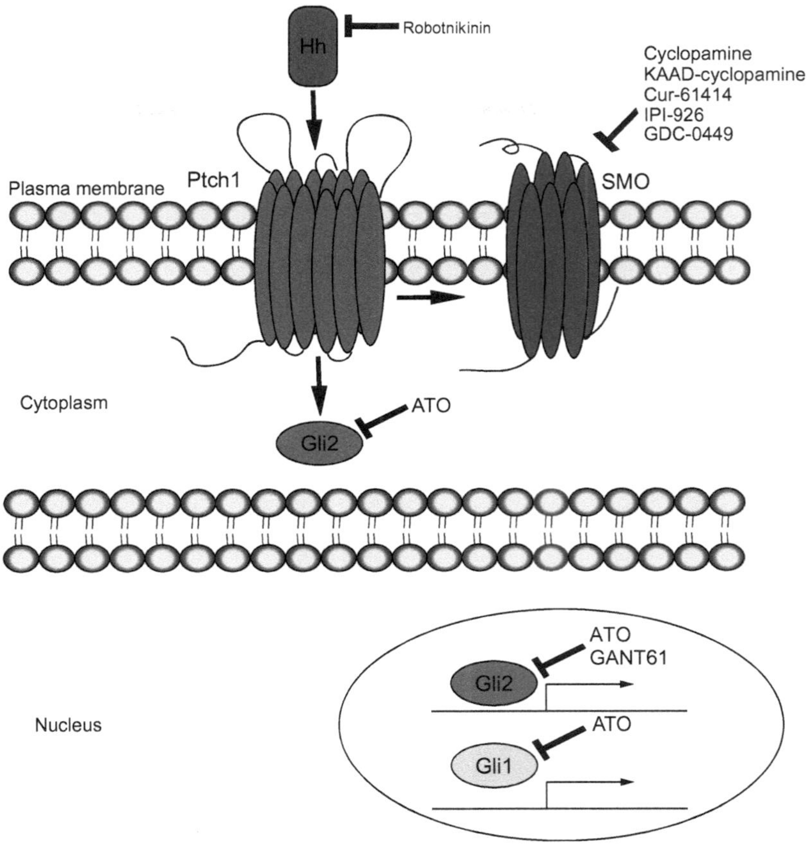

Figure 23.2 Hedgehog signaling pathway and small molecule antagonists. Schematic depicts several key components of the Hh signaling pathway and targets of several classes of Hh pathway inhibitors. Binding of Hh ligand to the PTCH1 receptor relieves a negative regulatory influence on the Smoothened (SMO) protein that through a poorly defined mechanism results in processing of GLI2 signal mediators to activator forms that translocate to the nucleus and turn on Hh dependent gene transcription. Most small molecule Hh antagonists in clinical trials bind the SMO protein; however, recently identified compounds bind either upstream or downstream of SMO at the level of Hh ligand or GLI proteins, respectively. (See Color Insert.)

testing should be performed before undertaking expensive clinical trials to better understand how to use specific Hh antagonists in the setting of cancers without genetic lesions in the Hh pathway. Additionally, preselection of patients with evidence of Hh signaling in their tumors will also likely result in better efficacy of these compounds in clinical trials. Finally, due to the molecular and cellular heterogeneity of most cancers, resistance mechanisms to Hh targeted therapies are likely to be a major hurdle in the use of Hh pathway inhibitors in cancer patients. A better understanding of

Table 23.1 Hedgehog pathway inhibitors in preclinical or clinical trials

Compound	Tumor type	Stage of development	Target
Cyclopamine	Medulloblastoma BCC Rhabdomycosarcoma	Preclinical	SMO
IPI-926	BCC Pancreatic Adenocarcinoma	Phase I	SMO
XL139/BMS-833923	Solid Tumors Gastric Myeloma Medulloblastoma Basal-cell nevus	Phase I	SMO
GDC-0449	Advanced BCC BCNS Breast CRC GI cancer GBM Medulloblastoma Ovarian Pancreatic SCLC	Phase II	SMO
Cur-61414	BCC	Phase I	SMO
Robotnikinin	Skin tumors	Preclinical	Shh
GANT-61	Prostate	Preclinical	GLI
ATO	Acute promyelocytic Leukemia Ewing sarcoma Medulloblastoma	Preclinical	GLI1,2

Abbreviations: BCC, basal-cell carcinoma; BCNS, basal-cell nevus syndrome; CRC, colorectal cancer; GBM, glioblastoma multiforme; SCLC, small-cell lung cancer; GI, gastrointestinal.

resistance mechanisms to the current and next generation of Hh pathway antagonists in preclinical studies will also be beneficial to patients as this may lead to the use of additional therapies upfront to minimize resistance or to treat patients at the time of initial relapse with rational therapies that make use of this preclinical information.

ACKNOWLEDGMENTS

G. P. R. was a Fellow of the Pediatric Scientist Development Program (NICHD K12-HD00850) and is currently supported by an NIH K08 Clinical Scientist Career Development Award (NINDS 5K08NS066083). The authors declare no competing financial interests.

REFERENCES

Athar, M., Li, C., Tang, X., Chi, S., Zhang, X., Kim, A. L., Tyring, S. K., Kopelovich, L., Hebert, J., Epstein, E. H., Jr., Bickers, D. R., and Xie, J. (2004). Inhibition of smoothened signaling prevents ultraviolet B-induced basal cell carcinomas through regulation of Fas expression and apoptosis. *Cancer Res.* **64,** 7545–7552.

Beauchamp, E. M., Ringer, L., Bulut, G., Sajwan, K. P., Hall, M. D., Lee, Y., Peaceman, D., Ozdemirli, M., Rodriguez, O., Macdonald, T. J., Albanese, C., Toretsky, J. A., *et al.* (2011). Arsenic trioxide inhibits human cancer cell growth and tumor development in mice by blocking Hedgehog/GLI pathway. *J. Clin. Invest.* **121**(1), 148–160.

Belloni, E., Muenke, M., Roessler, E., Traverso, G., Siegel-Bartelt, J., Frumkin, A., Mitchell, H. F., Doni-Keller, H., Helms, C., Hing, A. V., Heng, H. H., Koop, B., *et al.* (1996). Identification of Sonic hedgehog as a candidate gene responsible for holoprosencephaly. *Nat. Genet.* **14**(3), 353–356.

Berman, D. M., Karhadkar, S. S., Maitra, A., Montes De Oca, R., Gerstenblith, M. R., Briggs, K., Parker, A. R., Shimada, Y., Eshleman, J. R., Watkins, D. N., and Beachy, P. A. (2003). Widespread requirement for Hedgehog ligand stimulation in growth of digestive tract tumours. *Nature* **425**(6960), 846–851.

Binns, W., James, L. F., Shupe, J. L., and Thacker, E. J. (1962). Cyclopian-type malformation in lambs. *Arch. Environ. Health* **5,** 106–108.

Bitgood, M. J., Shen, L., and McMahon, A. P. (1996). Sertoli cell signaling by Desert hedgehog regulates the male germline. *Curr. Biol.* **6,** 298–304.

Brechbiel, J. L., Ng, J. M., and Curran, T. (2011). PTHrP treatment fails to rescue bone defects caused by Hedgehog pathway inhibition in young mice. *Toxicol. Pathol.* **39**(3), 478–485.

Buonamici, S., Williams, J., Morrissey, M., Wang, A., Guo, R., Vattay, A., Hsiao, K., Yuan, J., Green, J., Ospina, B., Yu, Q., Ostrom, L., *et al.* (2010). Interfering with resistance to smoothened antagonists by inhibition of the PI3K pathway in medulloblastoma. *Sci. Transl. Med.* **2**(51), 51ra70.

Chen, J. K., Taipale, J., Cooper, M. K., and Beachy, P. A. (2002a). Inhibition of Hedgehog signaling by direct binding of cyclopamine to Smoothened. *Genes Dev.* **16**(21), 2743–2748.

Chen, J. K., Taipale, J., Young, K. E., Maiti, T., and Beachy, P. A. (2002b). Small molecule modulation of smoothened activity. *Proc. Natl. Acad. Sci. USA* **99,** 14071–14076.

Chiang, C., Litingtung, Y., Lee, E., Young, K. E., Corden, J. L., Westphal, H., and Beachy, P. A. (1996). Cyclopia and defective axial patterning in mice lacking Sonic hedgehog gene function. *Nature* **383**(6599), 407–413.

Cooper, M. K., Porter, J. A., Young, K. E., and Beachy, P. A. (1998). Teratogen-mediated inhibition of target tissue response to Shh signaling. *Science* **280**(5369), 1603–1607.

Dahlen, A., Fletcher, C. D., Mertens, F., Fletcher, J. A., Perez-Atayde, A. R., Hicks, M. J., Debiec-Rychter, M., Sciot, R., Wejde, J., Wedin, R., Mandahl, N., and Panagopoulos, I. (2004). Activation of the GLI oncogene through fusion with the beta-actin gene (ACTB) in a group of distinctive pericytic neoplasms: Pericytoma with t(7;12). *Am. J. Pathol.* **164**(5), 1645–1653.

Dijkgraaf, G. J., Alicke, B., Weinmann, L., Januario, T., West, K., Modrusan, Z., Burdick, D., Goldsmith, R., Robarge, K., Sutherlin, D., Scales, S. J., Gould, S. E., *et al.* (2011). Small molecule inhibition of GDC-0449 refractory smoothened mutants and downstream mechanisms of drug resistance. *Cancer Res.* **71**(2), 435–444.

Hahn, H., Wicking, C., Zaphiropoulous, P. G., Gailani, M. R., Shanley, S., Chidambaram, A., Vorechovsky, I., Holmberg, E., Unden, A. B., Gillies, S., Negus, K., Smyth, I., *et al.* (1996). Mutations of the human homolog of Drosophila patched in the nevoid basal cell carcinoma syndrome. *Cell* **85**(6), 841–851.

Hellemans, J., Coucke, P. J., Giedion, A., De Paepe, A., Kramer, P., Beemer, F., and Mortier, G. R. (2003). Homozygous mutations in IHH cause acrocapitofemoral dysplasia, an autosomal recessive disorder with cone-shaped epiphyses in hands and hips. *Am. J. Hum. Genet.* **72,** 1040–1046.

Heretsch, P., Tzagkaroulaki, L., and Giannis, A. (2010). Modulators of the hedgehog signaling pathway. *Bioorg. Med. Chem.* **18,** 6613–6624.

Hyman, J. M., Firestone, A. J., Heine, V. M., Zhao, Y., Ocasio, C. A., Han, K., Sun, M., Rack, P. G., Sinha, S., Wu, J. J., Slow-Cordero, D. E., Jiang, J., *et al.* (2009). Small-molecule inhibitors reveal multiple strategies for Hedgehog pathway blockade. *Proc. Natl. Acad. Sci. USA* **106**(33), 14132–14137.

Incardona, J. P., Gaffield, W., Kapur, R. P., and Roelink, H. (1998). The teratogenic Veratrum alkaloid cyclopamine inhibits sonic hedgehog signal transduction. *Development* **125,** 3553–3562.

Ingham, P. W., Nakano, Y., and Seger, C. (2011). Mechanisms and functions of Hedgehog signaling across the metazoa. *Nat. Rev. Genet.* **12**(6), 393–406.

Johnson, R. L., Rothman, A. L., Xie, J., Goodrich, L. V., Bare, J. W., Bonifas, J. M., Quinn, A. G., Myers, R. M., Cox, D. R., Epstein, E. H., Jr., and Scott, M. P. (1996). Human homolog of patched, a candidate gene for the basal cell nevus syndrome. *Science* **272**(5268), 1668–1671.

Kim, J., Lee, J. J., Kim, J., Gardner, D., and Beachy, P. A. (2010). Arsenic antagonizes the Hedgehog pathway by preventing ciliary accumulation and reducing stability of the Gli2 transcriptional effector. *Proc. Natl. Acad. Sci. USA* **107**(30), 13432–13437.

Kimura, H., Ng, J. M., and Curran, T. (2008). Transient inhibition of the Hedgehog pathway in young mice causes permanent defects in bone structure. *Cancer Cell* **13**(3), 249–260.

Kinzler, K. W., Bigner, S. H., Bigner, D. D., Trent, J. M., Law, M. L., O'Brien, S. J., Wong, A. J., and Vogelstein, B. (1987). Identification of an amplified, highly expressed gene in a human glioma. *Science* **236**(4797), 70–73.

Lauth, M., Bergstrom, A., Shimokawa, T., and Toftgard, R. (2007). Inhibition of GLI-mediated transcription and tumor cell growth by small molecule antagonists. *Proc. Natl. Acad. Sci. USA* **104,** 8455–8460.

Lench, N. J., Telford, E. A., High, A. S., Markham, A. F., Wicking, C., and Wainwright, B. J. (1997). Characterisation of human patched germline mutations in naevoid basal cell carcinoma syndrome. *Hum. Genet.* **100**(5–6), 497–502.

Lipinski, R. J., Hutson, P. R., Hannam, P. W., Nydza, R. J., Washington, I. M., Moore, R. W., Girdaukas, G. G., Peterson, R. E., and Bushman, W. (2008). Dose- and route-dependent teratogenicity, toxicity, and pharmacokinetic profiles of the hedgehog signaling antagonist cyclopamine in the mouse. *Toxicol. Sci.* **104**(1), 189–197.

LoRusso, P. M., Rudin, C. M., Reddy, J. C., Tibes, R., Weiss, G. J., Borad, M. J., Hann, C. L., Brahmer, J. R., Chang, I., Darbonne, W. C., Graham, R. A., Zerivitz, K. L., *et al.* (2011). Phase I trial of hedgehog pathway inhibitor vismodegib (GDC-0449) in patients with refractory, locally advanced or metastatic solid tumors. *Clin. Cancer Res.* **17**(8), 2502–2511.

Maeda, Y., Nakamura, E., Nguyen, M. T., Suva, L. J., Swain, F. L., Razzaque, M. S., Mackem, S., and Lanske, B. (2007). Indian hedgehog produced by postnatal chondrocytes is essential for maintaining a growth plate and trabecular bone. *Proc. Natl. Acad. Sci. USA* **104**(15), 6382–6387.

Mahindroo, N., Punchihewa, C., and Fujii, N. (2009). Hedgehog-Gli signaling pathway inhibitors as anticancer agents. *J. Med. Chem.* **52**(13), 3829–3845.

Mas, C., and Ruiz i Altaba, A. (2010). Small molecule inhibition of HH-GLI signaling: Current leads, trials, and tribulations. *Biochem. Pharmacol.* **80**(5), 712–723.

McMahon, A. P., Ingham, P. W., and Tabin, C. J. (2003). Developmental roles and clinical significance of hedgehog signaling. *Curr. Top. Dev. Biol.* **53,** 1–114.

Merchant, A. A., and Matsui, W. (2010). Targeting Hedgehog—A cancer stem cell pathway. *Clin. Cancer Res.* **16**(12), 3130–3140.

Ng, J. M., and Curran, T. (2011). The Hedgehog's tale: Developing strategies for targeting cancer. *Nat. Rev. Cancer* **11**(7), 493–501.

Nusslein-Volhard, C., and Wieschaus, E. (1980). Mutations affecting segment number and polarity in Drosophila. *Nature* **287**(5785), 795–801.

Olive, K. P., Jacobetz, M. A., Davidson, C. J., Gopinathan, A., McIntyre, D., Honess, D., Madhu, B., Goldgraben, M. A., Caldwell, M. E., Allard, D., Frese, K. K., Denicola, G., *et al.* (2009). Inhibition of Hedgehog signaling enhances delivery of chemotherapy in a mouse model of pancreatic cancer. *Science* **324**(5933), 1457–1461.

Raffel, C., Jenkins, R. B., Frederick, L., Hebrink, D., Alderete, B., Fults, D. W., and James, C. D. (1997). Sporadic medulloblastomas contain PTCH mutations. *Cancer Res.* **57**(5), 842–845.

Raju, G. P. (2011). Arsenic: A potentially useful poison for Hedgehog-driven cancers. *J. Clin. Invest.* **121**(1), 14–16.

Reifenberger, J., Wolter, M., Knobbe, C. B., Kohler, B., Schonicke, A., Scharwachter, C., Kumar, K., Blaschke, B., Ruzicka, T., and Reifenberger, G. (2005). Somatic mutations in the PTCH, SMOH, SUFUH, and TP53 genes in sporadic basal cell carcinomas. *Br. J. Dermatol.* **152**(1), 43–51.

Robarge, K. D., Brunton, S. A., Castnedo, G. M., Cui, Y., Dina, M. S., Goldsmith, R., Gould, S. E., Guichert, O., Gunzner, J. L., Halladay, J., Jia, W., Khojasteh, C., *et al.* (2009). GDC-0449, a potent inhibitor of the hedgehog pathway. *Bioorg. Med. Chem. Lett.* **19**(19), 5576–5581.

Roberts, W. M., Douglass, E. C., Peiper, S. C., Houghton, P. J., and Look, A. T. (1989). Amplification of the gli gene in childhood sarcomas. *Cancer Res.* **49**(19), 5407–5413.

Roessler, E., Belloni, E., Gaudenz, K., Jay, P., Berta, P., Scherer, S. W., Tsui, L. C., and Muenke, M. (1996). Mutations in the human Sonic hedgehog gene cause holoprosencephaly. *Nat. Genet.* **14**(3), 357–360.

Romer, J. T., Kimura, H., Magdaleno, S., Sasai, K., Fuller, C., Baines, H., Connelly, M., Stewart, C. F., Gould, S., Rubin, L. L., and Curran, T. (2004). Suppression of the Shh pathway using a small molecule inhibitor eliminates medulloblastoma in Ptc1(+/-)p53 (-/-) mice. *Cancer Cell* **6**(3), 229–240.

Rudin, C. M., Hann, C. L., Laterra, J., Yauch, R. L., Callahan, C. A., Fu, L., Holcomb, T., Stinson, J., Gould, S. E., Coleman, B., LoRusso, P. M., Von Hoff, D. D., de Sauvage, F. J., and Low, J. A., (2009). Treatment of medulloblastoma with hedgehog pathway inhibitor GDC-0449. *N. Engl. J. Med.* **361**(12), 1173–1178.

Sanchez, P., Ruiz, I., and Altaba, A. (2005). In vivo inhibition of endogeneous brain tumors through systemic interference of Hedgehog signaling in mice. *Mech. Dev.* **122,** 223–230.

Scales, S. J., and de Sauvage, F. J. (2009). Mechanisms of Hedgehog pathway activation in cancer and implications for therapy. *Trends Pharmacol. Sci.* **30**(6), 303–312.

Stanton, B. Z., Peng, L. F., Maloof, N., Nakai, K., Wang, X., Duffner, J. L., Taveras, K. M., Hyman, J. M., Lee, S. W., Koehler, A. N., Chen, J. K., Fox, J. L., *et al.* (2009). A small molecule that binds Hedgehog and blocks its signaling in human cells. *Nat. Chem. Biol.* **5,** 154–156.

Taylor, M. D., Liu, L., Raffel, C., Hui, C. C., Mainprize, T. G., Zhang, X., Agatep, R., Chiappa, S., Gao, L., Lowrance, A., Hao, A., Goldstein, A. M., *et al.* (2002). Mutations in SUFU predispose to medulloblastoma. *Nat. Genet.* **31**(3), 306–310.

Theunissen, J. W., and de Sauvage, F. J. (2009). Paracrine Hedgehog signaling in cancer. *Cancer Res.* **69**(15), 6007–6010.

Tremblay, M. R., Lescarbeau, A., Grogan, M. J., Tan, E., Lin, G., Austad, B. C., Yu, L. C., Behnke, M. L., Nair, S. J., Hagel, M., White, K., Conley, J., *et al.* (2009). Discovery of a potent and orally active hedgehog pathway antagonist (IPI-926). *J. Med. Chem.* **52,** 4400–4418.

Tremblay, M. R., McGovern, K., Read, M. A., and Castro, A. C. (2010). New developments in the discovery of small molecule Hedgehog antagonists. *Curr. Opin. Chem. Biol.* **14**(3), 428–435.

Varjosalo, M., and Taipale, J. (2008). Hedgehog: Functions and mechanisms. *Genes Dev.* **22** (18), 2454–2472.

Von Hoff, D. D., LoRusso, P. M., Rudin, C. M., Reddy, J. C., Yauch, R. L., Tibes, R., Weiss, G. J., Borad, M. J., Hann, C. L., Brahmer, J. R., Mackey, H. M., Lum, B. L., *et al.* (2009). Inhibition of the hedgehog pathway in advanced basal-cell carcinoma. *N. Engl. J. Med.* **361**(12), 1164–1172.

Watkins, D. N., Berman, D. M., Burkholder, S. G., Wang, B., Beachy, P., and Baylin, S. B. (2003). Hedgehog signaling within airway epithelial progenitors and in small-cell lung cancer. *Nature* **422**(6929), 313–317.

Williams, J. A., Guicherit, O. M., Zaharian, B. I., Xu, Y., Chai, L., Wichterle, H., Kon, C., Gatchalian, C., Porter, J. A., Rubin, L. L., and Fang, F. Y. (2003). Identification of a small molecule inhibitor of the hedgehog signaling pathway: Effects on basal cell carcinoma-like lesions. *Proc. Natl. Acad. Sci. USA* **100**(8), 4616–4621.

Xie, J., Johnson, R. L., Zhang, X., Bare, J. W., Waldman, F. M., Cogen, P. H., Menon, A. G., Warren, R. S., Chen, L. C., Scott, M. P., and Epstein, E. H., Jr. (1997a). Mutations of the PATCHED gene in several types of sporadic extracutaneous tumors. *Cancer Res.* **57**(12), 2369–2372.

Xie, J., Murone, M., Luoh, S. M., Ryan, A., Gu, Q., Zhang, C., Bonifas, J. M., Lam, C. W., Hynes, M., Goddard, A., Rosenthal, A., Epstein, E. H., Jr., *et al.* (1997b). Activating Smoothened mutations in sporadic basal-cell carcinoma. *Nature* **391**(6662), 90–92.

Yang, L., Xie, G., Fan, Q., and Xie, J. (2010). Activation of the hedgehog-signaling pathway in human cancer and the clinical implications. *Oncogene* **29,** 469–481.

Yauch, R. L., Gould, S. E., Scales, S. J., Tang, T., Tian, H., Ahn, C. P., Marshall, D., Fu, L., Januario, T., Kallop, D., Nannini-Pepe, M., Kotkow, K., *et al.* (2008). A paracrine requirement for hedgehog signalling in cancer. *Nature* **455**(7211), 406–410.

Yauch, R. L., Dijkgraaf, G. J. P., Alicke, B., Januario, T., Ahn, C. P., Holcomb, T., Pujara, K., Stinson, J., Callahan, C. A., Tang, T., Bazan, F. J., Kan, Z., *et al.* (2009). Smoothened mutation confers resistance to a hedgehog pathway inhibitor in medulloblastoma. *Science* **326,** 572–574.

Index

Note: Page numbers followed by "*f*" indicate figures, and "*t*" indicate tables.

P

R

S

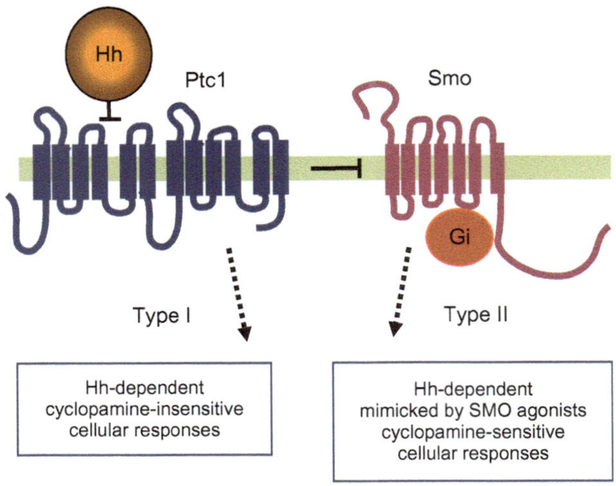

Donna Brennan *et al.*, Figure 3.1 Schematic representation of the two types of noncanonical Hh signaling. Type I requires only binding of an Hh isoform to Ptc1 and is mediated by novel functions of Ptc1 unrelated to Smo repression, and it is by definition insensitive to Smo modulators. Type II is dependent on Smo and in some cases it has been shown to rely on signaling through G_i proteins, and it is both mimicked by Smo agonists and inhibited by Smo antagonists.

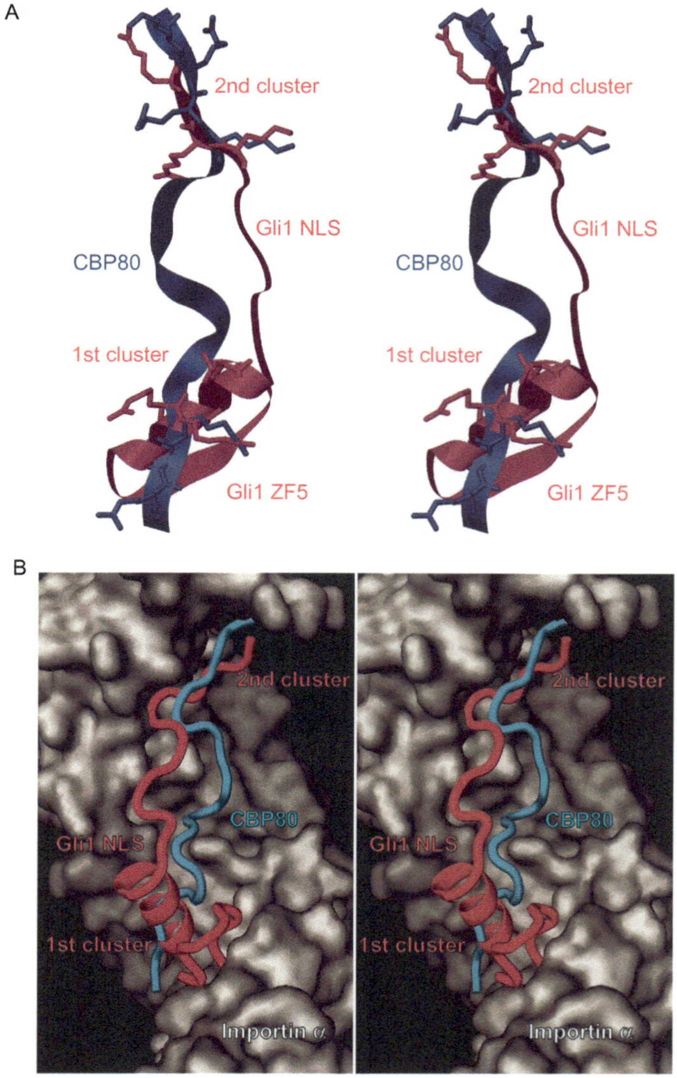

Minoru Hatayama and Jun Aruga, Figure 4.4 Three-dimensional model of Gli1 NLS. The results are shown as stereograms. (A) Superimposition of the Gli NLS structure (thin ribbon) and a known bipartite NLS (thick ribbon) structure from CBP80 (PDB ID, 3FEX). We calculated and created 10 models of the Gli1 NLS, referring to known bipartite NLS structures. NLS-forming side chains are shown as a stick model and the C-terminus is at the top. (B) Superimposition of the Gli NLS and Importin α. Critical steric hindrance was not found. Importin α is shown as a surface model, and the Gli NLS and CBP80 are shown as thick wire models.

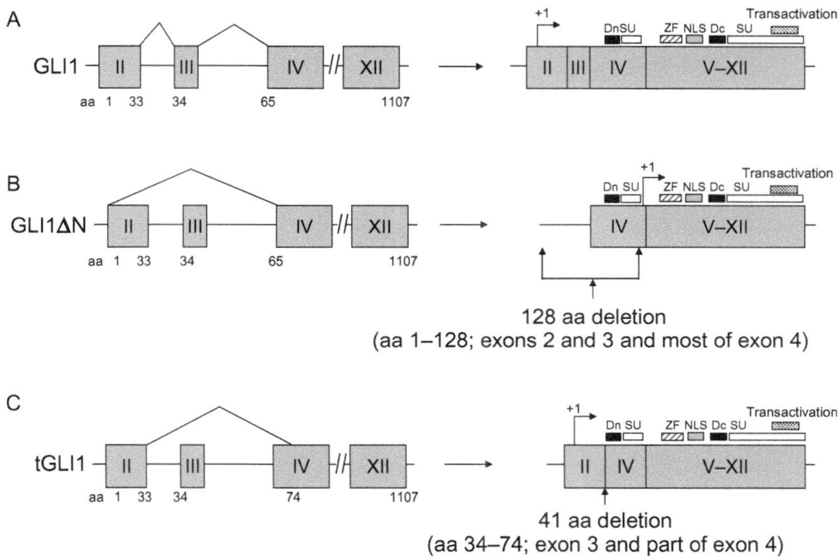

Richard L. Carpenter and Hui-Wen Lo, Figure 6.1 Structures of the human GLI1 gene and the encoded full-length GLI1, GLI1ΔN and tGLI1 isoforms. (A) The full-length GLI1 gene comprises 12 exons, including the 5′-untranslated exon 1. The GLI1 coding region spans nt +79 to +3399 with the initiating methionine codon, ATG, at +79 in exon 2 (arrows). Exons are indicated as gray boxes while introns are shown by lines. The known functional domains of full-length GLI1 include the degron degradation signals (Dn and Dc; aa 77–116; 464–469), SUFU-binding domains (SU; aa 111–125 and C-terminus) (Dunaeva *et al.*, 2003), zinc finger domains (ZF; aa 235–387), the nuclear localization signal (NLS; aa 380–420), and the transactivation domain (aa 1020–1091). (B) Alternative splicing of GLI1 RNA can lead to the deletion of exons 1–3 totaling 128 amino acids in the N-terminus, forming the GLI1ΔN variant. (C) The deletion of the entire exon 3 and part of exon 4 totaling 41 amino acids yields the tGLI1 isoform. Notably, tGLI1 retains all the known functional domains of the full-length GLI1.

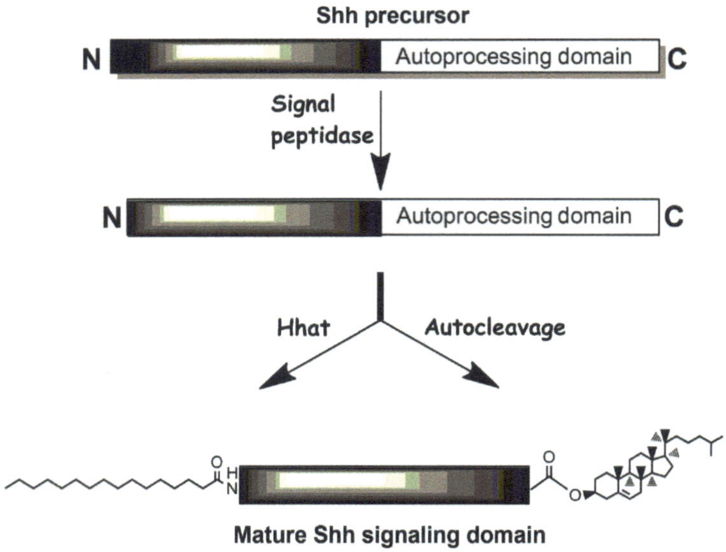

John A. Buglino and Marilyn D. Resh, Figure 10.1 Multiple processing events generate the mature Shh signaling protein. Signal peptidase removes the N-terminal signal sequence. The autoprocessing domain mediates autocleavage and attachment of cholesterol to the C-terminus of the Shh signaling domain, while Hhat catalyzes attachment of palmitate via amide linkage to the N-terminus of the Shh signaling domain.

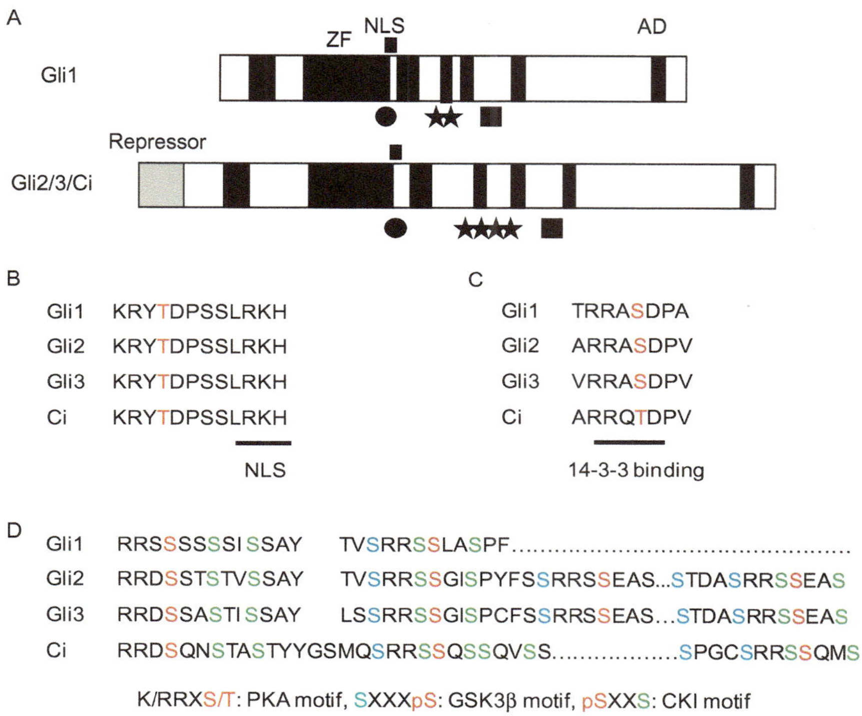

Yoshinari Asaoka, Figure 13.1 (A) Schematic drawing of Gli transcription factors. Repressor, N-terminal repressor domain; ZF, zinc finger; NLS, nuclear localization signal; AD, transactivation domain. Other conserved regions are indicated in black. (B–D) Conservation of amino acid sequences of Gli transcription factors (human Gli1, human Gli2, human Gli3, and Cubitus interruptus of *Drosophila melanogaster*). These regions are related to intracellular localization (B, black circle in (A)), 14-3-3 binding (C, black rectangle in (A)), and proteolytic cleavage (D, star in (A)). Each serine/threonine is phosphorylated by PKA, GSK3β, and CKI. The respective sites are color coded in red, blue, and green.

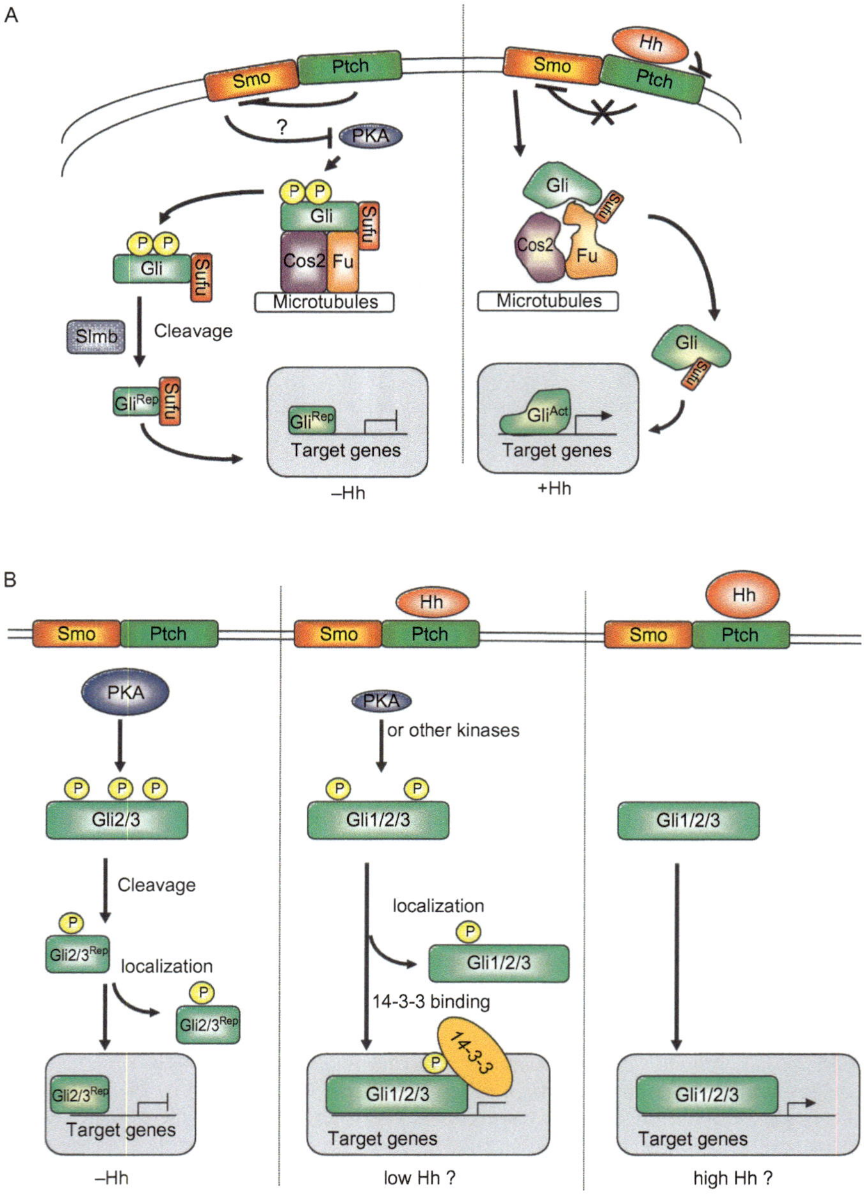

Yoshinari Asaoka, Figure 13.2 (A) The Hh pathway in the absence (left panel) or presence (right panel) of the Hh ligand. In the absence of Hh, Ptch prevents the activity of Smo, and Gli forms a complex with several components, which targets Gli for proteolytic cleavage into the repressor form. In the presence of Hh ligand, Ptch inhibition is relieved; Smo activates Gli. (B) Putative mechanism regulating Hh signaling by multiple phosphorylation of Gli by PKA. Inhibition of nuclear localization and 14-3-3 binding following PKA phosphorylation may limit the activity of Gli (middle panel).

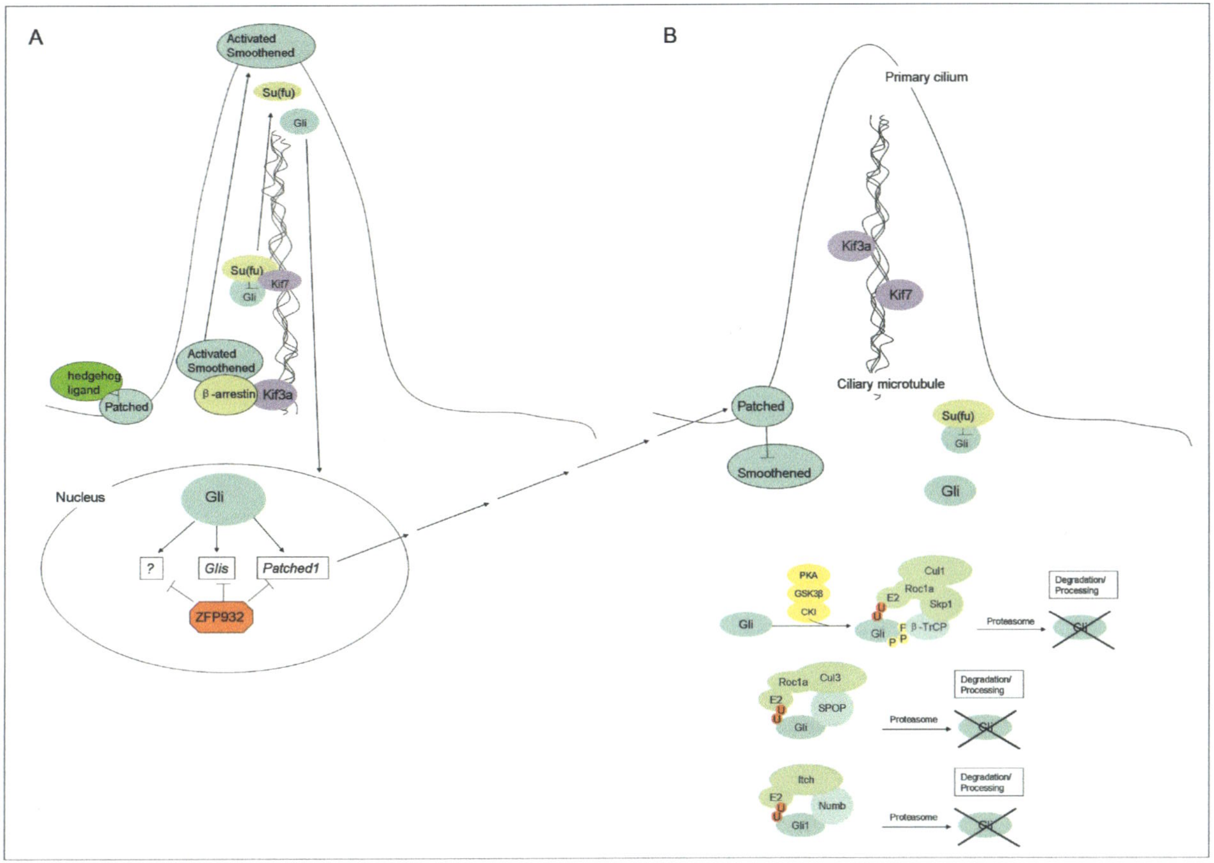

G. Jason Huang *et al.*, Figure 14.1 A simplified representation of the hedgehog-signaling pathway and ZFP932's role. (A) Hh pathway is activated by the Hh ligand. (B) When ligands are absent, Hh pathway is inactivated.

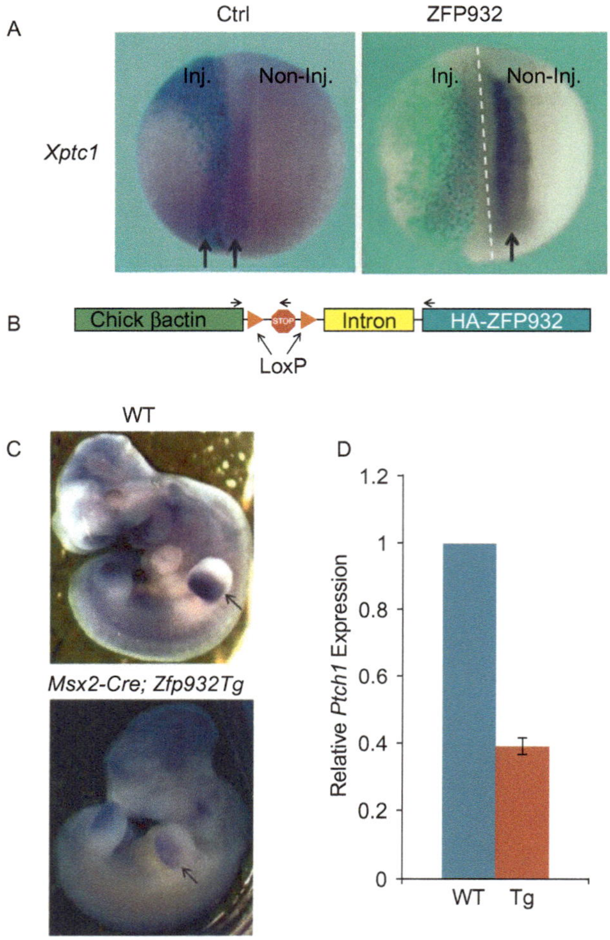

G. Jason Huang *et al.*, Figure 14.5 ZFP932 overexpression repressed *Ptch1* expression in *Xenopus* and transgenic mouse embryos. (A) *Xenopus* embryos showing the injected (Inj.) halves as indicated by X-gal staining (blue) and Non-injected (Non-Inj.) halves. Xptc1 staining by *in situ* hybridization is shown in purple. (B) HA-tagged ZFP932 was driven by the chicken beta-actin promoter preceded by a LoxP-flanked transcriptional stop cassette and a beta-globin intron. This construct was used to create transgenic mouse lines. Relative positions of genotyping primers are indicated by arrows. (C) Wild-type embryos (upper); HA-ZFP932 transgenic embryos (lower). *Ptch1* staining in the limb bud is indicated by arrows. (D) qPCR quantification of *Ptch1* expression in wild-type or HA-ZFP932 transgenic embryos is shown. The results are shown as relative *Ptch1* expression. This figure was originally published in the *Journal of Biological Chemistry*. He *et al.* (2011); © the American Society for Biochemistry and Molecular Biology.

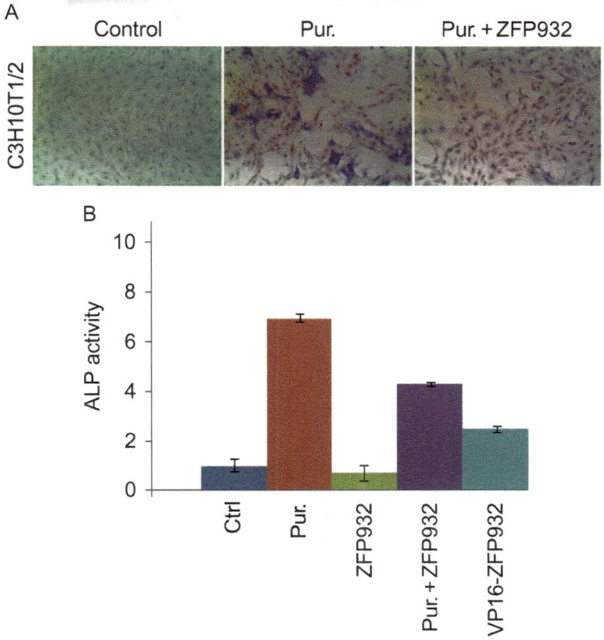

G. Jason Huang *et al.*, Figure 14.7 ZFP932 interferes with Hh-induced osteoblast differentiation. (A) Purmorphamine (Pur.) treatment induced osteoblast differentiation in C3H10T1/2 cells shown by ALP staining. Overexpression of ZFP932 partially blocked Hh induction shown by reduced ALP staining. (B) The level of osteoblast differentiation was quantitatively measured by the ALP activity. Purmorphamine induced a sevenfold increase in ALP activity which was much reduced when ZFP932 was overexpressed. The activator form (VP16-ZFP932ΔN) induced ALP activity in the absence of purmorphamine, mimicking Hh activation.

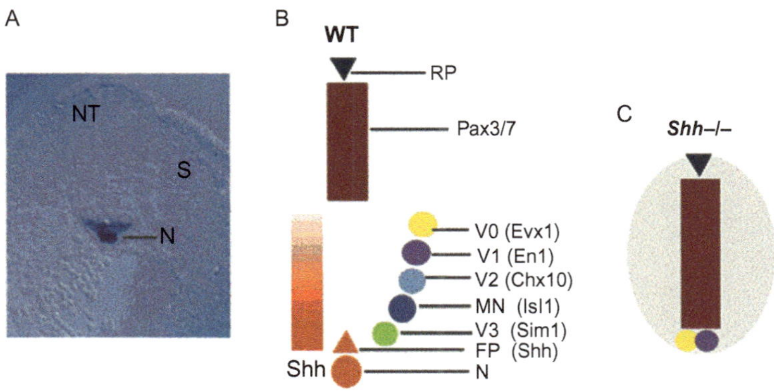

Nicholas C. Bambakidis and Kaine Onwuzulike, Figure 17.1 Neural patterning in spinal cords. Reprinted with permission from Litingtung and Chiang (2000).

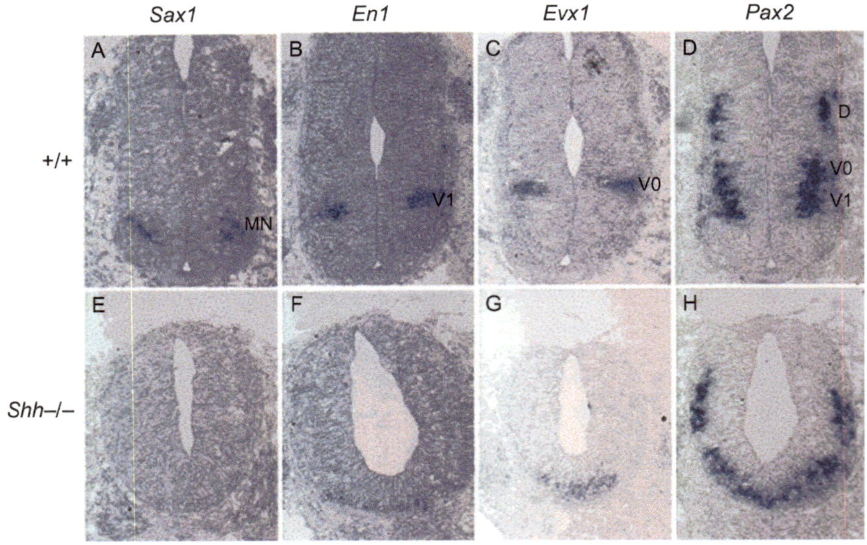

Nicholas C. Bambakidis and Kaine Onwuzulike, Figure 17.2 Absence of motor neurons and disrupted dorsal-ventral patterning. Reprinted with permission from Litingtung and Chiang (2000).

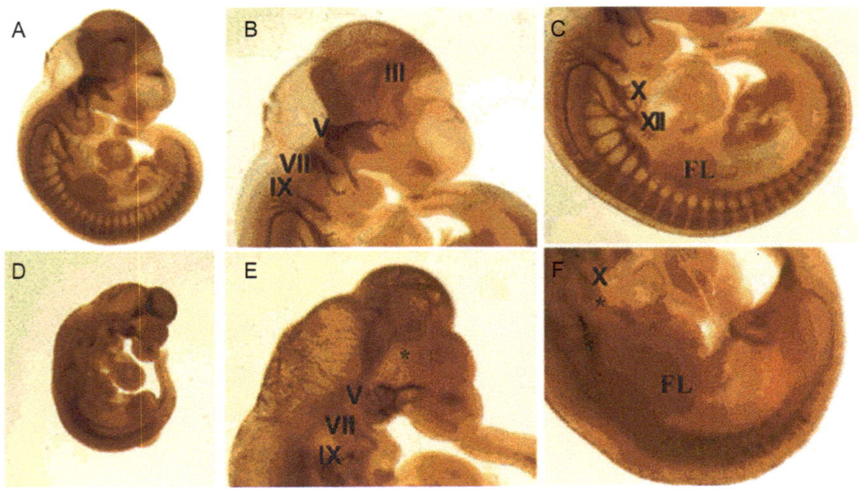

Nicholas C. Bambakidis and Kaine Onwuzulike, Figure 17.3 Lack of cranial motor axons in Shh. Reprinted with permission from Litingtung and Chiang (2000).

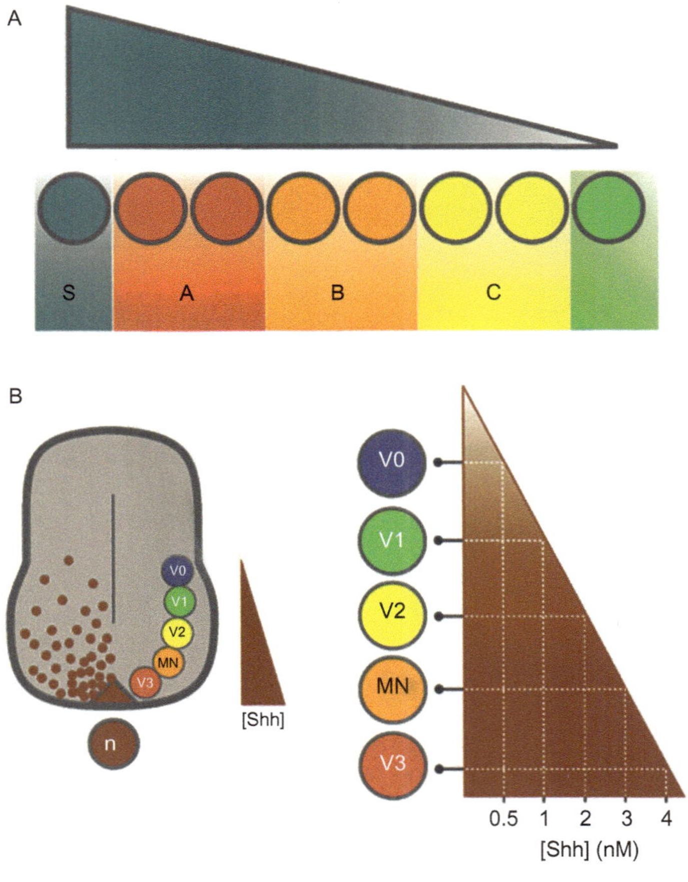

Nicholas C. Bambakidis and Kaine Onwuzulike, Figure 17.4 Shh acts in a graded manner to pattern the ventral neural tube. Reprinted with permission from Briscoe (2009).

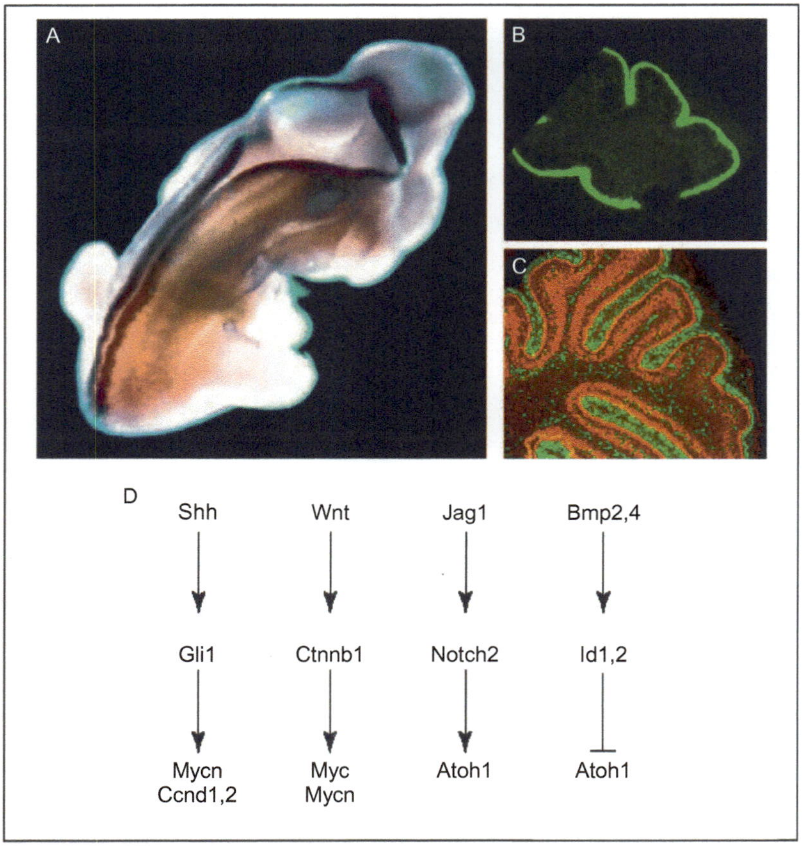

Nicholas C. Bambakidis and Kaine Onwuzulike, Figure 17.5 Changing patterns of gene expression and signaling pathways in cerebellar granule cell progenitor expansion. Reprinted with permission from Hatten and Roussel (2011).

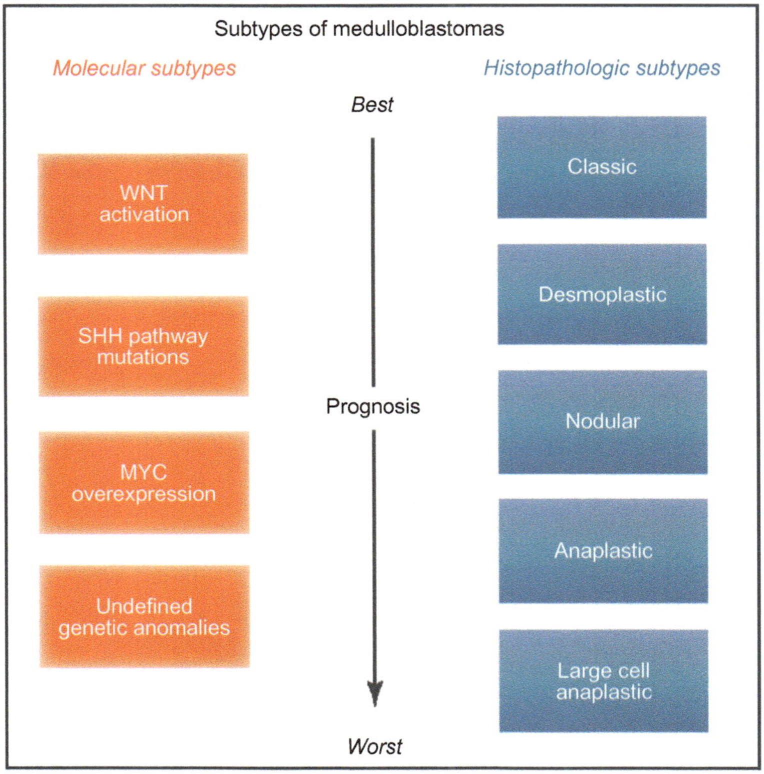

Nicholas C. Bambakidis and Kaine Onwuzulike, Figure 17.6 Subtypes of medulloblastomas. Reprinted with permission from Hatten and Roussel (2011).

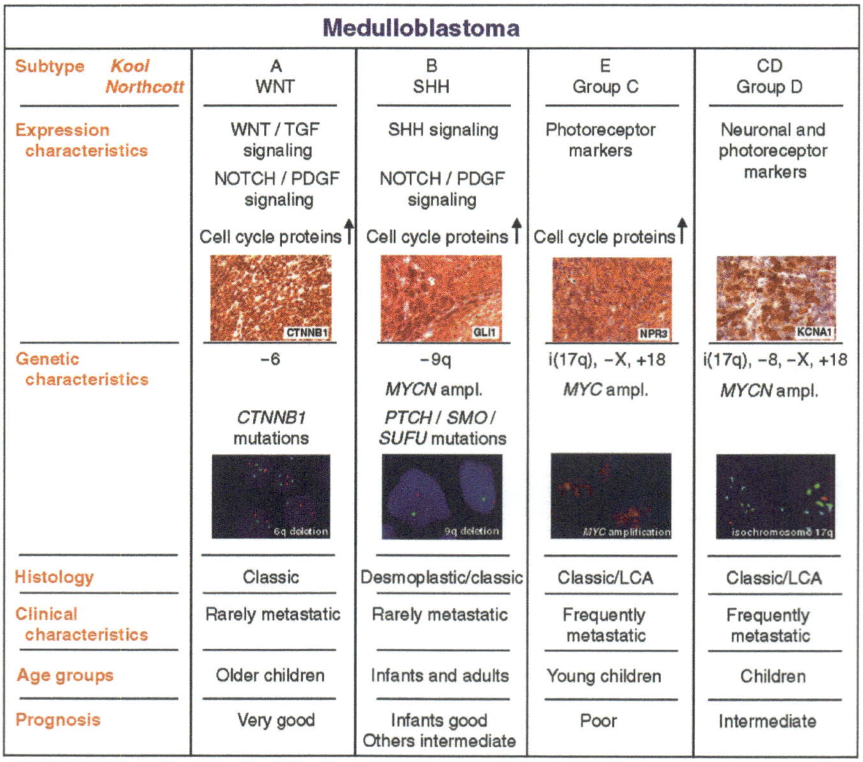

Nicholas C. Bambakidis and Kaine Onwuzulike, Figure 17.7 Medulloblastoma diagnostics. Reprinted with permission from Pfister *et al.* (2010).

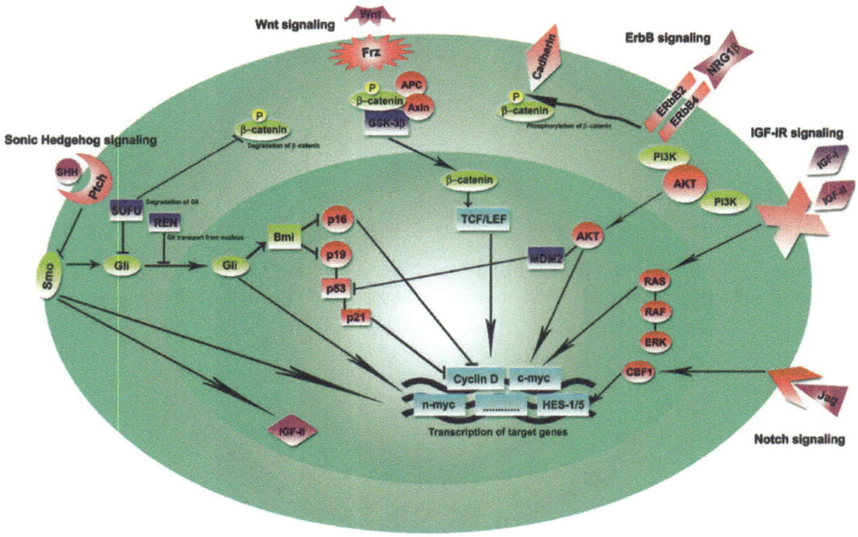

Nicholas C. Bambakidis and Kaine Onwuzulike, Figure 17.8 Signaling pathways involved in the development of the brain and pathogenesis of medulloblastoma and ependymoma. Reprinted with permission from de Bont *et al.* (2008).

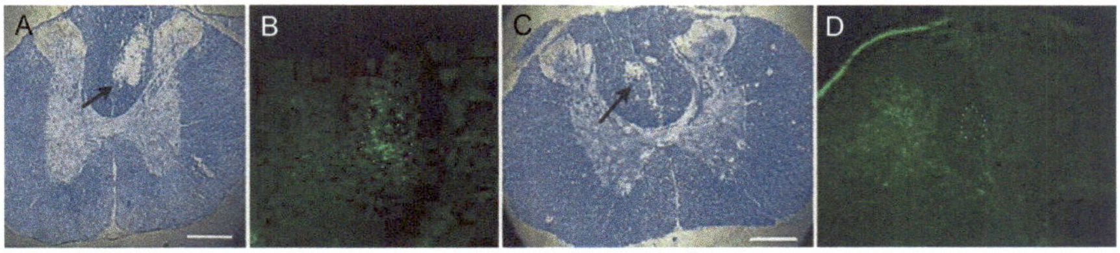

Nicholas C. Bambakidis and Kaine Onwuzulike, Figure 17.9 Photomicrographs showing midthoracic spinal cord cross-sections of a rat receiving spinal cord injury without Shh treatment (A and B) and following Shh injection (C and D). Reprinted with permission from Bambakidis *et al.* (2003). (See Color Insert.)

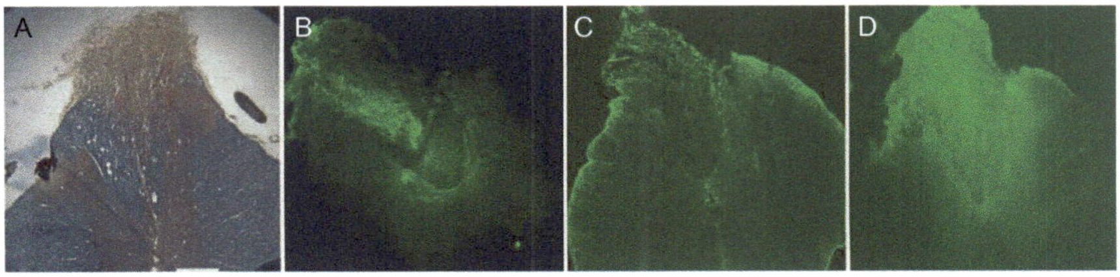

Nicholas C. Bambakidis and Kaine Onwuzulike, Figure 17.10 Photomicrographs showing posterior outgrowth characteristically observed in animals treated with a lesion and Shh. Reprinted with permission from Bambakidis *et al.* (2003).

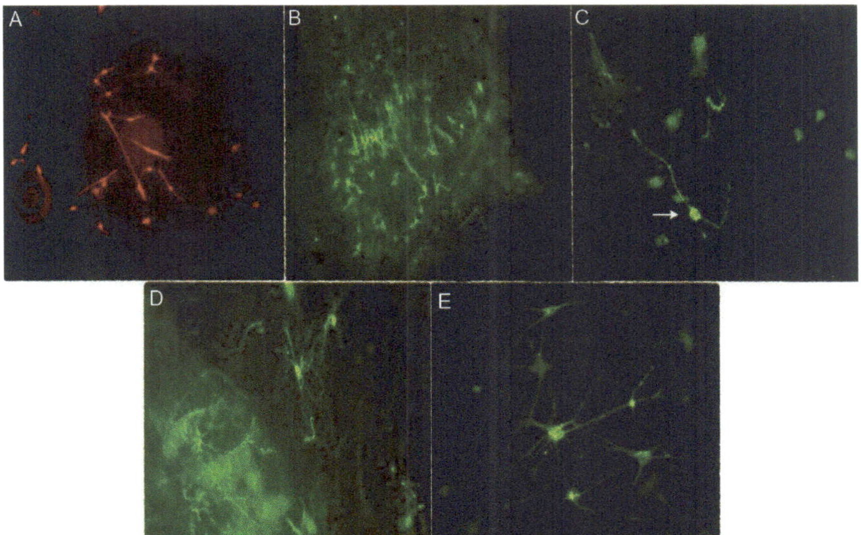

Nicholas C. Bambakidis and Kaine Onwuzulike, Figure 17.11 Photomicrographs showing that Shh treatment increases the number of neural precursors in posterior spinal cord cultures. Reprinted with permission from Bambakidis *et al.* (2003).

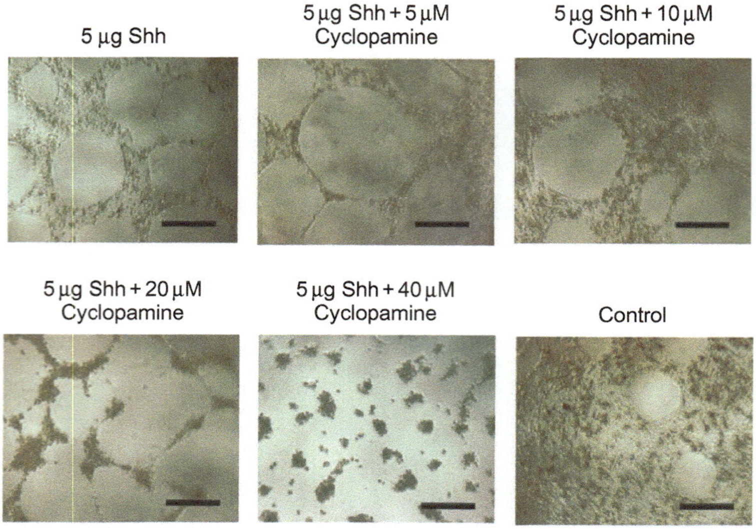

Sabine Fuchs *et al.*, Figure 22.2 OEC seeded on Matrigel® Basement Membrane Matrix: Simultaneous treatment of OEC monocultures with recombinant human sonic hedgehog and different concentrations of the Shh inhibitor cyclopamine (5–40 µM) for 24 h. n=3. Scale bar: 200 µm.

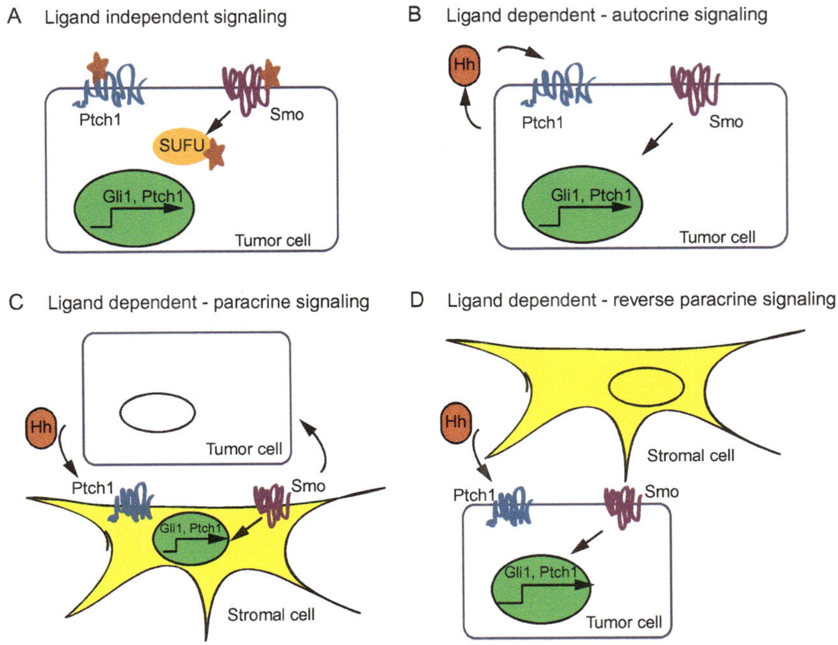

G. Praveen Raju and Diane Pham, Figure 23.1 Mechanisms of Hedgehog pathway activation in cancer. (A) Ligand-independent signaling typically occurs as a consequence of mutations (denoted by star) in pathway components resulting in cell autonomous constitutive Hh signaling. (B) Ligand-dependent autocrine signaling occurs secondary to overproduction of Hh ligands that result in cell autonomous Hh signaling. (C) Ligand-dependent paracrine signaling typically found in epithelial tumors that produce Hh ligands that signal to neighboring stromal cells that in turn provide reciprocal trophic support to the tumor cell. (D) Ligand-dependent reverse paracrine signaling where neighboring stromal cells of the tumor microenvironment produce Hh ligands that signal to tumor cells for growth and survival. Activation of hedgehog (Hh) target gene (Gli1 and PTCH1) expression is depicted in nuclei of either tumor or stromal cells depending on the mode of Hh signaling.

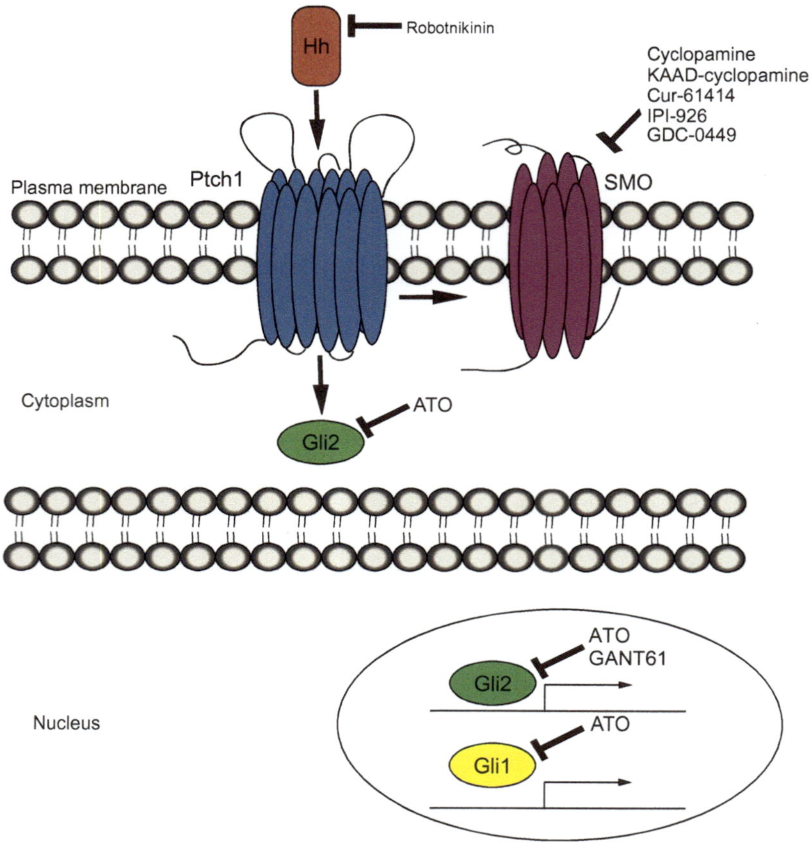

G. Praveen Raju and Diane Pham, Figure 23.2 Hedgehog signaling pathway and small molecule antagonists. Schematic depicts several key components of the Hh signaling pathway and targets of several classes of Hh pathway inhibitors. Binding of Hh ligand to the PTCH1 receptor relieves a negative regulatory influence on the Smoothened (SMO) protein that through a poorly defined mechanism results in processing of GLI2 signal mediators to activator forms that translocate to the nucleus and turn on Hh dependent gene transcription. Most small molecule Hh antagonists in clinical trials bind the SMO protein; however, recently identified compounds bind either upstream or downstream of SMO at the level of Hh ligand or GLI proteins, respectively.

Kolibri 360 GmbH
Ferdinand-Jühlke-Straße 7
99095 Erfurt, Deutschland
produktsicherheit@kolibri360.de